# FREE AND INTERACTING QUANTUM FIELDS

*First Order Processes on Electromagnetic and Gravitational Backgrounds*

# FREE AND INTERACTING QUANTUM FIELDS

## *First Order Processes on Electromagnetic and Gravitational Backgrounds*

Daniel Radu
Ioan Merches
Dorian Tatomir
Alexandru Ioan Cuza University, Romania

World Scientific

NEW JERSEY • LONDON • SINGAPORE • BEIJING • SHANGHAI • HONG KONG • TAIPEI • CHENNAI • TOKYO

*Published by*

World Scientific Publishing Co. Pte. Ltd.
5 Toh Tuck Link, Singapore 596224
*USA office:* 27 Warren Street, Suite 401-402, Hackensack, NJ 07601
*UK office:* 57 Shelton Street, Covent Garden, London WC2H 9HE

**Library of Congress Cataloging-in-Publication Data**
Names: Radu, Daniel, author. | Merches, Ioan, author. | Tatomir, Dorian, author.
Title: Free and interacting quantum fields / Daniel Radu (Alexandru Ioan Cuza University, Romania), Ioan Merches (Alexandru Ioan Cuza University, Romania), Dorian Tatomir (Alexandru Ioan Cuza University, Romania).
Description: Singapore ; Hackensack, NJ : World Scientific, [2017] | Includes bibliographical references and index.
Identifiers: LCCN 2016058646| ISBN 9789813145467 (hardcover ; alk. paper) | ISBN 9813145463 (hardcover ; alk. paper)
Subjects: LCSH: Quantum field theory. | Relativity (Physics) | Particles (Nuclear physics)
Classification: LCC QC174.45 .R335 2017 | DDC 530.14/3--dc23
LC record available at https://lccn.loc.gov/2016058646

**British Library Cataloguing-in-Publication Data**
A catalogue record for this book is available from the British Library.



Desk Editor: Christopher Teo

Printed in Singapore

*"There is nothing more practical*
*than a good theory"*
Kurt Lewin

# Preface

The study of interactions between various categories of particles belonging to the atomic and sub-atomic world is of quantum nature *par excellence*; it became possible due to several fundamental theoretical discoveries in modern physics. Usually, by "modern physics" one understands the period of physics' development beginning with the revolutionary Plank's and Einstein's theories and continuing up to today. Making abstraction for the progress performed by the empirical investigation, modern physics is characterized by two great theories, which make it fundamentally distinguishable from the whole of its previous development: *quantum theory* and *theory of relativity.*

In this book we want to bring to the reader's attention several solutions to problems connected to the quantum-relativistic interaction of particles. Remarkably, such solutions furnished rigorous and pertinent explanations of a large set of phenomena, both in microscopic universe and galactic world.

One of the most important targets of today's physics is to give a dynamic model capable to explain the complex properties of particles, their interactions, and symmetry relations existent in the microcosmos. Basically, the procedure consists in a simultaneous approach – both quantum and relativistic – of these interactions. The quantum nature of interactions emerges from similitude between the structure of particle systems and atomic structures. But the energies characterizing processes that involve subatomic particles are very high, so that they cannot be investigated only from the quantum point of view. Under these circumstances, only a quantum-relativistic model proved to be able to correctly explain the particles' properties and their interaction in the microcosmos.

The discovery of such models began with the Quantum Field Theory. The first step was performed by Planck (1900) by quantization of the energy. The theory was then developed by Jordan and Wigner, while in 1927 P.A.M. Dirac formulated the first quantum theory of the electromagnetic field. Quantum electrodynamics soon became a theory with impressive experimental checks (for example, the fine-structure constant shows an exact concordance between its experimental and theoretical values up to the eighth decimal digit). Despite of these initial performances, some difficulties appeared connected to the physical interpretation of some divergent expressions obtained for several concrete processes. The problem was solved by Heisenberg who, in 1943, elaborated the $S$-matrix theory. Subsequently, this theory became a mathematical complex structure, ideal for description of quantum interactions between particles. According to the $S$-matrix theory, the investigator's attention is displaced from objects to events; it mostly deals with the processes of particle interactions, rather than the particles themselves.

Investigation is oriented towards events (processes), this being demanded by both quantum and relativistic theories. As it was shown by quantum theory, a subatomic particle must be understood as an interaction between various measurement processes. Instead of an isolated object, this time we deal with an event responsible for the correlation of some other events. As Heisenberg [202] used to say, "Modern physics divides the world not in different groups of objects, but in various groups of correlations... What we can distinguish is the type of connection, which is of the first importance for a phenomenon. This way, the world appears as a complicated net of events, with alternating, overlapping, and combining connections of various types, determining the appearance of the whole".

On the other side, the theory of relativity displayed the space-time four-dimensional characteristics of microparticles as processes rather than as objects. Approaching these matters within the $S$-matrix formalism unifies both points of view. Using the mathematical tools of the Theory of Relativity, this formalism describes all the properties of particles in terms of interactions (*i.e.* in terms of transition probabilities between various quantum states), setting up a relationship between particles and processes. Any interaction implies particles connecting that interaction with some other interacting processes; this way, an entire system of interdependent processes is configured.

Among the multitude of processes and particles implied in these processes, in the book we investigate only a narrow (but demonstrating an increasing interest) domain, namely the class of high (*i.e.* superior to the conventional one) spin particles. Most recent investigations showed that a great amount of still unsolved problems of the contemporary physics could receive an answer by using high-spin quantum fields.

A considerable part of this book is dedicated to a set of research papers concerning interaction between various types of particles (mostly those of high-spin) and the gravitational and electromagnetic fields, considered as background fields within the modern formalism of the Quantum Field Theory. Our endeavor concerning both the study of high-spin fields, in general, and the scattering and photogeneration of several kinds of particles (both fermions and bosons, massive and zero-rest mass), in particular, is justified by necessity of clarifying some aspects of the fundamental theory of vector and tensor fields, especially those implying the mechanisms of interaction with some other fields (with applications in Quantum Chromodynamics and Cosmology). It is also determined by a considerable number of concrete practical applications, and, last but not least, by the permanent subjective necessity of the scientist to clarify any "mystery" of theoretical physics.

The importance of our investigation is proved by the great amount of research papers on this subject which appeared during the last three decades and still continue to appear in prestigious periodicals like Classical and Quantum Gravity, Physical Review, Nuclear Physics, Physics Letters, etc.

This book is mostly addressed to the senior undergraduates, MSc, and PhD students of the Faculties of Physics, their instructors, but it can also be used by those studying mathematics, physics researchers, and the high school teachers.

This volume is dedicated to the memory of our mentor and adviser Professor Emeritus Ioan Gottlieb, the founder of the theoretical physics school in our faculty, Alexandru Ioan Cuza University of Iasi, and profound investigator in the Quantum Mechanics, Quantum Field Theory and Theory of General Relativity. We also want to express our deep gratitude to Professor Gheorghe Zet, from the Technical University Gheorghe Asachi of Iasi, who read the manuscript and made several pertinent and useful observations.

Last but not least, we are grateful to our colleague, Dr. Ioana-Laura Velicu, for imagining and drawing up the nice design of the book cover.

*The Authors*
Iasi, June 2016

# Introduction

The research results exposed in this volume fall in the general international effort of investigation of the interaction processes of particles in gravitational and electromagnetic fields. They also concern the transformation process of particles on various space-time structures, the applicative chapters being mostly dedicated to processes implying the *high-spin particles.*

One must first explain the significance of the expression "high-spin" associated to a particle. There is no definite convention in this respect, but most frequently the values 0 and 1/2 (in $\hbar$ units) of the spin of a particle are considered small (or *low*), while the values 3/2, 2, 5/2, 3, etc. are considered large (or *high*). These two domains are separated by the value 1 of the spin, which value is being considered as *intermediate.* Therefore, according to this classification, any quantum field whose quanta have at least value 3/2 (in $\hbar$ units) for its spin is considered as being a *high-spin field.*

The importance and necessity of the study of quantum fields (in general) and those with high spin (in particular) is justified by numerous reasons, and here are only some of them:

• First, the spin 3/2, 2, and 5/2 fields play a fundamental role in the theory of quantum gravity and supergravity.

• Second, somewhat recently have been discovered several quasiparticles (in fact, nuclear resonances) whose spin exceeds the value 2 (getting even value 7/2). These quasiparticles can be described by high-spin fields, even if they are non-fundamental fields (like the multiplet or supermultiplet fields). In addition, it has been recently shown that for the massless fields with spin higher than 2 is possible to give an interpretation in terms of gauge-type potentials; such an interpretation was not obvious within previous investigations involving local (both massive and non-massive) fields having spin smaller or equal to 3 and massive local fields of an arbitrary spin as well. These new interpretations stand for advanced steps in understanding the high-spin fields, being capable to lead to the usage of these fields in phenomenological investigations.

• Third, there exists a more or less subjective reason: even if stable particles[1] having the spin higher than 7/2 have not been experimentally discovered yet, the inquisitive spirit of the human being cannot leave unexplored any domain of Reality. Furthermore, at the present we already have a complete and consequent theory – both in Lagrangian and Hamiltonian formalisms – involving classical and quantum fields of arbitrary spin, but deep investigations were performed only up to value 4 of spin, inclusive.

• Fourth, the string and superstring theories show that, at least up to the levels where the generation of particles is still well-defined, one must be considered massive excitations *of any spin.*

• Fifth, even if until now have not been observed particles with no internal structure having spin higher[2] than 2, particles whose

---

[1]Being aware of the fact that even the life of proton, within the Standard Model, has been predicted as being finite, the notion of *stable particle* became today at least questionable. That is why, in the present volume, by "stable particle" we shall understand that particle which has an average lifetime long enough to make it able to effectively participate in at least one observable and measurable physical process.

[2]Related to this value of the spin, *i.e.* 2, Paul van Nieuwenhuizen states in his notorious study published in Physics Reports, *Supergravity* [287], that "*It seems, at this moment, that Nature stops at spin 2*". Nevertheless, even within the extended supergravity theory with at least 8 gravitino (*i.e.* for the supergravity theories with $N > 8$) one appears the necessity of introduction of a 5/2 spin field and consideration of more than one graviton. The main argument of van Nieuwenhuizen is that only up to value 2 of the spin of a particle the mass discontinuity appearing in the Lagrangian of the theory (the so-called *van Dam–Veltman discontinuity*) is finite, for any other value superior to 2 this discontinuity being insurmountable. Nevertheless, the papers published in this field during the last decades determined the authors to approach such a problem. Recently [194] a procedure of construction of a gauge theory for the massive spin-2 tensor field, having a smooth limit for $m \to 0$ has been elaborated by Shinji Hamamoto, by using the Batalin–Fradkin algorithm for the massive spin-2 tensor field theory, with straight application to renormalization theory of quantum gravitation. This way, the solutions to infrared divergences met in quantum gravitation can be obtained in a manner similar

description demand fundamental fields of superior spin, within the frame of ToE (Theory of Everything) type theories, there are clear theoretical evidences showing that the already known interactions (including gravitation) can be incorporated into a supersymmetric scheme only by taking into account the *high-spin fields*. Thus, it is well-known the fact that an irreducible supermultiplet having spin lower than or equal to 2 cannot unify a gauge theory based on local symmetry $SU(3)_{color} \times SU(2) \times U(1)_{electroweak}$ with the fundamental fields of the necessary vector bosons. The simplest extension to a single supermultiplet able to incorporate both the gauge electromagnetic and strong standard interactions, together with the isospin doublets of the weak nuclear interactions, requires the use of both spin-5/2 and spin-3 fields.

• Sixth, there is an increasing importance of the non-massive Rarita–Schwinger field within the modern cosmological models, due to the role played in various scenarios by its quanta, *i.e.* the *gravitino*.

• Seventh – last but not least – the importance of the research involving classical and quantum high-spin fields comes also from some concrete applications in quantum chromodynamics. So, according to numerous studies performed by researches from CERN (Geneva) and DESY (Hamburg), the vector and tensor mesons play an essential role in various theoretical models on interaction involving strongly interacting particles.

The problem of gravitational-electromagnetic coupling (which is an important subject of study in this volume) has been widely investigated from the perspective of the interaction of various types of particles on backgrounds with various symmetries, the approach being done in both linear and nonlinear theories. The obtained results, even in some inchoate phases, were proved to be of a real interest in clarification of some important aspects in certain physical branches, such as cosmology and astrophysics. The great number of applications based on the results obtained in this field by the world scientists (results emphasized by awarding of the Nobel Prize in 1993

---

to the renormalization theory of quantum electrodynamics; here, as one knowns, these divergences are eliminated by description of the photon by means of a massive vector field, subsequently taking the limit when the photon mass goes to zero.

to physicists Russel A. Hulse and Joseph H. Taylor from Princeton University, New Jersey, USA, who evidentiated for the first time in 1974, even if indirectly, the highly controversial – at least regarding their concrete physical existence – gravitational waves) stands for an important reason which fully justified the initiation of several researches on properties of the interaction processes involving the two fundamental physical fields (the electromagnetic and gravitational fields). From this perspective, since the approached subjects have both theoretical and practical finalization, the present volume is intended to fill up some "gaps" in scientific research. So, among the applications of the exposed results, we mention the possibility of using a scattered field in order to be able to make distinction between various types of cosmic objects (for instance, between the black holes and compact cosmic objects with large and very large spatial extension), when the observation process is performed at big distances from the source of the scattering gravitational field with axial symmetry, as well as the possibility of explaining the phenomenon of intensification of the radiation – both in visible and invisible spectra – emmited by the arms of spiral galaxies with an intense own magnetic field, on account of the photon-graviton (and vice-versa) conversion process, within the framework of various configurations of electromagnetic background fields. One of the most discouraging feature of the theories on high-spin fields, even if these fields are considered to be free, rests in the very complicated formalism required by the description of such fields. This formalism becomes even more complicated if, for example, we intend to couple the high-spin fermions with gravitation, in the frame of a local theory with interaction. This is one of the main reasons why until recently all the issues related to the use of high-spin fields has been least addressed regarding concrete calculations. Nevertheless, we have to mention that during the last three decades the electronic calculus technique was intensively developed, in strong connection with emergence of several very performant software packages, capable to process information not only numerically, but especially analytically; we recall, in this respect, the software products such as Reduce, Maple, Mathematica, etc., as being the best known and currently used. Nowadays, we are in possession of very powerful computer equipments, not only at working offices, but also at our home private facilities, and this situation has revolutionized the research work. This way, the study of several complex processes of scattering, transformation and cosmological generation of particles became possible. These processes are very important in understanding of certain physical processes, not only at atomic and sub-atomic scale, but also at cosmic scale. A somewhat exotic phenomenon in this respect is furnished by the process of transformation/conversion of the photons into gravitons and vice-versa (or, classically, transformation/conversion of the electromagnetic into gravitational waves and vice-versa); at first sight this process seems less plausible, but it is perfectly possible from the quantum point of view, having important applications in astrophysics. In this connection it is important to mention that the existence of the gravitational waves is not anymore a controversial problem since, in 1974, Russel A. Hulse and Joseph H. Taylor from Princeton University put into evidence, in an indirect way, the existence of the gravitational waves and more recently, in September 2015, the two detectors of LIGO simultaneously registered a gravitational wave signal – the first direct detection ever achieved[3]. Due to the smallness of the gravitational coupling constant, the study of the photon-graviton conversion processes leads to very small scattering cross-sections. This was, in fact, the main reason why physicists oriented their attention mostly to the cosmic space, instead of laboratory experiments. In astrophysics there certainly exist the possibility that, under some special circumstances, the graviton generation rate in quantum processes, be as big as that corresponding to classical ones. However, recent results concerning this process show that the scattering cross-sections can receive detectable (measurable) values, even in laboratory experiments.

Before beginning a short presentation of the book, it is worth mentioning an important fact regarding the massless spinorial field and its corresponding particles/quanta. This remark concerns the

[3]The observation was reported in B.P. Abbott *et al.* (LIGO Scientific Collaboration and Virgo Collaboration), *Phys. Rev. Lett.* 116, 061102 (2016), http://journals.aps.org/prl/abstract/10.1103/PhysRevLett.116.061102.

fact that, so far, neutrinos have been considered as quanta of the massless spinorial field. Nevertheless, several recently performed experiments (Super-Kamiokande I – 1996 Apr., Super-Kamiokande II – 2002 Oct., Super-Kamiokande III – 2006 Jul., Super-Kamiokande IV – 2008 Sep.; Sudbury Neutrino Observatory (SNO) – 1999 May–2006 Nov.) contradict this theory, showing that the neutrino rest-mass is in fact non-zero. This experimental result has been theoretically suspected much time ago (as a proof, in the Appendix H are given the upper mass limits of the electron-neutrino and muon-neutrino), supposition turning into a certitude.

The experimental discovery of *neutrino oscillation*, and thus *neutrino mass*, by the *Super-Kamiokande Observatory* and the *Sudbury Neutrino Observatory* was rewarded with the 2015 Nobel Prize in Physics that was shared by Japanese physicist Takaaki Kajita (b. 1959) from Institute for Cosmic Ray Research, University of Tokyo, and Canadian physicist Arthur B. McDonald (b. 1943) from Princeton University and Queen's University, for their early pioneering observations of these oscillations.

Neutrino oscillation is a quantum phenomenon by which a neutrino created with a specific lepton flavour (electron, muon or tauon) can subsequently change its flavour, turning – for example – from an electron-neutrino into a muon-neutrino. The basic physics behind neutrino oscillation can be found in any system of coupled harmonic oscillators. The most simple example is a system of two mechanical pendulums connected by a weak spring (having a small spring constant).

The probability of determination of a particular flavour for a certain neutrino varies periodically together with its spatial displacement. The neutrino oscillation was first predicted by the Italian nuclear physicist Bruno Pontecorvo (1913–1993) in 1957, since then being observed in various contexts, such as: *solar neutrino oscillation* (the Ray Davis's Homestake Experiment in the late 1960s; the Sudbury Neutrino Observatory provided clear evidence of neutrino flavor change in 2001. Notably, the existence of neutrino oscillation resolved the long-standing solar neutrino problem – Arthur B. McDonald), *atmospheric neutrino oscillation* (the large detectors such as IMB, MACRO, and Kamiokande II have observed a deficit in the ratio of the flux of muon to electron flavor atmospheric neutrinos; the Super-Kamiokande experiment provided a very precise measurement of neutrino oscillation in an energy range of hundreds of MeV to a few TeV), *reactor neutrino oscillation* (the Daya Bay experiment, in 2012), *beam neutrino oscillation* (the MINOS, K2K, Super-K, LSND, MiniBooNE, Gran Sasso, T2K, Super-Kamiokande, NOvA experiments).

Neutrino oscillation is of great theoretical and experimental interest, as the precise properties of the process can shed light on several properties of the neutrino. In particular, it implies that the *neutrino has a non-zero rest mass*, which requires a modification to the Standard Model of particle physics. However, the question of how neutrino masses arise has not been answered conclusively. There exist several attempts to explain how neutrinos acquire their mass (such as the Majorana mass term, the seesaw mechanism, the leptogenesis, etc.), but none of them has prevailed so far.

*

* *

The book is organized in two parts: the first has a monographic character (based on reference books on the subject and consisting in basic elements necessary to understand the investigations to follow), and the second has a pronounced applicative feature (being composed mostly by authors' original research). The volume also contains eight appendices with relations and formulas necessary to understand the main text as well as certain auxiliary elements/information included for a better understanding of several notions introduced during the exposure, together with a paragraph with some useful formulas met in quantum field theory.

The first chapter is called *Classical and quantum free fields* and contains a short description of classical and quantum theory of free fields with the spin 0, 1/2, 1, 3/2, and 2. Although this subject was exhaustively approached by many famous scientists in several excellent books, the authors decided to include these fundamental notions/results for two reasons: first, all further applications presented in the book (and even all further theoretical considerations designated to prepare an appropriate understanding of the interaction processes studied in the second part of the book) directly appeal a lot of elements belonging to quantum field theory, and second, the authors wanted to ease – and even eliminate – the reader's effort of searching elsewhere for fundamentals of quantum field theory.

Being a fundamental and well-adjusted subject, the authors preferred to use the manner found in masters' investigations. Thus, in this chapter are briefly exposed the main characteristics of the fields having spins between 0 and 2, following as closely as possible the way of presentation used by these consecrated authors. This has been done extensively, keeping even the metric signature and the notation conventions used in their original works. As such masters the authors chose N.N. Bogoliubov and D.V. Shirkov (Ref. [41]) for quantization of the scalar, spinorial and electromagnetic fields, and D. Lurie (Ref. [252]) for quantization of the Rarita–Schwinger field. Regarding the free gravitational field, the authors preferred to expose a new variant, somewhat more natural (see subchapter I.3.7.4) as compared to procedures exposed in references 181, 182, 219, and 275.

The second chapter of the volume, named *The gravitational transmutations hypothesis* can be considered as an introduction to the second part of the book. It contains a survey on the main aspects of the quantum study of particles' interaction in gravitational (and electromagnetic) fields with applications in astrophysics, as they appear in scientific literature: particle diffusion in external gravitational field, transformation of the photon into graviton in an external electromagnetic field, the gravitational/electromagnetic radiation of deceleration of particles in electromagnetic/gravitational external fields (*bremsstrahlung*), the gravitational/electromagnetic annihilation/ creation of elementary particles, the gravitational Compton effect, the diffusion of elementary particles through virtual gravitons, and the gravitational radiation of nuclei.

The third chapter is titled *Tomonaga–Schwinger representation of dynamics of a quantum physical system. Matrix elements of the field operators and Feynman–Dyson-type rules for the high-spin particles.* It comprises two subchapters, presenting the two problems mentioned in the title. The first subchapter introduces the *evolution operator*, by means of which one can describe the time evolution of quantum physical systems – in principle of any nature. In addition, there have also been presented the three fundamental possibilities of describing the time behavior of quantum systems, namely *Schrödinger picture* (or *representation*), *Heisenberg picture*, and *Tomonaga–Schwinger picture* (or *Dirac picture* or, still, *interaction picture*), the last one underlying the study of all interaction processes approached in the next chapters. The second subchapter briefly introduces the essential elements necessary in the study of concrete processes approached in the volume, namely: the *operator/matrix $S$*, the *transition probability* between various quantum states of a system (strongly connected to the differential cross-section of the interaction processes), as well as the *matrix elements* of the field operators and the *Feynman–Dyson type rules* for the particles (bosons and fermions) having the spin between 0 and 2. It is worth mentioning that, in order to determine the matrix elements and the corresponding Feynman–Dyson rules, the authors have elaborated and shown an algorithm composed by two stages, each stage with four steps, very useful in order to give a solution to this problem in the general case. The proposed algorithm has been applied to the massive field with spin 3/2, considered in interaction with the gravitational field, within the external field approximation.

In the fourth chapter, called *Elements of gauge theories. The minimal coupling principle*, are presented the basic elements of the gauge theory of classical fields. On the basis of requirement of the local invariance of the Lagrangian with respect to the transformations of a certain symmetry group, this theory introduces the notion of *minimal substitution*, which is the cornerstone of the so-called *principle of minimal coupling*. This principle (formulated within the theory of quantum gravity) is then utilized on a large scale in chapters V–VIII on purpose to determine the first-order (in the corresponding coupling constant) interaction Lagrangians; these Lagrangians are specific to various interaction processes studied in the above mentioned chapters. Nowadays, there exist an impressive number of excellent books regarding gauge theories and their mathematical foundations, our presentation following the way of exposure of these subjects (with the adequate adjustments and frequent detailing of intermediate calculation) given in Ref. [66] – concerning the fundamentals of gauge theories – as well as the mathematical basis of the theory of symmetry groups and their representations – as being exposed in Ref. [349].

Chapter V is titled *The gravitational field interacting with other fields* and presents the interaction between gravitational field and the matter, starting with the minimal coupling principle applied in quantum gravity. Here by *matter* one understands the scalar, spinorial (electron-positron and neutrinic), and vectorial (electromagnetic and vectorial mesonic) fields. In this chapter it is also approached the external field (gravitational and electromagnetic) problem, being exposed a hypothesis regarding a possible influence of the gravitational field on fundamental interactions (electromagnetic, strong and weak) between particles, as well as the manner in which the gravitational field is capable to interact – irrespective of supergravity and supersymmetry reasons – with the spin-vectorial (Rarita–Schwinger) field.

In the sixth chapter, called *Interaction of scalar, spinorial, vectorial, spin-vectorial and tensorial particles, and the gravitational field described by the Schwarzschild metric*, are generalized some results obtained by various authors in the study of the gravitational scattering process (on a background described by the Schwarzschild geometry) of vectorial massive particles, by considering – within Proca and Stueckelberg formalisms – the first-order corrections in the gravitational coupling constant, due to covariant derivative. This chapter also deals with the study of scattering process of massive spin 3/2 and 2 particles on a Newtonian potential. Here are determined, in a closed form, the corresponding differential scattering cross-sections and, using a cut-off type procedure based on the Leibnitz–Newton formula in order to eliminate the divergent quantities, the integral scattering sections of these processes are calculated. The authors also established a general procedure of validation of the results regarding the gravitational differential scattering cross-sections on a

Schwarzschild-type background of any type of particle, based on the study of behavior of these cross-sections in the domain of small and very small scattering angles, with an additional approach in the ultrarelativistic domain for massive particles.

Chapter VII is named *Scattering of electrons and photogeneration of gravitons in external electromagnetic fields.* Using the external field formalism, in this chapter is discussed – in a unitary and consequent manner – the electron scattering process in four types of static electromagnetic fields: electrostatic and magnetostatic, together with electric and magnetic dipole fields, in each case being determined the corresponding differential cross-section. Also, within the same formalism, certain particular cases found in the literature about the process of transformation of photons into gravitons (for the above mentioned types of static fields) are generalized. It is also studied the graviton photogeneration process in an electric dipole field – considered as an exterior field – being determined the corresponding differential cross-section. This last process was not studied in such a context in the literature, but we have shown that it is perfectly possible, appearing as a purely quantum effect.

The eighth chapter named *Interaction of scalar, spinorial, vectorial, spin-vectorial and tensorial particles, with an axially-symmetric gravitational field* is dedicated to the study of gravitational scattering processes on an axially symmetric gravitational background (described by the Kerr metric) of scalar, spinorial, vectorial, spin-vectorial and tensorial massive and non-massive particles, within the quantum theory of weak gravitation. The basic working procedure is offered by the minimal coupling principle applied in quantum gravity, in strong connection with Feynman diagrams and the $S$-matrix theory. For the massive scalar field has also been used a non-minimal coupling introduced through the medium of the scalar curvature $R$, the same treatment being utilized for the non-massive scalar and vector fields. Calculations have been performed in the first-order Born approximation, but also taking into account the covariant derivative contribution, in agreement with the minimal coupling theory of quantum gravity. Consequently, the obtained results stand for a higher level of generality about the scale/range of values taken by the scattering angle which is not forced anymore to take only small and/or very small values. As in the previous chapters, the expressions of the differential scattering cross-sections for massive and non-massive particles with spin 0, 1/2, 1, 3/2, and 2 have been determined in a closed form. Besides, the particular limit case of backscattering process has also been considered; this could be done because the obtained results keep their validity for any value of the scattering angle. In this approach are also determined the total scattering cross-sections for all five types of massive and non-massive particles (scalar, spinorial, vectorial, spin-vectorial and tensorial). In addition, in this paragraph the authors have studied the invariance of the obtained results showing, for example, that the transition amplitude for the non-massive vector field is invariant under a transformation of $U(1)$-gauge group, while within the same formalism the transition amplitude for the massless tensorial field (corresponding to the graviton) is not invariant under analogous gauge transformations.

Finally, in chapter IX titled *Software package for analytical calculation of differential cross-sections of gravitational scattering of high-spin particles* are presented five examples of programs of analytical calculation, conceived and utilized in order to determine the differential scattering cross-sections of massive and non-massive vectorial and spin-vectorial particles, as well as of massive tensorial particles. The utilized software was Mathematica, which is high specialized in analytical computations.

# Contents

## Chapter I

# Classical and Quantum Free Fields

The historical evolution of the natural sciences, in general, and of the physics, in particular, took place over a considerable time interval – more than two millennia – culminating in the nowadays understanding of the Universe, essentially based on the concept of the *physical field*. The bases of the modern field theory were established during the last century, with its newest version – the *quantum field theory*. This theory met an extraordinary rate of development, many of its aspects being intensively studied even at present time [52, 59, 85, 90, 122, 172, 190, 232, 254, 257, 297, 342, 347, 351, 394, 406, 411].

Even if the structure of matter – as it is actually known – proves to be basically simple, the phenomena produced by the four fundamental forces known so far acting on the matter are extremely diverse and complex, being studied by all possible scientific procedures. A quasi-exhaustive image on this subject can be fulfilled by surveying the specialized papers, recently published in this domain [1, 42, 45, 67, 139, 143, 155, 187, 220, 259, 177, 325, 348].

The most comprehensive and, one can say, the most evolved theory of the Universe (conceived as a micro- and macrocosmos association) elaborated so far by the human mind is the so-called *Standard Model*. Even if this is an effective theory, depending on certain parameters from outside, it stands as the masterpiece of the mankind intellectual effort in this respect known so far. From the start and up to the present, this theory has been continuously adjusted, verified, enriched and confirmed, these processes being always actual [17, 18, 186, 256, 292, 326]. Starting with the discovery of the vector bosons $W^{\pm}$ (carriers/mediators of the weak nuclear interaction) and culminating in identification of the Higgs boson – the last one being subject of several adjacent studies [65, 81, 205, 285, 290, 298, 343] – (both discoveries being rewarded with Nobel Prize in 1984 and, respectively, 2013), the Standard Model has proved its validity. The most recent discoveries, such those connected to dark matter and dark energy [62, 250] imposed exiting from the already too narrow framework of this model and going further to models like *BSM* (Beyond Standard Model), being mostly investigated at present.

**Notations and working conventions**

Except for certain specified cases, *in this chapter* all the four-vector components are considered real. Following customary notation, the four-vector $a = (a^0, \vec{a})$ with the zero-component $a^0$ and the spatial vector $\vec{a}$ shall be called *contravariant four-vector* and shall be denoted by upper indices,

$$a^{\nu} = (a^0, a^1, a^2, a^3).$$

The product of the four-vectors $a$ and $b$ is given by

$$ab = a^0 b^0 - \vec{a} \cdot \vec{b} = a^0 b^0 - a^1 b^1 - a^2 b^2 - a^3 b^3.$$

A more convenient way of writing is

$$ab = \sum_{\mu,\nu} g_{\mu\nu} a^{\mu} b^{\nu}, \quad (\mu, \nu = 0, 1, 2, 3),$$

where $g_{\mu\nu}$ is the diagonal *metric tensor*

$$g_{\mu\nu} = 0 \;\text{ if }\; \mu \neq \nu, \quad g_{00} = 1, \quad g_{11} = g_{22} = g_{33} = -1,$$

and differs only by a minus sign from Minkowski tensor. Passing from contravariant components $a^{\nu}$ to covariant components $a_{\mu}$ and vice-versa (index lowering/raising) is performed by means of the metric tensor,

$$a_{\mu} = g_{\mu\nu} a^{\nu}, \quad a^{\mu} = g^{\mu\nu} a_{\nu}, \quad g_{\mu\nu} = g^{\mu\nu},$$

that is

$$a_0 = a^0, \quad a_k = -a^k \; (k = 1, 2, 3), \quad a_{\nu} = (a^0, -\vec{a}).$$

Here and hereafter Einstein's convention over repeated indices shall be used (*i.e.* if an index is repeated in a product of vectors or tensors, summation is implied over the repeated index and the summation symbol is omitted). The summation over spatial components are denoted by Latin letters,

$$\vec{a} \cdot \vec{b} = a_n b_n = a^k b^k = a_1 b_1 + a_2 b_2 + a_3 b_3,$$

while the summation over all four components $(0, 1, 2, 3)$ is marked by Greek letters,

$$ab = a_\nu b^\nu = a^\mu b_\mu = \sum_{\mu,\nu} g^{\mu\nu} a_\mu b_\nu.$$

Sometimes, in order to simplify writing of a complex expression, both Greek indices shall be placed only in upper/lower position, *e.g.*,

$$A_\nu B_\nu = A^\mu B^\mu \equiv AB = A_\nu B^\nu,$$

$$F_{\mu\nu} F_{\mu\nu} \equiv F_{\mu\nu} F^{\mu\nu} = \sum_{\mu,\nu} F_{\mu\nu} F^{\mu\nu}.$$

Thus, the presence of two identical Greek indices in various factors shall always signify a covariant summation, regardless of the place of indices. The quantities corresponding to certain groups on internal symmetry (such as isotopic indices) shall be denoted by Latin letters $(a, b, c, ...)$.

The contraction of a Dirac matrix with a four vector $a^\nu$ is conventionally abbreviated as

$$\hat{a} = a_\nu \gamma^\nu = a^\nu \gamma_\nu.$$

In order to give a simplified writing for the partial derivatives, the following notations shall be used:

$$\frac{\partial u}{\partial x_\nu} = \partial^\nu u = u^{,\nu}, \quad \frac{\partial \varphi_a}{\partial x^\nu} = \partial_\nu \varphi_a = \varphi_{a,\nu}.$$

Obviously,

$$\varphi_a^{,\mu} = g^{\mu\nu} \varphi_{a,\nu}.$$

With these notations, the D'Alembert operator

$$\Box = \nabla^2 - \partial_0^2 = \Delta - \partial_0^2$$

writes

$$\square = -\partial_\nu \partial^\nu.$$

In this chapter, the natural unit system ($c = 1$, $\hbar = 1$) shall be used. In this system, the dimension of energy and momentum is mass or inverse length, while the dimension of time is length or inverse mass (see Appendix H),

$$[E] = [p] = \mathrm{m} = \mathrm{l}^{-1}; \quad [x_0] = [\vec{x}] = \mathrm{l} = \mathrm{m}^{-1}.$$

The Fourier transforms of the functions of four-dimensional variables are given by

$$f(x) = \frac{1}{4\pi^2} \int e^{-ipx} \tilde{f}(p) dp; \quad \tilde{f}(p) = \frac{1}{4\pi^2} \int e^{ipx} f(x) dx.$$

The sign of the exponential index is chosen in such a way so as to be in concordance with the quantum-mechanical formula

$$f(x^0, \vec{x}) = f(t, \vec{x}) = \frac{1}{\sqrt{2\pi}} \int e^{-iEt} \tilde{f}(E, \vec{x}) dE.$$

The three-dimensional Fourier transform is of the form

$$\varphi(\vec{x}) = \frac{1}{(2\pi)^{3/2}} \int e^{i\vec{p}\cdot\vec{x}} \tilde{\varphi}(\vec{p})\, d\vec{p}; \quad \tilde{\varphi}(\vec{p}) = \frac{1}{(2\pi)^{3/2}} \int e^{-i\vec{p}\cdot\vec{x}} \varphi(\vec{x})\, d\vec{x}.$$

## I.1. Particles and fields

### I.1.1. Particles. Fundamental properties

The quantum field theory deals with the physical study of elementary particles and their interactions. This theory is based on the connection between relativistic particles and quantum fields emerged as a result of classical fields quantization. The quantum fields properties and those of the corresponding particles are in strong connection with each other. In view of this fact, one shall first enumerate the main properties of the particles.

An important attribute of relativistic particles is their *rest mass* $m$. According to the relativistic theory, there exists the following relationship between the particle energy $E$ and its momentum $p$:

$$E^2 - c^2 p^2 = m^2 c^4.$$

In natural unit system, this relationship writes

$$E^2 - p^2 = m^2.$$

In this unit system the mass can be evaluated in energetic quantities. For example, the electron mass is

$$m_e = 9.109534(47) \times 10^{-28}\mathrm{g} = 0.5110034(14)\,\mathrm{MeV},$$

and the proton mass writes

$$m_p = 938.2796(27)\,\mathrm{MeV} \cong 0.94\,\mathrm{GeV}.$$

The second important characteristic of the particles is their *spin*. According to the general theorems of quantum mechanics, the particle spin is quantized, its absolute value being an integer – or half-integer – multiple of

$$\hbar = \frac{h}{2\pi} = 1.0545887(57) \times 10^{-27}\mathrm{erg}\cdot\mathrm{s} = 6.582173(17) \times 10^{-22}\mathrm{MeV}\cdot\mathrm{s},$$

where $h$ is the *Planck constant*. For instance, in natural unit system, the spin values of the electron, proton and photon are, respectively,

$$s_{e^-} = s_p = \frac{1}{2}; \quad s_\gamma = 1.$$

Another important property of the particles, the third, is the *electric charge*, which is also quantifiable. Unlike the quantum nature of the spin, which is known, the discrete nature of the electric charge is not yet entirely comprehended.

The electric charge of the particles are integer multiples of the elementary (electron) charge[1]

$$e = 4.803242(14) \times 10^{-10}\mathrm{u.CGS} = 1.6021892(46) \times 10^{-19}\mathrm{C}.$$

Finally, a significant attribute of the particles is their lifetime $\tau$. Only a few particles at the free state "live" (exist) for a very long time, namely: the electron $e$, the proton $p$, the photon $\gamma$, and the neutrino[2] $\nu$, as well as their antiparticles: the positron $e^+$, the antiproton $\overline{p}$ and the antineutrino $\overline{\nu}$. The remaining particles are unstable and spontaneously disintegrate in some other particles according to an

[1] Except for the quarks, whose electric charge is a fraction of $e$ (*e.g.* $\pm(1/3)e$, or $\pm(2/3)e$).

[2] There are three neutrino families at the present: electron-neutrino $\nu_e$, muon-neutrino $\nu_\mu$, and tau-neutrino $\nu_\tau$.

exponential law of the form $e^{-t/\tau}$, where $\tau$ is the average lifetime of the corresponding particles. For example, the neutron decay (known as *beta decay*) is performed according to the scheme

$$n \to p + e^- + \overline{\nu}_e.$$

The neutron average lifetime is $\tau_n = 918(14)$ s $\cong 15.3\,\text{min}$. As another example, the charged meson $\pi^+$ decays into a muon and a muon-neutrino,

$$\pi^+ \to \mu^+ + \nu_\mu,$$

with the average lifetime $\tau_{\pi^+} = 2.6030(23) \times 10^{-8}$s.

The instability represents an important property of microcosmos – the property of reciprocal transformation of particles – which is a direct consequence of permanent interactions between them. The fundamental interactions between particles can be grouped into four large categories:

1. Strong nuclear interactions;
2. Electromagnetic interactions;
3. Weak nuclear interactions;
4. Gravitational interactions.

### I.1.2. Conservation laws

In general, the conservation laws are consequences of some symmetries which, in their turn, display the properties of unobservability of some characteristics of physical objects.

One knows, for example, that the energy conservation law in a conservative system stands for a consequence of the operation of time-translation. The covariance with respect to time translations, in its turn, is equivalent to unobservability of the *absolute time*. A similar example is offered by the parity conservation law, expressing the covariance with respect to mirror reflection and conventional notions of *right* and *left* (in other words, there is neither *absolute right*, nor *absolute left*).

The laws of conservation of electric charge and baryonic number also have a universal character. These laws correspond to the symmetry with respect to continuous phase transformations, but they do not have an obvious physical basis and are not connected with the space-time structure. Such symmetries are called *internal symmetries*. In this category fall the *isotopic symmetry*, and the *unitary symmetries* as well.

The isotopic invariance (the law of conservation of isotopic spin) characterizes only strong interactions and cannot be attached to electromagnetic and/or weak interactions.

The parity conservation law is connected to the symmetry of the wave function with respect to reflection of spatial axes. This law is valid in the case of strong nuclear and electromagnetic interactions, but it is not respected by weak nuclear interactions. The laws of conservation of "strangeness" and "charm" behave in the same way and, together with the isotopic invariance, stay at the basis of unitary symmetries. These properties are schematically presented in the Table 1 below.

**Table 1.** Symmetry properties (of interactions) and conservation features (of physical quantities).

| ***Physical quantity*** | ***Type of interaction (fundamental force)*** | | |
|---|---|---|---|
| | ***Strong*** | ***Electromagnetic*** | ***Weak*** |
| *Electric charge* | + | + | + |
| *Barionic number* | + | + | + |
| *Parity* | + | + | + |
| *Isospin* | + | − | − |
| *Strangeness ("eta-charge")* | + | + | − |

**Legend**: (+) → the quantity conserves; (−) → the quantity does not conserve.

### I.1.3. Particle – field correspondence

Any physical field represents a system with an infinite number of degrees of freedom. The field concept appears when wanted to give up the instantaneous reciprocal actions between particles (situated at an arbitrary mutual distance). Supposing that the space between particles is "filled-up" with field, it is assumed that the interactions between particles propagate with a finite velocity through the field.

This way, the introduction of notion of classical field in physics is motivated by the relativistic covariance necessity. In the microparticle theory, the leading role belongs to the relativistic quantum fields.

The fundamental concept of quantum physical field plays an essential role in formulation of particle dynamics and particle interactions theories. By means of a single physical object – the quantized field – one can describe various states of almost any particle systems. This concept is obtained by quantization of the classical field, while the field function becomes an operator, usually expressed through creation and annihilation particle operators. This way becomes possible description of the most important property of elementary particles, namely their reciprocal transformation.

A well-known example of classical field is the electromagnetic field. The classical investigation of this field, based on Maxwell's equations, leads to a purely wave image of electromagnetism.

Description of the corpuscular properties of light can be performed by means of the quantization procedure, through which the field is endowed with discrete quanta of energy, corresponding to various possible energy states of the field oscillators. The electromagnetic field quanta – the photons – have a zero rest mass, are charge-free and their spin is 1. The last characteristic corresponds to polarization property of classical electromagnetic field, being reflected by the fact that this field has more than one component, so that it can be described by a "totality" of field functions: the components of electric and magnetic field intensities or, equivalently, the components of the potential four-vector $A_\nu$.

The properties of the field functions express the discrete characteristics (such as spin, charge, etc.) of the corresponding particles. As soon as a field is quantized, its quanta are identified with the particles themselves.

### I.1.4. The Lorentz group

In this paragraph, the transformation laws of the field functions with respect to the relativistic transformation of coordinates shall be described, starting with a survey of some general properties of these transformations.

The *homogeneous Lorentz group*[3] is composed by the multitude of all linear transformations of the position four-vector $x^\nu$ which leaves covariant the four-dimensional form

$$x^2 = x_\nu x^\nu = (x^0)^2 - \vec{x}^2$$

and does not change the time orientation. This group contains the common rotations, the Lorentz boosts (*i.e.* the "usual" Lorentz transformations, which are in fact rotations in the planes $x^0x^1, x^0x^2, x^0x^3$), and reflection of spatial axes. If the reflection of the time axis is added, then the group is called *Lorentz general.*

In physics, an important role is played by the group formed by the transformations of the Lorentz homogeneous group, together with transformations representing translations along all four axes. This group is called the *inhomogeneous Lorentz group* or *Poincaré group.* The covariance with respect to this group shall be called *relativistic covariance.*

The coordinate transformations of the Poincaré group are written in a compact form as follows,

$$x \to x' = P(\omega; x), \tag{1.1}$$

where $\omega = (L, a)$ denotes the totality of parameters, both those representing translations $(a)$ and those designating rotations $(L)$,

$$x'_\nu = L_{\nu\mu}x^\mu + a_\nu; \quad L_{\nu\sigma}g^{\sigma\rho}L_{\mu\rho} = g_{\nu\mu}. \tag{1.2}$$

The field function $u(x)$ stands for one or several functions of $x^\nu$ (depending on one-component or several-component functions, respectively), given in each initial system. Transition from the initial system $x$ to the system $x'$ according to transformation (1.2) corresponds to the homogeneous linear transformation of the field functions components

$$u(x) \to u'(x') = \Lambda(\omega)u(x), \tag{1.3}$$

where $\Lambda$ is the transformation matrix.

[3]The Lorentz and Poincaré groups shall be more broadly considered in chapter IV.

One should specify that transformation (1.3) expresses more than a change of arguments ($x \to x'$); it describes transformation of the field function with respect to a given transformation of the coordinate system, and not the operation of spatial translation from one point to another.

To each Lorentz transformation $L$ corresponds a matrix $\Lambda_L$, to the unit element of the group $L$ corresponds the unit matrix $\Lambda = 1$, and the product of two elements $L_1$ and $L_2$ of the Lorentz group corresponds to the product

$$\Lambda_{L_1 L_2} = \Lambda_{L_1}\Lambda_{L_2}.$$

Within the group theory, a system of matrices with such properties form a linear representation of the group. The finite-rank matrices $\Lambda$ give a finite representation of the Lorentz group. The representation rank is given by the number of components of the field function $u(x)$.

The types of the field functions and the corresponding transformation laws can be determined by analysis of the finite representations of the Lorentz group. (Such an investigation stands for a distinct chapter in the theory of group representations, and shall not be analyzed in this approach). The Lorentz group representations can be monovalent or bivalent; in other words, the correspondence $L \to \Lambda_L$ can be monovalent or bivalent.

The operators $\Lambda_L$ must be continuous functions of the parameters $L$ of the Lorentz transformation; in other terms, to an infinitesimal transformation of the coordinate system $x$ must correspond a similar (infinitesimal) transformation of the field function $u(x)$.

The totality of these requirements lead to division of representations of the Lorentz group into two categories. The first category is characterized by the unitary correspondence $L \to \Lambda_L$ and contains vectorial and pseudo-vectorial representations. The field functions that change according to the tensorial representations are called *tensors* and, sometimes, they can be directly observed (*e.g.* the electromagnetic field). In the second case, this correspondence proves to be bivalent: $L \to \pm\Lambda_L$.

The law of transformation of the $N$-rank tensor $T^{\nu_1,\nu_2,...,\nu_N}$ under a continuous coordinate transformation writes

$$T'_{\nu_1,\nu_2,...,\nu_N}(x') = \frac{\partial x'_{\nu_1}}{\partial x_{\mu_1}} ... \frac{\partial x'_{\nu_N}}{\partial x_{\mu_N}} T_{\mu_1,...,\mu_N}(x), \tag{1.4}$$

or, if formula (1.2) is used,

$$T'^{\nu_1,\nu_2,...,\nu_N}(x') = L^{\nu_1}_{\mu_1} ... L^{\nu_N}_{\mu_N} T^{\mu_1,...,\mu_N}(x).$$

The bivalent representations are called *spinorial* and the corresponding quantities are called *spinors*. The transformation law of the spinorial quantities is complicated and several most simple cases are given in Appendix B. One deserves to be mentioned here, however, that according to formula (1.4) the following transformation for tensor quantities,

$$u(x) \to u'(x') = u(x), \tag{1.5}$$

under a coordinate translation

$$x^\nu \to x'^\nu = x^\nu + a^\nu,$$

is also valid for the spinors.

Here are two examples of tensor representations and their corresponding quantities:

• *The zero-rank tensor.* If under a continuous coordinate transformation (1.2) a quantity changes according to (1.5), it is an invariant called *scalar* (*pseudoscalar*);

• *The first-rank tensor.* If under a coordinate rotation a quantity changes according to the law

$$u'^\nu(x') = L^\nu_\mu u^\mu(x) = L^{\nu\mu} u_\mu(x), \tag{1.6}$$

then it is called *contravariant vector.* The associated covariant vector

$$u_\nu(x) = g_{\nu\mu} u^\mu(x)$$

changes according to the law

$$u'_\nu(x') = L^\mu_\nu u_\mu(x). \tag{1.7}$$

In a similar way can be written the transformation formulas for tensors of higher ranks and different variances.

One can also obtain the *inverse spatial transformation* $P$, that is the reflection of the three spatial axes,

$$x \to x' = Px; \quad x'_0 = x_0; \quad \vec{x}\,' = -\vec{x}. \tag{1.8}$$

In this case, the transformation laws of the field functions cannot be determined by means of the formula (1.4), so that they must be separately formulated. Since $P^2 = 1$, these laws associated with the tensor components $T^{\cdots}(x)$ can have two forms:

$$PT(x) = T'(x') = T'(Px) = \pm T(x).$$

The zeroth-rank tensor of the unchangeable sign under inversion

$$Pu(x) = +u(x) \tag{1.9}$$

is called a *scalar*, while in the second case the quantity satisfying the relation

$$Pu(x) = -u(x) \tag{1.10}$$

is a *pseudoscalar*.

A first-rank tensor whose spatial components change their sign under transformation $P$, but the sign of the temporal component remains unchanged,

$$PV^\mu(x) = V_\mu(x),$$

that is

$$PV^0(x) = V^0(x); \quad P\vec{V}(x) = -\vec{V}(x), \tag{1.11}$$

is called a *vector*.

If the change of sign affects only the temporal component (the sign of the spatial components remain unchanged),

$$PV^\mu(x) = -V_\mu(x),$$

that is

$$PV^0(x) = -V^0(x); \quad P\vec{V}(x) = \vec{V}(x), \tag{1.12}$$

then the first-rank tensor is called *pseudovector* (*axial vector*).

The parity property of the field functions plays an important role in determination of possible interaction forms of various fields.

## I.2. Free classical fields

### 1.2.1. Dynamical invariants

#### I.2.1.1. The Lagrangian

In order to determine the field motion equations, as well as the quantities corresponding to the invariance properties with respect to the transformations of the continuous symmetry groups, below we shall use the Lagrangian formalism.

In general, the Lagrange function (the Lagrangian) is a function of space-time variables $x^0, \vec{x}$, the dynamical variables of the studied system $u_a(x)$, and their derivatives $u_{a,\nu}(x)$ as well. In mechanics, the Lagrangian satisfies the superposition principle (it is written as a sum of functions for all material points of the system). In the case of a continuous system (such as a wave field) the sum is replaced by the spatial integral of the Lagrangian density $\mathcal{L}$,

$$L = \int \mathcal{L}\; d\vec{x}.$$

As a starting point in the development of the Lagrangian formalism deserves the *action* $\mathcal{A}$ of the studied system (field), defined as the time integral of the Lagrangian $L$,

$$\mathcal{A} = \int L\; dt = \int \mathcal{L}\; dx^0 d\vec{x} = \int \mathcal{L}\; dx. \tag{1.13}$$

It then follows that in the field theory the Lagrangian $L$ plays an intermediary role, the essential role being played by the Lagrangian density $\mathcal{L}$.

The equations of motion can be obtained by means of the principle of stationary action (Hamilton's principle), which states that the real motion between any two points in a conservative dynamical system is such that the action has an extreme value with respect to all paths between the points that correspond to the same energy. In other words, the real motion corresponds to a stationary action. Using the condition

$$\delta\mathcal{A} = 0,$$

supplemented by the supposition that variation of the field function $\delta u_a$ cancels on the spatial boundary of the integration domain, the

Euler–Lagrange equations are obtained,

$$\frac{\delta \mathcal{A}}{\delta u_a(x)} = \frac{\partial \mathcal{L}}{\partial u_a(x)} - \frac{\partial}{\partial x^\nu}\left(\frac{\partial \mathcal{L}}{\partial u_{a,\nu}}\right) = 0, \quad a = 1, 2, ..., n; \ \nu = 0, 1, 2, 3. \tag{1.14}$$

Here $\frac{\delta \mathcal{A}}{\delta u_a(x)}$ denotes the *functional derivative* of the action with respect to the field functions.

In quantum field theory, the Lagrangian density $\mathcal{L}$ has to obey the following main requirements:

— It is a function only of the dynamical variables, *i.e.* the components of the field functions $u_a(x)$ and their derivatives;

— It does not explicitly depend on the space-time coordinates, since such a dependence would violate the relativistic covariance. In addition, since the field is considered free (without sources), the Lagrangian density cannot explicitly depend on $x$;

— The local character of the theory is assured by the fact that the value of $\mathcal{L}$ at the point $x$ is determined by the values of the field function $u_a(x)$ and a finite number of partial derivatives at the same point $x$. Such a Lagrangian density is called *localized.*

— In order to obtain second-order differential equations (all known physical equations are of the second order at the most), the Lagrangian density must depend on the field function $u_a(x)$ and its first-order (at most) derivatives,

$$\mathcal{L}(x) = \Phi\Big(u_a(x), u_{a,\nu}(x)\Big). \tag{1.15}$$

It is worth mentioning that, since the physical properties of the system are determined by the action $\mathcal{A}$, containing the Lagrangian under the integral sign, the correlation $\mathcal{A} \to \mathcal{L}$ is not strictly univocal. Two Lagrangian densities differing only by a four-divergence of an arbitrary four-vector,

$$\mathcal{L}' = \mathcal{L} + \partial_\nu F^\nu(x) \tag{1.16}$$

are, from the physical point of view, *equivalent.* Indeed, according to definition (1.13) and applying the Green–Gauss theorem, the integral of $F^\nu_{,\nu}$ reduces to the spatial integral of $F^\nu$ over the three-dimensional boundary of the four-dimensional integration volume. Since the variations $\delta u_a$ of the field function cancel on this boundary, it follows that the term $F^{\nu}_{,\nu}$ has no influence on the physical quantities.

The most important properties required for a Lagrangian density are its real character (Hermiticity, in quantum terms) and its relativistic covariance. The property of being real leads to a similar character for the dynamical invariants (the observables) of the system such as: energy, momentum, current, etc., as well as to the $S$-matrix unitarity. The relativistic covariance of the Lagrangian density,

$$\mathcal{L}'(x') = \Phi\left(u'_a(x'), u'_{a,\nu}(x')\right) = \Phi\left(u_a(x), u_{a,\nu}(x)\right) = \mathcal{L}(x), \quad (1.17)$$

signifies that $\mathcal{L}$ behaves like a (pseudo)scalar under the Poincaré group transformations. Since the infinitesimal volume element in the four-dimensional integral (1.13), $dx = dx^0\, d\vec{x}$, is also a relativistic invariant, it follows that the magnitude of the action $\mathcal{A}$ does not change under the Poincaré group transformations. The fact that $\mathcal{L}$ is a scalar, assures the relativistic covariance of the action $\mathcal{A}$.

### I.2.1.2. The field energy-momentum four-vector

The time-independent physical quantities, such as: energy, momentum, angular momentum, electric charge, etc., are called *dynamical invariants*. For example, the energy-momentum four-vector of the field, $P^\nu$, can be represented under the form of the spatial integral

$$P^\nu = \int d\vec{x}\; T^{\nu 0}(x^0, \vec{x}), \quad (1.18)$$

where $T^{\nu\mu}$ is the field energy-momentum four-tensor defined as

$$T^{\nu\mu} = \frac{\partial \mathcal{L}}{\partial u_{a,\mu}} u_a^{,\nu} - g^{\nu\mu}\mathcal{L}(x). \quad (1.19)$$

The time-independence of the integral (1.18) is a consequence of the fact that the tensor $T^{\nu\mu}$ satisfies the equation of continuity

$$\frac{\partial T^{\nu\mu}}{\partial x^\mu} = 0. \quad (1.20)$$

To prove this statement, one considers the integral

$$\int dx \, \frac{\partial T^{\nu\mu}}{\partial x^\mu}$$

and, if the Green–Gauss theorem is applied,

$$\int dx \, \frac{\partial T^{\nu\mu}}{\partial x^\mu} = \int_\sigma d\sigma_\mu T^{\nu\mu}(x), \tag{1.21}$$

where $\sigma$ is the surface bounding the integration domain, and $d\sigma_\mu$ an infinitesimal surface element orthogonal on $x^\mu$-axis. Taking into account (1.20), one integrates (1.21) between two three-dimensional spatial varieties $\sigma_1(x^0 = t_1)$ and $\sigma_2(x^0 = t_2)$. Since on the boundary situated at infinity both the field function $u_a$ and its derivatives equal zero (which also involves cancellation of the energy-momentum tensor at the same points), one obtains

$$\int_{\sigma_1} d\vec{x} \; T^{\nu 0}(x) - \int_{\sigma_2} d\vec{x} \; T^{\nu 0}(x)$$

$$= \int d\vec{x} \; T^{\nu 0}(t_1, \vec{x}) - \int d\vec{x} \; T^{\nu 0}(t_2, \vec{x}) = 0,$$

which proves the statement.

It can also be shown that the equation (1.20) is a consequence of the equation of motion. To this end, one calculate the divergence of the r.h.s. of equation (1.19), that is

$$\partial_\mu T^{\nu\mu} = \frac{\partial}{\partial x^\mu}\left(\frac{\partial \mathcal{L}}{\partial u_{a,\mu}}\right) u_a^{;\nu} + \frac{\partial \mathcal{L}}{\partial u_{a,\mu}} \partial_\mu \left(u_a^{;\nu}\right) - \partial^\nu \mathcal{L}.$$

Changing the derivation order in the second derivative and using (1.14), one finally gets

$$\partial_\mu T^{\nu\mu} = \frac{\partial \mathcal{L}}{\partial u_a}\frac{\partial u_a}{\partial x_\nu} + \frac{\partial \mathcal{L}}{\partial u_{a,\mu}}\frac{\partial u_{a,\mu}}{\partial x_\nu} - \frac{\partial \mathcal{L}}{\partial x_\nu} = 0.$$

## Conclusions

Each component of the four-vector $P^\nu$ ($\nu = 0, 1, 2, 3$) remains constant over time because the spatial density $T^{\nu 0}$ is the null-component of the "four-vector" $\theta^\mu(\nu) = T^{\nu\mu}$ and satisfies the equation of continuity (1.20). The last property comes from the Euler–Lagrange equations, and also from the fact that the Lagrangian density $\mathcal{L}$ depends on coordinate $x$ only through the medium of the field function $u_a$ and its first-order derivatives $u_{a,\nu}$. As a result,

$$\frac{\partial \mathcal{L}}{\partial x_\nu} = \frac{\partial \mathcal{L}}{\partial u_a} u_a^{;\nu} + \frac{\partial \mathcal{L}}{\partial u_{a,\mu}} \frac{\partial}{\partial x_\nu} \left( \frac{\partial u_a}{\partial x^\mu} \right).$$

Conservation of the projection of the four-vector $P^\nu$ on the $x^\nu$-axis is a consequence of invariance of the Lagrangian density with respect to translations along the $x^\nu$-axis.

## Noether's theorem

Consider an arbitrary continuous transformation of the coordinates and/or field functions, which depends on $s$ parameters $\omega_k$ ($k = 1, 2, ..., s$),

$$x_\nu \to x'_\nu = f_\nu(x;\omega), \tag{1.22}$$

$$u_a(x) \to u'_a(x') = U_a\Big(u_b(x);\omega\Big), \tag{1.23}$$

and for which the variation of the action $\mathcal{A} = \int \mathcal{L}(x)dx$ cancels: $\delta\mathcal{A} = 0$. Under these conditions, there exist $s$ *dynamical invariants*, $C_k$, defined as the spatial integrals

$$C_k = \int d\vec{x}\ \theta^0_{(k)}(x), \tag{1.24}$$

where $\theta^0_{(k)}(x)$ are the temporal components of the four-vectors

$$\theta^\nu_{(k)} = \frac{\partial \mathcal{L}}{\partial u_{a,\nu}} \left( u_{a,\mu} X^\mu_k - \Psi_{a,k} \right) - X^\nu_k \mathcal{L}(x), \tag{1.25}$$

where

$$X^\mu_k = \left.\frac{\partial f^\mu(x,\omega)}{\partial \omega_k}\right|_{\omega=0}; \qquad \Psi_{a,k} = \left.\frac{\partial u_a}{\partial \omega_k}\right|_{\omega=0}. \tag{1.26}$$

This theorem shall not be proved here. As one can easily observe, formulas (1.18) and (1.19) are particular cases of the expressions (1.24) and (1.25) for

$$X^\nu_l = \delta^\nu_l; \quad \Psi_{a,k} = 0. \tag{1.27}$$

The quantities (1.27) correspond to the transformations

$$x'_\nu = x_\nu + \omega_\nu,$$

and

$$u'_a(x') = u_a(x),$$

*i.e.* to translations along the four coordinate axes. The last relation stands for the transformation law of the field functions in the case of a coordinate translation, being identical for all field functions, irrespective of their tensor rank. Introducing (1.27) into (1.25), it follows the equation (1.19) for the field energy-momentum tensor. Using (1.24), one can obtain formula (1.18) for the energy-momentum four-vector $P^\nu = const.$ (constant with respect to time).

### I.2.1.3. The angular momentum and the spin

The purpose of this section is to determine the dynamical invariants corresponding to the rotational Lorentz transformations (*i.e.* rotations in the six coordinate planes $x_1Ox_2$, $x_1Ox_3$, $x_2Ox_3$, $x_1Ox_4$, $x_2Ox_4$, $x_3Ox_4$). According to (1.26), it is sufficient to consider only the infinitesimal form of this transformation, namely

$$x'_\nu = x_\nu + x^\mu \delta L_{\nu\mu}, \tag{1.28}$$

where $\delta L_{\nu\mu}$ are the infinitesimal parameters of the transformations (rotations).

Since the quantities $\delta L_{\nu\mu}$ are antisymmetric, as independent parameters can be chosen the following six of them,

$$\delta\omega_{(\mu\nu)} = \delta L_{\mu\nu}, \quad \mu < \nu, \tag{1.29}$$

representing infinitely small rotation angles in the planes $x_\mu Ox_\nu$. Under these circumstances, the index $k$ appearing in formulas (1.24)–(1.27) is going to be a double-index $(\mu\nu)$. If variation $\delta x$ is represented in the form

$$\delta x_\nu = X^k_\nu \delta\omega_k = \sum_{\rho<\sigma} X^{(\rho\sigma)}_\nu \delta\omega_{(\rho\sigma)}, \tag{1.30}$$

by means of (1.28) and (1.29) one obtains

$$\delta x_\nu = x^\mu \delta L_{\nu\mu} = \sum_{\rho<\sigma} (x^\sigma \delta^\rho_\nu - x^\rho \delta^\sigma_\nu) \delta\omega_{(\rho\sigma)},$$

that is

$$X_\nu^{(\rho\sigma)} = x^\sigma \delta_\nu^\rho - x^\rho \delta_\nu^\sigma. \tag{1.31}$$

The infinitesimal variation of the field function, $u'_a(x') = u_a(x) + \delta u_a(x)$, can be written as

$$\delta u_a(x) = \sum_{b,\rho<\sigma} A_a^{b(\rho\sigma)} u_b(x) \delta\omega_{(\rho\sigma)}. \tag{1.32}$$

According to conditions (1.5) and (1.6), the scalar fields are characterized by

$$A_{\cdots}^{\cdots} = 0,$$

and the vector fields by

$$A_\mu^{\nu(\rho\sigma)} = \delta_\mu^\rho g^{\nu\sigma} - \delta_\mu^\sigma g^{\nu\rho}, \ \rho < \sigma.$$

Therefore, in the case of a vector field one can write

$$\Psi_\nu^{(\rho\sigma)} = A_\nu^{\mu(\rho\sigma)} u_\mu(x) = u^\sigma(x)\delta_\nu^\rho - u^\rho(x)\delta_\nu^\sigma. \tag{1.33}$$

Introducing (1.33) and (1.31) into (1.25), the angular momentum tensor is obtained,

$$M^{\tau(\rho\sigma)} = \frac{\partial \mathcal{L}}{\partial u_{\nu,\tau}} \left( u_\nu^{'\rho} x^\sigma - u_\nu^{'\sigma} x^\rho \right)$$

$$+ \mathcal{L}\left(x^\rho g^{\sigma\tau} - x^\sigma g^{\rho\tau}\right) - \frac{\partial \mathcal{L}}{\partial u_{\nu,\tau}} A_\nu^{\mu(\rho\sigma)} u_\mu(x). \tag{1.34}$$

If the first two terms of the r.h.s. are compared with (1.19), they can be written as

$$x^\sigma T^{\rho\tau} - x^\rho T^{\sigma\tau} = M_0^{\tau(\rho\sigma)}. \tag{1.35}$$

This relation expresses connection between the angular momentum tensor and the energy-momentum tensor in the material point dynamics. Due to this fact, the quantity (1.35) has to be identified with the orbital angular momentum of the field. In the case of a one-component field (scalar or pseudoscalar), the total angular momentum reduces to the orbital angular momentum. If the field has more than one component, the last term of the r.h.s. of (1.34) is different from zero. This term characterizes the polarization property of the field and corresponds to the *intrinsic angular momentum* (the *spin*) of the field. Therefore,

$$M^{\tau(\rho\sigma)} = M_0^{\tau(\rho\sigma)} + S^{\tau(\rho\sigma)},$$

where

$$S^{\tau(\rho\sigma)} = \frac{\partial \mathcal{L}}{\partial u_{a,\tau}} A_a^{b(\rho\sigma)} u_b(x). \tag{1.36}$$

For example, in the case of a vector field one finds

$$S^{(\rho\sigma)} = \int d\vec{x}\ S^{0(\rho\sigma)} = \int d\vec{x} \left[ u^\rho(x)\frac{\partial \mathcal{L}}{\partial \dot{u}_\sigma} - u^\sigma(x)\frac{\partial \mathcal{L}}{\partial \dot{u}_\rho} \right], \tag{1.37}$$

where $\dot{u}_a$ denotes the partial derivative of the field function $u_a$ with respect to time. The quantities $S^{(23)}$, $S^{(31)}$, and $S^{(12)}$ are the components of the spatial pseudovector

$$S^n = \varepsilon^{nml} S^{(ml)}, \quad (n, m, l = 1, 2, 3), \tag{1.38}$$

which is the *spin pseudovector.*

### I.2.1.4. The current vector and the charge

In the case of complex fields, corresponding to charged fields, the Lagrangian density is covariant with respect to the phase transformations of the field functions.

For a better understanding, consider a physical system described by a complex field. Since the Lagrangian density has to be a real quantity, it can depend on a complex field only through the medium of the quadratic forms $u_a^* u_a$ and/or $u_{a,\nu}^* u_a^{;\nu}$, where $u_a$ and $u_a^*$ are reciprocally complex conjugate field functions. It then follows that the field functions $u_a$ can be multiplied by a unitary phase factor of the type $\exp(i\alpha), \alpha = const.$, without changing the quadratic form $u_a^* u_a$ and/or $u_{a,\nu}^* u_a^{;\nu}$ or leading to some other effects.

Supposing that $u_a$ and $u_a^*$ are linearly independent functions, the first-order gradient transformation can be written as

$$u_a \to u_a' = e^{i\alpha} u_a, \quad u_b^* \to u_b^{*\prime} = e^{-i\alpha} u_b^*. \tag{1.39}$$

Assuming that $\alpha$ is a real infinitely small quantity, the first two terms of the series expansion give

$$u_a \to u_a' = u_a + i\alpha u_a; \quad u_b^* \to u_b^{*\prime} = u_b^* - i\alpha u_b^*.$$

According to (1.26), it follows that

$$\Psi_a = iu_a, \quad \text{for all } u_a,$$

and

$$\Psi_b = -iu_b^*, \quad \text{for all } u_b^*,$$

together with the fact that $X = 0$. Introducing these relations into (1.25), one finds a quantity whose dimensions are those of a four-dimensional vector,

$$J^\nu(x) = i\left(\frac{\partial\mathcal{L}}{\partial u_{a,\nu}^*}u_a^* - \frac{\partial\mathcal{L}}{\partial u_{b,\nu}}u_b\right). \tag{1.40}$$

Since this four-vector satisfies the equation of continuity

$$\partial_\nu J^\nu = 0, \tag{1.41}$$

it can be identified with the *current four-vector*. The spatial integral of the temporal component of this four-vector is

$$Q = \int J^0(x)\,d\vec{x} = i\int d\vec{x}\left(\frac{\partial\mathcal{L}}{\partial u_{a,0}^*}u_a^* - \frac{\partial\mathcal{L}}{\partial u_{b,0}}u_b\right). \tag{1.42}$$

This time-independent quantity is a constant characteristic of the field and can be identified with *charge*. Through the medium of this first-integral of the motion, the quantity $Q$ can signify both electric charge and some other characteristics specific to the internal field symmetries, such as baryonic charge, strangeness, charm, etc.

The systems characterized by such integrals of motion are described by means of complex fields. It is worth mentioning that the above analyzed transformations (with $\psi \neq 0,\ X = 0$) are called *internal symmetry transformations*. In physics, the most significant examples in this respect are the isotopic and unitary symmetries.

### I.2.2. The simplest fields

#### I.2.2.1. The scalar field

One of the simplest fields described by the field function $\varphi(x)$ and characterizing the spinless particles is the *real scalar field*. The Lagrangian density of the free real scalar field is chosen as

$$\mathcal{L}\Big(\varphi(x), \varphi_{,\nu}(x)\Big) = \frac{1}{2}\varphi_{,\nu}(x)\varphi^{,\nu}(x) - \frac{m^2}{2}\varphi^2(x). \tag{1.43}$$

This choice is convenient because the resulting equation of motion (1.14) is of a Klein–Gordon type,

$$\frac{\partial \mathcal{L}}{\partial \varphi} - \frac{\partial}{\partial x^\nu}\left(\frac{\partial \mathcal{L}}{\partial \varphi_{,\nu}}\right) = -m^2\varphi - \frac{\partial^2 \varphi}{\partial x^\nu \partial x_\nu} = \left(\Box - m^2\right)\varphi(x) = 0, \quad (1.44)$$

where $\Box = -\partial_\nu \partial^\nu = \Delta - \frac{\partial^2}{\partial t^2}$ is the D'Alembertian operator.

In view of (1.43) and using (1.19), one finds the energy-momentum tensor of the scalar field,

$$T^{\mu\nu}(x) = \varphi^{,\nu}(x)\varphi^{,\mu}(x) - g^{\mu\nu}\mathcal{L}\Big(\varphi(x), \varphi_{,\nu}(x)\Big). \quad (1.45)$$

By means of this expression, the corresponding angular momentum tensor (given by the general formula (1.35)) can be determined. It should be noted that the intrinsic angular momentum (the spin) of this field equals zero. If the Lagrangian density (1.43) is used in (1.45), one finds the energy density

$$T^{00} = \frac{1}{2}\left[\dot{\varphi}^2 + (\nabla\varphi)^2 + m^2\varphi^2\right], \quad (1.46)$$

as well as the momentum vector density

$$T^{0k} = \varphi^{,0}\varphi^{,k}; \quad (k = 1, 2, 3).$$

## Momentum representation

The notion of *momentum representation* is very frequently utilized in the field theory. First, because it is suitable to the enunciation of the physical content of the problem, when particles are characterized by well-defined energies and momenta. Second, because in this representation the dynamic variables have a more compact form and concrete structure.

Suppose, for example, that the field function $\varphi(x)$ is represented in terms of the Fourier integral

$$\varphi(x) = (2\pi)^{-3/2}\int e^{ipx}\tilde{\varphi}(p)dp, \quad dp = dp^0 d\vec{p}.$$

Since the field function is real, $\varphi^*(x) = \varphi(x)$, this leads to

$$\tilde{\varphi}^*(p) = \tilde{\varphi}(-p).$$

According to (1.44), the Fourier amplitude satisfies the relation

$$(p^2 - m^2)\tilde{\varphi}(p) = 0,$$

so that it can be represented in the form

$$\tilde{\varphi}(p) = \sqrt{2\pi}\,\delta(p^2 - m^2)\varphi(p).$$

The Dirac delta function accomplishes the connection

$$(p^0)^2 - \vec{p}^2 = m^2, \tag{1.47}$$

between the 0-component $p^0$ of the four-vector $p$, the three-vector $\vec{p}$, and the parameter $m$. As shall be shown further on, the quantity $p^0$ has to be identified with the energy, and the three-vector $\vec{p}$ with the momentum vector. Therefore, the relation (1.47) is well-known in particle relativistic mechanics, with significance of mass for the parameter $m$. The Fourier representation then becomes

$$\varphi(x) = (2\pi)^{-3/2} \int \delta(p^2 - m^2) e^{ipx} \tilde{\varphi}(p) dp.$$

Since the delta function is under the integral sign, the integration does not perform over the whole four-dimensional space of the four-vector $p$, but only for the two poles of the three-dimensional hyperboloid

$$p^0 = \pm\sqrt{\vec{p}^2 + m^2},$$

one pole being situated inside the superior light cone, and the other inside the inferior cone. Since each of the two poles are Lorentz invariant, the field function $\varphi(x)$ can be written as a sum of two relativistic-covariant parts, corresponding to the two poles, namely

$$\varphi(x) = \varphi^+(x) + \varphi^-(x), \tag{1.48}$$

where

$$\varphi^\pm(x) = (2\pi)^{-3/2} \int e^{+ipx} \delta(p^2 - m^2) \tilde{\varphi}^\pm(p) dp. \tag{1.49}$$

Here

$$\tilde{\varphi}^\pm(p) = \theta(p^0)\varphi(\pm p),$$

which contains the discontinuous *unit step function* (*Heaviside function*),

$$\theta(x) = \begin{cases} 1, & x > 0, \\ 0, & x < 0. \end{cases}$$

The two components $\varphi^+$ and $\varphi^-$ of the field function $\varphi$ are called *positive frequency part* and *negative frequency part*, respectively, of this function. In quantum field theory, the positive frequency part $\varphi^+$ corresponds to the particle creation operator, and the negative frequency part to the particle annihilation operator. The symbols $(+)$ and $(-)$ have been chosen for an appropriate and suggestive physical significance of the corresponding quantum operators.

As shall further be shown in this section, the above performed decomposition is very useful for writing the dynamical quantities in momentum representation, due to the fact that these quantities are expressed as bilinear forms in terms of $\varphi^+(p)$ and $\varphi^-(p)$.

According to (1.49), the complex conjugation rule for $\varphi^\pm(p)$ can be written as

$$\left(\varphi^\pm(p)\right)^* = \varphi^\mp(p).$$

If integration over $p^0$ is performed in (1.49), the result is

$$\varphi^\pm(x) = \frac{1}{(2\pi)^{3/2}} \int \frac{d\vec{p}}{\sqrt{2p^0}} e^{\pm ipx} \varphi^\pm(\vec{p}), \tag{1.50}$$

where the following notations have been used:

$$\varphi^\pm(\vec{p}) = \frac{\varphi^\pm(p)}{\sqrt{2p^0}} = \frac{\varphi^\pm(p^0, \vec{p})}{\sqrt{2p^0}}; \quad p^0 = +\sqrt{\vec{p}^2 + m^2}.$$

Substituting (1.48) and (1.50) into (1.46), one finds

$$\begin{aligned} P^0 &= \int T^{00}\, d\vec{x} = \frac{1}{2} \int \left[ (\partial_\nu \varphi(x))^2 + m^2 \varphi^2(x) \right] d\vec{x} \\ &= \frac{1}{2} \int \Big\{ \partial_\nu \varphi^+ \partial_\nu \varphi^+ + 2\partial_\nu \varphi^+ \partial_\nu \varphi^- + \partial_\nu \varphi^- \partial_\nu \varphi^- \\ &\quad + m^2 \Big[ \varphi^+(x)\varphi^+(x) + 2\varphi^+(x)\varphi^-(x) + \varphi^-(x)\varphi^-(x) \Big] \Big\} d\vec{x}. \end{aligned}$$

It is worth mentioning that all summations over $\nu$ in this formula are non-covariant. It is not difficult to show that the components containing products of the field functions $\varphi^\pm(x)$ of the same sign do not change anything in the dynamical invariant $P^0$. For example,

$$\int \left[ \left(\partial_\nu \varphi^+(x)\right)^2 + m^2 \left(\varphi^+(x)\right)^2 \right] d\vec{x}$$

$$= \frac{1}{(2\pi)^3} \int\int \frac{d\vec{p}\, d\vec{p}'}{\sqrt{2p^0 p^{0\prime}}} \varphi^+(\vec{p})\varphi^+(\vec{p}')e^{i(p^0+p^{0\prime})x_0}(m^2 - p_\nu p'_\nu)$$

$$\times \int e^{-i(\vec{p}+\vec{p}')\cdot\vec{x}} d\vec{x} = \int \frac{d\vec{p}}{2p^0} \varphi^+(\vec{p})\varphi^+(-\vec{p})e^{2ip^0 x_0}(m^2 - p^0p^0 - \vec{p}^2),$$

and, since $m^2 - p^0p^0 + \vec{p}^2 = 0$, this yields

$$\int \left[ \left(\partial_\nu \varphi^+(x)\right)^2 + m^2 \left(\varphi^+(x)\right)^2 \right] d\vec{x} = 0.$$

A similar expression is obtained for $\varphi^-(x)$, so that one can write

$$P^0 = \int \left[\partial_\nu \varphi^+(x) \partial_\nu \varphi^-(x) + m^2 \varphi^+(x)\varphi^-(x)\right] d\vec{x}.$$

Using formula (1.50), one finds

$$P^0 = \int p^0 \varphi^+(\vec{p})\varphi^-(\vec{p}) d\vec{p}.$$

The components of the three-dimensional momentum vector are

$$P^n = \int T^{0n} d\vec{x} = \int p^n \varphi^+(\vec{p})\varphi^-(\vec{p}) d\vec{p}; \quad (n = 1, 2, 3).$$

If the functions $\varphi^+(\vec{p})$ and $\varphi^-(\vec{p})$ were non-commutative, then the above two expressions could be combined and written in the form

$$P^\nu = \frac{1}{2} \int p^\nu \left[\varphi^+(\vec{p})\varphi^-(\vec{p}) + \varphi^-(\vec{p})\varphi^+(\vec{p})\right] d\vec{p}. \tag{1.51}$$

In order to realize the significance of the normalization factor appearing in equation (1.50), consider the quantity

$$n(\vec{p}) = \varphi^+(\vec{p})\varphi^-(\vec{p}) = |\varphi^+(\vec{p})|^2,$$

which plays the role of the average density of spinless particle number, characterized by mass $m$, momentum $\vec{p}$, and energy $\sqrt{m^2 + \vec{p}^2}$. As soon as quantization is performed, these products become operators having as eigenvalues these (integer) numbers of particles.

The four-dimensional amplitudes $\varphi^+(p)$ and $\varphi^-(p)$ enable us to write the energy-momentum four-vector $P^\nu$ in the manifestly covariant form,

$$P^\nu = \frac{1}{2}\int p^\nu \theta(p^0)\delta(p^2 - m^2)\Big(\varphi^+(p)\varphi^-(p) + \varphi^-(p)\varphi^+(p)\Big)\,dp.$$

## The complex scalar field

The theory of the complex scalar field can be constructed by analogy with the theory of the real scalar field. The main difference is that the complex scalar field is described not by one – as in the case of the real scalar field – but by two independent real functions $\varphi_1(x)$ and $\varphi_2(x)$. However, the formalism becomes more convenient if instead of $\varphi_1(x)$ and $\varphi_2(x)$ their complex combinations $\varphi(x) = \varphi_1(x) + i\varphi_2(x)$ and $\varphi^*(x) = \varphi_1(x) - i\varphi_2(x)$ are used (hence the name "complex scalar field").

The Lagrangian density of the complex scalar field is given by

$$\mathcal{L}\Big(\varphi(x), \varphi^*(x), \varphi_{,\nu}(x), \varphi^*_{,\nu}(x)\Big) = \varphi^*_{,\nu}(x)\varphi^{,\nu}(x) - m^2\varphi^*(x)\varphi(x).$$

With the aid of the general formulas given in the paragraph I.2.1, from this expression can be obtained:

— the complex scalar field equations,

$$\left(\Box - m^2\right)\varphi(x) = 0, \quad \left(\Box - m^2\right)\varphi^*(x) = 0;$$

— the energy-momentum tensor,

$$T_{\mu\nu} = {\varphi^*}_{,\mu}(x)\varphi_{,\nu}(x) + {\varphi^*}_{,\nu}(x)\varphi_{,\mu}(x) - g_{\mu\nu}\mathcal{L};$$

— the current four-vector,

$$J_\nu = i\Big(\varphi^*(x)\varphi_{,\nu}(x) - \varphi^*_{,\nu}(x)\varphi(x)\Big).$$

As it is expected, the intrinsic angular momentum (the spin) of the complex scalar field is found to be zero, as in the case of the real scalar field.

The analogous of the expressions (1.49) are, in this case,

$$\varphi^\pm(x) = (2\pi)^{-3/2}\int e^{\pm ipx}\delta(p^2 - m^2)\tilde{\varphi}^\pm(p)\,dp,$$

$$\varphi^*\pm(x) = (2\pi)^{-3/2}\int e^{\pm ipx}\delta(p^2 - m^2)\tilde{\varphi}^{*\pm}(p)\,dp.$$

After integration with respect to the variable $p^0$ one finds

$$\varphi^{\pm}(x) = \frac{1}{(2\pi)^{3/2}}\int \frac{d\vec{p}}{\sqrt{2p^0}} e^{\pm ipx}\varphi^{\pm}(\vec{p}),$$

$$\varphi^{*\pm}(x) = \frac{1}{(2\pi)^{3/2}}\int \frac{d\vec{p}}{\sqrt{2p^0}} e^{\pm ipx}\varphi^{*\pm}(\vec{p}),$$

where

$$\varphi^{\pm}(\vec{p}) = \frac{\varphi^{\pm}(p)}{\sqrt{2p^0}}, \quad \varphi^{*\pm}(\vec{p}) = \frac{\varphi^{*\pm}(p)}{\sqrt{2p^0}}, \quad p^0 = \sqrt{\vec{p}^2 + m^2}.$$

Here, care should be taken to the fact that $\varphi^{*+}(\vec{p})$ denotes not the function which is the complex conjugate of $\varphi^{+}(\vec{p})$, but the positive-frequency part of $\varphi^{*}(\vec{p})$. In this case, in the momentum representation we have

$$\left(\varphi^{+}(p)\right)^{*} = \varphi^{*-}(p), \quad \left(\varphi^{-}(p)\right)^{*} = \varphi^{*+}(p).$$

For the energy-momentum four-vector $P^{\nu}$ and for the charge $Q$, the following expressions can be found:

$$P^{\nu} = \int p^{\nu}\left[\varphi^{*+}(\vec{p})\varphi^{-}(\vec{p}) + \varphi^{*-}(\vec{p})\varphi^{+}(\vec{p})\right] d\vec{p},$$

$$(\nu = 0, 1, 2, 3; \quad p^0 = \sqrt{\vec{p}^2 + m^2});$$

$$Q = \int \left[\varphi^{*+}(\vec{p})\varphi^{-}(\vec{p}) - \varphi^{*-}(\vec{p})\varphi^{+}(\vec{p})\right] d\vec{p}.$$

From the above expressions it follows that the product $\varphi^{*-}(\vec{p})\varphi^{+}(\vec{p})$ can be interpreted as being the density of the average number of particles of mass $m$, energy $p^0$, momentum $\vec{p}$, and charge $-1$, while the product $\varphi^{*+}(\vec{p})\varphi^{-}(\vec{p})$ can be interpreted as the density of the average number of particles of the same energy and momentum, but of charge $+1$.

Obviously, after quantization, $\varphi^{*+}(\vec{p})$ will describe the creation of a scalar particle of mass $m$, energy $p^0$, momentum $\vec{p}$, and charge $+1$, while $\varphi^{-}(\vec{p})$ will describe its annihilation. The same is the situation with $\varphi^{+}$ and $\varphi^{*-}$, respectively.

In conclusion, the complex scalar field describes positively and negatively charged spinless particles.

In practice, the most important role is played by the field corresponding to $\pi$ mesons (pions). This is a pseudoscalar field. The mesons $\pi^+, \pi^0$, and $\pi^-$, which distinguish from each other by the electric charge (equal, respectively, to $+1, 0$ and $-1$) form an isotopic triplet. To this system corresponds the three-component field function $\vec{\varphi} = \{\varphi_1, \varphi_2, \varphi_3\}$, which components form a vector in the three-dimensional isotopic space. The Lagrangian density of the free field writes

$$\mathcal{L} = \frac{1}{2}\left(\vec{\varphi}^*_{,\nu} \cdot \vec{\varphi}^{,\nu}\right) - \frac{m^2}{2}(\vec{\varphi}^* \cdot \vec{\varphi}) \qquad (1.52)$$

and stands for the sum of the Lagrangian densities for each component of the vector defined in isotopic space. Due to this fact, the dynamical quantities corresponding to the Poincaré group transformations are represented by a sum of expressions corresponding to each isotopic component.

The new conservable quantity satisfies the transformation relations in the isotopic space. It is called the *isotopic spin vector.*

### I.2.2.2. The massive vectorial field

This vectorial field describes the massive particles (whose rest-mass is different from zero) with spin $s = 1$. Denoting the field function by $U_\nu$, $\nu = 0, 1, 2, 3$, its components form a covariant four-vector if, under a coordinate transformation $x'_\nu = x_\nu + x^\mu \delta L_{\nu\mu}$ (a Lorentz rotation), transform according to the law

$$U_\mu(x) \to U'_\mu(x') = U_\mu(x) + \delta U_\mu(x); \quad \delta U_\mu(x) = \delta\Omega_{\mu\nu} U^\nu(x).$$

There are different ways that the Lagrangian density of the vectorial field can be formed, since the expression given by formula (1.43),

$$\mathcal{L} = -\frac{1}{2} U_{\mu,\nu} U^{\mu,\nu} + \frac{m^2}{2} U_\mu U^\mu \qquad (1.53)$$

is not the only possible Lorentz-invariant choice. For example, a term proportional to

$$U_{\mu,\nu} U^{\nu,\mu} \qquad (1.54)$$

can be added, which, according to (1.16), is equivalent to the following expression:

$$U_{\mu}^{,\mu} U_{,\nu}^{\nu} = (\partial U/\partial x)^2.$$

The composition (1.53) of the Lagrangian density is superior to other choices because all quantities are covariant sums (over $\mu$) of the corresponding expressions of (1.43), which is valid for one-component field. Since the term related to the $U_0$ component appears in these sums with minus sign, its presence in the energy expression is undesirable. To solve this difficulty, one more condition is imposed to the component $U_\mu$, namely

$$\partial^\mu U_\mu = \frac{\partial}{\partial x_\mu} U_\mu(x) \equiv (\partial U) = 0. \qquad (1.55)$$

This condition diminishes from four to three the number of linear components of the field, but assures the positive sign of energy of the massive vectorial field. These three independent components correspond to the three possible values of the spin projections onto an arbitrary axis, namely $(1, 0, -1)$. Therefore, the field function describes the massive spin-1 particles. The inclusion of the above supplementary condition is equivalent to the exclusion of spin-0 particles (which possibility would lead to negative values of energy).

If the term (1.54) is added to the Lagrangian density (1.53), one obtains

$$\mathcal{L} = -\frac{1}{4} H_{\mu\nu} H^{\mu\nu} + \frac{m^2}{2} U_\mu U^\mu,$$

where

$$H_{\mu\nu} = \partial_\mu U_\nu - \partial_\nu U_\mu. \qquad (1.56)$$

The condition (1.55) can also be written as a consequence of the equation of motion

$$\partial_\nu H^{\nu\mu} - m^2 U^\mu = (\Box - m^2) U^\mu + \partial^\mu \partial_\nu U^\nu. \qquad (1.57)$$

These equations are called *Proca's equations*. Taking the derivative of (1.57) with respect to $x^\mu$ and performing some elementary transformations, the condition (1.55) is obtained. Therefore, Proca's equations are equivalent to Klein–Gordon equations, plus the subsidiary condition (1.55).

In order to describe the charged massive vectorial particles, one must consider a complex massive vector field. In this case, the dynamical quantities corresponding to the Lagrangian density of the complex massive vector field,

$$\mathcal{L} = -\frac{1}{2} H^*_{\mu\nu} H^{\mu\nu} + m^2 U^*_\mu U^\mu \tag{1.58}$$

are as follows:

— the energy-momentum tensor,

$$T_{\mu\nu} = H^*_{\mu\sigma} U^\sigma_{,\nu} + U^{*\sigma}_{,\nu} H_{\mu\sigma} - g_{\mu\nu}\mathcal{L}, \tag{1.59$a$}$$

— the current four-vector,

$$J_\nu = i\left(U^{*\sigma} H_{\sigma\nu} - H^*_{\sigma\nu} U^\sigma\right), \tag{1.59$b$}$$

— the spin angular momentum tensor,

$$S^{\nu(\mu\sigma)} = U^{*\mu} H^{\sigma\nu} - H^{*\mu\nu} U^\sigma + H^{*\sigma\nu} U^\mu - U^{*\sigma} H^{\mu\nu}. \tag{1.59$c$}$$

In order to perform further calculations it is necessary to decompose the potentials into positive and negative density parts and pass to the three-dimensional momentum representation. Thus we have

$$U_\nu(x) = U^+_\nu(x) + U^-_\nu(x), \quad U^*_\nu(x) = U^{*+}_\nu(x) + U^{*-}_\nu(x),$$

with

$$U^\pm_\nu(x) = \frac{1}{(2\pi)^{3/2}} \int \frac{d\vec{p}}{\sqrt{2p^0}} e^{\pm ipx} U^\pm_\nu(\vec{p}), \tag{1.60$a$}$$

$$U^{*\pm}_\nu(x) = \frac{1}{(2\pi)^{3/2}} \int \frac{d\vec{p}}{\sqrt{2p^0}} e^{\pm ipx} U^{*\pm}_\nu(\vec{p}), \tag{1.60$b$}$$

where, as usual,

$$U^\pm_\nu(\vec{p}) = \frac{U^\pm_\nu(p)}{\sqrt{2p^0}}, \quad U^{*\pm}_\nu(\vec{p}) = \frac{U^{*\pm}_\nu(p)}{\sqrt{2p^0}}, \quad p^0 = \sqrt{\vec{p}^2 + m^2},$$

and

$$U^\pm_\nu(p) = \theta(p^0) U_\nu(\pm p), \quad U^{*\pm}_\nu(p) = \theta(p^0) U^*_\nu(\pm p).$$

Substituting (1.60) into the temporal ($\nu = 0$) components of the quantities (1.59), and performing integration over $\vec{x}$, the following dynamical quantities of the complex massive vectorial field are obtained:

$$P^\nu = -\int d\vec{p}\, p^\nu \left[U^{*-}_\mu(\vec{p}) U^{+,\mu}(\vec{p}) + U^{*+}_\mu(\vec{p}) U^{-,\mu}(\vec{p})\right], \tag{1.61}$$

$$Q = \int d\vec{p}\left[U^{*-}_{\mu}(\vec{p})U^{+,\mu}(\vec{p}) - U^{*+}_{\mu}(\vec{p})U^{-,\mu}(\vec{p})\right], \qquad (1.62)$$

$$\vec{S} = i\int d\vec{p}\left\{\left[\vec{U}^{*+}(\vec{p}) \times \vec{U}^{-}(\vec{p})\right] - \left[\vec{U}^{*-}(\vec{p}) \times \vec{U}^{+}(\vec{p})\right]\right\}. \qquad (1.63)$$

Taking into account the complex conjugation condition resulting from (1.60),

$$\left(U^{\pm}_{\nu}(\vec{p})\right)^{*} = U^{*\mp}_{\nu}(\vec{p}),$$

as one can observe, the expression (1.61) is negative for $\mu = 0$, and the sing of $P^0$ (the energy) is undetermined. To eliminate this situation, one introduces some subsidiary conditions which, in momentum representation, have the form

$$p^{\nu}U^{\pm}_{\nu}(\vec{p}) = 0, \quad p^{\nu}U^{*\pm}_{\nu}(\vec{p}) = 0. \qquad (1.64)$$

Under these conditions, the components $U_{\nu}$ are not independent anymore. Using (1.64), the components $U^{\pm}_{0}$ can be expressed in terms of the other components as

$$U^{\pm}_{0}(\vec{p}) = \frac{1}{p^0}p^n U^{\pm}_{n}(\vec{p}), \quad U^{*\pm}_{0}(\vec{p}) = \frac{1}{p^0}p^n U^{*\pm}_{n}(\vec{p}), \quad (n = 1, 2, 3),$$

in which case the quadratic form under the integral (1.61) can be expressed in terms of the potential spatial components only,

$$-U^{*\pm}_{\nu}U^{\pm,\nu} = \vec{U}^{*\pm}(\vec{p}) \cdot \vec{U}^{\mp}(\vec{p}) - \frac{1}{p^0p^0}\left(\vec{p} \cdot \vec{U}^{*\pm}(\vec{p})\right)\left(\vec{p} \cdot \vec{U}^{\mp}(\vec{p})\right). \qquad (1.65)$$

**The local reference frame**

The last expression can be diagonalized by means of the linear representation

$$\vec{U}(\vec{p}) = \vec{e}_1 a_1(\vec{p}) + \vec{e}_2 a_2(\vec{p}) + \frac{\vec{p}}{|\vec{p}|}\frac{p^0}{m}a_3(\vec{p}), \qquad (1.66)$$

where $\vec{e}_1$ and $\vec{e}_2$ are two unit vectors perpendicular to the wave vector $\vec{p}$ and reciprocally orthogonal,

$$\vec{e}_i \cdot \vec{e}_j = \delta_{ij}; \quad \vec{e}_3 = \frac{\vec{p}}{|\vec{p}|} \quad (i, j = 1, 2, 3),$$

*i.e.* the versors of the transversal directions of polarization.

The representation (1.66) expresses the decomposition of the vector potential in longitudinal and transversal components with respect to momentum, that is transition to the *local frame* in momentum space. By means of (1.66) one obtains

$$-U_\nu^{*\pm}(\vec{p})U^{\mp,\nu}(\vec{p}) = a_n^{*\pm}(\vec{p})a_n^{\mp}(\vec{p}). \tag{1.67}$$

Substitution of (1.67) into (1.61) and (1.62) yields diagonal expressions for both the energy-momentum four-vector and the charge. In addition, the energy is positively defined in these new variables,

$$P^\nu = \int d\vec{p}\; p^\nu \left[a_n^{*+}(\vec{p})a_n^-(\vec{p}) + a_n^{*-}(\vec{p})a_n^+(\vec{p})\right];$$

$$Q = \int d\vec{p}\; \left[a_n^{*+}(\vec{p})a_n^-(\vec{p}) - a_n^{*-}(\vec{p})a_n^+(\vec{p})\right].$$

For the spin projection to the direction of momentum vector, one obtains

$$\begin{aligned} S_3 = i\int d\vec{p}\Big[&a_1^{*+}(\vec{p})a_2^-(\vec{p}) - a_2^{*+}(\varphi p)a_1^-(\vec{p}) \\ &+ a_2^{*-}(\vec{p})a_1^+(\vec{p}) - a_1^{*-}(\vec{p})a_2^+(\vec{p})\Big], \end{aligned} \tag{1.68}$$

which is non-diagonal. In order to diagonalize the spin projection (1.68) one uses the linear substitution

$$a_1^\pm = \frac{b_1^\pm + b_2^\pm}{\sqrt{2}}; \quad a_2^\pm = \frac{b_1^\pm - b_2^\pm}{i\sqrt{2}}; \quad a_3^\pm = b_3^\pm, \tag{1.69}$$

(together with analogous relations for the complex amplitudes $a_n^{*\pm}$) which leave unchanged the diagonal form of $P_\nu$ and $Q$. Indeed,

$$P^\nu = \int d\vec{p}\; p^\nu \left[b_n^{*+}(\vec{p})b_n^-(\vec{p}) + b_n^{*}\;(p')b_n^+(\vec{p})\right], \tag{1.70}$$

$$Q = \int d\vec{p}\; \left[b_n^{*+}(\vec{p})b_n^-(\vec{p}) - b_n^{*-}(\vec{p})b_n^+(\vec{p})\right], \tag{1.71}$$

$$\begin{aligned} S_3 = \int d\vec{p}\Big[&b_1^{*+}(\vec{p})b_1^-(\vec{p}) - b_1^{*-}(\vec{p})b_1^+(\vec{p}) \\ &+ b_2^{*-}(\vec{p})b_2^+(\vec{p}) - b_2^{*+}(\vec{p})b_2^-(\vec{p})\Big]. \end{aligned} \tag{1.72}$$

It then follows that the quadratic combinations of the amplitudes $b^{*\pm}$ and $b^{\mp}$ can be interpreted as densities of an average number of particles, possessing a certain energy, a given momentum, a certain charge, and a given spin projection along the direction of motion. For example, the quantity $b_2^{*-}(\vec{p})b_2^{+}(\vec{p})$ represents the particle number density with momentum $\vec{p}$, energy $p^0$, charge $-1$, and the spin projection along the direction of motion equal to $+1$. In the same way, $b_3^{*-}(\vec{p})b_3^{+}(\vec{p})$ stands for the particle number density with momentum $\vec{p}$, energy $p^0$, charge $-1$, and the spin projection along the direction of motion equal to zero. As shall further be shown, after the quantization procedure, the amplitude $b_2^{*+}(\vec{p})$ describes the creation of a particle having energy $p^0$, momentum $\vec{p}$, charge $+1$, and the spin projection along the direction of motion equal to $+1$, while the amplitude $b_2^{-}(\vec{p})$ defines the annihilation of such a particle, and so on.

According to formulas (1.66) and (1.69), the amplitudes $a_1$ and $a_2$ correspond to linearly polarized waves, and the amplitudes $b_1$ and $b_2$ to the circular polarization of these waves.

In conclusion, the complex massive vectorial field describes the charged particles, having a non-zero rest mass, and three possible values of the spin projection along the direction of motion.

### I.2.2.3. The electromagnetic field

### The electromagnetic field potential

The free electromagnetic field satisfies *Maxwell's equations*

$$\nabla \times \vec{E} = -\dot{\vec{H}}; \quad \nabla \times \vec{H} = \dot{\vec{E}},$$

$$\nabla \cdot \vec{E} = 0; \qquad \nabla \cdot \vec{H} = 0.$$

For a symmetric description of the electromagnetic field one defines the real covariant potential four-vector

$$A_\nu = (A_0, -\vec{A}),$$

so that

$$\vec{E} = -\nabla A_0 - \dot{\vec{A}}; \quad \vec{H} = \nabla \times \vec{A}.$$

The components of the four-dimensional curl of the potential $A_\nu$ form the well-known antisymmetric electromagnetic field tensor

$$F_{\mu\nu} = \partial_\mu A_\nu - \partial_\nu A_\mu, \tag{1.73}$$

whose components are expressed in terms of the electric and magnetic fields intensities as

$$E^n = -F_{n0}; \quad H^n = -\frac{1}{2}\varepsilon_{nml} F_{ml}, \quad (n, m, l = 1, 2, 3). \tag{1.74}$$

This connection can also be written as

$$F^{\mu\nu} = \begin{pmatrix} 0 & -E^1 & -E^2 & -E^3 \\ E^1 & 0 & H^3 & -H^2 \\ E^2 & -H^3 & 0 & H^1 \\ E^3 & H^2 & -H^1 & 0 \end{pmatrix}.$$

Written in terms of $F_{\mu\nu}$, Maxwell's equations are

$$\left.\begin{aligned} \partial^\nu F_{\nu\mu} &= 0, \\ \partial_\sigma F_{\mu\nu} + \partial_\nu F_{\sigma\mu} + \partial_\mu F_{\nu\sigma} &= 0. \end{aligned}\right\} \tag{1.75}$$

Passing from $F_{\mu\nu}$ to the four-potential $A_\nu$, one can observe that the second equation of (1.75) is a result of the definition (1.73), without furnishing any equation for $A_\nu$, while the first equation (1.75) yields

$$\Box A_\nu + \partial_\nu \partial^\mu A_\mu = 0, \tag{1.76}$$

which is analogous to Proca's equations (1.57).

To conclude, the formulation of the electromagnetic field equations in terms of the four-potential $A_\nu$ is really simple, symmetric and explicitly covariant.

### The covariant gauge and the Lorenz condition

The purpose of this section is to discuss the procedure of introduction of the potential four-vector, $A_\nu$. In this respect, it is not difficult to verify that the quantities $\vec{E}$, $\vec{H}$, $F_{\mu\nu}$, and Maxwell's equations remain unchanged under the so-called second species gradient transformation of the four-potential $A_\nu$,

$$A_\nu(x) \to A'_\nu(x) = A_\nu(x) + \partial_\nu f(x), \tag{1.77}$$

which is usually called the *gauge condition.* This way the four-potential $A_\nu$, which is not an observable quantity, can have different values for the same values of the observable quantities $\vec{E}$ and $\vec{H}$. For the moment, the function $f$ appearing on the r.h.s. of (1.77) is arbitrary. It usually has to obey the smoothness condition. Since the four-potential $A_\nu$ can take different values for the same values of the physical observables, this signifies the fact that description of the electromagnetic field in terms of $A_\nu$ contains a non-physical degree of freedom, associated with the four-divergence

$$\partial^\nu A_\nu(x) \equiv \partial A(x) = \chi(x).$$

The condition that this quantity vanishes,

$$\partial A(x) = 0 \tag{1.78}$$

is known in the theory of the electromagnetic field as the *Lorenz gauge condition.* Since – if we wish to keep the relativistic covariance of the theory – the Lorenz condition cannot be maintained in this form when quantizing the electromagnetic field, one must know that degree of freedom expressed by the scalar arbitrary function $\chi(x)$, and its physical significance as well.

In this respect, one first supposes that the four-potential $A_\nu(x)$ does not satisfy equation (1.76), but it describes an arbitrary field, produced by some external sources, and satisfies – for example – the equation

$$\Box\, A_\nu + \partial_\nu \left(\partial A(x)\right) = J_\nu(x),$$

where $J_\nu(x)$ is the current produced by those external sources.

Second, one passes to the momentum representation

$$A_\nu(x) = \int dk\; e^{ikx} A_\nu(k),$$

where the same symbol $A$ has been used for both the functions and their Fourier transforms. This way, under the change of representation, the derivation operations are going to be written in algebraic form. In this case, the transformation (1.77) writes

$$A'_\nu(k) = A_\nu(k) + k_\nu \tilde{f}(k), \tag{1.79}$$

where the function $\tilde{f}(k)$ differs from the Fourier transform of $f(x)$ by the imaginary unit.

At this stage of investigation, it is necessary to introduce the projection operators

$$P^{tr}_{\mu\nu}(k) = g_{\mu\nu} - \frac{k_\mu k_\nu}{k^2}, \quad P^{long}_{\mu\nu}(k) = \frac{k_\mu k_\nu}{k^2}. \tag{1.80}$$

In coordinate representation, the operation $(k^2)^{-1}$ is performed by integration, and can be written as

$$\frac{1}{k^2}A_\nu \sim \int G(x-y)A_\nu(y)\,dy,$$

where $G$ is the Green function for the D'Alembert equation

$$\Box\, G(x) = -\delta(x). \tag{1.81}$$

Using (1.80), the four-potential $A_\nu$ can be decomposed in its transversal and longitudinal components,

$$A_\nu = A^{tr}_\nu + A^{long}_\nu; \quad A^{tr}_\nu = P^{tr}_{\nu\mu}A^\mu; \quad A^{long}_\nu = P^{long}_{\nu\mu}A^\mu. \tag{1.82}$$

These components satisfy the relations

$$k^\nu A^{tr}_\nu(k) = 0; \quad k^\nu A^{long}_\nu(k) = kA(k). \tag{1.83}$$

It is now convenient to introduce the special notation

$$kA(k) = \xi(k),$$

for the quantity that identifies (up to the imaginary unit) with the Fourier image of the function $\chi(x)$.

According to (1.82) and (1.83), the parallel (longitudinal) component of the electromagnetic field is given by the function $\xi(x)$, hence

$$A^{long}_\nu(k) = \frac{k_\nu}{k^2}\xi(k).$$

When the four-potential $A_\nu$ transforms according to the gauge condition (1.79), the function $\xi(k)$ transforms according to

$$\xi'(k) = \xi(k) + k^2\tilde{f}(k),$$

but the transversal component $A^{tr}_\nu$ remains unchanged when the four-potential $A_\nu$ changes according to (1.79). It is not difficult to show that the covariant gauge tensor $F_{\mu\nu}$ does not depend on $A^{long}_\nu$ (which depends on $\xi$). This is the reason for appearance of the degree of freedom described by the scalar function $\xi(k)$ (that brings with it the arbitrariness in what concerns the non-univocity of the four-potential $A_\nu$), which degree of freedom thus proves to be of a non-physical nature.

**The generalized Lagrangian**

The equations of motion (1.76) can be also obtained by means of a variational principle, in terms of the Lagrangian density

$$\mathcal{L}_{tr} = -\frac{1}{4} F_{\mu\nu} F^{\mu\nu}, \tag{1.84}$$

that results from the well-known expression $\frac{1}{2}(E^2 - H^2)$, written in terms of the electromagnetic field tensor $F_{\mu\nu}$. The lower index "tr" shows that the Lagrangian density (1.84) is gauge-covariant, and does not contain any dependence on the parallel component $A_\nu^{long}$ of $A_\nu$.

The differential operator of the equation of motion (1.76),

$$K_{\mu\nu}^{tr} = \partial_\mu \partial_\nu - g_{\mu\nu} \partial^2$$

is proportional to the transversal projection operator $P^{tr}$ and, since it cannot be inverted, it makes more difficult to solve the corresponding non-homogeneous equation of motion.

But, within the Hamiltonian formalism, the transversal Lagrangian density $\mathcal{L}_{tr}$ possesses a certain useful particularity. In this respect, it is worth recalling that in this formalism an important role is played by the canonically conjugate momenta, obtained as partial derivative with respect to the generalized velocities $v_i = \dot{q}_i$, where $q_i$ are the generalized coordinates of the system. In the case of a system with an infinite number of degrees of freedom (physical fields), as generalized coordinates are chosen the field functions $\varphi(\vec{x}, t) = \varphi(x)$. Consequently, the generalized momenta of the classical fields are determined by

$$\pi_i(x) = \frac{\partial \mathcal{L}}{\partial \dot{\varphi}_i(x)}.$$

The Hamiltonian formalism lies at the basis of the so-called *canonical quantization*. Choosing as generalized coordinates the components of the four-potential $A_\nu$, one can easily prove that the associate momenta corresponding to the temporal component $A_0$ are zero:

$$\pi_0(x) = F_{00} = 0.$$

In other words, the Lagrangian density $\mathcal{L}_{tr}$ does not depend on the "velocity" $\dot{A}_0$. The Lagrangians possessing such a property are sometimes called *singular*, and correspond to *constrained physical systems*. In order to describe the constrained systems, one makes use of the generalized Hamiltonian dynamics formalism, elaborated by Dirac in 1968. In the present case, transition to generalized Hamiltonian dynamics is equivalent to the following modification of the Lagrangian density,

$$\mathcal{L}_{tr} \to \mathcal{L} = -\frac{1}{4}F_{\mu\nu}F^{\mu\nu} + \frac{a}{2}(\partial A)^2, \tag{1.85}$$

where $a$ is an arbitrary non-null coefficient. This Lagrangian density leads to the following equations of motion:

$$\partial^\nu F_{\mu\nu} + a\,\partial_\mu(\partial A) = \Box A_\mu + (1+a)\partial_\mu(\partial A) = 0. \tag{1.86}$$

Hence, it follows that the parallel component of the field satisfies the D'Alembert equation

$$\Box\partial^\nu A_\nu(x) = \chi(x) = 0. \tag{1.87}$$

This property of the parallel component is maintained even if the electromagnetic field interacts with some other field, in which case on the r.h.s. of the equation of motion (1.86) appear the components of the external current, $J_\nu$, that satisfy the equation of continuity $\partial^\nu J_\nu = 0$. This way, the degree of freedom associated with the parallel component $\chi(x)$ interact neither with the transversal components $A_\nu^{tr}$, nor with other fields.

As one can see, the Lagrangian density (1.85) and the equations of motion (1.86) are not anymore covariant with respect to transformation (1.77). Therefore, one can say that the introduction of the term $\sim(\partial A)^2$ into the Lagrangian fixes the gauge. At the same time, the equations (1.85) and (1.86) do not change under the specific gauge transformations

$$A'_\nu(x) = A_\nu(x) + \partial_\nu f_0(x), \tag{1.88}$$

where the gauge function $f_0(x)$ satisfies the D'Alembert equation

$$\Box f_0(x) = 0. \tag{1.89}$$

The insertion of the gauge (fixing) term into the Lagrangian density leads to the fact that the operator of the equation of motion,

$$K_{\mu\nu} = g_{\mu\nu}\Box + (1+a)\partial_\mu\partial_\nu$$

$$\to g_{\mu\nu}k^2 - (1+a)k_\mu k_\nu = k^2 P^{tr}_{\mu\nu}(k) + ak^2 P^{long}_{\mu\nu}(k)$$

stops being singular, therefore possessing the inverse operator

$$K^{-1}_{\mu\nu} = \frac{1}{k^2}P^{tr}_{\mu\nu}(k) + \frac{d_l}{k^2}P^{long}_{\mu\nu}(k); \quad d_l = \frac{1}{a}.$$

**The diagonal gauge**

If the coefficient $d_l$ which fixes the gauge equals unity, then $K$ and $K^{-1}$ shall have the following diagonal expressions:

$$d_l = 1; \quad K_{\mu\nu} = g_{\mu\nu}k^2; \quad K^{-1}_{\mu\nu} = g_{\mu\nu}(k^2)^{-1}.$$

This gauge is called *diagonal*. Under this gauge the Lagrangian density

$$\mathcal{L}_{diag} = -\frac{1}{4}F_{\mu\nu}F^{\mu\nu} + \frac{1}{2}(\partial A)^2 \tag{1.90}$$

can be transformed by means of integration by parts, that gives

$$\mathcal{L}_{diag} = -\frac{1}{2}\partial^\nu A_\mu(x)\partial_\nu A^\mu(x), \tag{1.91}$$

which represents the covariant sum of the Lagrangian densities for each component $A_\mu$. The Lagrangian density (1.91) leads to the equation of motion under the D'Alembert form

$$\Box A_\mu(x) = 0. \tag{1.92}$$

This Lagrangian density is the most convenient for quantization. From the physical point of view, the Lagrangian density (1.91), similarly to the more general expression (1.85), is equivalent to the initial "transversal" Lagrangian density (1.84). One can also observe that the equations of motion (1.92) and the Lagrangian density (1.91) represent a particular case of the equations (1.85) and (1.86), being covariant with respect to transformation (1.88). In any fixed reference system, the function $f_0$ can be chosen in such a way that the temporal component $A_0$ of the four-potential $A_\nu$, which satisfies the equation (1.92), is zero. The Lorenz condition then writes

$$\nabla \cdot \vec{A} = 0. \tag{1.93}$$

The corresponding gauge is called *Coulombian* (*Coulomb gauge*).

In momentum representation, equation (1.93) writes

$$\vec{k} \cdot \vec{A}(k) = 0.$$

As one can see, in the case of Coulomb gauge the electromagnetic field satisfies the transversality condition. Since the electromagnetic field is described by the four-component potential $A_\nu$, only two of them, linearly independent and orthogonal to the wave vector $\vec{k}$, have a physical significance.

Diminution of the fields components from four to two, as a result of the gradient covariance, is strongly connected to the vanishing rest-mass of the field particle – the photon. Due to this property, only the derivatives of $A_\nu$ appear in the field equations, and thus the gauge invariance property of the electromagnetic field is coming to light.

By using (1.91), one easily obtains:

— the field energy-momentum tensor,

$$T^{\mu\nu} = -A_\sigma^{;\mu} A^{\sigma,\nu} - g^{\mu\nu}\mathcal{L};$$

— the volumic density of the four-vector $P^\nu$,

$$T^{00} = -\frac{1}{2}\left(\dot{A}^\nu \dot{A}_\nu + \partial_n A^\nu \partial_n A_\nu\right), \quad T^{0k} = \dot{A}_\nu \partial^k A^\nu;$$

— the spin angular momentum,

$$S^{\tau(\nu\mu)} = A^\mu A^{\nu,\tau} - A^\nu A^{\mu,\tau};$$

— the volumic density of the spin vector,

$$\vec{S} = \left[\vec{A}(x) \times \dot{\vec{A}}(x)\right].$$

In order to determine the dynamical invariants of the system, we shall decompose the four-vector $A_\nu$ in terms of positive and negative frequency, then passing to the momentum representation for the free field,

$$A_\nu^\pm(x) = \frac{1}{(2\pi)^{3/2}} \int \frac{d\vec{k}}{\sqrt{2k_0}} e^{\pm ikx} A_\nu^\pm(\vec{k}). \tag{1.94}$$

## Transition to the local frame

In order to perform the diagonalization in momentum representation of the free electromagnetic field (1.94), it is necessary to decompose the field into components connected to the local frame. Since the photon rest-mass is zero ($m = 0$), it is necessary to modify the relation (1.66). To do this, one writes the potential $A(\vec{k})$ as a sum of the transversal, longitudinal, and temporal components,

$$A^{\pm}_{\nu}(\vec{k}) = e^{0}_{\nu} a^{\pm}_{0}(\vec{k}) + e^{1}_{\nu} a^{\pm}_{1}(\vec{k}) + e^{2}_{\nu} a^{\pm}_{2}(\vec{k}) + e^{3}_{\nu} a^{\pm}_{3}(\vec{k}), \tag{1.95}$$

where $\vec{e}^{\,1}$ and $\vec{e}^{\,2}$ are spatial polarization versors (similar to those introduced in (1.66)) orthogonal to each other and to the versor of the longitudinal direction (the direction of the momentum vector $\vec{k}$), $\vec{e}^{\,3} = \vec{k}/|\vec{k}|$, with

$$e^{i}_{0} = 0; \quad \vec{e}^{\,i} \cdot \vec{e}^{\,j} = \delta_{ij}; \quad \vec{e}^{\,i} \times \vec{e}^{\,j} = \varepsilon_{ijk} \vec{e}^{\,k}; \quad (i, j, k = 1, 2, 3),$$

while $\vec{e}^{\,0}$ is a time unit vector with $e^{0}_{\nu} = \delta_{\nu 0}$. The transformation (1.95) maintains the diagonal configuration of the main quadratic form

$$A^{+}_{\nu}(\vec{k}) A^{-}_{\nu}(\vec{k}) = g^{\sigma\rho} a^{+}_{\sigma}(\vec{k}) a^{-}_{\rho}(\vec{k}). \tag{1.96}$$

Substituting (1.95) into the Lorenz condition, written separately for the positive and negative components, one obtains

$$|\vec{k}| a^{\pm}_{3}(\vec{k}) - k_0 a^{\pm}_{0}(\vec{k}) = 0.$$

Since the photon rest-mass is zero, $|\vec{k}| = k_0$, and therefore

$$a^{+}_{3}(\vec{k}) a^{-}_{3}(\vec{k}) = a^{+}_{0}(\vec{k}) a^{-}_{0}(\vec{k}). \tag{1.97}$$

It then follows that, according to the Lorenz condition, the average numerical densities of the *longitudinal* and *temporal* photons, *i.e.* $a^{+}_{3} a^{-}_{3}$ and $a^{+}_{0} a^{-}_{0}$, respectively, are equal. Consequently, the longitudinal and temporal photons reciprocally compensate each other. Introducing (1.97) into (1.96), one obtains

$$-A^{+}_{\nu}(\vec{k}) A^{-}_{\nu}(\vec{k}) = a^{+}_{1}(\vec{k}) a^{-}_{1}(\vec{k}) + a^{+}_{2}(\vec{k}) a^{-}_{2}(\vec{k}).$$

Next, calculating the energy-momentum four-vector

$$P^{\nu} = \int d\vec{k}\; k^{\nu} a^{+}_{s}(\vec{k}) a^{-}_{s}(\vec{k}), \quad (s = 1, 2),$$

one finds that, as for the case of the massive vector field, the energy is positively defined, as a consequence of the Lorenz condition.

Using the formula that gives the vector $\vec{S}$ (the spin),

$$\vec{S} = i \int d\vec{k} \, \left[\vec{A}^{+}(\vec{k}) \times \vec{A}^{-}(\vec{k})\right] = i \int d\vec{k} \, \varepsilon_{lmn} \vec{e}^{\,l}(\vec{k}) a^{+}_{m}(\vec{k}) a^{-}_{n}(\vec{k}),$$

one obtains its projection on the direction of $\vec{k}$, as

$$S_3 = \int d\vec{k} \left[a^{+}_{1}(\vec{k}) a^{-}_{2}(\vec{k}) - a^{+}_{2}(\vec{k}) a^{-}_{1}(\vec{k})\right].$$

Introducing the amplitudes $b_1$ and $b_2$ as in the previous paragraph, one finds the following diagonal expressions for $P_\nu$ and $S_3$:

$$P^\nu = \int d\vec{k} \, k^\nu \left[b^{+}_{1}(\vec{k}) b^{-}_{1}(\vec{k}) + b^{+}_{2}(\vec{k}) b^{-}_{2}(\vec{k})\right], \tag{1.98}$$

$$S_3 = \int d\vec{k} \left[b^{+}_{1}(\vec{k}) b^{-}_{1}(\vec{k}) - b^{+}_{2}(\vec{k}) b^{-}_{2}(\vec{k})\right]. \tag{1.99}$$

According to (1.98) and (1.99), it follows that the products

$$n_s(\vec{k}) = b^{+}_{s}(\vec{k}) b^{-}_{s}(\vec{k}), \quad (s = 1, 2) \tag{1.100}$$

can be interpreted as the average densities of particles with zero rest-mass, characterized by momentum $\vec{k}$ and energy $|\vec{k}|$. The projections of the spin angular momentum of these particles on the direction of $\vec{k}$ are $+1$ for $s = 1$, and $-1$ for $s = 2$. The particles with these properties are the photons. Transition to the momentum representation allows one to realize that the electromagnetic field describes the transversal photons, characterized by two possible values of the spin projection on the direction of motion.

### I.2.2.4. The gravitational field

#### Overview

Einstein developed his ideas about the relativistic approach to gravity over many years, culminating with his 1915 papers presented to the Prussian Academy of Science.

Soon after the elaboration of special relativity, Einstein started pondering about the introduction of accelerated motion, in particular the gravity, in his theory. In 1907 a conceptual breakthrough took place, when Einstein realized that Galilei's law of free fall was the key element for expanding the principle of relativity to systems moving non-uniformly relative to each other. The mathematical description of these ideas was still to follow. Its development was partly facilitated by Einstein's personal friendship and collaboration with the Swiss mathematician Marcel Grossmann (1879–1936), who introduced Einstein to fundamental concepts in differential geometry and tensor calculus, developed by Elwin Bruno Christoffel (1829–1900), Gregorio Ricci-Curbastro (1853–1925) and Tullio Levi-Civita (1873–1941), as well as Riemannian geometry, initiated in 1854 by the German mathematician Bernhard Riemann (1826–1866). During the following eight years, Einstein worked out the conceptual and most of the technical details of general relativity, whose epitome are the second-order nonlinear differential equations known as *Einstein's field equations* which connect the curvature of space-time to the energy-momentum tensor of matter.

General relativity is a *geometric theory of gravitation*. In this theory, gravity is not seen as a force, but as an intrinsic distortion of space-time. According to Newtonian gravity, particles' trajectories deviate from the straight line due to the gravitational force. In general relativity, matter and energy deform the space-time into a curved manifold, which encodes all the gravitational effects. Particles still move on the straightest possible lines of the distorted space-time.

Newton's theory of gravitation gave a correct description of the majority of gravitational phenomena, but there were some effects which could not be explained within the Newtonian framework. Instead, the Einstein's theory was brilliantly confirmed by the astrophysical and experimental observations.

General relativity and the modern theories of gravitation have developed tremendously during the past century, influencing and being influenced by the progress in mathematics, astrophysics, and cosmology.

Even if, chronologically speaking, gravity was the first studied and – up to some extent – understood interaction, due to genius of some great scientists like Copernicus, Kepler, Galilei, Newton (and others), in fact it proved to be the most difficult to apprehend.

Indeed, there are several still unsolved problems regarding gravity, such as: the nature of the gravitational field, the enormous difference between strength of the gravitational interaction and of the other three fundamental forces known so far (more than 30 orders of magnitude), the way of action of the gravitational force at a scale smaller than $10^{-4}$ m, the absence of a generally valid theory of gravitational field quantization (only the weak gravitational field has been nowadays quantized), the difficulties arising from the gravitational field singularities, the huge difference (approximately 120 orders of magnitude) between the energy density corresponding to the cosmological constant (determined by astronomical observations) and the value predicted by the quantum field theory, and the list can be continued.

As previously mentioned, the most actual theory on gravitation is the general theory of relativity, based on *Einstein's equations* (in this investigation, Einstein's equations shall be written without the cosmological constant[4]),

$$R_{\mu\nu} - \frac{1}{2} R\, g_{\mu\nu} = \chi T_{\mu\nu}, \tag{1.101}$$

where

$$R_{\mu\nu} - \frac{1}{2} R\, g_{\mu\nu} \equiv G_{\mu\nu} \tag{1.102}$$

is the *Einstein's tensor*,

$$R_{\mu\nu} = R^{\lambda}_{\mu\lambda\nu} = \frac{\partial \Gamma^{\lambda}_{\mu\nu}}{\partial x^{\lambda}} - \frac{\partial \Gamma^{\lambda}_{\mu\lambda}}{\partial x^{\nu}} + \Gamma^{\rho}_{\mu\nu}\Gamma^{\lambda}_{\rho\lambda} - \Gamma^{\rho}_{\mu\lambda}\Gamma^{\lambda}_{\rho\nu} \tag{1.103}$$

is the *Ricci* (*curvature*) *tensor*, $g_{\mu\nu}$ is the (*fundamental*) *metric tensor*, $ds^2 = g_{\mu\nu} dx^{\mu} dx^{\nu}$ is the *metric* of the Riemann space-time, $R$ is the *scalar curvature* (or *curvature invariant*, or *Ricci scalar*) which is given by the contracted product

$$R = g^{\mu\nu} R_{\mu\nu}, \tag{1.104}$$

$\chi$ is an universal constant (related to both the Newtonian constant of gravitation $G$ and the speed of light in vacuum $c$), and $T_{\mu\nu}$ is the *energy-momentum tensor of the matter fields*. In definition of

[4]Einstein's equations with cosmological constant $\Lambda$ are $R_{\mu\nu} - \frac{1}{2} R\, g_{\mu\nu} + \Lambda\, g_{\mu\nu} (\equiv G_{\mu\nu} + \Lambda\, g_{\mu\nu}) = \chi T_{\mu\nu}$.

the Ricci tensor appears the *Riemann tensor* (or the *Riemann–Christoffel curvature tensor*) given by

$$R^{\rho}_{\lambda\mu\nu} = \frac{\partial \Gamma^{\rho}_{\lambda\nu}}{\partial x^{\mu}} - \frac{\partial \Gamma^{\rho}_{\lambda\mu}}{\partial x^{\nu}} + \Gamma^{\sigma}_{\lambda\nu}\Gamma^{\rho}_{\sigma\mu} - \Gamma^{\sigma}_{\lambda\mu}\Gamma^{\rho}_{\sigma\nu}, \tag{1.105}$$

where

$$\Gamma^{\rho}_{\mu\nu} = \frac{1}{2} g^{\lambda\rho} \left( \frac{\partial g_{\lambda\nu}}{\partial x^{\mu}} + \frac{\partial g_{\mu\lambda}}{\partial x^{\nu}} - \frac{\partial g_{\mu\nu}}{\partial x^{\lambda}} \right) \tag{1.106}$$

are the *Christoffel's symbols of the second kind*, also called *connection coefficients*.

Einstein's equations are a set of ten coupled *nonlinear* partial differential equations for the metric components. Nonlinearity distinguishes general relativity from many other physical theories (Maxwell's equations are linear in the electric and magnetic fields, Schrödinger's equation is linear in the wave function, etc.) and makes it so difficult to quantize, as the superposition principle used customarily in quantum mechanics and the theory of quantized fields is no more valid.

In the case of *weak gravitational fields*, Einstein's equations can be linearized to the first approximation. Since the field is weak, the space is only slightly curved, which means that $g_{\mu\nu}$ differs by a small amount from the metric tensor of Minkowski space, which is denoted by $\eta_{\mu\nu}$. Mathematically transposed, these considerations lead to the linear approximation of the general relativity, which allows one the quantization of a weak gravitational field.

If $T'^{\nu}_{\ \mu}$ is the generalized energy-momentum tensor of the matter together with the gravitational field, and the momentum covariant four-vector is defined (in the natural unit system $c = 1$; $\hbar = 1$) as

$$P_{\mu} = \int T'^{\nu}_{\ \mu}\, dS_{\nu}, \tag{1.107}$$

then the corresponding conservation equation writes

$$\frac{\partial}{\partial x^{\nu}} \left( \sqrt{-g}\, T'^{\nu}_{\ \mu} \right) = 0, \tag{1.108}$$

where

$$dS_0 = \sqrt{-g}\, dx^1 dx^2 dx^3, ..., dS_3 = \sqrt{-g}\, dx^0 dx^1 dx^2, \tag{1.109}$$

and $g = \det g_{\mu\nu}$.

It can be shown that $\nabla_\nu G^\nu_\mu \left(\equiv G^\nu_{\mu;\nu}\right) = 0$, where $\nabla_\mu$ represents the *covariant derivative operator* with respect to coordinate $x^\mu$, sometimes denoted by $;\mu$ as a subscript index. This leads to

$$\nabla_\nu T^\nu_\mu \left(\equiv T^\nu_{\mu;\nu}\right) = 0, \tag{1.110}$$

expressing the fact that the energy-momentum tensor of the matter fields $T_{\mu\nu}$ (interfering in Einstein's equations) does not satisfy the conservation law

$$\frac{\partial}{\partial x^\nu}\left(\sqrt{-g}\,T'^\nu_{\ \mu}\right) = 0, \tag{1.111}$$

so that, in addition to $T_{\mu\nu}$, one must consider an energy-momentum pseudotensor $t_{\mu\nu}$ – the energy-momentum pseudotensor of the gravitational field – which has to be chosen in such a way that

$$T'^\nu_{\ \mu} = T^\nu_\mu + t^\nu_\mu \tag{1.112}$$

satisfies the above mentioned law.

Quantization of the gravitational field presents major difficulties, mostly due to the fact that its theory is nonlinear. For this reason, the following investigation shall only refer to the linear approximation of the general theory of relativity, and quantization of the gravitational field within this approximation as well (see [181]).

### The linear approximation. Gravitational waves

Consider the Minkowski metric in the form

$$ds^2 = dt^2 - dx^2 - dy^2 - dz^2,$$

or

$$ds^2 = \eta_{\mu\nu} dx^\mu dx^\nu, \tag{1.113}$$

where

$$\eta_{\mu\nu} = \begin{pmatrix} 1 & 0 & 0 & 0 \\ 0 & -1 & 0 & 0 \\ 0 & 0 & -1 & 0 \\ 0 & 0 & 0 & -1 \end{pmatrix}, \tag{1.114}$$

the coordinates being

$$x^0 = t,\ x^1 = x,\ x^2 = y,\ x^3 = z \quad (c = 1). \tag{1.115}$$

In the presence of a weak gravitational field the space is only slightly curved, therefore $g_{\mu\nu}$ is not much different from $\eta_{\mu\nu}$ and one can write

$$g_{\mu\nu} = \eta_{\mu\nu} + \kappa h_{\mu\nu}, \tag{1.116}$$

where $\kappa$ is a parameter small enough so as to keep only terms in its first power. Since $g_{\mu\nu}g^{\nu\lambda} = \delta^{\lambda}_{\mu}$, one also has

$$g^{\mu\nu} = \eta^{\mu\nu} - \kappa h^{\mu\nu}. \tag{1.117}$$

Under these conditions, the Christoffel's symbols of the second kind become

$$\Gamma^{\lambda}_{\mu\nu} = \frac{1}{2}\kappa\eta^{\lambda\rho}\left(\frac{\partial h_{\nu\rho}}{\partial x^{\mu}} + \frac{\partial h_{\rho\mu}}{\partial x^{\nu}} - \frac{\partial h_{\mu\nu}}{\partial x^{\rho}}\right), \tag{1.118}$$

and the Ricci tensor, in the same approximation, writes

$$R_{\mu\nu} = \frac{\partial \Gamma^{\lambda}_{\mu\nu}}{\partial x^{\lambda}} - \frac{\partial \Gamma^{\lambda}_{\mu\lambda}}{\partial x^{\nu}}, \tag{1.119}$$

or, in view of the previous equation,

$$R_{\mu\nu} = \frac{1}{2}\kappa\eta^{\lambda\rho}\left(\frac{\partial^2 h_{\mu\lambda}}{\partial x^{\nu}\partial x^{\rho}} + \frac{\partial^2 h_{\nu\rho}}{\partial x^{\lambda}\partial x^{\mu}} - \frac{\partial^2 h_{\mu\nu}}{\partial x^{\lambda}\partial x^{\rho}} - \frac{\partial^2 h_{\lambda\rho}}{\partial x^{\nu}\partial x^{\mu}}\right). \tag{1.120}$$

Denoting

$$\eta^{\mu\nu}h_{\mu\nu} = h^{\nu}_{\nu} = h, \tag{1.121}$$

and admitting that the conditions for equality of the second-order derivatives are fulfilled, the Ricci tensor becomes

$$R_{\mu\nu} = \frac{1}{2}\kappa\left(\frac{\partial^2 h^{\rho}_{\mu}}{\partial x^{\nu}\partial x^{\rho}} + \frac{\partial^2 h^{\lambda}_{\nu}}{\partial x^{\lambda}\partial x^{\mu}} - \eta^{\lambda\rho}\frac{\partial^2 h_{\mu\nu}}{\partial x^{\lambda}\partial x^{\rho}} - \frac{\partial^2 h}{\partial x^{\mu}\partial x^{\nu}}\right), \tag{1.122}$$

while the curvature invariant $R$ writes

$$\begin{aligned} R = \eta^{\mu\nu}R_{\mu\nu} &= \frac{1}{2}\kappa\left(\frac{\partial^2 h^{\rho\nu}}{\partial x^{\nu}\partial x^{\rho}} + \frac{\partial^2 h^{\lambda\mu}}{\partial x^{\lambda}\partial x^{\mu}} - \eta^{\lambda\rho}\frac{\partial^2 h}{\partial x^{\lambda}\partial x^{\rho}} - \eta^{\mu\nu}\frac{\partial^2 h}{\partial x^{\mu}\partial x^{\nu}}\right) \\ &= \kappa\left(\frac{\partial^2 h^{\lambda\rho}}{\partial x^{\lambda}\partial x^{\rho}} - \eta^{\lambda\rho}\frac{\partial^2 h}{\partial x^{\lambda}\partial x^{\rho}}\right). \end{aligned} \tag{1.123}$$

By means of $R_{\mu\nu}$ and $R$ written above, and also using (1.116), the Einstein's tensor $G_{\mu\nu} = R_{\mu\nu} - \frac{1}{2} g_{\mu\nu} R$ can be written as follows,

$$\begin{aligned} G_{\mu\nu} = \frac{1}{2}\kappa \Bigg[ & \frac{\partial^2 h^\rho_\mu}{\partial x^\nu \partial x^\rho} + \frac{\partial^2 h^\lambda_\nu}{\partial x^\lambda \partial x^\mu} - \frac{\partial^2 h}{\partial x^\mu \partial x^\nu} \\ & - \Box \Big( h_{\mu\nu} - h\eta_{\mu\nu} \Big) - \eta_{\mu\nu} \frac{\partial^2 h^{\lambda\rho}}{\partial x^\lambda \partial x^\rho} \Bigg], \end{aligned} \tag{1.124}$$

where $\Box = \eta^{\lambda\rho} \frac{\partial^2}{\partial x^\lambda \partial x^\rho}$ is the D'Alembert operator.

As in the case of the electromagnetic field, where the four-potential $A_\mu$ has to satisfy the Lorenz-gauge condition (which is a constraint condition for the four-vector $A_\mu$, $\mu = 0, 1, 2, 3$), in the case of gravitational field there also exists a subsidiary condition (*Einstein's condition*, resulting from the analysis of the tensor $G_{\mu\nu}$),

$$\frac{\partial h^\nu_\mu}{\partial x^\nu} - \frac{1}{2}\frac{\partial h}{\partial x^\mu} = 0. \tag{1.125}$$

By virtue of Einstein's condition and taking advantage of the equality of mixed partial derivatives, the first three terms of $G_{\mu\nu}$ yield

$$\frac{\partial}{\partial x^\nu}\left( \frac{\partial h^\rho_\mu}{\partial x^\rho} - \frac{1}{2}\frac{\partial h}{\partial x^\mu} \right) + \frac{\partial}{\partial x^\mu}\left( \frac{\partial h^\lambda_\nu}{\partial x^\lambda} - \frac{1}{2}\frac{\partial h}{\partial x^\nu} \right) = 0,$$

while the last term of the same expression of $G_{\mu\nu}$ writes

$$\begin{aligned} -\eta_{\mu\nu} \frac{\partial^2 h^{\lambda\rho}}{\partial x^\lambda \partial x^\rho} &= -\eta_{\mu\nu}\eta^{\rho\sigma} \frac{\partial^2 h^\lambda_\sigma}{\partial x^\lambda \partial x^\rho} = -\eta_{\mu\nu}\eta^{\rho\sigma} \frac{\partial}{\partial x^\rho}\left( \frac{\partial h^\lambda_\sigma}{\partial x^\lambda} \right) \\ &= -\frac{1}{2}\eta_{\mu\nu}\eta^{\rho\sigma} \frac{\partial}{\partial x^\rho}\left( \frac{\partial h}{\partial x^\sigma} \right) \\ &= -\frac{1}{2}\eta_{\mu\nu}\eta^{\rho\sigma} \frac{\partial^2 h}{\partial x^\rho \partial x^\sigma} - -\frac{1}{2}\eta_{\mu\nu}\Box h. \end{aligned}$$

By means of the preceding two relations, the Einstein's tensor $G_{\mu\nu}$ becomes

$$G_{\mu\nu} = -\frac{1}{2}\kappa \, \Box \left( h_{\mu\nu} - \frac{1}{2} h \, \eta_{\mu\nu} \right) = -\frac{1}{2}\kappa \, \Box \tilde{g}_{\mu\nu}, \tag{1.126}$$

where the following notation has been used:

$$\tilde{g}_{\mu\nu} = h_{\mu\nu} - \frac{1}{2} h \, \eta_{\mu\nu}. \tag{1.127}$$

Multiplying this equation by $\eta^{\nu\lambda}$, one obtains

$$\tilde{g}^{\lambda}_{\mu} = h^{\lambda}_{\mu} - \frac{1}{2} h\, \delta^{\lambda}_{\mu}. \tag{1.128}$$

Setting $\mu = \lambda$, performing summation over $\mu$ and recalling that $h^{\mu}_{\mu} = h$, one has

$$\tilde{g}^{\mu}_{\mu} \equiv \tilde{g} = h - \frac{1}{2} 4h = -h,$$

and (1.127) becomes

$$h_{\mu\nu} = \tilde{g}_{\mu\nu} - \frac{1}{2} \tilde{g}\, \eta_{\mu\nu}. \tag{1.129}$$

Taking now the derivative of (1.128) with respect to $x^{\lambda}$, one obtains

$$\frac{\partial \tilde{g}^{\lambda}_{\mu}}{\partial x^{\lambda}} = \frac{\partial h^{\lambda}_{\mu}}{\partial x^{\lambda}} - \frac{1}{2} \delta^{\lambda}_{\ \mu} \frac{\partial h}{\partial x^{\lambda}} = \frac{\partial h^{\lambda}_{\mu}}{\partial x^{\lambda}} - \frac{1}{2} \frac{\partial h}{\partial x^{\mu}} = 0,$$

where Einstein's condition has been used. Therefore, Einstein's condition can be written in a simpler form, namely

$$\frac{\partial \tilde{g}^{\lambda}_{\mu}}{\partial x^{\lambda}} = 0. \tag{1.130}$$

In the case of the electromagnetic field, the analogous (Lorenz) condition, which expresses the transversality of the free electromagnetic field, is satisfied only for average values. The theory leads to four types of photons (corresponding to the four components of the field function – the four-potential $A_{\mu}$, $\mu = 0, 1, 2, 3$), but only two types correspond to the real photons, namely the transverse photons. Regarding the weak gravitational field, calculations give 11 types of gravitons, corresponding to the 11 independent quantities describing this field (10 independent components of the symmetric tensor $\tilde{g}_{\mu\nu}$, $\mu, \nu = 0, 1, 2, 3$, and the trace $\tilde{g}^{\mu}_{\mu} \equiv \tilde{g}$, considered as an independent variable). The relation $\tilde{g}^{\mu}_{\mu} = \tilde{g}$, together with Einstein's condition (1.130) remain also valid for average quantities, therefore one can write

$$\langle \tilde{g}^{\mu}_{\mu} \rangle = \langle \tilde{g} \rangle; \quad \left\langle \frac{\partial \tilde{g}^{\lambda}_{\mu}}{\partial x^{\lambda}} \right\rangle = 0. \tag{1.131}$$

These conditions eliminate nine types of gravitons out of 11, and the only remaining two types correspond to the real ones, namely the transverse gravitons.

Consider now Einstein's equations (without the cosmological constant) and choose the universal constant $\chi$ as: $\chi = -\frac{1}{2}\kappa^2$, so that

$$R_{\mu\nu} - \frac{1}{2}Rg_{\mu\nu} = -\frac{1}{2}\kappa^2 T_{\mu\nu}. \tag{1.132}$$

Comparing this equation with (1.126), one obtains

$$\Box\tilde{g}_{\mu\nu} = \kappa T_{\mu\nu}, \tag{1.133}$$

which is a wave-type equation for $\tilde{g}_{\mu\nu}$. The existence of gravitational waves was predicted by Einstein 100 years ago, during developing his ideas on the theory of general relativity. Several years later, a team led by J. Weber performed an experiment intended to put into evidence the gravitational waves, but the results were not concludent, because the effects of the gravitational waves and the experimental determination errors were of the same order of magnitude.

As it has been specified in Introduction, and shall also be mentioned in chapter VIII, the gravitational waves were indirectly put into evidence in 1974 by Russel A. Hulse and Joseph H. Taylor from the Princeton University, New Jersey, USA (Nobel Prize, 1993), and directly by the LIGO experiment (September, 2015).

### The energy-momentum pseudotensor and the Lagrangian formalism

In order to determine the expression of the energy-momentum pseudotensor, $t^{\nu}_{\mu}$, the equation (1.119) – showing that the four-divergence of the energy-momentum tensor of the matter fields (interfering in Einstein's equations) is null – shall be previously explicitated.

Taking into account the operator form of a four-divergence of a second-rank mixed tensor, one can write

$$\nabla_\nu T^{\nu}_{\mu} \equiv T^{\nu}_{\mu;\nu} = \frac{\partial T^{\nu}_{\mu}}{\partial x^{\nu}} + \Gamma^{\nu}_{\nu\rho}T^{\rho}_{\mu} - \Gamma^{\rho}_{\nu\mu}T^{\nu}_{\rho} = 0. \tag{1.134}$$

In view of (1.118), equation (1.134) still writes

$$\begin{aligned}\nabla_\nu T^\nu_\mu \equiv T^\nu_{\mu;\nu} = \frac{\partial T^\nu_\mu}{\partial x^\nu} + \frac{1}{2}\kappa\eta^{\nu\lambda}\left(\frac{\partial h_{\rho\lambda}}{\partial x^\nu} + \frac{\partial h_{\lambda\nu}}{\partial x^\rho} - \frac{\partial h_{\nu\rho}}{\partial x^\lambda}\right)T^\rho_\mu \\ -\frac{1}{2}\kappa\eta^{\rho\lambda}\left(\frac{\partial h_{\mu\lambda}}{\partial x^\nu} + \frac{\partial h_{\lambda\nu}}{\partial x^\mu} - \frac{\partial h_{\nu\mu}}{\partial x^\lambda}\right)T^\nu_\rho = 0.\end{aligned} \tag{1.135}$$

But

$$\begin{aligned}\eta^{\nu\lambda}\left(\frac{\partial h_{\rho\lambda}}{\partial x^\nu} - \frac{\partial h_{\nu\rho}}{\partial x^\lambda}\right) &= \eta^{\nu\lambda}\frac{\partial h_{\rho\lambda}}{\partial x^\nu} - \eta^{\nu\lambda}\frac{\partial h_{\nu\rho}}{\partial x^\lambda} = \eta^{\nu\lambda}\frac{\partial h_{\rho\lambda}}{\partial x^\nu} - \eta^{\lambda\nu}\frac{\partial h_{\lambda\rho}}{\partial x^\nu} \\ &= \eta^{\nu\lambda}\frac{\partial h_{\rho\lambda}}{\partial x^\nu} - \eta^{\nu\lambda}\frac{\partial h_{\rho\lambda}}{\partial x^\nu} = 0,\end{aligned}$$

where in the second term the summation indices $\lambda$ and $\nu$ have been interchanged, and the symmetry property of the tensors $\eta^{\lambda\nu}$ and $h_{\rho\nu}$ has been used. Similarly,

$$\begin{aligned}\eta^{\rho\lambda}\left(\frac{\partial h_{\mu\lambda}}{\partial x^\nu} - \frac{\partial h_{\nu\mu}}{\partial x^\lambda}\right)T^\nu_\rho &= \frac{\partial h_{\mu\lambda}}{\partial x^\nu}T^{\lambda\nu} - \frac{\partial h_{\nu\mu}}{\partial x^\lambda}T^{\lambda\nu} \\ &= \frac{\partial h_{\mu\lambda}}{\partial x^\nu}T^{\lambda\nu} - \frac{\partial h_{\lambda\mu}}{\partial x^\nu}T^{\nu\lambda} \\ &= \frac{\partial h_{\mu\lambda}}{\partial x^\nu}T^{\lambda\nu} - \frac{\partial h_{\mu\lambda}}{\partial x^\nu}T^{\lambda\nu} = 0,\end{aligned}$$

since the tensor $T^{\lambda\nu}$ is also symmetric.

By means of the last two equations, the four-divergence (1.135) becomes

$$\nabla_\nu T^\nu_\mu \equiv T^\nu_{\mu;\nu} = \frac{\partial T^\nu_\mu}{\partial x^\nu} + \frac{1}{2}\kappa\eta^{\nu\lambda}\frac{\partial h_{\lambda\nu}}{\partial x^\rho}T^\rho_\mu - \frac{1}{2}\kappa\eta^{\rho\lambda}\frac{\partial h_{\lambda\nu}}{\partial x^\mu}T^\nu_\rho = 0,$$

or

$$\frac{\partial T^\nu_\mu}{\partial x^\nu} + \frac{1}{2}\kappa\left(\eta^{\nu\lambda}\frac{\partial h_{\lambda\nu}}{\partial x^\rho}T^\rho_\mu - \frac{\partial h_{\lambda\nu}}{\partial x^\mu}T^{\nu\lambda}\right) = 0. \tag{1.136}$$

In order to explicitate the equation of conservation (1.111), it is necessary to know $g = \det g_{\mu\nu}$. Using the linearization relation

(1.116), one may write

$$g = \begin{vmatrix} \eta_{00} + \kappa h_{00} & \kappa h_{01} & \kappa h_{02} & \kappa h_{03} \\ \kappa h_{10} & \eta_{11} + \kappa h_{11} & \kappa h_{12} & \kappa h_{13} \\ \kappa h_{20} & \kappa h_{21} & \eta_{22} + \kappa h_{22} & \kappa h_{23} \\ \kappa h_{30} & \kappa h_{31} & \kappa h_{32} & \eta_{33} + \kappa h_{33} \end{vmatrix}$$

$$= \eta_{00}\eta_{11}\eta_{22}\eta_{33} + \kappa\left(h_{00}\eta_{11}\eta_{22}\eta_{33} + h_{11}\eta_{22}\eta_{33}\eta_{00} + h_{22}\eta_{33}\eta_{00}\eta_{11} + h_{33}\eta_{00}\eta_{11}\eta_{22}\right) + \mathcal{O}(\kappa^2),$$

or, in view of (1.114),

$$g = -1 - \kappa h + \mathcal{O}(\kappa^2),$$

therefore

$$-g \cong 1 + \kappa h, \tag{1.137}$$

since according to the considered approximation $\mathcal{O}(\kappa^2) = 0$. The equation of conservation (1.111) then writes

$$\sqrt{1+\kappa h}\,\frac{\partial T'^{\nu}_{\mu}}{\partial x^\nu} + \frac{\kappa}{2\sqrt{1+\kappa h}}\frac{\partial h}{\partial x^\nu}\,T'^{\nu}_{\mu} = 0,$$

or, dividing by $\sqrt{1+\kappa h}$, neglecting the terms in $\kappa^2$, $\kappa^3$, etc., and using (1.112),

$$\frac{\partial T^\nu_\mu}{\partial x^\nu} + \frac{\partial t^\nu_\mu}{\partial x^\nu} + \frac{1}{2}\kappa\frac{\partial h}{\partial x^\nu}\,T^\nu_\mu + \frac{1}{2}\kappa\frac{\partial h}{\partial x^\nu}\,t^\nu_\mu = 0.$$

Using (1.121) and (1.136), the last equation becomes

$$\frac{\partial t^\nu_\mu}{\partial x^\nu} + \frac{1}{2}\kappa\frac{\partial h}{\partial x^\nu}\,t^\nu_\mu = -\frac{1}{2}\kappa\frac{\partial h_{\nu\rho}}{\partial x^\mu}\,T^{\nu\rho}. \tag{1.138}$$

This equation shows that $t^\nu_\mu$ is multiplied by $\kappa$ and, according to the initial convention $\big(\mathcal{O}(k^2) = 0\big)$ the second term of the l.h.s. can be neglected and it follows that the pseudotensor $t^\nu_\mu$ must satisfy the condition

$$\frac{\partial t^\nu_\mu}{\partial x^\nu} = -\frac{1}{2}\kappa\frac{\partial h_{\nu\rho}}{\partial x^\mu}\,T^{\nu\rho}. \tag{1.139}$$

Using (1.129) and (1.133), one still has

$$\begin{aligned}\frac{\partial t^{\nu}_{\mu}}{\partial x^{\nu}} &= -\frac{1}{2}\kappa\frac{\partial \tilde{g}_{\nu\rho}}{\partial x^{\mu}}\,T^{\nu\rho} + \frac{1}{4}\kappa\eta_{\nu\rho}\frac{\partial \tilde{g}}{\partial x^{\mu}}\,T^{\nu\rho} \\ &= -\frac{1}{2}\frac{\partial \tilde{g}_{\nu\rho}}{\partial x^{\mu}}\,\Box\tilde{g}^{\nu\rho} + \frac{1}{4}\frac{\partial \tilde{g}}{\partial x^{\mu}}\,\Box\tilde{g}\,,\end{aligned}$$

or

$$\frac{\partial t^{\nu}_{\mu}}{\partial x^{\nu}} = -\frac{1}{2}\frac{\partial \tilde{g}_{\nu\rho}}{\partial x^{\mu}}\eta^{\sigma\theta}\frac{\partial^2 \tilde{g}^{\nu\rho}}{\partial x^{\sigma}\partial x^{\theta}} + \frac{1}{4}\frac{\partial \tilde{g}}{\partial x^{\mu}}\eta^{\sigma\theta}\frac{\partial^2 \tilde{g}}{\partial x^{\sigma}\partial x^{\theta}}.$$

On the r.h.s. one makes now a change of summation indices, namely: $\nu \to \lambda$ and $\theta \to \nu$. The result is

$$\begin{aligned}\frac{\partial t^{\nu}_{\mu}}{\partial x^{\nu}} &= -\frac{1}{2}\frac{\partial \tilde{g}_{\lambda\rho}}{\partial x^{\mu}}\eta^{\sigma\nu}\frac{\partial^2 \tilde{g}^{\lambda\rho}}{\partial x^{\sigma}\partial x^{\nu}} + \frac{1}{4}\frac{\partial \tilde{g}}{\partial x^{\mu}}\eta^{\sigma\nu}\frac{\partial^2 \tilde{g}}{\partial x^{\sigma}\partial x^{\nu}} \\ &= \frac{\partial}{\partial x^{\nu}}\left(-\frac{1}{2}\frac{\partial \tilde{g}_{\lambda\rho}}{\partial x^{\mu}}\eta^{\sigma\nu}\frac{\partial \tilde{g}^{\lambda\rho}}{\partial x^{\sigma}}\right) + \frac{1}{2}\frac{\partial^2 \tilde{g}_{\lambda\rho}}{\partial x^{\mu}\partial x^{\nu}}\eta^{\sigma\nu}\frac{\partial \tilde{g}^{\lambda\rho}}{\partial x^{\sigma}} \\ &\quad + \frac{\partial}{\partial x^{\nu}}\left(\frac{1}{4}\frac{\partial \tilde{g}}{\partial x^{\mu}}\eta^{\sigma\nu}\frac{\partial \tilde{g}}{\partial x^{\sigma}}\right) - \frac{1}{4}\frac{\partial^2 \tilde{g}}{\partial x^{\mu}\partial x^{\nu}}\eta^{\sigma\nu}\frac{\partial \tilde{g}}{\partial x^{\sigma}}.\end{aligned}$$

The second term of the last equation can be transformed as follows,

$$\begin{aligned}\frac{1}{2}\frac{\partial^2 \tilde{g}_{\lambda\rho}}{\partial x^{\mu}\partial x^{\nu}}\eta^{\sigma\nu}\frac{\partial \tilde{g}^{\lambda\rho}}{\partial x^{\sigma}} &= \frac{1}{4}\left(\frac{\partial^2 \tilde{g}_{\lambda\rho}}{\partial x^{\mu}\partial x^{\nu}}\eta^{\sigma\nu}\frac{\partial \tilde{g}^{\lambda\rho}}{\partial x^{\sigma}} + \frac{\partial^2 \tilde{g}_{\lambda\rho}}{\partial x^{\mu}\partial x^{\nu}}\eta^{\sigma\nu}\frac{\partial \tilde{g}^{\lambda\rho}}{\partial x^{\sigma}}\right) \\ &= \frac{1}{4}\left(\frac{\partial^2 \tilde{g}_{\lambda\rho}}{\partial x^{\mu}\partial x^{\nu}}\eta^{\sigma\nu}\frac{\partial \tilde{g}^{\lambda\rho}}{\partial x^{\sigma}} + \frac{\partial^2 \tilde{g}_{\lambda\rho}}{\partial x^{\mu}\partial x^{\sigma}}\eta^{\nu\sigma}\frac{\partial \tilde{g}^{\lambda\rho}}{\partial x^{\nu}}\right) \\ &= \frac{1}{4}\left(\frac{\partial^2 \tilde{g}_{\lambda\rho}}{\partial x^{\mu}\partial x^{\nu}}\eta^{\sigma\nu}\frac{\partial \tilde{g}^{\lambda\rho}}{\partial x^{\sigma}} + \frac{\partial \tilde{g}_{\lambda\rho}}{\partial x^{\nu}}\eta^{\sigma\nu}\frac{\partial^2 \tilde{g}^{\lambda\rho}}{\partial x^{\mu}\partial x^{\sigma}}\right) \\ &= \frac{\partial}{\partial x^{\mu}}\left(\frac{1}{4}\frac{\partial \tilde{g}_{\lambda\rho}}{\partial x^{\nu}}\eta^{\sigma\nu}\frac{\partial \tilde{g}^{\lambda\rho}}{\partial x^{\sigma}}\right),\end{aligned}$$

where the summation indices $\nu$ and $\sigma$ have been interchanged, the tensors $\tilde{g}^{\lambda\rho}$ and $\tilde{g}_{\lambda\rho}$ have been replaced with each other, and the symmetry property of $\eta^{\sigma\nu}$ has been used. In a similar manner, one obtains

$$\frac{1}{4}\frac{\partial^2\tilde{g}}{\partial x^\mu \partial x^\nu}\eta^{\sigma\nu}\frac{\partial\tilde{g}}{\partial x^\sigma} = \frac{\partial}{\partial x^\mu}\left(\frac{1}{8}\frac{\partial\tilde{g}}{\partial x^\nu}\eta^{\sigma\nu}\frac{\partial\tilde{g}}{\partial x^\sigma}\right).$$

In view of the above results, the equation for $\partial t^\nu_\mu/\partial x^\nu$ becomes

$$\begin{aligned}\frac{\partial t^\nu_\mu}{\partial x^\nu} &= \frac{\partial}{\partial x^\nu}\left(-\frac{1}{2}\eta^{\sigma\nu}\frac{\partial\tilde{g}_{\lambda\rho}}{\partial x^\mu}\frac{\partial\tilde{g}^{\lambda\rho}}{\partial x^\sigma} + \frac{1}{4}\eta^{\sigma\nu}\frac{\partial\tilde{g}}{\partial x^\mu}\frac{\partial\tilde{g}}{\partial x^\sigma}\right)\\ &\quad + \frac{\partial}{\partial x^\mu}\left(\frac{1}{4}\frac{\partial\tilde{g}_{\lambda\rho}}{\partial x^\nu}\eta^{\sigma\nu}\frac{\partial\tilde{g}^{\lambda\rho}}{\partial x^\sigma} - \frac{1}{8}\frac{\partial\tilde{g}}{\partial x^\nu}\eta^{\sigma\nu}\frac{\partial\tilde{g}}{\partial x^\sigma}\right).\end{aligned}$$

But the second term of the r.h.s. can also be written as

$$\begin{aligned}&\frac{\partial}{\partial x^\mu}\left(\frac{1}{4}\frac{\partial\tilde{g}_{\lambda\rho}}{\partial x^\nu}\eta^{\sigma\nu}\frac{\partial\tilde{g}^{\lambda\rho}}{\partial x^\sigma} - \frac{1}{8}\frac{\partial\tilde{g}}{\partial x^\nu}\eta^{\sigma\nu}\frac{\partial\tilde{g}}{\partial x^\sigma}\right)\\ &\qquad = \frac{\partial}{\partial x^\nu}\left(\frac{1}{4}\frac{\partial\tilde{g}_{\lambda\rho}}{\partial x^\theta}\eta^{\sigma\theta}\frac{\partial\tilde{g}^{\lambda\rho}}{\partial x^\sigma} - \frac{1}{8}\frac{\partial\tilde{g}}{\partial x^\theta}\eta^{\sigma\theta}\frac{\partial\tilde{g}}{\partial x^\sigma}\right)\delta^\nu_\mu,\end{aligned}$$

so that

$$\begin{aligned}\frac{\partial t^\nu_\mu}{\partial x^\nu} &= \frac{\partial}{\partial x^\nu}\left[-\frac{1}{2}\eta^{\sigma\nu}\frac{\partial\tilde{g}_{\lambda\rho}}{\partial x^\mu}\frac{\partial\tilde{g}^{\lambda\rho}}{\partial x^\sigma} + \frac{1}{4}\eta^{\sigma\nu}\frac{\partial\tilde{g}}{\partial x^\mu}\frac{\partial\tilde{g}}{\partial x^\sigma}\right.\\ &\quad \left. + \left(\frac{1}{4}\frac{\partial\tilde{g}_{\lambda\rho}}{\partial x^\theta}\eta^{\sigma\theta}\frac{\partial\tilde{g}^{\lambda\rho}}{\partial x^\sigma} - \frac{1}{8}\frac{\partial\tilde{g}}{\partial x^\theta}\eta^{\sigma\theta}\frac{\partial\tilde{g}}{\partial x^\sigma}\right)\delta^\nu_\mu\right].\end{aligned}$$

Consequently, the energy-momentum pseudotensor of the free gravitational field, $t^\nu_\mu$, has the following form:

$$\begin{aligned}t^\nu_\mu &= -\frac{1}{2}\eta^{\sigma\nu}\left(\frac{\partial\tilde{g}_{\lambda\rho}}{\partial x^\mu}\frac{\partial\tilde{g}^{\lambda\rho}}{\partial x^\sigma} - \frac{1}{2}\frac{\partial\tilde{g}}{\partial x^\mu}\frac{\partial\tilde{g}}{\partial x^\sigma}\right)\\ &\quad + \frac{1}{4}\eta^{\sigma\theta}\left(\frac{\partial\tilde{g}_{\lambda\rho}}{\partial x^\theta}\frac{\partial\tilde{g}^{\lambda\rho}}{\partial x^\sigma} - \frac{1}{2}\frac{\partial\tilde{g}}{\partial x^\theta}\frac{\partial\tilde{g}}{\partial x^\sigma}\right)\delta^\nu_\mu.\end{aligned} \tag{1.140}$$

Let $\mathcal{H}$ be the Hamiltonian density of the free gravitational field in the linear approximation (weak field). According to (1.140) then one can write

$$
\begin{aligned}
-\mathcal{H} = t_0^0 &= -\frac{1}{2}\eta^{\sigma 0}\left(\frac{\partial \tilde{g}_{\lambda\rho}}{\partial x^0}\frac{\partial \tilde{g}^{\lambda\rho}}{\partial x^\sigma} - \frac{1}{2}\frac{\partial \tilde{g}}{\partial x^0}\frac{\partial \tilde{g}}{\partial x^\sigma}\right) \\
&\quad + \frac{1}{4}\eta^{\sigma\theta}\left(\frac{\partial \tilde{g}_{\lambda\rho}}{\partial x^\theta}\frac{\partial \tilde{g}^{\lambda\rho}}{\partial x^\sigma} - \frac{1}{2}\frac{\partial \tilde{g}}{\partial x^\theta}\frac{\partial \tilde{g}}{\partial x^\sigma}\right) \\
&= -\frac{1}{2}\frac{\partial \tilde{g}_{\lambda\rho}}{\partial t}\frac{\partial \tilde{g}^{\lambda\rho}}{\partial t} + \frac{1}{4}\left(\frac{\partial \tilde{g}}{\partial t}\right)^2 + \frac{1}{4}\frac{\partial \tilde{g}_{\lambda\rho}}{\partial x^i}\frac{\partial \tilde{g}^{\lambda\rho}}{\partial x^i} \\
&\quad - \frac{1}{8}\frac{\partial \tilde{g}}{\partial x^i}\frac{\partial \tilde{g}}{\partial x^i} + \frac{1}{4}\frac{\partial \tilde{g}_{\lambda\rho}}{\partial t}\frac{\partial \tilde{g}^{\lambda\rho}}{\partial t} - \frac{1}{8}\left(\frac{\partial \tilde{g}}{\partial t}\right)^2, \quad (i = 1, 2, 3),
\end{aligned}
$$

where the properties of the metric tensor of the Minkowski space, $\eta^{\mu\nu}$, $(\mu, \nu = 0, 1, 2, 3)$ has been considered. Grouping the terms, one finally obtains the Hamiltonian density of the free gravitational field as being

$$
\mathcal{H} = \frac{1}{4}\frac{\partial \tilde{g}_{\lambda\rho}}{\partial t}\frac{\partial \tilde{g}^{\lambda\rho}}{\partial t} - \frac{1}{8}\left(\frac{\partial \tilde{g}}{\partial t}\right)^2 - \frac{1}{4}\frac{\partial \tilde{g}_{\lambda\rho}}{\partial x^i}\frac{\partial \tilde{g}^{\lambda\rho}}{\partial x^i} + \frac{1}{8}\frac{\partial \tilde{g}}{\partial x^i}\frac{\partial \tilde{g}}{\partial x^i}, \quad (i = 1, 2, 3). \tag{1.141}
$$

If summation is performed, one can realize that $\mathcal{H}$ is positively defined. Taking into account that the gravitational field is a real field and choosing as the field function $U_{(r)} \equiv (\tilde{g}_{\lambda\rho},\ \tilde{g})$, where $(r)$ represents a group of two indices (the gravitational field is a second-rank tensor field), for the first ten components $\tilde{g}_{\lambda\rho}$ of the field function, and for the eleventh component $\tilde{g}$ as well, one obtains

$$
t_\mu^\nu = \Pi^{\nu(r)}U_{(r),\mu} - \mathcal{L}\delta_\mu^\nu = \frac{\partial \mathcal{L}}{\partial \tilde{g}_{\lambda\rho,\nu}}\tilde{g}_{\lambda\rho,\mu} + \frac{\partial \mathcal{L}}{\partial \tilde{g}_{,\nu}}\tilde{g}_{,\mu} - \mathcal{L}\delta_\mu^\nu, \tag{1.142}
$$

where $\Pi^{\nu(r)} \equiv \Pi^{\nu\lambda\rho} = \dfrac{\partial \mathcal{L}}{\partial \tilde{g}_{\lambda\rho,\nu}} \equiv \dfrac{\partial \mathcal{L}}{\partial\left(\dfrac{\partial \tilde{g}_{\lambda\rho}}{\partial x^\nu}\right)}$ and $\Pi^\nu = \dfrac{\partial \mathcal{L}}{\partial \tilde{g}_{,\nu}} \equiv$ $\dfrac{\partial \mathcal{L}}{\partial\left(\dfrac{\partial \tilde{g}}{\partial x^\nu}\right)}$ are the generalized momentum densities of the free (weak) gravitational field.

Comparing (1.140) and (1.142), one deduces the Lagrangian density of the free weak gravitational field

$$\mathcal{L} = -\frac{1}{4}\eta^{\mu\nu}\left(\frac{\partial \tilde{g}_{\lambda\rho}}{\partial x^\nu}\frac{\partial \tilde{g}^{\lambda\rho}}{\partial x^\mu} - \frac{1}{2}\frac{\partial \tilde{g}}{\partial x^\nu}\frac{\partial \tilde{g}}{\partial x^\mu}\right). \tag{1.143}$$

Quantization of the free gravitational field is performed by means of Heisenberg's commutation relation

$$\left[\Pi^+_{0(k)}(\vec{x}\,', t),\ U_{(r)}(\vec{x}, t)\right] = \frac{\hbar}{i}\delta_{(k)(r)}\delta(\vec{x}\,' - \vec{x}), \tag{1.144}$$

or, if the natural unit system is used,

$$\left[\Pi^+_{0(k)}(\vec{x}\,', t),\ U_{(r)}(\vec{x}, t)\right] = -i\delta_{(k)(r)}\delta(\vec{x}\,' - \vec{x}), \tag{1.144$'$}$$

and also by decomposition of the field function in positive and negative frequency parts in the framework of the momentum representation. Elimination of the 9 types of virtual gravitons (out of the total of 11) can be done by means of the microscopic form of the supplementary conditions obeyed by the field function (analogous to the Lorenz–Fermi gauge condition encountered in the theory of electromagnetic field). These conditions were given by Einstein for both $\tilde{g}_{\mu\nu}$ and $\tilde{g}$:

$$\frac{\partial \tilde{g}_\mu^{\lambda(+)}}{\partial x^\lambda}|\psi\rangle = 0, \quad \text{(four relations)}, \tag{1.145}$$

$$\left(\eta^{\mu\lambda}\tilde{g}^{(+)}_{\mu\lambda} - \tilde{g}^{(+)}\right)|\psi\rangle = 0, \quad \text{(one relation)}, \tag{1.146}$$

where (+) denotes the positive frequency part of the field function. Obviously, by averaging, the microscopic condition (1.145) leads to macroscopic condition $\dfrac{\partial \tilde{g}_\mu^\lambda}{\partial x^\lambda} = 0$, with a similar observation for the component $\tilde{g}$ of the field function.

### I.2.2.5. The spinor (Dirac) field

#### The Dirac equation and the Dirac matrices

Those fields that transform according to the spinorial representations of the Poincaré group are called *spinorial fields*. The quantized spinorial fields describe the fractional-spin particles (1/2, 3/2,

5/2,...). The simplest spinorial field is the spin-1/2 field. Such a field, associated with non-zero rest mass particles, was introduced by Dirac in order to describe the electron, but it can also be used to describe particles such muons, protons, neutrons and some other hyperons. In the investigation to follow, this field shall be called the *Dirac field*. The equation describing this field can be obtained by "factorizing" the Klein–Gordon operator

$$\Box - m^2 = P_\nu P^\nu - m^2,$$

where

$$P_\nu = i\frac{\partial}{\partial x^\nu} = i\partial_\nu.$$

The operator $\Box - m^2$ is quadratic in the derivatives $\partial_\nu$ and cannot be written as a product of two factors linear in $P^\nu$ with numerical coefficients. Indeed, if we try to write the operator $\Box - m^2$ under the form

$$\Box - m^2 = (P + m)(P - m),$$

where $P$ is an arbitrary linear combination of the operators $P_\nu$ with the coefficients $\gamma^\nu$:

$$P = P_\nu \gamma^\nu,$$

then the condition

$$P_\nu P^\nu = (P_\mu \gamma^\mu)^2$$

is also necessary to be fulfilled. Decomposing the r.h.s. of this last relation, one finds that the coefficients $\gamma^\nu$ must obey the following condition:

$$\gamma^\nu \gamma^\mu + \gamma^\mu \gamma^\nu = 2g^{\mu\nu}. \tag{1.147}$$

According to this condition, the quantities $\gamma^\nu$ satisfy an anticommutation relation. Therefore, these quantities are not numbers and can be chosen as matrices. For example, the Klein–Gordon operator can be represented as a product of two matrix operators,

$$\Box - m^2 = (i\gamma^\nu \partial_\nu + m)(i\gamma^\mu \partial_\mu - m).$$

In order that the field function satisfies the Klein–Gordon equation

$$(\Box - m^2)\psi = 0,$$

one can require that the field function satisfies one of the two first-order equations

$$(i\gamma^\nu \partial_\nu - m)\psi(x) = 0, \qquad \text{or} \qquad (i\gamma^\nu \partial_\nu + m)\psi(x) = 0.$$

For the sake of tradition, the first equation shall be chosen as the equation of motion,

$$(i\hat{\partial} - m)\psi(x) = 0; \quad \hat{\partial} \equiv \gamma^\nu \partial_\nu. \tag{1.148}$$

By means of equation (1.148), Dirac succeeded for the first time to describe the electron spin, which is 1/2. This equation is called the *Dirac equation*, and the matrices $\gamma^\nu$ which satisfy equation (1.147) are named *Dirac matrices.*

The Dirac field function is a spinor with four components and can be represented as a $4 \times 1$ matrix,

$$\psi = \begin{pmatrix} \psi_1 \\ \psi_2 \\ \psi_3 \\ \psi_4 \end{pmatrix},$$

which, for economy of space, can also be written as

$$\psi = (\psi_1, \psi_2, \psi_3, \psi_4)^T.$$

Thus, the condensed matrix equation (1.148) is equivalent to the following four components (equations):

$$(i\gamma^\nu \partial_\nu - m\, I)_{\alpha\beta}\, \psi_\beta(x) = 0; \quad (\alpha, \beta = 1, 2, 3, 4).$$

The matrices $\gamma^\nu$ are square matrices of rank four, with the matrix elements $\gamma^\nu_{\alpha\beta}$, and $I$ is the unit matrix of rank four, $I_{\alpha\beta} = \delta_{\alpha\beta}$, which shall further be omitted.

According to (1.147), the Dirac matrices can be chosen as unitary matrices,

$$\gamma^\nu (\gamma^+)^\nu = (\gamma^+)^\nu \gamma^\nu = I; \quad (\text{no summation over } \nu),$$

if the Hermitian conjugation operation can be written as

$$(\gamma^+)^\nu = g_{\nu\mu}\gamma^\mu \equiv \gamma_\nu. \tag{1.149}$$

The Hermitian conjugate spinor $\psi^+$ can be written as a matrix $1 \times 4$, whose components are obtained by the complex conjugation of $\psi$,

$$\psi^+ = (\psi_1^*, \psi_2^*, \psi_3^*, \psi_4^*).$$

Multiplying this spinor on the right by $\gamma^0$, one obtains the so-called *Dirac conjugated spinor*

$$\overline{\psi}(x) = \psi^+(x)\gamma^0.$$

Using (1.149), one can show that the Dirac conjugated spinor satisfies the equation

$$i\partial_\nu \overline{\psi}(x)\gamma^\nu + m\,\overline{\psi}(x) = 0. \tag{1.150}$$

It is worth observing that each component of the spinor $\psi(x)$ satisfies the Klein–Gordon equation. This property can be easily evidentiated by multiplying the Dirac equation (on the left) by the operator $(i\partial\!\!\!/ + m)$, and then making use of the anticommutation relation (1.147).

One can also observe that in the product $\overline{\psi}O\psi$, where $O$ is an arbitrary (but suitable as dimension) matrix, the summation is performed over spinorial indices, that is

$$\overline{\psi}O\psi = \overline{\psi}_\alpha O_{\alpha\beta}\psi_\beta.$$

The result of this operation is a number, while an expression like the product $\psi\overline{\psi}$ represents a matrix,

$$\psi\overline{\psi} = A; \quad A_{\alpha\beta} = \psi_\alpha \overline{\psi}_\beta.$$

### The Lagrangian of the Dirac formalism

Both the Dirac equation (1.148) and its conjugate (1.150) can be obtained through the medium of the Lagrangian density

$$\mathcal{L} = \frac{i}{2}\left[\overline{\psi}(x)\gamma^\nu\partial_\nu\psi(x) - \partial_\nu\overline{\psi}(x)\gamma^\nu\psi(x)\right] - m\overline{\psi}(x)\psi(x), \tag{1.151}$$

by means of the variational principle. The main feature of this Lagrangian density is that it becomes identically null if $\overline{\psi}$ and $\psi$ satisfy the corresponding equations of motion.

Using Noether's theorem and the equations of motion, one finds the energy-momentum tensor

$$T^{\mu\nu} = \frac{i}{2}\left[\overline{\psi}(x)\gamma^\mu\partial^\nu\psi(x) - \partial^\nu\overline{\psi}(x)\gamma^\mu\psi(x)\right], \tag{1.152}$$

and the current vector as well,

$$J^\mu(x) = \overline{\psi}(x)\gamma^\mu\psi(x). \tag{1.153}$$

In order to find the spin tensor, after performing summation over spinorial indices, one uses formula (1.36), and the result is

$$S^{\tau(\mu\nu)} = -\frac{\partial \mathcal{L}}{\partial \psi_{,\tau}} A^{\psi(\mu\nu)} \psi(x) - \overline{\psi}(x) A^{\overline{\psi}(\mu\nu)} \frac{\partial \mathcal{L}}{\partial \overline{\psi}_{,\tau}}.$$

Here, by means of the infinitesimal Lorentz transformations

$$\psi'(x') = \left(1 - \frac{i}{2}\sigma^{\mu\nu}\varphi\right)\psi(x); \quad \overline{\psi}'(x') = \overline{\psi}(x)\left(1 + \frac{i}{2}\sigma^{\mu\nu}\varphi\right)$$

the infinitesimal generators $A^{\psi}$ and $A^{\overline{\psi}}$ are written as (see Appendix B),

$$A^{\psi(\mu\nu)} = -\frac{i}{2}\sigma^{\mu\nu}; \quad A^{\overline{\psi}(\mu\nu)} = \frac{i}{2}\sigma^{\mu\nu},$$

where $\sigma^{\mu\nu}$ is the *spin matrix tensor*. Therefore,

$$S^{\tau(\mu\nu)} = \frac{1}{4}\overline{\psi}(x)\left(\gamma^{\tau}\sigma^{\mu\nu} + \sigma^{\mu\nu}\gamma^{\tau}\right)\psi(x). \tag{1.154}$$

**Momentum representation**

Since the components of the Dirac spinor $\psi_{\alpha}(x), (\alpha = 1, 2, 3, 4)$ satisfy the Klein–Gordon equation, the momentum representation of the Dirac field function shall be writen as follows,

$$\psi(x) = (2\pi)^{-3/2} \int dp\; e^{ipx}\delta(p^2 - m^2)\tilde{\psi}(p),$$

where the amplitude $\tilde{\psi}(p)$ satisfies the Dirac equation written in the momentum representation,

$$(\hat{p} + m)\tilde{\psi}(p)|_{p^2=m^2} = 0.$$

Decomposing the field function in its components $\psi = \psi^{+} + \psi^{-}$ and integrating over $p^0$, one obtains

$$\psi^{\pm}(x) = (2\pi)^{-3/2} \int d\vec{p}\; \psi^{\pm}(\vec{p})e^{\pm ipx}, \tag{1.155}$$

where the following notations have been used:

$$\psi^{\pm}(\vec{p}) = \frac{\theta(p^0)\psi(\pm p)}{2p^0}, \quad p^0 = \sqrt{\vec{p}^2 + m^2}.$$

The spinors $\psi^{\pm}(\vec{p})$ satisfy the matrix equations

$$(m \pm \hat{p})\psi^{\pm}(\vec{p}) = 0, \tag{1.156}$$

while the equations corresponding to the conjugate spinors are

$$\overline{\psi}^{\pm}(\vec{p})(m \mp \hat{p}) = 0. \tag{1.157}$$

The actual form of the components $\psi^{\pm}_{\alpha}$ and $\psi^{\pm}_{\beta}$ depends on the Dirac matrices representation and can be determined as follows. Since the equation (1.156) is covariant, it can be written in a reference frame at rest, given the fact that passing to another reference frame can be easily performed using the transformations shown in Appendix B. Choosing such a convenient frame (in which $\vec{p} = 0$), the formula (1.156) yields

$$(\gamma^0 \pm I)\psi^{\pm}(0) = 0.$$

From here one obtains in the standard representation (see Appendix B),

$$\psi^{-}_{\alpha}(0) = c_1\delta_{1\alpha} + c_2\delta_{2\alpha}; \quad \psi^{+}_{\beta}(0) = c_3\delta_{3\beta} + c_4\delta_{4\beta}. \tag{1.158}$$

Here $\alpha$ and $\beta$ are spinorial indices, and $\delta_{\alpha\beta}$ is the Kronecker symbol. For an arbitrary $\vec{p}$, different from zero, the solution can be obtained by means of (1.158) and the corresponding Lorentz transformations. The equations satisfied by $\psi^{+}$ and $\psi^{-}$ can be written as

$$\left(\pm\gamma^0 p_0 - \vec{\gamma}\cdot\vec{p} + m\right)\psi^{\pm}(\pm\vec{p}) = 0,$$

which differ from each other only by the sign of $p^0$. To each equation correspond two linearly independent solutions. It then follows that the Dirac equation for each given four-vector $p$ (the component $p_0$ has a fixed sign) possesses only two linearly-independent solutions. The same property has the Dirac equation for $\overline{\psi}$ (only two linearly-independent solutions, $p_0$ being fixed). Due to the complex character of solutions of the Dirac equation, it follows that they describe particles with different charges (particles and antiparticles). The existence of two linearly-independent solutions for each spinor shows the fact that these two particles can exist in two different states, distinguishable – as shall further be shown – by the spin projection on the direction of motion.

### Decomposition in terms of the spin states

Denoting the normalized linearly-independent solutions for $p^0 > 0$ (*i.e.* for the first equation (1.156)) by $v^{s,+}(\vec{p})$, and for $p^0 < 0$ (*i.e.* for the second equation (1.156)) by $v^{s,-}(\vec{p})$, $(s = 1, 2)$, the decomposition of the function $\psi^{\pm}(\vec{p})$ in terms of the spin states can be written as

$$\psi_\alpha^{\pm}(\vec{p}) = \sum_{s=1,2} a_s^{\pm}(\vec{p}) v_\alpha^{s,\pm}(\vec{p}) = a_s^{\pm}(\vec{p}) v_\alpha^{s,\pm}(\vec{p}), \tag{1.159}$$

while for the conjugate spinor we have

$$\overline{\psi}_\alpha^{\pm}(\vec{p}) = a_s^{*\pm}(\vec{p}) \overline{v}_\alpha^{s,\pm}(\vec{p}) \equiv a_s^{*\pm}(\vec{p}) \overset{*}{v}_\alpha^{s,\pm}(\vec{p}) \gamma^0. \tag{1.160}$$

Since $\overline{\psi}^+$ and $\overline{\psi}^-$ represent the positive and negative frequency parts of the field function $\overline{\psi}$, the Hermitian conjugate conditions for the normalized spinors $v$ are of the form

$$\left(v^{s,\pm}(\vec{p})\right)^* = v^{*s,\mp}(\vec{p}), \tag{1.161}$$

because the conditions for the orthonormality of the spinors $v$ are

$$\overset{*}{v}^{s,\pm}(\vec{p}) v^{r,\mp}(\vec{p}) = \sum_{\alpha=1}^{4} \overset{*}{v}_\alpha^{s,\pm}(\vec{p}) v_\alpha^{r,\mp}(\vec{p}) = \delta^{sr}. \tag{1.162}$$

Using (1.161), (1.162), and Dirac equations, by means of some purely algebraic transformations, a number of relations for the quadratic forms of the spinors $v$ and $\overline{v} = v^+\gamma^0$ can be obtained. Here are the most important of them:

— the orthogonality condition for the conjugate Dirac spinors,

$$\overline{v}^{s,\pm}(\vec{p}) v^{r,\mp}(\vec{p}) = \pm \frac{m}{p^0} \delta^{sr}; \tag{1.163}$$

— the condition of reciprocal orthogonality of the spinors $v$, having arguments with different relative signs,

$$v^{*s,\pm}(\vec{p}) v^{r,\pm}(-\vec{p}) = 0; \tag{1.164}$$

— the relations involving $\gamma$-matrices,

$$v^{*s,\pm}(\vec{p}) \left(p^n \gamma^l - p^l \gamma^n - m \gamma^n \gamma^l\right) v^{r,\pm}(-\vec{p}) = 0, \tag{1.165a}$$

$$\sum_n p_n v^{*s,\pm}(\vec{p}) \left(\gamma^n \gamma^l - \gamma^l \gamma^n\right) v^{r,\mp}(-\vec{p}) = 0, \quad (n, l = 1, 2, 3); \tag{1.165b}$$

— the formulas of summation over spinorial indices,

$$v_\alpha^{s,\pm}(\vec{p})\overline{v}_\beta^{s,\mp}(\vec{p}) = \frac{(\hat{p} \mp m)_{\alpha\beta}}{2p^0}. \tag{1.166}$$

## Dynamical invariants

Substituting (1.155), (1.159), and (1.160) into (1.152), then integrating $T^{0\nu}$ over variable $x$, and taking into account the orthonormality conditions (1.162), one obtains the energy-momentum four-vector

$$P^\nu = \int d\vec{p}\, p^\nu \left[a_s^{*+}(\vec{p})a_s^-(\vec{p}) - a_s^{*-}(\vec{p})a_s^+(\vec{p})\right]. \tag{1.167}$$

In view of the previous relations, and using the fact that the complex conjugation of the amplitudes $a_s^\pm(\vec{p})$, $(s = 1, 2)$ in momentum representation writes $(a_s^\pm)^* = a_s^{*\mp}$, one can easily observe that the energy expression resulting from (1.167),

$$P^0 = \int d\vec{p}\, p^0 \left[a_s^{*+}(\vec{p})a_s^-(\vec{p}) - a_s^{*-}(\vec{p})a_s^+(\vec{p})\right] \tag{1.168}$$

is not positively defined. A positive determination of the spinorial field energy can only be done in the framework of quantum theory, in terms of the Fermi–Dirac quantization.

To calculate the spin vector, one remarks that in the standard representation (see Appendix B) the spin tensor components $\sigma^{\mu\nu}$ can be expressed in terms of matrices $\alpha_i, (i = 1, 2, 3)$ and $\sigma_j, (j = 1, 2, 3)$ as follows:

$$\sigma^{\mu\nu} = \begin{pmatrix} \sigma^{00} & \sigma^{01} & \sigma^{02} & \sigma^{03} \\ \sigma^{10} & \sigma^{11} & \sigma^{12} & \sigma^{13} \\ \sigma^{20} & \sigma^{21} & \sigma^{22} & \sigma^{23} \\ \sigma^{30} & \sigma^{31} & \sigma^{32} & \sigma^{33} \end{pmatrix} = \begin{pmatrix} 0 & i\alpha_1 & i\alpha_2 & i\alpha_3 \\ -i\alpha_1 & 0 & \sigma_3 & -\sigma_2 \\ -i\alpha_2 & -\sigma_3 & 0 & \sigma_1 \\ -i\alpha_3 & \sigma_2 & -\sigma_1 & 0 \end{pmatrix}. \tag{1.169}$$

Thus, taking $\tau = 0$ and $\mu, \nu = 1, 2, 3$ in (1.154), one observes that the spatial density of the spin vector is expressed in terms of the matrix vector $\vec{\sigma}$,

$$\vec{S} = \frac{1}{2}\int \psi^*(x)\vec{\sigma}\,\psi(x)\,dx. \tag{1.170}$$

Unlike the massive vectorial and electromagnetic fields, the spin vector (1.170) of the Dirac field is not conserved in time (due to the absence of symmetry of the energy-momentum four-vector). But if the field functions $\psi$ and $\overline{\psi}$ do not depend on some components $x^0, x^1, ...,$ the equation of continuity can be satisfied by certain components of the tensor $S^{\tau(\mu\nu)}$, meaning that the corresponding integrals are also conserved over time. For instance, supposing that $\frac{\partial}{\partial x_1} = \frac{\partial}{\partial x_2} = 0$, one obtains

$$\partial_\nu S^{\nu(12)} = 0.$$

It then follows that the projection of the spin vector on $Ox_3$-axis,

$$S_3 = \int S^{0(12)} \, d\vec{x}$$

is conserved over time. In momentum representation, this case corresponds to the projection of the spin vector along the direction of motion. Going to the three-dimensional momentum representation in (1.170) and performing the corresponding integration (over the three-dimensional configuration space), one finds

$$\vec{S} = \frac{1}{2} \int d\vec{p} \Big\{ \psi^{*+}(\vec{p})\vec{\sigma}\psi^-(\vec{p}) + \psi^{*-}(\vec{p})\vec{\sigma}\psi^+(\vec{p}) \\ + e^{2ip^0 x_0}\psi^{*+}(\vec{p})\vec{\sigma}\psi^+(\vec{p}) + e^{-2ip^0 x_0}\psi^{*-}(\vec{p})\vec{\sigma}\psi^-(\vec{p}) \Big\}.$$

Let analyze only the $S_3$-component, as is usually done; for this purpose the equation (1.165$a$) should be considered, but written in the form

$$v^{*\pm}(\vec{p})\vec{\sigma}v^\pm(-\vec{p}) = -\frac{i}{m} v^{*\pm}(\vec{p}) \, [\vec{p} \times \vec{\gamma}] \, v^\pm(-\vec{p}).$$

According to this equation, the time-dependent terms of $S_3$ disappear in the reference frame characterized by $p_1 = p_2 = 0$, and one can write

$$S_3 = \frac{1}{2} \int d\vec{p} \Big[ a_s^{*+}(\vec{p}) a_r^-(\vec{p}) v^{*s,+}(\vec{p}) \sigma_3 v^{r,-}(\vec{p}) \\ + a_s^{*-}(\vec{p}) a_r^+(\vec{p}) v^{*s,-}(\vec{p}) \sigma_3 v^{r,+}(\vec{p}) \Big]. \tag{1.171}$$

For a more concreteness of this relation, it is necessary to adopt some kind of particular representation of the Dirac matrices. To put this expression into a more detailed form, it is necessary to choose some particular concrete representation of the Dirac matrices. In the usual standard representation, the matrix $\sigma_3$ is diagonal, having the form

$$\sigma_3 = \begin{pmatrix} 1 & . & . & . \\ . & -1 & . & . \\ . & . & 1 & . \\ . & . & . & -1 \end{pmatrix}.$$

In this case, choosing the normalized spinors in the system characterized by $p_1 = p_2 = 0$ as

$$\left.\begin{aligned} v^{1,-} &= N^{-1}\left(1,\ 0,\ \tfrac{p^3}{p^0+m},\ 0\right)^T \\ v^{2,-} &= N^{-1}\left(0,\ 1,\ 0,\ \tfrac{-p^3}{p^0+m}\right)^T \\ v^{1,+} &= N^{-1}\left(\tfrac{p^3}{p^0+m},\ 0,\ 1,\ 0\right)^T \\ v^{2,+} &= N^{-1}\left(0,\ \tfrac{-p^3}{p^0+m},\ 0,\ 1\right)^T \end{aligned}\right\} \tag{1.172}$$

where $N$ is a normalization factor being equal to

$$N = \left[1 + \left(\frac{p^3}{p^0+m}\right)^2\right]^{1/2} = \left(\frac{2p^0}{p^0+m}\right)^{1/2},$$

the formula (1.171) becomes

$$\begin{aligned} S_3 = \frac{1}{2}\int d\vec{p}\Big[& a_1^{*+}(\vec{p})a_1^-(\vec{p}) - a_2^{*+}(\vec{p})a_2^-(\vec{p}) \\ &+ a_1^{*-}(\vec{p})a_1^+(\vec{p}) - a_2^{*-}(\vec{p})a_2^+(\vec{p})\Big]. \end{aligned} \tag{1.173}$$

Comparing expressions for the energy-momentum four vector (1.167), for the spin projection vector (1.173), and for the charge $Q$ resulting from (1.153),

$$Q = \int \psi^*(x)\psi(x)\, d\vec{x} = \int d\vec{p}\ \left[a_s^{*+}(\vec{p})a_s^-(\vec{p}) + a_s^{*-}(\vec{p})a_s^+(\vec{p})\right], \tag{1.174}$$

one can conclude that the Dirac (spinorial) field describes charged particles, whose possible spin projections on a given axis are $\pm 1/2$. A more detailed analysis of these fundamental quantities shall be given as soon as the quantization is performed.

### I.2.2.6. The spin-vector (Rarita–Schwinger) field

In this paragraph (as well as in the paragraph I.3.8.3, dedicated to the quantization of the massive spin-vectorial field) the following notations shall be used: Roman indices for the components of the three-vectors, Greek indices for the components of the four-vectors, the fourth component of the four-vectors is immaginary, the norm of the four-vector $q = (\vec{q}, iq_0)$ is $q^2 = q_\mu q_\mu = \vec{q}^2 - q_0^2 = -m^2$, the Dirac matrices $\gamma_\mu$, $(\mu = 1, 2, 3, 4)$ are Hermitian, $\gamma_5 = \gamma_1\gamma_2\gamma_3\gamma_4$, $\gamma_\mu\gamma_\nu + \gamma_\nu\gamma_\mu = 2\delta_{\mu\nu}$, $(\mu, \nu = 1, 2, 3, 4)$, $\gamma_\mu^2 = 1$, $(\mu = 1, ..., 5)$, and $\gamma_\mu q_\mu \equiv \gamma \cdot q$, $(\mu = 1, 2, 3, 4)$, $\partial_\mu = \dfrac{\partial}{\partial x_\mu} = \left(\vec{\nabla}, -i\dfrac{\partial}{\partial x_0}\right)$, $\partial_\mu^* = \dfrac{\partial}{\partial x_\mu} = \left(\vec{\nabla}, i\dfrac{\partial}{\partial x_0}\right)$, while if $a_\mu = (\vec{a}, ia_0)$ with $\vec{a} \neq \vec{a}^*$ and $a_0 \neq a_0^*$, then a distinction shall be made between $a^*{}_\mu = (\vec{a}^*, ia_0^*)$ and $a_\mu{}^* = (\vec{a}^*, -a_0{}^*)$, *i.e.* in dealing with complex four-vectors $a_\mu = (\vec{a}, ia_0)$ where $\vec{a} \neq \vec{a}^*$ and $a_0 \neq a_0^*$, it is convenient to work with the conjugate four-vector $a^*{}_\mu = (\vec{a}^*, ia_0^*)$ in which the metric $i$ is left unconjugated.

#### The Bargmann–Wigner equations

Below shall be presented a general system of relativistic wave equations for an arbitrary spin field. In the Bargmann and Wigner's formulation, the field of mass $m$ and spin $s \geq \frac{1}{2}$ is represented by a *completely symmetric multi-spinor* of rank $2s$, $\psi_{\alpha\beta\gamma\ldots\tau}(x)$, that satisfies the Dirac-type equation in all indices,

$$\begin{aligned}
&(\gamma \cdot \partial + m)_{\alpha\alpha'}\psi_{\alpha'\beta\gamma\ldots\tau}(x) = 0, \\
&(\gamma \cdot \partial + m)_{\beta\beta'}\psi_{\alpha\beta'\gamma\ldots\tau}(x) = 0, \qquad (1.175)\\
&\vdots
\end{aligned}$$

Obviously, for $s = \frac{1}{2}$ the system (1.175) reduces to a single equation, namely the Dirac equation for the spin-1/2 massive field. Multiplying the first equation (1.175) by $(\gamma \cdot \partial - m)_{\rho\alpha}$ and performing summation over $\alpha$, one obtains

$$(\partial^2 - m^2)\psi_{\rho\beta\ldots}(x) = 0,$$

which proves the validity of the relation between momentum and energy,

$$E^2 - \vec{p}^{\,2} = m^2,$$

showing that, indeed, the system (1.175) describes massive relativistic particles with a non-zero rest mass $m$. From the above equation it simply results that $E = \pm\sqrt{\vec{p}^{\,2} + m^2}$. The positive solution shall be denoted in what follows by[5] $\varepsilon_{\vec{p}}$, *i.e.* $\varepsilon_{\vec{p}} = \sqrt{\vec{p}^{\,2} + m^2}$.

It can be proved that the completely symmetric spinor of rank $2s$ that satisfies the system (1.175) describes a particle with the spin $s$. To this end, one construct the positive energy plane wave solutions. By setting

$$\psi_{\alpha\beta\gamma\ldots}(x) = w_{\alpha\beta\gamma\ldots}(\vec{p}, \varepsilon_{\vec{p}})e^{i\vec{p}\cdot\vec{x} - i\varepsilon_{\vec{p}}t},$$

then we have

$$\begin{aligned}
&(\gamma \cdot p - im)_{\alpha\alpha'} w_{\alpha'\beta\gamma\ldots\tau}(\vec{p}, \varepsilon_{\vec{p}}) = 0, \\
&(\gamma \cdot p - im)_{\beta\beta'} w_{\alpha\beta'\gamma\ldots\tau}(\vec{p}, \varepsilon_{\vec{p}}) = 0, \qquad (1.176) \\
&\vdots
\end{aligned}$$

which proves the statement.

### The non-zero rest mass case

Consider $m \neq 0$ and take the reference system in which the particle is at rest: $\vec{p} = 0$. Adopting the Dirac–Pauli representation of the Dirac matrices $\gamma_k$, $\gamma_4$ being diagonal with the elements $(+1, +1, -1, -1)$, that is

$$\gamma_i = \begin{pmatrix} 0 & -i\sigma_i \\ i\sigma_i & 0 \end{pmatrix}; \quad \gamma_4 = \begin{pmatrix} I & 0 \\ 0 & -I \end{pmatrix}; \quad I = \begin{pmatrix} 1 & 0 \\ 0 & 1 \end{pmatrix};$$

$$\gamma_5 = \gamma_1\gamma_2\gamma_3\gamma_4 = \begin{pmatrix} 0 & -I \\ -I & 0 \end{pmatrix}; \quad \{\gamma_\mu, \gamma_\nu\} = 2\delta_{\mu\nu},$$

one observes that $w_{\alpha\beta\gamma\ldots}(0, m) \neq 0$ if all indices take only the values 1 and 2. Since $w_{\alpha\beta\gamma\ldots}$ must be symmetric, one can construct the

[5] Unlike this paragraph, in the previous ones the positive energy was denoted by $p^0$.

following complete set of linearly independent solutions for $\varepsilon_{\vec{p}} = m$:

$$\begin{aligned} w^{(0)}_{\alpha\beta\gamma\ldots\tau} &= \delta_{\alpha 1}\delta_{\beta 1}\delta_{\gamma 1}\ldots\delta_{\tau 1}, \\ w^{(1)}_{\alpha\beta\gamma\ldots\tau} &= \delta_{\alpha 2}\delta_{\beta 1}\delta_{\gamma 1}\ldots\delta_{\tau 1} \\ &\quad + \delta_{\alpha 1}\delta_{\beta 2}\delta_{\gamma 1}\ldots\delta_{\tau 1} \\ &\quad + \delta_{\alpha 1}\delta_{\beta 1}\delta_{\gamma 2}\ldots\delta_{\tau 1} \\ &\quad + \ldots \\ w^{(2)}_{\alpha\beta\gamma\ldots\tau} &= \delta_{\alpha 2}\delta_{\beta 2}\delta_{\gamma 1}\ldots\delta_{\tau 1} \\ &\quad + \delta_{\alpha 2}\delta_{\beta 1}\delta_{\gamma 2}\ldots\delta_{\tau 1} \\ &\quad + \ldots \\ w^{(2s)}_{\alpha\beta\gamma\ldots\tau} &= \delta_{\alpha 2}\delta_{\beta 2}\delta_{\gamma 2}\ldots\delta_{\tau 2}. \end{aligned} \tag{1.177}$$

Since the number of the linearly independent states for the angular momentum $s$ is $2s+1$, it is expected that the solutions (1.177) refer to different polarization states of a spin-$s$ particle at rest. To show that this is true indeed, one appeals to the covariant Pauli–Lubanski spin operator,

$$(\omega_\rho)_{\alpha\alpha',\beta\beta',\ldots} = -\frac{1}{2}\varepsilon_{\rho\mu\nu\lambda}\,(\Sigma_{\mu\nu})_{\alpha\alpha',\beta\beta',\ldots}\,\partial_\lambda,$$

which generalizes the covariant Pauli–Lubanski spin-vector

$$\omega_\alpha = -\frac{1}{2}\varepsilon_{\alpha\mu\nu\lambda}\Sigma_{\mu\nu}\partial_\lambda,$$

where the spin tensor $\Sigma_{\mu\nu}$ is given by $\Sigma_{\mu\nu} = \frac{1}{2i}(\gamma_\mu\gamma_\nu - \gamma_\nu\gamma_\mu)$. The operator $\omega_\alpha$ reduces to the spin of a rest particle,

$$(\vec{\omega} = m\vec{\Sigma}),$$

or

$$\omega_\alpha = (\pm m\vec{\Sigma}, 0), \quad \vec{\Sigma} = (\Sigma_1, \Sigma_2, \Sigma_3).$$

$$\Sigma_1 = \Sigma_{23}, \quad \Sigma_2 = \Sigma_{31}, \quad \Sigma_3 = \Sigma_{12},$$

and transforms as a four-vector under Lorentz transformations. In the present case, $\Sigma_{\mu\nu}$ is replaced by the generalized spin-tensor

$$(\Sigma_{\mu\nu})_{\alpha\alpha',\beta\beta',\ldots} = \frac{1}{2i}\Big\{(\gamma_\mu\gamma_\nu - \gamma_\nu\gamma_\mu)_{\alpha\alpha'} + (\gamma_\mu\gamma_\nu - \gamma_\nu\gamma_\mu)_{\beta\beta'} + \ldots\Big\}. \tag{1.178}$$

For the particles with positive energy at rest, $\omega_\alpha$, $(\alpha = 1, 2, 3, 4)$ writes

$$\vec{\omega} = m\vec{\Sigma}, \quad \omega_4 = 0,$$

with the proper values of $\frac{1}{2}(\Sigma_z)_{\alpha\alpha',\beta\beta'\ldots}$ giving the spin projection along the $z$-axis.

Let us now act with the operator $\Sigma_z$ on the solutions (1.177). In the Dirac–Pauli representation the matrix $-i\gamma_1\gamma_2$ is diagonal, with the eigenvalues $+1, -1, +1, -1$, so that, if we act with $\frac{1}{2}\Sigma_z$ upon a solution $w^i_{\alpha\beta\gamma\ldots}$ with $i$ indices equal to 1 and $2s - i$ indices equal to 2, one obtains the eigenvalue $\frac{1}{2}i - \frac{1}{2}(2s - i) = i - s$. But, since the integer number $i$ takes values from 0 to $2s$, it follows that the solutions (1.303) form a set of eigenstates of the operator $\frac{1}{2}\Sigma_z$, with integer eigenvalues between $-s$ and $+s$.

### The zero-rest mass case

In this situation, it is not possible to appeal to a reference frame in which the particle is at rest, for the simple reason that in this frame the particle does not exist. One can choose instead a frame in which the particle moves along a single axis, for example the $z$-axis: $p_\mu = p^0_\mu = (0, 0, p, ip)$. In this case, the system (1.177) reduces to

$$\begin{aligned}
&(\gamma_3\gamma_4)_{\alpha\alpha'} w_{\alpha'\beta\gamma\ldots}(p^0) = i w_{\alpha\beta\gamma\ldots}(p^0), \\
&(\gamma_3\gamma_4)_{\beta\beta'} w_{\alpha\beta'\gamma\ldots}(p^0) = i w_{\alpha\beta\gamma\ldots}(p^0), \\
&\vdots
\end{aligned} \tag{1.179}$$

after a previous multiplication by $\gamma_3$. It is also convenient to choose a representation in which the matrix $\gamma_3\gamma_4$ is diagonal. This is the Weyl's representation of the Dirac's $\gamma_k$ matrices, in which

$$\gamma_j = \begin{pmatrix} 0 & -i\sigma_j \\ i\sigma_j & 0 \end{pmatrix}; \quad \gamma_4 = \begin{pmatrix} 0 & I \\ I & 0 \end{pmatrix}; \quad I = \begin{pmatrix} 1 & 0 \\ 0 & 1 \end{pmatrix};$$

$$\gamma_5 = \gamma_1\gamma_2\gamma_3\gamma_4 = \begin{pmatrix} I & 0 \\ 0 & -I \end{pmatrix}.$$

The diagonal elements of the matrix $\gamma_3\gamma_4$ then are $-i, +i, +i, -1$, so that $w_{\alpha\beta\gamma\ldots}(\vec{p}, \varepsilon_{\vec{p}}) \neq 0$ in this frame if all the indices $\alpha, \beta, \gamma\ldots$

take only the values 2 and 3. If the symmetry requirement is also considered, it leads again to a $(2s+1)$ states of the type (1.177), where all indices 1 and 2 must be replaced everywhere by 2 and, respectively, 3. If the rest mass is null, only two of these states remain as solutions, namely

$$w^{(0)}_{\alpha\beta\gamma\ldots\tau}(p^0) = \delta_{\alpha 2}\delta_{\beta 2}\ldots\delta_{\tau 2}, \tag{1.180$a$}$$

and

$$w^{(2s)}_{\alpha\beta\gamma\ldots\tau}(p^0) = \delta_{\alpha 3}\delta_{\beta 3}\ldots\delta_{\tau 3}, \tag{1.180$b$}$$

since, if $m = 0$, according to (1.176), the solutions of the equations Bargmann–Wigner can be multiplied by any of the following operators,

$$\begin{aligned} (\gamma_5)_{\alpha\alpha'} &= (\gamma_1\gamma_2\gamma_3\gamma_4)_{\alpha\alpha'}, \\ (\gamma_5)_{\beta\beta'} &= (\gamma_1\gamma_2\gamma_3\gamma_4)_{\beta\beta'}, \\ &\vdots \end{aligned} \tag{1.181}$$

without modifying the system. But, since in the Weyl representation the matrix $\gamma_5$ is diagonal, with the eigenvalues $(+1, +1, -1, -1)$, it follows that the only solutions satisfying the above requirement are (1.180$a$) and (1.180$b$).

Therefore, there are only two spin states for $m = 0$ (as compared with $2s+1$ states for the particles with non-zero rest mass), for each energy state (positive or negative). To identify the spin values of the two states one appeals again to the covariant Pauli–Lubanski spin operator (with $-i\partial_\mu \equiv p^0_\mu = (0, 0, p, ip)$),

$$\begin{aligned} \omega_1(p^0) &= (\Sigma_{23} + i\Sigma_{24})p, \\ \omega_2(p^0) &= -(\Sigma_{13} + i\Sigma_{14})p, \\ \omega_3(p^0) &= \Sigma_{12}p, \\ \omega_4(p^0) &= i\Sigma_{12}p. \end{aligned} \tag{1.182}$$

Acting with the first two components of this operator onto solution (1.179), one obtains zero:

$$\omega_1(p^0)_{\alpha\alpha',\beta\beta',\ldots} w^{(i)}_{\alpha'\beta'\ldots}(p^0) = 0, \tag{1.183$a$}$$

and

$$\omega_2(p^0)_{\alpha\alpha',\beta\beta',\ldots} w^{(i)}_{\alpha'\beta'\ldots}(p^0) = 0. \tag{1.183b}$$

Indeed, referring for example to (1.183$a$), this is a consequence of the fact that the operators $\gamma_2\gamma_3$ and $-i\gamma_2\gamma_4$ lead to the same eigenvalue when acting on $w^{(i)}$. This result follows by multiplying each equation of (1.306) by the matrix $\gamma_2\gamma_3$.

For $\omega_3$ one obtains

$$\begin{aligned} \omega_3(p^0)_{\alpha\alpha',\beta\beta',\ldots} w^{(0)}_{\alpha'\beta'\ldots}(p^0) &= -2s\, p\, w^{(0)}_{\alpha\beta\ldots}(p^0), \\ \omega_3(p^0)_{\alpha\alpha',\beta\beta',\ldots} w^{(2s)}_{\alpha'\beta'\ldots}(p^0) &= +2s\, p\, w^{(2s)}_{\alpha\beta\ldots}(p^0). \end{aligned} \tag{1.184}$$

It is easy to show this property because, multiplying each equation of (1.179) by $-i\gamma_1\gamma_2$, one obtains the same eigenvalue as $-\gamma_5$, when acting upon $w$. The result expressed by the equations (1.184) is achieved as a result of the fact that all the $\gamma_5$'s yield eigenvalues $+1$ and $-1$ when acting upon $w^{(0)}$ and $w^{(2s)}$, respectively. The result obtained for $\omega_4$ is the same as for $\omega_3$, except for the factor $i$. Denoting by $s_\mu$ the eigenvalues of the operator $\frac{1}{2}\omega_\mu(p)$, one therefore obtains

$$s_\mu = \pm s\, p_\mu,$$

which generalizes the result obtained for the spin-1/2 particle.

Thus, the spin $s$ of a massless particle is oriented along the direction of motion and can take only two extreme values: $\pm s$. Regarding the spin significance, the main difference between massive and massless particles is that in the first case the particle can be attached to a frame and its spin can have any orientation, while in the second such a reference frame cannot be conceived and the spin can be oriented only along the direction of motion. That is why the massless particles are characterized by helicity states, rather than spin states. It is important to mention that along the direction of motion the orbital angular momentum is null, so that the spin projection on the momentum direction can be identified with the projection of the total angular momentum on that direction.

**Observation**. In spite of their high level of generality, due to technical difficulties arising from their use (especially within the frame of a Lagrangian formalism), the Bargmann–Wigner equations are not suitable for general investigations. Therefore, one usually studies certain particular cases, separately for each value of the spin, so that the relativistic equations of motion of the system become more workable. Below, the special case of spin-3/2 particles shall be discussed.

Thus, for $s = 3/2$, the Bargmann–Wigner equations reduce to the following set of there equations for the symmetric spinor of the third rank $\psi_{\alpha\beta\gamma}$:

$$(\gamma \cdot \partial + m)_{\alpha\alpha'} \psi_{\alpha'\beta\gamma}(x) = 0, \tag{1.185a}$$

$$(\gamma \cdot \partial + m)_{\beta\beta'} \psi_{\alpha\beta'\gamma}(x) = 0, \tag{1.185b}$$

$$(\gamma \cdot \partial + m)_{\gamma\gamma'} \psi_{\alpha\beta\gamma'}(x) = 0. \tag{1.185c}$$

Following Abdus Salam's idea, the spinor $\psi_{\alpha\beta\gamma}(x)$ can be written in a form similar to that of the massive vectorial field, namely

$$\psi_{\alpha\beta\gamma}(x) = (\gamma_\mu C)_{\alpha\beta}\, \psi^\mu_\gamma(x) + \frac{1}{2}(\Sigma_{\mu\nu} C)_{\alpha\beta}\, \psi^{\mu\nu}_\gamma(x), \tag{1.186}$$

where $\psi^\mu_\gamma$ is a *spin-vector*, that is a quantity that obeys the transformation rule of a product of a four-vector and a Dirac spinor, $\psi^{\mu\nu}_\gamma$ is a *spin-2 tensor* that transforms as the product of a second-rank antisymmetric tensor and a Dirac spinor, and $C$ is the charge conjugation matrix defined in the frame of the Dirac–Pauli representation as

$$C^{-1}\gamma_\mu C = -\gamma^T_\mu, \quad C^+ = C^{-1}, \quad C^T = -C, \quad C^{-1}\gamma_5 C = \gamma^T_5,$$

which properties are satisfied by

$$C = \gamma_2\gamma_4 = \begin{pmatrix} 0 & i\sigma_2 \\ i\sigma_2 & 0 \end{pmatrix}.$$

The charge conjugate Dirac spinor is defined as $\psi^C = C\overline{\psi}^T$. The complete symmetry of $\psi_{\alpha\beta\gamma}(x)$ with respect to the three indices is assured by requirement that, as a result of its contraction with the second-rank independent antisymmetric spinors $C^{-1}_{\beta\gamma}$, $(iC^{-1}\gamma_5)_{\beta\gamma}$, and $(iC^{-1}\gamma_5\gamma_\lambda)_{\beta\gamma}$, one obviously obtains zero. This requirement leads to the following three conditions,

$$\gamma_\mu \psi_\mu + \frac{1}{2}\Sigma_{\mu\nu}\psi_{\mu\nu} = 0, \tag{1.187a}$$

$$\gamma_\mu \gamma_5 \psi_\mu + \frac{1}{2}\Sigma_{\mu\nu}\gamma_5\psi_{\mu\nu} = 0, \tag{1.187b}$$

$$\gamma_\mu \gamma_5 \gamma_\lambda \psi_\mu + \frac{1}{2}\Sigma_{\mu\nu}\gamma_5\gamma_\lambda\psi_{\mu\nu} = 0, \tag{1.187c}$$

where, for the sake of simplicity, in each term a spinorial index has been omitted (for example, $\gamma_\mu \psi_\mu$ means in fact $(\gamma_\mu)_{\alpha\gamma}\psi^{\mu}_{\gamma}$). The first two conditions can be replaced by the equivalent set

$$\gamma_\mu \psi_\mu = 0, \tag{1.188a}$$

$$\Sigma_{\mu\nu}\psi_{\mu\nu} = 0, \tag{1.188b}$$

obtained as a result of multiplication of (1.187$c$) by $\gamma_5$, then taking the sum and the difference with (1.187$b$). Using (1.188$a$), the third condition can be transformed into

$$4\psi_\lambda - \Sigma_{\mu\nu}\gamma_\lambda\psi_{\mu\nu} = 0.$$

Using the equation (which can be easily verified),

$$[\Sigma_{\mu\nu}, \gamma_\lambda] = 2i\delta_{\mu\lambda}\gamma_\nu - 2i\delta_{\nu\lambda}\gamma_\mu$$

and the condition (1.188$b$), the third condition also writes

$$\psi_\lambda + i\gamma_\mu\psi_{\mu\lambda} = 0. \tag{1.188c}$$

But, as one can see, (1.188$b$) is a consequence of (1.188$a$) and (1.188$c$). Therefore, in order that $\psi_{\alpha\beta\gamma}$ given by (1.186) to represent a third rank completely symmetric spinor, only two conditions must be fulfilled, namely

$$\gamma_\mu \psi_\mu = 0, \tag{1.189a}$$

$$\gamma_\mu \psi_{\mu\nu} - i\psi_\nu = 0. \tag{1.189b}$$

Consequently, the $16 + 24 = 40$ quantities $\psi_\mu$ and $\psi_{\lambda\nu}$ are subject to $4 + 16 = 20$ constraints, meaning that $\psi_{\alpha\beta\gamma}(x)$ has 20 independent components, as it must have. Indeed, a completely symmetric, third rank tensor, in a four-dimensional vector space, has $(C_4^3)4^{3+0-3} = 20$ components.

### The Rarita–Schwinger equations

In order to apply the Bargmann–Wigner equations to the spin-tensor $\psi_{\alpha\beta\gamma}(x)$ given by (1.186), one proceeds as follows. The equation (1.185$a$) can be written as

$$(\gamma\cdot\partial+m)_{\alpha\alpha'}(\gamma_\mu C)_{\alpha'\beta}\,\psi^\mu_\gamma(x)+\frac{1}{2}(\gamma\cdot\partial+m)_{\alpha\alpha'}(\Sigma_{\mu\nu}C)_{\alpha'\beta}\,\psi^{\mu\nu}_\gamma=0.$$

Multiply now this equation by $(C^{-1}\Sigma_{\lambda\rho})_{\beta\alpha}$. To evaluate the appearing traces, one uses the equations

$$\mathrm{Tr}\,\gamma_\nu\gamma_\mu\Sigma_{\lambda\rho}=4i(\delta_{\nu\lambda}\gamma_{\mu\rho}-\delta_{\mu\lambda}\delta_{\nu\rho}),$$

$$\mathrm{Tr}\,\Sigma_{\mu\nu}\Sigma_{\lambda\rho}=4(\delta_{\mu\lambda}\delta_{\nu\rho}-\delta_{\nu\lambda}\delta_{\mu\rho}),$$

and the result is

$$im\psi_{\lambda\rho}=\partial_\lambda\psi_\rho-\partial_\rho\psi_\lambda, \tag{1.190}$$

which determines $\psi_{\lambda\rho}$ in terms of $\psi_\mu$. In order that (1.190) is compatible with the symmetry conditions (1.189$a$) and (1.189$b$), one must have

$$(\gamma\cdot\partial+m)\psi_\lambda(x)=0. \tag{1.191}$$

This way, it was shown that each component $\psi_\lambda$ must satisfy the Dirac equation. It is worth mentioning that the same result can be obtained by applying the Bargmann–Wigner equation (1.185$c$) to (1.186). The result is

$$(\gamma_\mu C)_{\alpha\beta}\Big[(\gamma\cdot\partial+m)\psi_\mu\Big]_\gamma+\frac{1}{2}(\Sigma_{\mu\nu}C)_{\alpha\beta}\Big[(\gamma\cdot\partial+m)\psi_{\mu\nu}\Big]_\gamma=0. \quad (*)$$

The equation (1.191) is obtained by multiplying (*) by $(C^{-1}\gamma_\lambda)_{\beta\alpha}$ and performing the corresponding calculations.

In conclusion, the components of the spin-tensor $\psi_{\mu\nu}$ can be regarded as dependent variables, defined in terms of $\psi_\lambda$ via (1.190). The spin-3/2 field can therefore be completely described by means of the spin-vector $\psi_\lambda$, which satisfies the Dirac equation and condition (1.189$a$). These two equations, rewritten below for convenience as

$$(\gamma \cdot \partial + m)\psi_\lambda(x) = 0, \tag{1.192a}$$

$$\gamma_\lambda \psi_\lambda(x) = 0, \tag{1.192b}$$

are known as the *Rarita–Schwinger equations.* They lead to the constraint

$$\partial_\lambda \psi_\lambda(x) = 0, \tag{1.193}$$

since, according to (1.192$a$) and (1.192$b$),

$$0 = \gamma_\lambda \gamma_\nu \partial_\nu \psi_\lambda = (2\delta_{\lambda\nu} - \gamma_\nu \gamma_\lambda)\partial_\nu \psi_\lambda = 2\partial_\lambda \psi_\lambda.$$

Even more than that, the equations (1.192$a$) and (1.182$b$) can be combined in a single *equation of motion*,

$$\left[(\gamma \cdot \partial + m)\delta_{\mu\lambda} - \frac{1}{3}(\gamma_\mu \partial_\lambda + \gamma_\lambda \partial_\mu) + \frac{1}{3}\gamma_\mu(\gamma \cdot \partial - m)\gamma_\lambda\right]\psi_\lambda(x) = 0. \tag{1.194}$$

Indeed, multiplying (1.194) on the left by $\gamma_\mu$ and then by $\partial_\mu$, one obtains

$$\frac{2}{3}\partial_\lambda \psi_\lambda - \frac{1}{3}m\gamma_\lambda \psi_\lambda = 0, \tag{1.195a}$$

and, respectively,

$$m\partial_\lambda \psi_\lambda + \gamma \cdot \partial \left(\frac{2}{3}\partial_\lambda \psi_\lambda - \frac{1}{3}\gamma_\lambda \psi_\lambda\right) = 0. \tag{1.195b}$$

It follows from the last two equations that $\partial_\lambda \psi_\lambda = 0$, $\gamma_\lambda \psi_\lambda = 0$, and equation (1.192$a$) follows immediately from (1.194).

### The Lorentz invariance of the Rarita–Schwinger equation

The invariance of the Rarita–Schwinger equation (1.194) with respect to a Lorentz transformation

$$x_\mu \to x'_\mu = x_\mu + \delta x_\mu = x_\mu + \omega_{\mu\nu} x_\nu$$

requires that the spin-vector $\psi_\mu(x)$ satisfies the following transformation relation:

$$\begin{aligned}\psi_\mu(x) \to \psi'_\mu(x') &= \psi_\mu(x) + \delta\psi_\mu(x) \\ &= \psi_\mu(x) + \omega_{\mu\nu}\psi_\nu(x) + (i/4)\omega_{\rho\sigma}\Sigma_{\rho\sigma}\psi_\mu(x).\end{aligned}$$

The notation can be conveniently simplified and the last equation writes in the more suggestive form

$$\psi(x) \to \psi'(x') = \psi(x) + \delta\psi(x)$$
$$= \psi(x) + (i/2)\omega_{\rho\sigma}\left(S_{\rho\sigma} + (1/2)\Sigma_{\rho\sigma}\right)\psi(x),$$

where $S_{\rho\sigma}$ and $\Sigma_{\rho\sigma}$ operate on the vectorial and spinorial indices of the spin-vector $\psi$, respectively. The quantity $S_{\rho\sigma} + \frac{1}{2}\Sigma_{\rho\sigma}$ is called the *spin-tensor of the Rarita–Schwinger theory.*

### Current conservation in the Rarita–Schwinger theory

Denoting

$$\psi_\lambda = (\psi_1,\ \psi_2,\ \psi_3,\ \psi_4) \equiv (\psi_1,\ \psi_2,\ \psi_3,\ i\psi_0), \tag{1.196a}$$

the Hermitian conjugate of the spin-vector $\psi_\lambda$ with respect to the spinorial index is defined as

$$\psi_\lambda^+ = (\psi_1^+,\ \psi_2^+,\ \psi_3^+,\ i\psi_0^+). \tag{1.196b}$$

The adjoint field is defined, as in the Dirac theory, by

$$\overline{\psi}_\lambda = \psi_\lambda^+\gamma_4, \tag{1.197}$$

and satisfies the following two equations,

$$\overline{\psi}_\lambda(x)(\gamma \cdot \overleftarrow{\partial} - m) = 0, \tag{1.198a}$$

$$\overline{\psi}_\lambda(x)\gamma_\lambda = 0, \tag{1.198b}$$

or, combined into a single equation,

$$\overline{\psi}_\lambda(x)\left[(\gamma \cdot \overleftarrow{\partial} - m)\delta_{\lambda\mu} - \frac{1}{3}(\gamma_\lambda\overleftarrow{\partial}_\mu + \gamma_\mu\overleftarrow{\partial}_\lambda) + \frac{1}{3}\gamma_\lambda(\gamma \cdot \overleftarrow{\partial} + m)\gamma_\mu\right] = 0. \tag{1.199}$$

Using (1.194) and (1.199), one can deduce the current conservation law

$$\partial_\mu j_\mu(x) = 0, \tag{1.200a}$$

with

$$\begin{aligned} j_\mu = i\Big(&\overline{\psi}_\lambda\gamma_\mu\psi_\lambda - \frac{1}{3}\overline{\psi}_\lambda\gamma_\lambda\psi_\mu - \frac{1}{3}\overline{\psi}_\mu\gamma_\lambda\psi_\lambda \\ &+ \frac{1}{3}\overline{\psi}_\lambda\gamma_\lambda\gamma_\mu\gamma_\nu\psi_\nu\Big) = i\overline{\psi}_\lambda\gamma_\mu\psi_\lambda. \end{aligned} \tag{1.200b}$$

The total charge then is

$$\begin{aligned} Q &= -i \int j_4(x)\, d^3x = \int \psi_\lambda^+(x)\psi_\lambda(x)\, d^3x \\ &= \int (\psi_1^+\psi_1 + \psi_2^+\psi_2 + \psi_3^+\psi_3 - \psi_0^+\psi_0)\, d^3x. \end{aligned} \tag{1.201}$$

Even if the charge density is not a positive-defined quantity, the charge is necessarily positive, since $\psi_0$ cancels out in the reference frame relative to which the particle is at rest.

## The plane-wave solutions of the Rarita–Schwinger equation

These solutions are searched under the form

$$\psi^\lambda(x) \sim w^\lambda(\vec{p}, \varepsilon_{\vec{p}})e^{i\vec{p}\cdot\vec{x} - i\varepsilon_{\vec{p}}t}.$$

There exist solutions for both positive and negative energies. For the positive energies $\varepsilon = +\varepsilon_{\vec{p}}$, denoting $w^\lambda(\vec{p}, \varepsilon_{\vec{p}}) = u_{\vec{p}}^\lambda$, one can write

$$(\gamma \cdot p - im)u_{\vec{p}}^\lambda = 0, \tag{1.202$a$}$$

and

$$\gamma^\lambda u_{\vec{p}}^\lambda = 0. \tag{1.202$b$}$$

Examining these solutions in the rest frame, one observes that for $\vec{p} = 0$, Eq. (1.202$a$) reduces to

$$\gamma_4 u^\lambda = u^\lambda, \quad (\lambda = \overline{1,4}), \tag{1.203$a$}$$

while the subsidiary condition (1.202$b$) writes

$$u^4 + \gamma^i u^i = 0. \quad (i = \overline{1,3}). \tag{1.203$b$}$$

On the other hand, multiplying (1.202$b$) on the left by $\gamma^4$ and using (1.203$a$), one obtains

$$u^4 - \gamma^i u^i = 0,$$

so that in the rest frame $u^0 = 0$ (which justifies taking $\psi^0 = 0$ in (1.201)) and also $\gamma^i u^i = 0$, that is

$$\vec{\sigma} \cdot \vec{u} = 0. \tag{1.204}$$

Therefore, in the rest frame it is necessary to consider only the spin-vector $\vec{u}$. The condition (1.204) furnishes two constraints for the six non-null components of the spin-vector $\vec{u}$, leaving four independent components (*i.e.* $2 \times 3/2 + 1$), as desired. A set of four orthogonal solutions can be constructed starting with the eigenstates of the operator $S_z$, with the eigenvalues $+1, -1$ and $0$:

$$\vec{e}_1 = \frac{1}{\sqrt{2}} \begin{pmatrix} 1 \\ i \\ 0 \end{pmatrix}; \quad \vec{e}_2 = \frac{1}{\sqrt{2}} \begin{pmatrix} 1 \\ -i \\ 0 \end{pmatrix}; \quad \vec{e}_3 = \frac{1}{\sqrt{2}} \begin{pmatrix} 0 \\ 0 \\ 1 \end{pmatrix}. \tag{1.205}$$

With these solutions one can construct four spin-vectors that are eigenstates of the operator $S_z + \frac{1}{2}\sigma_z$, corresponding to the eigenvalues $+3/2, +1/2, -1/2$ and $-3/2$, respectively,

$$\vec{u}_1 = \vec{e}_1 u_+, \tag{1.206$a$}$$

$$\vec{u}_2 = \frac{1}{\sqrt{3}} \vec{e}_1 u_- - \sqrt{\frac{2}{3}}\, \vec{e}_3 u_+, \tag{1.206$b$}$$

$$\vec{u}_3 = \frac{1}{\sqrt{3}} \vec{e}_2 u_+ + \sqrt{\frac{2}{3}}\, \vec{e}_3 u_-, \tag{1.206$c$}$$

$$\vec{u}_4 = \vec{e}_2 u_-. \tag{1.206$d$}$$

These solutions are orthogonal, normalized, and satisfy the subsidiary condition (1.204). The coefficients appearing in (1.206) are precisely the *Clebsch–Gordan coefficients* for coupling spin 1 and 1/2 to obtain spin 3/2, and the subsidiary condition (1.204), while the condition $\vec{\sigma} \cdot \vec{u} = 0$ assures the property that $\vec{u}$ is an eigenstate of the total spin operator $\left(\vec{S} + \frac{1}{2}\vec{\sigma}\right)^2$ with eigenvalue 3/2(3/2+1). In general, the subsidiary condition (1.202*b*) serves to project the part of the spin-vector $u^\lambda$ that refers to the 3/2 spin (*i.e.* plays the role of a projector).

The same principle can be used to construct the solutions for the finite and non-null values of $\vec{p}$, except for the fact that in this case one must use the polarization four-vectors that satisfy the condition $p \cdot e_{\vec{p}} = 0$.

If $\vec{p}$ is oriented along the $z$-axis, the four independent solutions for the positive energy are given by

$$\vec{u}_{\vec{p}1} = \vec{e}_1 u_{\vec{p}1}, \quad u^4_{\vec{p}1} = 0, \tag{1.207$a$}$$

$$\vec{u}_{\vec{p}2} = \frac{1}{\sqrt{3}}\vec{e}_1 u_{\vec{p}2} - \sqrt{\frac{2}{3}\frac{\varepsilon_{\vec{p}}}{m}}\vec{e}_3 u_{\vec{p}1}, \quad u^4_{\vec{p}2} = -i\sqrt{\frac{2}{3}\frac{|\vec{p}|}{m}}u_{\vec{p}1}, \tag{1.207b}$$

$$\vec{u}_{\vec{p}3} = \frac{1}{\sqrt{3}}\vec{e}_2 u_{\vec{p}1} + \sqrt{\frac{2}{3}\frac{\varepsilon_{\vec{p}}}{m}}\vec{e}_3 u_{\vec{p}2}, \quad u^4_{\vec{p}3} = -i\sqrt{\frac{2}{3}\frac{|\vec{p}|}{m}}u_{\vec{p}2}, \tag{1.207c}$$

$$\vec{u}_{\vec{p}4} = \vec{e}_2 u_{\vec{p}2}, \quad u^4_{\vec{p}4} = 0, \tag{1.207d}$$

where $\vec{e}_1$, $\vec{e}_2$ and $\vec{e}_3$ are given by (1.205), and $u_{\vec{p}1}$, $u_{\vec{p}2}$ are the Dirac spinors

$$u_{\vec{p}\sigma} = w_\sigma(\vec{p}, \varepsilon_{\vec{p}}) = \sqrt{\frac{\varepsilon_{\vec{p}} + m}{2m}} \begin{pmatrix} \xi_\sigma \\ \dfrac{\vec{\sigma}\cdot\vec{p}}{\varepsilon_{\vec{p}} + m}\xi_\sigma, \end{pmatrix},$$

$$(\sigma = 1, 2), \quad \xi_1 = \begin{pmatrix} 1 \\ 0 \end{pmatrix}, \quad \xi_2 = \begin{pmatrix} 0 \\ 1 \end{pmatrix},$$

for $\vec{p} = (0, 0, |\vec{p}|)$. The result is also valid for an arbitrary $\vec{p}$, if $u_{\vec{p}1}$ and $u_{\vec{p}2}$ are replaced by the corresponding eigenstates of the helicity operator $\dfrac{\vec{\Sigma}\cdot\vec{p}}{|\vec{p}|}$, and $\vec{e}_i$ $(i = \overline{1,3})$ by the corresponding eigenstates of the operator $\dfrac{\vec{S}\cdot\vec{p}}{|\vec{p}|}$. Thus, the following complete set of solutions can be constructed,

$$\frac{1}{\sqrt{V}}\sqrt{\frac{m}{\varepsilon_{\vec{p}}}}u^\lambda_{\vec{p}\sigma}e^{ip\cdot x}, \quad (\sigma = \overline{1,4}), \tag{1.208a}$$

and

$$\frac{1}{\sqrt{V}}\sqrt{\frac{m}{\varepsilon_{\vec{p}}}}v^\lambda_{\vec{p}\sigma}e^{-ip\cdot x}, \quad (\sigma = \overline{1,4}), \tag{1.208b}$$

where $V$ is the volume of a large box in which the above plane waves are normalized with periodic boundary conditions, while the spin-vectors $u^\lambda_{\vec{p}\sigma}$ and $v^\lambda_{\vec{p}\sigma}$ satisfy the following orthogonality relations and normalization conditions:

$$\sum_\lambda \overline{u}^\lambda_{\vec{p}\sigma}u^\lambda_{\vec{p}\sigma'} = \delta_{\sigma\sigma'}, \tag{1.209a}$$

$$\sum_\lambda \overline{v}^\lambda_{\vec{p}\sigma}v^\lambda_{\vec{p}\sigma'} = -\delta_{\sigma\sigma'}, \tag{1.209b}$$

$$\sum_\lambda \overline{u}^\lambda_{\vec{p}\sigma} v^\lambda_{\vec{p}\sigma'} = \sum_\lambda \overline{v}^\lambda_{\vec{p}\sigma} u^\lambda_{\vec{p}\sigma'} = 0, \tag{1.209c}$$

$$\sum_\lambda \overline{u}^\lambda_{\vec{p}\sigma} \gamma_4 v^\lambda_{-\vec{p}\sigma'} = \sum_\lambda \overline{v}^\lambda_{\vec{p}\sigma} \gamma_4 u^\lambda_{-\vec{p}\sigma'} = 0. \tag{1.209d}$$

To evaluate the sums over spins, namely $P^{\mu\nu}_+(\vec{p}) = \sum_\sigma u^\mu_{\vec{p}\sigma}\overline{u}^\nu_{\vec{p}\sigma}$ and $P^{\mu\nu}_-(\vec{p}) = -\sum_\sigma v^\mu_{\vec{p}\sigma}\overline{v}^\nu_{\vec{p}\sigma}$, one must take into account that, on the basis of Lorentz invariance, they have to be linear combinations of the following ten tensors,

$$\begin{array}{ccccc} \delta_{\mu\nu}, & \gamma_\mu\gamma_\nu, & \dfrac{1}{m}\gamma_\mu p_\nu, & \dfrac{1}{m}\gamma_\nu p_\mu, & \dfrac{1}{m^2}p_\mu p_\nu, \\ \dfrac{\gamma\cdot p}{m}\delta_{\mu\nu}, & \dfrac{\gamma\cdot p}{m}\gamma_\mu\gamma_\nu, & \dfrac{\gamma\cdot p}{m^2}\gamma_\mu p_\nu, & \dfrac{\gamma\cdot p}{m^2}\gamma_\nu p_\mu, & \dfrac{\gamma\cdot p}{m^3}p_\mu p_\nu, \end{array} \tag{1.210}$$

with the coefficients depending only on $p^2 = -m^2$. There are no other terms that can be constructed with the required tensor properties. More than that, $P^{\mu\nu}_\pm$ must satisfy the following conditions:

$$p_\mu P^{\mu\nu}_\pm = P^{\mu\nu}_\pm p_\nu = 0, \tag{1.211a}$$

$$\gamma_\mu P^{\mu\nu}_\pm = P^{\mu\nu}_\pm \gamma_\nu = 0, \tag{1.211b}$$

$$P^{\mu\lambda}_\pm P^{\lambda\nu}_\pm = P^{\mu\nu}_\pm, \tag{1.211c}$$

$$P^{\mu\lambda}_\pm P^{\lambda\nu}_\mp = 0. \tag{1.211d}$$

In the rest frame $p_\mu = (0,0,0,im)$ and (1.211$a$) yields $mP^{4\nu}_\pm = mP^{\mu 4}_\pm = 0$, so that $P^{\mu\nu}_\pm$ reduces to a three-dimensional tensor $P^{ij}_\pm$. According to (1.337), $P^{ij}_\pm$ must have the following general form,

$$P^{ij}_\pm = a\delta_{ij} + b\gamma_i\gamma_j + c\delta_{ij}\gamma_4 p_4 + d\gamma_i\gamma_j\gamma_4 p_4,$$

where $p_4 = im$. The coefficients are determined by virtue of (1.211$b$)–(1.211$d$), and finally one finds

$$P^{ij}_\pm = \pm\frac{\gamma_4 p_4 \pm im}{2im}\left(\delta^{ij} - \frac{1}{3}\gamma^i\gamma^j\right). \tag{1.212}$$

Since in the rest frame

$$(\vec{\gamma}, 0) = \left(\delta_{\mu\nu} - \frac{p_\mu p_\nu}{p^2}\right)\gamma_\nu, \quad \begin{pmatrix} 1 & 0 & 0 & 0 \\ 0 & 1 & 0 & 0 \\ 0 & 0 & 1 & 0 \\ 0 & 0 & 0 & 0 \end{pmatrix} = \left(\delta_{\mu\nu} - \frac{p_\mu p_\nu}{p^2}\right),$$

and

$$\gamma_4 p_4 = \gamma \cdot p,$$

the equation (1.212) can be generalized to an arbitrary reference frame.

To conclude, the covariant form of the projection operators (projectors) is

$$P_+^{\mu\nu}(\vec{p}) = \frac{\gamma \cdot p + im}{2im}\left[\delta^{\mu\nu} - \frac{p^\mu p^\nu}{p^2}\right.$$

$$\left. - \frac{1}{3}\left(\delta^{\mu\rho} - \frac{p^\mu p^\rho}{p^2}\right)\left(\delta^{\nu\sigma} - \frac{p^\nu p^\sigma}{p^2}\right)\gamma^\rho\gamma^\sigma\right]$$

$$= \frac{\gamma \cdot p + im}{2im}\left[\delta^{\mu\nu} - \frac{1}{3}\gamma^\mu\gamma^\nu + \frac{i}{3m}\left(\gamma^\mu p^\nu - \gamma^\nu p^\mu\right) + \frac{2}{3m^2}p^\mu p^\nu\right], \tag{1.213a}$$

$$P_-^{\mu\nu}(\vec{p}) = -\frac{\gamma \cdot p - im}{2im}\left[\delta^{\mu\nu} - \frac{1}{3}\gamma^\mu\gamma^\nu\right.$$

$$\left. - \frac{i}{3m}\left(\gamma^\mu p^\nu - \gamma^\nu p^\mu\right) + \frac{2}{3m^2}p^\mu p^\nu\right]. \tag{1.213b}$$

## I.3. Quantization of free fields

### I.3.1. Field quantization. The sense of quantization

The quantization of the wave fields is analogous (as procedure) to that of non-relativistic quantum mechanics. The specific features of this quantization rests in the fact that, unlike the physical systems with a finite number of degrees of freedom (usually studied by quantum mechanics), the wave fields are systems with an infinite number of degrees of freedom. Thus, an wave field can be conceived as a system with a large number of degrees of freedom, $N$, with $N \to \infty$. Such a system in quantum mechanics is, for example, a structure with a very large number of identical particles. As a particular case, one can consider a system composed of $N$ oscillators.

An important technical feature of quantization procedure is the representation of the occupation numbers of the possible quantum states (occupation number representation). The easiest way to introduce this representation is to analyze the quantum-mechanical oscillator, widely used to study the systems with large numbers of identical particles. Let us remind in short the logic of the reasoning.

The Hamiltonian of the mono-dimensional oscillator with $m = 1$ writes

$$H = \frac{1}{2}\left(p^2 + \omega^2 q^2\right). \tag{1.214}$$

The wave functions $\psi_n$ associated with different oscillator states satisfy the equation

$$H\psi_n = E_n\psi_n; \quad E_n = \omega\left(n + \frac{1}{2}\right), \ (\hbar = 1)$$

and can be determined one from the other by means of the operators $a$ and $a^+$:

$$a^+\psi_n = \sqrt{n+1}\ \psi_{n+1}; \quad a\psi_n = \sqrt{n}\ \psi_{n-1}. \tag{1.215}$$

These operators emerge as certain combinations of the differentiation operators and multiplication by coordinates, in $p$ and/or $q$ variables ($\hbar = 1$):

$$a = \sqrt{\frac{\omega}{2}}\left(\hat{q} + \frac{i\hat{p}}{\omega}\right); \quad a^+ = \sqrt{\frac{\omega}{2}}\left(\hat{q} - \frac{i\hat{p}}{\omega}\right),$$

where $\hat{p} = -i\dfrac{\partial}{\partial q}$, $\hat{q} = q$ in coordinate representation, and $\hat{p} = p$, $\hat{q} = i\dfrac{\partial}{\partial p}$ in momentum representation, while the symbol "ˆ" indicates an operator. The operator $a^+$ increases the quantum state index by one (and the energy by quantity $\omega$); it is called *raising operator*. Correspondingly, the operator $a$ is named *lowering operator*. They are collectively known as *ladder operators*. The basic state satisfies the equation

$$a\psi_0 = 0. \tag{1.216}$$

The numerical factors $\sqrt{n+1}$ and $\sqrt{n}$ are introduced into (1.215) for the state normalization reasons (see further (1.218)). According to (1.215), the quantum states $\psi_n$ are the eigenfunctions of the products $aa^+$ and $a^+a$, with the eigenvalues $n+1$ and $n$, respectively. This way, one can write

$$aa^+ = \hat{n} + 1; \quad a^+a = \hat{n}; \quad \hat{n}\psi_n = n\psi_n,$$

as well as

$$\left.\begin{aligned} [a, a^+] = aa^+ - a^+a = 1, \\ [a, a] = [a^+, a^+] = 0. \end{aligned}\right\} \tag{1.217}$$

The operator $\hat{n}$ is called, naturally, the *state index operator*.

The above considerations allow one to write the Hamiltonian as

$$H = \omega\left(\hat{n} + \frac{1}{2}\right) = \frac{\omega}{2}\left(aa^+ + a^+a\right). \tag{1.218}$$

### I.3.2. Corpuscular interpretation of the occupation number representation

Transition to the occupation numbers consists in the fact that, instead of considering $\psi$ as a function of coordinates (coordinate/configuration representation) or momenta (momentum representation), the state of the system is characterized by means of the number $n$ – the *number of excited states*.

The linear dependence of the Hamiltonian (1.218) on the operator $\hat{n}$, in the harmonic oscillator problem, serves as a basis for the corpuscular interpretation.

Consider an oscillator in an excited state $\psi_n$ $(n \geq 1)$ and suppose that it can be considered as a system of $n$ excited quanta, each of them having the energy $\omega$ $(\hbar = 1)$. The linear dependence of the energy on $n$ leads to the important property of *indiscernibility of excited quanta*. Under such an interpretation, the raising and lowering index operators of excited state number, $a^+$ and $a$, receive the significance of *creation* and *annihilation operators* of the excited quanta. In this case, the index $n$ equals the number of excited quanta being in the state $\psi_n$, and $\hat{n}$ becomes an *operator of the quanta number*.

The state $\psi_n$, characterized by $n$ quanta (particles) can be obtained from the basic/fundamental state $\psi_0$, containing no particles (also called *vacuum state*), by applying $n$ times the creation operator

$$\psi_n = \frac{(a^+)^n}{\sqrt{n!}}\psi_0. \tag{1.219}$$

The numerical factor $1/\sqrt{n!}$ brings the last formula in agreement with formulas (1.215). The relations (1.215)–(1.219) describe the quantum-mechanical oscillator problem within the formalism of the occupation number representation. It should be mentioned that the particles do not correspond to the oscillator itself, but to the quanta of its excitation state. These particles appeared as a result of a simple quantization. In order to investigate systems with multiple degrees of freedom, $N > 1$, one considers $N$ oscillators with different frequencies $\omega_p$ $(p = 1, 2, ..., N)$. In this case the Hamiltonian writes

$$H = \sum_p H_p = \sum_p \omega_p \left(a_p^+ a_p + \frac{1}{2}\right).$$

The operators $a_p^+$, $a_q$ satisfy the commutation relations (1.217) for $p = q$, and commute with each other for $p \neq q$:

$$[a_p, a_q^+] = \delta_{pq}; \quad [a_p, a_q] = [a_p^+, a_q^+] = 0.$$

An arbitrary state of the system is characterized by a row of occupation numbers $n_1, n_2, ..., n_N$. The state

$$\psi(...n_p...) = \prod_{1 \leq p \leq N} \left\{ \frac{(a_p^+)^{n_p}}{\sqrt{n_p!}} \right\} \psi_0 \tag{1.220}$$

contains $n_1$ particles of type 1 with energy $\omega_1$, $n_2$ particles of type 2 with energy $\omega_2$, and so on.

It can be shown that the material scalar field function $u(x)$ that satisfies the Klein–Gordon equation

$$(\Box - m^2)u(x) = 0$$

can be represented as a row (set) of harmonic oscillators. To this end, one passes from the infinite configuration space to a large cube of volume $V = L^3$, and one utilizes the periodic boundary conditions for the solutions of the Klein–Gordon equation,

$$u(x) = u(t, x_1, x_2, x_3) = u(t, x_1 + L, x_2, x_3)$$

$$= u(t, x_1, x_2 + L, x_3) = u(t, x_1, x_2, x_3 + L).$$

Expressing the field function $u(x)$ as a Fourier series,

$$u(t, \vec{x}) = \sum_{p_1, p_2, p_3} \left(\frac{2\pi}{V\omega_p}\right)^{1/2} \left\{a(t, \vec{p})e^{ipx} + a^*(t, \vec{p})e^{-ipx}\right\},$$

one obtains the equation of motion for the amplitude $a(t, \vec{p})$,

$$\ddot{a}(t, \vec{p}) + \omega_p^2\, a(t, \vec{p}) = 0,$$

where $\omega_p^2 = \vec{p}^2 + m^2$. This is the well-known equation of the linear oscillator. Therefore, the amplitude $a(t, \vec{p})$ corresponds to the linear oscillator whose frequency is $\omega_p = \sqrt{\vec{p}^2 + m^2}$.

The periodic boundary condition written in the coordinate space leads to discrete values for momenta,

$$\vec{p}(n_1, n_2, n_3) = \left\{\frac{2\pi}{L}n_1, \frac{2\pi}{L}n_2, \frac{2\pi}{L}n_3\right\}.$$

To each value $\vec{p}(n)$ corresponds an oscillator whose energy is

$$\omega(n) = \left[m^2 + \frac{4\pi^2}{L^2}(n_1^2 + n_2^2 + n_3^2)\right]^{1/2}.$$

This way, to the relativistic scalar field that satisfies the Klein–Gordon equation one associates a Hamiltonian of the linear harmonic oscillator type, the oscillators contributions being denoted by three integers $n_1, n_2, n_3$.

The transition to the real and infinite ($L \to \infty$) configuration space leads to the necessity of definition of a continuous (but bounded) momentum space. In this case, all sums have to be replaced by the corresponding integrals, according to

$$\left(\frac{2\pi}{L}\right)^3 \sum_n \to \int d\vec{p}.$$

Similarly, the Kronecker symbol has to be replaced by the corresponding delta function,

$$\left(\frac{L}{2\pi}\right)^3 \delta_{pp'} \to \delta(\vec{p}-\vec{p}'),$$

as well as the substitution

$$a_p \to \left(\frac{2\pi}{L}\right)^{3/2} a(\vec{p}).$$

In view of all these conditions, the following relations are obtained:

$$\begin{aligned} [a(\vec{p}), a^+(\vec{q})] &= \delta(\vec{p}-\vec{q}), \\ [a(\vec{p}), a(\vec{q})] &= [a^+(\vec{p}), a^+(\vec{q})] = 0. \end{aligned} \tag{1.221}$$

When going from discrete to continuous states, the equation (1.220) keeps its form. The corresponding states shall contain quanta with well-defined (fixed) values for momenta. The wave functions of these states, in the configuration representation, are of plane wave type. Due to this fact, the respective states do not have finite norms.

It is worth mentioning that the field function $u(x) = u(t, \vec{x})$, which satisfies the Klein–Gordon equation, can be represented as an integral over the three-dimensional momentum, in a linear form, by means of the operators $a(\vec{p})$ and $a^+(\vec{p})$, this way becoming itself an operator.

### I.3.3. Canonical quantization

Within the framework of canonical formalism, the basic variables are the generalized coordinates, $q$, and the conjugate generalized momenta, $p$, given by

$$p = \frac{\partial \mathcal{L}}{\partial \dot{q}}; \quad \left(\dot{q} \equiv \frac{\partial q}{\partial t}\right).$$

Written in these variables, the Hamiltonian is given by (1.214). The equation of motion for some dynamical variable $A = A(q,p)$ (which does not explicitly depend on the time), is

$$\frac{dA(q,p)}{dt} = \{A, H\}, \tag{1.222}$$

where the braces are used to define the classical Poisson bracket

$$\{a, b\} = \frac{\partial a}{\partial q}\frac{\partial b}{\partial p} - \frac{\partial a}{\partial p}\frac{\partial b}{\partial q}. \tag{1.223}$$

In particular,

$$\{q, p\} = 1, \tag{1.224}$$

and

$$\dot{q} = \{q, H\} = p; \quad \dot{p} = \{p, H\} = -\omega^2 q.$$

The last relations are the *canonical equations of motion*. Their solutions can be arranged in positive and negative frequency parts,

$$q(t) = \frac{1}{\sqrt{2\omega}}\left[a^{(+)}(t) + a^{(-)}(t)\right]; \quad p(t) = i\sqrt{\frac{\omega}{2}}\left[a^{(+)}(t) - a^{(-)}(t)\right], \tag{1.225}$$

under condition that

$$\dot{a}^{(\pm)}(t) = \pm i\omega a^{(\pm)}(t); \quad a^{(\pm)}(t) = a^{(\pm)}(0)\exp(\pm i\omega t).$$

Inverting relations (1.225) and using (1.214), one obtains

$$\{a^{(-)}(t),\, a^{(+)}(t)\} = -i. \tag{1.226}$$

Substituting (1.225) into the Hamiltonian (1.214), one finds

$$H = \frac{1}{2}\left[a^{(+)}(t)a^{(-)}(t) + a^{(-)}(t)a^{(+)}(t)\right] = \frac{1}{2}(a^+a + aa^+),$$

where the following notations have been used:

$$a^{(-)} \equiv a; \quad a^{(+)} \equiv a^+.$$

Now is the right moment to formulate the quantization procedure, by means of the following *postulate of canonical quantization*:

*One supposes that both the dynamical variables $q, p, a, a^+$, and the functions dependent on these variables (such as the Hamiltonian $H$) are operators acting on the state function $\psi$. The commutation relations for these operators are settled according to the correspondence principle: the classical Poisson bracket appearing in canonical formalism is replaced by the "quantum brucket" (commutator) given by*

$$\{a,\, b\}_{quantum} = \frac{1}{i}[a,\, b] = \frac{1}{i}(ab - ba),$$

*that is*

$$\{a,\, b\}_{classical} \to \frac{1}{i}[a,\, b]. \tag{1.227}$$

Under these circumstances, equation (1.222) for the quantum quantity $A$ writes

$$i\frac{dA}{dt} = [A,\, H], \tag{1.228}$$

and the formulas (1.224) and (1.226) transform into the following commutation relations:

$$[q,\, p] = i; \quad [a,\, a^{+}] = 1. \tag{1.229}$$

In view of the last commutation relation, the Hamiltonian $H$ can be written under the form (1.216). This way, it becomes obvious that the above introduced postulate of canonical quantization for the harmonic oscillator is identically equivalent with the usual formulation of the quantization developed in paragraph I.3.1. Being applied to a physical field, this postulate can stand at the basis of quantization of any field (at least, in principle). Thus, there now exists a quantization mechanism in terms of the occupation numbers.

In order to comparatively analyze various formulations of quantization of systems possessing a great number of degrees of freedom, it would be useful to remind the reader several elementary notions of representations theory in quantum mechanics[6].

## I.3.4. Schrödinger and Heisenberg representations for a quantum physical system

The most widespread representation of a physical system in quantum mechanics is the Schrödinger representation/picture. In this representation, the time-evolution of the system is given through the medium o the wave function $\psi(t)$, as a solution of the Schrödinger's equation

$$i\frac{\partial\psi(t)}{\partial t} = H\psi(t).$$

Here the Hamilton operator $H$ corresponds to the total energy of the system and does not explicitly depend on the time for the conservative systems. In this representation, the dynamical variables

[6] These notions shall be presented in more detail in chapter III.

are characterized by means of certain time-independent operators $B$. Nevertheless, the average value of these operators

$$\overline{B}_t = \psi^*(t) B \, \psi(t) \tag{1.230}$$

can depend on the time by means of the wave function $\psi(t)$.

It therefore follows from (1.230) that the time-evolution of the system, expressed by the dynamical variable $B$, is fulfilled through the medium of the wave function $\psi(t)$. This time-dependence of the system can be entirely expressed by the variable $B$. To this end, a formal integration of the Schrödinger equation yields

$$\psi(t) = U(t)\psi, \tag{1.231}$$

where

$$U(t) = e^{-iHt}. \tag{1.232}$$

By means of (1.231), the r.h.s. of (1.230) can be written as

$$\psi^*(t) B \, \psi(t) = \psi^* U^*(t) B U(t) \psi = \psi^* B_H(t) \psi,$$

under condition

$$B_H(t) = U^*(t) B U(t) = e^{iHt} B e^{-iHt}. \tag{1.233}$$

This way, the average value $\overline{B}$ of the dynamical variable $B$ is expressed as the average value in terms of $\psi$ (which are time-independent) of the operator (1.233) which explicitly depends on the time. Therefore, the transition to Heisenberg representation – with the time-dependent operator instead of time-dependent wave function – has been performed. This fact explains the use of the index $H$ for the operator $B_H$, associated to dynamical variable $B$ (namely, to show that it belongs to Heisenberg representation). Taking the time derivative of (1.233), one obtains the equation of motion for the operator $B_H(t)$:

$$i \frac{dB_H(t)}{dt} = [B_H(t), H]. \tag{1.234}$$

It is worthwhile to mention, therefore, that the time evolution of a physical system can be described in two different ways. In order to pass from the initial moment $t_0 = 0$ to an arbitrary moment $t$ it is necessary either to transform the wave function according to (1.231), or to transform the operator $B$ following relation (1.233). Comparing relations (1.234) and (1.228), one observes that the canonical quantization scheme described in the section I.3.2 leads to the Heisenberg representation for the quantized wave fields.

In the framework of canonical formalism the time $t$ plays a special role as compared to the space coordinates. This fact could lead to the suspicion of absence of the relativistic covariance of the canonical quantization formalism. But, globally seeing the problem, this does not happen: the expressions obtained as a result of the canonical quantization are relativistically-covariant.

There also exists a quantization scheme characterized by the fact that the relativistic covariance maintains, step by step, along the whole quantization procedure. This is the *relativistic scheme of the fields quantization*. This procedure is based on the representation theory or, more exactly, on the requirement of correspondence of the laws of transformation of quantum and classical fields with respect to their transformations, and coordinate transformations as well. A sketch of the steps (levels) of this quantization shall be presented in the following paragraph.

### I.3.5. The relativistic scheme of the field quantization

As a result of quantization, the field functions become operators which are linearly expressed by the particle creation and annihilation operators, satisfying corresponding commutation relations. The operators act on the field function $\Phi$ (a common notations for all fields approached in the considerations to follow).

In agreement with the usual quantum mechanics, the wave function $\Phi$ completely characterizes the physical state of an arbitrary system described by the corresponding quantized field. Similar to the usual field function $\psi$, the quantity $\Phi$ can be considered as a vector of a linear space. Due to this property, the function $\Phi$ is called the *state amplitude* (the *state vector*).

As in quantum mechanics, not all the $\Phi$ vectors possess a finite norm (for example, those corresponding to the plane waves). But one can always choose the norms of the real physical states as being finite. The average values of the dynamical quantities, as well as the transition probability are expressed in terms of quadratic forms involving $\Phi$.

Now, one can determine the transformation properties of the state vectors $\Phi$ in the case of coordinate and field function transformations which have been introduced in I.1.4 and I.2.1.2 (see relations (1.1), (1.3), and (1.22)):

$$x \to x' = P(\omega; x), \quad u(x) \to u'(x') = \Lambda(\omega)u(x).$$

According to superposition principle, to such a transformation one must also correspond a linear transformation of the state vector, namely

$$\Phi \to \Phi' = U(\omega)\Phi, \tag{1.235}$$

which is also unitary by virtue of the norm invariance,

$$U^*(\omega)U(\omega) = 1. \tag{1.236}$$

In the simplest case offered by translation, $u'(x) = u(x - \omega)$, the operator $U$ has the form

$$U(\omega) = \exp\left(iP_\mu\omega^\mu\right). \tag{1.237}$$

The exponential behavior of the dependence on the parameter $\omega$ appears as a result of the group property of the transformation. Due to the unitarity condition (1.236), the translation operators have to be Hermitian, $P_\mu^+ = P_\mu$. In the particular case of the temporal translation, $\omega_\nu = -\delta_{\nu 0}t$, the formula (1.237) corresponds to relation (1.232). This way, the operator $P_\mu$ has the significance of a four-momentum. The relation analogous to (1.233) for the field function operator writes

$$u'(x) = u(x - \omega) = U^{-1}(\omega)u(x)U(\omega). \tag{1.238}$$

For infinitesimal values of $\omega$, the state amplitude transforms according to

$$\Phi' = (1 + iP\omega)\Phi,$$

whose form coincides with the infinitesimal transformation of the field function,

$$u'(x) = (1 + ip\omega)u(x). \tag{1.239}$$

Admitting that $u$ is an ordinary field function, then the coefficient $p_\nu$ can be considered as quanto-mechanical four-momentum operator

$$p_\nu = i\partial_\nu = i\frac{\partial}{\partial x^\nu}.$$

One must emphasize that $p_\nu$ is the translation operator in the space of the field functions $u(x)$, while $P_\nu$ is the translations' generator in the space of the state vectors $\Phi$, *i.e.* in the occupation number (or double quantization) representation.

The above relations can be generalized for the general transformations of the Poincaré group $x' = Lx + a$:

$$\begin{aligned} U &= \exp\left[i\left(P_\nu a^\nu + \frac{1}{2}M_{\mu\nu}\omega^{\mu\nu}\right)\right], \\ \Phi' &= \left(1 + iPa + \frac{i}{2}M\omega\right)\Phi, \end{aligned} \tag{1.240}$$

$$\begin{aligned} u'(x) &= \left(1 + ipa + \frac{i}{2}m\omega\right)u(x), \\ m_{\mu\nu} &= i\left(x_\mu\partial_\nu - x_\nu\partial_\mu\right), \end{aligned} \tag{1.241}$$

where $m_{\mu\nu}$ are the generators of the four-dimensional rotations.

The correspondence reasons allow one to interpret the quantities $M$ and $P$ as being the angular momentum four-tensor and energy-momentum four-vector, respectively.

Similarly, for the gradient transformation of the field functions,

$$\varphi \to \varphi' = i^{i\alpha}\varphi; \quad \varphi^* \to \varphi^{*\prime} = e^{-i\alpha}\varphi^*, \tag{1.242}$$

the unitary operator $U$ of the transformation

$$\Phi' = U_\alpha\Phi$$

has the exponential form

$$U_\alpha = \exp(i\alpha Q), \tag{1.243}$$

the operator $Q$ being interpreted as the charge operator.

As the basic quantization postulate of the wave fields one considers that the Hermitian operators of the energy-momentum four-vector, $P$, of the angular momentum four-tensor, $M$, of the charge,

$Q$, etc. which represent the generators of the infinitesimal transformations of the state vectors (see (1.240) and (1.243)), are expressed in terms of the field function operators by means of the same relations as in the classical field theory (of course, by establishing an appropriate rule of multiplication of these operators). This postulate allows the application of the *correspondence principle* in the following investigation and determines the transformation law of the double-quantization state amplitudes (the double-quantization formalism).

## I.3.6. Commutation relations

### I.3.6.1. Physical significance of the frequency components

Consider the relation (1.238) in the case of an infinitesimal translation. Performing the series expansion in terms of $\omega_\mu$ and keeping only the linear terms, one obtains

$$i\frac{\partial u(x)}{\partial x^\mu} = \left[u(x),\, P_\mu\right], \tag{1.244}$$

which are four equations representing the equation of motion (1.234). These equations prove to be very useful and convenient in the analysis of the frequency components of the field functions in momentum representation. Substituting the Fourier decompositions of the type (1.50), (1.85), and (1.155) into (1.244), one obtains the following algebraic operator equations,

$$q^\mu u^\pm(q) = \mp\left[u^\pm(q),\, P^\mu\right], \tag{1.245}$$

where the following notations have been used: $u^\pm(q) = u(\pm q)|_{q^0>0}$.

Consider now a state characterized by a well-determined value of the energy-momentum four-vector $P^\nu$, described by the amplitude $\Phi_p$,

$$P^\nu\Phi_p = p^\nu\Phi_p. \tag{1.246}$$

Multiplying the first of relations (1.245), on the wright, by $\Phi_p$ and making use of (1.246), one finds

$$P^\nu u^+(q)\Phi_p = (p^\nu + q^\nu)u^+(q)\Phi_p.$$

Proceeding in the same way with the second equation (1.245), one obtains

$$P^{\nu}u^{-}(q)\Phi_p = (p^{\nu} - q^{\nu})u^{-}(q)\Phi_p,$$

in both cases being valid the relation $q^0 = \sqrt{\vec{q}^2 + m^2} > 0$.

It then follows that the expression $u^{+}(q)\Phi_p$ is equal either to zero, or to the amplitude of the state with energy-momentum $p+q$. Similarly, the expression $u^{-}(q)\Phi_p$ is equal either to the amplitude of the state with energy-momentum $p-q$, or to zero. Since in our case the condition $q^2 = m^2$ is fulfilled, it follows that the operator $u^{+}(q)$ describes the creation of a particle of mass $m$ and four-momentum $q$, while the operator $u^{-}(q)$ corresponds to annihilation of the same particle.

It is worth mentioning that this property of the operators corresponding to positive and negative frequency parts of the field is most general, characterizing fields of any tensorial dimension, both real and complex, and does not depend on the type of commutation relations (Bose–Einstein or Fermi–Dirac).

One also observes that, using formula (1.248) for the gradient transformations (1.242) and (1.243),

$$e^{i\alpha}u = U^{-1}(\alpha)uU(\alpha); \quad e^{-i\alpha}u^* = U^{-1}(\alpha)u^*U(\alpha)$$

and equating the coefficients of the terms linear in $\alpha$, one obtains

$$u(x) = [u(x), Q]; \quad u^* = -[u^*(x), Q]. \tag{1.247}$$

The equations (1.247) complete formulas (1.245) and show that the operator $u$ (together with its positive and negative frequency parts $u^{\pm}$) is a lowering operator with respect to the system's charge (more exactly, this operator diminishes by one the eigenvalue of the charge operator $Q$). The corresponding conjugate operator $u^*$ raises by one the eigenvalue of the charge operator. Thus, in the case of a complex field, the operator $u^{*+}$ creates a particle with the charge $+1$, the operator $u^{-}$ annihilates a particle with the charge $+1$, and so on.

### I.3.6.2. The vacuum state amplitude and the Fock representation

The purpose of this paragraph is to determine the *vacuum state* and the *rules for constructing the amplitudes corresponding to the states with a well-defined number of different particles.*

To this end, let us consider a dynamical system composed by several non-interacting quantum fields, characterized by the operator functions $u_1(x), ..., u_n(x)$. For the sake of simplicity, the corresponding conjugate functions (if $u_i^*$ and $u_i$ differ from each other) are also included in this sequence.

One of the basic characteristics of the studied system is the *vacuum state amplitude.* Since the vacuum state contains no physical particles, the vacuum state momentum is null, and its energy is a minimum (even tending to zero). Since the operators of negative frequency, $u_i^-$ $(i = 1, 2, ..., n)$, decrease the energy of the system, while the vacuum state energy $\Phi_0$ is already a minimum, it follows that

$$u_i^-(x)\Phi_0 = 0, \quad (i = 1, ..., n), \tag{1.248}$$

for any value of $x$. For the same reason,

$$\Phi_0^* u_i^+(x) = 0.$$

In momentum representation, the relation (1.248) writes

$$u_i^-(p)\Phi_0 = 0. \tag{1.249}$$

The relation (1.249), together with its conjugate $\Phi_0^* u_i^+(p) = 0$, and the normalization condition

$$\Phi_0^*\Phi_0 = 1, \tag{1.250}$$

can be taken as a *definition of the vacuum state* for the free fields.

The amplitude of any state of a dynamical system can be represented by means of the vacuum state amplitude, and the creation operators of the respective particles. For example, the amplitude of the state containing $s$ particles of types $j_1, j_2, ..., j_s$ can be represented as

$$\int F_s^{(j_1,...,j_s)}(\vec{p}_1, ..., \vec{p}_s) u_{j_1}^+(\vec{p}_1)...u_{j_s}^+(\vec{p}_s) d\vec{p}_1...d\vec{p}_s\, \Phi_0,$$

where, as usual,

$$u^+(\vec{p}) = \frac{\theta(p^0)u(p)}{\sqrt{2p^0}}, \quad (p^0 = \sqrt{\vec{p}^2 + m^2}).$$

The functions $F_s^{(...j...)}$ under the integral signify the usual wave functions associated with a system of $s$ particles, written in momentum representation. Such a function describes the amplitude of the state strictly corresponding to $s$ particles with the given characteristics $j_1, ..., j_s$.

In the general case of a variable number of particles, the state amplitude is characterized by a sequence of functions $F_s$, with various values of $s$. Performing summation over discrete indices $j_1, ..., j_s$ and over $s$, one obtains

$$\Phi = \sum_{(j,s\geq 0)} \int F_s^{(...j...)}(...\vec{p}...)u_{j_1}^+(\vec{p}_1)...u_{j_s}^+(\vec{p}_s)\, d\vec{p}_1...d\vec{p}_s\ \Phi_0. \quad (1.251)$$

This is the *Fock representation* of the state amplitude. Passing from the momentum to coordinate/configuration representation by means of relations

$$F_s(...\vec{p}...) = \frac{1}{(2\pi)^{3s/2}} \int e^{i\sum_j p_j x_j} \varphi_s(...\vec{x}...)\, d\vec{x}_1...d\vec{x}_s,$$

$$u^+(\vec{p}) = \frac{1}{(2\pi)^{3/2}} \int e^{ipx} u^+(0,\vec{x})\, d\vec{x}; \quad u^+(0,\vec{x}) = u^+(x)|_{x^0=0},$$

instead of (1.251) one obtains the following relation for the state amplitude:

$$\Phi = \sum_{(j,s\geq 0)} \int \varphi_s(\vec{x}_1, ..., \vec{x}_s) \prod_{1\leq\nu\leq s} \left\{u_{j_\nu}^+(0, x_\nu)d\vec{x}_\nu\right\} \Phi_0. \quad (1.252)$$

In the non-relativistic case, the functions $\varphi_s$ appearing in (1.252) play the role of the wave functions in the configuration space.

As one can see, the time-dependence of the quantity $\Phi$ has disappeared. This fact has been expected, since in our representation characterized by no interactions, the state amplitude remains constant.

### I.3.6.3. Types of commutation relations

At this stage of investigation, it is necessary to settle the commutation relation for the operators acting on the wave functions. In the classical theory of the free fields with a quadratic Lagrangian, the wave functions are linearly expressed by means of the canonically conjugated coordinates and momenta. The classical Poisson brackets for the field functions, $\{u(x), u(y)\}$, prove to be functions of $x$ and $y$ (more precisely, of $x - y$) and do not depend on $u$. Due to this fact and following the correspondence principle, in the quantum theory of the free fields, the commutation relations for the operators associated with the field functions are usually written as

$$\begin{aligned}\{u_a(x), u_b(y)\}_- &\equiv [u_a(x), u_b(y)] \\ &= u_a(x)u_b(y) - u_b(y)u_a(x) = \Delta_{ab}(x-y).\end{aligned} \tag{1.253B}$$

This commutation relation is too severe (strong) and, besides, does not contain several important cases. Thus, as an alternative to (1.253$B$), one also considers the anticommutator of the two operators, that is,

$$\{u_a(x), u_b(y)\}_+ \equiv u_a(x)u_b(y) + u_b(y)u_a(x) = \Delta_{ab}(x-y). \tag{1.253F}$$

In the second case, the correspondence principle has to be understood as follows: the anticommutativity of the field operators, expressed by the relation (1.253$F$), leads to the fact that the operators of the dynamical quantities which are bilinear forms of the field operators, are expressed in terms of the commutation functions $\Delta$.

The field quanta that satisfy the commutation relation (1.253$B$) obey the Bose–Einstein statistics, and the associated particles are called *bosons*. Similarly, the field quanta that satisfy the relation (1.253$F$) obey the Fermi–Dirac statistics, and the associated particles are called *fermions*.

The exact form of the commutation functions $\Delta$ for any field is determined by means of relations (1.244), (1.247), and according to the structure of the energy operator for the considered field, as well. But, irrespective of the concrete form of the commutation relations, it can be proved that the commutation function of the free fields depends only on the difference $x - y$, that is,

$$\{u_a(x), u_b(y)\} = \Delta_{ab}(x-y). \tag{1.254}$$

To prove this property, it is necessary to express the commutation relations (1.253) in the momentum representation. Due to linearity of the Fourier transformation formulas, the commutators or anti-commutators of the positive and negative frequency components in momentum representation, $u^{\pm}(\vec{q})$, must also be $c$-numbers (real or complex classical numbers). First, one must show that the operators of the same frequency strictly commute or anticommute, that is,

$$\left\{u_a^{\pm}(\vec{q}),\, u_b^{\pm}(\vec{k})\right\} = 0. \tag{1.255}$$

Here, as well as in (1.254), the symbol $\{...,...\}$ designates either the commutator, or the anticommutator of the respective operators.

To determine the amplitude of the state function $\Phi_p$ with a certain value of the four-momentum and satisfying the equation (1.246), one conveniently applies the operators $u_a^+(\vec{q})$ and $u_b^+(\vec{k})$ on this state, which yields two new amplitudes,

$$\Phi_1 = u_a^+(\vec{q})u_b^+(\vec{k})\Phi_p; \quad \Phi_2 = u_b^+(\vec{k})u_a^+(\vec{q})\Phi_p,$$

satisfying the equations

$$P_0\Phi_{1,2} = (p_0 + q_0 + k_0)\Phi_{1,2}.$$

Adding and subtracting these equations, one obtains

$$\begin{aligned} P_0(\Phi_1 \pm \Phi_2) &= P_0\left\{u_a^+(\vec{q}), u_b^+(\vec{k})\right\}\Phi_p \\ &= (p_0 + q_0 + k_0)\left\{u_a^+(\vec{q}), u_b^+(\vec{k})\right\}\Phi_p. \end{aligned}$$

Admitting that $\{u_a^+, u_b^+\} = c \neq 0$, where $c$ is a number, and dividing the last equation by $c$, one arrives at

$$P_0\Phi_p = (p_0 + q_0 + k_0)\Phi_p,$$

which contradicts the relation (1.246). Therefore, the formula (1.255) is proved. Following the same procedure, one can also show that

$$\left\{u_a^{\pm}(\vec{q}), u_b^{\mp}(\vec{k})\right\} = 0, \quad \text{for } \vec{q} \neq \vec{k}. \tag{1.256}$$

The physical significance of the relations (1.255)–(1.256) is simple. They show that the processes of creation and annihilation of the particles of any field do not interfere with each other (do not reciprocally influence each other). The processes of creation and annihilation of the particles with different momenta have the same property.

If the r.h.s. (*i.e.* zero) of the formula (1.256) is replaced by $\delta(\vec{q}-\vec{k})$, then the formula (1.254) in configuration representation is regained. It should be mentioned that this formula expresses the covariance of the commutation relations with respect to translations. To demonstrate this formula, the properties of the energy-momentum four-vector – as generator of translations – have been used.

In the case of complex fields, it is easy to show that, in addition to relations (1.179), the commutators of the operators corresponding to oppositely charged particles are always zero:

$$\{u^{+}, u^{-}\} = \{u^{*+}, u^{*-}\} = 0. \tag{1.257}$$

In this respect, it is necessary to analyze the wave function amplitude, $\Phi_q$, corresponding to a certain charge value, and use the same procedure as that applied to obtain formula (1.255). The physical significance of the formula (1.257) is connected to the creation and annihilation processes of oppositely charged particles.

Therefore, it has been shown that only two $c$-numbers, $\{u_a^{-}(\vec{p}), u_a^{*+}(\vec{p})\}$ and $\{u_a^{*-}(\vec{p}), u_a^{+}(\vec{p})\}$ can exist for any non-null complex field. These are the commutators/anticommutators of the creation and annihilation operators corresponding to particles with the same energy-momentum four-vector and the same charge. In the coordinate/configuration representation, these expressions are translationally invariant,

$$\{u^{\pm}(x), u^{*\pm}(y)\} = \Delta^{\pm}(x-y),$$

while their sum gives the complete commutation function (the commutator),

$$\{u(x), u^{*}(y)\} = \Delta^{+}(x-y) + \Delta^{-}(x-y) = \Delta(x-y).$$

In order to determine the commutator, it is necessary to use the concrete form of the energy-momentum four-vector, and the equations (1.244) and (1.247) as well.

### I.3.6.4. The Fermi–Dirac and Bose–Einstein quantization rules

In order to find the concrete form of the commutation relations, one makes use of equations (1.244) and (1.245), as well as the energy-momentum four-vector operator written as

$$P^\nu = \sum_s \int d\vec{p}\, p^\nu \left[a_s^{*+}(\vec{p})a_s^-(\vec{p}) \pm a_s^{*-}(\vec{p})a_s^+(\vec{p})\right]. \tag{1.258}$$

In agreement with the classical expressions (1.51), (1.79), (1.98), and (1.167), the operator given by (1.258) has been expressed in terms of the independent amplitudes $a^\pm, a^{*\pm}$, connected with the quantities $u^\pm$ and $u^{*\pm}$ through the linear relations

$$u_a^\pm(\vec{p}) = \sum_s v_a^{s,\pm}(\vec{p})a_s^\pm(\vec{p}),$$

where the coefficients $v^\pm$ are $c$-numbers.

Since the operators $a^\pm$, $a^{*\pm}$ do not commute, their order in equation (1.258) corresponds to that of the functions $u$ and $u^*$ in the Lagrangian density. We remind the reader that the operators $a^\pm$, $a^{*\pm}$ are connected through the complex conjugate conditions

$$\left(a^\pm(\vec{p})\right)^* = a^{*\mp}(\vec{p}).$$

The sign $(+)$ in front of the second term of (1.258) refers to the integer-spin fields (scalar, massive vectorial, electromagnetic, etc.), while the sign $(-)$ concerns the half-integer spin (spinorial) fields.

To obtain the searched commutation relations, one observes that the commutators of the operators $a^\pm$ and the bilinear combinations $a^{*\pm}(\vec{p})a^\pm(\vec{p})$ of the r.h.s. of (1.258), can be written as

$$\begin{aligned}&[a^\pm(\vec{q}), a^{*+}(\vec{p})a^-(\vec{p})]\\ &= \{a^\pm(\vec{q}), a^{*+}(\vec{p})\}\, a^-(\vec{p}) - a^{*+}(\vec{p})\,\{a^-(\vec{p}), a^\pm(\vec{q})\}.\end{aligned} \tag{1.259}$$

As previously mentioned, the symbol $\{a, b\}$ stands for both the commutator and anticommutator of the operators $a$ and $b$, depending on the nature of the described particles. Analogously, one finds

$$\begin{aligned}&[a^\pm(\vec{q}), a^{*-}(\vec{p})a^+(\vec{p})]\\ &= \{a^\pm(\vec{q}), a^{*-}(\vec{p})\}\, a^+(\vec{p}) - a^{*-}(\vec{p})\,\{a^+(\vec{p}), a^\pm(\vec{q})\}.\end{aligned} \tag{1.260}$$

In view of (1.255) and (1.257), one obtains

$$\begin{aligned}\left[a^+(\vec{k}), a^{*+}(\vec{q})a^-(\vec{q})\right] &= \left[a^-(\vec{k}), a^{*-}(\vec{q})a^+(\vec{q})\right] = 0,\\ \left[a^-(\vec{k}), a^{*+}(\vec{q})a^-(\vec{q})\right] &= \left\{a^-(\vec{k}), a^{*+}(\vec{q})\right\} a^-(\vec{q}),\\ \left[a^+(\vec{k}), a^{*-}(\vec{q})a^+(\vec{q})\right] &= \left\{a^+(\vec{k}), a^{*-}(\vec{q})\right\} a^+(\vec{q}),\end{aligned}$$

Substituting (1.258) into (1.245), one also finds

$$q^\nu a_a^-(\vec{q}) = \int d\vec{p}\, q^\nu \sum_b \left\{a_a^-(\vec{q}), a_b^{*+}(\vec{p})\right\} a_b^-(\vec{p}),$$

$$\mp q^\nu a_a^+(\vec{q}) = \int d\vec{p}\, q^\nu \sum_b \left\{a_a^+(\vec{q}), a_b^{*-}(\vec{p})\right\} a_b^+(\vec{p}),$$

and hence

$$\left\{a_a^-(\vec{q}), a_b^{*+}(\vec{p})\right\} = \delta_{ab}\delta(\vec{q}-\vec{p}), \tag{1.261}$$

$$\left\{a_a^+(\vec{q}), a_b^{*-}(\vec{p})\right\} = \mp\delta_{ab}\delta(\vec{q}-\vec{p}), \tag{1.262}$$

where the indeterminacy upon the sign ($\pm$) appears from (1.258).

This way, two variants of the commutation relations have been found for each type of field. The requirement that these relations must be symmetric in the sign of the electric charge (or, equivalently, the replacement of particles by their antiparticles),

$$a_a^\pm(\vec{p}) \leftrightarrow a_a^{*\pm}(\vec{p}) \tag{1.263}$$

shall uniquely define the quantization recipe in each case. The symmetry displayed by the formula (1.263) shows the fact that the choice between the "main" field function $u$ and its complex conjugate $u^*$ is only a matter of convention, the inverse choice (*i.e.* $w(x) = u^*(x)$, $w^*(x) = u(x)$) leading only to the replacement of the "main" particles by their antiparticles. Such a transformation affects the charge operator, which is not covariant with respect to (1.263), but it must not influence the equation of motion (1.244), and the expression of the energy-momentum four-vector as well. The symmetry condition with respect to (1.263) shall also ensure the correct transition from the complex to the real field,

$$u^*(x) = u(x); \quad a^{*\pm}(\vec{p}) = a^\pm(\vec{p}).$$

Transformation of the type (1.263) is called *charge conjugation*, and the corresponding symmetry – the *charge symmetry*. If the $(-)$ sign is taken on the r.h.s. of the formula (1.262), then the symmetry of (1.261) and (1.262) with respect to transformation (1.263) leads to the fact that the symbol $\{a, b\}$ must be understood as a commutator.

In the other case (if the sign (+) is considered), the symbol $\{a,b\}$ signifies an anticommutator,

$$\{a,b\} = \{a,b\}_+ \equiv ab + ba.$$

Therefore, the quantization of the integer-spin fields is done by the Bose–Einstein rules. For this kind of fields the following commutation relations should be satisfied:

$$\left[a_a^-(\vec{p}), a_b^{*+}(\vec{q})\right] = \left[a_a^{*-}(\vec{p}), a_b^+(\vec{q})\right] = \delta_{ab}\delta(\vec{p}-\vec{q}). \tag{1.264}$$

In the case of half-integer spin fields, the quantization is done by the Fermi–Dirac rules, in this case being satisfied the following anticommutation relations:

$$\left\{a_a^-(\vec{p}), a_b^{*+}(\vec{q})\right\}_+ = \left\{a_a^{*-}(\vec{p}), a_b^+(\vec{q})\right\}_+ = \delta_{ab}\delta(\vec{p}-\vec{q}). \tag{1.265}$$

It is important to specify that the normalization of the operator field functions is uniquely defined by the commutation relations (1.264) and (1.265).

### I.3.7. Integer spin fields

The following investigation concerns the quantum theory of the scalar, massive complex vectorial, electromagnetic and weak gravitational fields.

#### I.3.7.1. The scalar field

In agreement with the results obtained in the paragraph I.3.6.4, the commutation relations for the scalar field in momentum representation are:

a) For the real field,

$$\left[\varphi^-(\vec{p}), \varphi^+(\vec{q})\right] = \delta(\vec{p}-\vec{q}), \quad \left[\varphi^\pm(\vec{p}), \varphi^\pm(\vec{q})\right] = 0; \tag{1.266}$$

b) For the complex field,

$$\left[\varphi^{*-}(\vec{p}), \varphi^+(\vec{q})\right] = \left[\varphi^-(\vec{p}), \varphi^{*+}(\vec{q})\right] = \delta(\vec{p}-\vec{q}), \tag{1.267}$$

all other commutators being equal to zero. Passing to the coordinate representation, on the r.h.s. of the previous relations instead of the delta symbol shall appear the functions

$$\frac{1}{(2\pi)^3}\int \frac{d\vec{p}}{\sqrt{2p^0}}\frac{d\vec{q}}{\sqrt{2q^0}} e^{i(qy-px)}\delta(\vec{p}-\vec{q})$$
$$= \frac{1}{(2\pi)^3}\int \frac{d\vec{p}}{2p^0} e^{ip(y-x)} = \frac{1}{i}D^-(x-y);$$

$$\frac{1}{(2\pi)^3}\int \frac{d\vec{p}}{2p^0} e^{ip(x-y)} = \frac{1}{i}D^-(y-x) = iD^+(x-y),$$

*i.e.* the frequency parts of the Pauli–Jordan commutation function

$$D(x) = D^+(x) + D^-(x) = i(2\pi)^{-3}\int e^{-ipx}\varepsilon(p^0)\delta(p^2-m^2)\,dp. \tag{1.268}$$

where

$$\varepsilon(p^0) = \theta(p^0) - \theta(-p^0) = \begin{cases} +1, & p^0 > 0, \\ -1, & p^0 < 0. \end{cases}$$

Considering only the real fields, the commutation relations for the field functions in coordinate representation are

$$\begin{aligned} [\varphi^-(x), \varphi^+(y)] &= -iD^-(x-y), \\ [\varphi^+(x), \varphi^-(y)] &= -iD^+(x-y), \\ [\varphi(x), \varphi(y)] &= -iD(x-y). \end{aligned} \tag{1.269}$$

The explicit form of the functions $D^\pm$ and $D$ is given in the Appendix D. The main property of the function $D$ is that it equals zero outside the light cone. Therefore,

$$[\varphi(x), \varphi(y)] = 0$$

for

$$(x-y)^2 < 0. \tag{1.270}$$

This property is called the *locality commutation property* and ensures the causality in space-time (more exactly, this important property ensures that events separated by space-like intervals are independent).

Next, several useful properties and definitions in the field quantization shall be presented.

## Normal product of operators

An operator whose application leads to such an arrangement of the creation and annihilation operators in which the creation operators $u^+$ are always on the left as compared to all annihilation operators $u^-$, is called the *normal ordering operator*. The process of putting a product into normal order is called *normal ordering* (also called *Wick ordering*).

In order to calculate the matrix element $\Phi^* A \Phi$ of any operator $A$ in the normal form, it is necessary to commute all the operators $u^-$ belonging to the operator $A$ with all $a^-$ of $\Phi$, and all the operators $u^+$ of $A$ with all $a^-$ of $\Phi^*$, until one of $u^-$ shall not act on $\Phi_0$ (or, one of $u^+$ shall not act on $\Phi_0^*$), in which case the result is zero.

As an example, consider the product of two bosonic operators $u^*(x)$ and $u(y)$. One can write

$$\begin{aligned} u^+(x)u(y) &= u^{*+}(x)u^+(y) + u^{*+}(x)u^-(y) \\ &\quad + u^{*-}(x)u^+(y) + u^{*-}(x)u^-(y) \\ &= u^{*+}(x)u^+(y) + u^{*+}(x)u^-(y) \\ &\quad + u^+(y)u^{*-}(x) + u^{*-}(x)u^-(y) - i\Delta^-(x-y). \end{aligned}$$

In a more general case, bringing to the normal form the product of several field functions (operators), $u$, one obtains the sum of products of the components $u^+$, $u^-$, and the *commutation function* $\Delta^-$. The resulting expression can be conventionally considered as a "polynomial" in powers of commutator $\Delta^-$ – functions. The sum of the terms that do not contain the commutation function $\Delta^-$ (the zero degree term of this polynomial) is called *normal product* of the original operator field functions. If, as a result of bringing to the normal form, all the commutation functions are zero, the obtained result is also a normal product. The normal product of the operators $u_1, u_2, ..., u_n$ is denoted by the symbol

$$N[u_1 ... u_n].$$

Using this notation, one can write

$$u^*(x)u(y) = N[u^*(x)u(y)] - i\Delta^-(x-y).$$

One agrees that all dynamic variables that depend quadratically on operators with the same arguments (the Lagrangian, current etc.) to be written in the form of normal products. For instance, the Lagrangian of the scalar complex field will be written as

$$\mathcal{L} = N[\varphi^*_{,\nu}\varphi^{,\nu}] - m^2 N[\varphi^*\varphi]. \tag{1.271}$$

Acording to the definition of the vacuum amplitude $\Phi_0$, one can write

$$\varphi^-(x)\Phi_0 = \varphi^{*-}(x)\Phi_0 = 0,$$

and

$$\Phi_0^*\varphi^+(x) = \Phi_0^*\varphi^{*+}(x) = 0.$$

It then follows that the expectation values of all the dynamic variables are zero for the vacuum state. In the investigation to follow, all the pseudo-physical quantities – such as the vacuum energy, the vacuum charge, etc. – shall be automatically excluded. It is worth specifying that all the conservation laws met in classical field theory remain valid in quantum terms (but written in the normal form), since the algebraic identities used to demonstrate the Noether's theorem do not change.

In the light of the above considerations, the basic dynamical quantities of the complex scalar field write

$$T^{\mu\nu} = N\left[\varphi^{*,\mu}\varphi^{,\nu} + \varphi^{*,\nu}\varphi^{,\mu}\right] - g^{\mu\nu}\mathcal{L}, \tag{1.272}$$

$$P^\nu = \int d\vec{p}\, p^\nu \left[\varphi^{*+}(\vec{p})\varphi^-(\vec{p}) + \varphi^+(\vec{p})\varphi^{*-}(\vec{p})\right], \tag{1.273}$$

$$Q = \int d\vec{p}\, \left[\varphi^{*+}(\vec{p})\varphi^-(\vec{p}) - \varphi^+(\vec{p})\varphi^{*-}(\vec{p})\right]. \tag{1.274}$$

Taking into account the structure of the operators $P^\nu$ and $Q$, it follows that $\varphi^{*+}(\vec{p})$ is the creation operator of the particle with energy-momentum four-vector $p$ and the charge $+1$, and $\varphi^-(\vec{p})$ is the annihilation operator of the same particle. Similarly, $\varphi^+(\vec{p})$ is the creation operator of the particle with energy-momentum four-vector $p$ and the charge $-1$, and $\varphi^{*-}(\vec{p})$ is the annihilation operator of the same particle.

### I.3.7.2. The massive complex vectorial field

In order to establish the quantization rules for the vectorial field (described by a four-potential), one observes that a simple generalization of the quantization rules used for the scalar field (in other words, the independent quantization of each component of the four-potential $U_\mu$, according to the scalar field model) is impossible, because such a procedure does not ensure the positively defined character for the expectation value of the energy. Moreover, such a procedure is incompatible with the subsidiary condition (1.55). Therefore, the quantization scheme of this field has to take into consideration that subsidiary condition which is able to automatically lead to the positively defined character of $P^0$.

When the problem of the massive vectorial field – from the classical point of view – has been approached, it has been shown that, by passing from the four-component field function $U_\mu$ to the transversal and longitudinal components – according to formula (1.66) – the energy-momentum four-vector can be expressed in terms of the usual linearly independent vectorial functions (amplitudes) $a_n(\vec{p})$ as

$$P^\nu = \int d\vec{p}\, p^\nu \left[a_n^{*+}(\vec{p})a_n^-(\vec{p}) + a_n^{*-}(\vec{p})a_n^+(\vec{p})\right], \quad (n = 1,2,3). \tag{1.275}$$

This form of writing down the classical four-momentum takes into account the subsidiary condition (1.55) and guarantees that the unquantized $P^0$ shall be positive. The Bose–Einstein quantization of the three independent amplitudes $a_n$, $(n = 1,2,3)$ will guarantee the positive-definite character of the expectation value of the energy operator. As a result of this, the operators $a_n$ must be made to obey the following requirements,

$$\left[a_m^-(\vec{p}), a_n^{*+}(\vec{q})\right]_- = \left[a_m^{*-}(\vec{p}), a_n^+(\vec{q})\right]_- = \delta_{mn}\delta(\vec{p} - \vec{q}), \tag{1.276}$$

all the other commutators being zero.

With the help of Eq. (1.69) the expressions for the energy-momentum four-vector (1.275), for the charge,

$$Q = \int d\vec{p}\, \left[a_n^{*+}(\vec{p})a_n^-(\vec{p}) - a_n^{*-}(\vec{p})a_n^+(\vec{p})\right],$$

and for the component of the spin along the direction of motion (1.68) can be expressed in terms of the amplitudes $b_n$. Then rewriting all these expressions in normal form one obtains the following formulas:

$$P^\nu = \int d\vec{p}\, p^\nu \left[b_n^{*+}(\vec{p})b_n^-(\vec{p}) + b_n^+(\vec{p})b_n^{*-}(\vec{p})\right], \tag{1.276'}$$

$$Q = \int d\vec{p}\, \left[b_n^{*+}(\vec{p})b_n^-(\vec{p}) - b_n^+(\vec{p})b_n^{*-}(\vec{p})\right], \tag{1.277}$$

$$S_3 = \int d\vec{p}\Big[b_1^{*+}(\vec{p})b_1^-(\vec{p}) - b_2^{*+}(\vec{p})b_2^-(\vec{p}) - b_1^+(\vec{p})b_1^{*-}(\vec{p}) + b_2^+(\vec{p})b_2^{*-}(\vec{p})\Big]. \tag{1.278}$$

By calculating the corresponding expectation values, one finds that $b_1^{*+}(\vec{p})$ and $b_1^-(\vec{p})$ are, respectively, the creation and annihilation operators for the particles of momentum $\vec{p}$, charge $+1$, and projection of the spin on the axis of motion $+1$, $b_2^{*+}(\vec{p})$ and $b_2^-(\vec{p})$ are, respectively, the creation and annihilation operators for the particles of momentum $\vec{p}$, charge $+1$, and projection of the spin $-1$, while $b_3^{*+}(\vec{p})$ and $b_3^-(\vec{p})$ are, respectively, the creation and annihilation operators for the particles of momentum $\vec{p}$, charge $+1$, and null spin projection.

The significance of the operators $b_n^+$ and $b_n^{*-}$ is evidentiated according to the following rule: *the transition from $b_n^\pm$ to $b_n^{*\pm}$ is "equivalent" to a change in the sign of the charge and in the sign of the spin projection/component.* One thus observes that the massive complex vectorial field describes the charged particles of mass $m$, and three possible projection values $(+1, 0, -1)$ of the spin on the given axis. The quanta of this field can be, for instance, the vectorial mesons (like the isospin triplet $\rho^+$, $\rho^0$, $\rho^-$ of vector rho-mesons).

Following the same reasoning, one can display the meaning of the amplitudes $a_n$. For example, the amplitude $a_3$ corresponds to particles of null spin projection along the direction of motion, while $a_1$ and $a_2$ describe combinations of states with $+1$ and $-1$ spin projections, associated with linear polarizations.

In order to determine the commutation relations for the four-potential (amplitudes) $U_\nu$, one first use (1.66) and write the commutator for the three-potential $U_n$, $(n = 1, 2, 3)$:

$$\left[U_n^{*-}(\vec{p}), U_m^{+}(\vec{q})\right] = \left[U_n^{-}(\vec{p}), U_m^{*+}(\vec{q})\right]$$
$$= \delta(\vec{p} - \vec{q})\left[\delta_{nm} + \frac{p_n p_m}{m^2}\right].$$

Using the equation $p^0 U_0(\vec{p}) + p^n U_n(\vec{p}) = 0$, $(p_0 = p^0,\ p_n = -p^n)$ one can find the commutators which contain $U_0$:

$$\left[U_0^{*-}(\vec{p}), U_n^{+}(\vec{q})\right] = \left[U_0^{-}(\vec{p}), U_n^{*+}(\vec{q})\right] = \frac{p_0 p_n}{m^2}\delta(\vec{p} - \vec{q}),$$

$$\left[U_0^{*-}(\vec{p}), U_0^{+}(\vec{q})\right] = \left[U_0^{-}(\vec{p}), U_0^{*+}(\vec{q})\right] = \left(\frac{p_0^2}{m^2} - 1\right)\delta(\vec{p} - \vec{q}),$$

whose combination yields the formula

$$\left[U_\nu^{*-}(\vec{p}), U_\mu^{+}(\vec{q})\right] = \left[U_\nu^{-}(\vec{p}), U_\mu^{*+}(\vec{q})\right]$$
$$= \left[\frac{p_\nu p_\mu}{m^2} - g_{\nu\mu}\right]\delta(\vec{p} - \vec{q}), \tag{1.279}$$

which is relativistic, symmetrical, and compatible with the subsidiary condition (1.64). Passing to the coordinate representation, one obtains

$$\left[U_\nu^{*-}(x), U_\mu^{+}(y)\right] = \left[U_\nu^{-}(x), U_\mu^{*+}(y)\right]$$
$$= \left(g_{\nu\mu} + \frac{1}{m^2}\frac{\partial^2}{\partial x^\nu \partial x^\mu}\right) iD^-(x - y), \tag{1.280}$$
$$\left[U_\nu^{*}(x), U_\mu(y)\right] = \left(g_{\nu\mu} + \frac{1}{m^2}\frac{\partial^2}{\partial x^\nu \partial x^\mu}\right) iD(x - y).$$

It is not difficult to observe that these commutation relations are compatible with both the field equations and the subsidiary conditions. Thus, acting on both sides of (1.280) with the Klein–Gordon operator, one obtains an identity $(0 = 0)$, since according to formula (1.268),

$$(\Box - m^2)D(x) = 0.$$

### I.3.7.3. The electromagnetic field: quantization difficulties

In order to quantize the electromagnetic field, the following conditions must simultaneously be fulfilled:

— the positive definite character of the energy density;
— the subsidiary Lorenz gauge condition;
— the transversality condition;
— the relativistic covariance of the theory.

A somehow similar situation has been met in the case of massive vectorial field. The difference lies in the fact that the massive vectorial particles can have three possible spin values, while the photons can have only two such states (due to their transversality). In addition, the photon rest mass equals zero.

Unlike the massive vectorial field, the electromagnetic field is characterized by a bigger number of variables (four components of the potential four-vector), while the real photons can only occur in two states. Then, the property $m = 0$ makes it impossible to apply the procedure used for quantization of the massive vectorial field in the case of electromagnetic field. Indeed, the quantization of a hypothetical massive vectorial field, having an infinitely small mass (which can disappear in the limit $m \to 0$), therefore of a field that differs from the electromagnetic field due to the absence of some gradient transformations (due to this, being a field with only three components) leads to essential difficulties. In such a quantization scheme several senseless expressions are obtained, first when relation (1.66) is used in order to diagonalize the energy-momentum four-vector, and second concerning the establishment of the commutation relations for the components $A_\nu$ of the four-potential, where difficulty comes from the presence of the mass $m$ in the denominators of the expressions (1.279) and (1.280).

To avoid such inconveniences, the quantization of the electromagnetic field is performed as follows. One considers the components of the four-vector potential as being independent quantities, this way neglecting the operatorial form of the Lorenz condition. Thus, according to the general quantization recipe emerging from the structure of the energy-momentum four-vector expressed in terms of the longitudinal, transversal and temporal components, $a_\mu(\vec{k})$,

$$P^\nu = \int d\vec{k}\, k^\nu \left(-a_\mu^+(\vec{k})a_\mu^-(\vec{k})\right), \tag{1.281}$$

one arrives at the commutation relations

$$\left[a_\nu^-(\vec{k}), a_\mu^+(\vec{q})\right] = -g_{\nu\mu}\delta(\vec{k}-\vec{q}). \tag{1.282}$$

The quantization condition (1.197) allows one to consider the operators $a_\nu^\pm$ as creation operators of four types of independent photons: *transversal* (two types), *longitudinal* and *temporal/"time-like"*. Proceeding in this manner, there appears another difficulty which has to be taken into account. Observing that the component $a_0$ satisfies the relation

$$\left[a_0^-(\vec{k}), a_0^+(\vec{q})\right] = -\delta(\vec{k}-\vec{q}),$$

and comparing it with (1.262), it looks like the creation and annihilation operators of the "time-like" photons have changed their places, leading to appearance of the sign $(-)$ in front of the term $a_0^+a_0^-$ in formula (1.281). But such a commutation relation contradicts the initial supposition on the positively defined metric of the Hilbert space of states.

To solve this difficulty, one uses the Bleuler–Gupta formal approach, based on the idea that this inconvenience is fully connected to the temporal photons. These photons do not actually exist, but they interfere in the intermediate levels of discussion on the transition from the observable quantities ($\vec{E}$ and $\vec{H}$) to the nonobservable components of the four-potential $A_\nu$ (used to ensure the properties of relativistic symmetry and covariance of the theory).

### Electromagnetic field. The Bleuler–Gupta quantization

The basic idea underlying the Bleuler–Gupta method is connected to the Lorenz condition. Recalling that the commutation relations (1.282) have been written under the supposition that the components of the four-potential $A_\nu$ are independent quantities, the Lorenz gauge condition cannot be used as such for the operators. In addition, it can be shown that the fulfillment of the Lorenz condition for the allowed states,

$$\partial A(x)\Phi = 0, \qquad (\partial A \equiv \partial^\nu A_\nu)$$

is not possible, since such a condition contradicts, for example, the vacuum definition. Indeed, taking $\Phi = \Phi_0$, one obtains

$$\partial A(x)\Phi_0 = \partial A^+(x)\Phi_0 = 0.$$

Multiplying on the left by $A_\mu^-(y)$, it follows that

$$\begin{aligned} A_\mu^-(y)\partial^\nu A_\nu^+(x)\Phi_0 &= \frac{\partial}{\partial x_\nu}\left(A_\mu^-(y)A_\nu^+(x)\right)\Phi_0 \\ &= \frac{\partial}{\partial x_\nu}\left[A_\mu^+(x)A_\nu^-(y) - ig_{\mu\nu}D_0^+(x-y)\right]\Phi_0 \\ &= -i\partial_\nu D_0^+(x-y)\Phi_0 \neq 0, \end{aligned}$$

where $D_0^+$ is the usual positive-frequency commutation function for the zero-rest mass.

Technically speaking, the Bleuler–Gupta approach implies to introduce a so-called *indefinite metric* in the space of the state amplitudes, through the medium of Hermitian operator $\eta$ defined by

$$\eta a_\nu = -a^\nu \eta, \quad \eta^2 = 1, \quad \eta\Phi_0 = \Phi_0.$$

With the help of this operator a new adjoint state amplitude can be defined by

$$\Phi^+ = \Phi^*\eta,$$

which can be used to define the expectation value as

$$\langle\varphi\rangle = \Phi^*\eta\varphi\Phi = \Phi^+\varphi\Phi,$$

so that the self-conjugate property for an operator $a_\nu$ writes

$$\left(\Phi^+ a_\nu \Phi\right)^* = \Phi^*\eta a_\nu\Phi = \Phi^+ a_\nu\Phi.$$

In the formal approach of Bleuler and Gupta, the quantum Lorenz condition is formulated as a condition for the allowed states but in a weaker form, *i.e.*,

$$\partial A^-(x)\Phi = \frac{\partial A_\nu^-(x)}{\partial x_\nu}\Phi = 0; \quad \Phi^*(\partial A^+) = 0. \tag{1.283}$$

The weaker condition (1.283) ensures the fulfillment of the Lorenz condition for expectation values,

$$\langle\partial A\rangle = \Phi^*\partial A(x)\Phi = 0,$$

which is strong enough to ensure the compliance of the correspondence principle (it is quite sufficient to establish correspondence with the classical field).

As it has been shown in paragraph I.2.2.3, the scalar function $\chi(x) = \partial A(x)$ in the actual diagonal gauge satisfies the equation (1.87), even in the case of interaction with sources. Therefore, the Lorenz condition (1.283) maintains its form if it refers to the interaction between the quantized electromagnetic field and the currents produced by charged particles.

In momentum representation, using the local frame variables and taking into account that $k_0 = |\vec{k}|$, equation (1.283) becomes

$$F^-(\vec{k})\Phi = 0; \quad \Phi^* F^+(\vec{k}) = 0, \tag{1.284}$$

where

$$F^\pm(\vec{k}) = a_0^\pm(\vec{k}) - a_3^\pm(\vec{k}).$$

It then follows that in the allowed states the combination $a_3^+ a_3^- - a_0^+ a_0^-$ plays no role since

$$\langle a_3^+ a_3^- - a_0^+ a_0^- \rangle = \Phi^*(a_3^+ - a_0^+) a_3^- \Phi = 0,$$

and, as a direct result, in the allowed states the total energy and momentum of the time-like and longitudinal pseudo-photons are null.

It is not difficult to show, by means of the equations (1.283) and (1.284), that the allowed states have the following structure,

$$\Phi = \Phi_{tr}^0 + \sum_n c_n \prod_{1\le m\le n} \left[a_3^+(\vec{k}_m) - a_0^+(\vec{k}_m)\right] \Phi_{tr}^n, \tag{1.285}$$

where $\Phi_{tr}^0$ and $\Phi_{tr}^n$ are the amplitudes of the real photon states (not containing pseudo-photons).

It is worth proving that for decomposition in the local frame (1.95) are essential only the transversal components

$$A_\nu^{tr}(\vec{k}) = e_\nu^1(\vec{k}) a_1(\vec{k}) + e_\nu^2(\vec{k}) a_2(\vec{k}).$$

To this end, one observes that the r.h.s. of the formula (1.95) can be written as

$$\begin{aligned} A_\nu^-(\vec{k}) &= A_\nu^{tr,-}(\vec{k}) + \left(\frac{k_\nu}{|\vec{k}|} - \delta_{\nu 0}\right) a_3^-(\vec{k}) + \delta_{\nu 0} a_0^-(\vec{k}) \\ &= A_\nu^{tr,-}(\vec{k}) + k_\nu \Lambda(\vec{k}) + \delta_{\nu 0} F^-(\vec{k}). \end{aligned}$$

Here the term $k_\nu \Lambda = k_\nu a_3^-(\vec{k})/|\vec{k}|$ has the structure of a four-gradient and can be omitted by using the gradient transform tool, while $F^-(\vec{k})\Phi = 0$, according to the first equation (1.284). Therefore,

$$A_\nu^-(\vec{k})\Phi = A_\nu^{tr,-}(\vec{k})\Phi,$$

as well as

$$\Phi^* A_\nu \Phi = \Phi^* A_\nu^{tr} \Phi.$$

It can be shown that a more general form is also possible,

$$\Phi^* K(A)\Phi = \Phi^* K(A^{tr})\Phi = \Phi_{tr}^{*0} K(A^{tr})\Phi_{tr}^0, \tag{1.286}$$

where $K(A)$ is an operator dependent on $A$, and $\Phi^0$ is the state function introduced by the relation (1.285).

Therefore, according to (1.286), the expectation value on the allowed states of any potential $A$ equals the expectation value of the transversal potential $A^{tr}$ on the real physical states.

To conclude, here is a set of formulas for the quantized electromagnetic field, subjected to the Lorenz gauge:

— the commutator of the field functions,

$$\Big[A_\mu(x), A_\nu(y)\Big] = ig_{\mu\nu} D_0(x-y), \tag{1.287}$$

where

$$D_0(x) = D(x)|_{m=0} = -i(2\pi)^{-3} \int e^{ikx} \delta(k^2)\varepsilon(k^0)\, dk;$$

— the field equation,

$$\Box A_\nu = 0; \tag{1.288}$$

— the Lorenz subsidiary condition,

$$\left(\frac{\partial A^-(x)}{\partial x}\right)\Phi = 0; \tag{1.289}$$

— the expectation value of the four-momentum over allowed states,

$$\Big\langle P^\nu \Big\rangle = \left\langle \int d\vec{k}\, k^\nu \sum_{\sigma=1,2} a_\sigma^+(\vec{k}) a_\sigma^-(\vec{k}) \right\rangle; \tag{1.290}$$

— the intrinsic angular momentum vector (the spin),

$$\vec{S} = i \int d\vec{k} \left[ \vec{a}^{+}(\vec{k}) \times \vec{a}^{-}(\vec{k}) \right] . \tag{1.291}$$

### I.3.7.4. The free gravitational field in the linear approximation

Even if the gravitational force was the first fundamental force (out of the four known at present) identified and studied as such, endowed with a universal action (affecting any type of particle) and being straightly observable, it still remains the lesser-known force. Significant efforts to understand its nature have been made during the last decades, and various studies on this subject have been elaborated [3, 20, 51, 58, 78, 94, 103, 120, 152, 127, 206, 233, 249, 278, 281, 282, 295, 344, 416].

Non-linearity of the Einstein's field equations leads to difficulties – proved so far to be insurmountable – regarding the quantization of the gravitational field, so that the gravity is still waiting for a viable quantization model, capable to include the gravitational force alongside the other three fundamental forces, within the framework of a *New Standard Model.* Nevertheless, the efforts devoted to solutioning of this major physical problem have never stopped, the scientific investigations being concretized in a considerable number of research papers and specific works [4, 28, 32, 93, 229, 302]. These efforts are also supported from another direction, namely that of gauge theories, knowing that the Standard Model is based on the concept of unitary gauge symmetry. In this respect, serious attempts to elaborate a number of gauge theories of gravitation have been recently performed [35, 36, 231, 244, 291, 315, 321,322], with the obvious purpose of including the gravity in a unitary scheme alongside the other fundamental interactions described by the Standard Model (or extended models, of the *BSM*-type).

As it is known, the weak gravitational field was already quantized (there are several procedures in this respect, and one of them is presented in this paragraph), but the quantization of an arbitrary (intense) gravitational field is still waiting for a solution. The massive cosmic bodies, generators of (very) intense gravitational fields –

such as the black-holes, the massive neutronic stars, the quasars and pulsars – play an essential role in the framework of modern cosmological models. Thus, it is worth mentioning several recent papers devotes to cosmological models [133, 167, 188, 218, 238, 280] and to the black-holes [8, 53, 61, 83, 87, 185, 203], as two research domains of great interest.

This paragraph is dedicated to a new variant of quantization of the free gravitational field in the linear approximation, starting with a generalized commutation relation of Heisenberg-type for the gravitational potentials $y_{\mu\nu}$. Using this procedure, the expressions for the energy-momentum four-vector and the spin tensor of the weak gravitational field shall be obtained. The results are in agreement with those first obtained by Gupta, and later by Just and Mitskevich but, in the authors opinion, the procedure adopted in this paragraph is more natural.

In the following investigation, the natural unit system ($c = 1$, $\hbar = 1$) and the notations corresponding to the flat space ($x_j$, $j = 1, 2, 3$; $x_4 = ix_0 = it$) shall be used. The Greek indices run from 1 to 4 (from 0 to 3 in the case of a curved space), while the Latin indices run from 1 to 3 (except for those specified by the letters $a, b, c, ...$) which run from 1 to 2. A bar under an index indicates no summation over repeated indices, and a comma stands for usual derivative. To express the positive and negative frequency parts for the creation and annihilation operators, the conventions used in the papers [181(p.161)] and [387] shall be adopted.

In his already classical paper [181], Gupta performs quantization of the free gravitational field in the linear approximation. In this formalism, the gravitational field has eleven components, corresponding to the ten independent components of the symmetric tensor $y_{\mu\nu}$ plus its trace $y$, considered as an independent variable. These results lead to eleven polarization states for the gravitons. If the virtual gravitons are eliminated, there remain only two types of (real, transversal) gravitons, interfering in the expressions of the Hamiltonian and the spin tensor of the free gravitational field.

In another paper [182(1968)], Gupta gives up the idea of independence of the quantities $y_{\mu\nu}$ and $y$, all the sixteen components of the tensor $y_{\mu\nu}$ being treated as independent variables. Thus, he constructs the *number operator* (the operator for the number of particles) starting with a combination of the Fourier coefficients of the plane-wave expansion of the gravitational potentials. Using this procedure he obtaines ten types of gravitons and, finally, two types of real (observable) gravitons. Similar results have been obtained by Just [219] and Mitskevich [275].

The procedure used in this paragraph eliminates the subsidiary variable $y = y_{\alpha\alpha}$ as independent variable and, starting with a generalized commutation relation, a new version of quantization of the free gravitational field is given, obtaining the main quantities (*e.g.* the energy-momentum four-vector, the graviton spin tensor), in agreement with [182(1968), 219, 275].

According to Einstein's theory, the gravitational field differs from the other fields in two respects. First, the investigation of such a field is performed by means of the Riemannian geometry and second, the gravitational field is non-linear.

In the case of a weak gravitational field, following Gupta's ideas [181], the covariant metric tensor of the gravitational field, $g_{\mu\nu}$, can be linearized according to

$$g_{\mu\nu} = \eta_{\mu\nu} + \kappa h_{\mu\nu}, \tag{1.292}$$

which is equivalent, in the first-order approximation (in terms of the powers of $\kappa$), to

$$\sqrt{-g}\, g^{\mu\nu} = \eta^{\mu\nu} - \kappa y^{\mu\nu}, \tag{1.293}$$

and

$$h_{\mu\nu} = y_{\mu\nu} - \frac{1}{2}\eta_{\mu\nu} y; \quad y = y_{\alpha}{}^{\alpha}, \tag{1.294}$$

where the Minkowski metric tensor $\eta^{\mu\nu}$ has the signature $(-1,-1,-1,+1)$, the tensor $y^{\mu\nu}$ corresponds to the gravitational field, $g = \det(g_{\mu\nu})$, and $\kappa = \sqrt{16\pi G}$ is the gravitational coupling constant[7] ($G$ being the Newtonian gravitational constant). Below,

[7] The constant $\kappa$ enters the Einstein's famous equations [181(p.161), 241] $R^{\mu\nu} - \frac{1}{2}g^{\mu\nu}R = -\frac{1}{2}\kappa^2 T^{\mu\nu}$, where $R^{\mu\nu}$ is the gravitational field tensor, $R$ – the curvature invariant, and $T^{\mu\nu}$ – the matter energy-momentum tensor (by "matter" is understood anything, except for gravitation).

the linearization variant given by (1.293) shall be used. In this case, the Minkowski space can be regarded as the zeroth-order approximation of the Riemannian space, and all the field quantities can be represented as series expansions in powers of $\kappa$ in this space.

When passing to the Minkowski space, the gravitational field separates into a linear and a non-linear parts, the first being considered as a free gravitational field, and the other as a direct interaction between gravitons (an interaction part[8]).

The consideration of the non-linearity of the gravitational field as a perturbation is justified, since the Einstein's linear approximation describes the experimental facts with high precision.

In order to quantize the free gravitational field in the frame of the linear approximation one starts with Einstein's Lagrangian density [181(p.608), 388(1968)]:

$$\mathcal{L}_g = \kappa^{-2}\sqrt{-g}\, g^{\mu\nu}\left(\Gamma^{\alpha}_{\mu\nu}\Gamma^{\beta}_{\alpha\beta} - \Gamma^{\alpha}_{\mu\beta}\Gamma^{\beta}_{\nu\alpha}\right), \tag{1.295}$$

where

$$\Gamma^{\alpha}_{\mu\nu} = \frac{1}{2}g^{\alpha\lambda}\left(g_{\mu\lambda,\nu} + g_{\nu\lambda,\mu} - g_{\mu\nu,\lambda}\right)$$

are the Christoffel's symbols of the second kind.

Expanding the Lagrangian density (1.210) in a series of powers of $\kappa$, and taking into account (1.208), the determinant of the metric tensor writes

$$g = \det\left(g^{\mu\nu}\right) = -1 + \kappa y + \frac{1}{2}\kappa^2\left(y^{\mu\nu}y_{\mu\nu} - y^2\right) + ..., \tag{1.296}$$

which yields

$$\sqrt{-g} = 1 - \frac{1}{2}\kappa y + \frac{1}{4}\kappa^2\left(\frac{1}{2}y^2 - y^{\mu\nu}y_{\mu\nu}\right) + ... \tag{1.297}$$

and

$$\frac{1}{\sqrt{-g}} = 1 + \frac{1}{2}\kappa y + \frac{1}{4}\kappa^2\left(\frac{1}{2}y^2 + y^{\mu\nu}y_{\mu\nu}\right) + ... \tag{1.298}$$

[8]Nowadays are already given gauge-type theories for the gravitational field.

The contravariant and covariant components of the metric tensor are given by

$$
\begin{aligned}
g^{\mu\nu} &= \eta^{\mu\nu} + \kappa\left(\frac{1}{2}y\eta^{\mu\nu} - y^{\mu\nu}\right) \\
&\quad + \frac{1}{2}\kappa^2\left(\frac{1}{4}y^2\eta^{\mu\nu} + \frac{1}{2}y^{\alpha\beta}y_{\alpha\beta}\eta^{\mu\nu} - yy^{\mu\nu}\right) + \ldots
\end{aligned}
\tag{1.299}
$$

and

$$
\begin{aligned}
g_{\mu\nu} &= -\frac{1}{\sqrt{-g}}\,\text{Minor}\,(g^{\mu\nu}) = \eta_{\mu\nu} + \kappa\left(-\frac{1}{2}y\eta_{\mu\nu} + y_{\mu\nu}\right) \\
&\quad + \kappa^2\left(\frac{1}{8}y^2\eta_{\mu\nu} - \frac{1}{4}y^{\alpha\beta}y_{\alpha\beta}\eta_{\mu\nu} - \frac{1}{2}yy_{\mu\nu} + y_\mu{}^\alpha y_{\alpha\nu}\right) + \ldots,
\end{aligned}
\tag{1.300}
$$

respectively. Passing to the flat space and using (1.293), (1.299), (1.300) as well as Hilbert's[9] [228] coordinate condition

$$y_{\mu\nu,\nu} = 0, \tag{1.301}$$

one obtains the linear part of the Lagrangian density (1.295) that describes the free gravitational field (omitting a four-divergence) (see [181(p.161), 219]) in the form

$$\mathcal{L} = -\frac{1}{4}y_{\mu\nu,\alpha}h_{\mu\nu,\alpha}\,, \tag{1.302}$$

where, in this case,

$$h_{\mu\nu} = y_{\mu\nu} - \frac{1}{2}\delta_{\mu\nu}y; \quad y = y_{\alpha\alpha}. \tag{1.302'}$$

Using the property $y_{\mu\nu} = y_{\nu\mu}$, explicitating (1.302), and giving up the hypothesis of independency of the variables $y_{\mu\nu}$ and $y$, the generalized momenta canonically conjugated to the variables $y_{\mu\nu}$ are[10]

$$
\begin{aligned}
\Pi_{\mu\nu} &= \frac{\partial\mathcal{L}}{\partial y_{\mu\nu,t}} = y_{\mu\nu,t}; \quad \mu \neq \nu, \\
\Pi_{\underline{\mu}\,\underline{\mu}} &= \frac{\partial\mathcal{L}}{\partial y_{\underline{\mu}\,\underline{\mu},t}} = \frac{1}{2} = h_{\underline{\mu}\,\underline{\mu},t}\,, \\
\Pi &= \Pi_{\mu\mu} = -\frac{1}{2}y_{,t}\,,
\end{aligned}
\tag{1.303}
$$

---

[9]This condition is obtained by linearizing the De Donder–Fock [148] harmonicity condition $g^{\mu\nu}{}_{,\nu} = 0$.

[10]Considering $y_{\mu\nu}$ and $y$ as independent variables, the quantities $y_{\mu\nu}$ appear instead of $h_{\mu\nu}$ in the equations (1.303') and (1.305) (see also [181(p.161)]).

or, in general,

$$\Pi_{\mu\nu} = \left(1 - \frac{1}{2}\delta_{\underline{\mu\nu}}\right) h_{\mu\nu,t} \, . \tag{1.303'}$$

This result justifies the use for quantization of a generalized commutation relation of Heisenberg-type,

$$\left[y_{\mu\nu}(\vec{x},t),\, \Pi_{\lambda\rho}(\vec{x}\,',t)\right]$$

$$= i\left(1 - \frac{1}{2}\delta_{\underline{\lambda\rho}}\right)(\delta_{\mu\lambda}\delta_{\nu\rho} + \delta_{\mu\rho}\delta_{\nu\lambda})\,\delta(\vec{x} - \vec{x}\,'), \tag{1.304}$$

or, by means of (1.218),

$$\left[y_{\mu\nu}(\vec{x},t),\, h_{\lambda\rho,t}(\vec{x}\,',t)\right] = i\,(\delta_{\mu\lambda}\delta_{\nu\rho} + \delta_{\mu\rho}\delta_{\nu\lambda})\,\delta(\vec{x} - \vec{x}\,'). \tag{1.305}$$

In the authors' opinion, the above postulated commutator is more general than those found in Gupta's papers [181(p.161), 182(1958)], and naturally follows from the expression of the generalized momenta $\Pi_{\mu\nu}$.

Since the gravitational potentials satisfy the vacuum wave equation [125, 207],

$$\Box y_{\mu\nu} = 0, \tag{1.306}$$

its solution can be written as a plane wave of the form

$$\begin{aligned} y_{\mu\nu}(x) &= \frac{1}{(2\pi)^{3/2}} \int \frac{d^3p}{\sqrt{2p_0}} \left[y^{(+)}_{\mu\nu}(\vec{p})e^{i(\vec{p}\cdot\vec{x}-p_0t)} + y^{(-)}_{\mu\nu}(\vec{p})e^{-i(\vec{p}\cdot\vec{x}-p_0t)}\right] \\ &= y^{(+)}_{\mu\nu}(x) + y^{(-)}_{\mu\nu}(x), \end{aligned} \tag{1.307}$$

where $\vec{p}$ is the gravitons' momentum, and $p_0 = |\vec{p}|$ is their energy. An analogous formula can be written for $h_{\mu\nu}$ as well.

Substituting these results into (1.305) and using the Fourier representation of the delta function, one obtains

$$\frac{i}{2(2\pi)^{3/2}} \int \frac{p'_0 d^3p\, d^3p'}{\sqrt{p_0 p'_0}} \Bigg\{ - \left[y^{(+)}_{\mu\nu}(\vec{p}),\, h^{(+)}_{\lambda\rho}(\vec{p}\,')\right] \exp\Big\{i\Big[\vec{p}\cdot\vec{x} + \vec{p}\,'\cdot\vec{x}\,'$$

$$- (p_0 + p'_0)t\Big]\Big\} + \left[y^{(-)}_{\mu\nu}(\vec{p}),\, h^{(-)}_{\lambda\rho}(\vec{p}\,')\right] \exp\Big\{ - i\Big[\vec{p}\cdot\vec{x} + \vec{p}\,'\cdot\vec{x}\,'$$

$$- (p_0 + p_0')t\Big]\Big\} + \Big[y^{(+)}_{\mu\nu}(\vec{p}),\, h^{(-)}_{\lambda\rho}(\vec{p}\,')\Big] \exp\Big\{i\Big[\vec{p}\cdot\vec{x} - \vec{p}\,'\cdot\vec{x}\,'$$

$$- (p_0 - p_0')t\Big]\Big\} - \Big[y^{(-)}_{\mu\nu}(\vec{p}),\, h^{(+)}_{\lambda\rho}(\vec{p}\,')\Big] \exp\Big\{-i\Big[\vec{p}\cdot\vec{x} - \vec{p}\,'\cdot\vec{x}\,'$$

$$- (p_0 - p_0')t\Big]\Big\}\Bigg\} = \frac{i}{(2\pi)^3}\,(\delta_{\mu\lambda}\delta_{\nu\rho} + \delta_{\mu\rho}\delta_{\nu\lambda})$$

$$\times \int \exp\Big[i\vec{p}\cdot(\vec{x} - \vec{x}\,')\Big]\, d^3p. \tag{1.308}$$

Eliminating the time-dependence in the exponentials $\exp[\pm i(p_0 + p_0')t]$ and using the delta Dirac function properties, one obtains by identification the following commutation relations:

$$\begin{aligned} &\Big[y^{(\pm)}_{\mu\nu}(\vec{p}),\, h^{(\pm)}_{\lambda\rho}(\vec{p}\,')\Big] = 0, \\ &\Big[y^{(\pm)}_{\mu\nu}(\vec{p}),\, h^{(\mp)}_{\lambda\rho}(\vec{p}\,')\Big] = \pm\,(\delta_{\mu\lambda}\delta_{\nu\rho} + \delta_{\mu\rho}\delta_{\nu\lambda})\,\delta(\vec{p} - \vec{p}\,'). \end{aligned} \tag{1.309}$$

Since $y = -h$, the last equation yields[11]

$$\Big[y^{(+)}_{\mu\nu}(\vec{p}),\, y^{(-)}(\vec{p}\,')\Big] = \Big[h^{(+)}_{\mu\nu}(\vec{p}),\, h^{(-)}(\vec{p}\,')\Big] = -2\delta_{\mu\nu}\delta(\vec{p} - \vec{p}\,'), \tag{1.310}$$

and, finally,

$$\Big[y^{(+)}_{\mu\nu}(\vec{p}),\, y^{(-)}_{\lambda\rho}(\vec{p}\,')\Big] = \Big[h^{(+)}_{\mu\nu}(\vec{p}),\, h^{(-)}_{\lambda\rho}(\vec{p}\,')\Big]$$

$$= (\delta_{\mu\lambda}\delta_{\nu\rho} + \delta_{\mu\rho}\delta_{\nu\lambda} - \delta_{\mu\nu}\delta_{\lambda\rho})\delta(\vec{p} - \vec{p}\,'). \tag{1.311}$$

As one can simply verify, unlike the corresponding result presented in the reference [181(p.161)], here the creation operators $y^{(-)}_{\mu\nu}(\vec{p})$, $h^{(-)}_{\mu\nu}(\vec{p})$ and the annihilation operators $y^{(+)}_{\mu\nu}(\vec{p})$, $h^{(+)}_{\mu\nu}(\vec{p})$ satisfy the same commutation relation. This relation has been obtained by Gupta [182(1868)], Just [219] and Mitskevich [275] in a different way, while here it naturally follows from (1.305).

In view of (1.307) and (1.311), for $t \neq t'$ one finds

$$\Big[y_{\mu\nu}(x), y_{\lambda\rho}(x')\Big] = \Big[h_{\mu\nu}(x), h_{\lambda\rho}(x')\Big]$$

$$= i(\delta_{\mu\lambda}\delta_{\nu\rho} + \delta_{\mu\rho}\delta_{\nu\lambda} - \delta_{\mu\nu}\delta_{\lambda\rho})D(x - x'), \tag{1.312}$$

---

[11] In the Ref. [181(p.161)] the commutator writes $\Big[y^{(+)}_{\mu\nu}(\vec{p}), y^{(-)}(\vec{p}\,')\Big] = 0$.

in agreement with [182(1968)], where

$$D(x) = -\frac{1}{(2\pi)^3} \int d^3p \; e^{i\vec{p}\cdot\vec{x}} \, \frac{\sin p_0 t}{p_0}, \tag{1.313}$$

is the *Schwinger function* [217].

Starting with (1.312) and using the standard procedure of calculation of the expectation value of the chronological product $T$ over the vacuum state [387], one obtains

$$\langle 0 \left| T \left[ y_{\mu\nu}(x) y_{\lambda\rho}(x') \right] \right| 0 \rangle = \langle 0 \left| T \left[ h_{\mu\nu}(x) h_{\lambda\rho}(x') \right] \right| 0 \rangle$$

$$= \frac{1}{2}(\delta_{\mu\lambda}\delta_{\nu\rho} + \delta_{\mu\rho}\delta_{\nu\lambda} - \delta_{\mu\nu}\delta_{\lambda\rho}) D_F(x - x'), \tag{1.314}$$

where

$$D_F(x) = -\frac{2i}{(2\pi)^3} \int d^4p \; e^{ipx} \frac{1}{p^2 - i\eta} \tag{1.315}$$

is the *Feynman causal function* [327].

As one can see, unlike Gupta's papers [181], the expressions of the graviton propagators $\langle y_{\mu\nu} y_{\lambda\rho} \rangle$ and $\langle h_{\mu\nu} h_{\lambda\rho} \rangle$ coincide, which is normal and natural[12].

In order to find the graviton momentum, energy, and spin in momentum representation, one starts with the well-known definitions from the quantum field theory [2, 105],

$$P_j = i \int d^3x \; t_{j4}, \quad H = \int d^3x \; t_{44}, \quad S_{kj} = -\int d^3x \; f_{kj,t}\,, \tag{1.316}$$

where

$$t_{\mu\nu} = -\frac{1}{2} y_{\alpha\beta,\mu} h_{\alpha\beta,\nu} + \frac{1}{4} \delta_{\mu\nu} y_{\alpha\beta,\lambda} h_{\alpha\beta,\lambda} \tag{1.317}$$

is the energy-momentum pseudotensor of the gravitational field, and

$$f_{\mu\nu,\alpha} = y_{\mu\beta,\alpha} y_{\nu\beta} - y_{\nu\beta,\alpha} y_{\mu\beta} \tag{1.318}$$

---

[12] This property also follows from the Lagrangian density (1.217), which is symmetric in the variables $y_{\mu\nu}$ and $h_{\mu\nu}$.

is the *Belinfante's tensor* (see also [181(p.161)]). Evidentiating the spatial and temporal parts, one finds

$$\begin{cases} P_j = -\frac{1}{2}\int d^3x \left[y_{\mu\nu,j}(x)y_{\mu\nu,t}(x) - \frac{1}{2}y_{,j}(x)y_{,t}(x)\right]; \\ H = \frac{1}{4}\int d^3x \Big\{y_{\mu\nu,j}(x)y_{\mu\nu,j}(x) + y_{\mu\nu,t}(x)y_{\mu\nu,t}(x) \\ \qquad - \frac{1}{2}\left[y_{,j}(x)y_{,j}(x) + y_{,t}(x)y_{,t}(x)\right]\Big\}; \\ S_{kj} = \int d^3x \left[y_{k\alpha}(x)y_{j\alpha,t}(x) - y_{j\alpha}(x)y_{k\alpha,t}(x)\right]. \end{cases} \tag{1.319}$$

Decomposing the operators of the gravitational field into positive and negative frequency parts – according to (1.307) – the momentum components write

$$P_j = \sum_{a=1}^{2} P_{aj}, \tag{1.320}$$

where

$$P_{aj} = P'^{+}_{aj} + P'^{-}_{aj} + P''^{+}_{aj} + P''^{-}_{aj}, \tag{1.321}$$

with

$$\begin{cases} P'^{(\pm)}_{1j} = -\frac{1}{2}\int d^3x\; y^{(\pm)}_{\mu\nu,j}(x)y^{(\pm)}_{\mu\nu,t}(x); \\ P''^{(\pm)}_{1j} = -\frac{1}{2}\int d^3x\; y^{(\pm)}_{\mu\nu,j}(x)y^{(\mp)}_{\mu\nu,t}(x); \\ P'^{(\pm)}_{2j} = \frac{1}{4}\int d^3x\; y^{(\pm)}_{,j}(x)y^{(\pm)}_{,t}(x); \\ P''^{(\pm)}_{2j} = \frac{1}{4}\int d^3x\; y^{(\pm)}_{,j}(x)y^{(\mp)}_{,t}(x). \end{cases} \tag{1.322}$$

To calculate $P'^{(\pm)}_{1j}$ one uses the plane wave representation (1.307) of the gravitational potentials, and a suitable form of the Fourier transform of the delta Dirac function,

$$\delta(\vec{p} + \vec{p}\,') = \frac{1}{(2\pi)^3}\int d^3x\; e^{i(\vec{p}+\vec{p}\,')\cdot\vec{x}}, \tag{1.323}$$

as well. Integrating over $\vec{p}'$, one obtains

$$P_{1j}^{'(\pm)} = -\frac{1}{4}\int d^3p\ p_j y_{\mu\nu}^{(\pm)}(\vec{p}) y_{\mu\nu}^{(\pm)}(-\vec{p}) e^{\mp 2ip_0 t}. \tag{1.324}$$

Interchanging $\vec{p}$ and $-\vec{p}$, it follows that

$$P_{1j}^{'(\pm)} = -P_{1j}^{'(\pm)} = 0. \tag{1.325}$$

Similarly, using (1.222) and the corresponding expression of $\delta(\vec{p}-\vec{p}')$, and integrating over $\vec{p}'$, one arrives at

$$P_{1j}^{''(\pm)} = \frac{1}{4}\int d^3p\ p_j y_{\mu\nu}^{(\pm)}(\vec{p}) y_{\mu\nu}^{(\mp)}(\vec{p}). \tag{1.326}$$

Following the same procedure, one also obtains

$$P_{2j}^{'(\pm)} = 0, \tag{1.327}$$

and

$$P_{2j}^{''(\pm)} = -\frac{1}{8}\int d^3p\ p_j y^{(\pm)}(\vec{p}) y^{(\mp)}(\vec{p}). \tag{1.328}$$

In view of the above results, using (1.311), (1.320)–(1.322) and (1.325)–(1.328), and omitting a null-term, the momentum of the free weak gravitational field finally writes

$$P_j = \frac{1}{2}\int d^3p\ p_j \left[ y_{\mu\nu}^{(-)}(\vec{p}) y_{\mu\nu}^{(+)}(\vec{p}) - \frac{1}{2} y^{(-)}(\vec{p}) y^{(+)}(\vec{p}) \right]. \tag{1.329}$$

The energy of the free weak gravitational field is determined in the same way. Thus, according to (1.319-2),

$$H = \sum_{\alpha=1}^{4} H_\alpha, \tag{1.330}$$

where

$$H_\alpha = H_\alpha^{'+} + H_\alpha^{'-} + H_\alpha^{''+} + H_\alpha^{''-}, \tag{1.331}$$

with

$$\begin{cases} H_1^{'(\pm)} = \frac{1}{4}\int d^3x\; y^{(\pm)}_{\mu\nu,j}(x)y^{(\pm)}_{\mu\nu,j}(x); \\ H_1^{''(\pm)} = \frac{1}{4}\int d^3x\; y^{(\pm)}_{\mu\nu,j}(x)y^{(\mp)}_{\mu\nu,j}(x); \\ H_2^{'(\pm)} = \frac{1}{4}\int d^3x\; y^{(\pm)}_{\mu\nu,t}(x)y^{(\pm)}_{\mu\nu,t}(x); \\ H_2^{''(\pm)} = \frac{1}{4}\int d^3x\; y^{(\pm)}_{\mu\nu,t}(x)y^{(\mp)}_{\mu\nu,t}(x), \end{cases} \tag{1.332$'$}$$

and

$$\begin{cases} H_3^{'(\pm)} = -\frac{1}{8}\int d^3x\; y^{(\pm)}_{,j}(x)y^{(\pm)}_{,j}(x); \\ H_3^{''(\pm)} = -\frac{1}{8}\int d^3x\; y^{(\pm)}_{,j}(x)y^{(\mp)}_{,j}(x); \\ H_4^{'(\pm)} = -\frac{1}{8}\int d^3x\; y^{(\pm)}_{,t}(x)y^{(\pm)}_{,t}(x); \\ H_4^{''(\pm)} = -\frac{1}{8}\int d^3x\; y^{(\pm)}_{,t}(x)y^{(\mp)}_{,t}(x). \end{cases} \tag{1.332$''$}$$

Since $|\vec{p}| = p_0$, these formulas finally lead to

$$\begin{cases} H_1^{'(\pm)} = -H_2^{'(\pm)} = \frac{1}{8}\int d^3p\; p_0 y^{(\pm)}_{\mu\nu}(\vec{p})y^{(\pm)}_{\mu\nu}(-\vec{p})e^{\mp 2ip_0t}; \\ H_1^{''(\pm)} = H_2^{''(\pm)} = \frac{1}{8}\int d^3p\; p_0 y^{(\pm)}_{\mu\nu}(\vec{p})y^{(\mp)}_{\mu\nu}(\vec{p}); \\ H_3^{'(\pm)} = -H_4^{'(\pm)} = -\frac{1}{16}\int d^3p\; p_0 y^{(\pm)}(\vec{p})y^{(\pm)}(-\vec{p})e^{\mp 2ip_0t}; \\ H_3^{''(\pm)} = H_4^{''(\pm)} = -\frac{1}{16}\int d^3p\; p_0 y^{(\pm)}(\vec{p})y^{(\mp)}(\vec{p}). \end{cases} \tag{1.333}$$

Introducing (1.333) into (1.330) and (1.331), and omitting a null-term, the expression of the searched energy is obtained in the form

$$H = \frac{1}{2}\int d^3p\; p_0\left[y^{-}_{\mu\nu}(\vec{p})y^{(+)}_{\mu\nu}(\vec{p}) - \frac{1}{2}y^{(-)}(\vec{p})y^{(+)}(\vec{p})\right]. \tag{1.334}$$

In an analogous manner, using the Fourier expansion (1.307), taking into account the properties of the delta Dirac function $\delta(\vec{p}\pm\vec{p}')$, interchanging $\vec{p}$ and $-\vec{p}$, and omitting a null-term, the graviton spin tensor given by (1.319-3) becomes

$$S_{kj} = i\int d^3p\ \left[y^{(-)}_{j\alpha}(\vec{p})y^{(+)}_{k\alpha}(\vec{p}) - y^{(-)}_{k\alpha}(\vec{p})y^{(+)}_{j\alpha}(\vec{p})\right]. \tag{1.335}$$

The energy-momentum four-vector follows by virtue of (1.329) and (1.334), being given by

$$P_\alpha = \frac{1}{2}\int d^3p\ p_\alpha y^{(-)}_{\mu\nu}(\vec{p})h^{(+)}_{\mu\nu}(\vec{p}), \tag{1.336}$$

and, if the contained quantities are explicitated[13], then it becomes

$$\begin{aligned} P_\alpha = \int d^3p\ p_\alpha \Big[ & y'^{(-)}_{11}(\vec{p})y'^{(+)}_{11}(\vec{p}) + y'^{(-)}_{00}(\vec{p})y'^{(+)}_{00}(\vec{p}) \\ & - y'^{(-)}_{22}(\vec{p})y'^{(+)}_{33}(\vec{p}) - y'^{(-)}_{33}(\vec{p})y'^{(+)}_{22}(\vec{p}) \\ & + y^{(-)}_{12}(\vec{p})y^{(+)}_{12}(\vec{p}) + y^{(-)}_{13}(\vec{p})y^{(+)}_{13}(\vec{p}) \\ & + y^{(-)}_{23}(\vec{p})y^{(+)}_{23}(\vec{p}) - y^{(-)}_{10}(\vec{p})y^{(+)}_{10}(\vec{p}) \\ & - y^{(-)}_{20}(\vec{p})y^{(+)}_{20}(\vec{p}) - y^{(-)}_{30}(\vec{p})y^{(+)}_{30}(\vec{p})\Big], \end{aligned} \tag{1.337}$$

where

$$\begin{cases} y'^{(\pm)}_{11}(\vec{p}) = \dfrac{1}{2}\left[y^{(\pm)}_{11}(\vec{p}) - y^{(\pm)}_{22}(\vec{p})\right]; \\ y'^{(\pm)}_{22}(\vec{p}) = \dfrac{1}{2}\left[y^{(\pm)}_{11}(\vec{p}) + y^{(\pm)}_{22}(\vec{p})\right]; \\ y'^{(\pm)}_{33}(\vec{p}) = \dfrac{1}{2}\left[y^{(\pm)}_{33}(\vec{p}) - y^{(\pm)}_{00}(\vec{p})\right]; \\ y'^{(\pm)}_{00}(\vec{p}) = \dfrac{1}{2}\left[y^{(\pm)}_{33}(\vec{p}) + y^{(\pm)}_{00}(\vec{p})\right]. \end{cases} \tag{1.338}$$

According to (1.311) and (1.338), the quantities $y'^{(\pm)}_{\underline{a}\,\underline{a}}(\vec{p})$ $(a=1,0)$, $y^{(\pm)}_{ij}(\vec{p})$ $(i \neq j)$, and $y^{(\pm)}_{i0}(\vec{p})$ are creation and annihilation operators,

[13] $y_{j4} = -y_{j0}$, and $y_{44} = -y_{00}$.

respectively, that satisfy the following commutation relations:

$$\begin{aligned}\left[y'^{(+)}_{\underline{a}\,\underline{a}}(\vec{p}),\, y'^{(-)}_{\underline{a}\,\underline{a}}(\vec{p}\,')\right] &= \left[y^{(+)}_{\underline{ij}}(\vec{p}),\, y^{(-)}_{\underline{ij}}(\vec{p}\,')\right] \\ &= -\left[y^{(+)}_{\underline{i0}}(\vec{p}),\, y^{(-)}_{\underline{i0}}(\vec{p}\,')\right] = \delta(\vec{p}-\vec{p}\,').\end{aligned} \tag{1.339}$$

This property is not valid for $y'^{(\pm)}_{\underline{b}\,\underline{b}}(\vec{p})$, $(b = 2,3)$, but their combinations

$$\begin{cases} y''^{(\pm)}_{22}(\vec{p}) = \dfrac{1}{\sqrt{2}}\left[y'^{(\pm)}_{22}(\vec{p}) - y'^{(\pm)}_{33}(\vec{p})\right], \\ y''^{(\pm)}_{33}(\vec{p}) = \dfrac{1}{\sqrt{2}}\left[y'^{(\pm)}_{22}(\vec{p}) + y'^{(\pm)}_{33}(\vec{p})\right], \end{cases} \tag{1.340}$$

are, also, creation and annihilation operators, respectively, that satisfy the following commutation relations:

$$\left[y'^{(+)}_{\underline{b}\,\underline{b}}(\vec{p}),\, y''^{(-)}_{\underline{b}\,\underline{b}}(\vec{p}\,')\right] = \begin{cases} \delta(\vec{p}-\vec{p}\,'), & b = 2, \\ -\delta(\vec{p}-\vec{p}\,'), & b = 3. \end{cases} \tag{1.341}$$

Adding and subtracting the term

$$P'_\alpha = \int d^3p\, p_\alpha \left[y'^{(-)}_{22}(\vec{p})y'^{(+)}_{22}(\vec{p}) + y'^{(-)}_{33}(\vec{p})y'^{(+)}_{33}(\vec{p})\right] \tag{1.342}$$

in the formula (1.337), this becomes

$$\begin{aligned} P_\alpha = \int d^3p\, p_\alpha \Big[& n'_{11}(\vec{p}) + n'_{00}(\vec{p}) + n''_{22}(\vec{p}) - n''_{33}(\vec{p}) \\ &+ n_{12}(\vec{p}) + n_{13}(\vec{p}) + n_{23}(\vec{p}) - n_{10}(\vec{p}) - n_{20}(\vec{p}) - n_{30}(\vec{p})\Big], \end{aligned} \tag{1.343}$$

where the quantities

$$n_{\underline{\mu\nu}}(\vec{p}) = y^{(-)}_{\underline{\mu\nu}}(\vec{p})y^{(+)}_{\underline{\mu\nu}}(\vec{p}) \tag{1.344}$$

are particle number operators.

As one can easily verify, unlike the corresponding results presented in Ref. [181(p.161)], the momentum representation of the energy-momentum four-vector of the free weak gravitational field contains ten types of gravitons, corresponding to the ten independent components of the symmetric tensor $y_{\mu\nu}$. The particle number operators have been obtained as linear combinations of the Fourier coefficients of the plane wave representation of the gravitational potentials. These results are in concordance with those obtained by Gupta [182(1968)], the only difference being the way of justifying these linear combinations in the present volume.

According to (1.339) and (1.341), the commutators of the quantities $y_{i0}^{(\pm)}(\vec{p})$ and $y_{33}^{\prime\prime(\pm)}(\vec{p})$ have a negative sign, and this fact imposes the use of the indefinite metric formal approach from quantum electrodynamics. Thus, introducing the Hermitian metric operator $\eta$ and defining the expectation value of some operator $F$ as

$$\langle F\rangle = \langle\psi|\eta F|\psi\rangle, \tag{1.345}$$

the condition for a Hermitian operator – in a generalized sense, *i.e.* defined by means of the operator $\eta$ – to have a real expectation value is not necessary anymore [227].

Suppose that – as in the case of the electromagnetic field – the operator $y_{\mu\nu}$ corresponding to the weak gravitational field, is Hermitian in a generalized sense. The requirements that the expectation values of $y_{ij}$ and $y_{44}$ be real, and those of $y_{i4}$ be imaginary (see [304]), yield[14]

$$\begin{cases} \Big[y_{ij}(x),\eta\Big] = \Big[y_{44}(x),\eta\Big] = 0; \\ \Big\{y_{i4}(x),\eta\Big\} = 0. \end{cases} \tag{1.346}$$

Consider, instead of (1.301), the weaker subsidiary condition of Bleuler–Gupta type [162, 181(p.161)],

$$y_{\mu\nu,\nu}^{(+)}(x)|\psi\rangle = 0, \tag{1.347}$$

or, more explicitly,

$$\begin{cases} \Big[y_{ij,j}^{(+)}(x) + y_{i4,4}^{(+)}(x)\Big]|\psi\rangle = 0, \\ \Big[y_{4j,j}^{(+)}(x) + y_{44,4}^{(+)}(x)\Big]|\psi\rangle = 0. \end{cases} \tag{1.347′}$$

---

[14] $\{A, B\} = AB + BA$.

Taking into account (1.346), by means of Hermitian conjugation and multiplication with $\eta$, one still has

$$\begin{cases} \langle\psi|\left[y^{(-)}_{ij,j}(x) - y^{(-)}_{i4,4}(x)\right]\eta = \langle\psi|\eta y^{(-)}_{i\nu,\nu}(x) = 0, \\ \langle\psi|\left[y^{(-)}_{4j,j}(x) - y^{(-)}_{44,4}(x)\right]\eta = -\langle\psi|\eta y^{(-)}_{4\nu,\nu}(x) = 0, \end{cases} \tag{1.348}$$

that is,

$$\langle\psi|\eta y^{(-)}_{\mu\nu,\nu}(x) = 0. \tag{1.348$'$}$$

In view of (1.347) and (1.348′), one finally gets

$$\langle\psi|\eta y_{\mu\nu,\nu}(x)|\psi\rangle = 0, \tag{1.349}$$

which ensures the validity of the expectation value for the subsidiary condition (1.347), and the connection with the classical theory.

The above considerations regarding the indefinite metric formal approach justify the use of this formalism for quantization of the free weak gravitational field, in concordance with Gupta's [181(p.161), 182(1968)] and Mitskevich's [275] results.

The subsidiary condition (1.347), together with its adjoint (in the generalized $\eta$ sense), eliminate eight types of gravitons (the virtual gravitons), thus remaining only two types of (real) gravitons, involved in the expressions of the expectation values of the quantities that characterize the free weak gravitational field [275, 391].

Passing to the momentum representation in (1.347),

$$p_\nu y^{(-)}_{\mu\nu}(\vec{p})|\psi\rangle = 0 \tag{1.350}$$

and choosing the coordinate system with $z$-axis oriented along the momentum direction [181(p.161)], $\vec{p} = (0, 0, p_0)$, it follows that

$$\left[y^{(+)}_{\mu 3}(\vec{p}) + iy^{(+)}_{\mu 4}(\vec{p})\right]|\psi\rangle = 0, \tag{1.351}$$

or, explicitly,

$$\begin{cases} \left[y^{(+)}_{13}(\vec{p}) - y^{(+)}_{10}(\vec{p})\right]|\psi\rangle = 0, \\ \left[y^{(+)}_{23}(\vec{p}) - y^{(+)}_{20}(\vec{p})\right]|\psi\rangle = 0, \\ \left[y^{(+)}_{33}(\vec{p}) - y^{(+)}_{30}(\vec{p})\right]|\psi\rangle = 0, \\ \left[y^{(+)}_{03}(\vec{p}) - y^{(+)}_{00}(\vec{p})\right]|\psi\rangle = 0. \end{cases} \tag{1.352}$$

Analogously, one also obtains

$$\begin{cases} \langle\psi|\eta\Big[y_{13}^{(-)}(\vec{p}) - y_{10}^{(-)}(\vec{p})\Big] = 0, \\ \langle\psi|\eta\Big[y_{23}^{(-)}(\vec{p}) - y_{20}^{(-)}(\vec{p})\Big] = 0, \\ \langle\psi|\eta\Big[y_{33}^{(-)}(\vec{p}) - y_{30}^{(-)}(\vec{p})\Big] = 0, \\ \langle\psi|\eta\Big[y_{03}^{(-)}(\vec{p}) - y_{00}^{(-)}(\vec{p})\Big] = 0. \end{cases} \tag{1.353}$$

Adding and subtracting the last two equations of (1.352), and also using (1.338) and (1.341), one finds

$$\begin{cases} \Big[y_{22}^{''(+)}(\vec{p}) - y_{33}^{''(+)}(\vec{p})\Big]\,|\psi\rangle = 0, \\ \Big[y_{00}^{'(+)}(\vec{p}) - y_{30}^{(+)}(\vec{p})\Big]\,|\psi\rangle = 0. \end{cases} \tag{1.352-1}$$

A similar calculation leads to

$$\begin{cases} \langle\psi|\eta\,\Big[y_{22}^{''(-)}(\vec{p}) - y_{33}^{''(-)}(\vec{p})\Big] = 0, \\ \langle\psi|\eta\,\Big[y_{00}^{'(-)}(\vec{p}) - y_{30}^{(-)}(\vec{p})\Big] = 0. \end{cases} \tag{1.353-1}$$

The equation (1.352-1), together with its Hermitian conjugate (1.353-1), show the absence of the gravitons of type $y_{13}$ and $y_{10}$ for the free weak gravitational field. Indeed, taking the combination

$$\begin{aligned} \langle\psi|\eta\Big\{ &\Big[y_{13}^{(-)}(\vec{p}) - y_{10}^{(-)}(\vec{p})\Big]\Big[y_{13}^{(+)}(\vec{p}) + y_{10}^{(+)}(\vec{p})\Big] \\ &+\Big[y_{13}^{(-)}(\vec{p}) + y_{10}^{(-)}(\vec{p})\Big]\Big[y_{13}^{(+)}(\vec{p}) - y_{10}^{(+)}(\vec{p})\Big]\Big\}|\psi\rangle = 0, \end{aligned} \tag{1.354}$$

and using (1.344), one gets

$$\langle n_{13}(\vec{p})\rangle - \langle n_{10}(\vec{p})\rangle = 0. \tag{1.355}$$

In a similar way, one also obtains

$$\begin{aligned} \langle n_{23}(\vec{p})\rangle - \langle n_{20}(\vec{p})\rangle &= \langle n_{22}^{''}(\vec{p})\rangle - \langle n_{33}^{''}(\vec{p})\rangle \\ &= \langle n_{00}^{'}(\vec{p})\rangle - \langle n_{30}(\vec{p})\rangle = 0. \end{aligned} \tag{1.355$'$}$$

Taking into account (1.355) and (1.355′), the expectation value of the energy-momentum four-vector of the free weak gravitational field given by (1.343) writes

$$\langle P_\alpha \rangle = \int d^3p\, p_\alpha \Big[ \langle n'_{11}(\vec{p}) \rangle + \langle n_{12}(\vec{p}) \rangle \Big]. \tag{1.356}$$

As one observes – in agreement with [275, 391] – only the transversal gravitons of the type $y'_{11} = \frac{1}{2}(y_{11} - y_{22})$ and $y_{12}$ contribute to the energy and momentum of the field. The quantity $y_{11} + y_{22}$, which is equal to the trace $y$ of the tensor $y_{\mu\nu}$, and the associated operator that describes the states of spin zero do not contribute to the energy and momentum of the free weak gravitational field. Therefore, unlike Gupta's results [181(p.161), 182(1954)], a subsidiary condition imposed to eliminate the gravitons of the $y$-type is not necessary.

Following Mitskevich ideas [275], the conditions satisfied by the real (observable) gravitons are

$$\begin{cases} y^{(+)}(x)|\psi\rangle = 0, \\ \langle\psi|\eta\, y^{(-)}(x) = 0, \end{cases} \tag{1.357}$$

so that

$$\langle y(x) \rangle = 0, \tag{1.358}$$

in agreement with [26(1969)].

Considering now the expectation value of the spin tensor given by (1.250), one realizes that only the component $S_z = S_{12}$ is connected to the real (transversal) gravitons[15],

$$\begin{aligned} \langle S_z \rangle = 2i \textstyle\int d^3p\ & \Big[ \Big\langle y_{12}^{(-)}(\vec{p}) y_{11}^{'(+)}(\vec{p}) \Big\rangle \\ & - \Big\langle y_{11}^{'(-)}(\vec{p}) y_{12}^{(+)}(\vec{p}) \Big\rangle \Big]. \end{aligned} \tag{1.359}$$

Since the terms between brackets do not represent numbers of particles, one must perform a separation of the gravitons in terms of their spin. This can be done by diagonalizing $S_z$. Introducing new creation and annihilation operators through [181(p.161)],

$$\begin{cases} y_\pm^{(-)}(\vec{p}) = \frac{1}{\sqrt{2}} \Big[ y_{11}^{'(-)}(\vec{p}) \pm i y_{12}^{(-)}(\vec{p}) \Big], \\ y_\pm^{(+)}(\vec{p}) = \frac{1}{\sqrt{2}} \Big[ y_{11}^{'(+)}(\vec{p}) \mp i y_{12}^{(+)}(\vec{p}) \Big], \end{cases} \tag{1.360}$$

[15]The rest of gravitons are eliminated in view of the subsidiary conditions.

which, in view of (1.311) and (1.338-1) satisfy the commutation relations

$$\begin{cases} \left[y_{\pm}^{(+)}(\vec{p}), y_{\pm}^{(-)}(\vec{p}\,')\right] = \delta(\vec{p}-\vec{p}\,'), \\ \left[y_{\pm}^{(+)}(\vec{p}), y_{\mp}^{(-)}(\vec{p}\,')\right] = 0, \end{cases} \tag{1.361}$$

the expectation value for the $S_z$ spin component of the free weak gravitational field's quanta becomes

$$\langle S_z \rangle = 2 \int d^3p \, \Big[ \langle n_+(\vec{p}) \rangle - \langle n_-(\vec{p}) \rangle \Big], \tag{1.362}$$

where

$$n_{\pm}(\vec{p}) = y_{\pm}^{(-)}(\vec{p}) y_{\pm}^{(+)}(\vec{p}) \tag{1.363}$$

are the particle number operators corresponding to the $y_{\pm}$-type gravitons.

Using a similar procedure, the expectation value for the energy-momentum four-vector of the free weak gravitational field (1.356) becomes

$$\langle P_\alpha \rangle = \int d^3p \; p_\alpha \Big[ \langle n_+(\vec{p}) \rangle + \langle n_-(\vec{p}) \rangle \Big]. \tag{1.364}$$

As one can see, the only operators corresponding to the observable gravitons whose spin projection on the direction of motion (helicity) have the eigenvalues $\pm 2$, are $y_{\pm}^{(\pm)}(\vec{p})$. These results are in agreement with those obtained by Gupta [181(p.161), 182(1968)], Just [219] and Mitskevich [275], except for the fact that in this volume the operators corresponding to the energy-momentum four-vector and the spin of the free weak gravitational field are obtained as expectation values. This fact ensures a natural connection with the corresponding classical theory.

## I.3.8. Spinorial fields

### I.3.8.1. Quantization of the massive Dirac field

Analyzing the structure of the energy-momentum four-vector of the massive spinorial field,

$$P^\nu = \int d\vec{p} \; p^\nu \Big[ a_s^{*+}(\vec{p}) a_s^{-}(\vec{p}) - a_s^{*-}(\vec{p}) a_s^{+}(\vec{p}) \Big], \; (s = 1, 2)$$

one realizes that only the independent amplitudes $a_s$ ($s = 1, 2$) can be subjected to the quantization procedure.

As it has been shown, the condition for the expectation value of the energy operator to be positive leads to the necessity of quantization of the massive spinorial field according to the Fermi–Dirac rule. Therefore, it is necessary to consider the Fourier amplitudes of the massive spinorial field, $a_a(\vec{k})$, as operators that satisfy the Fermi–Dirac anticommutation relations

$$\{a_s^{*-}(\vec{p}), a_r^+(\vec{q})\} = \{a_s^-(\vec{p}), a_r^{*+}(\vec{q})\} = \delta_{sr}\delta(\vec{p} - \vec{q}), \tag{1.365}$$

all the other (anti)commutators being null.

The anticommutation relations for the operator $\psi(x)$ are obtained by means of (1.155), (1.159), and (1.160), which give the Fourier representation of the positive and negative parts of the field function,

$$\psi^\pm(x) = (2\pi)^{-3/2} \int d\vec{p}\; e^{\pm ipx} a_s^\pm(\vec{p}) v^{s,\pm}(\vec{p}), \tag{1.366}$$

$$\overline{\psi}^\pm(x) = (2\pi)^{-3/2} \int d\vec{p}\; e^{\pm ipx} a_s^{*\pm}(\vec{p}) \overline{v}^{s,\pm}(\vec{p}), \tag{1.367}$$

as well as with the help of the formulas for summation over a spinorial index (1.166). One finds

$$\begin{aligned}\left[\psi_\alpha^-(x), \overline{\psi}_\beta^+(y)\right]_+ &= (2\pi)^{-3} \int d\vec{p}\; e^{ik(y-x)} \sum_\sigma v_\alpha^{\sigma,-}(\vec{p}) \overline{v}_\beta^{\sigma,+}(\vec{p}) \\ &= (2\pi)^{-3} \int d\vec{p}\; e^{ip(y-x)} \frac{(\hat{p} + m)_{\alpha\beta}}{2p_0} = (2\pi)^{-3} \int dp\, (\hat{k} + m)_{\alpha\beta} \\ &\times \delta(p^2 - m^2)\theta(p^0) e^{ip(y-x)} = \left(i\gamma^\nu \frac{\partial}{\partial x^\nu} + m\right)_{\alpha\beta} \frac{1}{i} D^-(x - y).\end{aligned}$$

Similarly, it can be shown that

$$\left\{\psi^+(x), \overline{\psi}^-(y)\right\} = (i\hat{\partial}_x + m)\frac{1}{i} D^+(x - y).$$

Introducing the notation

$$S_{\alpha\beta}^\pm(x) = (i\hat{\partial} + m)_{\alpha\beta} D^\pm(x), \quad S = S^+ + S^-,$$

one can write

$$\left[\psi(x), \overline{\psi}(y)\right] = \frac{1}{i} S(x - y), \tag{1.368}$$

where

$$S(x) = (i\hat{\partial} + m)D(x)$$
$$= \frac{i}{(2\pi)^3}\int e^{-ipx}\delta(p^2 - m^2)\varepsilon(p^0)(\hat{p} + m)dp. \tag{1.369}$$

The anticommutation relations are very tightly related to the field equations, because

$$(i\hat{\partial} - m)S(x) = (\Box - m^2)D(x) = 0.$$

If the Lagrangian density is written in terms of the normal product,

$$\mathcal{L} = \frac{i}{2}N\left[\overline{\psi}(x)\gamma^\nu\partial_\nu\psi - (\partial_\nu\overline{\psi})\gamma^\nu\psi(x)\right] - mN\left[\overline{\psi}\psi\right], \tag{1.370}$$

one finds the energy-momentum four-tensor,

$$T^{\mu\nu} = \frac{i}{2}N\left[\overline{\psi}\gamma^\mu\frac{\partial\psi}{\partial x_\nu} - \frac{\partial\overline{\psi}}{\partial x_\nu}\gamma^\mu\psi\right], \tag{1.371}$$

the spin tensor density,

$$S^{\lambda(\mu\nu)} = \frac{1}{4}N\left[\overline{\psi}\gamma^\lambda\sigma^{\mu\nu}\psi\right] + \frac{1}{4}N\left[\overline{\psi}\sigma^{\mu\nu}\gamma^\lambda\psi\right], \tag{1.372}$$

and the current four-vector as well,

$$J^\nu(x) = N\left[\overline{\psi}(x)\gamma^\nu\psi(x)\right]. \tag{1.373}$$

Passing to the momentum representation by means of (1.366) and (1.367), and according to (1.167), (1.173), and (1.174), the corresponding integrals of the motion write:

— the energy-momentum four-vector,

$$P^\nu = \int d\vec{p}\; p^\nu\left[a_s^{*+}(\vec{p})a_s^-(\vec{p}) + a_s^+(\vec{p})a_s^{*-}(\vec{p})\right], \tag{1.374}$$

— the charge,

$$Q = \int d\vec{p}\left[a_s^{*+}(\vec{p})a_s^-(\vec{p}) - a_s^+(\vec{p})a_s^{*-}(\vec{p})\right]; \tag{1.375}$$

— the spin projection on the direction of motion,

$$S_3 = \frac{1}{2}\int d\vec{p}\Big[a_1^{*+}(\vec{p})a_1^-(\vec{p}) - a_2^{*+}(\vec{p})a_2^-(\vec{p})$$
$$- a_1^+(\vec{p})a_1^{*-}(\vec{p}) + a_2^+(\vec{p})a_2^{*-}(\vec{p})\Big]. \tag{1.376}$$

According to (1.374)–(1.376), the operators $a_s^{*+}(\vec{p})$ and $a_s^-(\vec{p})$ are, respectively, the creation and annihilation operators for the particles with momentum $\vec{p}$, energy $p^0 = \sqrt{\vec{p}^2 + m^2}$, charge $+1$, and the spin projection on the $\vec{p}$-direction equal to $+1/2$ (for $s = 1$), or $-1/2$ (for $s = 2$).

The operators $a_s^+(\vec{p})$ and $a_s^{*-}(\vec{p})$ correspond to the above described particles, except for the charge $-1$ and the spin projection on the $\vec{p}$-direction equal to $+1/2$ (for $s = 2$), or $-1/2$ (for $s = 1$).

### I.3.8.2 The massless spinorial field

If one takes $m = 0$ in the Dirac equation, the result is

$$i\gamma^\nu \partial_\nu \psi(x) = 0.$$

This equation can be decomposed into two independent equations for two-component field functions. To prove this statement one uses the fact that, if the mass term is absent, the Dirac equation operator anticommutes with matrix $\gamma_5$.

Introducing the *projection operators* (*projectors*),

$$P_\pm = \frac{1 \pm \gamma_5}{2},$$

one obtains

$$P_\pm \hat{\partial}\psi = P_\pm \gamma^\nu \partial_\nu \psi = \gamma^\nu P_\mp \partial_\nu \psi,$$

that is two separated equations for $\psi_\pm = P_\pm \psi$,

$$i\hat{\partial}\psi_+ = 0, \quad i\hat{\partial}\psi_- = 0. \tag{1.377}$$

In the standard representation, one can write

$$\psi_\pm = \frac{1}{2}\begin{pmatrix} +\psi_1 \mp \psi_3 \\ +\psi_2 \mp \psi_4 \\ \mp\psi_1 + \psi_3 \\ \mp\psi_2 + \psi_4 \end{pmatrix}.$$

This way, each of the functions $\psi_+$, $\psi_-$ contains two independent components and, under a decomposed form,

$$\psi_+ = \begin{pmatrix} +\varphi_{(+)} \\ -\varphi_{(+)} \end{pmatrix}, \qquad \psi_- = \begin{pmatrix} +\varphi_{(-)} \\ -\varphi_{(-)} \end{pmatrix},$$

can be expressed as two-component spinors,

$$\varphi_{(+)} = \frac{1}{2}\begin{pmatrix} \psi_1 - \psi_3 \\ \psi_2 - \psi_4 \end{pmatrix}, \qquad \varphi_{(-)} = \frac{1}{2}\begin{pmatrix} \psi_1 + \psi_3 \\ \psi_2 + \psi_4 \end{pmatrix}.$$

Using Appendix B for the Dirac matrices, the equations (1.377) can be formally written as

$$\left(\frac{\partial}{\partial x^0} \pm \vec{\sigma} \cdot \frac{\partial}{\partial \vec{x}}\right) \varphi_{\pm}(x) = 0. \tag{1.378}$$

This type of equations has been proposed by Weyl. In order to clarify the physical significance of the two-component function $\varphi_\alpha$ $[\alpha = (+), (-)]$ one passes to the momentum representation,

$$\varphi_\alpha(x) = \varphi^+_\alpha(x) + \varphi^-_\alpha(x); \quad \varphi^\pm_\alpha(x) = (2\pi)^{-3/2} \int d\vec{k}\ \tilde{\varphi}^\pm_\alpha(\vec{k}) e^{\pm ikx},$$

in which case the Weyl equation becomes

$$(k^0 \mp \vec{\sigma} \cdot \vec{k})\tilde{\varphi}_{(\pm)}(\vec{k}) = 0. \tag{1.379}$$

In the paragraph I.2.2.4 (see (1.170)) it has been shown that the vector matrix $\vec{\sigma}$ describes the particles with half-integer spin (*i.e.* fermions). Following Eqs. (1.379), the spin of a massless spinor particle can be oriented along the momentum direction, or in opposite direction. Therefore, each component of the four-component momentum amplitude, $\psi_\pm$, contains only one spin state.

The projection of the spin onto the direction of momentum, $(\vec{\sigma} \cdot \vec{k})/|\vec{k}|$, is called *helicity*. It has been experimentally proved that the spin and momentum of a massless spinor particle are oriented antiparallel, which means that the helicity of such a particle is $(-1/2)$ (this property is also valid for the muon-neutrino, $\nu_\mu$, and the electron-neutrino, $\nu_e$). The function $\psi_-$ describes the particles with negative helicity, and $\psi_+$ those with positive helicity.

The field operator for a massless spinor particle, introduced as

$$\nu(x) = P_- \psi(x)$$

satisfies the massless Dirac equation

$$i\hat{\partial}\nu(x) = 0,$$

as well as the subsidiary condition

$$P_+\nu(x) = \frac{1+\gamma_5}{2}\nu(x) = 0. \tag{1.380}$$

The Dirac conjugated field function $\overline{\nu} = \nu^{+}\gamma^{0}$ satisfy the conjugated Dirac equation

$$i\partial_{\mu}\overline{\nu}(x)\gamma^{\mu} = 0,$$

and the subsidiary condition

$$\overline{\nu}(x)P_{-} = \overline{\nu}(x)\frac{1-\gamma_5}{2} = 0.$$

In terms of the normal product, the Lagrangian density of the massless spinorial field writes

$$\mathcal{L}(x) = \frac{i}{2}N\Big[\overline{\nu}(x)\gamma^{\mu}\nu_{,\mu}(x) - \overline{\nu}_{,\mu}(x)\gamma^{\mu}\nu(x)\Big]. \tag{1.381}$$

The expressions for the energy-momentum tensor and the spin tensor density written in the configuration representation are obtained from (1.371) and (1.372), by setting $\nu(x)$ instead of $\psi(x)$. In momentum representation, the expression for $P^{\mu}$ follows from (1.374), except for summation over the spin indices,

$$P^{\mu} = \int d\vec{k}\; k^{\mu}\Big[a^{*+}(\vec{k})a^{-}(\vec{k}) + a^{+}(\vec{k})a^{*-}(\vec{k})\Big]. \tag{1.382}$$

In order to calculate the spin of the massless spinor field, one starts from expression (1.171) for the spin vector density,

$$\begin{aligned}\frac{1}{2}N\Big[a^{*+}(\vec{k})a^{-}(\vec{k})u^{*+}(\vec{k})\sigma_3 u^{-}(\vec{k})&\\ + a^{*-}(\vec{k})a^{+}(\vec{k})u^{*-}(\vec{k})\sigma_3 u^{+}(\vec{k})\Big]&,\end{aligned} \tag{1.383}$$

which is valid in the reference system with $k_1 = k_2 = 0$, *i.e.* the system in which the $Ox_3$-axis is oriented along the momentum direction. In this reference frame, the normalized spinors satisfy the equations

$$\hat{p}u^{\pm}(\vec{k}) = u^{*\pm}(\vec{k})\hat{k} = 0,$$

and the subsidiary conditions

$$(1+\gamma^5)u^{\pm}(\vec{k}) = u^{*\pm}(1+\gamma^5) = 0,$$

and can be chosen as

$$u^{+}(\vec{k}) = \frac{1}{\sqrt{2}}\begin{pmatrix} w(\vec{k}) \\ w(\vec{k}) \end{pmatrix} = u(\vec{k}); \quad u^{*}(\vec{k}) = \frac{1}{\sqrt{2}}\Big(w^{*}(\vec{k}), w^{*}(\vec{k})\Big).$$

Here $w(\vec{k})$ is the two-component spinor satisfying the Weyl equation (1.379),

$$(k^0 + \sigma_3 k^3) w(\vec{k}) = 0,$$

and it has the form

$$w(\vec{k}) = \begin{pmatrix} 0 \\ 1 \end{pmatrix}.$$

Above we have considered that for massless spinor particle ($m = 0$), $k^0 = k^3$. Obviously,

$$w^*(\vec{k}) = (0, 1)$$

and therefore,

$$u^{*\pm}(\vec{k}) \sigma_3 u^{\mp}(\vec{k}) = 1.$$

Substituting this result into (1.383), one obtains

$$S_3 = -\frac{1}{2} \int d\vec{k} \left[ a^{*+}(\vec{k}) a^{-}(\vec{k}) - a^{+}(\vec{k}) a^{*-}(\vec{k}) \right]. \qquad (1.384)$$

The first term of the r.h.s. of Eq. (1.384) corresponds to the massless spinor particle with *left-handed chirality* and negative helicity ($-1$), and the second to the corresponding antiparticle, with *right-handed chirality* and positive helicity ($+1$). This equation corresponds to (1.376) for $a_1 = 0$, $a_2 = a$. If instead of $\nu = \psi_-$ one uses $\psi_+$, then one obtains the analogous of Eq. (1.277) for $a_2 = 0$, describing the particles with positive helicity and antiparticles with negative helicity.

It is worth mentioning that the study of the weak interactions, involving *left-handed massless spinor particle*, sometime appeals to the following representation of the Dirac matrices ($n = 1, 2, 3$),

$$\Gamma_0 = \Gamma^0 = \begin{pmatrix} 0 & I \\ I & 0 \end{pmatrix} \; ; \quad \Gamma^n = -\Gamma_n = \begin{pmatrix} 0 & \sigma_n \\ -\sigma_n & 0 \end{pmatrix},$$

$$\Gamma_5 = \Gamma^5 = \begin{pmatrix} -I & 0 \\ 0 & I \end{pmatrix} = -i\Gamma^0\Gamma^1\Gamma^2\Gamma^3 = i\Gamma_0\Gamma_1\Gamma_2\Gamma_3, \qquad (1.385)$$

where 0 and $I$ are, respectively, the $2 \times 2$ null and unit matrices, while $\sigma_n$ ($n = 1, 2, 3$) are the Pauli matrices.

In this representation, the projection operators $P_{\pm}$ are diagonal and, as a result, the negative helicity spinor has a simpler form,

$$\nu(x) = P_{-}\psi(x) = \begin{pmatrix} \psi_1 \\ \psi_2 \\ 0 \\ 0 \end{pmatrix}.$$

The representation (1.385) is sometimes called *chiral*, being connected to the standard representation of the Dirac matrices through the transformation

$$\Gamma = O^{-1}\gamma O, \quad O = \frac{1}{\sqrt{2}}\begin{pmatrix} I & I \\ -I & I \end{pmatrix}.$$

### I.3.8.3. Canonical quantization of the free massive spin-3/2 (Rarita–Schwinger) field

Among the books approaching this subject, one of the best presentation is offered by D. Lurie in Ref. [252][16] and the considerations regarding the canonical method of quantization of the massive Rarita–Schwinger field exposed in this paragraph shall closely follow Lurie's investigation. A similar exposure can also be found, for example, in the paper written by N.S. Baaklini and M. Tuite (Ref. [14]).

In the case of the spin-3/2 field (and, in general, of any field with spin higher than 1) the procedure of canonical quantization becomes particularly difficult, mostly due to the difficulties connected with the isolation of the independent degrees of freedom or, equivalently, the elimination of the non-physical degrees of freedom from the system. As an actual example, the spin-vector $\psi_\mu$ associated with the Rarita–Schwinger field must previously be decomposed in its irreducible parts of spin 1/2 and 3/2, respectively, and canonically quantized only afterwords.

In order to avoid calculation difficulties, the creation and annihilation operators of the normal modes shall be directly utilized.

[16] In this book the notations corresponding to the complex Euclidean space are used: $x_1, x_2, x_3, x_4 = it$.

A consistent quantization scheme for the Rarita–Schwinger field must assure the equivalence between the quantum equation of motion

$$\dot{\psi}_\mu = \frac{1}{i}\Big[\psi_\mu, H\Big], \tag{1.386}$$

and the Rarita–Schwinger equations (1.192$a$), (1.192$b$), or (1.194). The field can be expanded in terms of the complete set of solutions (1.208$a$) and (1.208$b$), as

$$\psi^\mu = \frac{1}{\sqrt{V}}\sum_{\vec{p}}\sqrt{\frac{m}{\varepsilon_{\vec{p}}}}\sum_{\sigma=1}^{4}\left(c_{\vec{p}\sigma}u^\mu_{\vec{p}\sigma}e^{ip\cdot x} + d^+_{\vec{p}\sigma}v^\mu_{\vec{p}\sigma}e^{-ip\cdot x}\right). \tag{1.387}$$

Within the usual procedure of canonical quantization, the commutation relations for $c_{\vec{p}\sigma}$ and $d_{\vec{p}\sigma}$ are obtained from the equal-time commutators of the fields. Here, for convenience, the procedure shall be somehow inverted, and the commutation relations for the creation and annihilation operators shall be directly imposed. In agreement with the connection between spin and statistics, one considers

$$\begin{aligned}\left\{c_{\vec{p}\sigma}, c^+_{\vec{p}'\sigma'}\right\} &= \delta_{\vec{p}\vec{p}'}\delta_{\sigma\sigma'},\\ \left\{d_{\vec{p}\sigma}, d^+_{\vec{p}'\sigma'}\right\} &= \delta_{\vec{p}\vec{p}'}\delta_{\sigma\sigma'},\end{aligned} \tag{1.388$a$}$$

and

$$\left\{c_{\vec{p}\sigma}, c_{\vec{p}'\sigma'}\right\} = \left\{d_{\vec{p}\sigma}, d_{\vec{p}'\sigma'}\right\} = \left\{c_{\vec{p}\sigma}, d^+_{\vec{p}'\sigma'}\right\} = \left\{c_{\vec{p}\sigma}, d_{\vec{p}'\sigma'}\right\} = 0. \tag{1.388$b$}$$

To verify the correctness of this method one must show that (1.386) is consistent with (1.387), (1.388$a$), and (1.388$b$). To this end, one substitutes (1.387) and its adjoint $\overline{\psi}^\mu(x)$,

$$\overline{\psi}^\mu(x) = \frac{1}{\sqrt{V}}\sum_{\vec{p}}\sqrt{\frac{m}{\varepsilon_p}}\sum_{\sigma=1}^{4}\left(c^+_{\vec{p}\sigma}\overline{u}^\mu_{\vec{p}\sigma}e^{-ip\cdot x} + d_{\vec{p}\sigma}\overline{v}^\mu_{\vec{p}\sigma}e^{ip\cdot x}\right) \tag{1.389}$$

into the classical Hamiltonian

$$H = \int \psi^+_\mu i\partial_t\psi_\mu\, d^3x = \int \overline{\psi}_\mu\gamma_4 i\partial_t\psi_\mu\, d^3x. \tag{1.390}$$

Using (1.209$a$)–(1.209$d$) and eliminating the zero point energy, one obtains the usual expression for the Hamiltonian

$$H = \sum_{\vec{p}}\sum_{\sigma=1}^{4}\varepsilon_{\vec{p}}\left(c^+_{\vec{p}\sigma}c_{\vec{p}\sigma} + d^+_{\vec{p}\sigma}d_{\vec{p}\sigma}\right), \tag{1.391}$$

expressing the field energy as the sum of energies of the field quanta. By means of (1.388$a$), (1.388$b$), and the formula

$$[A, BC] = \{A, B\}C - B\{A, C\}, \tag{1.392}$$

one can now calculate the commutator $[\psi_\mu, H]$. One finds

$$[\psi_\mu, H] = \frac{1}{\sqrt{V}} \sum_{\vec{p}} \sqrt{\frac{m}{\varepsilon_{\vec{p}}}} \sum_{\sigma=1}^{4} \left( c_{\vec{p}\sigma} u^\mu_{\vec{p}\sigma} e^{ip\cdot x} - d^+_{\vec{p}\sigma} v^\mu_{\vec{p}\sigma} e^{-ip\cdot x} \right) = i\dot{\psi}_\mu,$$

in agreement with (1.386).

### Commutation rules. The Feynman propagator

At this stage of the investigation it is necessary to express the commutator of the field functions $\psi^\mu(x)$ and $\overline{\psi}(x')$ for the arbitrary moments of time $t$ and $t'$. Using (1.388$a$) and (1.388$b$), one can write

$$\left\{\psi^\mu(x), \overline{\psi}^\nu(x')\right\}$$

$$= \frac{1}{V} \sum_{\vec{p}} \frac{m}{\varepsilon_{\vec{p}}} \left\{ \sum_{\sigma=1}^{4} u^\mu_{\vec{p}\sigma} \overline{u}^\nu_{\vec{p}\sigma} e^{ip\cdot(x-x')} + \sum_{\sigma=1}^{4} v^\mu_{\vec{p}\sigma} \overline{v}^\nu_{\vec{p}\sigma} e^{-ip\cdot(x-x')} \right\}.$$

In view of (1.213$a$) and (1.213$b$), one therefore obtains

$$\left\{\psi^\mu(x), \overline{\psi}^\nu(x')\right\} = i\partial_{\mu\nu}\Delta(x - x'), \tag{1.393a}$$

where the differential operator $\partial_{\mu\nu}$ is defined as

$$\partial_{\mu\nu} = -(\gamma\cdot\partial - m)\left[\delta_{\mu\nu} - \frac{1}{3}\gamma_\mu\gamma_\nu + \frac{1}{3m}(\gamma_\mu\partial_\nu - \gamma_\nu\partial_\mu) - \frac{2}{3m^2}\partial_\mu\partial_\nu\right]. \tag{1.393b}$$

Using the well-known formalism of the canonical quantization, for the Feynman propagator given by the formula

$$S_{F\mu\nu}(x_1 - x_2) = \langle 0|T\left(\psi_\mu(x_1)\overline{\psi}_\nu(x_2)\right)|0\rangle, \tag{1.394}$$

the following expression is obtained,

$$S_{F\mu\nu}(x_1 - x_2) = \theta(t_1 - t_2)\partial_{\mu\nu} i\Delta^{(+)}(x_1 - x_2) - \theta(t_1 - t_2)$$
$$\times \partial_{\mu\nu} i\Delta^{(-)}(x_1 - x_2) = \partial_{\mu\nu}\Delta_F(x_1 - x_2) + \dots, \tag{1.395}$$

where the dots signify the non-covariant supplementary terms, proportional to $\delta^4(x_1 - x_2)$ and its derivatives, appearing when the $\theta$ functions are commuted past the differential operator $\partial_{\mu\nu}$. These supplementary terms can be omitted in concrete calculations regarding interacting fields. Practically, the effect of these additional terms is canceled by the so-called *normal-dependent terms*, appearing in the interaction Hamiltonian which can be written as

$$H_I = -L_I + N.D.T. \tag{1.396}$$

Nevertheless, in the case of spin-3/2 field an extra complication appears as compared to the massive vectorial field, since cancellation of these non-covariant terms introduces a supplementary covariant term, proportional to $\delta^4(x_1 - x_2)$, together with its derivatives. The concrete calculations show that the effective propagator is given by

$$\partial_{\mu\nu}\Delta_F(x_1 - x_2)$$
$$- \frac{2i}{3m^2}\Big[(\gamma_\mu\partial_\nu - \gamma_\nu\partial_\mu) + (\gamma\cdot\partial - m)\,\gamma_\mu\gamma_\nu\Big]\delta^{(4)}(x_1 - x_2),$$

which satisfies the non-homogeneous equation for the Green function corresponding to (1.194).

It is worth mentioning the following properties regarding the interaction of the spin-3/2 particles:

— the simplest trilinear coupling of the massive spin-3/2 field, $\psi_\mu$, with the neutral pseudoscalar field is

$$ic_1\overline{\psi}_\mu\gamma_5\psi_\mu\Phi;$$

— the simplest trilinear coupling of the massive spin-3/2 field, $\psi_\mu$, with the vectorial field $A_\mu$ is

$$ic_2\overline{\psi}_\mu\gamma_\lambda\psi_\mu A_\lambda;$$

— the coupling describing the disintegration of a spin-3/2 barion into a spin-1/2 barion and a neutral vectorial meson is

$$c_3\overline{\psi}\psi_\mu A_\mu + h.c.;$$

— the coupling describing the disintegration of a spin-3/2 barion into a spin-1/2 barion and a pseudoscalar meson is given by

$$ic_4\overline{\psi}\gamma_5\psi_\mu\partial_\mu\Phi + h.c.$$

### I.3.8.4. Quantization of the free massless spin-3/2 field

The massless spin-3/2 field plays an essential role in the supergravity theory, which is a supersymmetric generalization of general relativity in higher dimensions. The simplest scheme of this theory consists in the coupling of two particular gauge fields, namely the Einstein field and the massless Rarita–Schwinger field.

There appear to be at least two important reasons for the massless spin-3/2 field to be studied: one of intrinsic nature, and the other based on the direct application of this field in the supergravity theory.

Being a gauge field, the most adequate method of its quantization is offered by the functional formalism of the *path integrals*. But within the framework of this formalism still are some different ways of approaching the problem, and two of them are given below:

— the classical quantization technique by means of the path integral of a gauge field, in direct connection with the Faddeev–Popov *ansatz* for the generating functional, which is in fact equivalent to the canonical procedure of quantization in a gauge analogous to the axial gauge of the Yang–Mills theory;

— the procedure developed later by Faddeev, which: 1) is more rigorous and systematic, 2) is explicitly canonical, 3) serves as a justification for the Faddeev–Popov *ansatz*.

This last method shall be hereinafter presented, closely following the exposure given in Ref. [332]. It is, in fact, a remarkable combination of the quantization technique by means of the path-integrals, and the Dirac method of the quantization of constrained Hamiltonian systems.

As it is known, the main difficulties accompanying the use of the functional integral formalism in the problems of field quantization (which formalism is very performant and, sometimes, even saving) first consist in the fact that the convergence of such integrals has not been generally proved yet, and second in difficulties connected with the choice of the integration measure in various concrete cases. Nevertheless, several recently published studies and synthesis papers on the subject [56, 176–178, 222] offer a consistent set of concrete examples of calculations on spaces of various symmetries, showing that the path-integral method of quantization is a promising tool of investigation.

The massless spin-3/2 field is described by the action

$$S = \int d^4x \; \mathcal{L}(x), \tag{1.397}$$

with

$$\mathcal{L}(x) = -\varepsilon^{\mu\nu\rho\sigma}\overline{\psi}_\mu \gamma_5 \gamma_\nu \partial_\rho \psi_\sigma, \tag{1.398}$$

where the complex spin-vector $\psi_\mu$ is the field function. Each vector component of $\psi_\mu$ is in fact a spinor with four complex components, totalizing 16 complex components for $\psi_\mu$ or, equivalently, 32 real components. The action integral (1.397) is invariant under the gauge transformation

$$\delta\psi_\rho(x) = \partial_\rho \lambda(x), \tag{1.399}$$

where $\lambda(x)$ is an arbitrary spinorial field. As previously mentioned, the field spin-vector $\psi_\sigma$ is complex, while the Dirac–Faddeev quantization procedure requires real variables. To obtain a description in terms of real variables, one must separate the spin-vectorial field $\psi_\sigma$ in its real and imaginary parts, and then utilize the Majorana representation for the Dirac $\gamma$ matrices. Thus, one writes

$$\psi_\sigma = \psi_{1\sigma} + i\psi_{2\sigma}, \tag{1.400}$$

where $\psi_{i\sigma}$ $(i = 1, 2)$ are real, and in the Majorana representation,

$$\gamma_\mu^* = -\gamma_\mu, \tag{1.401}$$

where the asterisk stands for complex conjugation.

Here are the working conventions used in this paragraph:

$$g^{\mu\nu} = \text{diag}\,(+1, -1, -1, -1); \quad \{\gamma^\mu, \gamma^\nu\} = 2g^{\mu\nu};$$

$$\gamma_5 = i\gamma_0\gamma_1\gamma_2\gamma_3; \quad \varepsilon^{0123} = -\varepsilon_{0123} = 1; \quad \gamma_\mu^+ = \gamma_0\gamma_\mu\gamma_0.$$

With these conventions, one easily obtains

$$\gamma_\mu^T = -\gamma_0\gamma_\mu\gamma_0. \tag{1.402}$$

Denoting $A_\mu = \gamma_0\gamma_5\gamma_\mu$ (with $A_\mu^T = -A_\mu$, according to (1.402)), the action shall be given by

$$S = \int d^4x\ \mathcal{L}(x) = -\int d^4x\ \varepsilon^{\mu\nu\rho\sigma} \sum_{i=1}^{2} \psi_{\mu i}^T A_\nu \partial_\rho \psi_{\sigma i}\,, \tag{1.403}$$

where the surface contribution has been eliminated.

As one observes, the variables $\psi_i$ are completely decoupled and their dynamics is identical. From the methodological point of view, this fact implies a considerable simplification of the Hamiltonian formulation of the problem, and of the quantization scheme as well. Under these circumstances, it is sufficient to consider only the component $\psi_1$ and remember that the inclusion of $\psi_2$ is trivial (taking into account the previous observation). Mare than that, since the most applications of the scheme of the spin-3/2 field quantization are in the theory of supergravitation (which theory demands a non-massive spin-3/2 field described by a Majorana spinor[17]) and taking into account the above mentioned decoupling, it emerges that $\psi_2$ can be simply set to zero, and the action (1.403) can be replaced by

$$\mathcal{L} = -\varepsilon^{\mu\nu\rho\sigma} \psi_\mu^T A_\nu \partial_\rho \psi_\sigma, \tag{1.404}$$

recalling that $\psi_\sigma$ is a real spinor and, in the light of the above considerations, the extension to the case of the complex field is immediate.

### A survey of the Dirac and Faddeev methods for the anticommuting variables

The purpose of this paragraph is to give a short presentation of the approaching methods used by Dirac and Faddeev regarding the quantization of a physical system.

Consider a physical system with $N$ degrees of freedom (extension over the field theory is immediate) described by the Lagrangian

$$L = L(q, \dot{q}), \tag{1.405}$$

so that

$$\det\left(\frac{\partial^2 L}{\partial \dot{q}_m \partial \dot{q}_n}\right) = 0. \tag{1.406}$$

[17]It is useful to mention that the Majorana spinors are real and have four components.

Such a Lagrangian is called *singular*. As is well-known, the generalized momentum conjugate to $q_n$ is

$$p_n = \frac{\partial L}{\partial \dot{q}_n}. \tag{1.407}$$

As one observes, (1.407) – taking also into account (1.496) – cannot be inverted for all $\dot{q}$. Therefore it is necessary to pass from the standard to the Hamiltonian description. This matter was first investigated by Dirac. By eliminating the quantities $\dot{q}$ from (1.407), one obtains a certain number of independent constraints,

$$\phi_m(p, q) = 0, \tag{1.408}$$

while some quantities $\dot{q}$ remain undetermined. Using (1.407), it can be shown that the Hamiltonian

$$H_c = \sum_n p_n \dot{q}_n - L \tag{1.409}$$

depends only on $q$ and $p$ (none of $\dot{q}$). In order that the constraints (1.408) are also considered, one defines the so-called *total Hamiltonian*

$$H = H_c + \sum_m \lambda_m \phi_m, \tag{1.410}$$

where $\lambda_m$ is a set of non-zero multipliers. The correct system of the equations of motion then writes

$$\begin{aligned} \dot{p}_n &= -\frac{\partial H_c}{\partial q_n} - \sum_m \lambda_m \frac{\partial \phi_m}{\partial q_n}, \\ \dot{q}_n &= \frac{\partial H_c}{\partial p_n} + \sum_m \lambda_m \frac{\partial \phi_m}{\partial p_n}. \end{aligned} \tag{1.411}$$

This implies, for some function $g$ that depends on dynamical variables $q$ and $p$, the validity of the following equation,

$$\dot{g} = \{g, H\},$$

where $\{..., ...\}$ designates the ordinary Poisson bracket. Since the constraints must remain constant at any moment of time, the following equation is also true:

$$\dot{\phi}_m = \{\phi_m, H\} = 0. \tag{1.412}$$

Sometimes, such equations serve to determine some of the generalized velocities, but most frequently they lead to new constraints (the so-called *secondary constraints*),

$$\rho_i(q,p) = 0. \tag{1.413}$$

This procedure is carried on until the generation of the secondary constraints has stopped. At this point one introduces an important classification of the constraints in two categories:

— the *first class constraints* (or *primary constraints*, or *first type constraints*);

— the *second class constraints* (or *secondary constraints*, or *second type constraints*).

If all the constraints are denoted by a common notation $\phi_k$ and $\{\varphi_a, \phi_k\}$ cancel for all $k$ as soon as the constraints are applied, then $\varphi_a$ is called *constraint of the first type*. Any other constraint is a *constraint of the second type*. To each constraint of the first class one can associate a gauge-type condition (Faddeev). Choosing a common notation $\Phi_i$ for the first-class constraints, $\varphi_a$, the gauge conditions $\chi_a$, and the second-type constraints $\theta_b$, that is,

$$\Phi_i \equiv (\varphi_a, \chi_a, \theta_b),$$

and defining the matrix $C_{ij} = \{\Phi_i, \Phi_j\}$, one can define the Dirac brackets as

$$\{A,B\}^D = \{A,B\} - \{A,\Phi_i\}C_{ij}^{-1}\{\Phi_j, B\}. \tag{1.414}$$

Obviously,

$$\{\Phi_i, B\}^D = 0 \tag{1.415}$$

for any arbitrary $B$. Under these conditions, the canonical quantization prescription is nothing else but the *correspondence principle*: if $A$ and $B$ are two operators attached to the observables $A$ and $B$ in the quantum description, then the Dirac bracket of these two operators is replaced by their commutator multiplied by $1/i$, that is,

$$\{A,B\}^D \to \frac{1}{i}[A,B].$$

Equation (1.415) then implies $[\Phi_i, B] = 0$, so that the constraints can now be implemented in a consistent manner as operatorial identities, namely $\Phi_i \equiv 0$. If the mechanical (or field) variables are anticommuting fermionic quantities, care has to be taken about definition of the usual Poisson brackets. If, for example, denote by $E_i$ and $O_i$ the dynamical variables which are even and, respectively, odd elements of their Grassmann algebra, then the Poisson brackets are

$$\{E_1, E_2\} = \sum_\alpha \left( \frac{\partial E_1}{\partial \psi_\alpha} \frac{\partial E_2}{\partial \pi_\alpha} - \frac{\partial E_2}{\partial \psi_\alpha} \frac{\partial E_1}{\partial \pi_\alpha} \right), \tag{1.416a}$$

$$\{O, E\} = -\sum_\alpha \left( \frac{\partial O}{\partial \psi_\alpha} \frac{\partial E}{\partial \pi_\alpha} + \frac{\partial E}{\partial \psi_\alpha} \frac{\partial O}{\partial \pi_\alpha} \right), \tag{1.416b}$$

$$\{O_1, O_2\} = -\sum_\alpha \left( \frac{\partial O_1}{\partial \psi_\alpha} \frac{\partial O_2}{\partial \pi_\alpha} + \frac{\partial O_2}{\partial \psi_\alpha} \frac{\partial O_1}{\partial \pi_\alpha} \right), \tag{1.416c}$$

where $\psi_\alpha$ and $\pi_\alpha$ are the generalized coordinate and the generalized momentum, respectively. The algebra of these Poisson brackets is a *graded Lie algebra*.

The quantization by the path-integral formalism in the Hamiltonian formulation with constraints has been performed for the special case of the first class constraints by Faddeev, and later generalized by Senjanovic and Yabuki.

The path-integral writes

$$\langle out|in\rangle = \int D_q D_p \square_i \delta(\Phi_i) \left(\det\{\Phi_i, \Phi_j\}\right)^{1/2} \exp\left[ i \int (p\dot{q} - H)\, dt \right]. \tag{1.417}$$

If the gauge conditions are in convolution, that is if $\{\chi_a, \chi_b\} = 0$, the path integral becomes

$$\begin{aligned} \langle out|in\rangle = \int & D_q D_p \square_a \delta(\varphi_a) \delta(\chi_a) \square_b \delta(\theta_b) \left(\det\{\chi_a, \varphi_b\}\right) \\ & \times \left(\det\{\theta_a, \theta_b\}\right)^{1/2} \exp\left[ i \int (p\dot{q} - H)\, dt \right]. \end{aligned} \tag{1.418}$$

The expressions (1.417) and (1.418) are written in a somewhat symbolic form, because they have to be completed with the adequate boundary conditions, so that a univocal correspondence with the states in the matrix elements appearing in these relations is fulfilled; differently speaking, certain adequate initial and final wave functions corresponding to these states have to be introduced. Thus, the general theory is prepared and the problem of quantization of the massless spin-3/2 field, by using the Dirac–Faddeev formalism, can be approached.

### The Hamiltonian formulation and quantization through the functional integral of the massless spin-3/2 field

The following investigation shall concern only the real massless spin-3/2 field. It is not difficult to include the electromagnetic interaction in this theory (at least in terms of developing formal calculations) but this would show a little practical interest since – as was shown by G. Velo and D. Zwanziger [385] – it possesses non-causal modes of propagation. An interesting – but complicated – example is offered by the interaction between the spin-3/2 field and the gravitational field, considered as a vierbein field. Complications are concerned with the quantization scheme of the vierbein gravitational field, but this problem is somewhat independent of the present approach. The research is still ongoing, but significant progress has been registered only for electromagnetic couplings so far [340].

Returning to the equation (1.404), one observes that the Lagrangian density does not contain time derivatives of $\psi_0$, so that according to (1.407) the corresponding conjugate momentum is $\pi^0 = 0$. This is a constraint, and so are the conjugate momenta of $\psi_k$,

$$\pi^k = \frac{\partial \mathcal{L}}{\partial \dot{\psi}_k} = -\varepsilon^{kij}\psi_i^T A_j = -\varepsilon^{kij} A_i \psi_j .$$

The fundamental Poisson brackets are

$$\{\pi_{\mu\alpha}(\vec{x},t), \psi_{\nu\beta}(\vec{y},t)\} = g_{\mu\nu}\delta_{\alpha\beta}\delta(\vec{x}-\vec{y}), \tag{1.419}$$

$$\{\pi_{\mu\alpha}(\vec{x},t), \pi_{\nu\beta}(\vec{y},t)\} = 0, \tag{1.420}$$

$$\{\psi_{\mu\alpha}(\vec{x},t), \psi_{\nu\beta}(\vec{y},t)\} = 0. \tag{1.421}$$

These brackets are symmetrical. This property immediately follows from the definition of the Poisson bracket of odd elements $O_i$ of the Grassmann algebra (1.416). Since it shall further be used, the commutation relation satisfied by the constraints $\theta^k \equiv \pi^k + \varepsilon^{kij} A_i \psi_j$ is also given as follows:

$$\left\{\theta^i_\alpha(\vec{x},t), \theta^j_\beta(\vec{y},t)\right\} = 2\varepsilon^{ijk}(A_k)_{\alpha\beta}\delta(\vec{x}-\vec{y}). \tag{1.422}$$

The next step is to find the Hamiltonian. Its canonical part can be written as

$$H_c = \int d^3x\ \mathcal{H}_c = \int d^3x\ \varepsilon^{\mu\nu i\sigma}\psi^T_\mu A_\nu \partial_i \psi_\sigma. \tag{1.423}$$

As one can see, there appears the secondary constraint

$$0 = \dot{\pi}_c(\vec{x}) = \{\pi_0(\vec{x},t), H\},$$

which gives

$$\varepsilon^{ijk}\partial_i A_j \psi_k = 0.$$

As can easily be verified, there are no other secondary constraints. It can also be proved that the constraint

$$\varphi \equiv \partial_k \pi^k - \varepsilon^{ijk}\partial_i A_j \psi_k$$

is of the first class.

Summarizing, this theory has two primary constraints,

$$\varphi^0 = \pi^0, \tag{1.424}$$

$$\varphi = \partial_k \pi^k - \varepsilon^{ijk}\partial_i A_j \psi_k, \tag{1.425}$$

and one secondary constraint,

$$\theta^k = \pi^k + \varepsilon^{kij} A_i \psi_j. \tag{1.426}$$

The gauge condition for the constraint (1.424) is simply

$$\chi^0 = \psi^0, \tag{1.427}$$

while in the case of the primary constraint given by (1.425) the gauge can be chosen in three different ways:

1)

$$\chi = \gamma^i \psi_i, \tag{1.428}$$

called *Coulomb gauge* by A. Das and D. Z. Freedman [89],

2)
$$\chi = \partial^i \psi_i, \tag{1.429}$$
named *transverse gauge*, and

3)
$$\chi = \psi_3, \tag{1.430}$$
known as the *axial gauge*.

Since the gauge conditions are in convolution, the equation (1.418) shall be used in order to write the path integral for the $S$-matrix elements. Due to the fact that the quantity $(\det\{\theta_a, \theta_b\})^{1/2}$ is a number, the integration measure is given in fact by $\det\{\chi_a, \chi_b\} \equiv \det M$. As one observes, except for considering the Coulomb gauge (in the sense specified by A. Das and D. Z. Freedman) the measure in the path integral of the $S$-matrix elements is essentially the same as for the Maxwell field. According to the gauge conditions (1.428)–(1.430), the expressions for $\det M$ are given, respectively by

1) $\det\ M = \det\ |\gamma^i \partial_i|$,
2) $\det\ M = \det\ \nabla^2$,
3) $\det\ M = \det\ |\partial_3|$.

Below are given the $S$-matrix elements in the Lagrangian formulation (by integration over momenta $\pi$). Using (1.418) and choosing the transverse gauge, one obtains

$$\begin{aligned}\langle out|in\rangle = &\int D\pi^i D\pi^0 D\psi^i D\psi^0 \delta\left(\psi^0\right)\delta\left(\pi^0\right)\delta\left(\partial_k \pi^k - \varepsilon^{ijk}\partial_i A_j \psi_k\right)\\ &\times \delta\left(\partial_k \psi^k\right)(\det M)\delta\left(\pi^k + \varepsilon^{ijk} A_i \psi_j\right)\\ &\times \exp\left[i\int d^3x\left(\pi^k \dot{\psi}_k + \pi^0 \dot{\psi}_0 - \varepsilon^{k0im}\psi_k^T A_0 \partial_i \psi_m\right)\right].\end{aligned}$$

Integration over $\pi^0$ and $\psi^0$ gives

$$\begin{aligned}\langle out|in\rangle = &\int D\pi^i D\psi^i \delta\left(\partial_k \psi^k\right)\delta\left(\partial_k \pi^k\right)(\det M)\delta\left(\pi^k + \varepsilon^{ijk} A_i \psi_j\right)\\ &\times \exp\left[i\int d^3x\left(-\varepsilon^{ijk}\psi_i^T A_j \dot{\psi}_k - \varepsilon^{k0im}\psi_k^T A_0 \partial_i \psi_m\right)\right].\end{aligned}$$

A new integration, this time over $\pi^i$, yields

$$\langle out|in\rangle = \int D\psi^i \delta\left(\partial_k \varepsilon^{ijk} \psi_i^T A_j\right) \delta\left(\partial_k \psi^k\right) (\det M)$$
$$\times \exp\left[i \int d^3x \left(-\varepsilon^{ijk} \psi_i^T A_j \dot{\psi}_k - \varepsilon^{k0im} \psi_k^T A_0 \partial_i \psi_m\right)\right].$$

This result can be put in a form completely analogous to that encountered in the case of the Maxwell field, that is,

$$\langle out|in\rangle = \int D\psi_\mu \delta\left(\partial_k \psi^k\right) (\det M) e^{iS}, \tag{1.431}$$

with

$$\delta(\partial_k \varepsilon^{ijk} \psi_i^T A_j) = \int D\psi_0 \det 2 \exp\left[i \int d^3x (-2) \left(\partial_k \varepsilon^{ijk} \psi_i^T A_j\right) \psi_0\right].$$

**Dirac's brackets for the transverse gauge**

The total number of constraints (including gauges) is considerably large, which fact leads to the necessity of reversing a relatively big matrix. Fortunately, one can appeal to an iterative procedure, based on the iterative property of the Dirac brackets. This property can be described as follows.

One first considers any subset of $\Phi$,

$$\{X_a,\ \{X\} \subset \{\Phi\},\ \det\{X_a, X_b\} \neq 0\ \}$$

and calculate a preliminary bracket

$$\{A, B\}_X = \{A, B\} - \{A, X_a\}(C_X^{-1})_{ab}\{X_b, B\}, \tag{1.432}$$

where $(C_X)_{ab} = \{X_a, C_b\}$. Consider, next, another subset, composed by the remaining quantities $\Phi$,

$$\{Y_c,\ \{Y\} \subset \{\Phi\}\backslash\{X\},\ \det\{Y_c, Y_d\} \neq 0\ \},$$

and calculate the bracket for the second step of the procedure,

$$\{A, B\}_Y = \{A, B\}_X - \{A, Y_c\}(C_Y^{-1})_{cd}\{Y_d, B\}, \tag{1.433}$$

where $(C_Y)_{cd} = \{Y_c, Y_d\}_X$. The procedure is applied until the number of the quantities $\Phi$ is finished. The idea is that the final result is independent of the calculation procedure, so that the result is the same if the Dirac bracket is calculated directly, in one step. Due to (1.415) (*i.e.* $\{\Phi_i, B\}^D = 0$), one immediately has $\{\pi^0, B\}^D = \{\psi_0, B\}^D = 0$, for any $B$. This means that only the brackets $\{\psi_i, \psi_j\}^D$, $\{\psi_i, \pi_j\}^D$, and $\{\pi_i, \pi_j\}^D$ have to be calculated, as being the fundamental brackets of the theory.

Therefore, in our case, as $X_a$ one shall consider $\theta^k_\alpha$ from the equation (1.426). This thing is possible, since the determinant of the matrix $\{\theta^k, \theta^j\}$ is non-zero, therefore it is invertible. Its inverse (in block-matrix notation) is

$$C_\theta^{-1}(\vec{x}, \vec{y}) = \frac{1}{4}\begin{vmatrix} -iI & A_3 & -A_2 \\ -A_3 & -iI & A_1 \\ A_2 & -A_1 & -iI \end{vmatrix}\delta(\vec{x} - \vec{y}). \tag{1.434}$$

For $Y_c$ one takes the remaining subset, $\rho_i : \rho_{1\alpha} = \varphi_\alpha$, from the equation (1.425), and $\rho_{2\alpha} = \chi_\alpha$, from the equation (1.429). The inverse of $\{\rho_i, \rho_j\}$ is

$$C_\rho^{-1}(\vec{x}, \vec{y}) = \begin{vmatrix} \frac{i}{4}(\vec{\nabla}^2)^{-1}\delta(\vec{x} - \vec{y}) & (\vec{\nabla}^2)^{-1}\delta(\vec{x} - \vec{y}) \\ (\vec{\nabla}^2)^{-1}\delta(\vec{x} - \vec{y}) & 0 \end{vmatrix}.$$

First, one calculates the Dirac bracket

$$\begin{aligned}\{\psi_{i\alpha}(\vec{x}), \psi_{j\beta}(\vec{y})\}^D &= \{\psi_{i\alpha}(\vec{x}), \psi_{j\beta}(\vec{y})\}_\theta \\ &\quad - \int d^3x' d^3y' \{\psi_{i\alpha}(\vec{x}), \rho_{k\gamma}(\vec{x}\,')\}_\theta \\ &\quad \times C^{-1}_{\rho,kl}(\vec{x}\,', \vec{y}\,')\{\rho_{l\gamma}(\vec{y}\,'), \psi_{j\beta}(\vec{y})\}_\theta.\end{aligned}$$

Then, using

$$\begin{aligned}\{\psi_{i\alpha}(\vec{x}), \psi_{j\beta}(\vec{y})\}_\theta &= \{\psi_{i\alpha}(\vec{x}), \psi_{j\beta}(\vec{y})\} - \int d^3x' d^3y' \{\psi_{i\alpha}(\vec{x}), \theta^k_\gamma(\vec{x}\,')\} \\ &\quad \times C^{-1}_{\theta kl,\gamma\delta}(\vec{x}\,', \vec{y}\,')\{\theta^l_\delta(\vec{x}\,'), \psi_{j\beta}(\vec{y})\} \\ &= -C^{-1}_{\theta ij,\alpha\beta}\delta(\vec{x} - \vec{y}),\end{aligned}$$

as well as

$$\{\psi_{i\alpha}(\vec{x}), \rho_{1\beta}(\vec{y})\}_\theta = -\delta_{\alpha\beta}\partial_i\delta(\vec{x} - \vec{y}),$$

$$\{\psi_{i\alpha}(\vec{x}), \rho_{2\beta}(\vec{y})\}_\theta = \partial^l C_{il,\alpha\beta}\delta(\vec{x} - \vec{y}),$$

one obtains

$$\begin{aligned}\{\psi_i(\vec{x}),\psi_j(\vec{y})\}^D &= -C^{-1}_{\theta ij}\delta(\vec{x}-\vec{y}) + \tfrac{i}{4}\partial_i\partial_j\tfrac{1}{\vec{\nabla}^2}\delta(\vec{x}-\vec{y})\\ &\quad - \left(\partial_j\partial^k C^{-1}_{\theta,ik} + \partial_i\partial^k C^{-1}_{\theta kj}\right)\tfrac{1}{\vec{\nabla}^2}\delta(\vec{x}-\vec{y}).\end{aligned} \tag{1.435}$$

Due to the constraints

$$\pi^k = -\varepsilon^{lmk}A_l\psi_m,$$

the equation (1.435) leads to the following Dirac bracket:

$$\begin{aligned}\{\psi_i(\vec{x}),\pi^j(\vec{y})\}^D &= \frac{1}{4}\left(g^j{}_i + \partial_i\partial^j\frac{1}{\vec{\nabla}^2}\right)\delta(\vec{x}-\vec{y})\\ &\quad - \frac{1}{4}(\partial^l A_l)\varepsilon^{ijk}\partial_k\frac{1}{\vec{\nabla}^2}\delta(\vec{x}-\vec{y}).\end{aligned} \tag{1.436}$$

Analogously, one finds

$$\begin{aligned}\{\pi^i(\vec{x}),\pi^j(\vec{y})\}^D &= \frac{i}{4}\left(g^{ij} + \partial^i\partial^j\frac{1}{\vec{\nabla}^2}\right)\delta(\vec{x}-\vec{y})\\ &\quad - \frac{1}{4}\varepsilon^{ijk}\partial_k(\partial^l A_l)\frac{1}{\vec{\nabla}^2}\delta(\vec{x}-\vec{y}).\end{aligned} \tag{1.437}$$

To complete the above considerations, below are given the fundamental Dirac brackets for a general complex field $\psi = \psi_1 + i\psi_2$. From

$$\{\psi,\psi\}^D = \{\psi_1,\psi_1\}^D + i\{\psi_1,\psi_2\}^D + i\{\psi_2,\psi_1\}^D - \{\psi_2,\psi_2\}^D,$$

one easily obtains

$$\{\psi_i(\vec{x}),\psi_j(\vec{y})\}^D = 0,$$

as well as

$$\{\pi_i(\vec{x}),\pi_j(\vec{y})\}^D = 0.$$

Also,

$$\begin{aligned}\{\psi^i(\vec{x}),\pi^j(\vec{y})\}^D &= \frac{1}{2}\left(g^{ij} + \partial^i\partial^j\frac{1}{\vec{\nabla}^2}\right)\delta(\vec{x}-\vec{y})\\ &\quad - \frac{i}{2}\varepsilon^{ijk}\partial_k(A^l\partial_l)\frac{1}{\vec{\nabla}^2}\delta(\vec{x}-\vec{y}).\end{aligned}$$

The factor 1/2 appearing in the last equation can be easily eliminated by recalibrating the Lagrangian by a factor of 1/2. One can also show that the investigated system is Poincaré-invariant. Indeed, it can be shown that the algebra of the Poincaré group is satisfied by the Dirac brackets. As a result of very long (but not difficult) calculations, one finds

$$\begin{aligned}\{P^\mu, P^\nu\}^D &= 0,\\ \{M^{\alpha\beta}, P^\mu\}^D &= g^{\mu\alpha}P^\beta - g^{\mu\beta}P^\alpha,\\ \{M^{\mu\nu}, M^{\alpha\beta}\}^D &= \Big( - g^\alpha_\sigma g^\mu_\rho g^{\nu\beta} + g^\alpha_\sigma g^\nu_\rho g^{\mu\beta}\\ &\qquad - g^\mu_\sigma g^\beta_\rho g^{\nu\alpha} + g^\nu_\sigma g^\beta_\rho g^{\mu\alpha}\Big) M^{\sigma\rho}.\end{aligned}$$

These results, together with the fact that $P^\mu$ and $M^{\mu\nu}$ are constants of motion, lead to the conclusion that the theory is Poincaré-invariant. In addition, one can verify that $\dot{\psi}_i = \{\psi_i, H\}$ generates the correct equation of motion.

One can also observes that the total number of degrees of freedom, given by the number of real variables $\psi_{\mu\alpha}$ and $\pi_{\nu\beta}$ is 32. On the other hand, the system is subject to 28 constraints, as follows: Eqs. (1.424) and (1.425) – primary constraints; Eqs. (1.426) – secondary constraints; the gauge conditions (1.327) and (1.429) – if the transverse gauge is considered. Therefore, the number of the independent degrees of freedom in phase space is 4, as it must be (obliviously, the dependence on coordinates has not been considered).

## I.3.9. The charge conjugation

Analyzing the results obtained in the second paragraph concerning the integer-spin complex fields, describing charged particles, one can realize that the transformation

$$\varphi_a(x) \to \varphi'_a(x) = \varphi^*_a(x), \tag{1.438a}$$

$$\varphi^*_a(x) \to \varphi^{*'}_a(x) = \varphi_a(x), \tag{1.438b}$$

keeps the form of the fundamental equations of dynamics, except for the current $J_\nu$ and the charge $Q$, whose expressions change their sign. Due to this reason, the above transformation is called the charge conjugation transformation or, simply, the *charge conjugation*.

The charge conjugation of the spinorial field is more complicated then the transformation (1.438) due to the matrix character of the spinorial field functions; this is a matrix transformation of the form

$$\psi'(x) = C\Big(\overline{\psi}(x)\Big)^T, \quad \overline{\psi}'(x) = \Big(\psi(x)\Big)^T (C^{-1})^T = \Big(C^{-1}\psi(x)\Big)^T, \tag{1.439}$$

with the subsidiary condition (for consistency in (1.439)),

$$C^T \gamma_0 C^+ \gamma_0 = 1. \tag{1.440}$$

When writing the transformation (1.439) it has been taken into account that $\psi$ and $\psi'$ are matrices of $4 \times 1$-type, while $\overline{\psi}$ and $\overline{\psi}'$ are matrices of $1 \times 4$-type. Like in the previous paragraph (no. I.3.8.4), in this paragraph the same notation $A^T$ is used for the transpose of the matrix $A$.

The inverse of the transformation (1.439) writes

$$\psi(x) = C\Big(\overline{\psi}'(x)\Big)^T, \quad \overline{\psi}(x) = \Big(\psi'(x)\Big)^T (C^{-1})^T = \Big(C^{-1}\psi'(x)\Big)^T.$$

One observes, therefore, that the charge conjugation transformation, determined by (1.439), in addition to the property that the twice-iterated transformation is quite the identity transformation, has also the property of reflectiveness (the "mirror" symmetry), meaning that the direct and inverse transformations have the same form.

The concrete form of the matrix $C$ is determined according to the following requirements:

— the Lagrangian density of the free field does not change under such a transformation,

$$\mathcal{L}(\psi) = \mathcal{L}(\psi');$$

— the energy-momentum four-vector does not change its form under such a transformation,

$$P^\nu(\psi) = P^\nu(\psi');$$

— the current four-vector changes its sign,

$$J^\nu(\psi) = -J^\nu(\psi').$$

All these requirements are fulfilled if the following two relations are simultaneously satisfied:

$$N\left[\overline{\psi}_1(x)\gamma^\nu\psi_2(x)\right] = -N\left[\overline{\psi}_2'(x)\gamma^\nu\psi_1'(x)\right], \tag{1.441}$$

and

$$N\left[\overline{\psi}(x)\psi(x)\right] = N\left[\overline{\psi}'(x)\psi'(x)\right], \tag{1.442}$$

where $\psi_1$ and $\psi_2$ are either equal to $\psi$ itself, or to its derivative $\partial_\mu\psi$, and transform according to (1.439).

To analyze the restrictions imposed on the matrix C by (1.441) and (1.442), one substitutes the inverse of (1.439) into (1.441) and obtains

$$N\left[\overline{\psi}_1(x)\gamma^\nu\psi_2(x)\right] = N\left[\psi_1'(x)(C^{-1})^T\gamma^\nu C\psi_2'(x)\right].$$

In view of the anticommutativity of the quantized spinors $\psi'$ and $\overline{\psi}'$, it follows that

$$N\left[\overline{\psi}_1(x)\gamma^\nu\psi_2(x)\right] = -N\left[\overline{\psi}_2'(x)C^T(\gamma^\nu)^T C^{-1}\psi_1'(x)\right].$$

Comparing this last equation with (1.441), one gets the first condition that must be fulfilled by the matrix $C$,

$$C^T(\gamma^\nu)^T C^{-1} = \gamma^\nu,$$

or, in transposed form,

$$(C^{-1})^T\gamma^\nu C = (\gamma^\nu)^T. \tag{1.443}$$

Analogously, substituting the inverse of (1.439) into (1.442), the following result is obtained,

$$N\left[\overline{\psi}(x)\psi(x)\right] = N\left[\psi'(x)(C^{-1})^T C\overline{\psi}'(x)\right] = -N\left[\overline{\psi}'(x)C^T C^{-1}\psi'(x)\right],$$

leading to the second condition that must be satisfied by the matrix $C$,

$$C^T C^{-1} = -1,$$

that is,

$$C^T = -C. \tag{1.444}$$

By using (1.444), Eq. (1.443) can be written as

$$C^{-1}\gamma^\nu C = -(\gamma^\nu)^T. \tag{1.445}$$

As one can easily observe, according to (1.444), (1.445), and (1.440), the charge conjugation matrix is unitary,

$$CC^{+} = 1. \tag{1.446}$$

Formulas (1.445) and (1.446) allow the representation of the exact form of the matrices $C$ in any fixed representation of the Dirac matrices. For example, it can be demonstrated that both in the *standard* and *chiral* (given by (1.385)) representations, the matrix $C$ can be expressed by means of the "zero" and "two" Dirac matrices,

$$C_{stand} = \gamma^0 \gamma^2, \quad C_{chir} = \Gamma^0 \Gamma^2.$$

It is worth mentioning that in going over from the standard representation of the Dirac matrices (in which the matrix $\gamma^0$ is diagonal) to some other representation $\tilde{\gamma}$ by means of the formula

$$\tilde{\gamma}^\nu = O \gamma^\nu O^{-1},$$

the new matrix of charge conjugation $\tilde{C}$ can be obtained from the relation

$$\tilde{C} = OCO^T.$$

The physical significance of the charge conjugation $C$ reduces to the replacement of particles by their antiparticles. That is why, for example, the invariance with respect to the $C$ operation (valid for strong and electromagnetic interactions) leads to the fact that, if some process $\Pi$ (representing an arbitrary combination of the strong and electromagnetic interactions) takes place, then the process $C\Pi$ takes place with the same probability, which differs from $\Pi$ by substituting particles with the corresponding antiparticles.

### I.3.10. The *CPT* theorem

The final paragraph of this investigation on the quantized free fields concerns an important element of the local quantum field theory, namely the *CPT theorem*, which formulates the invariance property with respect to the three discrete transformations:

— the *charge conjugation*, $C$,
— the *space inversion*, $P$, and
— the *time reversal*, $T$.

The $C$-operation has been briefly analyzed in the previous paragraph, the $P$-operation has been met when dealing with spatial inversion of the coordinate axes (see, *e.g.*, subchapter I.1 and Appendix B), while the third operation is given by

$$x \to x', \quad x_0' = -x_0, \quad \vec{x}\,' = \vec{x}.$$

Unlike the continuous transformations, that belong to the Poincaré group, the $P$ and $T$ operations do not correspond to some real motions of a physical system, but – as well as the $C$ operation – they connect various physical processes.

In order to formally express these transformations, let us consider the example offered by the scalar complex field, described by the operators

$$\varphi(x) = \varphi^+(x) + \varphi^-(x), \quad \varphi^*(x) = \varphi^{*+}(x) + \varphi^{*-}(x).$$

The $P$ and $T$ operations change the sign of the argument in the configuration representation. For example,

$$PT\varphi^\pm(x) = PT\varphi^\pm(x_0, \vec{x}) = P\varphi^\pm(-x_0, \vec{x}) = \varphi^\pm(-x_0, -\vec{x}) = \varphi^\pm(-x),$$

while the $C$ operation changes particles in their corresponding antiparticles,

$$C\varphi^\pm(x) = \varphi^{*\pm}(x), \quad C\varphi^{*\pm}(x) = \varphi^\pm(x).$$

Therefore,

$$CPT\varphi^\pm(x) = \varphi^{*\pm}(-x), \quad CPT\varphi^{*\pm}(x) = \varphi^\pm(-x).$$

As in the case of the Lorentz group, where to its specific transformations there correspond the unitary operators $U_L$ (which transform the field operators and the state vectors), in the case of the quantization of the discrete transformations $C$, $P$ and $T$, to those transformations will correspond the unitary operators $U_C$, $U_P$ and $U_T$, respectively. For example, in the case of the scalar complex field one has

$$\varphi_C(x) = U_C^{-1}\varphi(x)U_C = \varphi^*(x),$$

$$\varphi_P(x) = U_P^{-1}\varphi(x)U_P = \varphi(x_0, -\vec{x}),$$

and so on.

According to the $CPT$ theorem, the local Lagrangian of a system of quantized fields, constructed by virtue of the "normal rules", is invariant under a simultaneous application of the three operations, that is,

$$\mathcal{L}_{CPT}(x') = U_C^{-1} U_P^{-1} U_T^{-1} \mathcal{L}(x') U_T U_P U_C = \mathcal{L}(x), \quad x' = -x.$$

The "normal rules" include the field quantization rules in agreement with the Pauli theorem, and the choice of the constraint constants consistent with the Lagrangian hermiticity condition.

Here are some physical consequences of the $CPT$ theorem. The $C$-operation replaces particles by their antiparticles. The $P$-operation inverts the sense of any direction in the three-dimensional space. In other words, the $P$-operation inverts the sense of the three-dimensional vectors (momentum, electric field intensity, etc.). The $T$-operation changes the initial states in the final states and vice-versa; it also reverts the sense of the momenta and spins. On that account, the $CPT$ theorem gives rise to the equality of the squared matrix elements in the $\Pi$ and $CPT\Pi$ processes, where the process $CPT\Pi$ are obtained from $\Pi$ by replacing the particles by their antiparticles, the spins by their inverts, and the initial states by the final states. In particular, the $CPT$ theorem explains both the mass and life-time equality for the particles and their corresponding antiparticles, as well as the sign difference between the magnetic momenta of particles and their corresponding antiparticles.

## Chapter II

# The Gravitational Transmutation Hypothesis

The purpose of this chapter is to make a survey on the main aspects connected to the quantum study of the particle interaction in the gravitational field, with astrophysical applications, according to the scientific literature on the subject. The problem of gravitational radiation is also approached.

In this chapter the natural unit system ($\hbar = 1,\ c = 1$) shall also be used, except for the experimental results, which are going to be given in CGS units. As in the previous chapter, the Greek indices run from 0 to 3 if the space is curved, and from 1 to 4 in the case of a flat space, the asterix symbol ($*$) denotes the complex conjugate, while a simple cross $+$ denotes the Hermitian conjugate.

The considerable interest regarding the quantum gravitational theory is based on the hope that the inclusion of gravitation in the quantum field theory scheme could lead to a consequent theory of elementary particles. In this respect, there exist mainly three orientations.

One of them is connected to the use of some specific astrophysical elements in the theory of elementary particles such those of cosmological nature, as has been shown by Wheeler and collab. [269, 403, 404], Misner [264–266] and Callum [57].

Another orientation is trying to get to a unitary theory of the elementary particles based on supersymmetry criteria – the well-known supergravitation theory (see *e.g.* the research papers of Nieuwenhuizen and collab. [151], or Deser and collab. [113, 114]).

Finally, there is a research orientation – as discussed in detail below – in that the gravitational field is investigated by the methods used in the quantum field theory, as variants of the gauge theories, the role of the gauge transformation being played by the coordinate transformations, subject to the gauge Poincaré group.

The connection to the elementary particles theory also emerges as a result of supposition that precisely the gravitational field plays the role of a natural physical regulator capable to eliminate the divergences met in the quantum field theory (*e.g.* the works of De Witt [104–106], Khriplovich [226], Salam and collab. [209], *etc.*).

The first canonical scheme of quantization of the gravitational field was given by Dirac [116, 117], by means of a Hamiltonian formulation. Similar procedures were employed by Arnowitt and collab. [9, 10], Schwinger [328], and by some other authors [5, 210]. The absence of covariance makes difficult the application of the perturbation theory. Due to big complications arising in the effective mathematical calculation, this scheme of quantization becomes practically useless in concrete processes.

The first covariant quantization of the gravitational field was performed by Gupta [181, 182(1954), 183], by means of the quantum electrodynamics procedures, but only in the free weak field approximation. To avoid the difficulties caused by the singularity of the Lagrangian of the free gravitational field, he used the well-known Fermi's method of electromagnetic field quantization.

One can show, unfortunately, that the application of this method to more complicated systems leads to breaking of unitarity of the theory. This fact was discovered by Feynman first in the framework of the Yang–Mills field theory and then in the theory of gravitation [141, 142]. He showed that the unitarity of the diagram containing a closed loop can be re-established by subtracting a diagram of the same form, describing the propagation of a fictive particle.

A solution for any diagram through the medium of a general method has been given by De Witt [105, 106], and by Faddeev and Popov [132], by means of the functional integration applied to gauge fields. In the case of non-Abelian gauge groups, the scheme contains the lines of the fictive particles and the vertices describing their interaction with the quanta of the gauge fields. The lines describing the fictive particles enter the diagram only as closed loops, while the minus sign corresponding to a loop shows that the fictive particles obey the Fermi statistics. Due to this fact, their direct interpretation for integer-spin particles (scalar particles for the Yang–Mills, and vectorial particles for the gravitational field) is very difficult.

These results have been verified by many authors [110, 124, 210] and some other solutions to the problem have been proposed. Thus, the so-called *path-dependent formalism* used by Mandelstam [253], and then the *S-matrix generating functional* constructed by Fradkin and Tiutin [149] – led to results equivalent to those already mentioned.

*

*　*

The particle interaction in the gravitational field and, implicitly, the *gravitational transmutations hypothesis* was and still remains one of the main research subjects of many authors.

In order to describe the interaction between the gravitational field and other fields, according to the minimal coupling principle in quantum gravity (see chapter IV), the Lagrangian densities of the corresponding fields expressed in curved space [25–27, 37, 38(1967, 1968), 306, 388–391] have to be added to the Lagrangian density of the Einstein's gravitational field [126, 181(p.608)]:

$$\begin{aligned}\mathcal{L} &= \kappa^{-2}\sqrt{-g}g^{\mu\nu}\left(\Gamma^{\rho}_{\mu\nu}\Gamma^{\sigma}_{\rho\sigma} - \Gamma^{\sigma}_{\mu\rho}\Gamma^{\rho}_{\nu\sigma}\right) - \frac{1}{4}\sqrt{-g}g^{\mu\lambda}g^{\nu\rho}F_{\mu\nu}F_{\lambda\rho} \\ &\quad - \frac{1}{2}\sqrt{-g}\Bigg\{i\frac{\partial\overline{\psi}}{\partial x^{\mu}}\gamma(\alpha)L^{\mu}(\alpha)\psi - i\overline{\psi}\gamma(\alpha)L^{\mu}(\alpha)\frac{\partial\psi}{\partial x^{\mu}} + 2M\overline{\psi}\psi \\ &\quad - 2e\overline{\psi}\gamma(\alpha)L^{\mu}(\alpha)\psi A_{\mu} + \frac{i}{4}\overline{\psi}\Big[\gamma(\lambda)\gamma(\alpha)\gamma(\beta) - \gamma(\beta)\gamma(\alpha)\gamma(\lambda)\Big] \\ &\quad \times L^{\mu}(\lambda)L^{\nu}(\alpha)L^{\sigma}(\beta)\Delta_{\mu,\nu\sigma}\psi\Bigg\} + \sqrt{-g}\left(g^{\mu\nu}\frac{\partial\varphi^{*}}{\partial x^{\mu}}\frac{\partial\varphi}{\partial x^{\nu}} - m^{2}\varphi^{*}\varphi\right. \\ &\quad \left. + ieg^{\mu\nu}\varphi^{*}\frac{\partial\varphi}{\partial x^{\mu}}A_{\nu} - ieg^{\mu\nu}\frac{\partial\varphi^{*}}{\partial x^{\mu}}\varphi A_{\nu} + g^{\mu\nu}e^{2}\varphi^{*}\varphi A_{\mu}A_{\nu}\right). \qquad (2.1)\end{aligned}$$

Here $A_\mu$ and $F_{\mu\nu}$ are the potential four-vector and antisymmetric four-tensor of the electromagnetic field, respectively, $\overline{\psi}$, $\psi$ and $M$ – the components and mass of the spinorial field, respectively, $\gamma(\alpha)$ and $L^\mu(\alpha)$ – the Dirac matrices and tetradic components (with the help of which the generalized Dirac matrices are defined), respectively, $\Delta_{\mu,\nu\sigma}$ – the Ricci rotation coefficients, $\varphi^*$, $\varphi$ and $M$ – the components and mass of the scalar complex field, respectively, $g^{\mu\nu}$ – the contravariant metric tensor of the gravitational field, $\Gamma^\rho_{\mu\nu}$ – the Christoffel symbols, $e$ – the electron charge, and $\kappa = \sqrt{16\pi G}$ – the gravitational coupling constant[1] (with $G$ – the Newtonian constant).

In the case of a weak gravitational field, following Gupta's ideas [181 (p.608), 182(1954), 183], one uses the linearization

$$\sqrt{-g}g^{\mu\nu} = \eta^{\mu\nu} - \kappa y^{\mu\nu}, \tag{2.2}$$

where the Minkowski metric tensor $\eta^{\mu\nu}$ has the signature $(-1,-1,-1,+1)$, and $\eta^{\mu\nu}$ is the tensor corresponding to the weak gravitational field. In this case the expression (2.1) can be expanded in power series in powers of $\kappa$ and, following the standard procedures used in the quantum field theory [2, 40, 216, 217, 252, 387] one can construct the $S$-matrix for the flat space ($x_j, x_4 = it,\ j = 1,2,3$) and, consequently, one can calculate concrete processes of particle reciprocal transformation, including the gravitons. This way, going up to the second-order of the coupling constant powers, the Lagrangian density describing the interactions of the above mentioned fields takes the form [26 (1969), 27, 306, 388 (1963)]:

$$\mathcal{L}_{int.} = -\frac{1}{2}\kappa y_{\mu\nu} \times \left(\frac{\partial y_{\mu\alpha}}{\partial x_\beta}\frac{\partial y_{\nu\beta}}{\partial x_\alpha} - \frac{\partial y_{\mu\beta}}{\partial x_\alpha}\frac{\partial y_{\nu\beta}}{\partial x_\alpha} + \frac{1}{2}\frac{\partial y_{\alpha\beta}}{\partial x_\mu}\frac{\partial y_{\alpha\beta}}{\partial x_\nu} + \frac{1}{2}\frac{\partial y_{\mu\nu}}{\partial x_\alpha}\frac{\partial y}{\partial x_\alpha} - \frac{1}{4}\frac{\partial y}{\partial x_\mu}\frac{\partial y}{\partial x_\nu}\right)$$

$$+ \cdots - \frac{1}{2}\kappa\left(F_{\mu\nu}F_{\mu\rho}y_{\nu\rho} - \frac{1}{4}F_{\mu\nu}F_{\mu\nu}y\right)$$

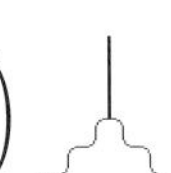

[1] The constant $\kappa$ enters Einstein's famous equations [181(p.161), 241] $R^{\mu\nu} - \frac{1}{2}g^{\mu\nu}R = -\frac{1}{2}\kappa^2T^{\mu\nu}$, where $R^{\mu\nu}$ is the Ricci curvature tensor, $R$ – the scalar curvature, and $T^{\mu\nu}$ – the stress-energy tensor (or the matter energy-momentum tensor, by "matter" being understood anything, except for gravitation).

$$-\frac{1}{4}\kappa^2\left[F_{\mu\nu}F_{\lambda\rho}y_{\mu\lambda}y_{\nu\rho} - F_{\mu\nu}F_{\mu\rho}y_{\nu\rho}y + \frac{1}{4}F_{\mu\nu}F_{\mu\nu}\left(y_{\lambda\rho}y_{\lambda\rho} + \frac{1}{2}y^2\right)\right]$$

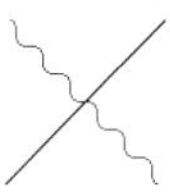

$$+ \cdots + \frac{1}{4}\kappa\left(y_{\mu\nu} + \frac{1}{2}\delta_{\mu\nu}y\right)\left(-i\overline{\psi}\gamma_\mu\frac{\partial\psi}{\partial x_\nu} + i\frac{\partial\overline{\psi}}{\partial x_\nu}\gamma_\mu\psi\right)$$

$$-\frac{1}{2}\kappa M\overline{\psi}\psi y$$

$$+ \frac{1}{16}\kappa^2\left(y_{\mu\alpha}y_{\nu\alpha} + \delta_{\mu\nu}y_{\alpha\beta}y_{\alpha\beta}\right)$$

$$\times\left(-i\overline{\psi}\gamma_\mu\frac{\partial\psi}{\partial x_\nu} + i\frac{\partial\overline{\psi}}{\partial x_\nu}\gamma_\mu\psi\right) + \frac{1}{4}\kappa^2 M\overline{\psi}\psi y_{\mu\nu}y_{\mu\nu}$$

$$-\frac{i}{16}\kappa^2\overline{\psi}\left(\gamma_\mu y_{\nu\alpha} - \gamma_\nu y_{\mu\alpha} - \gamma_\nu\gamma_\beta\gamma_\mu y_{\alpha\beta}\right)\frac{\partial y_{\nu\alpha}}{\partial x_\mu}\psi + \kappa^2 f(y^2)$$

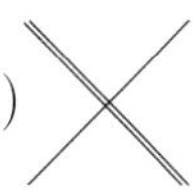

$$+ \cdots + e\overline{\psi}\gamma_\mu\psi A_\mu$$

$$-\frac{1}{2}e\kappa\left(\overline{\psi}\gamma_\mu\psi A_\mu y_{\mu\nu} + \frac{1}{2}\overline{\psi}\gamma_\mu\psi A_\mu y\right)$$

$$+ \cdots - \kappa\left(\frac{\partial\varphi^*}{\partial x_\mu}\frac{\partial\varphi}{\partial x_\nu}y_{\mu\nu} + \frac{1}{2}m^2\varphi^*\varphi y\right)$$

$$+ \frac{1}{4}\kappa^2 m^2\varphi^*\varphi$$

$$\times\left(y_{\mu\nu}y_{\mu\nu} - \frac{1}{2}y^2\right)$$

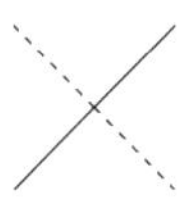

$$+ \cdots + ie\left(\frac{\partial\varphi^*}{\partial x_\mu}\varphi - \varphi^*\frac{\partial\varphi}{\partial x_\mu}\right)A_\mu$$

$$+ ie\kappa\left(\frac{\partial\varphi^*}{\partial x_\mu}\varphi - \varphi^*\frac{\partial\varphi}{\partial x_\mu}\right)A_\nu y_{\mu\nu}$$

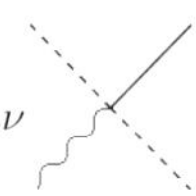

$$+ \cdots - e^2\varphi^*\varphi A_\alpha A_\alpha$$

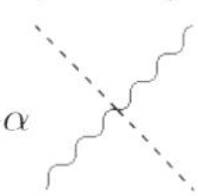

$$+ \cdots \qquad (2.3)$$

where $y = y_{\alpha\alpha}$. In Eq. (2.3), the solid lines, curly lines, double lines and dotted lines of the Feynman diagrams illustrating various type of couplings correspond to the gravitons, photons, spinorial particles and scalar particles, respectively.

The first hypothesis about possibility of reciprocal transformation of the gravitons and material quanta was suggested by Ivanenko in 1947 [211]. In the same year, Ivanenko and Sokolov [212] evaluated the probability of transformation of a scalar particle pair into two gravitons, showing that in the ultrarelativistic case the scattering cross-section increases proportional to the squared energy of the colliding particles. Later on, both the interaction between the gravitational field and other fields, and gravitational transformations were investigated by many researchers, such as: Gupta [181], Kimura [230], Piir [306], Brill and Wheeler [49], Mitskevich [270–275], Feynman [142], Vladimirov [388–391], Halpern [192], Weinberg [400], De Sabbata and collab. [37–39], B. De Witt [105], Gupta and collab. [25–27], Poznanin [307], Aslanian [11], C. De Witt and Wesley [107], Tourrenc [353, 354], and others [199, 243, 414].

The following presentation is dedicated to the main applications of the gravitational transformations with participation of the photons, scalar, spinorial, and vectorial particles, as shown in most of the mentioned references.

## II.1. Particle scattering in an external gravitational field

The first-order process of particle diffusion – photons, for example – in a static gravitational field (classical) with spherical symmetry described by the Schwarzschild metric is important, among other things, due to the fact that the quantum theory of this process leads, in the limit of small scattering angles, to the Einstein's classical law regarding the deviation (bending) of the light rays by the Sun's gravitational field.

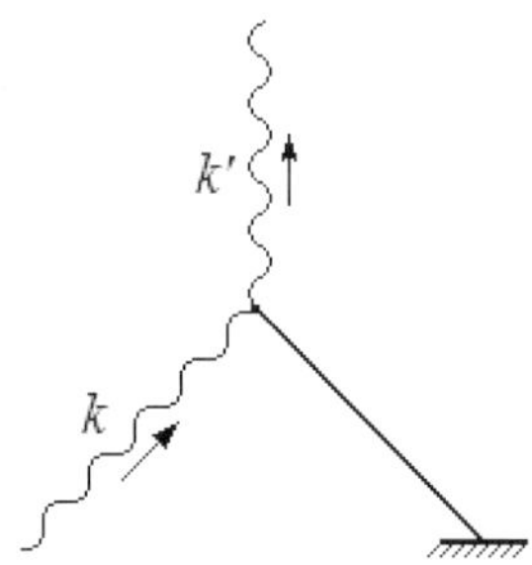

Fig. 2.1

### II.1.1. Photon scattering in Schwarzschild's static gravitational field

This diffusion/scattering process was studied by Piir [306], Mitskevich [271, 275], Halpern [192], and De Sabbata and collab. [38(1967)]. The corresponding Feynmna diagram[2] is shown in Fig. 2.1, and the differential scattering cross-section is given by

$$d\sigma = \left(\frac{\kappa^2 M_\odot}{16\pi}\right) \cot^4 \frac{\theta}{2}\, d\Omega, \tag{2.4}$$

[2]The time axis is oriented vertically upwards, with particles in its sense and antiparticles in its opposite sense. The hatched line at the bottom (in the right side of the diagram) represents the external gravitational field, while $k$ and $k'$ are the four-momenta of the initial/incident photon and final/emergent photon, respectively.

where $M_{\odot}$ is the mass of the central body that creates the static gravitational field (*e.g.* the Sun), and $\theta$ is the scattering angle.

### II.1.2. Scattering of scalar particles in the gravitational field described by Schwarzschild's metric

This case was analyzed by Mitskevich [271, 275]. The corresponding Feynman diagram is similar to that shown in Fig. 2.1, and the differential cross-section is given by

$$d\sigma = \left(\frac{\kappa^2 M_{\odot}}{16\pi}\right)^2 \left(\frac{k_0^2 - \frac{m^2}{2}}{\vec{k}^2}\right)^2 \frac{d\Omega}{\sin^4\frac{\theta}{2}}, \tag{2.5}$$

where $k_0$ and $\vec{k}$ are the energy and momentum of the scattered particles, respectively.

### II.1.3. Scattering of massive spinorial and vectorial particles in the gravitational field described by Schwarzschild's metric

Diffusion of the massive spinorial and vectorial particles was studied by Mitskevich [271, 275] and Vladimirov [388, 389]. The corresponding diagrams are similar to that shown in Fig. 2.1, and the expressions of the differential cross-sections – in the limit of small scattering angles – coincide with that given by the equation (2.5).

### II.1.4. Scattering of massless spinorial particles in the gravitational field described by Schwarzschild's metric

The diffusion of massless spinor particles in a static gravitational field was studied by De Sabbata and collab. [38(1972)]. The differential cross-section of the process – described by a diagram analogous with that previously shown – is of the form

$$d\sigma = \left(\frac{\kappa^2 M_{\odot}}{16\pi}\right)^2 \frac{\cos^2\frac{\theta}{2}}{\sin^4\frac{\theta}{2}}\, d\Omega, \tag{2.6}$$

showing that there is no difference (regarding the final result) between cases in which the massless spinor particles are described by a two-component or four-component wave function.

### II.1.5. Scattering of gravitons in the gravitational field described by Schwarzschild's metric

The effects considered so far were linear. Unlike these effects, the graviton scattering in the classical Schwarzschild's field is a nonlinear effect and demands the quantization of the gravitational field itself. This process – whose Feynman diagram is depicted in Fig. 2.2 – is investigated in Mitskevich's papers [271, 275], the differential cross-section being given by

$$d\sigma = \left(\frac{\chi^2 M_\odot}{16\pi}\right)^2 \frac{\cos^2\theta}{\sin^4\frac{\theta}{2}}\, d\Omega\,. \tag{2.7}$$

According to Vladimirov's [388(1964)] and Mitskevich's [275] ideas, from the coincidence of the differential cross-sections in the limit case of small scattering angles and, in addition – in the case of the massive particles – in the ultrarelativistic limit, it follows that in this limiting case the scattering of the particles in an external gravitational field does not depend on the spin of the particles.

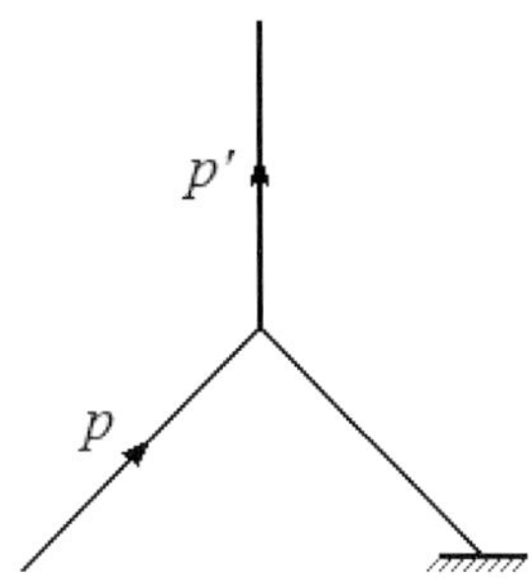


Fig. 2.2

One can also observe that the above mentioned differential cross-sections – in the limit of small scattering angles – are proportional to $d\theta/\theta^3$, corresponding to classical diffusion on a sphere. This fact was first shown for photons by Piir [306]. In his research paper, he also obtained one of the fundamental tests of Einstein's theory of relativity, namely the light ray deviation in the Sun's gravitational field.

All the processes mentioned above have also been analyzed by the authors of this book (see. *e.g.* [361] and [362]), by means of different procedures, the result of their investigations being presented in chapter VI.

## II.2. Conversion of photons into gravitons in external electromagnetic field

This first-order effect is very important among the multitude of possibilities of photon transmutation, being studied by many researchers. In the framework of classical theory, the interaction between the gravitational waves (see Ref. [125]) and the electromagnetic field was first analyzed by Gertzenstein [165] and Lupanov [251] who studied the possibilities of graviton creation (detection) in magnetic/electrostatic external field.

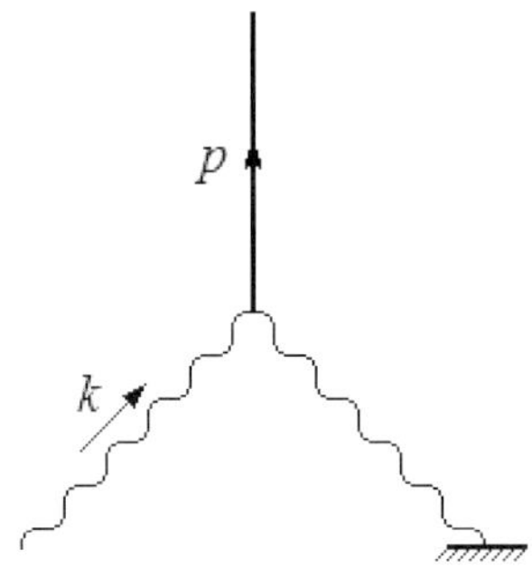

Fig. 2.3

In the framework of quantum theory, the conversion of photons into gravitons and vice versa and – in connection – the possibility of detecting the gravitational waves were studied by Halpern [192], Weber [395], Weber and Hinds [396], and De Sabbata and collab. [38(1970)]. The quantum study of the process of transformation of a photon into a graviton – described by the Feynman diagram shown in Fig. 2.3 – has been approached for various concrete types of classical external electromagnetic field.

### II.2.1. Conversion of photons into gravitons in Coulombian field

The case of a nucleus' Coulombian field of charge $Ze$ was analyzed by Mitskevich [275]. The differential cross-section of the process is given by

$$d\sigma = \frac{1}{2}\left(\frac{Ze\kappa}{16\pi}\right)^2 \cot^2\frac{\theta}{2}\left(1+\cos^2\theta\right) d\Omega, \tag{2.8}$$

where $\theta$ is the angle between the photon and graviton moment vectors. According to Mitskevich's [275] and Vladimirov's [391] evaluations, one can expect that, due to this effect, in the Sun appear about $10^{28}\,\mathrm{s}^{-1}$ gravitons, the global gravitational radiation power of the Sun being of about $P = 10^6$ erg/s. One can also observe that, owing to the independence of the differential cross-section of the photon energy, the spectrum of gravitons emitted by the Sun must totally reproduce the spectrum of the photons inside the Sun.

### II.2.2. Conversion of photons into gravitons in homogeneous electrostatic field

The case of the photons propagating in the homogeneous electrostatic field of capacitor (or a homogeneous magnetic field existent inside a rectangular domain of dimensions $A \times B \times C$) was studied by Weber and Hinds [396], Mitskevich [275], and Poznanin [307]. The differential cross-section of the process is given by

$$d\sigma = \frac{\kappa^2 F_\perp^2 A^2}{2(2\pi)^2 k_0^2} \sin^2 \frac{Bk_0}{2} \sin^2 \frac{Ck_0}{2} \, d\Omega, \tag{2.9}$$

where $k_0$ is the photons' energy, and $F_\perp$ is the intensity of the field oriented perpendicular to the photon direction of motion (which is $A$). If the linear dimensions of the scattering field considerably exceed the photon wavelength $\left(\lambda \ll V^{1/3}\right)$, the total cross-section becomes [165, 251, 307, 396],

$$\sigma = \frac{1}{8}\kappa^2 L F_\perp^2 V, \tag{2.10}$$

where $L$ is the length of the path's photon in the field existent in the volume $V$. If the photon propagates along the field, the cross-section of the process becomes null.

According to Mitskevich's [275] and Vladimirov's [391] evaluations, there exist laboratory conditions for the total scattering cross-section to be of the order of $\sigma \simeq 10^{-30}\text{cm}^2$, if the field intensity reaches the value $F_\perp = 10^{10}/A\lambda$, where $\lambda$ is the wavelength of the scattered photons.

This effect can also take place in the cosmic space, when the photons are passing through the magnetic fields of the galaxies. In order that the specified cross-section is reached, the intensity of the magnetic field has to be of about $10^{-5}$Oe, which is very possible. In this case – following Ref. [275] – the inverse phenomenon, *i.e.* the transformation of the gravitational waves into electromagnetic waves when passing through a region of magnetic field, is also possible. In this respect, one must observe an intensification of brightness of the galaxy arms within the long-wave domain, due to their irradiation with gravitational waves.

### II.2.3. Conversion of photons into gravitons in a magnetic dipole field

The case of photon diffusion in a dipole magnetic field was approached by Poznanin [307], and Papini and Valluri [301]. If the photon moves along the direction of the dipole magnetic moment $\vec{\mathcal{M}}$, the total cross-section of the process writes [307],

$$\sigma = \frac{\pi}{5}\kappa^2 \vec{\mathcal{M}}^2 k_0^2, \tag{2.11}$$

or, if the photon velocity is perpendicular to the direction of $\vec{\mathcal{M}}$,

$$\sigma = \frac{29}{120}\kappa^2 \vec{\mathcal{M}}^2 k_0^2. \tag{2.12}$$

According to Poznanin [307] and Vladimirov [391], the total scattering cross-section corresponding to diffusion of the photons of energy $k_0 \simeq 10^9$ eV in the magnetic field of an electron has the approximate value $\sigma \simeq 10^{-60}\mathrm{cm}^2$. Supposing that in laboratory conditions one can create a magnetic dipole that has the moment $\mathcal{M} = 10^4$ (in CGS e.m. unit system), then for the visible light ($k_0 \simeq 1$eV) the total cross-section is about $\sigma \simeq 10^{-32}\mathrm{cm}^2$.

As was shown by De Sabbata and collab., [37, 39, 100, 101], in astrophysics is also important the inverse phenomenon, namely emission of the electromagnetic waves as a result of interaction between the gravitational waves and the dipole magnetic field produced by some astrophysical objects, such as black holes and/or neutron stars [198, 269, 401, 413].

Thus, if the center of our galaxy is modeled by a neutron star cluster, by applying the above considerations and supposing that during the neutron star "collisions" arise gravitational waves, then their interaction with the magnetic field of the cluster gives rise to an emission of electromagnetic waves, the radiation power being of about $P \simeq 10^{31} - 10^{32}$erg/s, corresponding to a frequency of $1.6 \times 10^3$ Hz.

Analogously, supposing that the gravitational waves emitted from the center of our galaxy – due less to the gravitational collapse of some massive objects (the gravitational waves having the same frequency, as Weber's estimations showed [398, 399]), than to the phenomenon of the neutron star collisions – interact with the magnetic field of the Earth, for the power of the resulting electromagnetic radiation one obtains a value of about $P \simeq 10^{-4} - 10^{-3}\,\mathrm{erg/s}$.

The above mentioned effects, together with some other new phenomena, were studied by the authors of this book (see, *e.g.*, [361, 362]) and shall be discussed in detail in chapter VII.

## II.3. Electromagnetic/gravitational bremsstrahlung of particles in external electromagnetic/gravitational field

Among the second-order gravitational effects, this type of processes is the most important. Due to the presence of the external fields, this type of processes take place with a greater probability, somewhat compensating the smallness of the gravitational coupling constant, $\kappa$.

### II.3.1. Gravitational bremsstrahlung of photons in a static gravitational field

The gravitational bremsstrahlung of photons in a static gravitational field was investigated in the research papers written by Piir [306], Mitskevich [272], Korkina [236], and De Sabbata and collab. [38(1967, 1968)]. The corresponding Feynman diagrams are shown in Fig. 2.4.

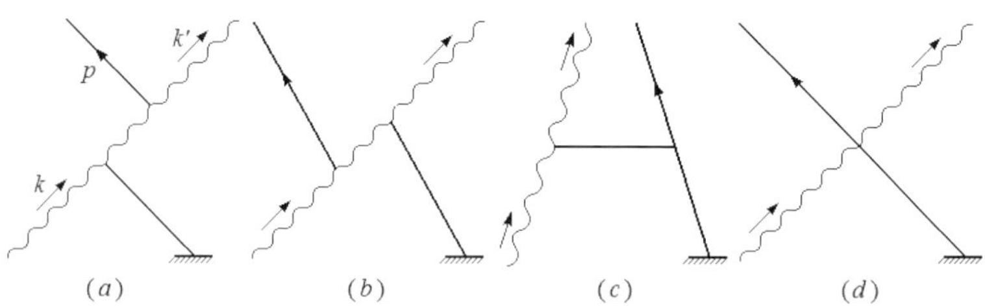


Fig. 2.4

In this case, the differential cross-section is proportional to $\kappa^6 M_\odot^2$, where $M_\odot$ is the mass of the body that creates the static gravitational field.

According to Piir's [306] evaluation, the total cross-section of the process is

$$\sigma \simeq \frac{\kappa^6 M_\odot^2}{(2\pi)^6} k_0^2, \tag{2.13}$$

where $k_0$ is the energy of the incident photons.

Under real conditions the process probability is small, but if supermassive bodies of small dimensions are present, the catalysing role of mass could be considered. Thus, this process was analyzed by De Sabbata and collab. [38(1967, 1968)] in the case of a hypothetical quasar with the following parameters: $M_\odot \sim 10^{41}$g, the internal temperature $T \sim 10^{70}$K, and the radius $R \sim 10^{16}$cm. The calculations show that the gravitational radiation power of the quasar has to be of the order of $P \sim 10^{19} - 10^{26}$erg/s in order to produce a photon radiation power of about $P \sim 10^{46}$erg/s (see also [391]). We mention that the total number of quasars existent in the visible part of the Universe, evaluated by the above mentioned authors, is of about $n \sim 10^7$.

The same authors performed an estimate calculation for the total scattering cross-section of photons of visible light, supposing a galaxy as the source of the static gravitational field ($M_\odot$ being in this case its average mass) and obtained the value $\sigma \sim 10^{-25} - 10^{-24}$cm$^2$.

The photons coming from the distant cosmic objects must obey the relativistic redshift effect, due to the gravitational bremsstrahlung in the gravitational field of the galaxies. The average photon energy lost by radiation due to the emission of a graviton, as shown in the papers [38(1968), 236], is given by

$$\Delta k_0 \simeq 10^4 \kappa^6 M_\odot^2 k_0^3 n L, \tag{2.14}$$

where $M_\odot \simeq 10^{44} g$ is the average mass of one galaxy, $n \simeq 10^{-69}$cm$^{-3}$ is the spatial density of galaxies in the observable part of the Universe, and $L$ is the distance traveled by the photon. For the visible light ($k_0 \simeq 10^{-12}$erg), one obtains [391],

$$\frac{\Delta k_0}{k_0} \sim 10^{-93} L, \tag{2.15}$$

and, taking a path of the order of the (observable) Universe radius $L \simeq 10^{28} - 10^{29}$cm, one finds $\Delta k_0 \sim 10^{-50}$eV [38(1968)], which is completely negligible.

On the other hand, one knows that the redshift determined by the galactic Friedmann expansion is obtained by means of

$$\frac{\Delta k_0}{k_0} \sim 10^{-28} L, \tag{2.16}$$

and does not depend on the photon initial energy [391].

To conclude, according to the above mentioned authors, the value of the scattering cross-section corresponding to the photon gravitational bremsstrahlung is too small to fully explain the redshift mechanism of the distant nebulae, this shift not being allowed to be considered as a result of "tired light" phenomenon. Similar results were found by the authors of this volume [358 (p.71)]. Based on semi-quantum reasons about photon energy loss in gravitational field, I. Gottlieb [168, 169] obtained the Hubble law in an original way. According to his ideas, both the Doppler–Fizeau effect – due to the relative movement of galaxies – and the "tired light" phenomenon – due to the enormous distances between galaxies – contribute to the shift of the spectral lines. Therefore, there would be no need of the Universe expansion hypothesis to explain the above discussed mechanism, but of the supposition that the Metagalaxy would be in a state of statistical equilibrium.

### II.3.2. Gravitational bremsstrahlung of massive scalar particles in a static gravitational field

The gravitational bremsstrahlung of massive scalar particles in a static gravitational field was studied by De Witt [105], Gupta and collab. [26(1969)], Weinberg [400], and Tourrenc [353(B69)]. The corresponding diagrams of the process are shown in Fig. 2.5.

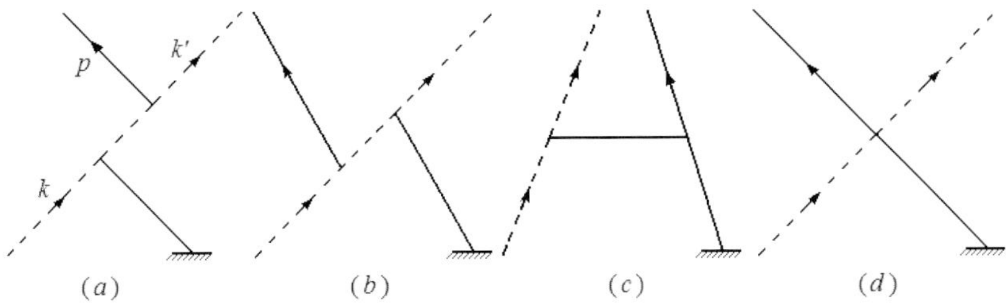


Fig. 2.5

In the non-relativistic approximation, the differential scattering cross-section is given by

$$d\sigma = \frac{\kappa^6 M_{\odot}^2 m^2}{15(4\pi)^3} \frac{dp_0}{p_0} \left[ \frac{|\vec{k}'|}{|\vec{k}|} + \frac{3}{2\vec{k}^2} \left( \vec{k}^2 + \vec{k}'^2 \right) \ln \frac{|\vec{k}| + |\vec{k}'|}{|\vec{k}| - |\vec{k}'|} \right], \tag{2.17}$$

where $p_0$ is the energy of the emitted gravitons, $\vec{k}$ and $\vec{k}'$ are the momenta of the scalar particles before and after the scattering process, respectively, while $m$ is the particles' mass. Due to the factor $\kappa^6$, the effect is extremely weak but, as in the previous case, one can rely on the catalysing role of the mass that creates the static gravitational field.

A similar process is the gravitational bremsstrahlung of electrons in an external gravitational field, studied by Aslanian [11]. The corresponding Feynman diagrams are analogous to those shown in Fig. 2.5, while the differential scattering cross-section in the non-relativistic approximation coincides (up to a factor) with the expression given by (2.17).

## II.3.3. Gravitational bremsstrahlung of electrons in an external electromagnetic field

The gravitational bremsstrahlung of electrons in an external electromagnetic field is also an important process that belongs to the category of gravitational transmutations. The effect – described by the Feynman diagrams shown in Fig. 2.6 – was studied by many researchers.

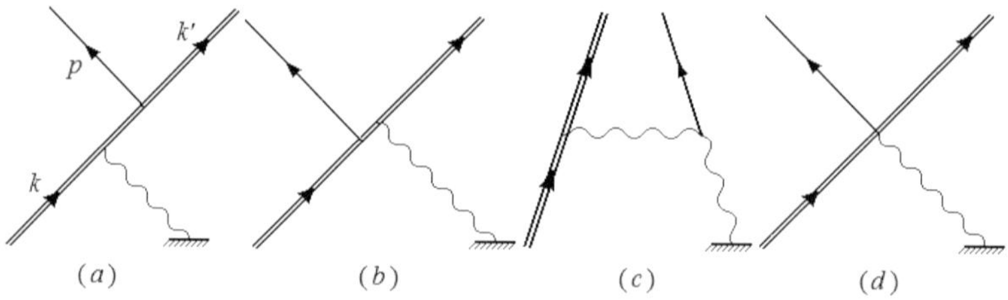


Fig. 2.6

So, according to Aslanian [11] and Vladimirov [391], the differential cross-section for the non-relativistic electrons diffusing in the Coulombian field of a nucleus with the charge $Ze$ is given by

$$d\sigma = \frac{\kappa^2 Z^2 e^4}{15\pi^3}\frac{dp_0}{p_0}\left[\frac{|\vec{k}'|}{|\vec{k}|} + \frac{3}{2\vec{k}^2}\left(\vec{k}^2 + \vec{k}'^2\right)\ln\frac{|\vec{k}| + |\vec{k}'|}{|\vec{k}| - |\vec{k}'|}\right], \tag{2.18}$$

which leads to a total cross-section of the order of $\sigma \simeq 10^{-68}\mathrm{cm}^2$. A detailed analysis of this effect can also be found in the papers written by Weinberg [400] and Carmeli [60].

Following Gandel'man's and Pinaev's [160] estimations, the power of the gravitational radiation of the Sun, determined by the electrons' bremsstrahlung on fully ionized atomic nuclei has the value $P \simeq 10^{15}\mathrm{erg/s}$, corresponding to a flux arrival on the Earth of the order of a few gravitons per $1\,\mathrm{cm}^2$, in one second. The main part of the gravitational radiation is situated in the interval $0 - 1\,\mathrm{KeV}$.

An effect similar to the gravitational bremsstrahlung must also appear in the case of motion of electrons in magnetic fields; this is the so-called *synchrotron gravitational radiation* [267, 268]. According to evaluations performed by Pustovoit and Gertzenstein [309], the power of this kind of radiation coming from the Sun is of the order of $P \simeq 10^9 - 10^{10}\mathrm{erg/s}$. In the ultrarelativistic limit, the same authors showed that the ratio between the electromagnetic power and that corresponding to gravitational bremsstrahlung is

$$\frac{P_{em}}{P_g} \simeq 10^{42}.$$

Another similar process is the gravitational bremsstrahlung of the charged scalar particles (of charge $e$) in a Coulombian field, analyzed by Gupta and collab. [27]. The Feynman diagrams are analogous to those shown in Fig. 2.6, while the differential cross-section of the process (in the non-relativistic approximation) is obtained by means of (2.17) – corresponding to the scattering in a static gravitational field – by replacing $GM_\odot m \to \frac{Ze^2}{4\pi}$. But, in the authors' opinion, this process has a very small probability due to the spin non-conservation (more precisely, due to the non-conservation of the spin projection along the direction of motion).

### II.3.4. Electromagnetic bremsstrahlung of charged scalar particles in an external gravitational field

The electromagnetic bremsstrahlung of charged scalar particles (of charge $e$) in an external gravitational field was studied by Halpern [192], and C. De Witt and Wesley [107]. The corresponding Feynman diagrams are shown in Fig. 2.7.

The differential scattering cross-section corresponding to this process has the following form,

$$d\sigma = \frac{\kappa^4 M_\odot^2 e^2}{3(4\pi)^3 v^2} \frac{dp_0}{p_0} \ln \frac{|\vec{k}| + |\vec{k}'|}{|\vec{k}| - |\vec{k}'|}, \tag{2.19}$$

where $p_0$ is the energy of the emitted photons, $v = |\vec{k}|/k_0$ is the velocity of the incident scalar particles, while $\vec{k}$ and $\vec{k}'$ are the momenta of the incident and emergent scalar particles, respectively.

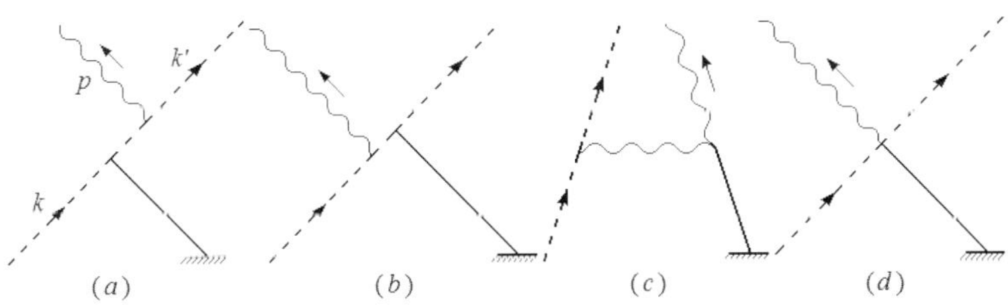


Fig. 2.7

A similar process is the electromagnetic electrons' bremsstrahlung in a static gravitational field, analyzed by Aslanian [11]. The Feynman diagrams describing the process are similar to those shown in

Fig. 2.7, and the differential cross-section of the process, corresponding to the non-relativistic electrons, has the following form,

$$d\sigma = \frac{\kappa^2 M_\odot^2 e^2 M^2}{3(4\pi)^3|\vec{k}|}\frac{dp_0}{p_0}\left(\frac{2|\vec{k}'|}{|\vec{k}|+|\vec{k}'|} - \frac{1}{|\vec{k}|}\ln\frac{|\vec{k}|+|\vec{k}'|}{|\vec{k}|-|\vec{k}'|}\right), \tag{2.20}$$

where $M$ is the mass of the non-relativistic electrons.

## II.4. Gravitational and/or electromagnetic annihilation of elementary particles

The gravitational and/or electromagnetic annihilation of the particles (and vice versa) represents another important process – with straight applications to astrophysics – being the first effect analyzed in the framework of hypothesis regarding the possibility of reciprocal transformation of the gravitons and material quanta.

### II.4.1. Annihilation of an electron-positron pair into two gravitons

Annihilation of an electron-positron pair into two gravitons was studied by Vladimirov [388], Mitskevich [275], Voronov [393] and Grossiord [180]. The corresponding Feynman diagrams are shown in Fig. 2.8.

The total cross-section of the process is very small. For example, in the non-relativistic approximation, one obtains the value $\sigma \simeq 10^{-110}\text{cm}^2$ for the electrons with velocity $v = c/100$. In the centre-of-mass reference frame, for the ultrarelativistic limit case ($E_k \sim p \gg M$), the differential and total scattering cross-sections are given by [388(1963), 391],

$$d\sigma = \frac{\kappa^4 E_k^2}{2(64\pi)^2}\left(3\sin^2 2\theta + 2\sin^4\theta\right) d\Omega, \tag{2.21}$$

and

$$\sigma \simeq r_g^2\left(\frac{E_k}{M}\right)^2, \tag{2.22}$$

respectively. In the above two relations $\theta$ is the angle between the direction of motion of the fermions and direction of graviton's momentum, $E_k$ is the energy of the spinorial particles, $M$ is their mass, and

$r_g = 2GM$ is the gravitational radius corresponding to the spinorial particles. One observes that the total cross-section increases proportional to the squared energy of the scattered particles, in agreement with the first results concerning the gravitational transmutations obtained by Ivanenko and Sokolov in the case of the scalar particles' annihilation [212].

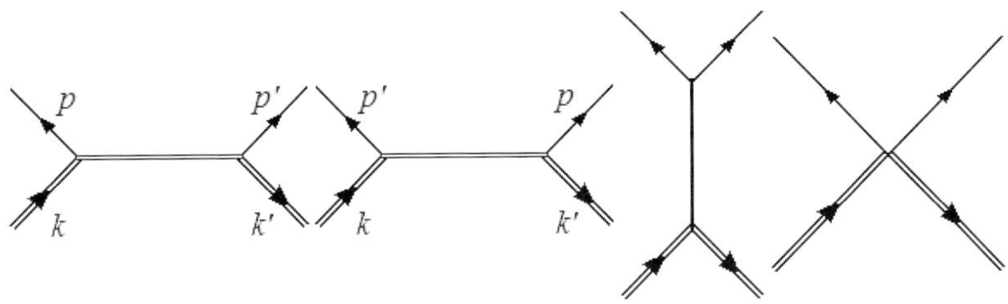


Fig. 2.8

We mention that the total scattering cross-section corresponding to the two-photon annihilation of the fermion pairs in the ultra-relativistic limit case diminishes when the energy of the spinorial particles increase, according to the law $\sigma_f \sim 1/E_k^2$ [2, 217, 240]. The Vladimirov's evaluations [388(1963), 391] show that – for energies of electrons and positrons of about $E_k \simeq 10^{21} M$ – the total cross-sections of the two-graviton and two-photon annihilation processes are of the same order of magnitude.

Even if in real physical conditions the analyzed effect is insignificant, it has to be considered when the "initial state" of the Universe (in the frame of the model in which the Universe is considered as being a sui generis "reactor" in equilibrium for the elementary particles) is constructed, or when the hypothesis of existence of collapsed objects is considered [391].

In a similar way takes place the two-graviton annihilation of the neutrinic pairs [391] (the neutrinos being described by a four-component wave function and considered as being massless particles); this process was first analyzed by Brill and Wheeler [49].

A suchlike process, *i.e.* the annihilation of a photonic pair into two gravitons, was studied by Piir [306], Korkina [236], Vladimirova

[392], and Tourrenc [353(B69)]. The Feynman diagrams describing this effect are shown in Fig. 2.9.

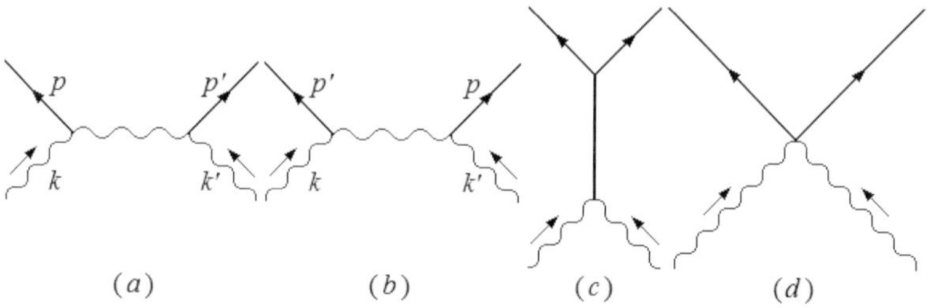


Fig. 2.9

The differential scattering cross-section of this process, in the centre-of-mass reference frame, is given by [353(B69)],

$$d\sigma = \frac{\kappa^4 k_0^2}{4(4\pi)^2}\left(\cos^8\frac{\theta}{2} + \sin^8\frac{\theta}{2}\right) d\Omega, \tag{2.23}$$

where $k_0$ is the photon energy, and $\theta$ stands for the angle between directions of motion of the photons and gravitons.

As in the previous case, the total cross-section varies proportional to the squared energy of the incident particles [236, 392]:

$$\sigma \simeq \frac{3\times 10^{-43}}{256}\kappa^4 k_0^2. \tag{2.24}$$

According to Vladimirova's estimations [392], for the visible spectrum ($\lambda \sim 500$ nm) one obtains $\sigma \simeq 10^{-124}\text{cm}^2$, but this value can be several order of magnitude higher in the case of laser fascicles [96]. Since the scattering cross-section is very small, this effect does not play an essential role in gravitational transmutations and presents a principial interest only.

Another similar process is the two-graviton annihilation of the charged or neutral particles, analyzed by Ivanenko and Sokolov [212], De Witt [105], and Gupta and collab. [26(1969)]. The corresponding Feynman diagrams are analogous to those shown in Figs. 2.8 and 2.9, while the total scattering cross-section corresponding to ultra-relativistic approximation is very small [26(1969)],

$$\sigma \simeq \frac{1}{480\pi}\kappa^4 k_0^2, \tag{2.25}$$

being proportional – as in the previous cases – to the squared energy of the colliding particles.

### II.4.2. Photon-graviton annihilation of the fermionic pairs

An important process is the photon-graviton annihilation of the fermionic pairs, investigated by Vladimirov [388(1963), 391] and Mitskevich [275]. The effect is described by the graphs shown in Fig. 2.10.

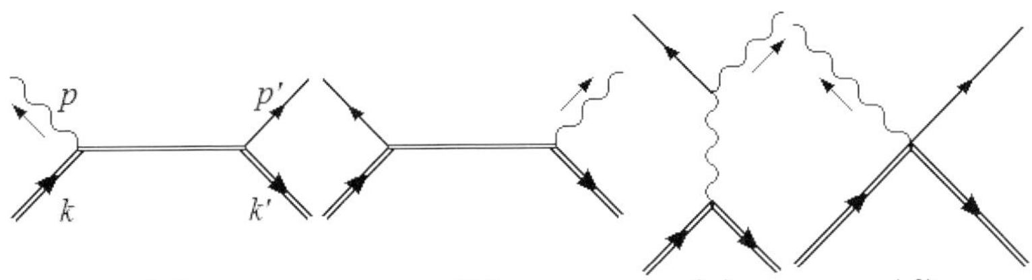


Fig. 2.10

The total scattering cross-sections in the non-relativistic and ultra-relativistic approximations are given, respectively by

$$\sigma_{N.R.} = \frac{13}{480\pi} r_e r_g v^3, \tag{2.26}$$

and

$$\sigma_{U.R.} = \frac{1}{48\pi} r_e r_g, \tag{2.26$'$}$$

where $v = |\vec{k}|/E_k$ is the velocity of the annihilating particles in the centre-of-mass reference frame, while $r_e = e^2/M$ and $r_g$ are the electromagnetic radius and the gravitational radius corresponding to the annihilating particles. Thus, for the fast particles colliding with velocity $v = c/100$ the calculations give [391],

$$\sigma_{N.R.} \simeq 10^{-75}\text{cm}^2,$$

in the non-relativistic approximation and

$$\sigma_{U.R.} \simeq 0.3 \cdot 10^{-68}\text{cm}^2.$$

in the ultrarelativistic limit [391].

Inside a supernova, where according to the model given by Chiu [74] the temperature exceeds the value $T \simeq 10^9$K, this effect can lead to generation of a remarkable number of hard gravitons. Following the research performed by De Sabbata and collab. [38(1968)], in such a superstar [334, 335, 413], in one second is formed a number of about $n \simeq 10^{22}$ gravitons whose average energy is 1 MeV ($\lambda \simeq 10^{-11}$cm), corresponding to a gravitational radiation of the power $P \simeq 10^{16}$erg/s. Since the estimated number of supernovae in the Universe is $N \simeq 10^9$, one obtains a value of the order of $10^{25}$erg/s for the total radiation power of hard gravitons, determined by the above mentioned process.

### II.4.3. Annihilation of elementary particles in the presence of external gravitational and/or electromagnetic fields

The probability of processes of elementary particles annihilation increases in the presence of gravitational/electromagnetic fields, that play a catalytic role. Such an effect is the transformation of two photons in one graviton in a static gravitational field. This process, analyzed by De Sabbata and collab. [38(1968)], is described by the Feynman diagrams shown in Fig. 2.11.

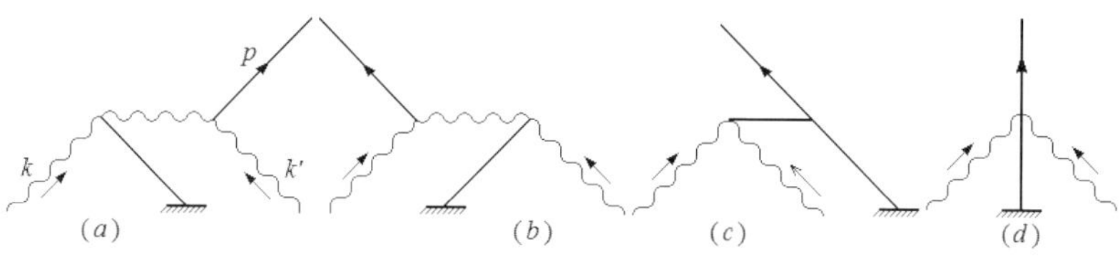


Fig. 2.11

The total scattering cross-section of the process – as in the case of the gravitational bremsstrahlung of the photons in external gravitational field – is proportional to $\kappa^6 M_\odot^2$, the mass that creates the static gravitational field playing again a catalytic role. The same authors evaluated the gravitational radiation of a quasar, determined by this effect. Thus, was estimated that in a quasar can take place $10^{34} - 10^{40}$ such processes in one second, corresponding to a power of the gravitational radiation of about $P \simeq 10^{28}$erg/s, in the energy interval of 1 eV – 1 KeV. Compared with this very strong emission, the number of gravitons produced in a "standard" star is negligible.

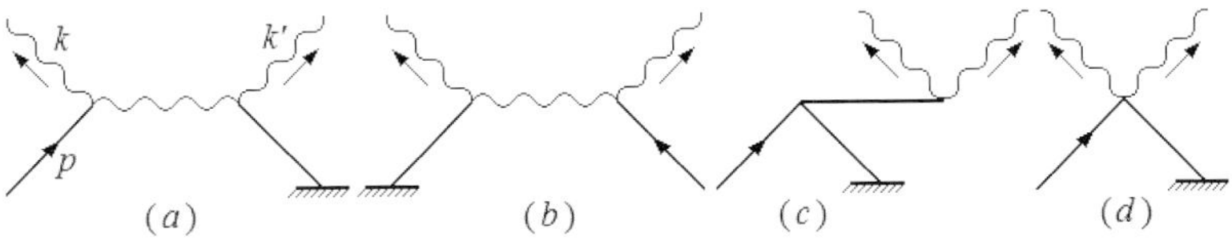


Fig. 2.12

The inverse process, namely the decay of a graviton into a photon pair in an external gravitational field, described by the Feynman diagrams shown in Fig. 2.12, is also interesting. It was studied by De Sabbata and collab. [38(1968)] and Poznanin [307], that gave the following expression for the total cross-section of the process,

$$\sigma \simeq 2\kappa^2 p_0^2 r_0^2, \tag{2.27}$$

where $p_0$ is the energy of the incident graviton, and $r_0$ is the gravitational radius of the external field source. Considering as the gravitons' source a rotating binary star system, and the galaxy as the catalytic external mass, De Sabbata and collab. obtained the value $\sigma \simeq 0.5 \times 10^{-47}\text{cm}^2$.

A similar process consists in the annihilation of an electron-positron pair into a static graviton, the corresponding Feynman diagrams being analogous to those given in Fig. 2.11. As shown by Aslanian [11], the total scattering cross-section in the ultrarelativistic approximation is proportional to $\kappa^6 M_\odot^2/E_k$, where $E_k$ is the fermion energy in the centre-of-mass reference system. The same author also studied the inverse effect: the graviton decomposition into a pair of spinorial particles in an external gravitational field, showing that, in the classical limit, the ratio between the differential cross-sections of the above mentioned process and that corresponding to the creation of an electron-positron pair out of a photon, in the Coulombian field generated by a nucleus of charge $Ze$, is of the order of magnitude

$$\frac{d\sigma(M_\odot, g)}{d\sigma(Ze, f)} \simeq 10^{-68}\frac{M_\odot^2}{Z^2}, \tag{2.28}$$

where the first letter between parentheses indicates the field source, and the second the respective boson.

Another important process is the one-graviton annihilation of the electron-positron pairs in external electromagnetic field. This effect was studied by Dobrev [118] and Aslanian [11], and the corresponding Feynman diagrams are shown in Fig. 2.13. As external electromagnetic field were considered the following concrete cases: the homogeneous electrostatic and magnetostatic fields [118], the magnetic dipole field [118], and the Coulombian field of a nucleus [11]. In the last case, the total scattering cross-section of the process in the ultrarelativistic approximation is given by

$$\sigma = \frac{\kappa^2 Z^2 e^4}{48 E_k^3}, \tag{2.29}$$

where $E_k$ is the fermions' energy in the centre-of-mass reference frame.

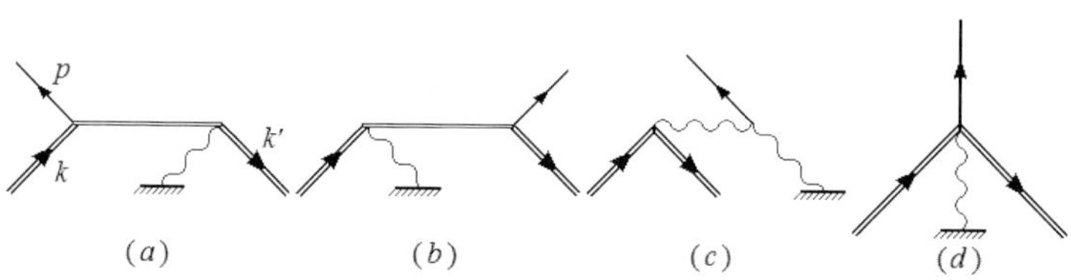


Fig. 2.13

Similarly, but with a smaller probability, one also takes place the annihilation of the electron-positron pair in a static gravitational field. This process was analyzed by Aslanian [11], being described by the Feynman diagrams shown in Fig. 2.14. The total scattering cross-section in the ultrarelativistic limit is obtained from (2.29) through the substitution $Z^2e^2 \to M_{\odot}^2\kappa/8$.

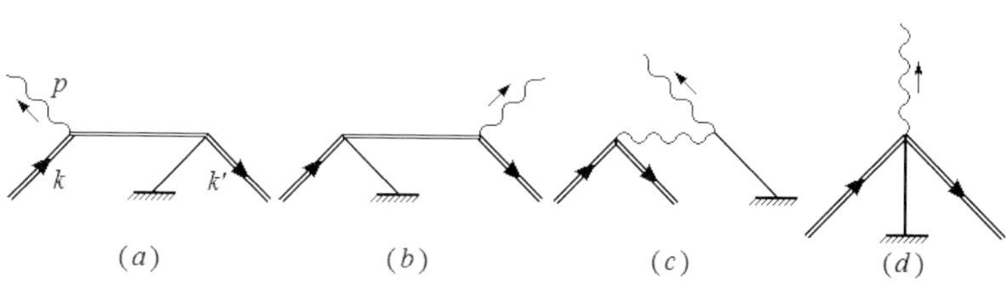


Fig. 2.14

In the paper [11] of Aslanian, the inverse processes – the graviton or photon decay into pairs in the external electromagnetic or gravitational field, respectively – are studied, and a comparison is made with the corresponding process of quantum electrodynamics, namely the creation of an electron-positron pair out of a photon in the Coulombian field of a nucleus. Thus, in the classical limit, the ratio between the differential scattering cross-sections corresponding to the pairs creation out of a single photon in the external gravitational field, and an external electromagnetic (Coulombian) field has the value

$$\frac{d\sigma(M_{\odot}, g)}{d\sigma(Ze, f)} = \frac{3}{4} \frac{M_{\odot}^2 \kappa^4 M^4}{Z^2 e^4 \left(\vec{k}^2 + \vec{k}'^2\right)}, \tag{2.30}$$

where $M$ is the mass of an electron/positron, $\vec{k}$ is the momentum of the electron, while $\vec{k}'$ is the momentum of the positron. If one chooses $M_{\odot} \simeq 10^{41}$g, and considering $|\vec{k}| = |\vec{k}'| \simeq 0.01\, M$, then the order of magnitude of this ratio equals one.

## II.5. Gravitational Compton effect

This process is similar to that studied by quantum electrodynamics and has a considerable importance in astrophysics.

### II.5.1. Scattering of gravitons on massive scalar particles

This case was investigated by Chester [73], Jackiw [213], Gross and Jackiw [179], Gupta and collab. [26(1969)], and De Witt [105] The corresponding Feynman diagrams are given in Fig. 2.15.

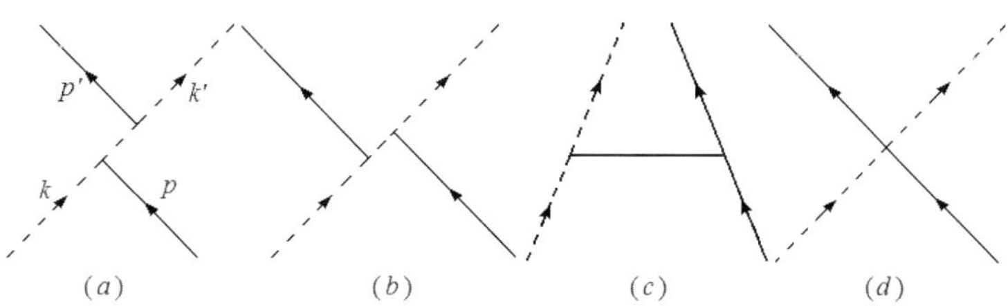


Fig. 2.15

The corresponding differential cross-section, in the non-relativistic approximation and in the laboratory system is given by [26(1969)],

$$d\sigma = \frac{\kappa^4 m^2}{(16\pi)^2}\left(\cot^4\frac{\theta}{2}\cos^4\frac{\theta}{2} + \sin^4\frac{\theta}{2}\right)d\Omega, \tag{2.31}$$

where $m$ is the mass of the scalar particles.

The Compton effect corresponding to the scattering of a graviton on an electron was studied by Voronov [393]. The Feynman diagrams of the process are analogous to those shown in Fig. 2.15, and the differential cross-section coincides with that given by relation (2.31).

The scattering of gravitons on photons was analyzed by Korkina [236], in the framework of Birkhoff's theory of gravitation. The Feynman diagrams are similar to the preceding ones, and the total scattering cross-section is proportional to $\kappa^4 k_0^2$, where $k_0$ is the photon energy in the centre-of-mass reference system.

## II.5.2. Photogeneration of gravitons with participation of spinorial particles

The process of the photoproduction of gravitons on spinorial particles proves to be somewhat more probable. It was studied by Vladimirov [388(1963)], Mironovski [262] and Voronov [393]. The corresponding Feynman diagrams are shown in Fig. 2.16.

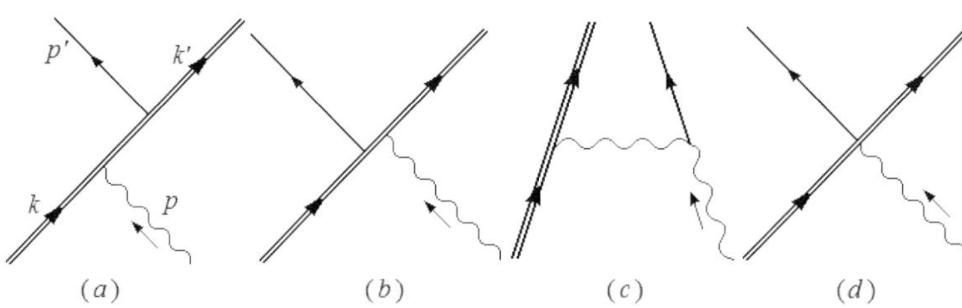


Fig. 2.16

The total scattering cross-section in the ultrarelativistic approximation is given by

$$\sigma \sim e^2\kappa^2\left(\frac{E_k}{p_0}\right)^2, \tag{2.32}$$

where $p_0$ and $E_k$ are the energies of the photon and incident fermion, respectively.

According to Mironovski's evaluations [263], the power of the Sun's gravitational radiation caused by this effect is about $P \simeq 10^8$erg/s, the frequency of the emitted gravitons being situated in the interval of the soft Roentgen radiation ($p_0' \sim 1\,\text{KeV}$).

But, as it was shown by Vladimirov [388(1963)], the inverse process, namely the Compton effect of the graviton on a fermion, with its transformation into a photon is much more important in astrophysics. Admitting that in the Universe exists a great number of gravitons, then by their interaction with the matter, the above mentioned process has to take place. It has to produce a supplementary photon radiation in cosmos, coming from the objects whose velocity is perpendicular to the direction connecting the Earth observer and the considered object (in which case the effect attains a maximum[3]). If this effect could be experimentally evidentiated, then – together with the Doppler (radial) redshift – it could allow a more precise evaluation of the velocity and direction of motion of the cosmic objects.

## II.6. Scattering of elementary particles via virtual gravitons

These processes take place without the effective participation of the gravitons, and they were studied by many authors. The first who investigated this type of effect were Gupta and collab. [25, 26], who showed that the gravitational interactions of the elementary particles through the exchange of a virtual graviton are independent on the spin of the respective particles.

### II.6.1. Photons-by-photons scattering

The photons-by-photons scattering was analyzed by Piir [306], Gupta and collab. [26], De Sabbata and collab. [38(1969)] and Tourrenc [353(B1)]. The Feynman diagrams describing the process are presented in Fig. 2.17.

---

[3]Due to the factor $\sin^4\theta$ in the differential scattering cross-section – see [388(1963)].

The differential scattering cross-section in the centre-of-mass reference frame is given by [38(1969)],

$$d\sigma = \frac{\kappa^4 k_0^2}{2(2\pi)^2}\left(1+\cos^{16}\frac{\theta}{2}+\sin^{16}\frac{\theta}{2}\right)\frac{d\Omega}{\sin^4\theta}, \tag{2.33}$$

where $k_0$ is the energy of the photon, and $\theta$ is the angle between momenta of the incident and emergent particles.

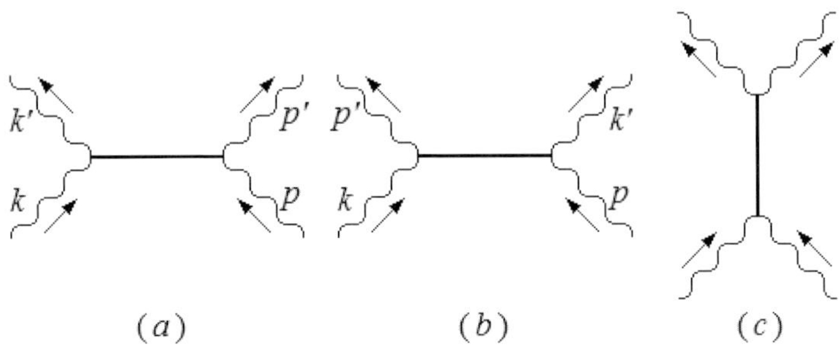


Fig. 2.17

The scalar particles' diffusion through the exchange of a virtual graviton was studied by Gupta and collab. [25], De Witt [105], Mitskevich [275], and Tourrenc [353(B69)], the corresponding Feynman diagrams being similar to those shown in Fig. 2.17[4]. The differential scattering cross-section for the neutral particles, in the ultrarelativistic approximation, is given by [105],

$$d\sigma = \frac{4\kappa^4 k_0^2}{(16\pi)^2}\left(\cot^2\frac{\theta}{2}+\tan^2\frac{\theta}{2}+\frac{1}{2}+\frac{1}{4}\sin^2\frac{\theta}{2}\right)^2 d\Omega, \tag{2.34}$$

where $k_0$ is the energy of these particles in the centre-of-mass reference frame.

The massless spinor particles by massless spinor particles diffusion was studied by De Sabbata and collab. [38(1972)]. The process is

[4]In the case of diffusion of neutral, scalar particles, the process is described by all the three Feynman diagrams. For the charged particles the contribution of the first two diagrams (Möller-type diffusion) is considered, while in the case of the diffusion of particles on antiparticles (of the Bhabha-type) the contributions are brought by the first and the last diagrams – the antiparticles being oriented, obviously, opposite to the time axis. The same observation is also valid for spinorial and vectorial particles; in these cases, the initial state is compounded by particles of the same type, and the corresponding Feynman diagrams describe the creation-annihilation processes.

described by the Feynman diagrams which are similar to the first two of Fig. 2.17, while the differential scattering cross-sections, in the limit of small angles, depending on the description of the massless spinor particles by a two- or four-component wave function, are given by

$$d\sigma = \frac{5\kappa^4}{4(2\pi)^2}\frac{k_0^2}{\theta^4}\,d\Omega, \tag{2.35}$$

and

$$d\sigma = \frac{\kappa^4}{(2\pi)^2}\frac{k_0^2}{\theta^4}\,d\Omega, \tag{2.35$'$}$$

respectively. Here $k_0$ is the particles' energy in the centre-of-mass reference frame, and $\theta$ is the scattering angle.

The massive spinorial and vectorial particles scattering takes place in an analogous way (see for example [25, 105]).

### II.6.2. Gravitational scattering of particles with different spins

The processes of gravitational diffusion of particles having different spins[5] are also important. Thus, the photons diffusion on scalar particles via virtual gravitons is studied by Gupta and collab. [25], Mitskevich [275], De Sabbata and collab. [38(1969)], and Tourrenc [353(B1)]. The corresponding Feynman diagram is given in Fig. 2.18.

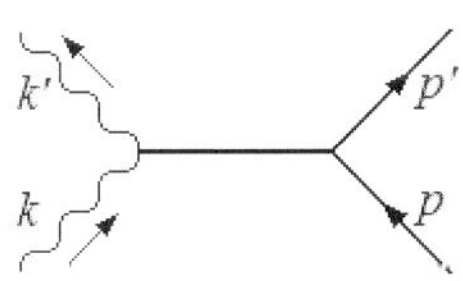


Fig. 2.18

The associated differential scattering cross-section is [38(1969), 353(B1)],

$$d\sigma = \frac{\kappa^4}{(16\pi)^2}\frac{E^4}{(E-k_0)^2}\cot^4\frac{\theta}{2}\,d\Omega, \tag{2.36}$$

[5] In this case, the initial state being composed by particles of different types, the corresonding Feynman diagram describes a process of elastic scattering type.

where $E$ is the total energy of the scalar particle and $k_0$ is the energy of the photon, both being considered in the centre-of-mass reference frame.

The photon-massless spinor particle scattering is of potential interest in astrophysical processes. This effect was studied by De Sabbata and his research group [38(1972)]. The Feynman diagram describing the process is similar to the previous one, while the differential scattering cross-section – in the limit case of small scattering angles and in the centre-of-mass reference system – has the same form as that of the previous process.

In the same manner takes place the photons scattering on massive spinorial particles [25].

The process of interaction between scalar particles and massless spinor particles through the exchange of virtual gravitons was analyzed by the same research group (led by De Sabbata) [38(1972)]. The corresponding Feynman diagram is shown in Fig. 2.19.

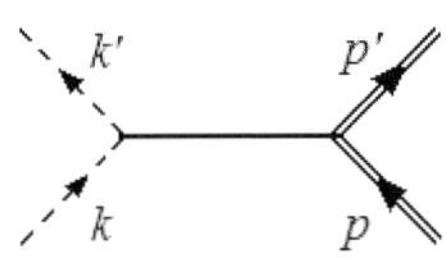


Fig. 2.19

The differential scattering cross-section, in the limit of small scattering angles, is

$$d\sigma = \frac{\kappa^4}{(4\pi)^2} \frac{E^2}{\theta^4} \, d\Omega, \tag{2.37}$$

where $E$ is the total energy of the scalar particle in the centre-of-mass system.

The scattering of scalar particles on massive spinorial particles follows a similar way of investigation.

## II.7. Gravitational radiation of nuclei

An important process is the generation of $\gamma$ (electromagnetic) radiation of nuclei, by their transition from a certain excited state in the fundamental state, or in an appropriate less excited state. One usually considers that the only way of dis-excitation is that fulfilled by emission of quanta of the electromagnetic field, with high and very high energy. However, one can also suppose that the excitation can be eliminated by the emission of hard gravitons.

Thus, in the paper written by Halpern and Laurent [193], the Sun's gravitational radiation determined by such processes is examined. As one knows, the internal temperature of the Sun is almost $T \simeq 10^7$K, in which case a part of the nuclei being capable to jump onto an excited state. For example, one can appear the excited state of the $^{57}$Fe nucleus, the excitation energy being of 14.4 keV. According to the estimates of the above mentioned authors, all the nuclei inside the Sun would emit about $10^{11}$ gravitons/s, corresponding to a gravitational radiation power of approximately $P \simeq 10^3$ erg/s, the flux created on the Earth being equal to 1 graviton/km$^2$ in 24 hours.

One supposes that in the centre of supernovas the temperature can reach the value $T \simeq 10^{10}$ K [334]. As it was shown by Vladimirov [391], at this temperature one excites the highest levels of the nuclei, such as the energy level of 16.1 MeV of the $^{12}$C. According to Vladimirov's estimates, the power of the gravitational radiation of the supernovas, due to the mentioned process, has the value $P \simeq 10^{25}$ erg/s. Taking an average distance from the Earth to the supernovae of about 300 parsecs, one can expect a flux of about 10 gravitons/km$^2$ in 24 hours.

The study of the above enumerated processes of graviton production and annihilation makes possible a rough evaluation of their density in the Universe. According to De Sabbata and collab. [38(1968)], due to the fact that the annihilation rate of gravitons is smaller than that of their production, in an evolutive model of the Universe (whose age is about $10^{18}$ s), the situation is far from equilibrium and the gravitons are continuously accumulated. The average energy density corresponding to all the gravitons in the actual Universe is estimated by the above mentioned authors at about $\overline{\rho}_g \simeq 10^{-19}$ erg/cm$^3$.

# Chapter III

# Tomonaga–Schwinger Representation of Dynamics of Quantum Physical Systems. Matrix Elements of the Field Operators and Feynman–Dyson-Type Rules for High-Spin Particles

## III.1. Tomonaga–Schwinger representation of dynamics of a quantum physical system

Since in the study of concrete processes approached in chapter VI, chapter VII, and chapter VIII, the *Tomonaga–Schwinger representation/picture* (also known as *Dirac representation* or *interaction representation*) has been used, in this subchapter the theoretical basis of this application shall be briefly exposed. The three basic/fundamental representations/pictures of a quantum system dynamics – namely the Schrödinger, the Heisenberg, and Tomonaga–Schwinger representations – are exposed in many papers which have started to appear together with "maturation" of quantum mechanics and its imposition as a stand-alone discipline. In this volume, the authors have preferred to expose this subject by means of Ref. [384], where the problem is presented in a very clear, systematic and unitary manner.

### III.1.1. Time evolution of quantum systems. The evolution operator

The assembly of $n$ coordinates $q_1, q_2, \ldots, q_n$ together with spin variables (if necessary) shall be denoted by $q$, while $\psi(t, q)$ will stand for the state vector describing the behavior (*i.e.* the space-time evolution) of the quantum system. The function $\psi(t,q)$ satisfies *Schrödinger's time-dependent equation*

$$\frac{\hbar}{i}\frac{\partial\psi(t,q)}{\partial t} + \hat{H}\psi(t,q) = 0. \tag{3.1}$$

Separation of the temporal variable is possible only if $\hat{H}$ does not explicitly depend on time. In this case, setting

$$\psi(t,q) = \varphi(q)\exp(-iEt/\hbar),$$

one obtains the *time-independent Schrödinger equation* for $\varphi(q)$: $\hat{H}\varphi = E\varphi$. This is the case when the external forces acting on the quantum system is constant in time. If the forces acting on the system are variable over the time, then the time separation is not possible anymore.

Eq. (3.1) allows one to determine $\psi$ at the moment $t$, if one knows its value at the moment $t'$ (anterior or posterior with respect to $t$). This operation can be sequentially done, step by step, with infinitesimal steps of "length" $\delta t$, since

$$\begin{aligned}\psi(t+\delta t,q) &= \psi(t,q) + \frac{\partial\psi(t,q)}{\partial t}\delta t = \psi(t,q) - \frac{i}{\hbar}\hat{H}\psi(t,q)\delta t\\ &= \left(1 - \frac{i}{\hbar}\hat{H}\delta t\right)\psi(t,q).\end{aligned}$$

Here it has been supposed that $\mathcal{O}(\delta t)^2 = 0$ and Eq. (3.1) has been used. Denoting the above round brackets by $T(t+\delta t,t) = 1 - \frac{i}{\hbar}\hat{H}\delta t$, (where 1 stands for the unity operator) the previous relation can be written as

$$\psi(t+\delta t,q) = T(t+\delta t,t)\psi(t,q),$$

or, by denoting $t+\delta t = t'$,

$$\psi(t',q) = T(t',t)\psi(t,q).$$

The operator $T(t',t) = 1 - \frac{i}{\hbar}\hat{H}(t)\delta t$ is called the *evolution operator*. Since Eq. (3.1) is linear, the operator $T$ is also linear: if $\psi(t,q) = a_1\psi_1(t,q) + a_2\psi_2(t,q)$, then

$$\begin{aligned}\psi(t_2,q) &= T\psi(t_1,q) = T[a_1\psi_1(t_1,q) + a_2\psi_2(t_1,q)]\\ &= a_1\psi_1(t_2,q) + a_2\psi_2(t_2,q) = a_1T\psi_1(t_1,q) + a_2T\psi_2(t_1,q).\end{aligned}$$

Leaving aside the dependence on spatial and/or spin variables, then one can write

$$\psi(t_2) = T(t_2, t_1)\psi(t_1)$$

and, by taking $t_1 \leftrightarrow t_2$,

$$\psi(t_1) = T(t_1, t_2)\psi(t_2).$$

Therefore, the operator $T(t_2, t_1)$ admits an inverse, which means $[T(t_2, t_1)]^{-1} = T(t_1, t_2)$. It also leaves unchanged the scalar products. This property follows from the fact that $\hat{H}$ is a self-adjoint operator. Indeed, if $\psi_1$ and $\psi_2$ are two state functions, then

$$\frac{\hbar}{i}\frac{\partial\psi_1}{\partial t} + H\psi_1 = 0, \tag{3.2}$$

and

$$\frac{\hbar}{i}\frac{\partial\psi_2}{\partial t} + H\psi_2 = 0. \tag{3.3}$$

(From now on, for the sake of convenience, $\hat{H}$ shall be replaced by $H$). According to (3.2),

$$\frac{\hbar}{-i}\frac{\partial\psi_1^*}{\partial t} + (H\psi_1)^* = 0,$$

or

$$\frac{\hbar}{i}\frac{\partial\psi_1^*}{\partial t} - (H\psi_1)^* = 0. \tag{3.4}$$

Multiplying (3.4) on the right by $\psi_2$ and (3.3) on the left by $\psi_1^*$, then adding the results, one obtains

$$\frac{\hbar}{i}\frac{\partial\psi_1^*}{\partial t}\psi_2 - (H\psi_1^*)\psi_2 + \frac{\hbar}{i}\psi_1^*\frac{\partial\psi_2}{\partial t} + \psi_1^*(H\psi_2) = 0,$$

or

$$\frac{\hbar}{i}\frac{\partial}{\partial t}(\psi_1^*\psi_2) + \psi_1^*(H\psi_2) - (H\psi_1^*)\psi_2 = 0. \tag{3.5}$$

Integrating (3.5) over spatial variables and summing over spins (these operations shall symbolically be denoted by $\int dq$), one gets

$$\frac{\hbar}{i}\frac{\partial}{\partial t}\int(\psi_1^*\psi_2)dq + \int\psi_1^*(H\psi_2)dq - \int(H\psi_1)^*\psi_2 dq = 0,$$

or, if the Dirac notation is used,

$$\frac{\hbar}{i}\frac{\partial}{\partial t}\langle\psi_1|\psi_2\rangle + \langle\psi_1|H\psi_2\rangle - \langle H\psi_1|\psi_2\rangle = 0.$$

Since $H$ is self-adjoint, it follows that

$$\langle\psi_1|H\psi_2\rangle = \langle H\psi_1|\psi_2\rangle,$$

and therefore, from the above relation it results that

$$\frac{\partial}{\partial t}\langle\psi_1|\psi_2\rangle = 0,$$

whence

$$\langle\psi_1(t_1)|\psi_2(t_1)\rangle = \langle\psi_1(t_2)|\psi_2(t_2)\rangle,$$

meaning that the operator $T$ leaves unchanged the scalar products:

$$\langle\psi_1|\psi_2\rangle = \langle T\psi_1|T\psi_2\rangle.$$

Since $T$ is also invertible, it follows that it is a *unitary operator*. It is also called the *time evolution operator*. Obviously,

$$T(t_1, t_1) = 1 \quad \text{(as unit operator)},$$

and

$$T(t_3, t_2)T(t_2, t_1) = T(t_3, t_1).$$

The operator $T$ satisfies a differential equation which is perfectly analogous to the Schrödinger equation. Indeed, one can write

$$\psi(t) = T(t, t_0)\psi(t_0).$$

Since $t_0$ is fixed, $\psi(t_0)$ is also fixed, and this yields

$$\frac{\partial\psi(t_0)}{\partial t} = 0.$$

In view of Eq. (3.1),

$$\frac{\partial\psi(t)}{\partial t} = -\frac{i}{\hbar}H\psi(t),$$

so that

$$-\frac{i}{\hbar}H\psi(t) = \frac{\partial T}{\partial t}\psi(t_0),$$

or

$$-\frac{i}{\hbar}HT(t,t_0)\psi(t_0) = \frac{\partial T}{\partial t}(t,t_0)\psi(t_0),$$

whence

$$\left[\frac{\hbar}{i}\frac{\partial T}{\partial t}(t,t_0) + HT(t,t_0)\right]\psi(t_0) = 0.$$

Since $t_0$ is arbitrary, it then follows that

$$\frac{\hbar}{i}\frac{\partial T}{\partial t}(t,t_0) + HT(t,t_0) = 0, \tag{3.6}$$

with the initial condition

$$T(t_0,t_0) = 1. \tag{3.7}$$

It can easily be shown that Eqs. (3.6) and (3.7) are equivalent to the integral equation

$$T(t,t_0) = 1 - \int_{t_0}^{t} HT(t_1,t_0)dt_1. \tag{3.8}$$

### III.1.2. Schrödinger representation

There are three possibilities to describe the time evolution of quantum systems, depending on the particularities of the considered system, namely:

a) *Schrödinger representation/picture*;
b) *Heisenberg representation/picture*;
c) *Tomonaga–Schwinger*, or *Dirac*, or *interaction representation/picture*.

Let $\psi$ be the state function of a quantum system, and $F$ an operator associated with an arbitrary physical quantity. Let us find the law of transformation of this operator as a result of changing of the state function when a unitary operator $U$ acts on it. We have

$$\psi \xrightarrow{\text{U}} \psi' = U\psi \quad \Rightarrow \quad \psi = U^{-1}\psi',$$

$$\psi_1 \xrightarrow{\text{F}} \psi_2 = F\psi_1.$$

In this case,

$$\psi_2' = U\psi_2 = UF\psi_1 = UFU^{-1}\psi_1' \equiv F'\psi_1'.$$

Therefore

$$\begin{cases} \psi \to^{\mathrm{U}} \psi' = U\psi\,, \\ F \to^{\mathrm{U}} F' = UFU^{-1}. \end{cases} \tag{3.9}$$

A specific property of the Schrödinger representation is the fact that the *unitary operators* $U$ *do not vary over time.* These operators perform, in fact, the transition from one representation of the studied quantum system to another one (for example, the transition from the configuration representation to momentum representation or vice versa).

To conclude, within the Schrödinger representation the unitary operators $U$ do not depend on time; this condition is necessary to ensure the time-constancy of the operators associated with physical quantities, expressed in terms of the fundamental operators $\hat{x}_r$ and $\hat{p}_r$, which also are time-independent. Therefore, the time evolution of quantum systems described by Schrödinger representation is fully included into the time variation of the state function/vector $\psi(t,q)$ which satisfies the Schrödinger equation, while the operators associated with physical quantities do not vary over time. In order to maintain this property as a result of unitary (3.9)-type transformations performed by the unitary operators $U$, these operators have to be explicitly time-independent.

The time evolution of the studied quantum system is wholly contained in $\psi(t)$. Since $\psi(t+\delta t) = T(t+\delta t, t)\psi(t)$, then also $\psi(t) = T(t, t-\delta t)\psi(t-\delta t)$. It is worthwhile proving – without using the Schrödinger time-dependent equation (3.1) satisfied by the state function $\psi(t,q)$ – that the evolution operator $T(t+\delta t, t) = 1 - \frac{i}{\hbar}H(t)\delta t$ satisfies the equation

$$\frac{\hbar}{i}\frac{\partial T}{\partial t} + HT = 0. \tag{3.10}$$

Indeed,

$$T(t,t_0) = T(t,t-\delta t)T(t-\delta t,t_0),$$

because

$$T(t_3, t_1) = T(t_3, t_2)T(t_2, t_1)$$

with

$$T(t, t - \delta t) = 1 - \frac{i}{\hbar}H(t)\delta t.$$

Therefore,

$$T(t, t_0) = T(t - \delta t, t_0) - \frac{i}{\hbar}H(t)T(t - \delta t, t_0)\delta t,$$

or

$$\frac{T(t, t_0) - T(t - \delta t, t_0)}{\delta t} = -\frac{i}{\hbar}H(t)T(t - \delta t, t_0).$$

Taking the limit $\delta t \to 0$, one finds

$$\frac{\partial T}{\partial t} = -\frac{i}{\hbar}H(t)T(t, t_0),$$

which completes the proof.

The fact that the function $\psi(t)$ satisfies the Schrödinger equation

$$\frac{\hbar}{i}\frac{\partial \psi}{\partial t} + H\psi = 0 \tag{3.11}$$

emerges directly from the equation satisfied by the evolution operator. Indeed, multiplying the equation $\frac{\hbar}{i}\frac{\partial T}{\partial t} + HT = 0$ on the right by $\psi(t_0)$, one obtains

$$\frac{\hbar}{i}\frac{\partial T(t, t_0)}{\partial t}\psi(t_0) + HT(t, t_0)\psi(t_0) = 0,$$

or

$$\frac{\hbar}{i}\frac{\partial \psi(t)}{\partial t} + H\psi(t) = 0\,, \quad (q.e.d.).$$

Under a unitary transformation $\psi \xrightarrow{U} \psi' = U\psi$, described by the time-independent operator $U$, one can write

$$\frac{\partial \psi'}{\partial t} = U\frac{\partial \psi}{\partial t},$$

because $\frac{\partial U}{\partial t} = 0$. Then

$$\frac{\hbar}{i}\frac{\partial \psi'}{\partial t} = \frac{\hbar}{i}U\frac{\partial \psi}{\partial t} = \frac{\hbar}{i}\left(-\frac{i}{\hbar}UH\psi\right) = -UH\psi = -UHU^{-1}\psi' = H'\psi',$$

so that

$$\frac{\hbar}{i}\frac{\partial \psi'}{\partial t} + H'\psi' = 0, \tag{3.12}$$

as is has to be.

### III.1.3. Heisenberg representation

Suppose now that the operator $U$ explicitly depends on time: $U = U(t)$. In this case, using the relation

$$\psi'(t) = U(t)\psi(t)$$

one can write

$$\begin{aligned}
\frac{\partial \psi'(t)}{\partial t} &= U(t)\frac{\partial \psi(t)}{\partial t} + \frac{\partial U(t)}{\partial t}\psi(t) \\
&= U(t)\left(-\frac{i}{\hbar}H\psi\right) + \frac{\partial U(t)}{\partial t}U^{-1}(t)\psi'(t) \\
&= -\frac{i}{\hbar}U(t)HU^{-1}(t)\psi'(t) + \frac{\partial U(t)}{\partial t}U^{-1}(t)\psi'(t),
\end{aligned}$$

that is

$$\frac{\hbar}{i}\frac{\partial \psi'}{\partial t} = -\left[U(t)HU^{-1}(t) - \frac{\hbar}{i}\frac{\partial U(t)}{\partial t}U^{-1}(t)\right]\psi'(t),$$

or

$$\frac{\hbar}{i}\frac{\partial \psi'}{\partial t} + H'(t)\psi'(t) = 0, \tag{3.13}$$

where

$$H'(t) \equiv U(t)HU^{-1}(t) - \frac{\hbar}{i}\frac{\partial U(t)}{\partial t}U^{-1}(t) \tag{3.14}$$

is only a notation ($H'(t)$ does not signify the energy).

The Heisenberg representation is obtained by imposing the condition $H'(t) = 0$. In other words, Heisenberg representation can be derived from Schrödinger picture by means of a time-dependent unitary operator $U = U(t)$, according to

$$\underset{\text{S.r.}}{F} \to \underset{\text{H.r.}}{F'} = U(t)\underset{\text{S.r.}}{F}U^{-1}(t) \equiv U(t)FU^{-1}(t), \tag{3.15}$$

where $F$ is considered in Schrödinger representation (and, therefore, does not change over time), while S.r. and H.r. stand for Schrödinger and Heisenberg representations, respectively. Since $H'(t) = 0$, it follows that $\left(\frac{\partial \psi'}{\partial t}\right)_{\text{H.r.}} = 0$, which says that $\psi'$ *remains unchanged* during the time evolution of the system.

As a result, in the Heisenberg picture the state vector remains unchanged, its constant value being equal to the initial one, namely

$$\psi'(t_0) = \text{ constant } .$$

The whole evolution of the quantum system is therefore transferred to the operators associated with physical quantities (*i.e.* somewhat opposite to Schrödinger representation). The unitary operator $U(t)$, responsible for transition from Schrödinger to Heisenberg pictures, is obtained as a result of condition $H'(t) = 0$, that is

$$H'(t) \equiv U(t)HU^{-1}(t) - \frac{\hbar}{i}\frac{\partial U(t)}{\partial t}U^{-1}(t) = 0,$$

which easily yields the equation satisfied by the operator $U(t)$:

$$\frac{\hbar}{i}\frac{\partial U(t)}{\partial t} - U(t)H = 0.$$

Taking the Hermitian conjugate of this equation and observing that the operator $H$ is Hermitian ($H^+ = H$), while $U(t)$ is unitary $[U^+(t) = U^{-1}(t)]$, one obtains

$$\frac{\hbar}{i}\frac{\partial U^{-1}(t)}{\partial t} + HU^{-1}(t) = 0, \tag{3.16}$$

which means that $U^{-1}(t)$ satisfies the differential equation of the evolution operator $T(t, t_0)$, so that one can write

$$U^{-1}(t) \equiv T(t, t_0)$$

that is

$$U(t) = T^{-1}(t, t_0) = T(t_0, t).$$

In this case, the equation

$$\frac{\hbar}{i}\frac{\partial U(t)}{\partial t} - U(t)H = 0$$

leads to

$$\frac{\hbar}{i}\frac{\partial T^{-1}(t,t_0)}{\partial t} - T^{-1}(t,t_0)H = 0,$$

while the equation

$$\frac{\hbar}{i}\frac{\partial U^{-1}(t)}{\partial t} + HU^{-1}(t) = 0$$

leads to

$$\frac{\hbar}{i}\frac{\partial T(t,t_0)}{\partial t} + HT(t,t_0) = 0.$$

Therefore,

$$\begin{aligned}\psi'(t) &= U(t)\psi(t) = T^{-1}(t,t_0)\psi(t)\\ &= T^{-1}(t,t_0)T(t,t_0)\psi(t_0) = \psi(t_0) \equiv const.|_{\text{over time}}\end{aligned} \tag{3.17}$$

while the operator $F'(t)$ varies over time according to

$$\frac{dF'}{dt} = \frac{\partial F'}{\partial t} + \frac{i}{\hbar}[H',F'] \; ; \quad F'(t) = U(t)FU^{-1}(t). \tag{3.18}$$

Here all the quantities denoted with "prime" (used as a notation) are considered in Heisenberg picture, and those without "prime" – in Schrödinger representation. Indeed, let suppose first that the operator $F$ does not explicitly depend on time ($\frac{\partial F}{\partial t} = 0$); then,

$$\begin{aligned}\frac{dF'}{dt} &= \frac{\partial U(t)}{\partial t}FU^{-1}(t) + U(t)F\frac{\partial U^{-1}(t)}{\partial t} = \frac{\partial T^{-1}(t,t_0)}{\partial t}FT(t,t_0)\\ &\quad + T^{-1}(t,t_0)F\frac{\partial T(t,t_0)}{\partial t} = \frac{i}{\hbar}T^{-1}(t,t_0)HFT(t,t_0)\\ &\quad + T^{-1}(t,t_0)F\left(-\frac{i}{\hbar}H\right)T(t,t_0)\\ &= \frac{i}{\hbar}\left(T^{-1}HTT^{-1}FT - T^{-1}FTT^{-1}HT\right) = \frac{i}{\hbar}[H',F'],\end{aligned}$$

since

$$T^{-1}HT = U(t)HU^{-1}(t) = H'(t),$$

and

$$T^{-1}FT = U(t)FU^{-1}(t) = F'(t).$$

If, in addition, $\frac{\partial F}{\partial t} \neq 0$, an extra term appears when taking the derivative, namely

$$U(t)\frac{\partial F}{\partial t}U^{-1}(t) = \frac{\partial F'}{\partial t},$$

in which case

$$\frac{dF'(t)}{dt} = \frac{\partial F'(t)}{\partial t} + \frac{i}{\hbar}[H', F'(t)]. \tag{3.19}$$

### III.1.4. Tomonaga–Schwinger representation

This description is utilized when the Hamiltonian $H$ of the quantum system can be written as $H = H_0 + H_I$, where $H_0$ corresponds to an exactly solvable problem (the so-called Hamiltonian of the *free problem*), while $H_I$ represents the *interaction* part. In the Heisenberg representation, one can write

$$\frac{\hbar}{i}\frac{\partial \psi'}{\partial t} + H'(t)\psi' = 0,$$

where

$$H'(t) \equiv U(t)HU^{-1}(t) - \frac{\hbar}{i}\frac{\partial U(t)}{\partial t}U^{-1}(t) = 0.$$

But in this case $H \to H = H_0 + H_I$, so that

$$H'(t) \to H'_D(t) = U(t)H_0U^{-1}(t) + U(t)H_IU^{-1}(t) - \frac{\hbar}{i}\frac{\partial U(t)}{\partial t}U^{-1}(t).$$

The transition from Schrödinger representation to Tomonaga–Schwinger picture is performed by the unitary operator $U(t)$ obtained from the condition

$$H'_D(t) = U(t)H_IU^{-1}(t) = 0,$$

so that

$$H'_D(t) = H'_I, \quad H'_I = UH_IU^{-1}.$$

Consequently,

$$U(t)H_0U^{-1}(t) - \frac{\hbar}{i}\frac{\partial U(t)}{\partial t}U^{-1}(t) = 0,$$

or

$$\frac{\hbar}{i}\frac{\partial U(t)}{\partial t}U^{-1}(t) - U(t)H_0U^{-1}(t) = 0.$$

Multiplying this equation on the right by $U(t)$, one obtains

$$\frac{\hbar}{i}\frac{\partial U(t)}{\partial t} - U(t)H_0 = 0, \tag{3.20}$$

and recalling that $U$ is a unitary operator, it finally follows that

$$\frac{\hbar}{i}\frac{\partial U^{-1}(t)}{\partial t} + H_0U^{-1}(t) = 0. \tag{3.21}$$

The last relation shows that $U^{-1}(t)$ satisfies an equation similar to that satisfied by the evolution operator $T(t, t_0)$ in Schrödinger picture, but only for the free part $H_0$ of the Hamiltonian $H$. Thus, if one introduces the notation

$$U^{-1}(t) \equiv T_0(t, t_0),$$

this leads to the following result:

$$U(t) = T_0^{-1}(t, t_0) = T_0(t_0, t).$$

The operator $F$ transforms according to

$$\begin{aligned} F_{\text{S.r.}} \to F'_{\text{D.r.}} &= U(t)\; F_{\text{S.r.}}\; U^{-1}(t) \\ &= T_0^{-1}(t, t_0)\; F_{\text{S.r.}}\; T_0(t, t_0) = T_0(t_0, t)\; F_{\text{S.r.}}\; T_0^{-1}(t_0, t). \end{aligned} \tag{3.22}$$

Since $U(t)$ and $U^{-1}(t)$ satisfy the same equations as in Heisenberg picture, but only for the free part $H_0$ of the Hamiltonian $H$, like in Heisenberg representation one can write

$$\frac{dF'(t)}{dt} = \frac{\partial F'(t)}{\partial t} + \frac{i}{\hbar}[H'_0, F'(t)], \tag{3.23}$$

which means that the time variation of the operators associated with physical quantities is determined by the free part of the Hamiltonian of Tomonaga–Schwinger representation. In this paragraph, all the quantities marked with "prime" symbol are considered in Tomonaga–Schwinger representation, while those without "prime" are considered in Schrödinger picture.

Therefore, within the interaction representation both the operators (through relations (3.20) and (3.21)) and the state function according to equation

$$\frac{\hbar}{i}\frac{\partial\psi'(t)}{\partial t}+U(t)H_IU^{-1}(t)\psi'(t)=0$$

are time-dependent. There is, nevertheless, one difference: the time-dependence of the operators is fulfilled through the medium of the free Hamiltonian $H_0'$, while the state function depends on time through the interaction Hamiltonian $H_I'$:

$$\frac{\hbar}{i}\frac{\partial\psi'(t)}{\partial t}+H_I'(t)\psi'(t)=0. \tag{3.24}$$

This equation lies at the basis of the scattering matrix theory (the *S-matrix theory*). The time-evolution of the operators is therefore given by the free part of the Hamiltonian (the so-called *kinematic problem*), and that of the wave function/state vector by the interaction part $H_I'$ (the *dynamical problem*), where

$$\begin{aligned} H'_{I\,\mathrm{D.r.}} &= U(t)H_IU^{-1}(t)=T_0^{-1}(t,t_0)\;H_{I\,\mathrm{S.r.}}\;T_0(t,t_0)\\ &= T_0(t_0,t)\;H_{I\,\mathrm{S.r.}}\;T_0^{-1}(t_0,t). \end{aligned} \tag{3.25}$$

Here one can also introduce an evolution operator, exclusively specific to the Tomonaga–Schwinger representation, by means of

$$\psi'(t)=V(t,t_0)\psi'(t_0), \tag{3.26}$$

with $V(t_0,t_0)=1$ (the unit operator). In this case,

$$\psi'(t)=U(t)\psi(t)=T_0^{-1}(t,t_0)\psi(t),$$

which yields $\psi'(t_0)=\psi(t_0)$. Also,

$$\begin{aligned} \psi(t) &= U^{-1}(t)\psi'(t)=T_0(t,t_0)\psi'(t)=T_0(t,t_0)V(t,t_0)\psi'(t_0)\\ &= T_0(t,t_0)V(t,t_0)\psi(t_0), \end{aligned}$$

that is

$$T(t,t_0)=T_0(t,t_0)V(t,t_0), \tag{3.27}$$

where $T(t,t_0)$ is the "true" evolution operator (belonging to Schrödinger picture), but in the presence of interaction (*i.e.* not for the "simple" Hamiltonian $H$, but for $H = H_0 + H_I$), since the evolution operator in the absence of interaction has been denoted by $T_0(t, t_0)$. Therefore, $V(t, t_0)$ can be interpreted as a contribution due to perturbation in the operator $T(t, t_0)$.

The great advantage of definition of the evolution operator $V(t, t_0)$ (found only in Tomonaga–Schwinger picture) can be put into evidence as follows: choosing the complete orthonormal system of eigenfunctions of the diagonal operator $H_0$, the operator $T_0(t, t_0)$ is, in its turn, diagonal, and its diagonal elements are

$$\exp\left[-\frac{i}{\hbar}E_m(t - t_0)\right], \quad m = 1, 2, 3, ...$$

because $T_0(t, t_0)$ satisfies the equation

$$\frac{\hbar}{i}\frac{\partial T_0}{\partial t} + H_0 T_0 = 0$$

or

$$\frac{\partial T_0}{\partial t} = -\frac{i}{\hbar}H_0 T_0,$$

which means that

$$T_0 = e^{-\frac{i}{\hbar}H_0(t-t_0)}$$

and then one can write

$$(T_0)_{mn} = \exp\left[-\frac{i}{\hbar}E_m(t - t_0)\right]\delta_{mn}.$$

Since $T(t, t_0) = T_0(t, t_0)V(t, t_0)$, the matrix elements of $T(t, t_0)$ are

$$T_{mn}(t, t_0) = V_{mn}(t, t_0)e^{-\frac{i}{\hbar}E_m(t-t_0)}, \tag{3.28}$$

giving the transition probability amplitudes from the states "$n$" to the states "$m$", both categories being stationary states of the free Hamiltonian $H_0$. The transition probability is then given by

$$|T_{mn}(t, t_0)|^2 = |V_{mn}(t, t_0)|^2, \tag{3.29}$$

hence *the transition probabilities can be calculated in a straight manner by means of the evolution operator* $V(t, t_0)$ *in the interaction picture.*

In the case of Tomonaga–Schwinger representation, it can be shown that $U(t) = T_0^{-1}(t, t_0)$ is unitary, by using the condition of invariance of the state vector norm:

$$\langle\psi'(t)|\psi'(t)\rangle = \langle U(t)\psi(t)|U(t)\psi(t)\rangle$$
$$= \langle\psi(t)|U^+(t)U(t)\psi(t)\rangle \overset{\text{impose}}{=} \langle\psi(t)|\psi(t)\rangle.$$

It then immediately follows that $U^+(t)U(t) = 1$, and therefore $U^+(t) = U^{-1}(t)$.

The average values of the operators associated with physical quantities are also invariant with respect to unitary transformations. Indeed,

$$\langle F'\rangle = \langle\psi'(t)|F'|\psi'(t)\rangle = \langle U(t)\psi(t)|U(t)FU^{-1}(t)|U(t)\psi(t)\rangle$$
$$= \langle\psi(t)|U^+(t)U(t)F|\psi(t)\rangle = \langle\psi(t)|F|\psi(t)\rangle = \langle F\rangle.$$

In order to introduce the scattering matrix $S$ (the so-called *$S$-matrix*) the *dynamical part* of the interaction picture shall be used:

$$\frac{\hbar}{i}\frac{\partial\psi'}{\partial t} + H_I'\psi' = 0.$$

Here, for convenience, the index "prime" shall be abandoned, and the natural unit system ($\hbar = 1,\ c = 1$) shall be used; therefore, one starts with the equation

$$\frac{\partial\psi}{\partial t} = -iH_I\psi. \tag{3.30}$$

**Observation.** It can easily be shown that, at the initial moment $t = t_0$, all these three ways to describe the dynamics of a physical system (all the three representations/pictures) coincide, that is: $\psi_S(t_0) = \psi_H(t_0) = \psi_D(t_0)$.

## III.2. The $S$-matrix. Matrix elements of the field operators and the Feynman–Dyson-type rules for high-spin particles

This subchapter shall make a survey of the main elements necessary in the study of concrete processes of gravitational interaction of high-spin particles which are going to be approached within next chapters.

More precisely, being situated within the Tomonaga–Schwinger picture, the $S$-matrix elements for high-spin fields shall be determined and the corresponding Feynman–Dyson-type rules will be obtained; these rules are necessary to calculate the differential scattering cross-sections of the processes analyzed in chapter VI and chapter VIII of the volume.

### III.2.1. The $S$-operator

As has already been mentioned, in order to introduce the scattering matrix (or $S$-matrix) one starts with the dynamical part of the problem within the Tomonaga–Schwinger picture,

$$\frac{\partial \psi}{\partial t} = (-i)H_I\psi.$$

To know the time evolution of the system it means to determine $\psi$ as a function of time. Equation (3.30), with the initial condition $\psi(t)|_{t=t_0} = \psi(t_0)$, is equivalent to the integral equation

$$\psi(t) = \psi(t_0) + (-i)\int_{t_0}^{t} H_I(t_1)\psi(t_1)dt_1, \tag{3.31}$$

which is an inhomogeneous Volterra-type equation, with the kernel $-iH_I(t_1)$. It can be solved by iterative procedures, such as the successive approximations method. The solution of equation (3.31) offered by this method is

$$\psi(t) = \Bigg[1+(-i)\int_{t_0}^{t} H_I(t_1)dt_1+(-i)^2\int_{t_0}^{t}\int_{t_0}^{t_1} H_I(t_1)H_I(t_2)dt_1dt_2$$
$$+ ... + (-i)^n \underbrace{\int_{t_0}^{t}\int_{t_0}^{t_1}...\int_{t_0}^{t_{n-1}}}_{\text{n integrals}} H_I(t_1)H_I(t_2)...H_I(t_n)dt_1dt_2...dt_n + ...\Bigg]$$
$$\times\, \psi(t_0), \tag{3.32}$$

with $t \geq t_1 \geq t_2 \geq ... \geq t_n \geq ...$ This solution is usually written under the form

$$\psi(t) = U(t, t_0)\psi(t_0), \tag{3.33}$$

where

$$U(t, t_0) = 1 + (-i)\int_{t_0}^{t} H_I(t_1)dt_1$$

$$+ (-i)^2 \int_{t_0}^{t}\int_{t_0}^{t_1} H_I(t_1)H_I(t_2)dt_1dt_2$$

$$+ ... + (-i)^n \underbrace{\int_{t_0}^{t}\int_{t_0}^{t_1}...\int_{t_0}^{t_{n-1}}}_{\text{n integrals}} H_I(t_1)H_I(t_2)...H_I(t_n)dt_1dt_2...dt_n + ...$$

$$= \sum_{n=0}^{\infty}(-i)^n \int_{t_0}^{t} H_I(t_1)dt_1 \int_{t_0}^{t_1} H_I(t_2)dt_2... \int_{t_0}^{t_{n-1}} H_I(t_n)dt_n$$

$$= \sum_{n=0}^{\infty} U_n(t, t_0) = 1 + \sum_{n=1}^{\infty} U_n(t, t_0), \tag{3.34}$$

with $t \geq t_1 \geq t_2 \geq ... \geq t_n \geq ...$, where

$$U_n(t, t_0) = (-i)^n \int_{t_0}^{t} H_I(t_1)dt_1 \int_{t_0}^{t_1} H_I(t_2)dt_2... \int_{t_0}^{t_{n-1}} H_I(t_n)dt_n, \tag{3.35}$$

with $t \geq t_1 \geq t_2 \geq ... \geq t_n \geq ...$.

Here $U(t, t_0)$ is a unitary operator. Indeed, one can write

$$\frac{\partial \psi(t)}{\partial t} = (-i)H_I(t)\psi(t)$$

and

$$\psi(t) = U(t, t_0)\psi(t_0) \equiv U\psi(t_0),$$

which yields

$$U(t_0, t_0) = 1.$$

According to relations $\frac{\partial \psi}{\partial t} = (-i)H_I\psi$ and Eq. (3.33), one can also write

$$\frac{\partial U}{\partial t} + iH_IU = 0, \tag{3.36}$$

as well as

$$\frac{\partial U^+}{\partial t} - iU^+H_I^+ = 0. \tag{3.37}$$

Multiplying Eq. (3.36) on the left by $U^+$ and Eq. (3.37) on the right by $U$, then adding the results, one finds

$$\frac{\partial}{\partial t}(U^+U) = 0,$$

since $H_I$ is Hermitian, that is $H_I = H_I^+$. So, it has been found that

$$const. = U^+(t, t_0)U(t, t_0) = U^+(t_0, t_0)U(t_0, t_0), \quad (\forall)\ t$$

and, since $U(t_0, t_0) = 1$, it follows that $U^+(t, t_0) = U^{-1}(t, t_0)$, therefore $U(t, t_0)$ is unitary (obviously, under necessary assumptions). As it is known, by using the time ordering operator, $U(t, t_0)$ can be written in a more symmetrical form (with regard to the integration limits). Using this definition, $U_n(t, t_0)$ can be written as

$$U_n(t, t_0) = \frac{(-i)^n}{n!} \underbrace{\int_{t_0}^{t} \int_{t_0}^{t} \dots \int_{t_0}^{t}}_{\text{n integrals}} T[H_I(t_1)H_I(t_2)\dots H_I(t_n)]$$
$$\times\ dt_1 dt_2 \dots dt_n. \tag{3.38}$$

By definition,

$$U(\infty, -\infty) \stackrel{\text{def}}{=} \lim_{\substack{t_0 \to -\infty \\ t \to +\infty}} U(t, t_0) \equiv S \tag{3.39}$$

is called the *S-operator*. Then

$$\psi(+\infty) = U(+\infty, -\infty)\psi(-\infty),$$

or, in Dirac notation,

$$|\psi(\infty)\rangle = U(+\infty, -\infty)|\psi(-\infty)\rangle. \tag{3.40}$$

The vector $|\psi(-\infty)\rangle$ plays the role of wave function of the initial state (before interaction), while $|\psi(\infty)\rangle$ represents the emergent wave function (*i.e.* of the final state, after interaction). Then one can write

$$S = \sum_{n=0}^{\infty} \frac{(-i)^n}{n!} \int_{-\infty}^{+\infty} dt_1 \int_{-\infty}^{+\infty} dt_2 \cdots \int_{-\infty}^{+\infty} dt_n$$
$$\times T[H_I(t_1)H_I(t_2)\cdots H_I(t_n)] = \sum_{n=0}^{\infty} S_n. \tag{3.41}$$

In view of relation between the unitary operator $U(t, t_0)$ and the $S$-operator, one can show that $S$ is also *unitary*.

It is worth mentioning that the above considerations have been presented in the simplest possible way (see, also, Ref. [258]); in general, there appear a lot of "small" problems connected to analyticity, unitarity, oscillant character and some possible divergences of the integrand, as well as some other properties of the $S$-operator. For a detailed documentation on this subject, we recommend to the interested reader several volumes, such as [216, 217], where the $S$-matrix subject is treated *in extenso*.

From now on, for the sake of convenience, the index "$I$" attached to the interaction Hamiltonian (*i.e.* that part of the total Hamiltonian which describes the interaction) shall be omitted, and the interaction Hamiltonian $H_I$ shall be simply denoted by $H$. Since

$$H(t) = \int H(x) d^3x = \int H(\vec{x}, t) d\vec{x},$$

it follows that

$$\int_{-\infty}^{+\infty} H(t) dt = \int H(\vec{x}, t) d\vec{x} dt = \frac{1}{i} \int H(x) dx,$$

where

$$dx = d\vec{x} dx_4 = d^3x dx_4 = d\vec{x}\, i\, dt = i\, d\vec{x}\, dt, \quad (c = 1)$$

so that

$$S = \frac{(-1)^n}{n!} \int dx_1 \int dx_2 ... \int dx_n\, T[H(x_1) H(x_2) ... H(x_n)]. \quad (3.42)$$

To avoid any difficulties connected to a possible introduction of various infinite quantities due to an inadequate interaction mechanism model, one takes

$$H(\vec{x}, t) \to H^{ad}(\vec{x}, t) = e^{-\varepsilon |t|} H(\vec{x}, t), \quad (3.43)$$

(the so-called *adiabatic hypothesis*), then passing to the limit for $\varepsilon \to 0$. Schematically, the image associated to the adiabatic hypothesis can be expressed as

$$t = -\infty \longrightarrow t = -T \longrightarrow$$

initial state — coupling coefficient increases starting from zero — an arbitrary finite value — coupling coefficient remains constant

$$t = +T \longrightarrow t = +\infty$$

coupling coefficient decreases to zero — final state

### III.2.2. Transition probability

Following reasoning exposed in Ref. [258], one can now determine the pure state vectors $\psi_m$ (corresponding to fixed and well-defined energies, momenta and spin projections/polarizations) denoted by $\psi_m \equiv |\psi_m\rangle$. The final state, described by $\psi(\infty)$, shall generally be given as a superposition of such (pure state) vectors,

$$\psi(\infty) \equiv |\psi(\infty)\rangle = \sum_{(m)} c_m |\psi_m\rangle = S|\psi(-\infty)\rangle,$$

so that

$$\sum_{(m)} \langle \psi_n | c_m | \psi_m \rangle = \langle \psi_n | S | \psi(-\infty) \rangle,$$

which gives

$$c_n = \langle \psi_n | S | \psi(-\infty) \rangle,$$

that is

$$|\psi(\infty)\rangle = \sum_{(m)} \langle \psi_m | S | \psi(-\infty) \rangle | \psi_m \rangle.$$

Supposing that the initial state is monoenergetic (a pure state, with well-defined energy and momentum): $|\psi(-\infty)\rangle = |\psi_n\rangle$, one has

$$|\psi(\infty)\rangle = \sum_{(m)} \langle \psi_m | S | \psi_n \rangle | \psi_m \rangle = \sum_{(m)} S_{mn} |\psi_m\rangle,$$

where

$$S_{mn} = \langle \psi_m | S | \psi_n \rangle \tag{3.44}$$

is the matrix form of the operator $S$ (the *S-matrix*).

Consider the pure state vectors $|i\rangle, |j\rangle, |k\rangle\dots$. Admitting a quasi-continuous spectrum in the final state (after interaction), one sets $\sum_{(m)} \to \int df$, where $f$ is a continuous parameter, and also $c_m \to c(f)$. Then $|\psi(-\infty)\rangle = |i\rangle$, and $|\psi(\infty)\rangle = \int c(f)|f\rangle\, df$, where $|f\rangle$ designates the pure final (after interaction) state. But

$$|\psi(\infty)\rangle = S|\psi(-\infty)\rangle,$$

so that

$$\int c(f)|f\rangle\, df = S|i\rangle.$$

One also has

$$\int \langle g|c(f)|f\rangle df = \langle g|S|i\rangle,$$

which yields

$$c(f) = \langle f|S|i\rangle,$$

since

$$\langle g|f\rangle = \delta(g - f)$$

and

$$\int \langle g|f\rangle c(f) df = \int c(f)\delta(g - f) df = c(g).$$

Then, the probability density of finding the system in the final state $|f\rangle$, if initially it was in the state $|i\rangle$, is

$$w(i \to f) = c^+(f)c(f) = |\langle f|S|i\rangle|^2. \tag{3.45}$$

Therefore, in order to calculate the transition probability densities (straightly connected to the differential scattering cross-sections), one actually has to evaluate the squared modulus of the matrix elements of the operator $S$. Since $S_0 = 1$, instead of $S_n$ one takes $R_n = S_n - I$, in which case

$$S_n^{(fi)} = \langle f|S_n|i\rangle = \langle f|(R_n + I)|i\rangle = \langle f|R_n|i\rangle + \langle f|i\rangle.$$

Since $\langle f|i\rangle$ describes transition without interaction, the only considered part is

$$w_n(i \to f) = |\langle f|R_n|i\rangle|^2, \quad n \geq 1.$$

For $n \geq 1$ one writes

$$\langle f|R_n|i\rangle \stackrel{\text{def}}{=} \langle f|M_n|i\rangle\, \delta\Big(\sum p_f - \sum p_i\Big), \tag{3.46}$$

where the factor $\delta\Big(\sum p_f - \sum p_i\Big)$ ensures the four-momentum conservation. Under these circumstances, one should have

$$w_n(i \to f) = \sum_{n=1}^{\infty} \Big[\delta\Big(\sum p_f - \sum p_i\Big)\Big]^2 |\langle f|M_n|i\rangle|^2,$$

meaning that when the integration is performed there would appear a $\delta^2(p) = \delta(p)\delta(p)$, which would give $\delta(0) = \infty$, and this is not admissible. The appearance of this divergence is natural and obvious, since – up to this point – the integration is performed over the whole space and the whole time interval (*i.e.* from $-\infty$ to $+\infty$), which covers an infinite number of processes. This divergence can be eliminated by calculating the transition probability per unit time and unit spatial volume, that is

$$\frac{|\langle f|S_n|i\rangle|^2}{VT},$$

by taking

$$w_n(i \to f) = \lim_{\substack{|V|\to+\infty \\ |T|\to+\infty}} \frac{|\langle f|S_n|i\rangle_{VT}|^2}{VT}. \tag{3.47}$$

### III.2.3. Matrix elements of the field operators and Feynman–Dyson-type rules for high-spin particles

The purpose of this paragraph is to derive the expressions of matrix elements of the field operators for high-spin particles, as well as the Feynman–Dyson-type rules written in the momentum space, necessary to facilitate writing of the matrix elements corresponding to a given interaction process.

To deduce the Feynman-type rules in momentum space one must first calculate the matrix elements of the normal products of operators, these matrix elements being calculated for a transition between states in which the numbers of free particles (as well as their characteristics: energy, momentum, spin projection or polarization state) are well-defined (are given). In this respect, it is necessary to decompose the normal product of the field operators into a sum of normal products of annihilation and creation operators, then commute the annihilation operators to the right until reaching the vector corresponding to the vacuum state $|0\rangle$, and creation operators to the left until reaching the adjoint vector $\langle 0|$. According to vacuum state definition, the terms obtained this way cancel, and the only contributions to the matrix element shall be given by the commutation functions. Concretely, here are the steps that must be followed to approach such a problem.

**A**. To calculate the matrix elements of the field operators:

1) One starts with expressions of Fourier series (or integral) expansions/representations (explicitly containing the annihilation and creation operators) of the field operators in the configuration representation;

2) One obtains the commutation relations satisfied by the annihilation and creation operators of the considered field, which easily follow from the equal-time commutators of that field;

3) The "standard" conditions satisfied by the vacuum state must be used;

4) Taking into account the preceding three points, a direct calculation shows that the matrix elements of the field operators are non-zero only between states whose number of particles differ one from the other by one; for example, between the vacuum state and the state which contains only one particle, with well-defined (fixed) energy, momentum, spin projection or polarization state.

**B**. To deduce the Feynman–Dyson-type rules in momentum space:

1) One obtains the interaction Lagrangian corresponding to the investigated process (*e.g.*, by using the principle of minimal coupling);

2) One determines the expression of the scattering matrix associated with the process of desired order (usually of the first and second-order, in which case all the Feynman-type rules are obtained; but, in regard with vertices, only those of the first and second-order

can be determined, the vertices of superior order being found only by considering the expression of the scattering matrix in the considered superior order);

3) One utilizes the matrix elements of the field operators obtained in section **A**;

4) Observing that integration over exponentials leads to Dirac-type functions, one obtains various kinds of vertices, propagators and free particles which can easily be turned into the Feynman-type rules (*i.e.* the appropriate correspondences between various parts of the terms of the $S$-matrix elements and "equivalent" mathematical expressions can be rigorously and unambiguously established).

Next, this algorithm shall be exemplified for a single (but representative) case that implies high-spin particles; in all other cases the calculations can be done by analogy. More exactly, the reader's attention shall be focused on the massive Rarita–Schwinger field (the massive spin-3/2 field), considered in interaction with the spin-2 non-massive field (the gravitational field), in the external field approximation.

**A.**

1) The Fourier integral expansion of the spin-vector $\psi_\mu(x)$ – the wave function of the massive Rarita–Schwinger field – writes

$$\begin{aligned}\psi_\mu(x) = \frac{1}{(2\pi)^{3/2}} \int d^3k \sqrt{\frac{M}{E_{\vec{k}}}} \sum_{r=1}^{4} \Big[ u_\mu^{(r)}(\vec{k}) d_r^{+}(\vec{k}) e^{ikx} \\ + v_\mu^{(r)}(\vec{k}) d_r^{-}(\vec{k}) e^{-ikx} \Big] = \psi_\mu^{(+)}(x) + \psi_\mu^{(-)}(x),\end{aligned} \tag{3.48}$$

while the Dirac conjugated spin-vector $\overline{\psi}_\mu(x)$ is

$$\begin{aligned}\overline{\psi}_\mu(x) = \frac{1}{(2\pi)^{3/2}} \int d^3k \sqrt{\frac{M}{E_{\vec{k}}}} \sum_{r=1}^{4} \Big[ \overline{v}_\mu^{(r)}(\vec{k}) d_r^{*(+)}(\vec{k}) e^{ikx} \\ + \overline{u}_\mu^{(r)}(\vec{k}) d_r^{*(-)}(\vec{k}) e^{-ikx} \Big] = \overline{\psi}_\mu^{(+)}(x) + \overline{\psi}_\mu^{(-)}(x),\end{aligned} \tag{3.49}$$

where $d_r^{+}(\vec{k})$ and $d_r^{*(+)}(\vec{k})$ are the annihilation operators of the massive spin-3/2 particle and antiparticle, respectively, $d_r^{*(-)}(\vec{k})$ and

$d_r^-(\vec{k})$ are the corresponding creation operators, $E_{\vec{k}}^2 = \vec{k}^2 + M^2$, while $M$ is the mass of the Rarita–Schwinger particles.

2) The commutation relations for the operators $d^{(+)}$ and $d^{(-)}$ can be deduced from the expressions of the equal-time commutators of the Rarita–Schwinger field. In order to shorten the calculations and having in view the connection between spin and statistics, we shall directly impose the commutation relations for the annihilation and creation operators:gindxOperator!creation

$$\left\{d_r^{(+)}(\vec{k}), d_{r'}^{*(-)}(\vec{k}')\right\} = \delta_{rr'}\delta(\vec{k} - \vec{k}');$$

$$\left\{d_r^{*(+)}(\vec{k}), d_{r'}^{(-)}(\vec{k}')\right\} = \delta_{rr'}\delta(\vec{k} - \vec{k}');$$

$$\left\{d_r^{(+)}(\vec{k}), d_{r'}^{(+)}(\vec{k}')\right\} = 0;$$

$$\left\{d_r^{*(+)}(\vec{k}), d_{r'}^{*(+)}(\vec{k}')\right\} = 0;$$

$$\left\{d_r^{(+)}(\vec{k}), d_{r'}^{(-)}(\vec{k}')\right\} = 0;$$

$$\left\{d_r^{(+)}(\vec{k}), d_{r'}^{*(+)}(\vec{k}')\right\} = 0.$$

3) By definition, the vacuum state satisfies the following relations:

$$d_r^{(+)}(\vec{k})|0\rangle = 0\ ; \quad d_r^{*(+)}(\vec{k})|0\rangle = 0;$$

$$\langle 0|d_r^{(-)}(\vec{k}) = 0\ ; \quad \langle 0|d_r^{*(-)}(\vec{k}) = 0\ ;$$

$$\langle 0|0\rangle = 1.$$

4) Let us calculate, for example, the matrix element corresponding to creation of a massive particle of spin 3/2, having the energy $E_{\vec{k}}$, the momentum $\vec{k}$ and the polarization state $(r)$. According to the prescriptions exposed above for this step of the algorithm, the expression $\langle 0|d_r^{(+)}(\vec{k})\overline{\psi}_\mu^{(-)}(x)|0\rangle$ should be calculated. In this regard we have

$$\langle 0|d_r^{(+)}(\vec{k})\overline{\psi}_\mu^{(-)}(x)|0\rangle = \langle 0|\left\{d_r^{(+)}(\vec{k}), \overline{\psi}_\mu^{(-)}(x)\right\}|0\rangle$$

$$- \underbrace{\langle 0|\overline{\psi}_\mu^{(-)}(x)}_{=0}\ \underbrace{d_r^{(+)}(\vec{k})|0\rangle}_{=0} = \langle 0|\left\{d_r^{(+)}(\vec{k}), \overline{\psi}_\mu^{(-)}(x)\right\}|0\rangle$$

$$= \langle 0| \left\{ d_r^{(+)}(\vec{k}), (2\pi)^{-\frac{3}{2}} \int d^3k' \sqrt{\frac{M}{E_{\vec{k}'}}} \sum_{r'=1}^{4} \overline{u}_\mu^{(r')}(\vec{k}') d_{r'}^{*(-)}(\vec{k}') e^{-ik'x} \right\} |0\rangle$$

$$= \langle 0| \frac{1}{(2\pi)^{3/2}} \int d^3k' \sqrt{\frac{M}{E_{\vec{k}'}}} \sum_{r'=1}^{4} \overline{u}_\mu^{(r')}(\vec{k}') \left\{ d_r^{(+)}(\vec{k}), d_{r'}^{*(-)}(\vec{k}') \right\} e^{-ik'x} |0\rangle$$

$$= \langle 0|0\rangle \frac{1}{(2\pi)^{3/2}} \int d^3k' \sqrt{\frac{M}{E_{\vec{k}'}}} \sum_{r'=1}^{4} \overline{u}_\mu^{(r')}(\vec{k}') \delta_{rr'} \delta(\vec{k} - \vec{k}') e^{-ik'x}$$

$$= \frac{1}{(2\pi)^{3/2}} \sqrt{\frac{M}{E_{\vec{k}}}} \overline{u}_\mu^{(r)}(\vec{k}) e^{-ikx},$$

which leads to the following correspondence:

$$\overline{\psi}_\mu^{(-)}(x) \to \frac{1}{(2\pi)^{3/2}} \sqrt{\frac{M}{E_{\vec{k}}}} \overline{u}_\mu^{(r)}(\vec{k}) e^{-ikx}. \tag{3.50}$$

In an analogous manner, the matrix elements corresponding to creation of the massive Rarita–Schwinger antiparticle, as well as to annihilation of the massive Rarita–Schwinger particle and antiparticle, respectively, can be calculated, all these particles and antiparticles being characterized by the same properties (energy, momentum and polarization state) as those associated with the creation of the massive Rarita–Schwinger particle. Here are the results:

$$\psi_\mu^{(-)}(x) \to \frac{1}{(2\pi)^{3/2}} \sqrt{\frac{M}{E_{\vec{k}}}} v_\mu^{(r)}(\vec{k}) e^{-ikx} \quad \text{(antiparticle creation);} \tag{3.51}$$

$$\psi_\mu^{(+)}(x) \to \frac{1}{(2\pi)^{3/2}} \sqrt{\frac{M}{E_{\vec{k}}}} u_\mu^{(r)}(\vec{k}) e^{ikx} \quad \text{(particle creation);} \tag{3.52}$$

$$\overline{\psi}_\mu^{(+)}(x) \to \frac{1}{(2\pi)^{3/2}} \sqrt{\frac{M}{E_{\vec{k}}}} \overline{v}_\mu^{(r)}(\vec{k}) e^{ikx} \quad \text{(antiparticle annihilation).} \tag{3.53}$$

**B.**

1) The first-order interaction Lagrangian (with $\kappa$ as the gravitational coupling constant) between the massive Rarita–Schwinger field and gravitation is

$$\mathcal{L}_{int}^{(1)}(\kappa) = \frac{1}{4}\kappa(\overline{\psi}_{\alpha,\nu}\gamma_\mu\psi_\alpha - \overline{\psi}_\alpha\gamma_\mu\psi_{\alpha,\nu})y_{\mu\nu} - \frac{1}{4}\kappa m\overline{\psi}_\alpha\psi_\alpha y. \qquad (3.54)$$

Written in the normal form it has the following expression:

$$N\left[\mathcal{L}_{RS-g}^{(1)}(x)\right] = \frac{1}{4}\kappa\left[\overline{\psi}_{\alpha,\nu}^{(-)}(x)\gamma_\mu\psi_\alpha^{(+)}(x)\right.$$

$$\left. - \overline{\psi}_\alpha^{(-)}(x)\gamma_\mu\psi_{\alpha,\nu}^{(+)}(x)\right] y_{\mu\nu}^{ext}(x) - \frac{1}{4}\kappa m\overline{\psi}_\alpha^{(-)}(x)\psi_\alpha^{(+)}(x)y^{ext}(x). \quad (3.55)$$

2) Being a first-order process, the expression of the corresponding scattering matrix is

$$S_1 = i\int d^4x N\left[\mathcal{L}_{RS-g}^{(1)}(x)\right]. \qquad (3.56)$$

3) To deduce the corresponding Feynman–Dyson-type rules, the following elements are needed:

*i*) the relations (3.50)–(3.53);

*ii*) the expression of the symmetric tensor $y_{\mu\nu}$ of the free weak gravitational field, written in the exterior field approximation (the only one which allows the existence of the first-order processes):

$$y_{\mu\nu}^{ext}(\vec{x}) = \delta_{\mu 4}\delta_{\nu 4}y^{ext}(\vec{x}), \qquad (3.57)$$

where, if the exterior gravitational field is static and has a central symmetry (which implies, for example, a Schwarzschild-type metric),

$$y^{ext}(\vec{x}) = \frac{\kappa M_\odot}{4\pi|\vec{x}|}, \qquad (3.58)$$

$M_\odot$ being the mass of the body that creates the gravitational field;

*iii*) the obvious relationship

$$\delta(k' - k \pm q) = \frac{1}{i(2\pi)^4}\int e^{i(k-k'\mp q)x}dx, \qquad (3.59)$$

where $\vec{q} = \vec{k}' - \vec{k}$ is the momentum transfer towards the exterior field (in this case the momentum is not conserved in the interaction process existing a non-null momentum transfer towards the exterior field), and

*iv*) the easily deductible relationship

$$\int \frac{1}{|\vec{x}|} e^{-iqx} d^4x = 2(2\pi)^2 \frac{\delta(q_0)}{|\vec{q}|^2}, \tag{3.60}$$

where $q_0 = E_{\vec{k}'} - E_{\vec{k}} = 0$ (the energy is conserved).

Based on all these considerations, one can write the following

4) *Feynman–Dyson-type rules*:

**Table 1**

| **Particle** | **Type** | **S-matrix factor** |
|---|---|---|
| Massive Rarita–Schwinger | $\bar{\psi}_\mu^{(-)}(\vec{k})$ | $\frac{1}{(2\pi)^{3/2}}\sqrt{\frac{M}{E_{\vec{k}}}}\bar{u}_\mu^{(r)}(\vec{k})$ |
| Massive Rarita–Schwinger | $\psi_\mu^{(-)}(\vec{k})$ | $\frac{1}{(2\pi)^{3/2}}\sqrt{\frac{M}{E_{\vec{k}}}}v_\mu^{(r)}(\vec{k})$ |
| Massive Rarita–Schwinger | $\psi_\mu^{(+)}(\vec{k})$ | $\frac{1}{(2\pi)^{3/2}}\sqrt{\frac{M}{E_{\vec{k}}}}u_\mu^{(r)}(\vec{k})$ |
| Massive Rarita–Schwinger | $\bar{\psi}_\mu^{(+)}(\vec{k})$ | $\frac{1}{(2\pi)^{3/2}}\sqrt{\frac{M}{E_{\vec{k}}}}\bar{v}_\mu^{(r)}(\vec{k})$ |

The other expressions of the matrix elements of the field operators and first-order vertices associated with gravitational interactions of particles, which shall be used in chapter VI, chapter VII, and chapter VIII, are presented in Appendix F.

**Table 2**

| **Gravitational field** | **Type** | **S-matrix factor** |
|---|---|---|
| Schwarzschild | $y^{ext}(\vec{q})$ | $\dfrac{1}{(2\pi)^{3/2}}\dfrac{\kappa M_\odot}{\lvert\vec{q}\rvert^2}$ |
| Schwarzschild | $y^{ext}_{\mu v}(\vec{q})$ | $\dfrac{1}{(2\pi)^{3/2}}\dfrac{\kappa M_\odot}{\lvert\vec{q}\rvert^2}\delta_{\mu 4}\delta_{v4}$ |
| Kerr | $y^{ext}_{ij}(\vec{q}),\ (i,j=\overline{1,3})$ | 0 |
| Kerr | $y^{ext}_{00}(\vec{q})=-y^{ext}_{44}(\vec{q})$ | $\dfrac{1}{(2\pi)^{3/2}}\dfrac{\kappa M_\odot}{\lvert\vec{q}\rvert^2}$ |
| Kerr | $y^{ext}_{0j}(\vec{q})=-iy^{ext}_{4j}(\vec{q})$ | $\dfrac{1}{(2\pi)^{3/2}}\dfrac{\kappa M_\odot}{2\lvert\vec{q}\rvert^2}\varepsilon_{jkl}a_k x_l;$<br>$(j,k,l=\overline{1,3})$ |

**Table 3**

| **Physical process** | **Type** | **S-matrix factor** |
|---|---|---|
| Particle scattering in external gravitational field | $V_{1,R-S-g}(k,k',q)$ | $-\dfrac{i}{4}\kappa(2\pi)^4$<br>$\times\,\delta(\vec{k}' \quad \vec{k}-\vec{q})\delta(q_0)$ |
| Particle scattering in external gravitational field | $V_{2,R-S-g}(k,k',q)$ | $-\dfrac{i}{4}\kappa M(2\pi)^4$<br>$\times\,\delta(\vec{k}'-\vec{k}-\vec{q})\delta(q_0)$ |
| Everywhere $q_0=E_{\vec{k}'}-E_{\vec{k}}=0$ and $\int \cdot d^3q$ | | |

Chapter IV

# Fundamentals of Gauge Theories. The Minimal Coupling Principle

The elaboration of the gauge theories played an essential role in conceiving and development of the Standard Model; at present, out of this concept, the unification of the fundamental forces is an impossible action. This way, to thoroughly study all the aspects connected to the gauge fields and their properties became more than necessary, as a condition of a correct comprehension of the theory. This fact is proved by the continuous increase of the number of scientific papers and publications in the above mentioned field of investigation [50, 85, 150, 204, 214, 350]. The special impetus taken by the High Energy Physics (HEP) as a result of the epochal discoveries produced at CERN, Geneva, as well as in some other experimental physics centres provided with particle accelerators, located in USA and Germany, led to an intensification of efforts to comprehend the Microcosmos phenomena from the theoretical physics point of view [16, 47, 170, 191, 200, 221, 261, 303, 355, 410]. On the other side, this impetus represented a premise for enlarging the research area in the domain of gauge theories, in the sense of discovering some new symmetries and/or supersymmetries capable to include in a unitary scheme the multitude of recently discovered particles, and to explain the nature and role of the matter known as *dark matter* and *dark energy*.

But, at this moment, talking about symmetry without resorting to the theory of symmetry groups is almost nonsensical. That is why the study of the theory of symmetry groups is at present more than a necessity for those who want to approach the fascinating field of HEP.

## IV.1. Generalities

The *group theory* appeared as a result of necessity of finding a mathematical tool capable to describe the *symmetry properties* of the world, since the knowledge of the group properties of some mathematical or physical objects furnishes important information about the structure of these objects.

The symmetry concept plays an essential role in physics. In order to know/comprehend a physical system, one must predict its space and time behavior in terms of certain *invariant quantities*, that satisfy certain *laws of conservation.*

But, according to Emmy Noether's theorem (1918), a law of conservation is "equivalent" to a symmetry. Indeed, if the variational principle leading to the equations of motion of a physical system is invariant with respect to certain transformations (called *symmetry transformations*), then there exist as many laws of conservation (*i.e.* invariant quantities) as the symmetry transformations.

Since these symmetry transformations form group structures – the so-called *symmetry groups* – there appears necessity to closely study the *group* and *group representation* notions.

Nowadays, the modern group theory stands as a well set theory with respect to its fundamental notions and includes several branches, such as:

1. The general group theory;
2. The finite group theory;
3. The topological (continuous) group theory;
4. The symplectic and metaplectic group theory;
5. The theory of discrete transformation groups;
6. The representation theory of groups;
7. The theory of groups' characters.

Here are several significant examples for the use of group theory in modern physics:

• The elementary particle classification, whose accomplishment is inconceivable without the group representation theory.

• Quantization of a physical system whose symmetry proprieties are known, without being necessary to solve the eigenvalue problem for various operators, not even to know the explicit form of these operators (one frequently happens that, due to complexity of these operators, solving their eigenvalue equations is practically impossible).

• Description within the framework of the gauge theory of the interaction mechanisms between particles/fields.

• Determination of the invariant quantities of a physical system with a space-time evolution by means of the laws of conservation offered by the Noether's theorem, etc.

One of the most important application of the group theory in Quantum Field Theory (QFT) is the quantum study of the interaction between particles, which interaction, from the quantum point of view, is equivalent to the interaction between the physical fields associated to the particles. As far as one knows, in order to determine – for example – the differential scattering cross sections (*i.e.* those quantities able to offer concrete and directly verifiable data), one starts with the interaction Lagrangian of the interacting fields and, by means of the $S$-matrix formalism, one can find the desired results.

Until now, the interaction Lagrangian $L_{int}$ used to be constructed (one can even say *postulated*) on the basis of some reasons raised to the status of *principles*, and it is worth mentioning some of them:

1) $L_{int}$ must be a scalar containing terms formed from factors that belong to each of the interacting fields;

2) $L_{int}$ must keep the symmetry properties off all the fields interfering as factors in its terms;

3) $L_{int}$ must be invariant with respect to the transformations associated with the nature of the developing theory. For example, in case of the study of interactions between particles whose velocity approaches the speed of light, $L_{int}$ must be a relativistic invariant;

4) $L_{int}$ must by as simple as possible, in order to avoid an inutile complication of the mathematical calculation;

5) The superposition principle should be also obeyed.

Since this way is somewhat empirical, one has been searched for a more elaborate and systematic procedure, capable to determine $L_{int}$ in terms of a smaller number of hypotheses which, eventually, could include the previously enumerated ones. So, one has arrived to the *minimal coupling principle*, elaborated on the basis of *minimal substitution*, which substitution is found under a rigorous manner in the framework of *gauge field theories*. But this is not the most general recipe; at present, in order to describe some more particular effects, in addition to this principle one also make use of certain non-minimal couplings, obtained by means of the above described empirical way. Therefore, one can speak about a harmonious symbiosis of the two methods.

The minimal coupling principle allows one to determine the interaction Lagrangian in all cases in which the gauge symmetry of the field that mediates the considered symmetry is known.

Here is a schematic overview of the procedure:

1) One starts with a field called *basic field*[1] (for example, the Dirac field – in which case the associated wave function is a Dirac spinor – that describes the electrons and positrons as quanta (particles) of the quantized spinorial field). One also starts from a certain type of interaction (out of the four fundamental types of presently known interactions, namely the gravitational, electromagnetic, weak-nuclear and strong-nuclear interactions), which interaction can take place with participation of the basic field quanta (*e.g.* since the electromagnetic interaction occurs only between charged particles, one must start with a charged basic field – and by no means with a neutral field describing, let say, the neutral mesons). The only universal interaction, able to couple any type of particle, but most difficult to be included in a quantum scheme, is the gravitational interaction.

2) The next step is to "discover" the symmetry of the field that mediates the interaction. From the point of view of the gauge theories, two particles can interact only through the medium of one (or more) of the four types of known fundamental interactions. From the quantum point of view, this interaction takes place through the so-called *exchange particles* – quanta of the field that mediates the

[1] *Basic field* is here a generic name which is frequently replaced by that of *matter field* or *material field*; this name is intentionally used in order to distinguish these fields from those mediating the interactions (the so-called *gauge fields*) between the quanta of the basic fields.

interaction. Thus, any gauge theory associates to each type of interaction a certain physical field, called *mediation field* or *gauge field.* Therefore, an interaction process between two particles (quanta of the basic field) takes always place through the medium of a second type of particles, which are precisely the gauge field quanta, that is of the field that mediates the interaction. For example, if the Dirac field is considered as the basic field (whose quanta are, for example, the electrons and positrons), in order to study the electromagnetic interaction between these particles (electrons and positrons) one must choose the electromagnetic field as the gauge field. As one knows, the electromagnetic field possesses the symmetry of the gauge group $U(1)$[2], that is the symmetry with respect to the infinitesimal phase transformations which, from the point of view of the gauge theories, are precisely the gauge transformations satisfied by the scalar potential $V$ and the vector potential $\vec{A}$; if considered together, they represent the field function of the electromagnetic field, which in this case plays the role of the gauge field. Therefore, here the gauge field function – or, simply called, the gauge field – is precisely the potential four-vector $A_\mu$ of the electromagnetic field. The great advantage of the gauge theories is that they offer a rigorous method for determining the interaction Lagrangian. And this method is the minimal coupling principle, whose mathematical and physical justification is given within the framework of these gauge[3] theories.

3) The third step consists in expressing the gauge covariant derivative of the field functions for the basic field and replacing the usual derivatives from the Lagrangian of the basic field by the covariant ones (this operation is called the *minimal substitution*), leading to the addition of the interaction Lagrangian to the old one;

4) Finally, the last step concerns the determination of the gauge field Lagrangian which, added to the Lagrangian of the basic field (with the minimal substitution already performed), furnishes

---

[2]In the case of electromagnetism, the transformations that leave the action unchanged form a gauge group known as the unitary group $U(1)$.

[3]The name of *gauge* theories, fields, etc. comes as a result of extension of the *phase transformations* name of the $U(1)$ group, which are specific to this field and are also called "gauge transformations".

the total Lagrangian of the system which, this time, is gauge invariant.

The necessity of being acquainted with the basic notions of the group theory emerges from the fact that, in order to write the covariant derivative of the basic field, it is necessary to know the generators of the gauge symmetry group. Indeed, as we shall see, the covariant derivative writes

$$\nabla_\mu u_i = \partial_\mu u_i - T^k_{ij} u_j A^k_\mu,$$

where

— $\nabla_\mu u_i$ is the covariant derivative of the basic field function, $u_i$;
— $\partial_\mu u_i$ is the usual partial derivative with respect to $x^\mu$ of $u_i$;
— $T^k_{ij} \equiv (T_{ij})^k$ are the $k$ generators of the gauge group, with $k$ — the dimension of the symmetry group of the gauge field;
— $A^k_\mu \equiv (A_\mu)^k$ are the gauge fields (with the same value for $k$).

Symbolically (using the operators), the minimal substitution can be written as

$$\partial_\mu \to \nabla_\mu = \partial_\mu - T_k A^k_\mu,$$

where the significance of $T_k$ and $A^k_\mu$ was mentioned above.

This whole mechanism is called the *minimal coupling principle* being used under this form – in the above considered example – in determination of the interaction Lagrangian when the gauge field is the electromagnetic field, $A_\mu$ (here the index $k$ was omitted because it takes only one value, since the symmetry transformation group for this field – the $U(1)$ group – has only one generator).

In addition, depending on the particularities of the analyzed physical system, the minimal coupling principle can signify more than the minimal substitution (as in the example above). Such a case is – for example – the linear quantum theory of gravitation. Here the interaction Lagrangian is also determined by means of the minimal coupling principle, with observation that in this particular case, in addition to the minimal substitution, one must perform the following two replacements:

1) The Minkowski metric is replaced by the fundamental metric tensor $g_{\mu\nu}$;

2) The four-volume element $d^4x$ is replaced by the invariant quantity $\sqrt{-g}\, d^4x$.

Therefore, in quantum gravitation, by minimal coupling principle one must understand all these three "substitutions".

In view of the above discussion it then follows that the interaction between particles, from the point of view of gauge theories, does not perform instantly, at the distance, by their direct/instantaneous reciprocal influence (as in the classical, Newtonian theory), but through the medium of the gauge field particles (the quanta of the gauge field). In the above considered example, the interaction itself (or, classically speaking, the *collision*) takes place between quanta/particles of the gauge field and the quanta/particles of the basic field. It is said that the interaction process is performed through a *particle exchange*, the quanta of the mediation field being called *exchange quanta* (see Fig. 4.1). In fact, any exchange between the two interacting particles (two electrons, or two positrons, or an electron and a positron, etc.), that can be an exchange of momentum, energy, angular momentum, etc., is accomplished by means of the gauge field quanta (here, the photons $\gamma$).

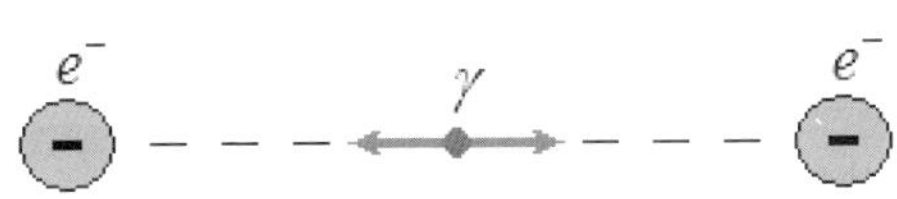


Fig. 4.1

In reality, the things are a little more complicated, since the gauge field itself can play the role of the basic field[4] (matter field). One can show that, in this case, the electromagnetic field quanta that mediate the interactions between fermions are the so-called *virtual photons* (the "longitudinal" and "time-like" photons, corresponding to the non-physical degrees of freedom of the electromagnetic field), while the quanta of the real (measurable) electromagnetic field are the *real photons* (the transversal photons, corresponding to the two physical degrees of freedom of the electromagnetic field).

[4]Such as, for example, the case of the electromagnetic field.

**Observation**. All the mediating (gauge) fields are bosonic fields, therefore their quanta (the particles that mediate the interactions) are *bosons*. The fields that effectively participate to the interaction processes (the so-called *matter fields*) can be both bosonic and fermionic fields[5].

*

* *

### Symmetry classification

The symmetry properties of the particles/fields are of two types:

1. *Space-time* (or *external*) which are those symmetries whose transformations directly affect only the coordinates of the space on which the field is given,

$$x \to x' \neq x, \quad u_{(r)} \to u'_{(r)} = u_{(r)}\,;$$

2. *Internal*, *i.e.* those symmetries that affects only the field (as

[5]Here the "beauty resulting from the simplicity" of a theory has not been considered, and the bosonic fields have also been included in the "matter fields". A "nicer" image is that which calls *matter* everything described only by fermionic fields, while the interaction between the matter fields are mediated by the (always bosonic) gauge fields. Since nowadays is known that the light (photons) can be scattered by photons, the inclusion of the bosonic fields in the category of *matter* fields was preferred. Moreover, as it is known, in the framework of the non-Abelian gauge theories necessarily appears the gauge fields self-interaction which, from the phenomenological point of view, also belong to the "material" domain. More than that, the experimental discovery of the Higgs boson at CERN, Geneva (announced for the first time on July 12, 2012) followed by the Nobel Prize in Physics, 2013, awarded jointly to François Englert and Peter W. Higgs who proposed the so-called "Higgs mechanism" of generation of the particles mass in Microcosmos bring a new complication to the correspondence "material field – fermionic quanta" and "gauge field – bosonic quanta" since, according to nowadays understanding of the Higgs boson, the corresponding field (called *Higgs field*) is omnipresent – it fills the entire Universe. This property could entitle us to consider it as a "matter field", even if its quanta are bosons. But the origin of the Higgs field remains an enigma for the contemporary physics. As the former director general of CERN, Rolf-Dieter Heuer, stated in a 2011 talk on the Higgs field, "All the matter particles are spin-1/2 fermions. All the force carriers are spin-1 bosons. Higgs particles are spin-0 bosons (scalars). The Higgs is neither matter nor force. The Higgs is just different. This would be the first fundamental scalar ever discovered. The Higgs field is thought to fill the entire universe. Could it give some handle of dark energy (scalar field)? Many modern theories predict other scalar particles like the Higgs. Why, after all, should the Higgs be the only one of its kind? [The] LHC can search for and study new scalars with precision."

an operatorial function) and has no effect upon coordinates,

$$x \to x' = x, \quad u_{(r)} \to u'_{(r)} \neq u_{(r)}.$$

In its turn, each symmetry can be either *global* or *local*, so that one can imagine the following classification scheme:

- Symmetries
  - external
    - global
    - local
  - internal
    - global
    - local

The *global* symmetry transformations are those transformations whose parameters do not change in space and time (do not vary from one "point" of space-time to another). The symmetry transformations are called *local* if their parameters depend on the space-time coordinates (vary continuously from one "point" of space-time to another).

It is worth noting that any gauge field appears (can be introduced) when passing from the global to the local symmetries. These two notions are so much connected, that the term *local* involves *gauge*, and conversely. Nowadays is deeply established the idea that any theory pretending to describe an interaction process in QFT must by all means be invariant with respect to the *local* symmetry transformations of the local symmetry group corresponding to the studied physical system.

The demand for local (gauge) invariance is stronger than that for the global invariance. At the first sight, the global symmetry requirement seems to be more general than the local symmetry, but in fact the situation is quite the opposite.[6] The condition for local invariance

[6]In the usual language, the term *global* implies a higher generality level than the term *local*, which is frequently understood as something limited (restricted, particular).

is more general than that for global invariance, since in the first case the form of the symmetry transformations differ from one "point" to another, while in the second the symmetry transformations have the same form everywhere (at all "points" of the space-time), this last case being particular as against the first, namely when the parameters of the symmetry transformations are the same in every "point" of the space-time.[7]

One can ask the following legitimate question: why is it necessary to introduce the local invariance instead of/in addition to the global one? Intuitively, this fact can be understood by appealing to the General Theory of Relativity (GTR). Thus, in the framework of this theory, due to the space-time curvature, it is necessary to define the covariant derivative of the co- and contra-variant vectors and tensors. Here we have to do with a somewhat similar situation: in principle, the space-time remains curved, even if the considered process is analyzed on a small scale. This suggests the fact that, in the most general case, the parameters of the symmetry transformation cannot be considered as being constant (the same in all "points" of the space-time), but dependent on coordinates. Since the space-time is curved and these parameters are imagined as some vectors attached to the "points" of the space-time, then at least their direction and sense must vary. This way, appears as natural the necessity of considering the local invariance (the parameters of the symmetry transformations are functions of coordinates) as being more general as compared to the global transformations (the parameters of the symmetry transformations are the same everywhere in space-time).

The necessity of studying the continuous group theory in QFT also emerges from the fundamental requirement that the variational principle (which stays at the basis of construction of any theory) be invariant with respect to the symmetry transformations of certain group. For example, in the case of a free field, this requirement writes

$$\delta \int_{(\Omega)} dx\ \mathcal{L}(u_i, \partial_\mu u_i) = 0,$$

[7]Denoting by $\varepsilon$ these parameters, then in the case of local symmetry one can write $\varepsilon = \varepsilon(x)$, while in the case of global symmetry $\varepsilon$ does not depend on $x$.

where $\Omega$ is an arbitrary four-domain, where the field is considered. The invariance condition of the variational principle (which is here equivalent to that of the Lagrangian of the considered problem) under the internal symmetry transformations

$$u_i(x) \to u_i'(x) = u_i(x) + \delta u_i(x)$$

is

$$\frac{\partial \mathcal{L}}{\partial u_i}\delta u_i + \frac{\partial \mathcal{L}}{\partial(\partial_\mu u_i)}\delta(\partial_\mu u_i) = 0,$$

where, for a global symmetry, the variation $\delta u_i(x)$ of the field function is given by

$$\delta u_i(x) = T_{ij}^k \varepsilon_k u_j(x).$$

Therefore, again appears the necessity of knowing the infinitesimal generators of the symmetry group of the studied system, $(T^k)_{ij}$. Under these circumstances, the next subchapter summarizes the main notions and results of group theory and representation theory of groups.

## IV.2. Notions and results in group theory and group representations

### IV.2.1. Definitions and fundamental theorems

The primary notion is that of a *set*. Let $M$ be a nonempty set, $M \neq \emptyset$.

**Definition**: We call *internal composition law* or *internal algebraic operation* on the set $M$ any application

$$f : M \times M \to M.$$

Let $f$ be an internal composition law on $M \neq \emptyset$.

♣ **The laws of composition properties**

— The law $f$ is *associative* if

$$\forall\ x, y, z \in M, \quad f\left(f(x,y), z\right) = f\left(x, f(y,z)\right);$$

— The law $f$ is *commutative* if

$$\forall\ x, y \in M, \quad f(x,y) = f(y,x).$$

Let $g$ be another internal law of composition on $M \neq \emptyset$.

**Definition**: It is said that the law $f$ is *right-distributive*, respectively *left-distributive* with respect to the law $g$ if

$$\forall\, x, y, z \in M, \quad f(g(x, y), z) = g(f(x, z), f(y, z)),$$

respectively

$$\forall\, x, y, z \in M, \quad f(z, g(x, y)) = g(f(z, x), f(z, y)).$$

The *bilateral distributivity* or, simply, the *distributivity* is equivalent to simultaneous distributivity in both senses.

If

$$\forall\, x, y \in M, \quad f(g(x, y), x) = x$$

and

$$g(f(x, y), x) = x,$$

then the property of *absorption* takes place.

**Definition**: An element $x \in M$ is called *idempotent* if

$$f(x, x) = x.$$

**Definition**: An element $e \in M$ is called *neutral element* for the law $f$ if

$$\forall\, x \in M, \quad f(x, e) = f(e, x) = x.$$

**Definition**: If the law of composition is additive, then the neutral element is called *null element*; if the law of composition is multiplicative, the neutral element is called *unit element*.

**Theorem**: If an internal law of composition admits a neutral element, then this element is *unique*.

**Definition**: It is said that an element $x' \in M$ is *symmetric* with respect to $x \in M$ if

$$f(x, x') = f(x', x) = e.$$

**Definition**: In this case, it is said that the element $x$ is *symmetrizable*. If the law of composition is additive, then the symmetric element is called *opposite element*; if the law of composition is multiplicative, the symmetric element is called *inverse element*.

**Theorem**: Let $f$ be an internal law of composition, associative, with neutral element, defined on $M \neq \emptyset$. If $x \in M$ has a symmetric element, then it is *unique*.

**Definition**: It is called *regular element* (or *simplifiable*) with respect to the law $f$ an element $a \in M$, so that

$$\forall\, x, y \in M, \quad f(a, x) = f(a, y) \text{ and } f(x, a) = f(y, a) \Leftrightarrow x = y.$$

**Theorem**: Any symmetrizable element is a regular element for an internal law of composition defined on $M \neq \emptyset$, which is associative and has a neutral element.

Consider two sets $M \neq \emptyset$ and $\Omega \neq \emptyset$.

**Definition**: An application

$$\varphi : \Omega \times M \to M, \quad \text{or} \quad \varphi : M \times \Omega \to M$$

is called *external composition law* on the set $M$, with the operators domain $\Omega$.

**Definition**: The element $\varphi(\alpha, x) \in M$ with $\alpha \in \Omega$ and $x \in M$ is called *composite of $\alpha$ with $x$ with respect to $\varphi$*.

### ♣ Algebraic structures

**Definition**: A set $M \neq \emptyset$ endowed with a single internal composition law is called *groupoid*.

**Definition**: The groupoid having an associative internal composition law is called *semigroup*.

**Definition**: A semigroup with an identity element is called *monoid*.

**Definition**: If the internal composition law is commutative, the algebraic structures defined with respect to it are called *Abelian*.

**Definition**: A *group* is a set $G \notin \emptyset$ provided with an internal composition law that verifies the following axioms:

$G_1$: it is *associative*;
$G_2$: it has a *neutral element*;
$G_3$: any element of $G$ is *symmetrizable* with respect to the law.

**Definition**: In other words, a group is any semigroup with the property that, if $\forall\, a, b \in G$, then the equations $f(a, x) = b$ and $f(y, a) = b$ have unique solutions in $G$.

**Definition**: The group is called *finite* if the set $G$ is finite.

**Definition**: The *rank* of a group is the cardinal of the set $G$, denoted *card* $G$. If the set $G$ is finite, then the group rank coincides with the number of elements of the group.

**Definition**: A *transformation* of the set $M$ is a biunivocal application of the set $M$ on itself. It can be shown that the set of transformations $T$ form a group.

Consider two groups $(F, f)$ and $(G, g)$.

**Definition**: An application $h : F \to G$ is called *homomorphism* or *group morphism* if

$$\forall\ x, y \in F, \quad h(f(x, y)) = g(h(a), h(y)).$$

**Definition**: A morphism is *injective* or *monomorphism* if $h$ is an injective or surjective application, respectively.

**Definition**: A *bijective morphism* (a *bimorphism*) is a morphism simultaneously injective and surjective.

**Definition**: If $G \equiv F$, then the morphism is called *endomorphism*, being denoted by $End\,(G)$.

**Theorem**: If $h$ is a groups homomorphism, then

$$h(e) = e',$$

where $e$ and $e'$ are the unit elements of $F$ and $G$, respectively.

**Theorem**: The image of the symmetric of an element through a group morphism is the symmetric of the image of that element.

**Theorem**: A group morphism $f : F \to G$ is a *group isomorphism*, if and only if it is bijective.

**Definition**: If $G \equiv F$, the isomorphism is called *automorphism*.

If $f : G \to G'$ and $g : G' \to G''$ are two morphisms, then $h(g, f)$ is also a group homomorphism. Using the multiplicative notations, one can write

$$(g \circ f)(xy) = (g \circ f)(x)(g \circ f)(y), \quad \forall\ x, y \in G.$$

**Definition**: In a group $G$ an element $g \in G$ is called *conjugated* to the element $h \in R$ if $\exists\ x \in G$ so that $xgx^{-1} = h$.

**Definition**: The application $f : G \to G$, through $f(g) = xgx^{-1}$, $\forall\, g \in G$ is called *inner automorphism.* Therefore, the conjugate elements are defined by the inner automorphisms of $G$.

**Definition**: The set of elements conjugated to a given element form a *class.* In an Abelian group, the number of classes equals the number of elements of the group (a class contains only one element), since $xgx^{-1} = gxx^{-1} = g$.

**Definition**: A set $H \subset G$, $H \neq \emptyset$ is called *subgroup* of the group $G$ if the composition law in $G$ induces on $H$ a composition law, together with which $H$ forms a group structure.

**Definition**: The subsets $G$ and $\{e\}$ of a group $G$ form the *improper subgroups* of this group. All the other subgroups are called *proper subgroups.*

Let $f : F \to G$ be a morphism. Then,

**Definition**: $f(F)$ is a subgroup of $G$, called the *image through $f$ of $F$*, (denoted by $Im\ f$);

**Definition**: If $H$ is a subgroup of $G$, then $f^{-1}(H)$ is a subgroup of $F$. In particular, $f^{-1}(e)$, where $e \in G$, is called the *nucleus of $f$*, (denoted by $Ker\ f$).

Let $G$ be a group and $X \subset G$, $X \neq \emptyset$. The subgroup of $G$ generated by $X$ (called *system of generators* of this subgroup) is represented by the intersection of all subgroups of $G$ that contain $X$.

**Definition**: A *cyclic* (*monogeneous*) *group* is a group that is generated by a single element.

**Definition**: A *finite group* (or *finite generated group*) is a group that is generated by a finite set of elements of $G$.

**Definition**: An *invariant subgroup* (*normal divisor* or *distinguished subgroup*) $H$ of a group $G$ is a subgroup with the property that if $\forall\, x \in G$ and $\forall\, h \in H$, then $xhx^{-1} \in H$.

Therefore, a group is invariant if, together with any of its own elements, it also contains all the conjugated elements (that is its transformations by any interior automorphism of $G$). In any group, the improper subgroups are invariant. Any subgroup of an Abelian group is invariant (the only inner automorphism being the identical one).

**Definition**: The *center of a group* $G$, denoted by $\mathcal{C}(G)$, is the set of all elements $x \in G$ with the property $\forall\ y \in G,\ xy = yx$.

The center of a group, $\mathcal{C}(G)$, is an invariant subgroup of $G$. The group of internal automorphisms of $G$ is an invariant subgroup in the group of all automorphisms of $G$.

**Theorem**: If $f : F \to G$ is a morphism and $H$ is an invariant subgroup in $G$, then $f^{-1}(H)$ is an invariant subgroup in $F$.

In particular, *Ker* $f$ is an invariant subgroup in $F$.

**Definition**: A *binary relation* between two sets $A$ and $B$, distinct or coincident, is a subset $R \subset A \times B$.

The elements associated through this relation are those elements $a \in A,\ b \in B$ for which $(a, b) \in R$. The notation is $bRa$.

**Definition**: The *equivalence relation* is a binary relation that satisfies the following properties:

1) it is *reflexive*: $\forall\ a \in A,\ aRa$;
2) it is *symmetric*: $\forall\ a \in A,\ \forall\ b \in B,\ bRa \Leftrightarrow aRb$;
3) it is *transitive*: $\forall\ a \in A,\ \forall\ b \in B, \forall\ c \in C,\ bRa$ and $cRb \Rightarrow cRa$.

**Definition**: A binary relation on a set where no element is related to itself (that is no $aRa,\ a \in A$) is called *antireflexive* or *irreflexive*. A relation is called *antisymmetric* if from $aRb$ and $bRa \Rightarrow a = b,\ a \in A,\ b \in B$.

**Definition**: By *partition* of a set $M$ one understands a class $C$ of nonempty subsets of $M$, so that,

1) the subsets of $C$ are disjoint two by two;
2) the reunion of subsets of $C$ is $M$.

Consider the set $M$, an equivalence relation $R$ on $M$, and an element $x \in M$.

**Definition**: By an *equivalence class* $C_x$ one understands the set of elements $y \in M$, so that $yRx$, that is

$$C_x = \{y \,|\, yRx;\ x, y \in M\}.$$

**Theorem**: Given an equivalence relation $R$ in $M$, the set of all equivalence classes forms a partition of $M$; reciprocally, given a partition in $M$, the relation defined by $yRx$, if $x$ and $y$ belong to the same subset, is an equivalence relation.

**Definition**: Given $M$ and $R$ on $M$, the set of equivalence classes is called *factor set* (*ratio set*) of $M$ with respect to $R$ (denoted by $M/R$).

**Theorem**: Let $R$ be an equivalence relation defined on the multiplicative group $G$. If the equivalence relation is left (right) compatible with the group law, then there exists a subgroup $H$ in $G$, so that $yRx \Leftrightarrow y^{-1}x \in H$ (respectively $yRx \Leftrightarrow xy^{-1} \in H$). Reciprocally, if $H$ is a subgroup, then the relation $xRy \Leftrightarrow x^{-1}y \in H$ (respectively $xRy \Leftrightarrow yx^{-1} \in H$) is an equivalence relation on $G$, left (right) compatible with the group law.

**Definition**: The cardinal of the set of equivalence classes on the left (right) is called *left* (*right*) *index* of the subgroup $H$ of $G$.

**Theorem**: Let $G$ be a group and $H \subset G$ one of its subgroups. Then the left and right indices of $H$ are equal and are called the *index of $H$ in $G$*.

**Theorem**: An equivalence relation $R$ defined on $G$ is compatible with the group structure of $G$ if and only if it is determined by an invariant subgroup of $G$.

**Theorem**: If $H$ (of rank $n$) $\subset G$ (of rank $N$), then $N = ni$, where $i$ is the index of $H$ in $G$.

**Definition**: A *simple group* is a group that has as normal divisors (invariant subgroups) only $G$ and $\{e\}$ (in other words, a group is called simple if it does not contain proper invariant subgroups). A finite group of odd rank is always simple.

**Definition**: A *semisimple group* is a group that does not contain Abelian proper invariant subgroups.

## ♣ Notions of representation theory of groups

In general, working with the objects of a group (especially when these elements are not usual mathematical objects) is quite difficult to perform and, on that account, to each object of a certain group

$G$ one associates by means of a certain law (application), $T$, another object, in the idea that working with this new object is simpler

$$g \in G \xrightarrow{\mathrm{T}} T(g) \in \mathcal{L},$$

where $\mathcal{L}$ is called the *representation space*. Practically, the new "object" associated with the element $g \in G$ is a linear operator and, since this is a pure mathematical object, it is easier and more useful to work with it.

In general, the group representation theory studies the morphisms of some arbitrary group $G$ on all possible groups of linear operators.

**Definition**: Let $G$ be a group and $\mathcal{L}$ a linear space. A *representation of the group* $G$ in/on the linear space $\mathcal{L}$ is any application $T$ with the property that to any element $g \in G$ corresponds a linear operator $T(g)$ of $\mathcal{L}$, so that

$R_1 : T(e) = 1$, where 1 is the identity operator in $\mathcal{L}$;

$R_2 : T(g_1 g_2) = T(g_1)T(g_2), \quad \forall\, g_1, g_2 \in G.$

**Definition**: Under the above conditions, $\mathcal{L}$ is called the *representation space*, and $T(g)$ are called the *representation operators*.

In fact, $T(g)$ is an application (more exactly, a bijection) of $\mathcal{L}$ on $\mathcal{L}$, because

$$T(g^{-1})T(g) = T(g^{-1}g) = T(e) = 1,$$

$$T(g)T(g^{-1}) = T(gg^{-1}) = T(e) = 1,$$

$$T(g^{-1}) = (T(g))^{-1}.$$

Let $G_{\mathcal{L}}$ denote the group of the linear operators defined on $\mathcal{L}$, which bijectively applies $\mathcal{L}$ on $\mathcal{L}$.

**Definition**: If $G$ is isomorphic to $G_{\mathcal{L}}$, then the representation is called *exact*, or *punctual*, or, still, *one to one* (in the sense that to each element $g$ of the group $G$ one associates a linear operator $T$ of the group $G_{\mathcal{L}}$ – and only one – and conversely).

- **Particular representations**

**1)** Suppose that to each element $g$ of the group $G$ one associates the same element, $g$. Obviously, in this case, $G$ is a representation of itself, therefore $G_{\mathcal{L}}$ coincides with $G$. This representation is called *identical* and it is always existent, being frequently of a great importance. Thus, one can give the following

**Definition**: The correspondence $g \to T(g) = g$ furnishes the so-called *identical representation* of $G$.

**2)** Let $G$ be a group and $\mathcal{L}$ a linear space of some dimension. Let associate to each element $g \in G$ the unit (identity) operator of $\mathcal{L}$, that is $g \to T(g) = 1 \in \mathcal{L}$. The obtained representation (that always exists) is called a *unit representation* of the group $G$ in $\mathcal{L}$. This leads to the following

**Definition**: If to each element $g \in G$ one associates the unit operator of the linear space $\mathcal{L}$, it is said that the *unit representation* has been defined.

**Definition**: The dimension of the space $\mathcal{L}$ is called the *representation dimension* (denoted by dim $\mathcal{L}$).

Consider $\dim \mathcal{L} = n$ and $\{\vec{e}_i\}_{i=\overline{1,n}}$ a basis in $\mathcal{L}$. Then the matrix representation $T(g)$ can be written as

$$T(g) \to \vec{t}(g) = \begin{pmatrix} t_{11}(g) & \cdots & t_{1n}(g) \\ \vdots & \ddots & \vdots \\ t_{n1}(g) & \cdots & t_{nn}(g) \end{pmatrix},$$

and

$$T(g)\vec{e}_k = \sum_{j=1}^{n} t_{jk}(g)\vec{e}_j, \quad (k = \overline{1,n}).$$

Therefore, a representation of finite dimension can be considered as a matrix function, $g \to \vec{t}(g)$ with the conditions $\vec{t}(e) = 1$ and $\vec{t}(g_1 g_2) = \vec{t}(g_1)\vec{t}(g_2)$, where

$$t_{jk}(e) = \begin{cases} 1, & j = k, \\ 0, & j \neq k \end{cases} \quad \text{and} \quad t_{jk}(g_1 g_2) = \sum_{s=1}^{n} t_{js}(g_1) t_{sk}(g_2).$$

In particular, a unidimensional representation can be assimilated with a numerical function $g \to t(g)$ with the same conditions.

For a closer investigation, consider the same linear space $\mathcal{L}$ as above, with dimension $\dim \mathcal{L} = n$. Then in $\mathcal{L}$ there exist $n$ linearly independent elements, $\vec{e}_k$ $(k = \overline{1,n})$, through which all the other elements can be linearly expressed. It is said that the system $\{\vec{e}_k\}_{k=\overline{1,n}}$ form a *basis* in the linear space $\mathcal{L}$, and then

$$\forall\, \vec{x} \in \mathcal{L}, \quad \text{one has} \quad \vec{x} = \sum_{k=1}^{n} x_k \vec{e}_k,$$

where $x_k$, $(k = \overline{1,n})$ are called the *components* of the vector $\vec{x}$ in the basis $\{\vec{e}_k\}_{k=\overline{1,n}}$.

Consider now another linear space $\mathcal{L}'$ of dimensions $\dim \mathcal{L}' = m$ and $T$ an operator from $\mathcal{L}$ to $\mathcal{L}'$:

$$T : \mathcal{L} \to \mathcal{L}', \quad \vec{x} \in \mathcal{L} \to \vec{x}\,' = T(\vec{x}) \in \mathcal{L}',$$

with

$$T(\vec{x} + \vec{y}) = T(\vec{x}) + T(\vec{y});$$
$$T(\alpha\vec{x}) = \alpha T(\vec{x}).$$

Consider now two bases $\{\vec{e}_i\}_{i=\overline{1,n}}$ and $\{\vec{e}\,'_k\}_{k=\overline{1,m}}$ in $\mathcal{L}$ and $\mathcal{L}'$ respectively. Then

$$\vec{x}\,' = \sum_{k=1}^{m} x'_k \vec{e}\,'_k = T(\vec{x}) = T\left(\sum_{i=1}^{n} x_i \vec{e}_i\right) = \sum_{i=1}^{n} x_i T(\vec{e}_i)$$
$$= \sum_{i=1}^{n} x_i \sum_{k=1}^{m} T_{ki} \vec{e}\,'_k,$$

that is

$$x'_k = \sum_{i=1}^{n} T_{ki} x_i.$$

Therefore, to the operator $T$ one associates the matrix $T_{ki}$, which is an $(m \times n)$ matrix. If $\mathcal{L}'$ coincides with $\mathcal{L}$, then the matrix associated to the operator $T$ is a square matrix, $(n \times n)$. We are now talking about a *matrix representation*. In general, one asks that the linear space on which a representation of a group $G$ is performed, be a *Hilbert space*.

**Definition**: A prehilbertian space which is complete as normed linear space is called *Hilbert space*.

**Definition**: A linear space on which is defined a scalar product is called *prehilbertian space*.

**Definition**: A *scalar product* on a linear space $\mathcal{L}$ is an operation that associates to any pair $(\vec{x}, \vec{y}) \in \mathcal{L} \times \mathcal{L}$ a complex number

$$\vec{x}, \vec{y} \in \mathcal{L} \to (\vec{x}, \vec{y}) \in \mathbb{C},$$

with the following properties:

$$SP_1 : (\alpha\vec{x} + \beta\vec{y}, \vec{z}) = \alpha(\vec{x}, \vec{z}) + \beta(\vec{y}, \vec{z}),\ \forall\ \alpha, \beta \in \mathbb{C},\ \forall\ \vec{x}, \vec{y}, \vec{z} \in \mathcal{L};$$

$$SP_2 : (\vec{x}, \vec{y}) = (\vec{y}, \vec{x})^* \equiv \overline{(\vec{y}, \vec{x})},\ \forall\ \vec{x}, \vec{y} \in \mathcal{L};$$

$$SP_3 : (\vec{x}, \lambda\vec{y}) = (\overline{\lambda}\vec{x}, \vec{y}) = (\lambda^*\vec{x}, \vec{y}) = \lambda^*(\vec{x}, \vec{y}),\ \forall\ \lambda \in \mathbb{C},\ \forall\ \vec{x}, \vec{y} \in \mathcal{L};$$

$$SP_4 : (\vec{x}, \vec{x}) \geq 0;\ (\vec{x}, \vec{x}) = 0 \text{ if and only if } \vec{x} = 0,\ \forall\ \vec{x} \in \mathcal{L}.$$

**Example**. Consider the linear space $\mathcal{L}$ of the dimension $\dim \mathcal{L} = n$ and $\{\vec{e}\}_{i=\overline{1,n}}$ a basis in $\mathcal{L}$. If $\vec{x} = \sum_{i=1}^{n} x_i\vec{e}_i$ and $\vec{y} = \sum_{j=1}^{n} y_j\vec{e}_j$, then the scalar product of $\vec{x}$ and $\vec{y}$ can be defined as

$$\vec{x}, \vec{y} \in \mathcal{L} \to (\vec{x}, \vec{y}) =^{\text{def}} \sum_{i=1}^{n}\sum_{j=1}^{n} x_i\overline{y}_j\vec{e}_i\overline{\vec{e}}_j = \sum_{i=1}^{n}\sum_{j=1}^{n} x_i\overline{y}_j\delta_{ij}$$

$$= \sum_{i=1}^{n} x_i\overline{y}_j \overset{\text{not}}{-} x\overline{y}.$$

**Definition**: A *complete linear space* is a linear space in which any Cauchy sequence converges to an element of the space.

**Definition**: A sequence $\{x_n\}_{n\in\mathbb{N}}$ of a metric space $(X, d)$ is called *Cauchy sequence* if $\forall\ \varepsilon > 0, \exists\ n_\varepsilon \in \mathbb{N}$, so that $d(x_n, x_m) < \varepsilon$, as soon as $n, m \geq n_\varepsilon$.

**Definition**: A *metric space* is any pair $(X, d)$, where $X$ is a nonempty set and $d$ a distance on $X$.

**Definition**: A *distance* on the nonempty set $X$ is any application $d$, defined on $X \times X$, with real positive values and satisfying the following axioms:

$D_1$: $d(x, y) = 0$ if and only if $x = y$;
$D_2$: $d(x, y) = d(y, x), \ \forall \ x, y \in X$;
$D_3$: $d(x, y) \geq d(x, z) + d(z, y), \ \forall \ x, y, z \in X$.

**Definition**: A *normed space* is a linear space on which a norm $p$ was given.

**Definition**: A *norm* is any seminorm that satisfies the axiom

$$p(x) = 0 \Rightarrow x = 0.$$

**Definition**: A *seminorm* on the linear space $\mathcal{L}$ is any real function $p : \mathcal{L} \to \mathbb{R}$ that satisfies the following axioms:

$SN_1$: $p(\vec{x} + \vec{y}) \leq p(\vec{x}) + p(\vec{y}), \ \forall \ \vec{x}, \vec{y} \in \mathcal{L}$;
$SN_2$: $p(\lambda \vec{x}) = |\lambda| p(\vec{x}), \ \forall \ \vec{x} \in \mathcal{L}$ and $\lambda$ a non-null scalar.

The most frequently used notation is $p(\vec{x}) = \|\vec{x}\|$, where by $\|\vec{x}\|$ one understands the square root of the scalar product of $\vec{x}$ by itself, $\|\vec{x}\| \stackrel{def}{=} \sqrt{(\vec{x}, \vec{x})}$.

**Definition**: An operator $T : \mathcal{L} \to \mathcal{L}$ is called *bounded* if $\exists \ c > 0$ and finite, so that $\forall \ \vec{x} \neq 0, \ \dfrac{\|T\vec{x}\|}{\|\vec{x}\|} \leq c$.

**Definition**: For any bounded operator, $T$, one can define its *adjoint* $T^+$, by

$$(T\vec{x}, \vec{y}) = (\vec{x}, T^+\vec{y}), \ \forall \ \vec{x}, \vec{y} \in \mathcal{L}.$$

**Definition**: If $T^+ = T$, the operator $T$ is called *Hermitian* or *self-adjoint*.

**Definition**: If $T^+ = T^{-1}$, the operator $T$ is called *unitary*.

**Definition**: If all the operators $T(g)$ of some representation are unitary, then the representation is called *unitary*.

Let $G$ be a group and $\mathcal{L}$ of dimension $\dim \mathcal{L} = n$, with the basis $\{\vec{e}\}_{i=\overline{1,n}}$, a linear space of some representation of the group $G$, the representation operators being $T$. In the given basis, the operator $T$ has the matrix form given by $T_{ki}$:

$$\forall\ \vec{x} \in \mathcal{L},\ \vec{x} = \sum_{i=1}^{n} \vec{e}_i x_i;$$

$$\begin{aligned}\vec{x}\,' &= \sum_{i=1}^{n} \vec{e}_i{}' x_i' = T(\vec{x}) = T\left(\sum_{i=1}^{n} \vec{e}_i x_i\right)\\ &= \sum_{i=1}^{n} x_i T(\vec{e}_i) = \sum_{i=1}^{n}\sum_{k=1}^{n} x_i T_{ki} \vec{e}_k{}' = \sum_{k=1}^{n}\left(\sum_{i=1}^{n} x_i T_{ki}\right)\vec{e}_k{}',\end{aligned}$$

so that

$$x_k' = \sum_{i=1}^{n} T_{ki} x_i \overset{\text{not}}{=} T_{ki} x_i.$$

Under a change of basis in $\mathcal{L}$, $\vec{e}_i{}' = \vec{e}_k M_{ki}$, $(i, k = \overline{1, n})$, a new matrix shall correspond to the operator $T$, connected to the matrix of the old basis, $T_{ki}$, through a *similarity transformation*,

$$T_{ki}' = \left(M^{-1} T M\right)_{ki}.$$

Indeed,

$$\vec{e}_i{}' = \vec{e}_k M_{ki}, \quad \vec{e}_i{}'(M^{-1})_{ik} = \vec{e}_k M_{ki} (M^{-1})_{ik} = \vec{e}_k,$$

so that

$$\vec{e}_i = \vec{e}_k{}'(M^{-1})_{ki},$$

and therefore

$$\begin{aligned}T\vec{e}_i{}' &= T\vec{e}_k M_{ki} = \vec{e}_l T_{lk} M_{ki} = \vec{e}_j{}'(M^{-1})_{jl} T_{lk} M_{ki}\\ &= \vec{e}_j{}'(M^{-1}TM)_{ji},\end{aligned}$$

that is

$$T\vec{e}_i{}' = \vec{e}_k{}'(M^{-1}TM)_{ki},$$

hence, in the new basis $\{\vec{e}_i{}'\}_{i=\overline{1,n}}$, the matrix $(M^{-1}TM)_{ki}$ corresponds to the operator $T$, so that one can write

$$T_{ki}' = (M^{-1}TM)_{ki}, \quad \text{(q.e.d.)}.$$

**Definition**: Two *representations*, $T$ and $M^{-1}TM$, connected by a similarity transformation, are called *equivalent*.

The general case. Let $T : \mathcal{L} \to \mathcal{L}$ and $S : \mathcal{M} \to \mathcal{M}$ be two representations of a group $G$ on the linear spaces $\mathcal{L}$ and $\mathcal{M}$ respectively.

**Definition**: If there exists a linear operator $A$ that bijectively applies $\mathcal{L}$ into $\mathcal{M}$, so as

$$AT(g) = S(g)A, \quad \forall\, g \in G,$$

then the two representations are called *equivalent*.

**Definition**: A subspace $\ell \subset \mathcal{L}$ is called *invariant* with respect to the representation $g \to T(g)$ if it is invariant with respect to all the operators $T(g)$ of this representation (*i.e.* if the vectors of the subspace $\ell$ transform between them under the action of the representation operators, $T(g)$).

**Definition**: A representation on $\mathcal{L}$ is called *irreducible* if $\mathcal{L}$ does not contain subspaces different from $\{0\}$ and $\mathcal{L}$, which are invariant under the action of the representation operators. In the opposite case, the representation is called *reducible*.

An example of reducible representation is that whose dimension is $(n+m)$ and whose matrix form is block-diagonal, that is

$$(T) = \begin{pmatrix} \overbrace{(T_1)}^{n} & \vdots & \overbrace{O}^{m} \\ \cdots & \vdots & \cdots \\ O & \vdots & (T_2) \end{pmatrix} \begin{matrix} \}n \\ \\ \}m \end{matrix}$$

where $(T_1)$ and $(T_2)$ are square matrices of $n$-dimensional and $m$-dimensional representations, of dimension $(n \times n)$ and $(m \times m)$, respectively. In this case, the space of the representation $\mathcal{L}$, of dimension $(n+m)$ can be written as a direct sum of two subspaces of dimensions $n$, respectively $m$, namely $\mathcal{L} = \mathcal{L}_1 \oplus \mathcal{L}_2$, and for the representation one writes $T = T_1 \oplus T_2$, which means (as shall further be seen) a *direct sum of representations*.

Thus, by separation/decomposition of the representation space into invariant subspaces, one practically obtains the irreducible representations of a group, this way the study of a reducible representation consisting in the study/determination of all irreducible representations of that group. This aspect has direct applications in the elementary particles classification and determination of the possible eigenstates (and their degrees of degeneration as well), of the quantum systems, using the representation theory of groups.

**Theorem**: If $T$ and $S$ are two equivalent representations and $T$ is irreducible, then $S$ is also irreducible.

**Schur's lemmas**

**1)** If $T$ and $S$ are two irreducible representations of $G$ in $\mathcal{L}$ and $\mathcal{M}$ respectively, and one supposes that $A$ is an operator from $\mathcal{L}$ to $\mathcal{M}$, that satisfies the relation $AT(g) = S(g)A,\ \forall\ g \in G$, then the operator $A$ applies bijectively $\mathcal{L}$ into $\mathcal{M}$ (therefore $T$ is equivalent to $S$) or $A = 0$.

**2)** Let $T$ be a finite representation of $G$ in $\mathcal{L}$. Then any linear operator $B$ in the space $\mathcal{L}$, which commutes with all the operators $T(g)$, $g \in G$, is of the form $B = \lambda \mathbf{1}$, where $\lambda$ is a number. Therefore, if

$$BT(g) = T(g)B,\ \forall\ g \in G,$$

then $B = \lambda \mathbf{1}$.

One can shown that all the irreducible representations of finite dimension of a commutative group are unidimensional.

**Definition**: Let $\mathcal{L}$ and $\mathcal{M}$ be two linear spaces. A *bilinear form* on the pair $(\mathcal{L},\ \mathcal{M})$ is the numerical function denoted $(\vec{x},\ \vec{y})$, defined on $\mathcal{L} \times \mathcal{M}$, that satisfies the properties:

$BF_1$: $(\alpha\vec{x}, \vec{y}) = \alpha(\vec{x}, \vec{y})$;

$BF_2$: $(\vec{x}, \alpha\vec{y}) = \alpha(\vec{x}, \vec{y})$;

$BF_3$: $(\vec{x}_1 + \vec{x}_2, \vec{y}) = (\vec{x}_1, \vec{y}) + (\vec{x}_2, \vec{y})$;

$BF_4$: $(\vec{x}, \vec{y}_1 + \vec{y}_2) = (\vec{x}, \vec{y}_1) + (\vec{x}, \vec{y}_2)$.

**Definition**: The representation $S$ is called the *adjoint* of the representation $T$ regarding the bilinear form $(\vec{x}, \vec{y})$ if

$$(T(g)\vec{x}, S(g)\vec{y}) = (\vec{x}, \vec{y}),\ \forall\ g \in G,\ \vec{x} \in \mathcal{L},\ \vec{y} \in \mathcal{M},$$

or, equivalently,

$$\left(T(g^{-1})\vec{x},\vec{y}\right) = (\vec{x}, S(g)\vec{y})\,,\ \forall\ g \in G,\ \vec{x} \in \mathcal{L},\ \vec{y} \in \mathcal{M}.$$

• **Direct sum and direct product of representations**

Consider $\mathcal{L} = \mathcal{L}_1 \oplus \mathcal{L}_2 \oplus ... \oplus \mathcal{L}_m$. In this case, $\vec{x} \in \mathcal{L}$ can be uniquely written as a linear combination of the form: $\vec{x} = \vec{x}_1 + \vec{x}_2 + ... + \vec{x}_m$, $\vec{x}_k \in \mathcal{L}_k$, $(k = \overline{1,m})$. Consider, also, on the space $\mathcal{L}_k$, the representation $T^{(k)}$ of $G$, $(k = \overline{1,m})$.

One defines

$$\begin{aligned} T(g)\vec{x} &= T(g)(\vec{x}_1 + \vec{x}_2 + ... + \vec{x}_m) \\ &= T^{(1)}(g)\vec{x}_1 + T^{(2)}(g)\vec{x}_2 + ... + T^{(m)}(g)\vec{x}_m. \end{aligned}$$

Then, one obtains $T(e) = 1$ and $T(g_1g_2) = T(g_1)T(g_2)$, $\forall\ g_1, g_2 \in G$.

**Definition**: In view of the above notions and notations, $g \to T(g)$ is a representation of $G$ on $\mathcal{L}$, called *direct sum representation* of the representations $T^{(1)}, T^{(2)}, ..., T^{(m)}$, being denoted as $T = T^{(1)} \oplus T^{(2)} \oplus ... \oplus T^{(m)}$.

Obviously, each $\mathcal{L}_k$, $(k = \overline{1,m})$ is invariant with respect to $T$, and the restriction of $T$ to $\mathcal{L}_k$ is $T^{(k)}$.

Let $T^{(1)}$ and $T^{(2)}$ be two representations of $G$ in $\mathcal{L}_1$ and $\mathcal{L}_2$, of dimensions $n_1$ and $n_2$ respectively, and also let $\{\vec{e}_i\}_{i=\overline{1,n_1}}$ and $\{\vec{f}_j\}_{j=\overline{1,n_2}}$ be two bases in these spaces. If $\mathcal{L} = \mathcal{L}_1 \otimes \mathcal{L}_2$ has the basis $\{\vec{e}_i \otimes \vec{f}_j\} = \{\vec{h}_{ij}\}$ of the dimension $n_1n_2$, then $\forall\ \vec{z} \in \mathcal{L}$ one has

$$\vec{z} = \sum_{i=1}^{n_1}\sum_{j=1}^{n_2}(\vec{e}_i \otimes \vec{f}_j)z_{ij} = \sum_{i=1}^{n_1}\sum_{j=1}^{n_2}\vec{h}_{ij}z_{ij}.$$

One also has

$$T^{(1)}(g)\vec{e}_i = \vec{e}_i{}' = \sum_{j=1}^{n_1} t_{ji}^{(1)}(g)\vec{e}_j, \quad (i = \overline{1,n_1})$$

and

$$T^{(2)}(g)\vec{f}_i = \vec{f}_i{}' = \sum_{j=1}^{n_2} t_{ji}^{(2)}(g)\vec{f}_j, \quad (i = \overline{1,n_2})$$

Then,

$$T(g)\vec{h}_{ij} = \vec{h}_{ij}{}' = \vec{e}_i{}' \otimes \vec{f}_j{}' = T^{(1)}(g)\vec{e}_i \otimes T^{(2)}(g)\vec{f}_j$$

$$= \sum_{k=1}^{n_1}\sum_{l=1}^{n_2} t_{ki}^{(1)}(g)t_{lj}^{(2)}(g)\vec{e}_k \otimes \vec{f}_l = \sum_{k=1}^{n_1}\sum_{l=1}^{n_2} t_{(kl)(ij)}\vec{h}_{kl},$$

$$z'_{ij} = \sum_{k=1}^{n_1}\sum_{l=1}^{n_2} t_{(ij)(kl)}(g)z_{kl} = \sum_{k=1}^{n_1}\sum_{l=1}^{n_2} t_{ik}^{(1)}(g)t_{jl}^{(2)}(g)z_{kl},$$

and

$$t_{(ij)(kl)}(g) = t_{ik}^{(1)}(g)t_{jl}^{(2)}(g).$$

Also,

$$t_{(ij)(kl)}(gu) = t_{ik}^{(1)}(gu)t_{jl}^{(2)}(gu) = \sum_{s=1}^{n_1} t_{is}^{(1)}(g)t_{sk}^{(1)}(u)\sum_{p=1}^{n_2} t_{jp}^{(2)}(g)t_{pl}^{(2)}(u)$$

$$= \sum_{s=1}^{n_1}\sum_{p=1}^{n_2} t_{(ij)(sp)}(g)t_{(sp)(kl)}(u).$$

**Definition**: Under the above conditions and notations, $g \to T(g)$ is a representation of $G$ on $\mathcal{L}$, called *direct product representation* of the representations $T^{(1)}$ and $T^{(2)}$, being denoted by $T = T^{(1)} \otimes T^{(2)}$. The generalization is immediate:

$$T(g)(\vec{x}_1 \otimes \vec{x}_2 \otimes ... \otimes \vec{x}_m) = T^{(1)}(g)\vec{x}_1 \otimes T^{(2)}(g)\vec{x}_2 \otimes ... \otimes T^{(m)}(g)\vec{x}_m,$$

where $T$ is a representation of $G$ on $\mathcal{L} = \mathcal{L}_1 \otimes \mathcal{L}_2 \otimes ... \otimes \mathcal{L}_m$.

**Definition**: The *character* of a representation of finite dimension $T$ is the trace (spur) of matrix of the operator $T$ of this representation

$$\chi(g) = \mathrm{Tr}\,\vec{t}(g) = t_{11}(g) + t_{22}(g) + ... + t_{nn}(g).$$

**Proprieties of the characters of finite dimension representations**

1) the characters of the equivalent representations coincide;
2) a character is a constant on every class of conjugate elements;
3) if $T$ and $S$ are adjoint, then $\chi_S(g) = \overline{\chi_T(g^{-1})}$;

4) the character of the direct sum of a finite number of representations equals the sum of the characters of these representations;

5) the character of the direct product of a finite number of representations equals the product of the characters of these representations.

### IV.2.2 Continuous groups

The concept of continuous group can be introduced in several ways, and two of them shall be given in this paragraph, for a better understanding of the group theory particularities and details.

One must specify from the beginning that, as shall further be seen, the study of the continuous groups and their representations is indispensable in the framework of the gauge theories. Since in order to introduce the notion of continuous (or topological) group it is absolutely necessary the notion of topological space (that plays here a fundamental role), here is the following.

**Definition**: A *topological space* is any nonempty set, $\mathcal{T}$ with a system of subsets (parts) called *vicinities*, that satisfy the following axioms:

$T_1$: $\forall\, x \in \mathcal{T}$, $x$ belongs to a vicinity denoted by $V(x)$;
$T_2$: any subset that includes $V(x)$ is, also, a vicinity of $x$;
$T_3$: the intersection of a finite number of vicinities of $x$ is also a vicinity of $x$;
$T_4$: a vicinity $V(x)$ of $x$ is also a vicinity of an element $y \in \mathcal{T}$, close enough to $x$.

Denoting by

$$\mathcal{V}(x) = \bigcup_i V_i(x)$$

the set of all vicinities of $x$, called *vicinity system* for $x \in \mathcal{L}$, then the first three axioms[8] can be rewritten as follows:

$T_1$:

$$\forall\, x \in \mathcal{T},\ \exists\, V(x) \in \mathcal{V}(x) \neq \emptyset;$$

[8]The fourth axiom remains unchanged.

$T_2$:

$$\forall\, V_1(x) \supset V(x) \ni x \;\Rightarrow V_1(x) \in \mathcal{V}(x);$$

$T_3$:

$$\forall\, V_\circ = \bigcap_i V_i(x) \text{ with } V_i(x) \in \mathcal{V}(x) \Rightarrow V_\circ \in \mathcal{V}(x).$$

Example of topological space: the basic nonempty set is the set of real numbers, $\mathcal{T} \equiv \mathbb{R}$, and the vicinities are given by the sets $V(x) = \{y \in \mathbb{R} |\ \|x - y\| \leq R\}$, where $R \in \mathbb{R}_+$ is a real positive number.

Let now consider two definitions for the continuous (or topological) group, one in the language of vicinities, and the other in the language $(\varepsilon, \delta(\varepsilon))$. But, before going further, it is worthy recalling (for comparison) the definition of a continuous function, in both languages.

**Definition**: Let $f : X \subset \mathbb{R} \to \mathbb{R}$ be a function defined on a subset $X$ of the set of real numbers, and $x_0 \in X$ an arbitrary point of this subset. The function $f$ is called *continuous at the point* $x_0$ if $\forall\, V\left(f(x_0)\right),\ \exists\, V(x_0)$, so that $\forall\, x \in X \cap V(x_0) \Rightarrow f(x) \in V\left(f(x_0)\right)$.

**Definition**: Let $f : X \subset \mathbb{R} \to \mathbb{R}$ be a function defined on a subset $X$ of the set of real numbers, and $x_0 \in X$ an arbitrary point of this subset. The function $f$ is called *continuous at the point* $x_0$ if $\forall\, \varepsilon > 0,\ \exists\, \delta(x) > 0$, so that $|f(x) - f(x_0)| < \varepsilon,\ \forall\, x \in X$, as soon as $|x - x_0| < \delta(x)$.

**Definition**: A *continuous (topological) group* is a nonempty set $G \neq \emptyset$ endowed with a law of internal composition that satisfies the group axioms and that can also be structured as a topological space, the group and topological structures being compatible according to the following conditions:

($i$) If $x, y \in G$, then $\forall\, W$ a vicinity of the element $x \circ y$ (where "$\circ$" denotes the internal composition law of the group), $\exists\, X$ a vicinity of $x$ and $Y$ a vicinity of $y$, so that $x \in X$ and $y \in Y$ must be followed by $x \circ y \in W$;

($ii$) If $x \in G$, then $\forall\, Z$ a vicinity of $x^{-1}$, $\exists\, X$ a vicinity of $x$, so that $x \in X$ must be followed by $x^{-1} \in Z$.

If one takes $z = x \circ y$ and $w = x^{-1}$, the above conditions can be rewritten as:

(*i'*) $\forall\ W(z),\ \exists\ X(x)$ and $Y(y)$, so that $x \in X$ and $y \in Y \Rightarrow z = x \circ y \in W(z)$;

(*ii'*) $\forall\ Z(w),\ \exists\ X(x)$ so that $x \in X(x) \Rightarrow w = x^{-1} \in Z(w)$.

In other words, $x \circ y$ is required to be a continuous function of $x$ and $y$, and $x^{-1}$ a continuous function of $x$.

Next, the $(\varepsilon, \delta(\varepsilon))$ language shall be used in order to define a topological group.

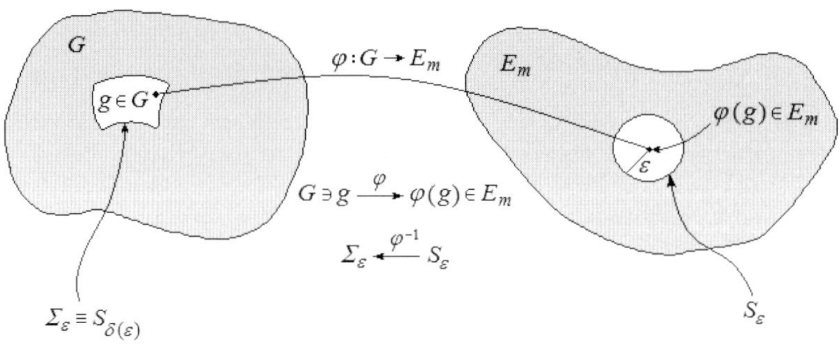


Fig. 4.2

Let $(G, \circ)$ be a group and $\varphi : G \rightarrow E_m$ a biunivocal correspondence between the elements of the group $G$ and the "points" (in fact, the $m$-dimensional vectors) in the $m$-dimensional Euclidean space $E_m$. The set $G$, which has already a group structure, can be simultaneously organized as a topological space. So, as vicinities of the "points" of the set $G$ shall be considered those sets of points whose image through the application $\varphi$ are in a vicinity of the points of the set $E_m$, which points are precisely the images through $\varphi$ of the initially considered points of $G$, *i.e.* the sets (see Fig. 4.2 for an easier understanding):

$$\Sigma_\varepsilon(g) \left(\equiv S_{\delta(\varepsilon)}\right) = \left\{g' \in G\right\} \mid \varphi(g') \in S_\varepsilon\left(\varphi(g)\right)\right\}.$$

Let a spherical vicinity, $S_\varepsilon$, be in $E_m$, of radius $\varepsilon$ and the centre at $\varphi(g)$,

$$S_\varepsilon \left(\equiv S_\varepsilon\left(\varphi(g)\right)\right) = \{f \in E_m | \ |f - \varphi(g)| \leq \varepsilon,\ \forall\ \varepsilon > 0\}.$$

The elements $f \in E_m$ have therefore the property that $\forall\ f \in S_\varepsilon \subset E_m$, $\exists\ g' \in G$ so that $f = \varphi(g')$. $\Sigma_\varepsilon(g) \equiv S_{\delta(\varepsilon)}$ shall be a vicinity of $g \in G$ if through $\varphi$ its elements are taken in the vicinity $S_\varepsilon$ of $\varphi(g)$. Thus, $\varphi^{-1}$ induces from $E_m$ into $G$ a topological structure, that is by means of $\varphi^{-1}$ one can construct vicinities in the set $G$ of the group. This way, the set $G$ becomes structurable as a topological space, so that one can speak about the notion of continuity on this set as follows:

Consider $g_1, g_2, g_3 \in G$ and $g_1 \circ g_2 = g_3$.

**Definition**: One says that the composition law of $G$ is *continuous in the element* $g_2$ if $\forall\ \varepsilon > 0$, $\exists\ \delta(\varepsilon) > 0$, so that (see Fig. 4.3) $\forall\ g \in \Sigma_{\delta(\varepsilon)}(g_2) \subset \Sigma(g_2) \subset G$, then $g_1 \circ g \in \Sigma_\varepsilon(g_3) \subset G$ or, formally, $g_1 \circ \Sigma_{\delta(\varepsilon)}(g_2) \subset \Sigma_\varepsilon(g_3)$.

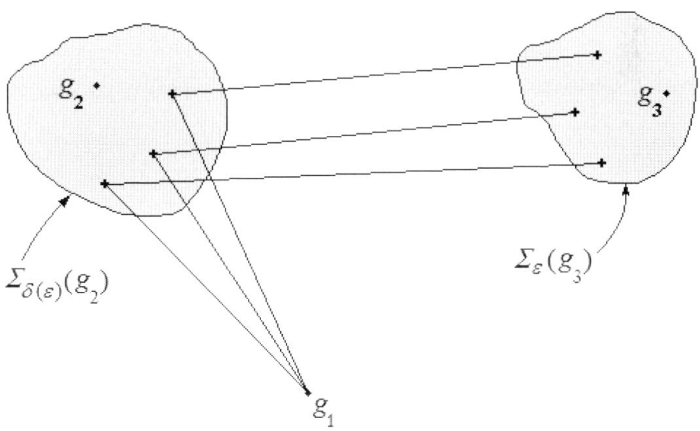


Fig. 4.3

Referring now to the space $E_m$, under this situation being called the *group space*, one can write: $\forall\ \varepsilon > 0$, $\exists\ \delta(\varepsilon) > 0$, so that $|\varphi(g_1 \circ g) - \varphi(g_3)| < \varepsilon$ as soon as $|\varphi(g) - \varphi(g_2)| < \delta(\varepsilon)$, $\forall\ g \in \Sigma_{\delta(\varepsilon)}(g_2)$. In other words, "multiplying" (composing) the elements of vicinity of $g_2$ with the fixed element $g_1$, one obtains elements of vicinity of $g_3$.

Under these circumstances, the compatibility between the group structure and the topological structure is ensured and one can define the notion of topological (or continuous) group as follows:

**Definition**: One calls a *topological* (*continuous*) *group*, that group whose composition law together with its inverse are continuous in the sense defined above.

**Definition**: A *topological group* is called *compact* if its image, $\varphi(G)$, in the group space $E_m$ is bounded, that is if

$$|\varphi(g) - \varphi(g')| < M < \infty, \quad \forall\, g, g' \in G.$$

In the reverse case, the group is called *non-compact.*

For example, the group whose elements are the translations along the real axis is non-compact (there can be translations of any value). In turn, the rotation group about a fixed axis is compact, since a rotation of any angle can be reduced to a rotation of an angle contained in the bounded interval $[0, 2\pi]$, once the zero-angle rotation (the origin or the null element of the group) is fixed.

Before defining the concept of a *Lie group*, one observes that the elements of a topological group can depend on certain number of parameters. The dimension of the group space is in fact determined by this number of parameters. In this regard, the things can be seen as follows.

Suppose that, in a one-to-one manner, to each element $G$ of a topological group $G$ one associates a set of parameters (called *essential*), $a = (a_1, a_2, ..., a_m)$, which are exactly the elements (vectors) of the group space (or the *parameter space*, as it is also called).

This way, it is said that a *parametrization* of the elements of the topological group $G$ (with the intention to obtain a Lie group) had been performed. This parametrization is always done in a way that makes it possible an arbitrary parameter transformation $a' = f(a)$, differentiable as many times $(n)$ as necessary.

### IV.2.3. Lie groups

**Definition**: A topological group $G$ is called a *group Lie of dimension* $m$ if there exists a vicinity $V_m$ of the unit element $e \in G$ with the following properties:

$L_1$: There exists a biunivocal correspondence between the elements of $V_m$ and the points of a "subset" of the group space, $E_m$, the parameters $a_k$ $(k = \overline{1,m})$ of $E_m$ being essential (*i.e.* none of them can be expressed in terms of the remaining $m-1$ parameters);

$L_2$: If $g_1 = g_1(a_1, a_2, ..., a_m)$ and $g_2 = g_2(a'_1, a'_2, ..., a'_m)$ belong to the vicinity $V_m$ and $g_1 g_2 = g_3(a''_1, a''_2, ..., a''_m)$ and $g_1^{-1} = g(a'''_1, a'''_2, ..., a'''_m)$, then one requires that $a''_i = a''_i(a_1, a_2, ..., a_m; a'_1, a'_2, ..., a'_m)$ and $a'''_i = a'''_i(a_1, a_2, ..., a_m)$, $(i = \overline{1,m})$ be analytic functions of their arguments.

**Definition**: A *function* $f : \Omega \to \mathbb{C}$, where $\Omega$ is an open set in $\mathbb{C}^n$ is called *analytic* (more exactly $\mathbb{C}$-analytic) if $\forall\ z_0 \in \Omega$, $\exists\ \{c_n\}_{n\geq 0}$ a sequence of complex numbers with properties:

**1)** The power series $\sum\limits_{n\geq 0} c_n z^n$ has a positive radius of convergence;

**2)** On a vicinity of $z_0$ the function $f(z)$ can be represented as $f(z) = \sum\limits_{n\geq 0} c_n (z - z_0)^n$, the coefficients $c_n$ being this way univocally determined by the function $f$.

(The radius of convergence is defined as follows:

$$R = \sup\left\{\rho \geq 0 \,\middle|\, \sum_{n\geq 0} |c_n| \rho^n < \infty\right\}).$$

**Observations**:

**1)** It has been chosen a vicinity of the unit element, $V_m$, and not another arbitrary one, because one usually studies the infinitesimal transformations, when due to the values of the parameters, the elements of $G$ are situated in the vicinity of the unit element of the group. Any finite transformation can be considered (if needed) as a succession of infinitesimal transformations;

**2)** It is useful to choose the parameters $a_1, a_2, ..., a_m$ so that the image of $e \in G$ be precisely the origin of $E_m$, that is $e - e\underbrace{(0, 0, ..., 0)}_{\text{m times}}$.

## ♣ The parameter group

A Lie group can be "constructed" as follows: to each element $g$ of a given topological group one associates in a one-to-one (biunivocal) manner a finite set of essential parameters $a = (a_1, a_2, ..., a_m)$. These sets of essential parameters form the *group space* or the *parameter space*. The parametrization of the elements of a topological group, in order to obtain a Lie group, is performed in a way that always makes possible an arbitrary parameter transformation $a' = f(a)$ that is differentiable as many times as necessary.

The one-to-one correspondences

$$\begin{aligned} g_1 &\leftrightarrow (a_1, a_2, ..., a_m) = a, \\ g_2 &\leftrightarrow (a'_1, a'_2, ..., a'_m) = a', \\ g_1 g_2 = g_3 &\leftrightarrow (a''_1, a''_2, ..., a''_m) = a'', \end{aligned}$$

imply the existence of a group composition law in the group space of the form

$$a''_i = \varphi_i(a, a'),$$

where $\varphi_i$ $(i = \overline{1, m})$ are analytic functions of $a$ and $a'$. Therefore the parameters of a Lie group form in their turn a group structure, called the *parameter group*, with the relation $a''_i = \varphi_i(a, a')$ as the composition law of the group.

The following investigation concerns the *Lie groups* (topological groups whose elements depend on a finite number of parameters, that continuously take values in a certain space, called the *group space*) only.

The necessity of considering the study of the Lie groups is implicitly understood, since the requirement for the Lagrangian of a physical system to possess some kind of invariance (*e.g.* relativistic invariance, or $SU(2)$-invariance, or $U(1)$-invariance, etc.) consists in the Lagrangian invariance with respect to the particular transformations (depending on the required type of invariance) of the wave functions which compose that Lagrangian. In fact, almost all the required types of invariance consist in some transformations of the wave functions *depending on a finite number of parameters*, that continuously take values in a certain domain. A banal example is the special relativistic invariance (that is the invariance with respect to the Lorentz group transformations), in which case the respective transformations are nothing else but "rotations" in the four-dimensional space (as far as the group elements are concerned), depending on six parameters (the rotation angles in the six distinct planes $x_1Ox_2$, $x_1Ox_3$, $x_1Ox_4$, $x_2Ox_3$, $x_2Ox_4$, $x_3Ox_4$) that can take any value (*i.e.* continuously) in the interval $[0, 2\pi]$.

In view of the above considerations, here is another (but equivalent) definition[9] of the Lie group.

**Definition**: Let $\vec{x} = (x^1, x^2, ..., x^n)$ be a vector in a $n$-dimensional space and consider the set of transformations

$$\overline{x}^i = f^i\left(x^1, x^2, ..., x^n; a^1, a^2, ..., a^m\right), \quad (i = \overline{1, n}), \tag{4.1}$$

that depend on $m$ independent, real and continuous parameters $a^1, a^2, ..., a^m$. One admits that $f^i$, $(i = \overline{1, n})$ are analytic functions of the parameters $a^j$, $(j = \overline{1, m})$, so that for $a^1 = a^2 = ... = a^m = 0$ one obtains the identical transformation:

$$x^i = f^i\left(x^1, x^2, ..., x^n; 0, 0, ..., 0\right), \quad (i = \overline{1, n}). \tag{4.2}$$

One also assumes that there exists a set of values of the parameters $a^j$, $(j = \overline{1, m})$, so that

$$x^i = f^i\left(\overline{x}^1, \overline{x}^2, ..., \overline{x}^n;\ \tilde{a}^1, \tilde{a}^2, ..., \tilde{a}^m\right), \quad (i = \overline{1, n}), \tag{4.2$'$}$$

meaning that there exists the inverse transformation of (4.1). Consider now the effect of following two successive transformations,

$$\begin{aligned}\overline{\overline{x}}^i &= f^i\left(\overline{x}^1, \overline{x}^2, ..., \overline{x}^n;\ \overline{a}^1, \overline{a}^2, ..., \overline{a}^m\right)\\ &= f^i\Big(f^1(x^1, x^2, ..., x^n;\ a^1, a^2, ..., a^m),\\ &\quad f^2(x^1, x^2, ..., x^n;\ a^1, a^2, ..., a^m), ...,\\ &\quad f^n(x^1, x^2, ..., x^n;\ a^1, a^2, ..., a^m); \overline{a}^1, \overline{a}^2, ..., \overline{a}^m\Big), \quad (i = \overline{1, n})\end{aligned} \tag{4.3}$$

and suppose that there exists a set of values of the parameters $a^j$, $(j = \overline{1, m})$ so that

$$\overline{\overline{x}}^i = f^i\left(x^1, x^2, ..., x^n;\ \overline{\overline{a}}^1, \overline{\overline{a}}^2, ..., \overline{\overline{a}}^m\right), \ (i = \overline{1, n}). \tag{4.3$'$}$$

[9] The presentation of this subject follows the exposure given in *Applications of the group theory in mechanics and physics* written by P.P. Teodorescu and N. Nicorovici-Porumbaru, Technical Publishing House, Bucharest, 1985.

Under these hypotheses, the set of transformations (4.1) form a topological finite group with $m$ parameters. Obviously, the following relations are also valid:

$$\overline{\overline{a}}^j = \varphi^j\left(a^1, a^2, ..., a^m;\ \overline{a}^1, \overline{a}^2, ..., \overline{a}^m\right),\ (j = \overline{1,m}). \qquad (4.4)$$

If $\varphi^j$, $(j = \overline{1,m})$ are analytic functions of their arguments ($a^k$, $k = \overline{1,m}$ and $\overline{a}^l$, $l = \overline{1,m}$), then it is said that the transformations (4.1) form a *Lie group*.

**Definition**: If $m$ is the minimum number of independent parameters necessary to characterize the elements of the Lie group, then these parameters are called *essential*, and $m$ is the *rank* or *dimension* of the group.

The relations (4.4) necessarily follow as a result of comparison of (4.3) and (4.3′), their form being very similar to that of the starting relations (4.1). Due to this fact, these relations can be seen as parameter transformations, rather than coordinate transformations, the parameters of the parameters' transformation being here $\overline{a}^1, \overline{a}^2, ..., \overline{a}^m$. If $\varphi^j$, $(j = \overline{1,m})$ are analytic functions, then the relations (4.4) define, in their turn, a Lie group isomorphic with that generated by the transformations (4.1), called *parameter group*. Due to the above mentioned isomorphism, the parameter Lie group – which is simpler – is more frequently studied. For the sake of convenience (simplicity), the transformation (4.1) shall be denoted by $\overline{x}^i = f^i(\vec{x}, \vec{a})$, $(i = \overline{1,n})$, where $\vec{x} = (x^1, x^2, ..., x^n)$ and $\vec{a} = (a^1, a^2, ..., a^m)$, or, under a vector form, $\vec{\overline{x}} = \vec{f}(\vec{x}, \vec{a})$, with $\vec{\overline{x}} = (\overline{x}^1, \overline{x}^2, ..., \overline{x}^n)$, $\vec{f} = (f^1, f^2, ..., f^n)$, and $\vec{\overline{\overline{a}}} = \vec{\varphi}(\vec{a}, \vec{\overline{a}})$. One also has $\vec{x} = \vec{f}(\vec{x}, \vec{0})$ and $\vec{a} = \vec{\varphi}(\vec{a}, \vec{0})$.

Since this is a matter of coordinate transformations, an infinitesimal displacement $d\vec{x}$ of $\vec{x}$ is a differential one and can be obtained in two ways:

**1)** as a succession of two transformations,

(i) $\vec{x}_0 \to \vec{x}$, $\qquad \vec{x} = \vec{f}(\vec{x} + 0, \vec{a})$, $\qquad \left(\vec{x}_0 = \vec{f}(\vec{x}_0, \vec{0})\right)$;

(ii) $\vec{x} \to \vec{x} + d\vec{x}$, $\qquad \vec{\overline{x}} = \vec{x} + d\vec{x} = \vec{f}(\vec{x}, \delta\vec{a})$, $\qquad (\vec{x} \to \vec{\overline{x}} = \vec{x} + d\vec{x})$;

**2)** directly, in which case one can write

$$\vec{x}_0 \to \vec{x} + d\vec{x}, \qquad \vec{x} + d\vec{x} = \vec{f}(\vec{x}_0, \vec{a} + d\vec{a}).$$

Due to the associative property, one can write $\vec{a} + d\vec{a} = \vec{\varphi}(\vec{a}, \delta\vec{a})$. Since the transformations are infinitesimal, one can express the differential as

$$d\vec{x} = \left(\frac{\partial \vec{f}(\vec{x}, \vec{a})}{\partial a^k}\right)_{\vec{a}=0} \delta a^k,$$

or

$$d\vec{x} = \frac{\partial \vec{f}(\vec{x}_0, \vec{a})}{\partial a^j} da^j,$$

since

$$\begin{aligned} \vec{x} + d\vec{x} &= \vec{f}(\vec{x}, \delta\vec{a}) \\ &= \vec{f}(\vec{x}, \vec{0}) + \left.\frac{\partial \vec{f}(\vec{x},\vec{a})}{\partial a^k}\right|_{\vec{a}=0} \delta a^k + \mathcal{O}(\delta^2 a) \simeq \vec{x} + \left.\frac{\partial \vec{f}(\vec{x},\vec{a})}{\partial a^k}\right|_{\vec{a}=0} \delta a^k \end{aligned}$$

and, respectively,

$$\begin{aligned} \vec{x} + d\vec{x} &= \vec{f}(\vec{x}_0, \vec{a} + d\vec{a}) \\ &= \vec{f}(\vec{x}_0, \vec{a}) + \left.\frac{\partial \vec{f}(\vec{x}_0, \vec{a})}{\partial a^j}\right|_{\vec{a}=\vec{a}} da^j + \mathcal{O}(d^2 a) \simeq \vec{x} + \frac{\partial \vec{f}(\vec{x}_0, \vec{a})}{\partial a^j} da^j. \end{aligned}$$

Denoting $u^i_k(\vec{x}) = \left.\frac{\partial f^i(\vec{x},\vec{a})}{\partial a^k}\right|_{\vec{a}=0}$, $(i = \overline{1,n},\ k = \overline{1,m})$, one can write

$$dx^i = u^i_k(x)\delta a^k, \quad (i = \overline{1,n}). \tag{4.5}$$

Consider, next, the relation $\vec{a} + d\vec{a} = \vec{\varphi}(\vec{a}, \delta\vec{a})$. Since $\varphi^j$ are analytic functions and $\vec{\varphi}(\vec{a}, \vec{0}) = \vec{a}$, by means of a Taylor expansion one obtains

$$\begin{aligned} \vec{a} + d\vec{a} &= \vec{\varphi}(\vec{a}, \vec{0}) + \left(\frac{\partial \vec{\varphi}(\vec{a}, \vec{\overline{a}})}{\partial \overline{a}^k}\right)_{\vec{\overline{a}}=0} \delta a^k \\ &= \vec{a} + \left.\frac{\partial \vec{\varphi}(\vec{a}, \vec{\overline{a}})}{\partial \overline{a}^k}\right|_{\vec{\overline{a}}=0} \delta a^k, \end{aligned}$$

therefore one can write

$$da^j = \mu^j_k(\vec{a})\delta a^k, \quad (j, k = \overline{1, m}), \tag{4.6}$$

where

$$\mu^j_k(\vec{a}) = \left.\frac{\partial \varphi^j(\vec{a}, \vec{\overline{a}})}{\partial \overline{a}^k}\right|_{\vec{\overline{a}}=0}, \quad (j, k = \overline{1, m}).$$

According to (4.6),

$$\delta a^k = \left(\mu^j_k(\vec{a})\right)^{-1} da^j \overset{\text{not}}{=} \lambda^k_j(\vec{a}) da^j, \quad (k = \overline{1, m}), \tag{4.7}$$

with $\lambda^l_j(\vec{a})\mu^j_k(\vec{a}) = \delta^l_k$ (Kronecker symbol). Then, in view of (4.5) and (4.7),

$$dx^i = u^i_k(\vec{x})\delta a^k = u^i_k(\vec{x})\lambda^k_j(\vec{a}) da^j,$$

so that

$$\frac{\partial x^i}{\partial a^j} = u^i_k(\vec{x})\lambda^k_j(\vec{a}), \quad (i = \overline{1, n};\ j, k = \overline{1, m}). \tag{4.8}$$

The equations (4.8) are called the *differential equations of the group*. In order to be completely integrable, the integrability conditions must be fulfilled:

$$\frac{\partial^2 x^i}{\partial a^j \partial a^l} = \frac{\partial^2 x^i}{\partial a^l \partial a^j}.$$

Since

$$\frac{\partial x^i}{\partial a^j} = u^i_k \lambda^k_j,$$

it follows that

$$\frac{\partial^2 x^i}{\partial a^l \partial a^j} = \frac{\partial u^i_k}{\partial x^s}\frac{\partial x^s}{\partial a^l}\lambda^k_j + u^i_k \frac{\partial \lambda^k_j}{\partial a^l} = \frac{\partial u^i_k}{\partial x^s} u^s_p \lambda^p_l \lambda^k_j + u^i_k \frac{\partial \lambda^k_j}{\partial a^l},$$

while according to

$$\frac{\partial x^i}{\partial a^l} = u^i_r \lambda^r_l,$$

one can write

$$\frac{\partial^2 x^i}{\partial a^j \partial a^l} = \frac{\partial u^i_r}{\partial x^s}\frac{\partial x^s}{\partial a^j}\lambda^r_l + u^i_r \frac{\partial \lambda^r_l}{\partial a^j} = \frac{\partial u^i_r}{\partial x^s} u^s_n \lambda^n_j \lambda^r_l + u^i_r \frac{\partial \lambda^r_l}{\partial a^j}.$$

The integrability condition then becomes

$$\frac{\partial u_k^i}{\partial x^s} u_p^s \lambda_l^p \lambda_j^k + u_k^i \frac{\partial \lambda_j^k}{\partial a^l} = \frac{\partial u_r^i}{\partial x^s} u_n^s \lambda_j^n \lambda_l^r + u_r^i \frac{\partial \lambda_l^r}{\partial a^j},$$

or, if the summation indices are conveniently changed ($k$ to $r$ on the l.h.s., and $r$ to $p$ and $n$ to $r$ in the first term on the r.h.s.),

$$\frac{\partial u_r^i}{\partial x^s} u_p^s \lambda_l^p \lambda_j^r + u_r^i \frac{\partial \lambda_j^r}{\partial a^l} = \frac{\partial u_p^i}{\partial x^s} u_r^s \lambda_j^r \lambda_l^p + u_r^i \frac{\partial \lambda_l^r}{\partial a^j},$$

one can still write

$$\left( u_p^s \frac{\partial u_r^i}{\partial x^s} - u_r^s \frac{\partial u_p^i}{\partial x^s} \right) \lambda_l^p \lambda_j^r + u_r^i \left( \frac{\partial \lambda_j^r}{\partial a^l} - \frac{\partial \lambda_l^r}{\partial a^j} \right) = 0,$$

and, if the quantities $\mu_k^j$ are used,

$$\left( u_p^s \frac{\partial u_r^i}{\partial x^s} - u_r^s \frac{\partial u_p^i}{\partial x^s} \right) = \left( \frac{\partial \lambda_l^k}{\partial a^j} - \frac{\partial \lambda_j^k}{\partial a^l} \right) \mu_p^l \mu_r^j u_k^i \stackrel{\text{not}}{=} \sum_{k=1}^m c_{pr}^k(\vec{a}) u_k^i, \quad (i = \overline{1,n};\ p,r = \overline{1,m}), \tag{4.9}$$

where

$$c_{pr}^k = \left( \frac{\partial \lambda_l^k}{\partial a^j} - \frac{\partial \lambda_j^k}{\partial a^l} \right) \mu_p^l \mu_r^j, \quad (j,l,k,p,r = \overline{1,m}). \tag{4.10}$$

Taking the derivative of (4.9) with respect to $a^j$, one gets

$$\frac{\partial c_{pr}^k}{\partial a^j} u_k^i = 0; \quad i = \overline{1,n}; \quad (k,p,r = \overline{1,m}), \tag{4.11}$$

because the functions $u_k^i(\vec{x})$ do not depend on the parameters $a^l$, $(l = \overline{1,m})$ by definition. But the parameters $a^j$, $(j = \overline{1,m})$ are essential, therefore the functions $u_k^i(\vec{x})$ are linearly independent. Consequently, the Eq. (4.11) yields $\frac{\partial c_{pr}^k}{\partial a^j} = 0$, so that the quantities $c_{pr}^k$, $(k,p,r = \overline{1,m})$ are constant, and the equations (4.9) and (4.10) can be rewritten as follows,

$$\left( u_p^s \frac{\partial u_r^i}{\partial x^s} - u_r^s \frac{\partial u_p^i}{\partial x^s} \right) = c_{pr}^k u_k^i, \quad (i = \overline{1,n};\ k,j,l,p,r = \overline{1,m}), \tag{4.9$'$}$$

and

$$\frac{\partial \lambda_l^k}{\partial a^j} - \frac{\partial \lambda_j^k}{\partial a^l} = c_{pr}^k \lambda_l^p \lambda_j^r, \quad (k,j,l,p,r = \overline{1,m}). \tag{4.10$'$}$$

## ♣ Generators of a Lie group

In order to introduce the generators of a Lie group one must refer to a group representation, more exactly, one must introduce the space of the group representation which, in this case, can be the space of the $\mathbb{C}^2$-class functions. When an elementary coordinate transformation of the form $\vec{x} \to \vec{\overline{x}} = \vec{x} + d\vec{x}$ (a group element) takes place, a transformation of the representation space elements under the action of the representation operators takes place too. Let $F(\vec{x})$ be an element of the representation space. Then,

$$dF(\vec{x}) = \frac{\partial F}{\partial x^i} dx^i = u^i_j \frac{\partial F}{\partial x^i} \delta a^j \overset{\text{not}}{=} X_j F(\vec{x}) \delta a^i,$$

where

$$x_j \overset{\text{def}}{=} u^i_j \frac{\partial}{\partial x^i}, \quad (i = \overline{1,n},\ j = \overline{1,m})$$

are $m$ operators called the *group generators*. Going now to the parameters group, one still can write

$$dF = \frac{\partial F}{\partial a^i} \delta a^i,$$

so that the generators can also be expresses as

$$X_j = \frac{\partial F}{\partial a^j}, \quad (j = \overline{1,m}).$$

Sometimes is more convenient to define the generators as

$$X_j = K \frac{\partial}{\partial a^j}, \quad (j = \overline{1,m}),$$

where $K$ is a complex constant.

Consider, now, the successive action of two generators

$$X_p X_r F(\vec{x}) = \left( u^s_p \frac{\partial}{\partial x^s} \right) u^i_r \frac{\partial}{\partial x^i} F(\vec{x})$$

$$= u^s_p \frac{\partial u^i_r}{\partial x^s} \frac{\partial F(\vec{x})}{\partial x^i} + u^s_p u^i_r \frac{\partial^2 F(\vec{x})}{\partial x^s \partial x^i}.$$

Analogously,

$$X_r X_p F(\vec{x}) = \left(u_r^i \frac{\partial}{\partial x^i}\right) u_p^s \frac{\partial}{\partial x^s} F(\vec{x})$$
$$= u_r^i \frac{\partial u_p^s}{\partial x^i} \frac{\partial F(\vec{x})}{\partial x^s} + u_r^i u_p^s \frac{\partial^2 F(\vec{x})}{\partial x^i \partial x^s}.$$

Then,

$$(X_p X_r - X_r X_p) F(\vec{x}) \overset{\text{not}}{=} [X_p, X_r] F(\vec{x})$$
$$= u_p^s \frac{\partial u_r^i}{\partial x^s} \frac{\partial F(\vec{x})}{\partial x^i} - u_r^i \frac{\partial u_p^s}{\partial x^i} \frac{\partial F(\vec{x})}{\partial x^s}$$
$$= \left(u_p^s \frac{\partial u_r^i}{\partial x^s} - u_r^s \frac{\partial u_p^i}{\partial x^s}\right) \frac{\partial F(\vec{x})}{\partial x^i} \overset{(4.11)}{=} c_{pr}^k u_k^i \frac{\partial F(\vec{x})}{\partial x^i}$$
$$= c_{pr}^k u_k^i \frac{\partial}{\partial x^i} F(\vec{x}) = c_{pr}^k X_k F(\vec{x}),$$

where Schwarz's theorem (that ensure the equality of the second-order mixt partial derivatives, if the required conditions are fulfilled – as is the case here), as well as the definition (4.11) of the constants $c_{pr}^k$ have been used. Therefore, it has been found that

$$[X_p, X_r] = c_{pr}^k X_k.$$

This is the *commutator of generators* of a Lie group, the constants $c_{pr}^k$ being called *structure constants* of the group. The antisymmetry property of the commutator

$$[X_p, X_r] = -[X_r, X_p]$$

is reflected in these constants as the antisymmetry condition of the lower indices

$$c_{pr}^k = -c_{rp}^k,$$

while Jacobi's identity

$$\Big[X_p, [X_r, X_k]\Big] + \Big[x_r, [X_k, X_p]\Big] + \Big[X_k, [X_p, X_r]\Big] = 0,$$

easily yields the relationship

$$c_{pl}^i c_{rs}^l + c_{rl}^i c_{sp}^l + c_{sl}^i c_{pr}^l = 0, \quad (l, i, p, r, s = \overline{1, m}).$$

The structure constants of each Lie group are univocally determined by the group structure.

Summarizing, it has been seen that, starting from a group of transformations of the form

$$\overline{x}^i = f^i(x^1, x^2, ..., x^n;\ a^1, a^2, ..., a^m), \quad (i = \overline{1, n}), \tag{4.12}$$

the group differential equations have been found,

$$\frac{\partial x^i}{\partial a^j} = u^i_k(\vec{x})\lambda^k_j(\vec{a}), \quad (i = \overline{1, n};\ j, k = \overline{1, m}), \tag{4.13}$$

together with the conditions

$$\left(u^s_p \frac{\partial u^i_r}{\partial x^s} - u^s_r \frac{\partial u^i_p}{\partial x^s}\right) = c^k_{pr} u^i_k, \quad (i = \overline{1, n};\ k, s, p, r = \overline{1, m}), \tag{4.14}$$

and

$$\frac{\partial \lambda^k_l}{\partial a^j} - \frac{\partial \lambda^k_j}{\partial a^l} = c^k_{pr} \lambda^p_l \lambda^r_j, \quad (i = \overline{1, n};\ j, l, k, p, r = \overline{1, m}), \tag{4.15}$$

and, finally,

$$c^k_{pr} = -c^k_{rp}, \quad (k, p, r = \overline{1, m}), \tag{4.16}$$

and

$$c^i_{pl} c^l_{rs} + c^i_{rl} c^l_{sp} + c^i_{sl} c^l_{pr} = 0, \quad (i, l, p, r, s = \overline{1, m}). \tag{4.17}$$

Lie demonstrated the remarkable fact that the way leading to this result can be reversed, that is: if some constants satisfying (4.16) and (4.17) are identified, one can find some "functions" $u$ and $\lambda$ that satisfy (4.14) and (4.15) and, finally, one can find the functions (4.12) as integrals of the system (4.13) that form a group structure.

**Theorem (Ado)**: Any Lie group admits a linear representation. According to this theorem (given here, as the other ones, without demonstration) any Lie group is isomorphic with a group of linear transformations on a vector space, each element of this group being able to be represented as a linear operator acting in such a space. Due to the above mentioned isomorphism, it is sufficient to deal with the group of linear transformations (coordinate transformations) that depend on $m$ parameters, of the type

$$x^i \to \overline{x}^i = f^i(x^j;\ a^k), \quad (i, j = \overline{1, n};\ k = \overline{1, m}),$$

previously used in the definition of the Lie group. These transformations can be considered as a set of linear transformations on a $n$-dimensional space $\mathcal{L}$, the space of the vectors with $n$ components, $\vec{x} = (x^1, x^2, ..., x^n)$. Then, a vector $\vec{x}$ passes to another vector $\vec{\overline{x}}$ as a result of the action of a linear operator $T$, defined on the space of these vectors,

$$T : \mathcal{L} \to \mathcal{L}, \quad \vec{x} \xrightarrow{\text{T}} \vec{\overline{x}} = T\vec{x},$$

meaning that, in fact, $f^i$, $i = (\overline{1,n})$ is a set of linear operators that depend on $m$ real parameters $a^k$, $(k = \overline{1,m})$, so that one can write

$$\overline{x}^i = f^i(x^j; a^k) \to g = T(a^k), \quad (k = \overline{1,m}),$$

where $g$ is an element of the operators $T$ group (in fact, the same group of transformations that has been used so far). The unit element of the group

$$x^i \to x^i = f^i(x^1, x^2, ..., x^n; \underbrace{0,\ 0, ..., 0}_{\text{m times}}),$$

under the new notation shall be written as

$$e = T(\underbrace{0,\ 0, ..., 0}_{\text{m times}}).$$

Defining a basis on the vector space $\mathcal{L}$ of the operators $T$, they shall be expressed in a matrix form. The action of an operator $T$ on a vector of the respective linear space writes, by components,

$$x^i \xrightarrow{\text{T}} \overline{x}^i = T^i_j(a^1, a^2, ..., a^m)x^j,$$

or, in a matrix form,

$$[x] \xrightarrow{\text{T}} [\overline{x}] = T\,[x],$$

where

$$[x] = \begin{pmatrix} x^1 \\ x^2 \\ \vdots \\ x^n \end{pmatrix} \text{ and } [\overline{x}] = \begin{pmatrix} \overline{x}^1 \\ \overline{x}^2 \\ \vdots \\ \overline{x}^n \end{pmatrix}.$$

Supposing that the transformation is infinitesimal, the operators $T(a^k)$ can be expanded into a Taylor series and, if only the first two terms are kept, one can write

$$T^i_j(a^k) \simeq T^i_j(a^k = 0) + \left.\frac{\partial T^i_j}{\partial a^k}\right|_{\vec{a}=0} \delta a^k.$$

The unit element $e$ of the group corresponds to the unit operator $e \to T(0, 0, ..., 0) = I$, or, expressed in terms of its components, $T^i_j(a^k = 0) = I(\equiv \delta^i_j)$ so that

$$\overline{x}^i = \left\{ \delta^i_j + \left( \frac{\partial T^i_j}{\partial a^k} \right)_{a^k=0} \delta a^k \right\} x^j = x^i + dx^i,$$

where

$$dx^i = \left( \frac{\partial T^i_j}{\partial a^k} \right)_{a^k=0} x^j \delta a^k,$$

or, in a matrix form,

$$d[x] = \left( \frac{\partial T}{\partial a^k} \right)_{a^k=0} [x] \delta a^k.$$

Comparing with the previous results, it follows that the matrices of the generators are of the form

$$X_k = \left( \frac{\partial T^i_j}{\partial a^k} \right)_{a^k=0}, \quad (k = \overline{1, m}).$$

**Observation**: As one can easily observe, the number of generators of a Lie group equals the number of essential parameters of the group.

Let us define the generators of an arbitrary compact Lie group $G$ in the most general case (in other words, not only for the particular case when the elements of the Lie group are the coordinate linear transformations, depending on $m$ real parameters).

But before doing that, one must specify that, when talking about generators of a Lie group, one really understands the generators of the identical representation of the considered Lie group, since the notion of generator is defined in strong connection with the notion of representation of a group (*i.e.* it cannot be defined without considering a representation of that group). That is why, in general, it is correct to talk about the generators of some representation of a group, instead of generators of that group (the last case being a particular one, namely when the identical representation of the group is considered).

A concept necessary to the following development of the investigation is that of topological representation.

**Definition**: A *representation* is called *topological* if

1) the representation space is topological;
2) the operators of the representation of the topological group $G$ on the topological space of the representation $\mathcal{L}$, satisfy the following properties:
   ($i$) $T(e) = 1$ (the unit operator);
   ($ii$) $T(g_1 g_2) = T(g_1)T(g_2)$;
   ($iii$) The application $\{x, g\} \in \mathcal{L} \times G \to T(g)\,x \in \mathcal{L}$ from $\mathcal{L} \times G$ to $\mathcal{L}$ is continuous, $\forall\ x \in \mathcal{L}$, $\forall\ g \in G$.

**Observation**: any topological representation is a representation.

**Definition**: Let $G$ be a compact Lie group and $g \to T(g)$ a finite-dimensional topological representation of this group on the Hilbert space $\mathcal{L}$. Since $G$ is a compact $m$-dimensional Lie group, one can parametrize the vicinity $V_m$ of the unit element $e \in G$, and therefore, $\forall\ g \in V_m$, $T(g) \equiv T(a_1, a_2, ..., a_m)$. Since the representation is topological (continuous), and the dependence $g = g(a_1, a_2, ..., a_m)$ is analytical, it follows that the operators of the representation $g \to T(g)$ are analytic functions of the parameters $a_k$, $(k = \overline{1,m})$ and there exist (can be defined) $m$ operators

$$I_k \overset{\text{def}}{=} \frac{\partial T(\vec{a})}{\partial a_k}\bigg|_{a_1=a_2=...=a_m=0}, \qquad (k = \overline{1,m}),$$

called the *infinitesimal generators* of the Lie group $G$.

The parametrization is carried out in such a way that

$$e = e(\underbrace{0,\,0,\,...,\,0}_{\text{m times}}).$$

The number of generators is equal to the number of parameters (according to their definition), as well as to the group dimension.

The operators of the representation $g \to T(g)$ are connected to the generators $I_k$ through a differential equation satisfied by the representation operators $T$. To determine this equation, let $f$ and $g$ be two elements of the Lie group $G$, and $x \in \mathcal{L}$ an element of the representation space. Let, also, introduce the notation

$$y(g) \overset{\text{not}}{=} T(g)x \in \mathcal{L}.$$

Since $g \to T(g)$ is a representation, one can write

$$T(fg) = T(f)T(g),$$

$$T(e) = 1_{\mathcal{L}},$$

and therefore

$$\begin{aligned} T(fg)y\left(g^{-1}\right) &= T(fg)T\left(g^{-1}\right)x = T(f)T(g)T\left(g^{-1}\right)x \\ &= T(f)T\left(gg^{-1}\right)x = T(f)T(e)x = T(f)x = y(f). \end{aligned}$$

Thus, it has been found that

$$y(f) = T(fg)y\left(g^{-1}\right).$$

Taking into account the above parametrization and limiting this investigation to the elements $f, g$ in the vicinity $V_m$ of the unit element of the group $G$, where the parametrization is valid, one can write

$$y\left(a_i(f)\right) = T\left(a_k(fg)\right) y\left(a_j(g^{-1})\right), \tag{4.18}$$

where the bijection existent between the elements of the group and those of the parameter space $E_m$, has been taken into account. Taking the derivative of (4.18) with respect to $a_l(f)$, one obtains

$$\begin{aligned} &\frac{\partial y\left(a_1(f),\, a_2(f), ..., a_m(f)\right)}{\partial a_l(f)} \\ &= \sum_{k=1}^{m} \frac{\partial T(fg)}{\partial a_k(fg)} \frac{\partial a_k(fg)}{\partial a_l(f)} y\left(a_j\left(g^{-1}\right)\right). \end{aligned} \tag{4.19}$$

Denote

$$s_{kl}(a_j) \stackrel{\text{def}}{=} \frac{\partial a_k(fg)}{\partial a_l(g)}.$$

These are real functions and depend on:

— the elements $f,\, g \in V_m \subset G$;
— the composition law of the group $G$;
— the parametrization of the vicinity $V_m$ of the unit element $e \in G$;
— but they *do not depend* on the considered representation, $g \to T(g)$ of the group.

The equation (4.19) is valid $\forall\, f,\, g \in V_m$. Taking $f = g^{-1} \in V_m$ (see the Lie group definition), one can write

$$\frac{\partial y(a_1,\, a_2, ...,\, a_m)}{\partial a_l} = I_k s_{kl}(a_1,\, a_2, ...,\, a_m) y(a_1,\, a_2, ...,\, a_m),$$

because the element $g \in G$ is arbitrary, with the "boundary condition"

$$y(0,\, 0, ...,\, 0) = x,$$

(since $T(0,\, 0, ... 0) = 1_{\mathcal{L}}$ and $y(g) = T(g)x$, it follows that $y(0,\, 0, ...,\, 0) = x$). Taking also into account the arbitrary character of $x \in \mathcal{L}$, one can finally write

$$\begin{cases} \dfrac{\partial T(a_1,\, a_2, ...,\, a_m)}{\partial a_l} = I_k s_{kl}(a_1,\, a_2, ...,\, a_m)\, T(a_1,\, a_2, ...,\, a_m), \\ T(0,\, 0, ...,\, 0) = 1_{\mathcal{L}}\,, \end{cases}$$

which is exactly the searched differential equation (in fact it is a system of partial differential equations, because $l = \overline{1, m}$).

The integrability conditions for this system write

$$\frac{\partial^2 T(f)}{\partial a_k \partial a_l} = \frac{\partial^2 T(f)}{\partial a_l \partial a_k}, \quad \left(l, k = \overline{1, m}\right),\ \forall\, f \in V_m \subset G.$$

But

$$\frac{\partial^2 T}{\partial a_k \partial a_l} = I_j \frac{\partial s_{jl}}{\partial a_k} T + I_j s_{jl} \frac{\partial T}{\partial a_k} = I_j \frac{\partial s_{jl}}{\partial a_k} T + I_j s_{jl} I_p s_{pk} T$$

and

$$\frac{\partial^2 T}{\partial a_l \partial a_k} = I_j \frac{\partial s_{jk}}{\partial a_l} T + I_j s_{jk} I_p s_{pl} T.$$

For $f = e$, $(a_1 = a_2 = ... = a_m = 0)$, one has $T = 1_{\mathcal{L}}$, hence

$$I_j \left( \frac{\partial s_{jl}}{\partial a_k} - \frac{\partial s_{jk}}{\partial a_l} \right)_{\vec{a}=0} = -I_j s_{jl}\Big|_{\vec{a}=0} I_p s_{pk}\Big|_{\vec{a}=0} + I_j s_{jk}\Big|_{\vec{a}=0} I_p s_{pl}\Big|_{\vec{a}=0}.$$

But

$$\frac{\partial T(\vec{a})}{\partial a_l} = I_k s_{kl}(\vec{a}) T(\vec{a})$$

and, since for $\vec{a} = 0$, $T(\vec{0}) = 1_{\mathcal{L}}$, one also has

$$I_k I_l - I_l I_k = [I_k, I_l] = I_j \left( \frac{\partial s_{jl}}{\partial a_k} - \frac{\partial s_{jk}}{\partial a_l} \right)_{\vec{a}=0} = c^j_{kl} I_j.$$

The quantities $c^j_{kl}$ are called the *structure constants* of the group and they have the following properties:

— are independent of the chosen representation;
— are real numbers;
— depend on the group structure (the composition law, etc.);
— depend on the group parametrization;
— satisfy the relations:

(*i*) $c^j_{kl} = -c^j_{lk}$,

(*ii*) $c^j_{kl} c^p_{ji} + c^j_{ik} c^p_{jl} + c^j_{li} c^p_{jk} = 0.$

These two last properties follow from those of the commutators:

(*i*) $[I_k, I_l] = c^j_{kl} I_j = -[I_l, I_k]$,

(*ii*) $\Big[I_i, [I_k, I_l]\Big] + \Big[I_k, [I_l, I_i]\Big] + \Big[I_l, [I_i, I_k]\Big] = 0,$

expressing the antisymmetry property and Jacobi's identity, respectively.

We conclude this section with two important theorems and also very useful in representation theory of groups:

**Theorem** (*Schur–Auerbach*): Any representation of a finite group is equivalent to a unitary representation.

**Theorem** (*Maschke*): Any reducible representation of a finite group is completely reducible.

These two theorems remain valid in the case of compact groups; the proof of this fact is based on the existence of the invariant integral for such groups. In particular, according to Maschke's theorem, any unitary representation is completely reducible (the representation decomposes in a sum of irreducible representations).

This way, an important result in the representation theory is obtained: any representation of a compact group is equivalent to a unitary representation that, it its turn, is completely reducible. Therefore, in such cases it is sufficient to study the irreducible unitary representations.

### IV.2.4. Algebras. Lie algebras

**Definition**: An *algebra* over a scalar corpus (usually, the corpus of complex numbers, $\mathbb{C}$) is a set of elements, $A$, with the properties:

1) $\forall\ \lambda \in \mathbb{C}, \forall\ x \in A$ one defines the element $\lambda x \in A$ called the *product* of the algebra elements with scalars (the *first operation* of the algebra);

2) $\forall\ x, y \in A$ is uniquely determined the element $z = x + y \in A$, called the *sum* of the considered elements (the *second operation* of the algebra);

3) $\forall\ x, y \in A$ there exists an operation called the *product of the algebra elements*, through $(x, y) \to z' = xy \in A$ (the *third operation* of the algebra).

4) The product of the elements and multiplication by scalars are distributive with respect to addition, and the multiplication by scalars is associative:

$$\lambda(x + y) = \lambda x + \lambda y, \quad \forall\ \lambda \in \mathbb{C}, \quad \forall\ x, y \in A;$$
$$x(y + x) = xy + xz, \quad (x + y)z = xz + yz, \quad \forall\ x, y, z \in A;$$
$$\lambda(xy) = (\lambda x)y = x(\lambda y), \quad \lambda \in \mathbb{C}, \quad \forall\ x, y \in A.$$

**Definition**: A *Lie algebra* over a corpus of scalars $\mathbb{C}$, is an algebra $L$, that satisfy the following properties:

$LA_1$: $xy + yx = 0$;

$LA_2$: $x(yz) + y(zx) + z(xy) = 0, \ \forall\ x, y, z \in L$.

If, in addition, $xy = 0, \ \forall\ x, y \in L$, then the Lie algebra is called *Abelian*.

**Observation**: Usually, the *product* of elements of a Lie algebra is defined as

$$xy \stackrel{\text{def}}{=} [x, y],$$

being called the *crochet* of the elements (or *commutator* of the elements).

**Definition**: A *finite-dimensional Lie algebra* is a Lie algebra $L$ endowed with an independent system of algebra elements, $\{X_i\}_{i=\overline{1,m}}$, ($m$ is called the *algebra dimension*) called the *algebra basis*, so that any algebra element can be represented as a linear combination of the basis elements,

$$\forall\ X \in L,\ \exists\ X_i \in L,\ (i = \overline{1,m}),$$

so that

$$X = \sum_{i=1}^{m} \alpha^i X_i\,, \quad (i = \overline{1,m}).$$

Any Lie group "gives rise" to a Lie algebra, whose elements are exactly the group generators (that is why the algebra elements have been denoted by capital letters), always satisfying – as was seen – an equation of the form

$$[X_i, X_j] = f_{ij}^k X_k,$$

that define the *product* operation of the Lie algebra elements (the Lie algebra *crochet*). The Lie algebra is *Abelian* if all the structure constants are null, since in this situation the generators commute: $X_i X_j = X_j X_i$.

**Definition**: A *Lie subalgebra*, $L'$, of a Lie algebra $L$ is a subset $L' \subset L$, that form in its turn an algebra with the same operations as those of the basic algebra.

**Definition**: An *invariant subalgebra* (or *ideal*), $I$, of a Lie algebra $L$ is a subalgebra that satisfies the property

$$\forall\, x \in I,\ \forall\, y \in L \Rightarrow xy \in I.$$

**Definition**: A Lie algebra is called *simple* if it has no proper ideals (a proper ideal is an ideal different from the improper ideals, which are $L$ and $\emptyset$).

**Definition**: A Lie algebra is called *semisimple* if it has no proper Abelian ideals.

There is a strong connection between the type of the Lie group and the type of the Lie algebra of that group, namely:

| | | |
|---|---|---|
| Lie group | $\longrightarrow$ | Lie algebra |
| Lie subgroup | $\longrightarrow$ | Lie subalgebra |
| Abelian subgroup | $\longrightarrow$ | Abelian subalgebra |
| Invariant subgroup | $\longrightarrow$ | Invariant subalgebra (ideal) |
| Simple group | $\longrightarrow$ | Simple algebra |
| Semisimple group | $\longrightarrow$ | Semisimple algebra |

**Definition**: The Abelian subalgebra of maximum dimension of a Lie algebra $L$ is called the *Cartan subalgebra*.

The basis elements of the Cartan subalgebra of a Lie algebra $L$ are usually denoted by $H_i$, ($i = \overline{1,l}$; $l$ is called the *dimension of the Cartan subalgebra* and, obviously, $[H_i, H_j] = 0$, $(i, j = \overline{1,l})$).

**Definition**: The *rank of a Lie algebra* $L$ is the dimension $l$ of the Cartan subalgebra of the Lie algebra $L$.

**Definition**: The *rank of a Lie group* $L$ is the rank of the Lie algebra of that group.

**Observations**

1) The generators of the Lie groups form Lie algebras in which the product of algebra elements is the crochet/commutator of the generators.

2) The Lie algebra of a Lie group is uniquely and completely determined by the respective Lie group. The reciprocal statement is valid up to an isomorphism, meaning that two or more isomorphic groups admit the same Lie algebra.

3) It is always possible the exponential parametrization of a Lie group, *i.e.* the parametrization in which the elements of the Lie group can be represented under the form $g = e^{\varepsilon_k I_k}$, where $\varepsilon_k$ are the group parameters in the exponential parametrization, and $I_k$ are the generators of the Lie group, because

$$\left(\frac{\partial g}{\partial \varepsilon_k}\right)_{\varepsilon_k=0} = \left[\frac{\partial}{\partial \varepsilon_k}\left(e^{\varepsilon_k I_k}\right)\right]_{\varepsilon_k=0} = I_k,$$

which is precisely the generator definition. Since $\sum_{k=1}^{m} \varepsilon_k I_k$ is a linear combination of generators (forming the Lie algebra of the considered Lie group), and since any algebra element writes as a linear combination of vectors of the algebra basis (which vectors actually are the generators of the Lie group), it follows that, obviously, $\varepsilon_k I_k$ is an element of the Lie group in question. Therefore, formally, one can write $G = e^A$, where $G$ is a Lie group, and $A$ is the algebra of that group.

## ♣ Representations of a Lie algebra

Let $L$ be a Lie algebra and $\mathcal{L}$ a $n$-dimensional linear space.

**Definition**: A *representation of the Lie algebra* $L$ on the linear space $\mathcal{L}$ is an application $T : \mathcal{L} \to \mathcal{L}$, $\forall\, x \in L \xrightarrow{\mathrm{T}} T(x) \in \mathcal{L}$, so that to each algebra element corresponds a linear operator on the space $\mathcal{L}$ with the following properties:

$AR_1$: $T(\alpha x + \beta y) = \alpha T(x) + \beta T(y), \quad \forall\, x, y \in L, \quad \forall\, \alpha, \beta \in \mathbb{C};$

$AR_2$: $T([x, y]) = [T(x), T(y)], \quad \forall\, x, y \in L.$

## ♣ The structure of the Lie algebras

As was previously seen, a Lie algebra is, basically, a set of elements $X_\alpha$ that form a linear space $L$, and satisfy the following relations:

$$[X_i, X_j] = -[X_j, X_i],$$
$$[X_i, X_j] = c_{ij}^k X_k,$$
$$\Big[X_i, [X_j, X_k]\Big] + \Big[X_j, [X_k, X_i]\Big] + \Big[X_k, [X_i, X_j]\Big] = 0.$$

As it has been seen, a system $X_i$, $(i = \overline{1,r})$ of $r$ linearly-independent vectors form an algebra ($r$-dimensional) basis, if $\forall\ X \in L$, $X = x_i X_i$, $(i = \overline{1,r})$, where $x_i$ are real numbers.

To find all the Lie algebras structures is a formidable mathematical task. The problem consists, in fact, in finding (for any value of $r$), all the real solutions of the equation

$$c^l_{ij}c^p_{lk} + c^l_{jk}c^p_{li} + c^l_{ki}c^p_{lj} = 0, \tag{4.20}$$

that satisfy the condition

$$c^s_{ij} = -c^s_{ji}, \tag{4.21}$$

since, as it has been already shown, any Lie algebra $L$ is univocally determined by the structure constants of the Lie group to which the algebra $L$ corresponds.

But the problem is difficult, because (4.20) represents a system of equations in squared unknowns. In addition, many solutions shall be equivalent to some others, since a change of the basis $\{X_i\}_{i=\overline{1,r}}$ leads to a new set of structure constants that satisfy the same equations (4.20) and (4.21).

If $X_i \to X'_j$ is a basis change expressed by

$$X'_j = a_{jk}X_k,$$

where $a_{jk}$ is the (nonsingular) matrix of transition from the old basis to the new one, then the new structure constants $c'^l_{jk}$ shall satisfy

$$c'^l_{js}X'_l = [X'_j, X'_s] = [a_{jk}X_k, a_{sm}X_m] = a_{jk}a_{sm}[X_k, X_m]$$
$$= a_{jk}a_{sm}c^p_{km}X_p = c'^l_{js}a_{lp}X_p,$$

which yields

$$c'^l_{js}a_{lp} = a_{jk}a_{sm}c^p_{km}.$$

Multiplying on the right by the inverse of the matrix $a$, one obtains

$$c'^l_{js} = a_{jk}a_{sm}c^p_{km}a^{-1}_{pl}$$

and one can easily verify that these new structure constants verify Eqs. (4.20) and (4.21).

**A)** For $r = 1$, all the elements of the Lie algebra are multiples of the same basis vector $X$, therefore all the commutators cancel and the Lie algebra is Abelian. The corresponding Lie group is uniparametric and Abelian (for example, the phase transformations group $U(1)$).

**B)** For $r = 2$, there are two vectors in the basis, $X_1$ and $X_2$, so that

$$[X_1, X_2] = aX_1 + bX_2.$$

**a)** If, for example, $a = b = 0$, then $[X_2, X_2] = 0$ and the algebra is Abelian, being given by the direct sum of the Lie algebras generated by $X_1$ and $X_2$, respectively:

$$L = L_1 \oplus L_2;$$

**b)** If $a \neq 0$, then one can find a basis in which $[X_1', X_2'] = X_1'$. Indeed, this basis is

$$X_1' = aX_1 + bX_2,$$
$$X_2' = \tfrac{1}{a}X_2,$$

because

$$[X_1', X_2'] = [aX_1 + bX_2, \frac{1}{a}X_2] = [aX_1, \frac{1}{a}X_2] + [bX_2, \frac{1}{a}X_2]$$
$$= [X_1, X_2] = aX_1 + bX_2 = X_1'.$$

The Lie subalgebra generated by $X_1'$ is invariant and Abelian, therefore the Lie algebra $L$ is not semisimple. Thus, even for $r = 2$ one obtains Lie algebras that are not semisimple.

**Example**: Consider a transformation group of the form $x' = f(x) = ax + b$. The generators then are

$$X_1 = u_{11}\frac{\partial}{\partial x_1} = u_{11}\frac{\partial}{\partial x} = x\frac{\partial}{\partial x},$$

$$X_2 = u_{12}\frac{\partial}{\partial x_1} = u_{12}\frac{\partial}{\partial x} = \frac{\partial}{\partial x},$$

because, according to the general theory, $X_k = u_{ik}\frac{\partial}{\partial x_i}$, $u_{ik}(\vec{x}) = \left(\frac{\partial f_i}{\partial a_k}\right)_{\vec{a}=0}$. In the present case $u_{11} = \left(\frac{\partial f(x)}{\partial a}\right)_{a=b=0} = x$, $u_{12} =$

$\left(\frac{\partial f(x)}{\partial b}\right)_{a=b=0} = 1$, and therefore one finds $[X_1, X_2] = \left[x\frac{\partial}{\partial x}, \frac{\partial}{\partial x}\right] = X_1$.

## ♣ Compact semisimple Lie groups

By definition, if the parameters of a Lie group vary in a finite domain, the group is called *compact*. The Lie algebra of a compact Lie group is also called compact.

Let $G$ be a semisimple compact Lie group and $L$ its associated Lie algebra. To each element $A \in L$ of the Lie algebra of the group $G$ one can associate a linear transformation (an operator) as follows: whatever other algebra element $S$ would be, one can define the commutator of $S$ and some fixed element $A$, which commutator is also an algebra element. This linear application is defined as follows:

$$p_A : \; L \to L, \quad p_A S = [A, S] \in L.$$

Therefore, with any fixed element $A \in L$ one associates the operator $p_A$ as previously defined (in fact, in this way it has been performed a representation of the Lie algebra $L$ on the space of the Lie algebra itself). If in $L$ a basis is considered, then the operator $p_A$ shall have a matrix form. Let $\{X_i\}_{i=\overline{1,n}}$ be the basis of the $n$-dimensional Lie algebra $L$; then, the two algebra elements can be written as

$$A = a_i X_i, \quad S = s_i X_i, \quad (i = \overline{1, n}).$$

Therefore, in the chosen basis, the matrix of the operator $p_A$ is $(p_A)_{ij} = c^i_{kj} a_k$. Indeed,

$$\begin{aligned} p_A S &= (p_A S)_i X_i = [A, S] = [a_k X_k, s_j X_j] \\ &= a_k s_j [X_k, X_j] = a_k s_j c^i_{kj} X_i, \end{aligned}$$

so that

$$(p_A S)_i = c^i_{kj} a_k s_j.$$

But $(p_A S)_i = (p_A)_{ij} s_j$, and thus the matrix elements of the operator $p_A$ write

$$(p_A)_{ij} = c^i_{kj} a_k \,, \quad \text{(q.e.d.)}.$$

Let now $B$ be another element of the Lie algebra $L$. To this element one associates the operator $p_B$ and one can write

$$p_B p_A S = p_B(p_A S) = p_B([A,S]) = \Big[B,[A,S]\Big] \in L.$$

One also has

$$\begin{aligned}(p_B p_A S)_i X_i &= \Big[B,[A,S]\Big] = [b_s X_s, a_k s_j c^p_{kj} X_p] \\ &= b_s a_k s_j c^p_{kj}[X_s, X_p] = b_s a_k s_j c^p_{kj} c^i_{sp} X_i,\end{aligned}$$

therefore

$$(p_B p_A S)_i = c^i_{sp} c^p_{kj} a_k b_s s_j = (p_B p_A)_{ij} s_j,$$

that is

$$(p_B p_A)_{ij} = c^i_{sp} c^p_{kj} a_k b_s = c^i_{sp} c^p_{kj} b_s a_k.$$

The trace of the matrix of the product application $p_B p_A$ is

$$\mathrm{Tr}(p_B p_A) = (p_B p_A)_{ii} = c^i_{sp} c^p_{ki} b_s a_k \overset{s\leftrightarrow p}{=} c^i_{ps} c^s_{ki} b_p a_k = g_{pk} b_p a_k, \quad (4.22)$$

where, by definition, the quantities $g_{pk}$ are given by

$$g_{pk} \overset{\text{def}}{=} c^i_{ps} c^s_{ki}$$

and satisfy the relationship

$$g_{kp} = c^i_{ks} c^s_{pi} \overset{i\leftrightarrow s}{=} c^s_{ki} c^i_{ps} = c^i_{ps} c^s_{ki} = g_{pk},$$

*i.e.* $g_{pk}$ are symmetric quantities.

Thus, the equation (4.22) allows one to associate with any two algebra elements a symmetric bilinear form called the *scalar product* of the elements $A$ and $B$, and denoted by $(A,B)$. A suggestive index change yields

$$(A,B) = g_{\mu\nu} b_\mu a_\nu.$$

Perform now a basis change, $X_k \to X'_n = a_{nk} X_k$. Under this operation, the structure constants transform according to the already known equation $c'^l_{ks} = a_{kn} a_{sm} c^p_{nm} a^{-1}_{pl}$. Then,

$$g_{\mu\nu} \to g'_{\mu\nu} = c'^{j}_{\mu i} c'^{i}_{\nu j} = a_{\mu u} a_{ip} c^{k}_{up} a^{-1}_{kj} a_{\nu m} a_{jn} c^{l}_{mn} a^{-1}_{li}$$
$$= a_{\mu u} a_{\nu m} c^{k}_{up} c^{l}_{mn} \delta_{lp} \delta_{nk} = a_{\mu u} a_{\nu m} c^{k}_{ul} c^{l}_{mk} = a_{\mu u} a_{\nu m} g_{um},$$

that is

$$g'_{\mu\nu} = a_{\mu\alpha} a_{\nu\beta} g_{\alpha\beta}.$$

Thus, the quantities $g_{\mu\nu} = g_{\nu\mu}$ transform as a second rank symmetric tensor. It can also be proved that the scalar product defined through the medium of $g_{\mu\nu}$ is an invariant. Indeed,

$$A = a_i X_i = a'_i X'_i = a'_i a_{ij} X_j = a'_j a_{ji} X_i,$$

showing that $a_i = a'_j a_{ji}$, that is $a'_j = a_i a'_{ij}$, so that

$$(A', B') = g'_{\mu\nu} b'_{\mu} a'_{\nu} = a_{\mu u} a_{\nu m} g_{um} b_{\alpha} a'_{\alpha\mu} a_{\beta} a^{-1}_{\beta\nu}$$
$$= g_{um} b_{\alpha} a_{\beta} \delta_{\alpha u} \delta_{\beta m} = g_{\alpha\beta} b_{\alpha} a_{\beta} = (A, B).$$

It follows that $g_{\mu\nu}$ acts as a metric tensor in the vectorial space of the Lie algebra $L$. Then one can define the new structure constants

$$c_{\lambda\mu\nu} = g_{\lambda\rho} c^{\rho}_{\mu\nu},$$

and one can be proved that they are completely antisymmetric. To this end, one uses the definition of the metric tensor $g_{\mu\nu} = c^{\alpha}_{\mu\beta} c^{\beta}_{\nu\alpha}$, its symmetry property, $g_{\mu\nu} = g_{\nu\mu}$, and the relation (resulting from Jacobi's identity for generators) satisfied by the structure constants, $c^{l}_{ij} c^{p}_{lk} + c^{l}_{jk} c^{p}_{li} + c^{l}_{ki} c^{p}_{lj} = 0$. Indeed, one can write

$$c_{lik} = g_{ls} c^{s}_{jk}, \quad \text{with} \quad g_{ls} = c^{\alpha}_{l\beta} c^{\beta}_{s\alpha},$$

so that

$$c_{ljk} = c^{\alpha}_{l\beta} c^{\beta}_{s\alpha} c^{s}_{jk} = -c^{\alpha}_{l\beta} \left( c^{\beta}_{sj} c^{s}_{k\alpha} + c^{\beta}_{sk} c^{s}_{\alpha j} \right),$$

because $c^{s}_{jk} c^{\beta}_{s\alpha} + c^{s}_{k\alpha} c^{\beta}_{sj} + c^{s}_{\alpha j} c^{\beta}_{sk} = 0$. Consequently,

$$c_{ljk} = -c^{\alpha}_{l\beta} c^{\beta}_{sj} c^{s}_{k\alpha} - c^{\alpha}_{l\beta} c^{\beta}_{sk} c^{s}_{\alpha j}.$$

Since the structure constants are antisymmetric in the last two indices, it remains to show their antisymmetry in the first two indices, all the other possibilities/combinations being obvious. Therefore, one has to show that $c_{jlk} = -c_{ljk}$. One can write

$$c_{jlk} = -c^{\alpha}_{j\beta}c^{\beta}_{sl}c^{s}_{k\alpha} - c^{\alpha}_{j\beta}c^{\beta}_{sk}c^{s}_{\alpha l} \overset{\alpha\to s\to\beta\to\alpha}{=} -c^{s}_{j\alpha}c^{\alpha}_{\beta l}c^{\beta}_{ks} - c^{\beta}_{js}c^{s}_{\alpha k}c^{\alpha}_{\beta l}$$

$$\overset{\alpha\to\beta\to s\to\alpha}{=} c^{\alpha}_{l\beta}c^{\beta}_{sk}c^{s}_{\alpha j} + c^{\alpha}_{l\beta}c^{\beta}_{sj}c^{s}_{k\alpha} = -c_{ljk},$$

which completes the proof.

*

* *

One of the most important groups in QFT, possessing the most important and numerous subgroups, is the group $GL(n,\mathbb{C})$.

Consider the set of square matrices of rank $n$, with elements as complex numbers, denoted $\mathcal{M}(n)$. If in this set one introduces the commutation operation defined by the relation

$$[A,B] = AB - BA; \quad A,B \in \mathcal{M}(n),$$

then the following relations can be easily verified:

$$[A,B] = -[B,A],$$
$$\Big[A,[B,C]\Big] + \Big[B,[C,A]\Big] + \Big[C,[A,B]\Big] = 0.$$

This way, the set $\mathcal{M}(n)$ forms a Lie algebra with respect to the commutation relation.

**Definition**: The *linear group of rank* $n$, denoted by $GL(n,\mathbb{C})$, is the set of square matrices, invertible, of rank $n$, with complex numbers as elements, the composition law of the group being the matrix product:

$$GL(n,\mathbb{C}) = \{g \in \mathcal{M}(n) |\ \det g \neq 0\}.$$

**Theorem**: The Lie algebra of the group $GL(n,\mathbb{C})$ is isomorphic to the Lie algebra $\mathcal{M}(n)$.

The generators of the group $GL(n,\mathbb{C})$ are real matrices, $X_{ij} \in \mathcal{M}(n)$, $(i,j = \overline{1,n})$, with all elements zero, except for the element

$(ij)$ which equals unity, so that $(X_{ij})_{kl} = \delta_{ik}\delta_{jl}$. These generators satisfy the commutation relations

$$[X_{ij}, X_{kl}] = \delta_{jk}X_{il} - \delta_{il}X_{kj}$$

and form the basis of the Lie algebra of the group $GL(n, \mathbb{C})$, the algebra elements being of the form $X = \alpha_{ij}X_{ij}$, $(\alpha_{ij} \in \mathbb{C})$. The above commutation relations are also satisfied by the differential operators

$$X_{ij} = x_i \frac{\partial}{\partial x_j}, \quad (x_i, x_j \in \mathbb{C}^n),$$

that can also be used as generators.

**Theorem**: The *exponential of the group* $GL(n, \mathbb{C})$ is the application $e^{tX} = \sum_{p=0}^{\infty} \frac{1}{p!}(tX)^p$, $(t \in \mathbb{R})$, with the following properties:

1) $t \to e^{tX}$ is a homomorphism of $\mathbb{R}$ in $GL(n, \mathbb{C})$;

2) $X \to e^{tX}$ is an isomorphism of a vicinity of the null matrix of $\mathcal{M}(n)$ (denoted by $U \subset \mathcal{M}(n)$), in a vicinity of the unit element of $GL(n, \mathbb{C})$, (denoted by $V \subset GL(n, \mathbb{C})$);

3) The same property is also true between a subalgebra of $\mathcal{M}(n)$ and the subgroup of $GL(n, \mathbb{C})$ corresponding to this subalgebra. By virtue of this theorem, one can show that

$$e^{gXg^{-1}} = ge^X g^{-1}, \quad e^{X+Y} = e^X e^Y,$$

where $g \in GL(n, \mathbb{C})$, $X, Y \in \mathcal{M}(n)$, and $[X, Y] = 0$.

**Theorem**: If $\{\lambda_i\}_{i=\overline{1,n}}$ are $n$ characteristic roots, distinct or not, of the matrix $X \in \mathcal{M}(n)$, then the characteristic roots of the exponential $e^X$ are $\{e^{\lambda_i}\}_{i=\overline{1,n}}$.

A consequence of this theorem is that

$$\det e^X = e^{\text{Tr}(X)}.$$

**Theorem**: Any unitary matrix $U$, or Hermitian matrix $H$, can be brought to a diagonal form by a similarity transformation, $U' = V^+UV$ and $H' = V^+HV$, respectively, where $V$ is a unitary matrix, and $U'$ and $H'$ are diagonal.

This result remains also valid for symmetric and orthogonal matrices with real numbers as elements, but not for those with complex numbers as elements, even they satisfy the same properties. Therefore, $\forall\ U \in \{U|U^+U = I\}$ and $\forall\ H \in \{H|H = H^+\}$, $\exists\ V \in \{V|V^+V = I\}$, so that $U' = V^+UV$ and $H' = V^+HV$ are diagonal.

Using these results, one can show that $\det e^H = e^{\mathrm{Tr}(H)}$, where $H$ is a Hermitian matrix. According to the above theorem, $H = (h_{ii}) = \mathrm{diag}(h_{ik})$. Then

$$e^H = I + \frac{1}{1!}H + \frac{1}{2!}H^2 + ... + \frac{1}{n!}H^n + ... = \left( \underbrace{\begin{bmatrix} 1 & 0 & \cdots & 0 \\ 1 & 0 & \cdots & 0 \\ 0 & 1 & \cdots & 0 \\ \vdots & \vdots & \vdots & \vdots \\ 0 & 0 & \cdots & 1 \end{bmatrix}}_{\text{n elements}} \right)$$

$$+ \frac{1}{1!}\begin{pmatrix} h_{11} & 0 & \cdots & 0 \\ 0 & h_{22} & \cdots & 0 \\ \vdots & \vdots & \vdots & \vdots \\ 0 & 0 & \cdots & h_{nn} \end{pmatrix} + \frac{1}{2!}\begin{pmatrix} h_{11}^2 & 0 & \cdots & 0 \\ 0 & h_{22}^2 & \cdots & 0 \\ \vdots & \vdots & \vdots & \vdots \\ 0 & 0 & \cdots & h_{nn}^2 \end{pmatrix} + ...$$

$$+ \frac{1}{n!}\begin{pmatrix} h_{11}^n & 0 & \cdots & 0 \\ 0 & h_{22}^n & \cdots & 0 \\ \vdots & \vdots & \vdots & \vdots \\ 0 & 0 & \cdots & h_{nn}^n \end{pmatrix} + ... = \begin{pmatrix} e^{h_{11}} & 0 & \cdots & 0 \\ 0 & e^{h_{22}} & \cdots & 0 \\ \vdots & \vdots & \vdots & \vdots \\ 0 & 0 & \cdots & e^{h_{nn}} \end{pmatrix},$$

therefore

$$\det\ e^H = \det \begin{pmatrix} e^{h_{11}} & 0 & \cdots & 0 \\ 0 & e^{h_{22}} & \cdots & 0 \\ \vdots & \vdots & \vdots & \vdots \\ 0 & 0 & \cdots & e^{h_{nn}} \end{pmatrix} = \prod_{i=1}^{n} e^{h_{ii}}$$

hence

$$\ln\left(\det e^H\right) = \ln \prod_{i=1}^{n} e^{h_{ii}} = \sum_{i=1}^{n} h_{ii} = \mathrm{Tr}(H),$$

that is

$$\det e^H = e^{\mathrm{Tr}(H)}, \quad \text{(q.e.d.)}.$$

Using this result, one can conclude that the determinant of a unitary matrix expressed under the form $u = e^{iH}$ (where $H$ is a Hermitian matrix) equals one (*i.e.* is a unimodular matrix), if the Hermitian matrix $H$ has a null trace.

Among the Lie subgroups of the group $GL(n, \mathbb{C})$ one can mention the following:

1) The special (unimodular) linear group

$$SL(n, \mathbb{C}) = \{g|g \in GL(n, \mathbb{C}),\ \det\ g = 1\};$$

2) The unitary group $U(n) = \{g|g \in GL(n, \mathbb{C}), g^+g = e\}$;

3) The special unitary group

$$\begin{aligned} SU(n) &= \{g|g \in SL(n, \mathbb{C}),\ g^+g = e\} \\ &= \{g|g \in U(n),\ \det g = 1\} = U(n) \cap SL(n, \mathbb{C}); \end{aligned}$$

4) The real linear group $GL(n, \mathbb{R}) = \{g|g \in GL(n, \mathbb{C}),\ \mathrm{Im}\ g = 0\}$;

5) The special real linear group

$$SL(n, \mathbb{R}) = \{g|g \in GL(n, \mathbb{R}),\ \det g = 1\};$$

6) The real orthogonal group

$$\begin{aligned} O(n, \mathbb{R}) &= \{g|g \in GL(n, \mathbb{R}),\ g^T g = e\} \\ &= U(n) \cap GL(n, \mathbb{R}); \end{aligned}$$

7) The special real orthogonal group

$$\begin{aligned} SO(n, \mathbb{R}) &= \{g|g \in SL(n, \mathbb{R}),\ g^T g = e\} \\ &= \{g|g \in O(n, \mathbb{R}),\ \det g = 1\} - O(n, \mathbb{R}) \cap SL(n, \mathbb{R}). \end{aligned}$$

One can prove that all these groups are compact.

**Definition**: An application $f : \mathcal{L} \to \mathcal{M}$ (with $\mathcal{L}$ and $\mathcal{M}$ topological spaces) is called *homeomorphism* or *topological application* is $f$ is a bijection, and the applications $f$ and $f^{-1}$ are continuous.

**Definition**: Two topological spaces $\mathcal{L}$ and $\mathcal{M}$ are called *homeomorphic* if there exists a homeomorphism between $\mathcal{L}$ and $\mathcal{M}$.

In the case of existence of a topological group structure, a homeomorphism induces a homomorphism.

**Theorem**: The groups $U(n)$ and $U(1) \times SU(n)$ are homeomorphic.

Since all three groups $U(1)$, $U(n)$ and $SU(n)$ are topological groups, from the above theorem it follows that the groups $U(n)$ and $U(1) \times SU(n)$ are also homomorphic.

**Theorem**: Let $L$ be a Lie algebra of matrices $X$. The application

$$X \in L \to u = e^X = I + \frac{1}{1!}X + \frac{1}{2!}X^2 + ... + \frac{1}{n!}X^n + ...$$

defines the matrices $u$ that form a Lie group structure as against the matrix multiplication.

**Theorem**: Any matrix $u \in U(n)$ can be written under the form $u = e^{ih}$, where $h$ is a Hermitian matrix. Reciprocally, any matrix $e^{ih}$, where $h$ is a Hermitian matrix, is a unitary matrix.

Therefore, the elements of the group $SU(n)$ can be represented through the matrices $u = e^{iX}$, where $X$ is a Hermitian matrix of null trace. Indeed,

1) since $u$ is unitary, this implies $u^+u = I,\ e^{-ih^+}e^{ih} = e^{i(h-h^+)} = I \Rightarrow h = h^+$;

2) since $u$ is unimodular, this means $\det u = 1 \Rightarrow \det e^{ih} = 1 \Rightarrow \mathrm{Tr}(h) = 0$.

**Theorem**: Let $H$ be a subgroup of the group $SU(n)$, composed of matrices of the form $h = \begin{pmatrix} 1 & 0 \\ 0 & g \end{pmatrix}$, where $g$ is a matrix of rank $(n-1)$. In this case the subgroup $H$ is isomorphic with the group $SU(n-1)$, and the quotient set $SU(n)/H$ is homeomorphic (or homomorphic) with the hypersphere of the $(2n-1)$-dimensional space.

One can also prove that it is possible a decomposition of the form $SU(n) \supset U(1) \times SU(n-1)$ called *canonical decomposition*. This decomposition leads to the canonical chain of subgroups

$$SU(n) \supset SU(n-1) \supset ... \supset SU(3) \supset SU(2).$$

*

*   *

Let us now return to the invariance condition of the variational principle[10] under the internal symmetry transformations

$$u_i(x) \to u_i'(x) = u_i(x) + \delta u_i(x).$$

For the first variation of the Lagrangian one gets

$$\frac{\partial \mathcal{L}}{\partial u_i}\delta u_i + \frac{\partial \mathcal{L}}{\partial(\partial_\mu u_i)}\delta(\partial_\mu u_i) = 0.$$

Therefore, in order to find this invariance condition, one must know the variation of the field function, $\delta u_i(x)$, and of its derivatives $\delta\Big(\partial_\mu u_i(x)\Big)$ as well.

Under a transformation of the symmetry group of a system, the wave functions transform according to a certain group representation. For a better understanding, it is worth considering a concrete example, namely the relativistic invariance. As it is known, any field theory must be relativistically invariant (more precisely, relativistically covariant). This condition follows from the necessity of giving a correct description to physical phenomena for high velocities, comparable to the velocity of light. In other words, the Lagrangian describing the studied physical system must be invariant with respect to the space-time transformations of the Minkowski space; these transformations form a group structure, called *Lorentz group*.

Consider the Minkowski space-time with coordinates $x_0, x_1, x_2, x_3$ and with the metric $ds^2 = dx_0^2 - d\vec{x}^2$. An arbitrary rotation about the origin in this space can be represented as a product of six successive rotations in the planes $x_1Ox_2, x_2Ox_3, x_3Ox_1, x_0Ox_1, x_0Ox_2, x_0Ox_3$. For example, a rotation of angle $\alpha$ in the plane $x_1Ox_2$ writes

$$\begin{cases} x_1' = +x_1\cos\alpha + x_2\sin\alpha, \\ x_2' = -x_1\sin\alpha + x_2\cos\alpha, \\ x_3' = x_3, \\ x_0' = x_0, \end{cases}$$

[10]This invariance, for the internal symmetries, coincides with that of the Lagrangian of our problem.

while a rotation of angle $w$ in the plane $x_0Ox_3$ writes

$$\begin{cases} x_1' = x_1, \\ x_2' = x_2, \\ x_3' = x_3 \cosh w + x_0 \sinh w, \\ x_0' = x_3 \sinh w + x_0 \cosh w. \end{cases}$$

These rotations can be expressed in the matrix form

$$a_{12} = \begin{pmatrix} +\cos\alpha & \sin\alpha & 0 & 0 \\ -\sin\alpha & \cos\alpha & 0 & 0 \\ 0 & 0 & 1 & 0 \\ 0 & 0 & 0 & 1 \end{pmatrix}, \quad a_{03} = \begin{pmatrix} 1 & 0 & 0 & 0 \\ 0 & 1 & 0 & 0 \\ 0 & 0 & \cosh w & \sinh w \\ 0 & 0 & \sinh w & \cosh w \end{pmatrix}.$$

Therefore, an arbitrary rotation can be represented by means of a matrix obtained as a product of six corresponding matrices. These rotations in the four-dimensional space have the following properties:

1) The result of two successive rotations is also a rotation; the product of two rotations (which corresponds to the two succesive rotations) is represented by the product of their matrices which is a matrix of the same type;

2) Among these rotations there is one that transforms the space-time into itself (referred to us the unit rotation, *i.e.* the rotation of angle zero); this rotation is represented by the unit matrix;

3) There is an inverse for any rotation (the rotation of opposite angle). For instance, the inverse rotation in the $x_1Ox_2$ plane is given by the angle $(-\alpha)$. The product of a given rotation and its inverse is equivalent to the unit rotation, hence the matrix corresponding to the inverse rotation is inverse to the matrix of the original rotation.

These three properties ensure the group structure of the rotations, and also of the representing matrices.

The rotation group in the four-dimensional Minkowski space-time is called the *Lorentz group*. In the case of Lorentz group, a general coordinate transformation writes

$$x_\mu \to x'_\mu = a_{\mu\nu} x_\nu, \quad (\mu, \nu = 0, 1, 2, 3),$$

where $a_{\mu\nu}$ represents (and also includes) the product of all matrices corresponding to the rotations in all possible planes. If the translations along the coordinate axes are also considered, then one obtains the coordinate transformations

$$x_\mu \to x'_\mu = a_{\mu\nu}x_\nu + a_\mu, \quad (\mu, \nu = 0, 1, 2, 3)$$

of the Poincaré group, where $a_\mu$ is an arbitrary constant four-vector.

The number of the independent parameters that determine the group transformations (six, in the case of Lorentz group, and ten, in the case of Poincaré group) gives the *group dimension*. If the parameters can vary continuously in a certain domain, the group is called *continuous*.

As it has been seen, the finite continuous groups (possessing a finite number of parameters) are called *Lie groups*. The Lorentz and Poincaré groups are examples of non-Abelian Lie groups.

A coordinate transformation $x_\mu \to x'_\mu$ implies – obviously – a transformation of the field function $u_{(r)}(x_\mu)$. A particle of mass $m$ and spin $s$ is described by a wave function with $2s + 1$ components or – differently speaking – by a $2s + 1$ - component multiplet.

For example, a four-component wave function transforms according to

$$\begin{aligned}
u_0(x) &\to u'_0(x') = A_{00}u_0(x) + A_{01}u_1(x) + A_{02}u_2(x) + A_{03}u_3(x);\\
u_1(x) &\to u'_1(x') = A_{10}u_0(x) + A_{11}u_1(x) + A_{12}u_2(x) + A_{13}u_3(x);\\
u_2(x) &\to u'_2(x') = A_{20}u_0(x) + A_{21}u_1(x) + A_{22}u_2(x) + A_{23}u_3(x);\\
u_3(x) &\to u'_3(x') = A_{30}u_0(x) + A_{31}u_1(x) + A_{32}u_2(x) + A_{33}u_3(x).
\end{aligned}$$

Like the coordinate transformation, the transformations of the field function are characterized by a matrix; in the analyzed case it is

$$(T_a) = \begin{pmatrix} A_{00} & A_{01} & A_{02} & A_{03} \\ A_{10} & A_{11} & A_{12} & A_{13} \\ A_{20} & A_{21} & A_{22} & A_{23} \\ A_{30} & A_{31} & A_{32} & A_{33} \end{pmatrix}.$$

The number of components of the field function (the multiplet dimension) gives the dimension of the matrix $(T_a)$. Therefore, under a rotation of some angle, the coordinates transform according to the matrices $a_{\mu\nu}$, and the wave functions according to the matrices $(T_a)$.

This way, to each space-time rotation one can associate a transformation matrix of the field function components, $(T_a)$, so that the product of two such matrices $(T_{a_1a_2}) = (T_{a_1}T_{a_2})$ corresponds to the product of the matrices $a_1$ and $a_2$, that is the matrix $a_1a_2$. This way, one obtains a representation of the Lorentz group on the vector space of the field functions,

$$a \to a_1a_2,$$

$$T_a \to T_{a_1a_2} = T_{a_1}T_{a_2}.$$

The vectors of the linear space of representation are just the field functions. The number of components of the field function multiplet equals the dimension of the transformation matrix of these components and, also, equals the dimension of the representation space.

Here are several examples of representations of the Lorentz group.

1) To each rotation in the four-dimensional space-time one can associate the unit operator of the linear space of the field functions (this operator acts on these functions). On rotating the four-dimensional space-time, a wave function then transforms into itself. In other words, under the action of the representation operators (in fact only one, namely the unit operator which acts in the representation space), the field function goes into itself. Such a field function is called *scalar* and describes the spin-zero particles, having a single component in the multiplet $(2s + 1 = 1)$. The representation space is unidimensional;
2) To each rotation in the four-dimensional space-time one can associate a matrix of the same type $(4 \times 4)$, acting on four-component field/wave functions: $u'_\alpha(x') = (T_a)_{\alpha\beta}u_\beta(x)$, $(\alpha, \beta = 0, 1, 2, 3)$. Such a field function, whose components $u_\alpha(x)$, $(\alpha = 0, 1, 2, 3)$ transform through the medium of the matrices of the same type as those of the Lorentz group $(4 \times 4)$ is called *vector function*. It describes the spin-1 particles, provided that a certain additional condition is imposed on its components (see Chap. I);

3) Consider, now, that each element of the Lorentz group corresponds to a matrix of the type $T_{\alpha\beta,\gamma\delta} = a_{\alpha\gamma}a_{\beta\delta}$. The field function that transforms through the medium of these matrices is of the form $u_{a\beta}(x)$ and one can write

$$u'_{\alpha\beta}(x') = T_{\alpha\beta,\gamma\delta}u_{\gamma\delta}(x).$$

Such a representation of the Lorentz group is called a *tensor* one (or *tensorial*. The field function $u_{\alpha\beta}(x)$ is a second-rank tensor and describes the spin-2 particles;

4) Suppose that to each element of the Lorentz group corresponds a matrix of the type

$$\sigma_{\mu\nu} = \frac{1}{2i}(\gamma_\mu\gamma_\nu + \gamma_\nu\gamma_\mu),$$

where $\gamma_\mu$ are the Dirac matrices,

$$\gamma_0 = \begin{pmatrix} \varepsilon & 0 \\ 0 & -\varepsilon \end{pmatrix}, \quad \gamma_1 = \begin{pmatrix} 0 & \sigma_1 \\ -\sigma_1 & 0 \end{pmatrix}, \quad \gamma_2 = \begin{pmatrix} 0 & \sigma_2 \\ -\sigma_2 & 0 \end{pmatrix},$$

$$\gamma_3 = \begin{pmatrix} 0 & \sigma_3 \\ -\sigma_3 & 0 \end{pmatrix}.$$

Here $\varepsilon = \begin{pmatrix} 1 & 0 \\ 0 & 1 \end{pmatrix}$, and $\sigma_i$, $(i = \overline{1,3})$ are the *Pauli matrices*:
$\sigma_1 = \begin{pmatrix} 0 & 1 \\ 1 & 0 \end{pmatrix}$, $\sigma_2 = \begin{pmatrix} 0 & -i \\ i & 0 \end{pmatrix}$, and $\sigma_3 = \begin{pmatrix} 1 & 0 \\ 0 & -1 \end{pmatrix}$.

The field function that transforms according to the matrices $\sigma_{\mu\nu}$ under rotations of the four-dimensional space-time is a *spinor* with four components, $\psi(x) = \begin{pmatrix} \psi_1(x) \\ \psi_2(x) \\ \psi_3(x) \\ \psi_4(x) \end{pmatrix}$ and describes the spin-1/2 particles. Under arbitrary rotations by the angles $\varepsilon_{\mu\nu}$ in the four-dimensional space-time, $\psi(x)$ transforms according to

$$\psi'(x') = e^{i\,\varepsilon_{\mu\nu}\sigma_{\mu\nu}}\psi(x),$$

and, if $\mathcal{O}(\varepsilon^2) = 0$, *i.e.* if the infinitesimal transformations are considered, then

$$\psi(x) \to \psi'(x') = \psi(x) + i\varepsilon_{\mu\nu}\sigma_{\mu\nu}\psi(x),$$

where $\varepsilon_{\mu\nu}$ stand for the rotation parameters in the four-dimensional space-time. The spinor

$$\overline{\psi}(x) = \psi^+(x)\gamma_0 = \Big(\psi_1^*(x),\, \psi_2^*(x),\, -\psi_3^*(x),\, -\psi_4^*(x)\Big)$$

is called the *Dirac conjugate spinor*. This representation of the Lorentz group is called the *spinorial representation*.

The number of independent components in the multiplet is referred to as the *dimension of the representation*.

**Definition**: A representation whose dimension is equal to the order (dimension) of the group is said to be *adjoint*.

Thus, an adjoint representation of the Lorentz group is a representation of dimension six.

**Definition**: The representation of minimum dimension (excluding the unit one) is called *fundamental*.

Any other representation of higher dimension can be obtained from the fundamental one by direct products of representations (by means of "multiplication").

Under infinitesimal rotations by the angles $\varepsilon_{\mu\nu}$, the field functions transform as follows:

1) $\varphi(x) \to \varphi'(x') = \varphi(x)$, (scalar field function);
2) $u_\mu(x) \to u'_\mu(x') = (\delta_{\mu\nu} + \varepsilon_{\mu\nu})u_\nu(x) + \mathcal{O}(\varepsilon)^2$, (vectorial field function);
3) $\psi(x) \to \psi'(x') = (1 + i\varepsilon_{\mu\nu}\sigma_{\mu\nu})\psi(x) + \mathcal{O}(\varepsilon)^2$, (spinorial field function).

Here are several examples of relativistic invariant quantities (Lorentz invariants):

*a*) $\varphi(x)\varphi(x)$ — scalar field;
*b*) $u_\mu(x)u_\mu(x)$ — vectorial field;
*c*) $\overline{\psi}(x)\gamma_\mu\partial_\mu\psi(x)$ — spinorial field;
*d*) $\overline{\psi}(x)\gamma_\mu\psi(x)u_\mu(x)$ — spinorial and vectorial fields;
*e*) $\overline{\psi}(x)\psi(x)\varphi(x)$ — spinorial and scalar fields.

### IV.2.5. Examples of external symmetry groups

This paragraph is dedicated to two of the best-known and utilized external symmetry groups, namely the *Lorentz group* and the *Poincaré group*.

### The Lorentz group

Consider the Minkowski space-time $M_4$ with $x^0 = t, x^1 = x, x^2 = y, x^3 = z$ and the metric of signature $(-2)$,

$$ds^2 = g_{\mu\nu} dx^\mu dx^\nu,$$

where $g_{\mu\nu} = \mathrm{diag}\,(+1, -1, -1, -1)$ is the fundamental metric tensor. The set of transformations of the form

$$x^\mu \to x'^\mu = \Lambda^\mu{}_\nu x^\nu, \quad (\mu, \nu = 0, 1, 2, 3) \tag{4.23}$$

or, in a matrix representation,

$$x' = \Lambda x$$

form a group structure called the *Lorentz group*, if these transformations conserve the metric of the Minkowski space-time (or, equivalently, if they conserve the vector norm in the Minkowski space-time). In fact, the transformations of the type (4.23) represent "rotations" in the considered four-dimensional space. As it is known, under a rotation the vectors norm remain unchanged,

$$x'_\mu x'^\mu = x_\nu x^\nu, \quad (\mu, \nu = 0, 1, 2, 3),$$

or

$$\|x'\|^2 = \|x\|^2.$$

But

$$\|x'\|^2 = g_{\mu\nu} x'^\mu x'^\nu = x'^T G x' = (\Lambda x)^T G (\Lambda x) = x^T (\Lambda^T G \Lambda) x$$

and

$$\|x\|^2 = g_{\mu\nu} x^\mu x^\nu = x^T G x,$$

therefore

$$G = \Lambda^T G \Lambda, \tag{4.24}$$

where

$$G = \begin{pmatrix} +1 & 0 & 0 & 0 \\ 0 & -1 & 0 & 0 \\ 0 & 0 & -1 & 0 \\ 0 & 0 & 0 & -1 \end{pmatrix} = (g_{\mu\nu}).$$

According to (4.24), $\det G = \det G(\det \Lambda)^2$, because $\det \Lambda^T = \det \Lambda$, that is $\det \Lambda = \pm 1$. Written by components, the equation (4.24) becomes

$$g_{\alpha\beta} = \Lambda^{\nu}{}_{\alpha} g_{\mu\nu} \Lambda^{\mu}{}_{\beta}.$$

Taking $\alpha = \beta = 0$, one obtains

$$\begin{aligned} g_{00} = \Lambda^{\nu}{}_{0} g_{\mu\nu} \Lambda^{\mu}{}_{0} &= \Lambda^{0}{}_{0} g_{00} \Lambda^{0}{}_{0} + \Lambda^{i}{}_{0} g_{ji} \Lambda^{j}{}_{0} \\ &= \left(\Lambda^{0}{}_{0}\right)^2 - \left(\Lambda^{1}{}_{0}\right)^2 - \left(\Lambda^{2}{}_{0}\right)^2 - \left(\Lambda^{3}{}_{0}\right)^2, \end{aligned}$$

which yields

$$\left(\Lambda^{0}{}_{0}\right)^2 = 1 + \left(\Lambda^{1}{}_{0}\right)^2 + \left(\Lambda^{2}{}_{0}\right)^2 + \left(\Lambda^{3}{}_{0}\right)^2,$$

so that

$$\left(\Lambda^{0}{}_{0}\right)^2 \geq 1,$$

hence $\Lambda^{0}{}_{0} \geq +1$, or $\Lambda^{0}{}_{0} \leq -1$, that is $\left|\Lambda^{0}{}_{0}\right| \geq 1$, and the Lorentz group can be written as

$$L = \left\{ \Lambda \in GL(4, \mathbb{R}) \middle| \Lambda^T G \Lambda = G, \ \det \Lambda = \pm 1, \ |\Lambda^{0}{}_{0}| \geq 1 \right\}.$$

The constraints $\det \Lambda = \pm 1$, $\left|\Lambda^{0}{}_{0}\right| \geq 1$ lead to a division of the Lorentz group into classes. Let $L_+$ be the Lorentz transformations subset with $\det \Lambda = +1$ (*proper Lorentz transformations*), and $L_-$ the Lorentz transformations subset with $\det \Lambda = -1$ (*improper Lorentz transformations*). Let also denote by $L^{\uparrow}$ the subset of the Lorentz transformations with $\Lambda^{0}{}_{0} \geq 1$ (*orthochronous Lorentz transformations*) and by $L^{\downarrow}$ the subset of the Lorentz transformations with $\Lambda^{0}{}_{0} \leq -1$ (*antichronous Lorentz transformations*). One can then distinguish the following four classes:

1) $L_+^\uparrow = \left\{\Lambda \in L \middle| \det\Lambda = +1,\ \Lambda^0{}_0 \geq +1\right\}$;

2) $L_+^\downarrow = \left\{\Lambda \in L \middle| \det\Lambda = +1,\ \Lambda^0{}_0 \leq -1\right\}$;

3) $L_-^\uparrow = \left\{\Lambda \in L \middle| \det\Lambda = -1,\ \Lambda^0{}_0 \geq +1\right\}$;

4) $L_-^\downarrow = \left\{\Lambda \in L \middle| \det\Lambda = -1,\ \Lambda^0{}_0 \leq -1\right\}$.

The proper Lorentz transformations $L_+ = L_+^\uparrow \cup L_+^\downarrow$ form an invariant subgroup of the Lorentz group $L$, called the *proper rotations subgroup*. One can prove that $L_+$ is a simple group. The factor group $L/L_+$ contains only two elements $\left\{E, G\right\}$, where $E$ is the unit matrix $4\times 4$. The improper Lorentz transformations $L_- = L_-^\uparrow \cup L_-^\downarrow$ do not form a group structure. The condition $\left|\Lambda^0{}_0\right| \geq 1$ allows separation of the Lorentz group into two subsets: $L^\uparrow = L_+^\uparrow \cup L_-^\uparrow$ and $L^\downarrow = L_+^\downarrow \cup L_-^\downarrow$ that contain the orthochronous and antichronous transformations, respectively. The orthochronous transformations form a group structure (the orthochronous Lorentz group, which is an invariant subgroup of the Lorentz group), the factor group $L/L^\uparrow$ containing the elements $\left\{E,\ -E\right\}$.

It can be proved that the group $SL(2,\,\mathbb{C})$ is homomorphic with $L_+$. As a result, any transformation of the Lorentz group can be obtained from two transformations of the group $SL(2,\,\mathbb{C})$, whose coefficients differ only in sign. The elements $\left\{E,\ -E\right\}$ form the invariant subgroup $H \subset SL(2,\,\mathbb{C})$, while the group $L_+$ is isomorphic with the complementary group of $H$.

In the case of infinitesimal transformations $x'^\alpha = x^\alpha + \omega^\alpha{}_\beta x^\beta$, one can write

$$x'^\alpha - \left(\delta^\alpha{}_\beta + \omega^\alpha{}_\beta\right)x^\beta,$$

that is

$$\Lambda^\alpha{}_\beta = \delta^\alpha{}_\beta + \omega^\alpha{}_\beta,$$

with $\mathcal{O}(\omega^2) = 0$. According to $g_{\alpha\beta} = \Lambda^\nu{}_\alpha g_{\mu\nu}\Lambda^\mu{}_\beta$ one then follows that

$$g_{\alpha\beta} = \left(\delta^{\nu}{}_{\alpha} + \omega^{\nu}{}_{\alpha}\right) g_{\mu\nu} \left(\delta^{\mu}{}_{\beta} + \omega^{\mu}{}_{\beta}\right)$$
$$= \left(g_{\mu\alpha} + \omega^{\nu}{}_{\alpha} g_{\mu\nu}\right)\left(\delta^{\mu}{}_{\beta} + \omega^{\mu}{}_{\beta}\right)$$
$$= g_{\beta\alpha}\omega^{\nu}{}_{\alpha} g_{\beta\nu} + g_{\mu\alpha}\omega^{\mu}{}_{\beta},$$

yielding

$$\omega^{\nu}{}_{\alpha} g_{\beta\nu} + g_{\mu\alpha}\omega^{\mu}{}_{\beta} = 0,$$

therefore

$$\omega_{\beta\alpha} + \omega_{\alpha\beta} = 0,$$

showing that $\omega_{\alpha\beta}$ is an antisymmetric quantity.

Suppose that $\Lambda \in L_{+}^{\uparrow}$, $\Lambda \equiv g$. Then,

$$g = \exp X = \exp\left(\sum_{j} \alpha^{j} X_{j}\right), \quad (j = \overline{1, 6}).$$

One usually takes $\alpha^{j} \to -\frac{i}{2}\omega^{\alpha\beta}$ and $X_{j} \to M_{\alpha\beta}$, where the quantities $M_{\alpha\beta}$ must be determined afterwards. Thus, using the exponential parametrization, one can write

$$\Lambda = \exp\left(-\frac{i}{2}\omega^{\alpha\beta} M_{\alpha\beta}\right), \tag{4.25}$$

with $M_{\alpha\beta} = -M_{\beta\alpha}$, since

$$\omega^{\alpha\beta} M_{\alpha\beta} = \frac{1}{2}\left(\omega^{\alpha\beta} M_{\alpha\beta} + \omega^{\alpha\beta} M_{\alpha\beta}\right) = \frac{1}{2}\omega^{\alpha\beta}\left(M_{\alpha\beta} - M_{\beta\alpha}\right),$$

so that

$$M_{\alpha\beta} = \frac{1}{2}\left(M_{\alpha\beta} - M_{\beta\alpha}\right) \Rightarrow M_{\alpha\beta} = -M_{\beta\alpha}.$$

The antisymmetric quantities $M_{\alpha\beta}$ are considered as generators; since no supplementary conditions act on them and keeping in mind that they are defined on a four-dimensional space, it follows that $M_{\alpha\beta}$, $(\alpha, \beta = 0,\, 1,\, 2,\, 3)$ has $C_{n}^{s} n^{p+q-s} = C_{4}^{2} 4^{2+0-2} = 6$ essentially distinct components ($n$ is the space dimension, $p$ – the number of covariant indices, $q$ – the number of contravariant indices, and $s$ – the number of symmetry indices).

In the case of infinitesimal transformations, the equation (4.25) becomes

$$\Lambda = I - \frac{i}{2}\omega^{\mu\nu} M_{\mu\nu},$$

or, if the indices are putting into evidence,

$$\Lambda^{\alpha}{}_{\beta} = \delta^{\alpha}{}_{\beta} - \frac{i}{2}\omega^{\mu\nu}\Big(M_{\mu\nu}\Big)^{\alpha}{}_{\beta}.$$

Comparing to $\Lambda^{\alpha}{}_{\beta} = \delta^{\alpha}{}_{\beta} + \omega^{\alpha}{}_{\beta}$, it follows that

$$\omega^{\alpha}{}_{\beta} = -\frac{i}{2}\omega^{\mu\nu}\Big(M_{\mu\nu}\Big)^{\alpha}{}_{\beta},$$

which yields

$$\Big(M_{\mu\nu}\Big)^{\alpha}{}_{\beta} = i\big(g_{\nu\beta}\delta^{\alpha}{}_{\mu} - g_{\mu\beta}\delta^{\alpha}{}_{\nu}\Big),$$

since

$$\omega^{\alpha}{}_{\beta} = g_{\gamma\beta}\omega^{\alpha\gamma} = \frac{1}{2}g_{\gamma\beta}\omega^{\alpha\gamma} + \frac{1}{2}g_{\gamma\beta}\omega^{\alpha\gamma} = \frac{1}{2}\delta^{\alpha}{}_{\mu}g_{\nu\beta}\omega^{\mu\nu} + \frac{1}{2}\delta^{\alpha}{}_{\nu}g_{\mu\beta}\omega^{\nu\mu} = \frac{1}{2}\omega^{\mu\nu}\Big(g_{\nu\beta}\delta^{\alpha}{}_{\mu} - g_{\mu\beta}\delta^{\alpha}{}_{\nu}\Big).$$

In total, there are six generators, namely $M_{01}, M_{02}, M_{03}, M_{12}, M_{23}, M_{31}$.

The group algebra is given by

$$\Big[M_{\alpha\beta}, M_{\mu\nu}\Big] = -i\Big(g_{\alpha\mu}M_{\beta\nu} + g_{\beta\nu}M_{\alpha\mu} - g_{\alpha\nu}M_{\beta\mu} - g_{\beta\mu}M_{\alpha\nu}\Big).$$

One usually takes

$$M_{\alpha\beta} = \varepsilon_{\alpha\beta i}J_i \quad \text{and} \quad M_{0j} = K_j,$$

in which case one can write the following commutation relations:

$$\Big[J_i, K_j\Big] = i\varepsilon_{ijl}K_l,$$

$$\Big[J_i, J_j\Big] = i\varepsilon_{ijk}J_k,$$

$$\Big[K_i, K_j\Big] = -i\varepsilon_{ijk}J_k.$$

Using the notations

$$S_i = \frac{1}{2}(J_i + iK_i),$$
$$T_i = \frac{1}{2}(J_i - iK_i),$$

the above commutation relations write in a simpler form

$$\Big[T_i, T_j\Big] = i\varepsilon_{ijk}T_k,$$
$$\Big[S_i, T_j\Big] = 0.$$

The six generators of the Lorentz group are

$$K_1 = \begin{pmatrix} 0 & -i & 0 & 0 \\ -i & 0 & 0 & 0 \\ 0 & 0 & 0 & 0 \\ 0 & 0 & 0 & 0 \end{pmatrix}, \quad K_2 = \begin{pmatrix} 0 & 0 & -i & 0 \\ 0 & 0 & 0 & 0 \\ -i & 0 & 0 & 0 \\ 0 & 0 & 0 & 0 \end{pmatrix},$$

$$K_3 = \begin{pmatrix} 0 & 0 & 0 & -i \\ 0 & 0 & 0 & 0 \\ 0 & 0 & 0 & 0 \\ -i & 0 & 0 & 0 \end{pmatrix}, \quad J_1 = \begin{pmatrix} 0 & 0 & 0 & 0 \\ 0 & 0 & 0 & 0 \\ 0 & 0 & 0 & -i \\ 0 & 0 & i & 0 \end{pmatrix},$$

$$J_2 = \begin{pmatrix} 0 & 0 & 0 & 0 \\ 0 & 0 & 0 & i \\ 0 & 0 & 0 & 0 \\ 0 & -i & 0 & 0 \end{pmatrix}, \quad J_3 = \begin{pmatrix} 0 & 0 & 0 & 0 \\ 0 & 0 & -i & 0 \\ 0 & i & 0 & 0 \\ 0 & 0 & 0 & 0 \end{pmatrix}.$$

Here are several usual representations of the Lorentz group. Taking $-\frac{i}{2}\omega^{\mu\nu} \to i\varepsilon^{\mu\nu}$ as parameters of the Lorentz group, one can write

$$\Lambda = e^{i\varepsilon^{\mu\nu}M_{\mu\nu}},$$

with $\varepsilon^{\mu\nu} = -\varepsilon^{\nu\mu}$ and $M_{\mu\nu} = -M_{\nu\mu}$. Generically, a representation of the Lorentz group writes

$$\Lambda \in G \xrightarrow{T:\mathcal{L}\to\mathcal{L}} T(\Lambda) \in \mathcal{L},$$

where $\mathcal{L}$ is the representation space. In the case of infinitesimal transformations, one has

$$\Big[T(\Lambda)\Big]_{\alpha\beta} = \delta_{\alpha\beta} + i\varepsilon^{\mu\nu}\Big(M_{\mu\nu}\Big)_{\alpha\beta} + \mathcal{O}(\varepsilon^2).$$

Here $\left(M_{\mu\nu}\right)_{\alpha\beta}$ are the group generators in the representation given by $T$ or, more exactly, the generators of the $T$-representation of the group. Formally, one can write

$$T(\Lambda) = e^{i\varepsilon^{\mu\nu}M_{\mu\nu}}.$$

Suppose that the operators $T(\Lambda)$ – vectors of the representation space $\mathcal{L}$ – act upon the fields denoted by $u_i(x)$. The most known and used four representations of the Lorentz group are the following:

a) Let $T(\Lambda) = I_n$, where $I_n$ is the unit (identity) matrix of rank $n$ ($n$ is the dimension of the representation space $\mathcal{L}$, and the representation dimension as well). Then,

$$T(\Lambda)\begin{pmatrix} u_1(x) \\ u_2(x) \\ \vdots \\ u_n(x) \end{pmatrix} = I_n \begin{pmatrix} u_1(x) \\ u_2(x) \\ \vdots \\ u_n(x) \end{pmatrix} = \begin{pmatrix} u'_1(x) \\ u'_2(x) \\ \vdots \\ u'_n(x) \end{pmatrix},$$

with

$$u'_1(x) = u_1(x),$$
$$u'_2(x) = u_2(x),$$
$$\vdots$$
$$u'_n(x) = u_n(x).$$

Therefore, each of the functions $u_i(x)$, $\left(i = \overline{1,n}\right)$ is a scalar $\left(e.g.\ u_1(x) = \phi(x),\ u_2(x) = \vartheta(x), ...,\ u_n(x) = \zeta(x)\right)$, meaning that each $u_i(x)$ has only one component. These functions describe the spin-zero particles, and the representation $T(\Lambda)$ is called a *scalar representation*.

b) Consider now $n = 4$. In this case $\left(T(\Lambda)\right)_{\alpha\beta}$ are $4 \times 4$ matrices, acting on the space $\mathcal{L}$ of the vectors

$$\begin{pmatrix} u_1(x) \\ u_2(x) \\ u_3(x) \\ u_4(x) \end{pmatrix} = u_\alpha(x), \quad (\alpha = 1,\ 2,\ 3,\ 4),$$

that is

$$T(\Lambda)\begin{pmatrix} u_1(x) \\ u_2(x) \\ u_3(x) \\ u_4(x) \end{pmatrix} = \begin{pmatrix} u'_1(x) \\ u'_2(x) \\ u'_3(x) \\ u'_4(x) \end{pmatrix}.$$

For example, $u_\alpha(x) = A_\alpha(x)$, where $A_\alpha(x)$ are the components of the four-potential of the electromagnetic field. The wave function (vector of $\mathcal{L}$) $u_\alpha(x)$ has four components, is called vectorial field, describes the spin-1 particles, and the representation is called *vectorial*. In this case one can write

$$u_\alpha \to u'_\alpha = \Big(I_4 + i\varepsilon^{\mu\nu}M_{\mu\nu}\Big)_{\alpha\beta} u_\beta(x),$$

$$(\alpha, \beta = 1, 2, 3, 4;\ \mu, \nu = 0, 1, 2, 3),$$

where $I_4$ is the $4 \times 4$ unit matrix.

c) *Tensorial representation.* In this case $n = 4 \times 4 = 16$, therefore $\mathcal{L}$ is the space of the vectors with 16 components (which means that the wave functions $u_{\alpha\beta}(x)$ have sixteen components). The operator $T_{\alpha\beta,\gamma\delta}$ acts according to

$$T_{\alpha\beta,\gamma\delta}u_{\gamma\delta}(x) = u'_{\alpha\beta}(x'),$$

while for representation one has

$$u'_{\alpha\beta}(x') = \Big(T(\Lambda)\Big)_{\alpha\gamma}\Big(T(\Lambda)\Big)_{\beta\delta}u_{\gamma\delta}(x),$$

which means that in this case the wave function is a second-rank tensor and describes the spin-2 particles. The tensorial representation $T_{\alpha\beta,\gamma\delta}$ is reducible; it decomposes into three irreducible representations: one of them is 1-dimensional and corresponds to the trace of the symmetric part $u^s_{\alpha\beta}(x)$ of the field/wave function $u_{\alpha\beta}(x) = u^s_{\alpha\beta}(x) + u^a_{\alpha\beta}(x)$, where $u^s_{\alpha\beta}(x) = \frac{1}{2}(u_{\alpha\beta}(x) + u_{\beta\alpha}(x))$ and $u^a_{\alpha\beta}(x) = \frac{1}{2}(u_{\alpha\beta}(x) - u_{\beta\alpha}(x))$, another one is 6-dimensional[11] and corresponds to

[11] As is well known, in a four-dimensional space a symmetric second-rank tensor has ten independent components, and an antisymmetric second-rank tensor has six independent components. If these components are submitted to some supplementary constraints, the total number of independent components of the tensor reduces by the number of independent supplementary constraints.

the antisymmetric part $u^a_{\alpha\beta}(x)$ of $u_{\alpha\beta}(x)$, and the last one is 9-dimensional and corresponds to the zero-trace $3 \times 3$ matrix whose elements are the nine remaining components of $u^s_{\alpha\beta}(x)$ (the tenth component, which is the trace of $u^s_{\alpha\beta}(x)$ has already been considered). This decomposition is allowed by the facts that, on the one hand, the components of the symmetric tensor $u^s_{\alpha\beta}(x)$ and the antisymmetric tensor $u^a_{\alpha\beta}(x)$ transform independently upon rotations of the four-dimensional space-time, without any mixing, and, on the other hand, the ten components of the symmetric tensor $u^s_{\alpha\beta}(x)$ decompose, in their turn, into two independent subsets; hence, the sixteen-component, second-rank tensor $u_{\alpha\beta}(x)$ decomposes under four-dimensional rotations into three invariant subspaces: a 1-dimensional, a 6-dimensional, and a 9-dimensional one.

d) The *spinorial* (*Dirac*) *representation*. Let $\sigma_{\mu\nu}$ be the matrices associated to the generators $M_{\mu\nu}$ of the Lorentz group, according to

$$M_{\mu\nu} \to \sigma_{\mu\nu} = \frac{1}{2i}\Big(\gamma_\mu\gamma_\nu - \gamma_\nu\gamma_\mu\Big), \quad (\mu, \nu = 0, 1, 2, 3),$$

where $\gamma_\mu$ are the Dirac matrices

$$\gamma_0 = \begin{pmatrix} 1 & 0 & 0 & 0 \\ 0 & 1 & 0 & 0 \\ 0 & 0 & -1 & 0 \\ 0 & 0 & 0 & -1 \end{pmatrix}, \quad \gamma_1 = \begin{pmatrix} 0 & 0 & 0 & 1 \\ 0 & 0 & 1 & 0 \\ 0 & -1 & 0 & 0 \\ -1 & 0 & 0 & 0 \end{pmatrix},$$

$$\gamma_2 = \begin{pmatrix} 0 & 0 & 0 & -i \\ 0 & 0 & i & 0 \\ 0 & i & 0 & 0 \\ -i & 0 & 0 & 0 \end{pmatrix}, \quad \gamma_3 = \begin{pmatrix} 0 & 0 & 1 & 0 \\ 0 & 0 & 0 & -1 \\ -1 & 0 & 0 & 0 \\ 0 & 1 & 0 & 0 \end{pmatrix}.$$

Under these conditions, the field function $u_i(x)$ is

$$u_i(x) \to \psi(x) = \begin{pmatrix} \psi_1(x) \\ \psi_2(x) \\ \psi_3(x) \\ \psi_4(x) \end{pmatrix},$$

that is, a Dirac spinor (with four components), $\psi(x)$, whose Dirac conjugate is

$$\overline{\psi}(x) = \psi^+(x)\gamma_0 = \Big(\psi_1^*(x),\ \psi_2^*(x),\ \psi_3^*(x),\ \psi_4^*(x)\Big)\gamma_0$$
$$= \Big(\psi_1^*(x),\ \psi_2^*(x),\ -\psi_3^*(x),\ -\psi_4^*(x)\Big).$$

In this case

$$T(\Lambda) = e^{i\varepsilon^{\mu\nu}\sigma_{\mu\nu}},$$

and the field function transforms in accordance with

$$\psi(x) \to \psi'(x') = T(\Lambda)\psi(x)$$
$$= e^{i\varepsilon^{\mu\nu}\sigma_{\mu\nu}}\psi(x) \overset{\substack{\text{infinitesim.}\\ \text{transf.}}}{=} \Big(I_4 + i\varepsilon^{\mu\nu}\sigma_{\mu\nu}\Big)\psi(x).$$

Since the order (dimension) of the Lorentz group is six, it follows that the dimension of an *adjoint representation* of the Lorentz group is also six. The *fundamental representation* of the Lorentz group (including the spatial inversion) is given by the matrices $\sigma_{\mu\nu}$.

We conclude this section by noting that the Lorentz group has representations of any dimension or, differently speaking, the wave functions – vectors of the representation space – can have any number of components. The transformation matrices for the components of the wave functions of a given dimension are also determined by the Lorentz group.

## 2. The Poincaré group

This group is represented by transformations of the form

$$x^\mu \to x'^\mu = \Lambda^\mu{}_\nu x^\nu + a^\mu,$$

expressing the Lorentz transformations complemented with the four-dimensional translations, given by the constant four-vector $a^\mu$, ($\mu = 0, 1, 2, 3$). Actually, this group is also called the *inhomogeneous Lorentz group*.

In order to determine the composition law of the group, consider two successive transformations, which are elements of the Poincaré group,

$$x' = \Lambda_1 x + a_1 \to (\Lambda_1, a_1) \in P \text{ and } x'' = \Lambda_2 x' + a_2 \to (\Lambda_2, a_2) \in P.$$

Then

$$x'' = \Lambda_2(\Lambda_1 x + a_1) + a_2 = (\Lambda_2\Lambda_1)x + (\Lambda_2 a_1 + a_2) = (\Lambda, a) \in P,$$

where

$$\Lambda = \Lambda_2\Lambda_1, \quad a = \Lambda_2 a_1 + a_2.$$

The neutral element is $(I, 0)$, since $(I, 0)(\Lambda, a) = (\Lambda, a)$. The inverse element is $(\Lambda, a)^{-1} = (\Lambda^{-1}, -\Lambda^{-1}a)$. Indeed, consider $(\Lambda, a)^{-1} = (\Lambda', a')$. Then,

$$(\Lambda', a')(\Lambda, a) = (\Lambda\Lambda', \Lambda a' + a) \overset{\text{impose}}{=} (I, 0),$$

which implies $\Lambda\Lambda' = I$ (that is $\Lambda' = \Lambda^{-1}$) and $\Lambda a' + a = 0$ (which results in $a' = -\Lambda^{-1}a$).

Here one also has $\det \Lambda = \pm 1$ and $|\Lambda^0{}_0| \geq 1$, leading to the same structure of subsets of the transformations set of the Poincaré group, as that valid for the Lorentz group. Some of these subsets (in fact, the same as in the case of the Lorentz group) are organized as subgroups, also keeping the names. For example, the proper orthochronous Poincaré group is

$$P_+^{\uparrow} = \left\{ (\Lambda, a) \in P \middle| \det \Lambda = +1, \ \Lambda^0{}_0 \geq 1 \right\}.$$

The Poincaré group has ten generators: six of them also belong to the Lorentz group, $M_{\mu\nu} = -M_{\nu\mu}$, which are the rotation generators, being connected to the angular momentum, and four generators $P_\mu$, $(\mu = 0, 1, 2, 3)$ coming from the four translations – one time-like, along the $x^0$-axis, and three space-like – which are the generators of translations, being straightly connected to the four-momentum.

The dimension of the group algebra is also ten. In addition, the ten generators satisfy the following relations:

a) $[M_{\alpha\beta}, M_{\mu\nu}] = -i\left(g_{\alpha\mu}M_{\beta\nu} + g_{\beta\nu}M_{\alpha\mu} - g_{\alpha\nu}M_{\beta\mu} - g_{\beta\mu}M_{\alpha\nu}\right);$

b) $\left.\begin{array}{l}[P_\mu, P_\mu] = 0, \\ [P_\mu, P_\nu] = 0,\end{array}\right\}$ since the translation group is Abelian; indeed,

$$(I, a_1)\,(I, a_2) = (I, a_1 + a_2) = (I, a_2) + (I, a_1) = (I, a_2 + a_1)\,,$$

c) $[M_{\mu\nu}, P_\alpha] = -i\,(g_{\mu\alpha} P_\nu - g_{\nu\alpha} P_\mu)\,.$

If the exponential parametrization is considered, then

$$(\Lambda, a) = \exp\left(\alpha^i X_i\right),$$

which, in the linear approximation (infinitesimal transformations), becomes

$$(\Lambda, a) = I - \frac{i}{2}\omega^{\mu\nu} M_{\mu\nu} + i\alpha^\mu P_\mu.$$

The element $(\Lambda, a)$ of the group can also be seen as a $5 \times 5$ matrix

$$(\Lambda, a) \to \begin{pmatrix} \Lambda & a \\ 0 & 1 \end{pmatrix} \equiv \overline{\Lambda},$$

acting in the 5-dimensional space-time of the 5-component vectors of the form

$$\begin{pmatrix} x^0 \\ x^1 \\ x^2 \\ x^3 \\ 1 \end{pmatrix}.$$

Since the composition law holds with this new notation for the group elements $(\overline{\Lambda})$ because

$$\overline{\Lambda}_2 \overline{\Lambda}_1 = \begin{pmatrix} \Lambda_2 \Lambda_1 & \Lambda_2 a_1 + a_2 \\ 0 & 1 \end{pmatrix},$$

one can introduce the fifth coordinate, $s$, as

$$(x^\mu, s) : \begin{cases} x'^\mu = \Lambda^\mu{}_\nu x^\nu + a^\nu s, \\ s' = s, \end{cases}$$

or

$$\begin{cases} x'^\mu = \Lambda^\mu{}_\nu x^\nu + a^\nu s \\ s' = 0 \cdot x^\nu + 1 \cdot s \end{cases} \to \begin{pmatrix} \Lambda & a \\ 0 & 1 \end{pmatrix},$$

with

$$\begin{pmatrix} x'^0 \\ x'^1 \\ x'^2 \\ x'^3 \\ s' \end{pmatrix} = \begin{pmatrix} \Lambda & a \\ 0 & 1 \end{pmatrix} \begin{pmatrix} x^0 \\ x^1 \\ x^2 \\ x^3 \\ s \end{pmatrix}.$$

In this notation, a translation of the four-vector $a^\mu$ writes

$$T_a = \begin{pmatrix} 1 & 0 & 0 & 0 & a^0 \\ 0 & 1 & 0 & 0 & a^1 \\ 0 & 0 & 1 & 0 & a^2 \\ 0 & 0 & 0 & 1 & a^3 \\ 0 & 0 & 0 & 0 & 1 \end{pmatrix}.$$

The translation generators then are $P'_\mu = \left(\frac{dT_a}{da^\mu}\right)_{a^\mu=0}$, $(\mu = 0, 1, 2, 3)$. For example,

$$P'_2 = \left(\frac{dT_a}{da^2}\right)_{a^2=0} = \begin{pmatrix} 0 & 0 & 0 & 0 & 0 \\ 0 & 0 & 0 & 0 & 0 \\ 0 & 0 & 0 & 0 & 1 \\ 0 & 0 & 0 & 0 & 0 \\ 0 & 0 & 0 & 0 & 0 \end{pmatrix}.$$

Under this notation, the Poincaré group generators are obtained by completing those of the Lorentz group (the actual rotations and the Lorentz boosts) with one more row and one more column of 0 (zero), also adding the translation generators, $P_\mu = iP'_\mu$; here the purpose of $i = \sqrt{-1}$ is to ensure the homogeneity in relation to the above considered Lorentz group[12]. The generators of the Poincaré

[12] To ensure the Hermiticity of the generators of the Lorentz group (required by the fact that the physical quantities corresponding to these generators are – from the quantum point of view – observables), they are multiplied by $i = \sqrt{-1}$, if directly deduced from the generators general definition, *i.e.* by taking the derivatives of the general matrix of the Lorentz transformation with respect to "rotation angles" (parameters of the Lorentz transformations) and evaluating these derivatives for the zero value of the derivation variable. In the present case, this procedure has been fulfilled in a simpler manner, by including the imaginary $i = \sqrt{-1}$, in the framework of exponential parametrization, in the expressions of the group parameters, $\alpha^j \to -\frac{i}{2}\omega^{\alpha\beta}$, the generators being given by $(M_{\mu\nu})^\alpha{}_\beta = i(g_{\nu\beta}\delta^\alpha{}_\mu - g_{\mu\beta}\delta^\alpha{}_\nu)$, as shown above.

group therefore are:

$$K_1 = \begin{pmatrix} 0 & -i & 0 & 0 & 0 \\ -i & 0 & 0 & 0 & 0 \\ 0 & 0 & 0 & 0 & 0 \\ 0 & 0 & 0 & 0 & 0 \\ 0 & 0 & 0 & 0 & 0 \end{pmatrix}, \quad K_2 = \begin{pmatrix} 0 & 0 & -i & 0 & 0 \\ 0 & 0 & 0 & 0 & 0 \\ -i & 0 & 0 & 0 & 0 \\ 0 & 0 & 0 & 0 & 0 \\ 0 & 0 & 0 & 0 & 0 \end{pmatrix},$$

$$K_3 = \begin{pmatrix} 0 & 0 & 0 & -i & 0 \\ 0 & 0 & 0 & 0 & 0 \\ 0 & 0 & 0 & 0 & 0 \\ -i & 0 & 0 & 0 & 0 \\ 0 & 0 & 0 & 0 & 0 \end{pmatrix}, \quad J_1 = \begin{pmatrix} 0 & 0 & 0 & 0 & 0 \\ 0 & 0 & 0 & 0 & 0 \\ 0 & 0 & 0 & -i & 0 \\ 0 & 0 & i & 0 & 0 \\ 0 & 0 & 0 & 0 & 0 \end{pmatrix},$$

$$J_2 = \begin{pmatrix} 0 & 0 & 0 & 0 & 0 \\ 0 & 0 & 0 & i & 0 \\ 0 & 0 & 0 & 0 & 0 \\ 0 & -i & 0 & 0 & 0 \\ 0 & 0 & 0 & 0 & 0 \end{pmatrix}, \quad J_3 = \begin{pmatrix} 0 & 0 & 0 & 0 & 0 \\ 0 & 0 & -i & 0 & 0 \\ 0 & i & 0 & 0 & 0 \\ 0 & 0 & 0 & 0 & 0 \\ 0 & 0 & 0 & 0 & 0 \end{pmatrix},$$

$$P_0 = \begin{pmatrix} 0 & 0 & 0 & 0 & i \\ 0 & 0 & 0 & 0 & 0 \\ 0 & 0 & 0 & 0 & 0 \\ 0 & 0 & 0 & 0 & 0 \\ 0 & 0 & 0 & 0 & 0 \end{pmatrix}, \quad P_1 = \begin{pmatrix} 0 & 0 & 0 & 0 & 0 \\ 0 & 0 & 0 & 0 & i \\ 0 & 0 & 0 & 0 & 0 \\ 0 & 0 & 0 & 0 & 0 \\ 0 & 0 & 0 & 0 & 0 \end{pmatrix},$$

$$P_2 = \begin{pmatrix} 0 & 0 & 0 & 0 & 0 \\ 0 & 0 & 0 & 0 & 0 \\ 0 & 0 & 0 & 0 & i \\ 0 & 0 & 0 & 0 & 0 \\ 0 & 0 & 0 & 0 & 0 \end{pmatrix}, \quad P_3 = \begin{pmatrix} 0 & 0 & 0 & 0 & 0 \\ 0 & 0 & 0 & 0 & 0 \\ 0 & 0 & 0 & 0 & 0 \\ 0 & 0 & 0 & 0 & i \\ 0 & 0 & 0 & 0 & 0 \end{pmatrix}.$$

To conclude, it is worth mentioning several important properties of the infinitesimal generators of the Poincaré group.

Thus, there are two operators that completely determine the representations of the Poincaré group; they are called *Casimir operators* and commute with all the group generators. One of these operators is the *squared mass operator*, $P^2 = P_\mu P^\mu$, since

$$\left[P_\mu, P^2\right] = 0,$$
$$\left[M_{\mu\nu}, P^2\right] = 0.$$

The operator $P^2$ is called the "squared mass operator" because in the centre-of-mass reference frame the eigenvalue problem for the operator $P_\mu$ writes $P_\mu|p\rangle = p_\mu|p\rangle$, with $p_\mu = \left(-\frac{i}{c}\mathcal{E}, \vec{p}\right)$ and, since in the COM frame $\vec{p} = 0$, and $\mathcal{E}$ is only the rest energy, one can write $P^2 = P_\mu P^\mu = p_\mu p^\mu = M^2$. Here $P^\mu = g^{\mu\nu} P_\nu$ and the natural unit system $(c = 1, \hbar = 1)$ was used.

The other Casimir operator is the *squared Pauli–Lubanski vector.* The Pauli–Lubanski vector is defined by

$$W_\mu = \frac{1}{2}\varepsilon_{\mu\nu\rho\sigma} P^\nu M^{\rho\sigma}.$$

One can easily be verified that

$$[P_\mu, W_\nu] = 0, \quad W_\mu P^\mu = 0,$$

and

$$[A, P^\mu] = W^\mu,$$

where

$$A = \frac{i}{8}\varepsilon_{\mu\nu\rho\sigma} M^{\mu\nu} M^{\rho\sigma}.$$

One also has

$$[M_{\mu\nu}, W_\rho] = -i\left(g_{\mu\rho} W_\nu - g_{\nu\rho} W_\mu\right).$$

The squared Pauli–Lubanski vector is

$$W^2 = W_\mu W^\mu = -\frac{1}{2} M_{\mu\nu} M^{\mu\nu} P^2 + M^{\rho\sigma} M_{\nu\sigma} P_\rho P^\nu.$$

It can also be shown that

$$\left[P_\mu, W^2\right] = 0,$$

and

$$\left[M_{\mu\nu}, W^2\right] = 0.$$

In COM system, the only component of $P_\mu$ is $P_0$, connected to the particle mass. Taking $\nu = 0$ in the definition relation of $W_\mu$, one has

$$W_\mu = \frac{1}{2}\varepsilon_{\mu 0\rho\sigma} P^0 M^{\rho\sigma}.$$

Then

$$W_\mu = P^0 \frac{1}{2}\varepsilon_{\mu\rho\sigma} M^{\rho\sigma} = P^0 S_\mu,$$

where

$$S_\mu = \frac{1}{2}\varepsilon_{\mu\rho\sigma} M^{\rho\sigma},$$

because, if one index is fixed, the space becomes three-dimensional. Thus, in the COM frame one can write

$$W^2 = W^0 W_0 + W^i W_i = -\vec{W}^2,$$

because in this frame $W^0 W_0 = 0$. Since $W_\mu = P^0 S_\mu$, it follows that

$$W^2 = W_\mu W^\mu = -\left(P^0\right)^2 \vec{S}^2,$$

since

$$W_\mu W^\mu = P^0 S_\mu P^0 S^\mu = \left(P^0\right)^2 \left(-\vec{S}^2\right).$$

Therefore, the eigenvalues of $W^2$ are $-M^2 S(S+1)$, where $S$ is the spin quantum number, showing that all particles of a multiplet have the same spin and the same mass.

**Conclusion**: The Casimir operators give the characteristics common to all particles of a multiplet (corresponding to a certain representation of the symmetry group).

If the particles' rest mass is zero, the COM reference system cannot be defined anymore and the above considerations do not remain valid. The state of a particle with zero rest mass can be specified by means of the energy-momentum four-vector, whose eigenvalue equation writes

$$P_\mu |p\rangle = p_\mu |p\rangle, \quad p^2 = 0.$$

Because both $W_\mu|p\rangle$ and $P_\mu|p\rangle$ are four-vectors of magnitude zero, it follows that they are proportional, that is, if one denotes by $\lambda$ the proportionality constant, one can write

$$W_\mu = \lambda P_\mu.$$

Taking $\mu = 0$ in the last equation, the constant $\lambda$ (called particle *helicity*) can be expressed as

$$\lambda = \frac{W_0}{P_0} = \frac{\frac{1}{2}\varepsilon_{0ijk}P^i M^{jk}}{P_0} = \frac{\vec{P}\cdot\vec{S}}{P_0} = \frac{\vec{p}\cdot\vec{S}}{p_0}.$$

But, since $M = 0$, one has $p_0^2 = \vec{p}^2$, *i.e.* $p_0 = |\vec{p}|$, so that

$$\lambda = \frac{\vec{p}\cdot\vec{S}}{|\vec{p}|},$$

showing that the helicity of a particle is nothing else but the projection of its spin angular momentum along the direction of motion. If the helicity differs from zero, then for a given value of $p_\mu$ there exist two independent states of the particle, corresponding to the values $\pm\lambda$ of helicity (the two circular polarization states – right and left). For $\lambda = 0$ the particle can have a single possible state. Therefore, a particle with null rest mass and spin quantum number $s \neq 0$ has only two polarization states ($|\lambda, p\rangle$ and $|-\lambda, p\rangle$), and not $2s + 1$ as the massive particles.

### IV.2.6. Examples of internal symmetry groups

To illustrate the general theory exposed so far, here are several examples of the most frequently used internal symmetry groups, met in Quantum Field Theory, namely the phase transformations group $U(1)$, the set of all two dimensional, complex unitary matrices with unit determinant ($SU(2)$ group), and the set of all three dimensional, complex unitary matrices with unit determinant ($SU(3)$ group).

#### 1. The $U(1)$ group. The $U(1)$ symmetry

According to the definition of the group $U(n)$, it follows that the elements of the group $U(1)$ are $1\times 1$ matrices, in other words, numbers[13]

[13]The $U(1)$ group is, therefore, an Abelian group.

that can be written (as it is known by virtue of a preceding theorem), under the form $e^{ih}$, where $h$ is a real number, usually expressed as $h = g\varepsilon$, where $\varepsilon$ is the group parameter, and $g$ is a *coupling constant* that depends on the nature of the considered physical field of $U(1)$ symmetry. For example, in the case of the electromagnetic field (whose Lagrangian is invariant with respect to the $U(1)$ group transformations), $g$ is just the electron charge $e$, entering into the fine structure constant; in this case, it is said that "$e$" is the coupling constant of the electromagnetic field.

Following the previously discussed subjects, the wave function $u(x)$ of some physical system transforms according to

$$u(x) \to u'(x) = e^{-ig\varepsilon} u(x).$$

If one wishes to describe charged particles, then the wave function must be complex. In this case, in addition to the above relation, one has to consider its complex conjugate,

$$u^*(x) \to u^{*\prime}(x) = e^{ig\varepsilon} u^*(x).$$

If the transformation is also infinitesimal, then $\mathcal{O}(\varepsilon^2) = 0$ and, in the first-order approximation, one can write $e^{\pm ig\varepsilon} \cong 1 \pm ig\varepsilon$. The above relations then become

$$u(x) \to u'(x) = u(x) - ig\varepsilon\, u(x) = u(x) + \delta u(x),$$

and

$$u^*(x) \to u^{*\prime}(x) = u^*(x) + ig\varepsilon\, u^*(x) = u^*(x) + \delta u^*(x),$$

which yields

$$\delta u(x) = -ig\varepsilon\, u(x),$$

and

$$\delta u^*(x) = ig\varepsilon\, u^*(x).$$

Comparing the last two relations with that giving the variation of the wave function under the global internal symmetry transformations of a group, $\delta u_i(x) = T^k_{ij}\varepsilon_k u_j(x)$, it follows that the group $U(1)$ is characterized by $\varepsilon_k \to \varepsilon$ (therefore it is *uniparametric*) and, if one denotes $u(x) = u_1(x)$ and $u^*(x) = u_2(x)$ then, for the generator matrix elements – which in this bidimensional representation can be written as $u(x) \equiv \begin{pmatrix} u_1(x) \\ u_2(x) \end{pmatrix} = \begin{pmatrix} u(x) \\ u^*(x) \end{pmatrix}$ – one has $T_{11} = -ig$, $T_{12} = 0$, $T_{21} = 0$, $T_{22} = ig$, or, in a more compact writing,

$$T_{ij}^k \to (T)_{ij} = \begin{pmatrix} -ig & 0 \\ 0 & ig \end{pmatrix} = -ig \begin{pmatrix} 1 & 0 \\ 0 & -1 \end{pmatrix} = -ig(\tau_3)_{ij},$$

where $\tau_3$ is the Pauli's third matrix:

$$\tau_1 = \begin{pmatrix} 0 & 1 \\ 1 & 0 \end{pmatrix}, \quad \tau_2 = \begin{pmatrix} 0 & -i \\ i & 0 \end{pmatrix}, \quad \tau_3 = \begin{pmatrix} 1 & 0 \\ 0 & -1 \end{pmatrix}.$$

Obviously, if the studied physical system is describable by a real scalar field (the field quanta are neutral particles), only one relation remains,

$$u(x) \to u'(x) = u(x) - ig\varepsilon\, u(x) = u(x) + \delta u(x),$$

meaning that, in this case, the generator is $(T)_{ij} \to -ig$ (identical representation, here of dimension one). Thus, depending on the dimension of the representation space, the group generator (in fact, the generator of the considered group representation) is a matrix of the same dimension.

As another example, let

$$L = \frac{i}{2}\left(\overline{\psi}\gamma_\mu \partial_\mu \psi - \partial_\mu \overline{\psi}\gamma_\mu \psi\right) - M\overline{\psi}\psi$$

be the Lagrangian density of a spinorial field of mass $M$. This expression is invariant with respect to the group of global symmetry transformations, $U(1)$ . Indeed, consider the transformations

$$\psi \to \psi' = e^{-ig\varepsilon}\psi,$$
$$\overline{\psi} \to \overline{\psi}' = \overline{\psi} e^{ig\varepsilon}.$$

If the transformations are infinitesimal $\left(\mathcal{O}(\varepsilon^2) = 0\right)$, then the above relations write

$$\begin{cases} \psi \to \psi' = \psi - ig\varepsilon\psi = \psi + \delta\psi, \\ \overline{\psi} \to \overline{\psi}' = \overline{\psi} + ig\varepsilon\overline{\psi} = \overline{\psi} + \delta\overline{\psi}, \end{cases}$$

showing that

$$\begin{cases} \delta\psi = -ig\varepsilon\psi, \\ \delta\overline{\psi} = +ig\varepsilon\overline{\psi}. \end{cases}$$

Comparing with the general relation $\delta u_i(x) = T_{ij}^k \varepsilon_k u_j(x)$, the generator matrix elements are

$$T_{11} = -ig, \quad T_{22} = ig, \quad T_{12} = T_{21} = 0.$$

In this case $u_i(x) \to \begin{pmatrix} \psi \\ \overline{\psi} \end{pmatrix}$, therefore the representation space is bidimensional and the representation generator is a $2 \times 2$ matrix (the same with the above example for the complex scalar field $u(x)$, namely

$$T_{ij}^k \to (T)_{ij} = \begin{pmatrix} -ig & 0 \\ 0 & ig \end{pmatrix} = -ig \begin{pmatrix} 1 & 0 \\ 0 & -1 \end{pmatrix} = -ig(\tau_3)_{ij}).$$

## 2. The $SU(2)$ group. The $SU(2)$ symmetry

As it has been exposed in the general theory chapter, $SU(2) = \{U \,|\, UU^+ = I,\ \det U = 1\}$, $U$ being a square matrix of the second rank. Supposing that the generic group element is of the form

$$U = \begin{pmatrix} a & b \\ c & d \end{pmatrix},$$

one has

$$U^+ = \begin{pmatrix} a^* & c^* \\ b^* & d^* \end{pmatrix},$$

where $a^*$ designates the complex conjugate of the matrix element $a$. The condition $UU^+ = I$ writes in this case as follows,

$$UU^+ = \begin{pmatrix} a & b \\ c & d \end{pmatrix} \begin{pmatrix} a^* & c^* \\ b^* & d^* \end{pmatrix}$$

$$= \begin{pmatrix} aa^* + bb^* & ac^* + bd^* \\ ca^* + db^* & cc^* + dd^* \end{pmatrix} = I = \begin{pmatrix} 1 & 0 \\ 0 & 1 \end{pmatrix},$$

while the unimodularity condition

$$\det U = 1$$

writes

$$\det U = ad - bc = 1.$$

One then obtains a system of five algebraic equations for the four unknown quantities (the four complex numbers $a$, $b$, $c$, $d$), namely,

$$\begin{cases} ad - bc = 1, \\ aa^* + bb^* = 1, \\ cc^* + dd^* = 1, \\ ac^* + bd^* = 0, \\ ca^* + db^* = 0. \end{cases}$$

For the sake of compatibility one must have $c = -b^*$ and $d = a^*$, so that

$$U = \begin{pmatrix} a & b \\ -b^* & a^* \end{pmatrix}, \tag{4.26}$$

with $|a|^2 + |b|^2 = 1$. The complex numbers $a$ and $b$ are called the *Cayley–Klein complex parameters*. According to the general theory, any Lie group admits a representation of the form

$$g \in G, \ g = \exp X = \exp \sum_i \alpha_i X_i,$$

where $\alpha_i$ are the group parameters, and $X_i$ its generators. In order to perform the exponential parametrization of the group, one takes the matrix $U$ of the form

$$\begin{aligned} U = a_0\sigma_0 - i\vec{a} \cdot \vec{\sigma} &= a_0 \begin{pmatrix} 1 & 0 \\ 0 & 1 \end{pmatrix} - i(a_1\sigma_1 + a_2\sigma_2 + a_3\sigma_3) \\ &= a_0 \begin{pmatrix} 1 & 0 \\ 0 & 1 \end{pmatrix} - i \left[ a_1 \begin{pmatrix} 0 & 1 \\ 1 & 0 \end{pmatrix} + a_2 \begin{pmatrix} 0 & -i \\ i & 0 \end{pmatrix} + a_3 \begin{pmatrix} 1 & 0 \\ 0 & -1 \end{pmatrix} \right] \\ &= \begin{pmatrix} a_0 & 0 \\ 0 & a_0 \end{pmatrix} - \begin{pmatrix} 0 & ia_1 \\ ia_1 & 0 \end{pmatrix} - \begin{pmatrix} 0 & a_2 \\ -a_2 & 0 \end{pmatrix} - \begin{pmatrix} ia_3 & 0 \\ 0 & -ia_3 \end{pmatrix} \\ &= \begin{pmatrix} a_0 - ia_3 & -a_2 - ia_1 \\ a_2 - ia_1 & a_0 + ia_3 \end{pmatrix}. \end{aligned}$$

Comparing the above equation with Eq. (4.26), one obtains the following relations,

$$\begin{cases} a = a_0 - ia_3, \\ b = -a_2 - ia_1, \\ -b^* = a_2 - ia_1, \\ a^* = a_0 + ia_3, \end{cases}$$

which relations are compatible (as one can easily be shown).

The condition $|a|^2 + |b|^2 = 1$ becomes $a_0^2 + a_1^2 + a_2^2 + a_3^2 = 1$. The numbers $a_0, a_1, a_2, a_3$ are referred to as the *real Cayley–Klein parameters.* Since the trace of a matrix operator does not depend on the basis choice, it follows that $a_0$ is an invariant with respect to the choice of the basis, because $\mathrm{Tr}\, U = 2a_0$. Since for any unitary (even Hermitian) matrix there exists another unitary (Hermitian) matrix so that by a similarity transformation the first matrix can be brought to the diagonal form ($U$ unitary and $H$ Hermitian, respectively $\Rightarrow \exists\, V$ unitary or Hermitian, respectively, so that $U \to U' = V^+ UV$ unitary and diagonal, and $H \to H' = V^+ HV$ Hermitian and diagonal, respectively), it follows that there exists a basis relative to which $U$ is diagonal. The basis with respect to which $U$ is diagonal must have $b = 0$, that is $b^* = 0$, and then $|a|^2 + |b|^2 = 1$ reduces to $|a|^2 = 1$, therefore one can take $a = e^{i\psi}$ and the matrix $U$ becomes

$$U = \begin{pmatrix} e^{i\psi} & 0 \\ 0 & e^{-i\psi} \end{pmatrix}.$$

Denoting $\psi = \frac{1}{2}\omega$, from $\mathrm{Tr}\, U = 2a_0 = e^{i\psi} + e^{-i\psi} = e^{i\frac{\omega}{2}} + e^{-i\frac{\omega}{2}}$ it follows that $a_0 = \frac{e^{i\frac{\omega}{2}} + e^{-i\frac{\omega}{2}}}{2} = \cos\frac{\omega}{2}$. Next, according to $a_0^2 + \vec{a}^2 = 1$ one has $\vec{a}^2 = 1 - a_0^2 = 1 - \cos^2\frac{\omega}{2} = \sin^2\frac{\omega}{2}$, that is $a = \sqrt{\vec{a}^2} = \sqrt{a_1^2 + a_2^2 + a_3^2} = \pm\sin\frac{\omega}{2}$. One usually chooses the positive sign. Since $\forall\ \vec{a}$, $\sqrt{\vec{a}^2} \geq 0$, it follows that $\omega$ must satisfy the condition $\sin\frac{\omega}{2} \geq 0$, that is $0 \leq \omega \leq 2\pi$. Thus, the matrix $U$ becomes

$$U = a_0\sigma_0 - i\vec{a}\cdot\vec{\sigma} = \sigma_0 \cos\frac{\omega}{2} - i\sin\frac{\omega}{2}\hat{a}\cdot\vec{\sigma},$$

with

$$\hat{a} \equiv \vec{u} = \frac{\vec{a}}{a} = \frac{\vec{a}}{|\vec{a}|}.$$

Consequently,

$$U = \sigma_0 \cos\frac{\omega}{2} - i\sin\frac{\omega}{2} n_j \sigma_j, \quad (j = \overline{1,3})$$

with

$$n_j = \frac{a_j}{a}.$$

Since one often deals with infinitesimal transformations, if $\omega$ is taken to be infinitesimal, then $\cos\frac{\omega}{2} \to 1$ and $\sin\frac{\omega}{2} \to \frac{\omega}{2}$, so that

$$U = \sigma_0 - i\frac{\omega}{2} n_j\sigma_j = I_2 - i\frac{\omega}{2} n_j \sigma_j, \quad (j = \overline{1,3}).$$

Denoting $\omega n_j \overset{\text{not}}{=} \varepsilon_j$, one still has

$$U = I_2 - \frac{i}{2}\varepsilon_j\sigma_j, \quad (j = \overline{1,3}),$$

representing the first two terms of the series expansion of the exponential

$$\exp\left(-\frac{i}{2}\varepsilon_j\sigma_j\right) (= U).$$

The group has three parameters, $\varepsilon_j$ $(n^2 - 1 = 2^2 - 1 = 3)$, while the generators of the identical representation of the group are

$$X_j = \left(\frac{\partial U}{\partial \varepsilon_j}\right)_{\varepsilon_j = 0} = -\frac{i}{2}\sigma_j, \quad (j = 1, 2, 3),$$

*i.e.* three $2 \times 2$ matrices. Since the matrices $\sigma_j$ are Hermitian, it follows that the matrix $U$ is unitary, the unitarity being ensured by the complex number $i = \sqrt{-1}$ in the exponent. The group generators form a Lie algebra, whose crochet is

$$\left[X_i, X_j\right] = \varepsilon_{ijk} X_k,$$

the basis of the Lie algebra being given by the very set $\{X_k\}_{k=\overline{1,3}}$. The group elements therefore write

$$\underset{\substack{\text{group}\\ \text{element}}}{U} = \exp\left(\sum_k \varepsilon_k X_k\right) = \exp \underset{\substack{\text{algebra}\\ \text{element}}}{X}.$$

As a general rule, the group generators in an arbitrary representation "acquire" the coupling constant $g$, so that, for example, in the identical representation (*i.e.* the generators of the representation, $T^k_{ij}$ are precisely the group generators $X^k_{ij}$) instead of

$$X^k_{ij} = -\frac{i}{2}(\sigma_k)_{ij}, \quad (k = \overline{1,3};\ i,j = 1,2),$$

one should write

$$T^k_{ij} = -\frac{i}{2}g(\sigma_k)_{ij}, \quad (k = \overline{1,3};\ i,j = 1,2).$$

Therefore, the generators of the *identical representation* are given by the above relation and, since the dimension of the identical representation is two (the generators are $2 \times 2$ matrices) it follows that the representation space is precisely the space of the two-components wave functions

$$u^a(x) = \begin{pmatrix} u^1(x) \\ u^2(x) \end{pmatrix}, \quad (a = 1,2),$$

and the transformation writes

$$u^a \to u^{a\prime}(x) = (T_A)_{ab}u^b(x), \quad (a,b = 1,2).$$

In the case of the infinitesimal transformations, according to the general procedure,

$$u^a(x) \to u^{a\prime}(x) = u^a(x) + \delta u^a(x) = u^a(x) + T^k_{ab}\varepsilon_k u^b(x),$$

one can write

$$u^a(x) \to u^{a\prime}(x) = u^a(x) + \delta u^a(x) = u^a(x) - \frac{i}{2}g\varepsilon_k(\sigma_k)_{ab}u^b(x).$$

Since the transformation is infinitesimal, one can assume $\mathcal{O}(\varepsilon^2) = 0$, and the last relation can also be written as

$$u^a(x) \to u^{a\prime}(x) = \left[\exp\left(-\frac{i}{2}g\varepsilon_k\sigma_k\right)\right]_{ab} u^b(x).$$

**Example.** Consider a spinorial field describing a doublet of spinorial particles,

$$\psi^a = \begin{pmatrix} \psi^1 \\ \psi^2 \end{pmatrix}, \quad (a = 1,2),$$

for example, a proton and a neutron. Having a half-integer spin (1/2), they are described by spinors, $N \equiv \begin{pmatrix} p \\ n \end{pmatrix} \to \begin{pmatrix} \psi^1 \\ \psi^2 \end{pmatrix}$.

The $SU(2)$ invariant Lagrangian density of this system writes

$$L = \frac{i}{2}\left(\overline{\psi}^a \gamma_\mu \partial_\mu \psi^a - \partial_\mu \overline{\psi}^a \gamma_\mu \psi^a\right) - m\overline{\psi}^a \psi^a.$$

The doublet $\psi^a$, $(a = 1, 2)$ is usually called *isodoublet*, because the wave function $\psi^a$ follows the transformations of the group $SU(2)$, also called *isospin group* (since its representations allow grouping the hadrons into isospin multiplets). Since $\left[T^k, T^l\right] = g\varepsilon_{klm}T^m$ it follows that the group $SU(2)$ is non-Abelian (the generators of the group do not commute; the generators of any of its representations have the same property), therefore in the Lie algebra of the group there certainly exist non-commutative elements (that is the group Lie algebra admits at least one non-commutative subalgebra).

The above Lagrangian density is invariant under the action of the global symmetry transformations of the isospin group $SU(2)$,

$$\psi^a \to \psi^{a\prime} = \left[\exp\left(-\frac{i}{2} g\varepsilon_k \sigma_k\right)\right]_{ab} \psi^b,$$

$$\overline{\psi}^a \to \overline{\psi}^{a\prime} = \overline{\psi}^b \left[\exp\left(\frac{i}{2} g\varepsilon_k \sigma_k\right)\right]_{ba}.$$

In the case of infinitesimal transformations $(\mathcal{O}(\varepsilon^2) = 0)$,

$$\psi^a \to \psi^{a\prime} = \psi^a + \delta\psi^a = \psi^a - \frac{i}{2} g\varepsilon_k (\sigma_k)_{ab} \psi^b,$$

$$\overline{\psi}^a \to \overline{\psi}^{a\prime} = \overline{\psi}^a + \delta\overline{\psi}^a = \overline{\psi}^a + \frac{i}{2} g\varepsilon_k \overline{\psi}^b (\sigma_k)_{ba}.$$

Using these preliminaries, one can easily find the total Lagrangian density of the system, invariant with respect to the *local* symmetry transformations.

The wave functions transform according to the $SU(2)$ group representations, that is according to the representations whose dimension equals the number of components of the wave function.

In the case of $SU(2)$ group, the *fundamental representation* has dimension two, its generators being

$$T^k_{ij} = -\frac{i}{2} g(\sigma_k)_{ij}, \quad (k = \overline{1,3};\ i,j = 1,2)\,.$$

The *vectorial* (*isovectorial*) *representation* of the group $SU(2)$ has dimension three; it is also called *adjoint representation*, since its dimension equals the group order. In this case, the matrices $\sigma$ are substituted by the rotation matrices in the three-dimensional space, $i\Sigma$ (it can be proved that the group $SU(2)$ is homomorphic with the rotation group in the three-dimensional space):

$$T^k_{ij} = -\frac{i}{2} g(\sigma_k)_{ij} \to T^k_{ij} = -ig\Sigma^k_{ij}, \quad (k = \overline{1,3};\ i,j = 1,2)\,,$$

where

$$\Sigma_1 = \begin{pmatrix} 0 & 0 & 0 \\ 0 & 0 & -i \\ 0 & i & 0 \end{pmatrix}, \quad \Sigma_2 = \begin{pmatrix} 0 & 0 & i \\ 0 & 0 & 0 \\ -i & 0 & 0 \end{pmatrix}, \quad \Sigma_3 = \begin{pmatrix} 0 & -i & 0 \\ i & 0 & 0 \\ 0 & 0 & 0 \end{pmatrix}.$$

Up to the factor "$i$", the three matrices $\Sigma_j$, $(j = 1,2,3)$ are precisely the matrices of the generators of rotations in the three-dimensional space. Then one has

$$u^a_\mu(x) \to u^{a\prime}_\mu(x) = \left[\delta^a_b - ig\varepsilon^k(\Sigma^k)^a_b\right] u^b_\mu(x) + \mathcal{O}(\varepsilon^2).$$

The *scalar representation* (or *isoscalar*, as usually called) is represented by the unit matrix: $T \to I$, therefore $\varphi(x) \to \varphi'(x) = \varphi(x)$ (the isospin singlet).

## 3. The group $SU(3)$. The $SU(3)$ symmetry

According to the definition, the elements of the $SU(3)$ group are $3 \times 3$ unitary and unimodular matrices, that can be considered as representing some matrix operators acting on a space whose vectors have three components. This space is precisely $\mathbb{C}^3$, that is the three-dimensional complex linear space, with the vectors

$$z = \begin{pmatrix} z_1 \\ z_2 \\ z_3 \end{pmatrix}, \quad z_i \in \mathbb{C}, \quad \vec{z} = z_1\vec{e}^{\,1} + z_2\vec{e}^{\,2} + z_3\vec{e}^{\,3},$$

where $\{\vec{e}^1, \vec{e}^2, \vec{e}^3\}$ form a basis in this space. Then, the group $SU(3)$ can be regarded as the group whose elements are the unitary and unimodular operators $U$ defined on $\mathbb{C}^3$ by

$$z \to z' = Uz,$$

where $U$ is a $3 \times 3$ matrix. In terms of the components, this writes

$$z_i \to z_i' = U_i^j z_j,$$

and

$$U^+U = I_3, \quad \det U = 1.$$

The set of matrices $U$ form the *fundamental representation* of the group, usually denoted by "3", also called *gradient representation* or *covariant triplet representation*, since the numbers $z_i$ that transform according to $z_i \to z_i' = U_i^j z_j$ form a covariant triplet under the $SU(3)$ transformations.

The vector $z^+$ (the Hermitian conjugate of $z$) writes

$$z^+ = (z_1^*, z_2^*, z_3^*)\,,$$

where $z_i^*$, $(i = \overline{1,3})$ denotes the complex conjugate of the element $z_i$. Since $z_i \to z_i' = U_i^j z_j$, $(i,j = \overline{1,3})$, one has

$$z_i^{*\prime} = U_i^{*j} z_j^* = \left(U^+\right)_j^i z^j,$$

where the notation $z_j^* \overset{\text{not}}{=} z^j$ has been used. Thus,

$$z^i \to z^{i\prime} = z^j \left(U^+\right)_j^i = z^j \left(U^{-1}\right)_j^i, \tag{4.27}$$

hence

$$z^+ \to z^{+\prime} = z^+U^+ = z^+U^{-1}.$$

The quantities $z^i$ appearing in (4.27) form a contravariant triplet under the transformations of the group $SU(3)$. The numbers $z_i$ transform by means of the matrix $U$, and $z^i$ by means of $U^*$. The set of matrices $U^*$ also form a representation of the group $SU(3)$ (like $U$) called *contragradient representation* or *contravariant triplet representation*, being usually denoted by $\overline{3}$ or $3^*$. These two representations are not equivalent, because there is no similarity transformation which can turn $U$ into $U^*$; in other words, there exists no matrix $C$ such that $U^* = CUC^{-1}$.

Since the most frequently used parametrization of the group $SU(3)$ is the exponential parametrization, the following presentation shall be concerned only with it. In fact, any Lie group of the type $SU(n)$ admits such a parametrization, which can be written as

$$\forall\, g \in G, \quad g = \exp(iX) = \exp\left(i\sum_i \alpha^i X_i\right),$$

where $i\alpha^i$ are the group parameters, and $X_i$ its generators. In addition, the matrix $X$ has to be Hermitian (in order for $g$ to be unitary) and having the null trace (in order for $g$ to be unimodular). These requirements are fulfilled by the matrix $X$, written in the form $X = \sum_{i=1}^{8} \varepsilon^i \lambda_i$ (because $SU(3)$ has eight generators, $n^2 - 1 = 3^2 - 1 = 8$), where $\lambda_i, (i = \overline{1,8})$ are called the *Gell-Mann matrices* and form the basis of the Lie algebra associated with the Lie group $SU(3)$. These matrices are

$$\lambda_1 = \begin{pmatrix} 0 & 1 & 0 \\ 1 & 0 & 0 \\ 0 & 0 & 0 \end{pmatrix}, \quad \lambda_2 = \begin{pmatrix} 0 & -i & 0 \\ i & 0 & 0 \\ 0 & 0 & 0 \end{pmatrix}, \quad \lambda_3 = \begin{pmatrix} 1 & 0 & 0 \\ 0 & -1 & 0 \\ 0 & 0 & 0 \end{pmatrix},$$

$$\lambda_4 = \begin{pmatrix} 0 & 0 & 1 \\ 0 & 0 & 0 \\ 1 & 0 & 0 \end{pmatrix}, \quad \lambda_5 = \begin{pmatrix} 0 & 0 & -i \\ 0 & 0 & 0 \\ i & 0 & 0 \end{pmatrix}, \quad \lambda_6 = \begin{pmatrix} 0 & 0 & 0 \\ 0 & 0 & 1 \\ 0 & 1 & 0 \end{pmatrix},$$

$$\lambda_7 = \begin{pmatrix} 0 & 0 & 0 \\ 0 & 0 & -i \\ 0 & i & 0 \end{pmatrix}, \quad \lambda_8 = \frac{1}{\sqrt{3}} \begin{pmatrix} 1 & 0 & 0 \\ 0 & 1 & 0 \\ 0 & 0 & -2 \end{pmatrix}.$$

The crochet of the Lie algebra is

$$[\lambda_k, \lambda_j] = 2i f_{kji} \lambda_i,$$

and also

$$\mathrm{Tr}\,(\lambda_k \lambda_j) = 2\delta_{kj}.$$

The structure constants $f_{kji}$ of the Lie algebra are completely antisymmetric, and the non-zero components are

$$f_{123} = 1, \; f_{147} = f_{246} = f_{345} = f_{516} = f_{257} = f_{637} = \frac{1}{2},$$

$$f_{458} = f_{678} = \frac{\sqrt{3}}{2}.$$

The rank of the $SU(3)$ group is 2. In the basis formed by the Gell-Mann matrices, the group elements write under an exponential form (the exponential parametrization of the group),

$$U = g = e^{-\frac{i}{2}\varepsilon^k \lambda_k},$$

as generators being the matrices

$$F_k = -\frac{i}{2}\lambda_k, \quad \left(k = \overline{1,8}\right).$$

When studying the interaction of a field that admits the $SU(3)$ symmetry, targeting the development of a corresponding gauge theory, one adds to $F_k$ the coupling constant $g$ as a factor, namely one takes

$$F_k = -\frac{i}{2} g\lambda_k, \quad \left(k = \overline{1,8}\right).$$

The field functions of various fields transform according to corresponding representations of the group $SU(3)$. Just as the $SU(2)$ group is called the *isospin group* because it describes the *isospin properties* of the particles, the group $SU(3)$ is named the *unitary symmetry group*, because it describes the *unitary symmetry properties* of the particles.

**Examples** of transformations of the field functions

(*i*) In the case of the *unit representation*, that is $\forall\, g \in SU(3)$, $g \equiv U \to T(g) \equiv T(U) = I$ (the unit matrix), one has $\varphi(x) \to \varphi'(x) = \varphi(x)$ (a $SU(3)$-scalar, or unitary spin singlet).

(*ii*) In the case of the *fundamental representation* (also called *triplet covariant representation* or *gradient representation*), the wave function has three components,

$$\psi^a(x) = \begin{pmatrix} \psi^{(1)}(x) \\ \psi^{(2)}(x) \\ \psi^{(3)}(x) \end{pmatrix}, \quad \left(a = \overline{1,3}\right),$$

(a triplet of particles, for example, the quarks $u$, $d$, and $s$). In this case,

$$\psi^a(x) \to \psi^{a\prime}(x) = \left[\delta^a_b - \frac{i}{2}\varepsilon^k (\lambda_k)^a_b\right] \psi^b(x) + \mathcal{O}(\varepsilon^2),$$
$$\left(a, b = \overline{1,3}\right),$$

or, if the transformations are infinitesimal,

$$\begin{aligned}\psi^a(x) \to \psi^{a\prime}(x) &= \psi^a(x) + \delta\psi^a(x) \\ &= \psi^a(x) + \left[-\frac{i}{2}\varepsilon^k (\lambda_k)^a_b \psi^b(x)\right],\end{aligned}$$

the representation generators being given by

$$(T_k)^a_b = -\frac{i}{2}(\lambda_k)^a_b, \quad \left(a, b = \overline{1,3}\right).$$

(*iii*) In the case of the *contragradient representation* (or *contravariant triplet representation*), the wave function is

$$\psi_a(x) = \Big(\psi_1(x),\, \psi_2(x),\, \psi_3(x)\Big), \quad \left(a = \overline{1,3}\right),$$

and transforms in accordance with

$$\psi_a(x) \to \psi^\prime_a(x) = \left[\delta^a_b + \frac{i}{2}\varepsilon^k (\lambda^T_k)^b_a\right] \psi_b(x) + \mathcal{O}(\varepsilon^2),$$

or, if the transformations are infinitesimal,

$$\begin{aligned}\psi_a(x) \to \psi^\prime_a(x) &= \psi_a(x) + \delta\psi_a(x) \\ &= \psi_a(x) + \frac{i}{2}\varepsilon^k (\lambda^T_k)^b_a\, \psi_b(x), \quad \left(a, b = \overline{1,3}\right),\end{aligned}$$

therefore the representation generators are

$$(T_k)^b_a = +\frac{i}{2}(\lambda^T_k)^b_a, \quad \left(a, b = \overline{1,3}\right).$$

*

* *

To have a concrete view on the necessity of introduction of *minimal substitution* (which lies on the basis of *minimal coupling principle*), in the followings the reader's attention shall be drawn on the basic elements of the gauge theories. This way, the general considerations presented at the beginning of the chapter (and based on some equations given without justification or demonstration) shall receive a rigorous mathematical support.

## IV.3. Fundamentals of gauge theories

### IV.3.1. Condition of global invariance of a Lagrangian

Let $G$ be a Lie group of global internal symmetry of dimension $n$ and the infinitesimal generators $T^k$, $\left(k=\overline{1,n}\right)$. In a certain representation, the generator matrices are of the form $\left(T^k_{ij}\right)=\left(T^k\right)_{ij}$, while the matter fields $u_i(x)$ transform (infinitesimally) according to

$$u_i(x)\to u'_i(x)=u_i(x)+\delta u_i(x), \tag{4.28}$$

where

$$\delta u_i(x)=T^k_{ij}\varepsilon_k u_j(x), \tag{4.29}$$

since, in view of the previously presented mathematical background, one has

$$\begin{aligned}u'_i(x)&=e^{T^k\varepsilon_k}u_i(x)\cong\left(I+T^k\varepsilon_k\right)_{ij}u_j(x)\\&=\left(\delta_{ij}+\varepsilon_k T^k_{ij}\right)u_j(x)=u_i(x)+\varepsilon_k T^k_{ij}u_j(x),\end{aligned}$$

where $\varepsilon_k$ are the (constant, here) parameters of the group $G$, and the generators $T^k$ satisfy the well-known commutation relation

$$\left[T^k,T^l\right]=f_{klm}T^m, \tag{4.30}$$

where $f_{klm}$ are the *structure constants* of the group. In a matrix form, the commutation relation (4.30) writes

$$T^k_{ij}T^l_{jn}-T^l_{ij}T^k_{jn}=f_{klm}T^m_{in}, \tag{4.31}$$

while the structure constants satisfy the relation (Jacobi's identity),

$$f_{ikl}f_{lmn}+f_{kml}f_{lin}+f_{mil}f_{lkn}=0, \tag{4.32}$$

as well as the antisymmetry property in the first two indices,

$$f_{ikl} = -f_{kil}. \tag{4.33}$$

Consider, now, the invariance condition of the variational principle, $\delta S = 0$. In this respect, let

$$\mathcal{L} = \mathcal{L}\left(u_i(x), \partial_\mu u_i(x)\right) \tag{4.34}$$

be the Lagrangian density of a certain physical system (free matter field) described by the field/wave function $u_i(x)$. The action of the system then is

$$S = \int\limits_{(\Omega)} dx\ \mathcal{L}\left(u_i(x), \partial_\mu u_i(x)\right), \tag{4.35}$$

and its first variation writes

$$\delta S = \delta \int\limits_{(\Omega)} dx\ \mathcal{L}\left(u_i(x), \partial_\mu u_i(x)\right) = \int\limits_{(\Omega)} dx\ \delta\mathcal{L}\left(u_i(x), \partial_\mu u_i(x)\right)$$
$$= \int\limits_{(\Omega)} dx\ \left[\frac{\partial \mathcal{L}}{\partial u_i}\delta u_i + \frac{\partial \mathcal{L}}{\partial(\partial_\mu u_i)}\delta(\partial_\mu u_i)\right],$$

where $\Omega$ is an *arbitrary* four-domain. The invariance condition of the variational principle then writes

$$\frac{\partial \mathcal{L}}{\partial u_i}\delta u_i + \frac{\partial \mathcal{L}}{\partial(\partial_\mu u_i)}\delta(\partial_\mu u_i) = 0. \tag{4.36}$$

Replacing here $\delta u_i(x)$ given by (4.29) and observing that $\delta\left(\partial_\mu u_i(x)\right) = \partial_\mu\left(\delta u_i(x)\right) = \partial_\mu\left(T_{ij}^k \varepsilon_k u_j(x)\right) = T_{ij}^k \varepsilon_k \partial_\mu u_j(x)$, one obtains

$$\left[\frac{\partial \mathcal{L}}{\partial u_i}T_{ij}^k u_j + \frac{\partial \mathcal{L}}{\partial(\partial_\mu u_i)}T_{ij}^k \partial_\mu u_j\right]\varepsilon_k = 0. \tag{4.37}$$

Since the parameters $\varepsilon_k$ are *arbitrary*, it follows that

$$\frac{\partial \mathcal{L}}{\partial u_i}T_{ij}^k u_j + \frac{\partial \mathcal{L}}{\partial(\partial_\mu u_i)}T_{ij}^k \partial_\mu u_j = 0, \tag{4.38}$$

which is the necessary and sufficient condition for the *global invariance* of the Lagrangian density $\mathcal{L}$ with respect to the transformations of the Lie group $G$. Here the index $k$ runs from 1 to $n$, taking only natural values, $n$ is the dimension of the Lie group $G$, while indices $i$ and $j$ run from 1 to $m$, taking also only natural values, $m$ being the dimension of the considered representation of the Lie group $G$.

### IV.3.2. Condition of gauge/local invariance of a Lagrangian. Minimal substitution. Minimal coupling principle

The transition from global to local invariance is performed by considering the space-time dependence of the group parameters, $\varepsilon_k \to \varepsilon_k(x^\mu) \equiv \varepsilon_k(x)$ (as it has been explained at the beginning of this chapter). The variation of the field function then writes

$$\delta u_i(x) = T^k_{ij}\varepsilon_k(x)u_j(x), \tag{4.39}$$

while the variation of its derivatives is given by

$$\begin{aligned}\delta\left(\partial_\mu u_i(x)\right) &= \partial_\mu\left(\delta u_i(x)\right) = \partial_\mu\left(T^k_{ij}\varepsilon_k(x)u_j(x)\right)\\ &= T^k_{ij}\varepsilon_k(x)\partial_\mu u_j(x) + T^k_{ij}u_j(x)\partial_\mu\varepsilon_k(x).\end{aligned} \tag{4.40}$$

The variation $\delta\mathcal{L}$ of the Lagrangian density then becomes

$$\begin{aligned}\delta\mathcal{L} = \frac{\partial\mathcal{L}}{\partial u_i}\delta u_i(x) + \frac{\partial\mathcal{L}}{\partial(\partial_\mu u_i)}\delta\left(\partial_\mu u_i(x)\right) = \frac{\partial\mathcal{L}}{\partial u_i}T^k_{ij}\varepsilon_k(x)u_j(x)\\ + \frac{\partial\mathcal{L}}{\partial(\partial_\mu u_i)}T^k_{ij}\varepsilon_k(x)\partial_\mu u_j(x) + \frac{\partial\mathcal{L}}{\partial(\partial_\mu u_i)}T^k_{ij}u_j(x)\partial_\mu\varepsilon_k(x).\end{aligned}$$

Taking now into account the global invariance condition (which is automatically satisfied, being weaker than that of the local invariance), it follows that

$$\delta\mathcal{L} = \frac{\partial\mathcal{L}}{\partial(\partial_\mu u_i)}T^k_{ij}u_j(x)\partial_\mu\varepsilon_k(x) \neq 0, \tag{4.41}$$

meaning that the Lagrangian density is not invariant with respect to the local symmetry transformations (4.39). In order to obtain, nevertheless, a Lagrangian density invariant with respect to the local transformations, one introduces some new fields, $A'_l(x)$, $\left(l = \overline{1,s}\right)$, called *compensating fields* (their aim being to compensate the appearance of the term $\frac{\partial\mathcal{L}}{\partial(\partial_\mu u_i)}T^k_{ij}u_j(x)\partial_\mu\varepsilon_k(x)$ that brakes the invariance), or *gauge fields*, this denomination being tightly connected to the particular case of the $U(1)$ symmetry (where the compensating field is just the electromagnetic field $A_\mu(x)$, which, as it is known, satisfies a *gauge* condition). The number of these fields (*i.e.* the maximum value of the index $s$) is for the moment undetermined; for the sake of simplicity, one presumes that the new Lagrangian density depends – in addition to the old variables – only on the gauge fields, $A'_l(x)$ (and not on its derivatives, $\partial_\mu A'_l(x)$), that is

$$\mathcal{L}_1 = \mathcal{L}_1\Big(u_i(x), \partial_\mu u_i(x), A'_l(x)\Big). \tag{4.42}$$

Since in the Lagrangian variation under the infinitesimal local transformations appear both the parameters of the group $G$, $\varepsilon_k(x)$, and their derivatives, $\partial_\mu \varepsilon_k(x)$, one postulates the variation of the compensating fields of the form

$$\delta A'_l(x) = P^k_{li} A'_i(x)\varepsilon_k(x) + R^k_{l\mu}\partial_\mu \varepsilon_k(x), \tag{4.43}$$

while the variation of the matter fields is maintained in the same form,

$$\delta u_i(x) = T^k_{ij}\varepsilon_k(x)u_j(x). \tag{4.44}$$

The quantities $P^k_{li} = (P_l)^k_i$ and $R^k_{l\mu} = (R_l)^k_\mu$ appearing in (4.43) are constant matrices, whose form shall be determined from the invariance condition of the new Lagrangian density (4.42) with respect to the local symmetry transformations.

Imposing the local invariance condition to the new Lagrangian density, $\delta\mathcal{L}_1 = 0$, one can write

$$\delta\mathcal{L}_1 = \frac{\partial \mathcal{L}_1}{\partial u_i}\delta u_i + \frac{\partial \mathcal{L}_1}{\partial(\partial_\mu u_i)}\delta\left(\partial_\mu u_i\right) + \frac{\partial \mathcal{L}_1}{\partial A'_l}\delta A'_l = 0. \tag{4.45}$$

Taking into account the variations of the implied functions, namely those given by (4.40), (4.43) and (4.44), the variation (4.45) becomes

$$\begin{aligned}
&\left(\frac{\partial \mathcal{L}_1}{\partial u_i}T^k_{ij}u_j + \frac{\partial \mathcal{L}_1}{\partial(\partial_\mu u_i)}T^k_{ij}\partial_\mu u_j + \frac{\partial \mathcal{L}_1}{\partial A'_l}P^k_{lm}A'_m\right)\varepsilon_k(x)\\
&\quad + \left(\frac{\partial \mathcal{L}_1}{\partial(\partial_\mu u_i)}T^k_{ij}u_j + \frac{\partial \mathcal{L}_1}{\partial A'_l}R^k_{l\mu}\right)\partial_\mu\varepsilon_k(x) = 0,
\end{aligned} \tag{4.46}$$

where $\varepsilon_k(x)$ and $\partial_\mu \varepsilon_k(x)$ are *arbitrary* quantities. The last equation yields the following two conditions:

$$\frac{\partial \mathcal{L}_1}{\partial u_i} T^k_{ij} u_j + \frac{\partial \mathcal{L}_1}{\partial(\partial_\mu u_i)} T^k_{ij} \partial_\mu u_j + \frac{\partial \mathcal{L}_1}{\partial A'_l} P^k_{lm} A'_m = 0, \tag{4.47}$$

and

$$\frac{\partial \mathcal{L}_1}{\partial(\partial_\mu u_i)} T^k_{ij} u_j + \frac{\partial \mathcal{L}_1}{\partial A'_l} R^k_{l\mu} = 0. \tag{4.48}$$

Taking into consideration the summation indices, eqs. (4.48) represents a system of $4n$ equations ($k = \overline{1,n}$ and $\mu = 0,1,2,3$). Thus, to univocally determine the compensating fields $A'_l(x)$, ($l = \overline{1,s}$), the index $s$ has to take $4n$ values.

Suppose that the matrices $R^k_{l\mu}$ exist, are nonsingular and are invertible, their inverses being given by

$$\left(R^k_{l\mu}\right)^{-1} R^k_{m\mu} = \delta_{lm}, \quad \left(R^k_{l\mu}\right)^{-1} R^i_{l\nu} = \delta_{ki}\eta_{\mu\nu}. \tag{4.49}$$

One can now define the "true" gauge fields by

$$A^k_\mu(x) = \left(R^k_{l\mu}\right)^{-1} A'_l(x). \tag{4.50}$$

According to their definition, under the Lorentz group transformations the new fields transform like the four-vectors; the index $k$ determines the number of components with respect to the gauge group. The way of transformation of these fields under a gauge group transformation shall be seen later on.

By means of (4.49) and (4.50), the system of equations (4.48) can be written as

$$\frac{\partial \mathcal{L}_1}{\partial(\partial_\mu u_i)} T^k_{ij} u_j + \frac{\partial \mathcal{L}_1}{\partial A^k_\mu} = 0. \tag{4.51}$$

Indeed,

$$\begin{aligned} \frac{\partial \mathcal{L}_1}{\partial A'_l} R^k_{l\mu} &= \frac{\partial \mathcal{L}_1}{\partial A^j_\nu} \frac{\partial A^j_\nu}{\partial A'_l} R^k_{l\mu} = \frac{\partial \mathcal{L}_1}{\partial A^j_\nu} \left(R^j_{l\nu}\right)^{-1} R^k_{l\mu} \\ &= \frac{\partial \mathcal{L}_1}{\partial A^j_\nu} \delta_{jk}\eta_{\nu\mu} = \frac{\partial \mathcal{L}_1}{\partial A^k_\mu}. \end{aligned}$$

The system of equations (4.51) is complete and Jacobian. In order for this system of equations be satisfied by the Lagrangian density $\mathcal{L}_1$, the gauge fields $A_\mu^k$ must enter into $\mathcal{L}_1$ as a combination of the form

$$\nabla_\mu u_i \equiv \partial_\mu u_i - T_{ij}^k u_j A_\mu^k. \tag{4.52}$$

Indeed, in order to have $\frac{\partial \mathcal{L}_1}{\partial(\partial_\mu u_i)} T_{ij}^k u_j = -\frac{\partial \mathcal{L}_1}{\partial A_\mu^k}$, it is necessary that the derivatives of $\mathcal{L}_1$ be performed with respect to the same variable, and their coefficients (of the two sides) be equal. Thus, $\mathcal{L}_1$ must depend on $\partial_\mu u_i$ and $A_\mu^k$ through the medium of another quantity, denoted by $\nabla_\mu u_i$. Therefore, one takes

$$\nabla_\mu u_i = a\partial_\mu u_i + b_{ik} A_\mu^k,$$

where the coefficients $a$ and $b_{ik}$ – unknown for the moment – shall be determined later on.

One has

$$\frac{\partial \mathcal{L}_1}{\partial(\partial_\mu u_i)} = \frac{\partial \mathcal{L}_1}{\partial(\nabla_\nu u_j)} \frac{\partial(\nabla_\nu u_j)}{\partial(\partial_\mu u_i)} = a\eta_{\nu\mu}\delta_{ij} \frac{\partial \mathcal{L}_1}{\partial(\nabla_\nu u_j)} = a\frac{\partial \mathcal{L}_1}{\partial(\nabla_\mu u_i)},$$

and, analogously,

$$\begin{aligned}\frac{\partial \mathcal{L}_1}{\partial A_\mu^k} &= \frac{\partial \mathcal{L}_1}{\partial(\nabla_\nu u_j)} \frac{\partial(\nabla_\nu u_j)}{\partial A_\mu^k} = b_{jk}\eta_{\nu\mu} \frac{\partial \mathcal{L}_1}{\partial(\nabla_\nu u_j)} \\ &= b_{jk}\frac{\partial \mathcal{L}_1}{\partial(\nabla_\mu u_j)} = b_{ik}\frac{\partial \mathcal{L}_1}{\partial(\nabla_\mu u_i)}.\end{aligned}$$

Substituting these results into (4.51), one finds

$$a\frac{\partial \mathcal{L}_1}{\partial(\nabla_\mu u_i)} T_{ij}^k u_j = -b_{ik}\frac{\partial \mathcal{L}_1}{\partial(\nabla_\mu u_i)},$$

which implies

$$aT_{ij}^k u_j = -b_{ik}.$$

Since there is no other condition for the coefficients $a$ and $b_{ik}$, for the sake of simplicity one can take $a = 1$, which yields $b_{ik} = -T_{ij}^k u_j$, this way justifying (4.52). The quantity $\nabla_\mu u_i$ is usually called *covariant derivative* (in the gauge sense) or *gauge covariant derivative*. It then follows that the locally invariant Lagrangian density must have the following functional dependence:

$$\mathcal{L}_1 = \mathcal{L}_1\left(u_i, \partial_\mu u_i, A'_l\right) = \mathcal{L}'_1\left(u_i, \nabla_\mu u_i\right).$$

Therefore,

$$\begin{aligned}\frac{\partial \mathcal{L}_1}{\partial u_i} &= \left.\frac{\partial \mathcal{L}'_1}{\partial u_i}\right|_{\nabla_\mu u_i = const.} + \left.\frac{\partial \mathcal{L}'_1}{\partial(\nabla_\mu u_j)}\right|_{u_i = const.} \frac{\partial(\nabla_\mu u_j)}{\partial u_i} \\ &= \left.\frac{\partial \mathcal{L}'_1}{\partial u_i}\right|_{\nabla_\mu u_i = const.} - T^k_{ji} A^k_\mu \left.\frac{\partial \mathcal{L}'_1}{\partial(\nabla_\mu u_j)}\right|_{u_i = const.},\end{aligned} \tag{4.53-1}$$

$$\frac{\partial \mathcal{L}_1}{\partial(\partial_\mu u_i)} = \left.\frac{\partial \mathcal{L}'_1}{\partial(\nabla_\mu u_j)}\right|_{u_i = const.} \frac{\partial(\nabla_\mu u_j)}{\partial(\partial_\mu u_i)} = \left.\frac{\partial \mathcal{L}'_1}{\partial(\nabla_\mu u_i)}\right|_{u_i = const.}, \tag{4.53-2}$$

$$\begin{aligned}\frac{\partial \mathcal{L}_1}{\partial A'_l} &= \left.\frac{\partial \mathcal{L}'_1}{\partial(\nabla_\mu u_j)}\right|_{u_i = const.} \frac{\partial(\nabla_\mu u_j)}{\partial A'_l} = -T^k_{ji} u_i \frac{\partial A^k_\mu}{\partial A'_l} \left.\frac{\partial \mathcal{L}'_1}{\partial(\nabla_\mu u_j)}\right|_{u_i = const.} \\ &= -\left.\frac{\partial \mathcal{L}'_1}{\partial(\nabla_\mu u_j)}\right|_{u_i = const.} T^k_{ji} u_i \left(R^k_{l\mu}\right)^{-1}.\end{aligned} \tag{4.53-3}$$

The requirement of the local invariance then consists in replacing the usual derivatives by the covariant ones in the globally invariant Lagrangian. This operation is called the *minimal substitution*. The second term in the gauge covariant derivative determines the interaction between the fields $u_i(x)$ – naturally called the *matter fields* – and the gauge fields $A^k_\mu(x)$. Thus, the interaction Lagrangian can be easily obtained by performing the minimal substitution. This procedure has been raised to the rank of principle, usually known as the *minimal coupling principle*. It is necessary to mention that the number of the gauge fields $A^k_\mu$ equals the number of generators of the *gauge group* $G$ (whose parameters $\varepsilon_k(x)$ are not constant anymore, in contrast to the initial Lie group of *global* internal symmetry which was denoted by the same letter $G$).

### IV.3.3. The gauge fields

The purpose of this paragraph is to present the most important properties of the gauge fields.

For the beginning, one shall deduce the transformation law of the gauge fields $A^k_\mu(x)$ with respect to the local (gauge) transformations of the internal symmetry group $G$. Thus,

$$\begin{aligned}\delta A^k_\mu(x) &= \delta\left[\left(R^k_{l\mu}\right)^{-1} A'_l(x)\right] = \left(R^k_{l\mu}\right)^{-1}\delta A'_l(x)\\ &= \left(R^k_{l\mu}\right)^{-1} P^j_{li} A'_i(x)\varepsilon_j(x) + \left(R^k_{l\mu}\right)^{-1} R^j_{l\nu}\partial_\nu\varepsilon_j(x)\\ &= \left(R^k_{l\mu}\right)^{-1} P^j_{li} R^m_{i\nu} A^m_\nu(x)\varepsilon_j(x) + \partial_\mu\varepsilon_k(x),\end{aligned}$$

or

$$\delta A^k_\mu(x) = \left(C^k_\mu\right)^{jm}_\nu A^m_\nu(x)\varepsilon_j(x) + \partial_\mu\varepsilon_k(x), \tag{4.54}$$

where

$$\left(C^k_\mu\right)^{jm}_\nu = \left(R^k_{l\mu}\right)^{-1} P^j_{li} R^m_{i\nu}$$

are – for the moment – unknown quantities. To determine these matrices, one uses the Eq. (4.47). In view of Eqs. (4.53) and the definition (4.50) of the gauge fields, Eq. (4.47) becomes

$$\begin{aligned}&\frac{\partial\mathcal{L}'_1}{\partial u_i}T^k_{ij}u_j - \frac{\partial\mathcal{L}'_1}{\partial(\nabla_\mu u_s)}T^m_{si}A^m_\mu T^k_{ij}\partial_\mu u_j + \frac{\partial\mathcal{L}'_1}{\partial(\nabla_\mu u_i)}T^k_{ij}\partial_\mu u_j\\ &\qquad - \frac{\partial\mathcal{L}'_1}{\partial(\nabla_\mu u_s)}T^p_{si}u_i \underbrace{\left(R^p_{l\mu}\right)^{-1} P^k_{lm}R^n_{m\nu}}_{\left(C^p_\mu\right)^{kn}_\nu} A^n_\nu = 0,\end{aligned}$$

or, if in the third term the common derivative is replaced by the covariant one,

$$\frac{\partial \mathcal{L}_1'}{\partial u_i} T^k_{ij} u_j - \underbrace{\frac{\partial \mathcal{L}_1'}{\partial(\nabla_\mu u_s)} T^m_{si} T^k_{ij} A^m_\mu \partial_\mu u_j}_{s\to i\to j\to s} + \frac{\partial \mathcal{L}_1'}{\partial(\nabla_\mu u_i)} \nabla_\mu u_j T^k_{ij}$$

$$+ \frac{\partial \mathcal{L}_1'}{\partial(\nabla_\mu u_i)} T^k_{ij} T^m_{js} A^m_\mu u_s - \underbrace{\frac{\partial \mathcal{L}_1'}{\partial(\nabla_\mu u_s)} T^p_{si} \left(C^p_\mu\right)^{kn}_\nu A^n_\nu u_i}_{s\to i\to j} = 0,$$

that is,

$$\frac{\partial \mathcal{L}_1'}{\partial(\nabla_\mu u_i)} \left[ \left( T^k_{ij} T^m_{js} - T^k_{js} T^m_{ij} \right) u_s A^m_\mu - T^p_{ij} \left(C^p_\mu\right)^{kn}_\nu A^n_\nu u_j \right] = 0.$$

Now using the commutation relation satisfied by the generators,

$$T^k_{ij} T^m_{js} - T^m_{ij} T^k_{js} = f_{kml} T^l_{is},$$

the previous relation becomes

$$\frac{\partial \mathcal{L}_1'}{\partial(\nabla_\mu u_i)} \left[ \underbrace{f_{kml} T^l_{is} A^m_\mu u_s}_{s\to j} - \underbrace{T^p_{ij} \left(C^p_\mu\right)^{kn}_\nu A^n_\nu u_j}_{p\to l,\, n\to m} \right] = 0,$$

or, if the indicated index replacements are performed,

$$\frac{\partial \mathcal{L}_1'}{\partial(\nabla_\mu u_i)} \left[ f_{kml} T^l_{ij} A^m_\mu u_j - T^l_{ij} \left(C^l_\mu\right)^{km}_\nu A^m_\nu u_j \right] = 0.$$

Forcing the quantity $T^l_{ij} u_j A^m_\nu$ to be taken out as a common factor, one gets

$$\frac{\partial \mathcal{L}_1'}{\partial(\nabla_\mu u_i)} \left[ f_{kml} \eta_{\mu\nu} - \left(C^l_\mu\right)^{km}_\nu \right] T^l_{ij} A^m_\nu u_j = 0,$$

which yields

$$\left(C^l_\mu\right)^{km}_\nu = f_{kml} \eta_{\mu\nu}.$$

Replacing these matrices into (4.54) (which relation expresses the variation of the gauge fields), one finds

$$\begin{aligned} \delta A^k_\mu(x) &= \left(C^k_\mu\right)^{jm}_\nu A^m_\nu(x) \varepsilon_j(x) + \partial_\mu \varepsilon_k(x) \\ &= f_{jmk} \eta_{\mu\nu} A^m_\nu(x) \varepsilon_j(x) + \partial_\mu \varepsilon_k(x) \\ &= f_{lmk} A^m_\mu(x) \varepsilon_l(x) + \partial_\mu \varepsilon_k(x). \end{aligned} \tag{4.55}$$

To summarize, the locally invariant Lagrangian density is obtained by replacing the ordinary derivative $(\partial_\mu)$ by the gauge covariant derivative $(\nabla_\mu)$ into the globally invariant Lagrangian density (the so-called *minimal substitution*). The gauge fields $A_\mu^k(x)$ transform with respect to the local (gauge) group according to (4.55).

One can now deduce the transformation law of the gauge covariant derivative, $\nabla_\mu u_i$, with respect to the gauge group. One has

$$\begin{aligned}
\delta\left(\nabla_\mu u_i\right) &= \delta\left(\partial_\mu u_i - T_{ij}^k u_j A_\mu^k\right) = \delta\left(\partial_\mu u_i\right) - T_{ij}^k \delta u_j A_\mu^k - T_{ij}^k u_j \delta A_\mu^k \\
&= T_{ij}^k \partial_\mu u_j \varepsilon_k(x) + T_{ij}^k u_j \partial_\mu \varepsilon_k(x) - T_{ij}^k T_{js}^m \varepsilon_m(x) u_s A_\mu^k \\
&\quad - T_{ij}^k u_j f_{lmk} A_\mu^m \varepsilon_l(x) - T_{ij}^k u_j \partial_\mu \varepsilon_k(x) = T_{ij}^k \partial_\mu u_j \varepsilon_k(x) \\
&\quad - \left(T_{ij}^k T_{js}^m + f_{mkl} T_{is}^l\right) \varepsilon_m(x) u_s A_\mu^k = T_{ij}^k \partial_\mu u_j \varepsilon_k(x) \\
&\quad - T_{ij}^m T_{js}^k \varepsilon_m(x) u_s A_\mu^k = T_{ij}^k \varepsilon_k(x) \left(\partial_\mu u_j - T_{js}^m u_s A_\mu^m\right),
\end{aligned}$$

that is

$$\delta\left(\nabla_\mu u_i\right) = T_{ij}^k \varepsilon_k(x) \nabla_\mu u_j. \tag{4.56}$$

This equation shows that $\nabla_\mu u_i$ transforms with respect to the gauge group as the field functions $u_i(x)$; due to this property, it is called *covariant* derivative.

### ♣ The Lagrangian of the gauge fields

The gauge invariant Lagrangian density $\mathcal{L}_1'\left(u_i, \nabla_\mu u_i\right)$ contains both the Lagrangian density of the free matter fields $u_i(x)$ and the Lagrangian density expressing the interaction between the free fields and the gauge fields $A_\mu^k(x)$. In order to determine the Lagrangian density of the gauge fields, $\mathcal{L}_0\left(A_\mu^k, \partial_\nu A_\mu^k\right) \equiv \mathcal{L}_{gauge}\left(A_\mu^k, \partial_\nu A_\mu^k\right)$, one must first study its invariance with respect to the transformations of the internal symmetry gauge group:

$$\mathcal{L}_0\left(A_\mu^k, \partial_\nu A_\mu^k\right) \to \delta\mathcal{L}_0 = \frac{\partial\mathcal{L}_0}{\partial A_\mu^k}\delta A_\mu^k + \frac{\partial\mathcal{L}_0}{\partial\left(\partial_\nu A_\mu^k\right)}\delta\left(\partial_\nu A_\mu^k\right) = 0.$$

Taking into account (4.55) and the fact that $\delta(\partial_\nu A^k_\mu) = \partial_\nu(\delta A^k_\mu)$, one can write

$$\left(\frac{\partial \mathcal{L}_0}{\partial A^k_\mu} f_{lmk} A^m_\mu + \frac{\partial \mathcal{L}_0}{\partial\left(\partial_\nu A^k_\mu\right)} f_{lmk} \partial_\nu A^m_\mu\right) \varepsilon_l(x)$$

$$+ \left(\frac{\partial \mathcal{L}_0}{\partial A^l_\nu} + \frac{\partial \mathcal{L}_0}{\partial\left(\partial_\nu A^k_\mu\right)} f_{lmk} A^m_\mu\right) \partial_\nu \varepsilon_l(x) + \frac{\partial \mathcal{L}_0}{\partial\left(\partial_\nu A^k_\mu\right)} \partial_\nu \partial_\mu \varepsilon_k(x) = 0.$$

Since the parameters $\varepsilon_k(x)$ are *arbitrary* quantities, there are three resulting equations:

$$1)\ \frac{\partial \mathcal{L}_0}{\partial A^k_\mu} f_{lmk} A^m_\mu + \frac{\partial \mathcal{L}_0}{\partial\left(\partial_\nu A^k_\mu\right)} f_{lmk} \partial_\nu A^m_\mu = 0,\ \ (l = \overline{1,n}), \tag{4.57}$$

$$2)\ \frac{\partial \mathcal{L}_0}{\partial A^l_\nu} + \frac{\partial \mathcal{L}_0}{\partial\left(\partial_\nu A^k_\mu\right)} f_{lmk} A^m_\mu = 0,\ \ (l = \overline{1,n}), \tag{4.58}$$

$$3)\ \frac{\partial \mathcal{L}_0}{\partial\left(\partial_\mu A^k_\nu\right)} + \frac{\partial \mathcal{L}_0}{\partial\left(\partial_\nu A^k_\mu\right)} = 0,\ \ (k = \overline{1,n}). \tag{4.59}$$

The last equation is obtained as follows:

$$\frac{\partial \mathcal{L}_0}{\partial(\partial_\nu A^k_\mu)} \partial_\nu \partial_\mu \varepsilon_k(x) = \frac{1}{2}\left[\frac{\partial \mathcal{L}_0}{\partial(\partial_\nu A^k_\mu)} \partial_\nu \partial_\mu \varepsilon_k(x) + \frac{\partial \mathcal{L}_0}{\partial\left(\partial_\nu A^k_\mu\right)} \partial_\nu \partial_\mu \varepsilon_k(x)\right]$$

$$= \frac{1}{2}\left[\frac{\partial \mathcal{L}_0}{\partial\left(\partial_\mu A^k_\nu\right)} \partial_\mu \partial_\nu \varepsilon_k(x) + \frac{\partial \mathcal{L}_0}{\partial\left(\partial_\nu A^k_\mu\right)} \partial_\nu \partial_\mu \varepsilon_k(x)\right]$$

$$= \frac{1}{2}\left[\frac{\partial \mathcal{L}_0}{\partial\left(\partial_\mu A^k_\nu\right)} + \frac{\partial \mathcal{L}_0}{\partial\left(\partial_\nu A^k_\mu\right)}\right] \partial_\mu \partial_\nu \varepsilon_k(x) = 0,$$

where the equality of the mixed second-order derivatives of $\varepsilon_k(x)$ (supposing that the Schwarz's theorem condition are fulfilled) has been considered.

According to condition 3), the derivatives of the gauge fields $\partial_\nu A^k_\mu$ can only enter the Lagrangian density $\mathcal{L}_0$ as the combination

$$A^k_{\mu\nu} = \partial_\mu A^k_\nu - \partial_\nu A^k_\mu,$$

with $A^k_{\mu\nu} = -A^k_{\nu\mu}$. Indeed,

$$\frac{\partial \mathcal{L}_0}{\partial\left(\partial_\nu A^k_\mu\right)} = \frac{\partial \mathcal{L}_0}{\partial A^m_{\rho\sigma}}\frac{\partial A^m_{\rho\sigma}}{\partial\left(\partial_\nu A^k_\mu\right)} = \frac{\partial \mathcal{L}_0}{\partial A^m_{\rho\sigma}}\left(\delta_{mk}\eta_{\rho\nu}\eta_{\sigma\mu} - \delta_{mk}\eta_{\rho\mu}\eta_{\sigma\nu}\right)$$
$$= \frac{\partial \mathcal{L}_0}{\partial A^k_{\nu\mu}} - \frac{\partial \mathcal{L}_0}{\partial A^k_{\mu\nu}} = -2\frac{\partial \mathcal{L}_0}{\partial A^k_{\mu\nu}}.$$

In an analogous way one shows that

$$\frac{\partial \mathcal{L}_0}{\partial\left(\partial_\mu A^k_\nu\right)} = -2\frac{\partial \mathcal{L}_0}{\partial A^k_{\nu\mu}} = 2\frac{\partial \mathcal{L}_0}{\partial A^k_{\mu\nu}},$$

and then, the condition 3) is immediately verified. With this result, 1) and 2) become

$$1')\ \frac{\partial \mathcal{L}_0}{\partial A^k_\mu} f_{lmk} A^m_\mu - 2\frac{\partial \mathcal{L}_0}{\partial A^k_{\mu\nu}} f_{lmk}\partial_\nu A^m_\mu = 0,\ (l = \overline{1,n}), \tag{4.57'}$$

$$2')\ \frac{\partial \mathcal{L}_0}{\partial A^l_\mu} - 2\frac{\partial \mathcal{L}_0}{\partial A^k_{\nu\mu}} f_{lmk} A^m_\nu = 0,\ (l = \overline{1,n}). \tag{4.58'}$$

The condition 2′) can be satisfied only if $\mathcal{L}_0$ depends on $A^k_\mu$ and $A^k_{\mu\nu}$ through the medium of some other quantity, denoted – let's say – by $F^k_{\mu\nu} = F^k_{\mu\nu}\left(A^k_\mu, A^k_{\mu\nu}\right)$. Then,

$$\frac{\partial \mathcal{L}_0}{\partial A^l_\mu} = \frac{\partial \mathcal{L}_0}{\partial F^m_{\rho\sigma}}\frac{\partial F^m_{\rho\sigma}}{\partial A^l_\mu},$$

and

$$\frac{\partial \mathcal{L}_0}{\partial A^k_{\mu\nu}} = \frac{\partial \mathcal{L}_0}{\partial F^m_{\rho\sigma}}\frac{\partial F^m_{\rho\sigma}}{\partial A^k_{\mu\nu}},$$

in which case the condition 2') becomes

$$\frac{\partial \mathcal{L}_0}{\partial F^m_{\rho\sigma}}\frac{\partial F^m_{\rho\sigma}}{\partial A^l_\mu} + 2\frac{\partial \mathcal{L}_0}{\partial F^m_{\rho\sigma}}\frac{\partial F^m_{\rho\sigma}}{\partial A^k_{\mu\nu}} f_{lsk} A^s_\nu = 0,$$

or

$$\frac{\partial \mathcal{L}_0}{\partial F^m_{\rho\sigma}}\left(\frac{\partial F^m_{\rho\sigma}}{\partial A^l_\mu} + 2 f_{lsk} A^s_\nu \frac{\partial F^m_{\rho\sigma}}{\partial A^k_{\mu\nu}}\right) = 0. \tag{4.60}$$

In order for this equation to be satisfied, the simplest choice is to have the following relation between $F^k_{\mu\nu}$, $A^k_{\mu\nu}$ and $A^k_\mu$:

$$F^k_{\mu\nu} = A^k_{\mu\nu} + \text{a quadratic term in the gauge fields } A^k_\mu. \qquad (4.61)$$

In addition, the quadratic term in $A^k_\mu$ must contain the structure constants of the group, $f_{lmk}$, with the minus sign, so that it has to be of the form $\left(-f_{lmk}A^l_\mu A^m_\nu\right)$. Therefore, $F^m_{\rho\sigma}$ must be of the form $F^m_{\rho\sigma} = A^m_{\rho\sigma} - f_{nsm}A^n_\rho A^s_\sigma$. Indeed, since

$$\frac{\partial F^m_{\rho\sigma}}{\partial A^l_\mu} = -f_{nsm}A^s_\sigma \delta_{nl}\eta_{\rho\mu} - f_{nsm}A^n_\rho \delta_{sl}\eta_{\sigma\mu} = -f_{lsm}A^s_\sigma\eta_{\rho\mu} - f_{nlm}A^n_\rho\eta_{\sigma\mu}$$

and

$$\frac{\partial F^m_{\rho\sigma}}{\partial A^k_{\mu\nu}} = \delta_{km}\eta_{\rho\mu}\eta_{\nu\sigma},$$

by substituting these two last expressions into (4.60) one has

$$\frac{\partial \mathcal{L}_0}{\partial F^m_{\rho\sigma}}\left(-f_{lsm}A^s_\sigma\eta_{\rho\mu} - f_{nlm}A^n_\rho\eta_{\sigma\mu}\right) + 2f_{lsk}A^s_\nu \frac{\partial \mathcal{L}_0}{\partial F^m_{\rho\sigma}}\delta_{km}\eta_{\rho\mu}\eta_{\nu\sigma}$$

$$= -f_{lsm}A^s_\sigma \frac{\partial \mathcal{L}_0}{\partial F^m_{\mu\sigma}} - f_{nlm}A^n_\rho \frac{\partial \mathcal{L}_0}{\partial F^m_{\rho\mu}} + 2f_{lsm}A^s_\nu \frac{\partial \mathcal{L}_0}{\partial F^m_{\mu\nu}}$$

$$= -f_{lsm}A^s_\nu \frac{\partial \mathcal{L}_0}{\partial F^m_{\mu\nu}} - f_{slm}A^s_\nu \frac{\partial \mathcal{L}_0}{\partial F^m_{\nu\mu}} + 2f_{lsm}A^s_\nu \frac{\partial \mathcal{L}_0}{\partial F^m_{\mu\nu}}$$

$$= -f_{lsm}A^s_\nu \frac{\partial \mathcal{L}_0}{\partial F^m_{\mu\nu}} - f_{lsm}A^s_\nu \frac{\partial \mathcal{L}_0}{\partial F^m_{\mu\nu}} + 2f_{lsm}A^s_\nu \frac{\partial \mathcal{L}_0}{\partial F^m_{\mu\nu}} = 0.$$

Without outbidding, one can consider that the quantity $F^k_{\mu\nu}$ can depend on $A^k_\mu$ and $A^k_{\mu\nu}$ under the general form

$$F^m_{\rho\sigma} - aA^m_{\rho\sigma} + b^m_{e\sigma}A^e_\rho, \qquad (4.62)$$

where $a$ is a true constant, and $b^m_{e\sigma} = b^m_{e\sigma}(A^s_\sigma)$ is a linear function of $A^s_\sigma$ which, for the sake of simplicity, is considered of the form $b^m_{e\sigma} = c^m_{pe}A^p_\sigma$, where $c^m_{pe}$ are true constants. In this case,

$$\frac{\partial F^m_{\rho\sigma}}{\partial A^l_\mu} = b^m_{e\sigma}\delta_{el}\eta_{\rho\mu} + A^e_\rho \frac{\partial b^m_{e\sigma}}{\partial A^l_\mu},$$

and

$$\frac{\partial F^m_{\rho\sigma}}{\partial A^k_{\mu\nu}} = a\delta_{mk}\eta_{\rho\mu}\eta_{\nu\sigma},$$

so that (4.60) becomes

$$\frac{\partial \mathcal{L}_0}{\partial F^m_{\rho\sigma}} \left( b^m_{l\sigma}\eta_{\rho\mu} + A^e_\rho \frac{\partial b^m_{e\sigma}}{\partial A^l_\mu} + 2af_{lsk}A^s_\nu \delta_{mk}\eta_{\rho\mu}\eta_{\nu\sigma} \right) = 0,$$

or,

$$\frac{\partial \mathcal{L}_0}{\partial F^k_{\mu\nu}} b^k_{l\nu} + \frac{\partial \mathcal{L}_0}{\partial F^m_{\rho\sigma}} A^e_\rho \frac{\partial b^m_{e\sigma}}{\partial A^l_\mu} + 2af_{lsk}A^s_\nu \frac{\partial \mathcal{L}_0}{\partial F^k_{\mu\nu}} = 0.$$

Since there are no other conditions acting upon $a$ and $b^m_{e\sigma} = b^m_{e\sigma}(A^s_\sigma)$, for the sake of simplicity one can take $a = 1$, in which case the previous equation becomes

$$\frac{\partial \mathcal{L}_0}{\partial F^k_{\mu\nu}} b^k_{l\nu} + \frac{\partial \mathcal{L}_0}{\partial F^m_{\rho\sigma}} A^e_\rho \frac{\partial b^m_{e\sigma}}{\partial A^l_\mu} + 2f_{lsk}A^s_\nu \frac{\partial \mathcal{L}_0}{\partial F^k_{\mu\nu}} = 0,$$

or

$$\frac{\partial \mathcal{L}_0}{\partial F^k_{\mu\nu}} b^k_{l\nu} + \frac{\partial \mathcal{L}_0}{\partial F^k_{\rho\nu}} A^e_\rho \frac{\partial b^k_{e\nu}}{\partial A^l_\mu} + 2f_{lsk}A^s_\nu \frac{\partial \mathcal{L}_0}{\partial F^k_{\mu\nu}} = 0. \tag{4.63}$$

In order to determine $b^m_{e\sigma} = b^m_{e\sigma}(A^s_\sigma)$ from this equation, one observes that the second term contains the derivative of $\mathcal{L}_0$ with respect to $F^k_{\rho\nu}$ (instead of the desired $F^k_{\mu\nu}$). In addition, since the significant index $\mu$ does not belong to $F^k_{\rho\nu}$, and $\rho$ is a summation index, the change $\rho \to \mu$ cannot be simply performed. Thus, the only solution for transforming $F^k_{\rho\nu}$ into $F^k_{\mu\nu}$ is to consider the derivative $\partial b^k_{e\nu}/\partial A^l_\mu$ as being equal to

$$\frac{\partial b^k_{e\nu}}{\partial A^l_\mu} = \frac{\partial \left( c^k_{pe} A^p_\nu \right)}{\partial A^l_\mu} = c^k_{pe}\delta_{lp}\eta_{\mu\nu} = c^k_{le}\eta_{\mu\nu},$$

where the tensor $\eta_{\mu\nu}$ ensures the desired change of the index. This way, the second term in (4.63) writes

$$\begin{aligned} \frac{\partial \mathcal{L}_0}{\partial F^k_{\rho\nu}} A^e_\rho \frac{\partial b^k_{e\nu}}{\partial A^l_\mu} &= \frac{\partial \mathcal{L}_0}{\partial F^k_{\rho\nu}} A^e_\rho c^k_{le}\eta_{\mu\nu} = \frac{\partial \mathcal{L}_0}{\partial F^k_{\rho\mu}} A^e_\rho c^k_{le} \\ &= \frac{\partial \mathcal{L}_0}{\partial F^k_{\nu\mu}} A^e_\nu c^k_{le} = -\frac{\partial \mathcal{L}_0}{\partial F^k_{\mu\nu}} A^s_\nu c^k_{ls}, \end{aligned}$$

and (4.63) becomes

$$\frac{\partial \mathcal{L}_0}{\partial F^k_{\mu\nu}} b^k_{l\nu} - \frac{\partial \mathcal{L}_0}{\partial F^k_{\mu\nu}} A^s_\nu c^k_{ls} + 2 f_{lsk} A^s_\nu \frac{\partial \mathcal{L}_0}{\partial F^k_{\mu\nu}} = 0,$$

which yields

$$b^k_{l\nu} - A^s_\nu c^k_{ls} + 2 f_{lsk} A^s_\nu = 0.$$

The constants $c^k_{ls}$ were completely arbitrary so far (but, obviously, non-null). The most convenient choice[14] for these constants is $c^k_{ls} = f_{lsk}$. In this case, the functions $b^k_{l\nu}$ write

$$b^k_{l\nu} = -f_{lsk} A^s_\nu.$$

Introducing this expression into (4.62) – and also considering the value of $a$ – one finally finds the correct expression of $F^m_{\rho\sigma}$, namely

$$F^m_{\rho\sigma} = A^m_{\rho\sigma} - f_{egm} A^e_\rho A^g_\sigma,$$

or,

$$F^k_{\mu\nu} = A^k_{\mu\nu} - f_{lmk} A^l_\mu A^m_\nu = A^k_{\mu\nu} - \frac{1}{2} f_{lmk} \left( A^l_\mu A^m_\nu - A^l_\nu A^m_\mu \right),$$

$$(k = \overline{1,n};\ \mu \neq \nu),$$

where the property of antisymmetry in the first two indices of the structure constants $f_{lmk}$ has been taken into account.

It is interesting (and, also, important) to observe that the quantity $F^k_{\mu\nu} = -F^k_{\nu\mu}$, called the *tensor of the gauge fields* (in fact, the tensor is $F_{\mu\nu} = F^k_{\mu\nu} T_k$) also follows from the commutator of the gauge covariant derivative operators. Indeed,

$$[\nabla_\mu, \nabla_\nu] u_i = \nabla_\mu (\nabla_\nu u_i) - \nabla_\nu (\nabla_\mu u_i) = \partial_\mu (\nabla_\nu u_i) - T^m_{is} (\nabla_\nu u_s) A^m_\mu$$
$$- \partial_\nu (\nabla_\mu u_i) + T^m_{is} (\nabla_\mu u_s) A^m_\nu = \partial_\mu \left( \partial_\nu u_i - T^k_{ij} u_j A^k_\nu \right)$$
$$- T^m_{is} \left( \partial_\nu u_s - T^k_{sp} u_p A^k_\nu \right) A^m_\mu - \partial_\nu \left( \partial_\mu u_i - T^k_{ij} u_j A^k_\mu \right)$$

[14] In fact, the only true (even real) constants interfering in the theory and having three indices are precisely the structure constants of the Lie group $G$, $f_{klm}$, so that the choice $c^k_{ls} = f_{lsk}$ is not only logical, but also the only natural and, at the same time, affordable alternative.

$$-T_{is}^m\left(\partial_\mu u_s - T_{sp}^k u_p A_\mu^k\right)A_\nu^m = \partial_\mu\partial_\nu u_i - T_{ij}^k\partial_\mu u_j A_\nu^k - T_{ij}^k u_j\partial_\mu A_\nu^k$$

$$-T_{is}^m\partial_\nu u_s A_\mu^m + T_{is}^m T_{sp}^k u_p A_\nu^k A_\mu^m - \partial_\nu\partial_\mu u_i + T_{ij}^k\partial_\nu u_j A_\mu^k + T_{ij}^k u_j\partial_\nu A_\mu^k$$

$$+T_{is}^m\partial_\mu u_s A_\nu^m - T_{is}^m T_{sp}^k u_p A_\mu^k A_\nu^m = -\left(\partial_\mu A_\nu^k - \partial_\nu A_\mu^k\right)T_{ij}^k u_j$$

$$+\left(T_{is}^m T_{sp}^k - T_{is}^k T_{sp}^m\right)u_p A_\mu^m A_\nu^k = -\left(\partial_\mu A_\nu^k - \partial_\nu A_\mu^k\right)T_{ij}^k u_j$$

$$+f_{mkl}T_{ip}^l u_p A_\mu^m A_\nu^k = -\left(\partial_\mu A_\nu^k - \partial_\nu A_\mu^k\right)T_{ij}^k u_j + f_{mlk}T_{ij}^k u_j A_\mu^m A_\nu^l$$

$$= -\left(\partial_\mu A_\nu^k - \partial_\nu A_\mu^k - f_{mlk}A_\mu^m A_\nu^l\right)T_{ij}^k u_j = -F_{\mu\nu}^k T_{ij}^k u_j.$$

The commutator $[\nabla_\mu, \nabla_\nu]$ is connected to the curvature of the connection $\nabla$; that is why $F_{\mu\nu}^k$ are also called the *components of the curvature 2-form* associated to this connection [415].

Admitting that the Lagrangian density of the (free) gauge field, $\mathcal{L}_0$, contains the gauge fields $A_\mu^k$ and their first derivatives, $\partial_\nu A_\mu^k$ only through the medium of the gauge field tensor $F_{\mu\nu}^k$,

$$\mathcal{L}_0 = \mathcal{L}_0\left(A_\mu^k, \partial_\nu A_\mu^k\right) = \mathcal{L}_0'\left(F_{\mu\nu}^k\right),$$

it follows that

$$\frac{\partial\mathcal{L}_0}{\partial(\partial_\mu A_\nu^k)} = 2\frac{\partial\mathcal{L}_0'}{\partial F_{\mu\nu}^k}, \qquad \frac{\partial\mathcal{L}_0}{\partial A_\mu^k} = 2\frac{\partial\mathcal{L}_0'}{\partial F_{\mu\nu}^l}f_{mkl}A_\nu^m. \tag{4.64}$$

Indeed, recalling that $F_{\rho\sigma}^l = \partial_\rho A_\sigma^l - \partial_\sigma A_\rho^l - f_{mnl}A_\rho^m A_\sigma^n$, one has

$$\frac{\partial\mathcal{L}_0}{\partial(\partial_\mu A_\nu^k)} = \frac{\partial\mathcal{L}_0'}{\partial F_{\rho\sigma}^l}\frac{\partial F_{\rho\sigma}^l}{\partial(\partial_\mu A_\nu^k)} = \frac{\partial\mathcal{L}_0'}{\partial F_{\rho\sigma}^l}\left(\delta_{\mu\rho}\delta_{kl}\delta_{\nu\sigma} - \delta_{\mu\sigma}\delta_{kl}\delta_{\nu\rho}\right)$$

$$= \frac{\partial\mathcal{L}_0'}{\partial F_{\mu\nu}^k} - \frac{\partial\mathcal{L}_0'}{\partial F_{\nu\mu}^k} = \frac{\partial\mathcal{L}_0'}{\partial F_{\mu\nu}^k} + \frac{\partial\mathcal{L}_0'}{\partial F_{\mu\nu}^k} = 2\frac{\partial\mathcal{L}_0'}{\partial F_{\mu\nu}^k},$$

$$\frac{\partial\mathcal{L}_0}{\partial A_\mu^k} = \frac{\partial\mathcal{L}_0'}{\partial F_{\rho\sigma}^l}\frac{\partial F_{\rho\sigma}^l}{\partial A_\mu^k} = \frac{\partial\mathcal{L}_0'}{\partial F_{\rho\sigma}^l}\left(-f_{mnl}\delta_{km}\delta_{\rho\mu}A_\sigma^n - f_{mnl}\delta_{nk}\delta_{\sigma\mu}A_\rho^m\right)$$

$$= -\frac{\partial\mathcal{L}_0'}{\partial F_{\mu\sigma}^l}\left(\underset{\substack{n\to m\\ \sigma\to\nu}}{f_{knl}A_\sigma^n}\right) - \frac{\partial\mathcal{L}_0'}{\partial F_{\rho\mu}^l}\left(\underset{\rho\to\nu}{f_{mkl}A_\rho^m}\right) = -\frac{\partial\mathcal{L}_0'}{\partial F_{\mu\nu}^l}f_{kml}A_\nu^m$$

$$-\frac{\partial \mathcal{L}'_0}{\partial F^l_{\nu\mu}} f_{mkl} A^m_\nu = \frac{\partial \mathcal{L}'_0}{\partial F^l_{\mu\nu}} f_{mkl} A^m_\nu + \frac{\partial \mathcal{L}'_0}{\partial F^l_{\mu\nu}} f_{mkl} A^m_\nu$$

$$= 2\frac{\partial \mathcal{L}'_0}{\partial F^l_{\mu\nu}} f_{mkl} A^m_\nu$$

and

$$\frac{\partial \mathcal{L}_0}{\partial A^k_{\mu\nu}} = \frac{\partial \mathcal{L}'_0}{\partial F^l_{\rho\sigma}} \frac{\partial F^l_{\rho\sigma}}{\partial A^k_{\mu\nu}} = \frac{\partial \mathcal{L}'_0}{\partial F^k_{\mu\nu}}. \tag{4.65}$$

So far, only two out of three relations resulting from the requirements that the gauge fields Lagrangian is invariant with respect to the internal symmetry gauge group have been used. The third equation, (4.57′), still unutilized, writes

$$\frac{\partial \mathcal{L}_0}{\partial A^k_\mu} f_{lmk} A^m_\mu - 2\frac{\partial \mathcal{L}_0}{\partial A^k_{\mu\nu}} f_{lmk} \partial_\nu A^m_\mu = 0.$$

In view of (4.64) and (4.65), this equation becomes

$$2\frac{\partial \mathcal{L}'_0}{\partial F^s_{\mu\nu}} f_{pks} \underset{s\leftrightarrow k}{A^p_\nu f_{lmk}} A^m_\mu - 2\frac{\partial \mathcal{L}'_0}{\partial F^k_{\mu\nu}} f_{lmk} \partial_\nu A^m_\mu = 0, \tag{4.66}$$

or

$$\frac{\partial \mathcal{L}'_0}{\partial F^k_{\mu\nu}} \Big( f_{psk} f_{lms} A^p_\nu A^m_\mu - f_{psk} \underset{p\leftrightarrow m}{f_{lms} A^p_\mu} A^m_\nu$$

$$- f_{lmk} \partial_\nu A^m_\mu + f_{lmk} \partial_\mu A^m_\nu \Big) = 0,$$

that is

$$\frac{\partial \mathcal{L}'_0}{\partial F^k_{\mu\nu}} \left[ \left( f_{psk} f_{lms} - f_{msk} f_{lps} \right) A^m_\mu A^p_\nu + f_{lmk} \left( \partial_\mu A^m_\nu - \partial_\nu A^m_\mu \right) \right] = 0. \tag{4.67}$$

Taking now into account the properties of the structure constants – the Jacobi's identity and the antisymmetry in the first two indices,

$$f_{ikl} f_{lmn} + f_{kml} f_{lin} + f_{mil} f_{lkn} = 0$$

and

$$f_{ikl} = -f_{kil},$$

respectively, in our case one can write

$$f_{lms}f_{spk} + f_{mps}f_{slk} + f_{pls}f_{smk} = 0,$$

hence

$$f_{psk}f_{lms} - f_{msk}f_{lps} = f_{mps}f_{slk}. \tag{4.68}$$

Introducing (4.68) into (4.67), one obtains

$$\frac{\partial \mathcal{L}_0'}{\partial F_{\mu\nu}^k}\left[\underset{s\leftrightarrow m}{f_{mps}\, f_{slk} A_\mu^m\ A_\nu^p} + f_{lmk}\left(\partial_\mu A_\nu^m - \partial_\nu A_\mu^m\right)\right] = 0,$$

or

$$\frac{\partial \mathcal{L}_0'}{\partial F_{\mu\nu}^k}\left[\underset{f_{mlk}=-f_{lmk}}{f_{mlk}f_{spm}A_\mu^s A_\nu^p} + f_{lmk}\left(\partial_\mu A_\nu^m - \partial_\nu A_\mu^m\right)\right] = 0, \tag{4.69}$$

which leads to

$$\frac{\partial \mathcal{L}_0'}{\partial F_{\mu\nu}^k} f_{lmk}\left[\left(\partial_\mu A_\nu^m - \partial_\nu A_\mu^m\right) - f_{spm}A_\mu^s A_\nu^p\right] = 0,$$

or

$$\frac{\partial \mathcal{L}_0'}{\partial F_{\mu\nu}^k} f_{lmk}F_{\mu\nu}^m = 0, \quad (l = \overline{1,n}), \tag{4.70}$$

and, still

$$-\frac{\partial \mathcal{L}_0'}{\partial F_{\mu\nu}^k} f_{mlk}F_{\mu\nu}^m = 0, \quad (l = \overline{1,n}). \tag{4.70$'$}$$

Therefore, the Lagrangian density of the gauge fields is a function of $F_{\mu\nu}^k$ (otherwise, the derivative $\partial\mathcal{L}_0'/\partial F_{\mu\nu}^k$ would be null) which must satisfy (4.70). But the choice of the gauge field Lagrangian density capable to satisfy these conditions is not unique. The simplest Lagrangian density quadratic in $F_{\mu\nu}^k$ was first proposed by Chen Ning Yang and Robert Mills in 1953. This is why it is called the *Yang–Mills Lagrangian density* and its form is analogous to that met in electrodynamics,

$$\mathcal{L}_{YM} = -\frac{1}{4}F_{\mu\nu}^k F_{\mu\nu}^k \equiv \mathcal{L}_0'\left(F_{\mu\nu}^k\right), \tag{4.71}$$

where

$$F^k_{\mu\nu} = \partial_\mu A^k_\nu - \partial_\nu A^k_\mu - \frac{1}{2} f_{lmk} \left( A^l_\mu A^m_\nu - A^l_\nu A^m_\mu \right). \qquad (4.72)$$

This Lagrangian density satisfies the conditions (4.70) if the properties (4.49) of the structure constants $f_{lmk}$ are considered. It can be shown that, in the case of compact semisimple Lie groups, the quantities $f_{lmk}$ are completely antisymmetric,

$$f_{lmk} = -f_{mlk} = -f_{lkm}. \qquad (4.73)$$

Indeed, the structure constants are also antisymmetric in the last two indices, because

$$\frac{\partial \mathcal{L}'_0}{\partial F^k_{\mu\nu}} f_{lmk} F^l_{\mu\nu} = \frac{\partial}{\partial F^k_{\mu\nu}} \left( -\frac{1}{4} F^s_{\rho\sigma} F^s_{\rho\sigma} \right) f_{lmk} F^l_{\mu\nu}$$

$$= -\frac{1}{4} \left( \delta_{ks}\delta_{\rho\mu}\delta_{\sigma\nu} F^s_{\rho\sigma} + F^s_{\rho\sigma} \delta_{ks}\delta_{\rho\mu}\delta_{\sigma\nu} \right) f_{lmk} F^l_{\mu\nu}$$

$$= -\frac{1}{2} f_{lmk} F^k_{\mu\nu} F^l_{\mu\nu} = -\frac{1}{4} \left[ f_{lmk} \left( F^k_{\mu\nu} F^l_{\mu\nu} \right) + f_{lkm} \left( F^l_{\mu\nu} F^k_{\mu\nu} \right) \right]$$

$$= -\frac{1}{4} \left( f_{lmk} + f_{lkm} \right) F^l_{\mu\nu} F^k_{\mu\nu} = 0.$$

The *full Lagrangian density* $\mathcal{L}$ of the system of the matter fields $u_i(x)$ and of gauge fields $A^k_\mu(x)$ is given by

$$\mathcal{L} = \mathcal{L}'_1 \left( u_i(x), \nabla_\mu u_i(x) \right) + \mathcal{L}_{YM} \left( F^k_{\mu\nu} \right). \qquad (4.74)$$

It is worth mentioning that in $\mathcal{L}'_1 \left( u_i, \nabla_\mu u_i \right)$ also appears the interaction Lagrangian density $\mathcal{L}_{int}$ between the matter fields $u_i(x)$ and gauge fields $A^k_\mu(x)$ due to the covariant derivative $\nabla_\mu u_i$ that contains both the matter fields (even under the derivative) and gauge fields, according to the definition (4.52):

$$\nabla_\mu u_i(x) = \partial_\mu u_i(x) - T^k_{ij} u_j(x) A^k_\mu(x).$$

It is important to establish the transformation law for the tensor of the gauge fields $F^k_{\mu\nu}$ with respect to the gauge transformations. To this end, one uses the definition

$$F^k_{\mu\nu} = \partial_\mu A^k_\nu - \partial_\nu A^k_\mu - \frac{1}{2} f_{lmk} A^l_\mu A^m_\nu$$

to obtain

$$\delta F^k_{\mu\nu} = \frac{\partial F^k_{\mu\nu}}{\partial(\partial_\rho A^s_\sigma)}\delta\left(\partial_\rho A^s_\sigma\right) + \frac{\partial F^k_{\mu\nu}}{\partial A^s_\rho}\delta A^s_\rho$$

$$= \frac{\partial F^k_{\mu\nu}}{\partial(\partial_\rho A^s_\sigma)}\partial_\rho\left(\delta A^s_\sigma\right) + \frac{\partial F^k_{\mu\nu}}{\partial A^s_\rho}\delta A^s_\rho,$$

or, if the gauge field variation

$$\delta A^k_\mu(x) = A'^k_\mu(x) - A^k_\mu(x) = f_{lmk}A^m_\mu(x)\varepsilon_l(x) + \partial_\mu\varepsilon_k(x),$$

is used, then we get

$$\delta F^k_{\mu\nu} = \left(\delta_{\mu\rho}\delta_{ks}\delta_{\nu\sigma} - \delta_{\nu\rho}\delta_{ks}\delta_{\mu\sigma}\right)\partial_\rho\left(f_{lms}A^m_\sigma\varepsilon_l + \partial_\sigma\varepsilon_s\right)$$

$$- f_{lmk}\left(\delta_{ls}\delta_{\mu\rho}A^m_s - \delta_{ms}\delta_{\nu\rho}A^l_\mu\right)\left(f_{pns}A^n_\rho\varepsilon_p + \partial_\rho\varepsilon_s\right)$$

$$= \partial_\mu\left(f_{lmk}A^m_\nu\varepsilon_l + \partial_\nu\varepsilon_k\right) - \partial_\nu\left(f_{lmk}A^m_\mu\varepsilon_l + \partial_\mu\varepsilon_k\right) - f_{lmk}f_{pnl}A^n_\mu A^m_\nu$$

$$\times\varepsilon_p - f_{lmk}\partial_\mu\varepsilon_l A^m_\nu - f_{lmk}f_{pnm}A^n_\nu A^l_\mu\varepsilon_p - f_{lmk}\partial_\nu\varepsilon_m A^l_\mu$$

$$= f_{lmk}\left(\partial_\mu A^m_\nu - \partial_\nu A^m_\mu\right)\varepsilon_l - \left(\underset{l\to m\to n\to l}{f_{lmk}f_{pnl}A^n_\mu A^m_\nu} + f_{lmk}f_{pnm}A^n_\nu A^l_\mu\right)\varepsilon_p$$

$$+ \partial_\mu\partial_\nu\varepsilon_k - \partial_\nu\partial_\mu\varepsilon_k + f_{lmk}A^m_\nu\partial_\mu\varepsilon_l - f_{lmk}A^m_\mu\partial_\nu\varepsilon_l - f_{lmk}\partial_\mu\varepsilon_l A^m_\nu$$

$$- \underset{l\leftrightarrow m}{f_{lmk}\partial_\nu\varepsilon_m A^l_\mu} = f_{lmk}\left(\partial_\mu A^m_\nu - \partial_\nu A^m_\mu\right)\varepsilon_l - \left(f_{mnk}f_{plm} + f_{lmk}f_{pnm}\right)$$

$$\times A^l_\mu A^n_\nu\varepsilon_p = f_{lmk}\left(\partial_\mu A^m_\nu - \partial_\nu A^m_\mu\right)\varepsilon_l - \left(f_{plm}f_{mnk} + f_{npm}f_{mlk}\right)$$

$$\times A^l_\mu A^n_\nu\varepsilon_p = f_{lmk}\left(\partial_\mu A^m_\nu - \partial_\nu A^m_\mu\right)\varepsilon_l - \underset{p\leftrightarrow l}{f_{nlm}f_{mpk}A^l_\mu A^n_\nu\varepsilon_p}$$

$$= f_{lmk}\left(\partial_\mu A^m_\nu - \partial_\nu A^m_\mu\right)\varepsilon_l + f_{mlk}f_{pnm}A^p_\mu A^n_\nu\varepsilon_l$$

$$= f_{lmk}\left(\partial_\mu A^m_\nu - \partial_\nu A^m_\mu - f_{pnm}A^p_\mu A^n_\nu\right)\varepsilon_l = f_{lmk}F^m_{\mu\nu}(x)\varepsilon_l(x).$$

Therefore,

$$\delta F^k_{\mu\nu}(x) = f_{lmk}\varepsilon_l(x)F^m_{\mu\nu}(x). \tag{4.75}$$

### IV.3.4. Field currents and their conservation

According to Noether's theorem, the Lagrangian density invariance with respect to a group of continuous transformations leads to the conservation of some (physical) quantity. In the case of local transformations, the conserved quantities are the *field currents* (or simply, the *currents*). In order to obtain their expressions, we start from the total Lagrangian density

$$\begin{aligned}\mathcal{L} &= \mathcal{L}_1'\Big(u_i(x), \nabla_\mu u_i(x)\Big) + \mathcal{L}_{YM}\left(F^k_{\mu\nu}\right)\\ &= \mathcal{L}\left(u_i(x), \nabla_\mu u_i(x),\ A^k_\mu,\ \partial_\nu A^k_\mu\right),\end{aligned}$$

and then we shall pass to the new variables $u_i(x), \nabla_\mu u_i(x),\ A^k_\mu(x)$, and $F^k_{\mu\nu}$.

The condition of local invariance of $\mathcal{L}$ writes

$$\begin{aligned}\delta\mathcal{L} = \frac{\partial\mathcal{L}}{\partial u_i}\delta u_i + \frac{\partial\mathcal{L}}{\partial(\partial_\mu u_i)}\delta(\partial_\mu u_i) + \frac{\partial\mathcal{L}}{\partial A^k_\mu}\delta A^k_\mu\\ + \frac{\partial\mathcal{L}}{\partial(\partial_\nu A^k_\mu)}\delta(\partial_\nu A^k_\mu) = 0,\end{aligned} \tag{4.76}$$

or, since $\delta(\partial_\mu u_i) = \partial_\mu(\delta u_i)$ and $\delta(\partial_\nu A^k_\mu) = \partial_\nu(\delta A^k_\mu)$,

$$\begin{aligned}\delta\mathcal{L} = \frac{\partial\mathcal{L}}{\partial u_i}\delta u_i - \partial_\mu\left(\frac{\partial\mathcal{L}}{\partial(\partial_\mu u_i)}\right)\delta u_i + \partial_\mu\left(\frac{\partial\mathcal{L}}{\partial(\partial_\mu u_i)}\delta u_i\right)\\ + \frac{\partial\mathcal{L}}{\partial A^k_\mu}\delta A^k_\mu - \partial_\nu\left(\frac{\partial\mathcal{L}}{\partial(\partial_\nu A^k_\mu)}\right)\delta A^k_\mu + \partial_\nu\left(\frac{\partial\mathcal{L}}{\partial(\partial_\nu A^k_\mu)}\delta A^k_\mu\right) = 0.\end{aligned} \tag{4.77}$$

Using the field equations for $u_i(x)$ and $A^k_\mu(x)$ in the form

$$\frac{\partial\mathcal{L}}{\partial u_i} - \partial_\mu\left(\frac{\partial\mathcal{L}}{\partial(\partial_\mu u_i)}\right) = 0;\quad \frac{\partial\mathcal{L}}{\partial A^k_\mu} - \partial_\nu\left(\frac{\partial\mathcal{L}}{\partial(\partial_\nu A^k_\mu)}\right) = 0, \tag{4.78}$$

the Eq. (4.77) becomes

$$\begin{aligned}\delta\mathcal{L} &= \partial_\mu\left(\frac{\partial\mathcal{L}}{\partial(\partial_\mu u_i)}\delta u_i\right) + \partial_\nu\left(\frac{\partial\mathcal{L}}{\partial(\partial_\nu A^k_\mu)}\delta A^k_\mu\right)\\ &= \partial_\mu\left(\frac{\partial\mathcal{L}}{\partial(\partial_\mu u_i)}\delta u_i + \frac{\partial\mathcal{L}}{\partial(\partial_\mu A^k_\nu)}\delta A^k_\nu\right) = 0.\end{aligned} \tag{4.79}$$

Using the gauge field variation $\delta A_\mu^k = f_{lmk} A_\mu^m \varepsilon_l + \partial_\mu \varepsilon_k$ and going from the derivatives with respect to $\partial_\mu u_i$ and $\partial_\mu A_\nu^k$ to those with respect to $\nabla_\mu u_i$ and $F_{\mu\nu}^k$, Eq. (4.79) becomes

$$\begin{aligned}
\delta\mathcal{L} &= \partial_\mu \left[ \frac{\partial \mathcal{L}}{\partial(\nabla_\mu u_i)} T_{ij}^k u_j(x) \varepsilon_k(x) + 2 \frac{\partial \mathcal{L}}{\partial F_{\mu\nu}^k} \left( \underset{l\to k\to m\to l}{f_{lmk} A_\nu^m \varepsilon_l} + \partial_\nu \varepsilon_k \right) \right] \\
&= \partial_\mu \left( \frac{\partial \mathcal{L}}{\partial(\nabla_\mu u_i)} T_{ij}^k u_j(x) + 2 \frac{\partial \mathcal{L}}{\partial F_{\mu\nu}^m} f_{klm} A_\nu^l \right) \varepsilon_k(x) \\
&\quad + \left[ \frac{\partial \mathcal{L}}{\partial(\nabla_\mu u_i)} T_{ij}^k u_j(x) + 2 \frac{\partial \mathcal{L}}{\partial F_{\mu\nu}^m} f_{klm} A_\nu^l - 2\partial_\nu \left( \frac{\partial \mathcal{L}}{\partial F_{\mu\nu}^k} \right) \right] \partial_\mu \varepsilon_k \\
&\quad + 2 \frac{\partial \mathcal{L}}{\partial F_{\mu\nu}^k} \partial_\mu \partial_\nu \varepsilon_k = 0.
\end{aligned} \tag{4.80}$$

But

$$\begin{aligned}
2 \frac{\partial \mathcal{L}}{\partial F_{\mu\nu}^k} \partial_\mu \partial_\nu \varepsilon_k &= \frac{\partial \mathcal{L}}{\partial F_{\mu\nu}^k} \partial_\mu \partial_\nu \varepsilon_k + \frac{\partial \mathcal{L}}{\partial F_{\mu\nu}^k} \partial_\mu \partial_\nu \varepsilon_k \\
&= \frac{\partial \mathcal{L}}{\partial F_{\mu\nu}^k} \partial_\mu \partial_\nu \varepsilon_k + \frac{\partial \mathcal{L}}{\partial F_{\nu\mu}^k} \partial_\nu \partial_\mu \varepsilon_k \\
&= \left( \frac{\partial \mathcal{L}}{\partial F_{\mu\nu}^k} + \frac{\partial \mathcal{L}}{\partial F_{\nu\mu}^k} \right) \partial_\mu \partial_\nu \varepsilon_k
\end{aligned}$$

and

$$\partial_\nu \left( \frac{\partial \mathcal{L}}{\partial F_{\nu\mu}^k} \right) = \frac{1}{2} \partial_\nu \left( \frac{\partial \mathcal{L}}{\partial(\partial_\nu A_\mu^k)} \right) \overset{\text{by means of the field eqs.}}{=} \frac{1}{2} \frac{\partial \mathcal{L}}{\partial A_\mu^k}.$$

The Eq. (4.80) then becomes

$$\begin{aligned}
\delta\mathcal{L} &= \partial_\mu \left( \frac{\partial \mathcal{L}}{\partial(\nabla_\mu u_i)} T_{ij}^k u_j(x) + 2 \frac{\partial \mathcal{L}}{\partial F_{\mu\nu}^m} f_{klm} A_\nu^l \right) \varepsilon_k(x) \\
&\quad + \left( \frac{\partial \mathcal{L}}{\partial(\nabla_\mu u_i)} T_{ij}^k u_j(x) + 2 \frac{\partial \mathcal{L}}{\partial F_{\mu\nu}^m} f_{klm} A_\nu^l + \frac{\partial \mathcal{L}}{\partial A_\mu^k} \right) \partial_\mu \varepsilon_k \\
&\quad + \left( \frac{\partial \mathcal{L}}{\partial F_{\mu\nu}^k} + \frac{\partial \mathcal{L}}{\partial F_{\nu\mu}^k} \right) \partial_\mu \partial_\nu \varepsilon_k = 0.
\end{aligned} \tag{4.81}$$

Since $\varepsilon_k(x)$ and $\partial_\mu \varepsilon_k(x)$ are *arbitrary*, (4.81) yields

$$\partial_\mu \left( \frac{\partial \mathcal{L}}{\partial(\nabla_\mu u_i)} T^k_{ij} u_j(x) + 2 \frac{\partial \mathcal{L}}{\partial F^m_{\mu\nu}} f_{klm} A^l_\nu \right) = 0, \tag{4.82}$$

and

$$\frac{\partial \mathcal{L}}{\partial(\nabla_\mu u_i)} T^k_{ij} u_j(x) + 2 \frac{\partial \mathcal{L}}{\partial F^m_{\mu\nu}} f_{klm} A^l_\nu + \frac{\partial \mathcal{L}}{\partial A^k_\mu} = 0. \tag{4.83}$$

Defining the *field currents* as

$$J^k_\mu(x) \stackrel{\text{def}}{=} \frac{\partial \mathcal{L}}{\partial A^k_\mu}, \quad (k = 1, 2, 3, ..., n),$$

where $n$ is the number of currents which is equal to the dimension of the gauge group $G$ (being also equal to the number of generators of the group), the Eq. (4.83) leads to

$$J^k_\mu(x) = -\frac{\partial \mathcal{L}}{\partial(\nabla_\mu u_i)} T^k_{ij} u_j(x) - 2 \frac{\partial \mathcal{L}}{\partial F^m_{\mu\nu}} f_{klm} A^l_\nu,$$

while Eq. (4.82) shows that

$$\partial_\mu J^k_\mu = 0,$$

expressing the *law of conservation of the charges*

$$Q^k = \int J^k_0(x) \; d^3x.$$

Indeed,

$$0 = \int\limits_{(V)} \partial_\mu J^k_\mu(x) \; d^3x = d_0 \int\limits_{(V)} J^k_0 \; d^3x + \int\limits_{(V)} \partial_i J^k_i \; d^3x$$

$$= d_0 Q^k + \underbrace{\oint\limits_{(\Sigma_V)} J^k_i \; dS_i}_{\xrightarrow{\Sigma_V \to \infty} 0}$$

so that $Q^k = const.$ and therefore, $Q = \sum_k Q^k = const.$

## Examples

### 1) The Abelian group $U(1)$. The electromagnetic field – a gauge field

One starts with the Lagrangian density of the free spinorial (Dirac) field $\psi$ of the particles of mass $M$, whose expression is well-known,

$$\mathcal{L} = \frac{i}{2}\left(\overline{\psi}\gamma_\mu\partial_\mu\psi - \partial_\mu\overline{\psi}\gamma_\mu\psi\right) - M\overline{\psi}\psi. \tag{4.84}$$

This Lagrangian is invariant under the global phase transformations (belonging to the group $U(1)$),

$$\psi \to \psi' = e^{-ig\varepsilon}\psi; \quad \overline{\psi} \to \overline{\psi}' = \overline{\psi}e^{ig\varepsilon}, \tag{4.85}$$

where $\varepsilon$ is the group parameter (here $\varepsilon$ is a constant), and $g$ is called *coupling constant.* One can immediately verify the invariance of the Lagrangian density $\mathcal{L}$ with respect to (4.85). If the parameter $\varepsilon$ is supposed to be infinitesimal (*i.e.* $\mathcal{O}(\varepsilon^2) = 0$), then it follows from (4.85) that

$$\delta\psi = -i\varepsilon\, g\,\psi, \qquad \delta\overline{\psi} = i\varepsilon\, g\,\overline{\psi}. \tag{4.86}$$

In order to use the general theory, one takes $u_1(x) = \psi(x)$ and $u_2(x) = \overline{\psi}(x)$ as independent fields. Since $\delta u_i(x) = T^k_{ij}\varepsilon_k u_j(x)$ and using (4.85), one can write

$$\delta u_i = T^1_{ij}\varepsilon_1 u_j \equiv T_{ij}\varepsilon u_j, \quad (j = 1, 2), \tag{4.87}$$

that is

$$T_{11} = -ig, \quad T_{12} = 0, \quad T_{22} = ig, \quad T_{21} = 0. \tag{4.88}$$

Since the group $U(1)$ is Abelian, all the structure constants are zero: $f_{klm} = 0$.

Imposing now the condition of local invariance with respect to the gauge transformations of the continuous group $U(1)$, that is considering $\varepsilon = \varepsilon(x)$, the Lagrangian density (4.84) is no longer invariant under the transformations (4.86) (in which the replacement $\varepsilon \to \varepsilon(x)$ was done). In order to obtain an invariant Lagrangian density one performs the *minimal substitution* (one introduces the *covariant derivative*),

$$\partial_\mu\psi \to \nabla_\mu\psi = \partial_\mu\psi + ig\psi A_\mu, \tag{4.89}$$

$$\partial_\mu \overline{\psi} \to \nabla_\mu \overline{\psi} = \partial_\mu \overline{\psi} - ig\overline{\psi} A_\mu, \tag{4.90}$$

where $A_\mu(x) \equiv A_\mu^1(x)$ is the *gauge field* introduced in agreement with the general theory. With this change, (4.84) becomes

$$\begin{aligned}
\mathcal{L}(\psi, \partial_\mu \psi) &\to \mathcal{L}'(\psi, \nabla_\mu \psi) \\
&= \frac{i}{2}\left[\overline{\psi}\gamma_\mu \left(\partial_\mu \psi + ig\psi A_\mu\right) - \left(\partial_\mu \overline{\psi} - ig\overline{\psi} A_\mu\right)\gamma_\mu \psi\right] - M\overline{\psi}\psi \\
&= \underbrace{(i/2)\left(\overline{\psi}\gamma_\mu \partial_\mu \psi - \partial_\mu \overline{\psi}\gamma_\mu \psi\right) - M\overline{\psi}\psi}_{\text{free field Lagrangian density}} \\
&\quad \underbrace{-g\overline{\psi}\gamma_\mu \psi A_\mu}_{\text{interaction Lagrangian density}} .
\end{aligned} \tag{4.91}$$

Using the definition of $F_{\mu\nu}^k$,

$$F_{\mu\nu}^k = A_{\mu\nu}^k - f_{lmk} A_\mu^l A_\nu^m = A_{\mu\nu}^k - \frac{1}{2} f_{lmk}\left(A_\mu^l A_\nu^m - A_\nu^l A_\mu^m\right),$$

$$\left(k = \overline{1, n}\right)$$

and using the fact that in this case $f_{lmk} = 0$, one can construct the *gauge field tensor*,

$$F_{\mu\nu}^1 \equiv F_{\mu\nu} = \partial_\mu A_\nu - \partial_\nu A_\mu. \tag{4.92}$$

The gauge (or Yang–Mills) Lagrangian density for the (free) gauge field $A_\mu(x)$ therefore is

$$\mathcal{L}'_0(F_{\mu\nu}) \equiv \mathcal{L}_{YM} = -\frac{1}{4} F_{\mu\nu} F_{\mu\nu}. \tag{4.93}$$

One can observe that the tensor (4.92) and that of the free electromagnetic field have the same form/expression. Therefore, in this interpretation, $A_\mu(x)$ represents the four-potential of the electromagnetic field, and $g$ (the coupling constant) is exactly the electron charge. According to the general expression $\delta A_\mu^k(x) = f_{lmk} A_\mu^m(x)\varepsilon_l(x) + \partial_\mu \varepsilon_k(x)$, in the case of gauge group $U(1)$, one has

$$\delta A_\mu(x) = \partial_\mu \varepsilon(x). \tag{4.94}$$

In its turn, the current (to be more precisely, the four-current) $J_\mu(x)$ has the form

$$J_\mu(x) = -g\,\overline{\psi}(x)\gamma_\mu\psi(x). \tag{4.95}$$

Comparing this expression with the one known from electrodynamics, $J_\mu(x) = -e\,\overline{\psi}(x)\gamma_\mu\psi(x)$, it follows that $g = e$. This way, the identification of the gauge field $A_\mu(x)$ with the electromagnetic field is totally justified.

**2) The non-Abelian group $SU(2)$ and the Yang–Mills field**

Consider an isodoublet ($SU(2)$-doublet) of spinorial fields,

$$\psi^a = \begin{pmatrix} \psi^1 \\ \psi^2 \end{pmatrix}, \quad (a = 1, 2), \tag{4.96}$$

where $\psi^1(x)$ and $\psi^2(x)$ are Dirac spinors.

For example, $\psi^1(x)$ can describe a proton of mass $M$ and $\psi^2(x)$ a neutron of the same mass. The Lagrangian density of this (iso)doublet is given by

$$\mathcal{L} = \frac{i}{2}\left(\overline{\psi}^a\gamma_\mu\partial_\mu\psi^a - \partial_\mu\overline{\psi}^a\gamma_\mu\psi^a\right) - M\overline{\psi}^a\psi^a, \quad (a = 1, 2). \tag{4.97}$$

This Lagrangian density is invariant under the global symmetry transformations of the Lie group $SU(2)$,

$$\psi^a \to \psi'^a = \left[\exp\left(-\frac{i}{2}g\varepsilon_k\tau_k\right)\right]_{ab}\psi^b, \quad \left(k = \overline{1,3};\; a, b = 1, 2\right), \tag{4.98}$$

$$\overline{\psi}^a \to \overline{\psi}'^a = \overline{\psi}^b\left[\exp\left(+\frac{i}{2}g\varepsilon_k\tau_k\right)\right]_{ba}, \quad \left(k = \overline{1,3};\; a, b = 1, 2\right), \tag{4.99}$$

where $\varepsilon_k$, $(k = \overline{1,3})$ denotes the parameters of the Lie group $SU(2)$, $\tau_k$, $(k = \overline{1,3})$ are the Pauli's matrices (regarded as isospin matrices), and $g$ is a constant, called *coupling constant.*

If $\varepsilon_k$ are infinitesimal quantities (*i.e.* $\mathcal{O}(\varepsilon^2) = 0$), the previous transformations become

$$\delta\psi^a = -\frac{i}{2}g\varepsilon_k(\tau_k)_{ab}\psi^b, \tag{4.100}$$

$$\delta\overline{\psi}^a = +\frac{i}{2} g \varepsilon_k \overline{\psi}^b (\tau_k)_{ba}. \tag{4.101}$$

It means that the generators $T^k$ of the Lie group $SU(2)$, in this bidimensional representation (isodoublet), have the form

$$T^k_{ab} = -\frac{i}{2} g (\tau_k)_{ab}, \text{ or simply } T^k = -\frac{i}{2} g \tau_k. \tag{4.102}$$

In this case,

$$\left[T^k, T^l\right] = -\frac{g^2}{4} \left[\tau_k, \tau_l\right] = -\frac{g^2}{4} 2i \varepsilon_{klm} \tau_m = g \varepsilon_{klm} T^m, \tag{4.103}$$

therefore the structure constants are

$$f_{klm} = g \varepsilon_{klm}, \tag{4.104}$$

where $\varepsilon_{klm}$ is the Levi-Civita symbol.

Consider now the $SU(2)$ local invariance. In this case, the space-time dependence of the group parameters $\varepsilon_k$ has to be considered, $\varepsilon_k \rightarrow \varepsilon_k(x)$, and the Lagrangian density (4.97) becomes locally (gauge) invariant if within it one performs the minimal substitution $\partial_\mu \rightarrow \nabla_\mu$:

$$\partial_\mu \psi^a \rightarrow \nabla_\mu \psi^a = \partial_\mu \psi^a + \frac{i}{2} g (\tau_k)_{ab} \psi^b A^k_\mu, \quad \left(k = \overline{1,3}\right), \tag{4.105}$$

where the gauge fields $A^k_\mu$, $(k = \overline{1,3})$ have been introduced in order to compensate the non-invariance of $\mathcal{L}$. The quantities $F^k_{\mu\nu}$ are given by

$$F^k_{\mu\nu} = \partial_\mu A^k_\nu - \partial_\nu A^k_\mu - \frac{1}{2} g \varepsilon_{klm} \left(A^l_\mu A^m_\nu - A^l_\nu A^m_\mu\right), \tag{4.106}$$

being used to construct the gauge Lagrangian density

$$\mathcal{L}'_0 \left(F^k_{\mu\nu}\right) \equiv \mathcal{L}_{YM} = -\frac{1}{4} F^k_{\mu\nu} F^k_{\mu\nu}. \tag{4.107}$$

The gauge fields transform according to the law

$$\delta A^k_\mu(x) = g \varepsilon_{lmk} A^m_\mu(x) \varepsilon_l(x) + \partial_\mu \varepsilon_k(x). \tag{4.108}$$

Substituting (4.105) into (4.97) and arranging the terms, one obtains the following expression for the total Lagrangian density:

$$\mathcal{L}_{total} = \underbrace{(i/2)\left(\overline{\psi}^a\gamma_\mu\partial_\mu\psi^a - \partial_\mu\overline{\psi}^a\gamma_\mu\psi^a\right) - M\overline{\psi}^a\psi^a}_{\mathcal{L}_{\text{free}}}$$

$$+ \underbrace{\mathcal{L}_{YM}}_{\mathcal{L}_{\text{gauge}}} \quad \underbrace{-(1/2)\, g\overline{\psi}^a\gamma_\mu(\tau_k)_{ab}\psi^b A^k_\mu}_{\mathcal{L}_{\text{interaction}}} . \tag{4.109}$$

The constant $g$ plays the role of a coupling constant between the gauge field $A^k_\mu(x)$ and the spinorial field $\psi^a(x)$, as well as with itself[15].

Using the definition of the field currents $J^k_\mu$, it follows that

$$J^k_\mu = -\frac{g}{2}\overline{\psi}^a\gamma_\mu(\tau_k)_{ab}\psi^b$$

$$-g\varepsilon_{klm}A^m_\nu \underbrace{\left[\partial_\mu A^l_\nu - \partial_\nu A^l_\mu - \frac{g}{2}\varepsilon_{lij}\left(A^i_\mu A^j_\nu - A^i_\nu A^j_\mu\right)\right]}_{F^l_{\mu\nu}} . \tag{4.110}$$

Indeed, since $F^m_{\sigma\nu} = \partial_\sigma A^m_\nu - \partial_\nu A^m_\sigma - g\varepsilon_{mlp}A^l_\sigma A^p_\nu$, one has

$$J^k_\mu = \frac{\partial\mathcal{L}}{\partial A^k_\mu} = \frac{\partial\mathcal{L}_{YM}}{\partial A^k_\mu} - \frac{g}{2}\overline{\psi}^a\gamma_\mu(\tau_k)_{ab}\psi^b = -\frac{1}{4}\frac{\partial}{\partial A^k_\mu}\left(F^m_{\sigma\nu}F^m_{\sigma\nu}\right)$$

$$-\frac{g}{2}\overline{\psi}^a\gamma_\mu(\tau_k)_{ab}\psi^b = -\frac{g}{2}\overline{\psi}^a\gamma_\mu(\tau_k)_{ab}\psi^b - \frac{1}{2}\frac{\partial F^m_{\sigma\nu}}{\partial A^k_\mu}F^m_{\sigma\nu}$$

$$= -\frac{g}{2}\overline{\psi}^a\gamma_\mu(\tau_k)_{ab}\psi^b + \frac{1}{2}g\varepsilon_{mlp}\left(\delta_{kl}\delta_{\sigma\mu}A^p_\nu - A^l_\sigma\delta_{kp}\delta_{\nu\mu}\right)F^m_{\sigma\nu}$$

$$= -\frac{g}{2}\overline{\psi}^a\gamma_\mu(\tau_k)_{ab}\psi^b + \frac{1}{2}g\left(\underbrace{\varepsilon_{mkp}A^p_\nu F^m_{\mu\nu}}_{p\to m\to l} - \underbrace{\varepsilon_{mlk}A^l_\sigma F^m_{\sigma\mu}}_{m\leftrightarrow l,\ \sigma\to\nu}\right)$$

$$= -\frac{g}{2}\overline{\psi}^a\gamma_\mu(\tau_k)_{ab}\psi^b + \frac{1}{2}g\left(\underset{-\varepsilon_{klm}}{\varepsilon_{lkm}}\ A^m_\nu F^l_{\mu\nu} - \underset{-\varepsilon_{klm}}{\varepsilon_{mlk}}\ A^m_\nu\ \underset{-F^l_{\mu\nu}}{F^l_{\nu\mu}}\right)$$

$$= -\frac{g}{2}\overline{\psi}^a\gamma_\mu(\tau_k)_{ab}\psi^b - g\varepsilon_{klm}A^m_\nu F^l_{\mu\nu}.$$

[15] Any non-Abelian gauge theory is a theory with self-interaction (in this case the gauge fields are self-interacting fields).

The three gauge fields $A_\mu^k(x)$ are tightly related (through the so-called *mixing angle* $\theta$ and the gauge field $A_\mu(x)$ of the electromagnetic interaction, in the frame of *spontaneous symmetry-breaking*) to the three fields describing the three vectorial bosons, $W^+, W^-$ and $Z^0$, that intermediate the (electro)weak interactions between the particles described by the fields $\psi^a(x)$ (for example, the protons and neutrons).

### 3) The $SU(3)$ symmetry and gluonic fields

Consider a $SU(3)$-triplet of spinorial fields,

$$\psi^a(x) = \begin{pmatrix} \psi^1(x) \\ \psi^2(x) \\ \psi^3(x) \end{pmatrix}, \quad \left(a = \overline{1,3}\right), \tag{4.111}$$

which can be, for example, the three fields describing the three initially known quarks $q_1, q_2, q_3$ denoted by[16] $(u, d, s)$. The Lagrangian density of this triplet of free fields writes

$$\mathcal{L} = \frac{i}{2}\left(\overline{\psi}^a \gamma_\mu \partial_\mu \psi^a - \partial_\mu \overline{\psi}^a \gamma_\mu \psi^a\right) - M\overline{\psi}^a \psi^a, \quad \left(a = \overline{1,3}\right). \tag{4.112}$$

It is invariant under the global symmetry transformations of the Lie group $SU(3)$,

$$\psi^a \to \psi'^a = \left[\exp\left(-i\frac{g}{2}\lambda_m \varepsilon_m\right)\right]_{ab} \psi^b, \quad \left(m = \overline{1,8};\ a, b = \overline{1,3}\right), \tag{4.113}$$

$$\overline{\psi}^a \to \overline{\psi}'^a = \overline{\psi}^b \left[\exp\left(i\frac{g}{2}\lambda_m \varepsilon_m\right)\right]_{ba}, \quad \left(m = \overline{1,8};\ a, b = \overline{1,3}\right). \tag{4.114}$$

Here $\varepsilon_m$ $(m = 1, 2, 3, ..., 8)$ are the eight parameters of the Lie group $SU(3)$, $\lambda_m$ $(m = \overline{1,8})$ are the *Gell-Mann matrices*, and $g$ is the coupling constant.

**Observation**: The elements of the Lie group $SU(3)$ are considered as acting in an eight-dimensional space, called *colour space*, and the model outlined here leads to the theory called *quantum chromodynamics*. If the parameters $\varepsilon_k$ of the Lie group $SU(3)$ are infinitesimal (*i.e.* $\mathcal{O}(\varepsilon^2) = 0$), then (4.113) and (4.114) become

$$\delta\psi^a(x) = -\frac{i}{2} g(\lambda_m)_{ab} \psi^b(x) \varepsilon_m, \tag{4.115}$$

[16] $u$ – "up", $d$ – "down", and $s$ – "strange".

$$\delta\overline{\psi}^a(x) = \frac{i}{2} g\overline{\psi}^b(x)(\lambda_m)_{ba}\varepsilon_m. \tag{4.116}$$

The infinitesimal generators of the considered representation (the fundamental or triplet covariant representation) of the Lie group $SU(3)$, therefore are

$$T^k_{ab} = -\frac{i}{2} g(\lambda_k)_{ab}, \tag{4.117a}$$

where

$$[\lambda_k, \lambda_l] = 2if_{klm}\lambda_m. \tag{4.117b}$$

Thus,

$$\left[T^k, T^l\right]_{ab} = -\frac{g^2}{4}[\lambda_k, \lambda_l]_{ab} = gf_{klm}T^m_{ab}. \tag{4.118}$$

Consider now the local (gauge) $SU(3)$ invariance. In this case, as it is known, the space-time dependence of the group parameters $\varepsilon_k$ has to be considered: $\varepsilon_m \to \varepsilon_m(x)$.

Let introduce/define eight gauge fields $V^m_\mu(x)$, $\left(m = \overline{1,8}\right)$ and write – by the help of these fields – the covariant derivative of the matter fields $\psi^a(x)$:

$$\nabla_\mu\psi^a(x) = \partial_\mu\psi^a(x) + \frac{i}{2} g(\lambda_m)_{ab}\psi^b(x)V^m_\mu(x). \tag{4.119}$$

The eight fields $V^m_\mu(x)$, $\left(m = \overline{1,8}\right)$ are called *gluonic fields* and their role is to intermediate the strong interactions between hadrons.

The gauge Lagrangian density for the (free) gluonic fields is given by

$$\mathcal{L}'_0\left(F^k_{\mu\nu}\right) \equiv \mathcal{L}_{YM} = -\frac{1}{4}F^k_{\mu\nu}F^k_{\mu\nu}, \tag{4.120}$$

where

$$F^k_{\mu\nu} = \partial_\mu V^k_\nu - \partial_\nu V^k_\mu - gf_{npk}\left(V^n_\mu V^p_\nu - V^n_\nu V^p_\mu\right)$$

is the tensor of the gluonic fields, which fields transform according to the law

$$\delta V^m_\mu(x) = g\, f_{npm}V^p_\mu(x)\varepsilon_n(x) + \partial_\mu\varepsilon_m(x). \tag{4.121}$$

The total Lagrangian density, invariant under the local (gauge) symmetry transformations of the Lie group $SU(3)$ is given by

$$\mathcal{L}_{total} = \frac{i}{2}\left(\overline{\psi}^a \gamma_\mu \partial_\mu \psi^a - \partial_\mu \overline{\psi}^a \gamma_\mu \psi^a\right) - M\overline{\psi}^a \psi^a$$
$$- \overline{\psi}^a \gamma_\mu (\lambda_m)_{ab} \psi^b V_\mu^m - \frac{1}{4} F_{\mu\nu}^m F_{\mu\nu}^m .$$

This is the *Lagrangian density of the quantum chromodynamics*. At the moment, quantum chromodynamics is considered to be the best quantum theory describing the strong interactions between subnuclear particles.

Taking into account the definition relation $J_\mu^k(x) \stackrel{\text{def}}{=} \partial\mathcal{L}/\partial A_\mu^k$, the corresponding four-currents $J_\mu^m(x)$ can be easily calculated; they are given by

$$J_\mu^m(x) = \frac{\partial\mathcal{L}}{\partial(\nabla_\mu \psi^a)} \frac{ig}{2} (\lambda_m)_{ab} \psi^b - \frac{\partial\mathcal{L}}{\partial(\nabla_\mu \overline{\psi}^a)} \frac{ig}{2} \overline{\psi}^b (\lambda_m)_{ba}$$
$$- 2g \frac{\partial\mathcal{L}}{\partial F_{\mu\nu}^n} f_{mpn} V_\nu^p, \quad \left(m = \overline{1,8}\right). \tag{4.122}$$

Examining the previously obtained Lagrangian densities, one can observe that the various gauge fields $\left(A_\mu(x),\ A_\mu^k(x),\ (k = 1,2,3);\ V_\mu^m,\ (m = \overline{1,8})\right)$ are massless, since there are no mass terms in the corresponding Lagrangian density $\mathcal{L}_{YM}$. Part of gauge fields (more precisely, those gauge fields that intermediate the short-range interactions between particles) can acquire mass for their quanta by means of spontaneous breaking of the local (gauge) symmetries and Higgs' mechanism, but this subject exceeds the content limits of this book and the authors advise the interested readers to consult, for instance, the Ref. [66].

## Chapter V

# The Gravitational Field Interacting with Other Fields

This chapter is dedicated to the study of interaction of the gravitational field with the matter, by *matter* being understood the scalar, vectorial and spinorial fields. The interaction Lagrangians were obtained in the frame of the formalism offered by the minimal coupling principle applied in quantum gravity. Using the $S$-matrix formalism, a general treatment of the first and second-order processes is approached, being obtained – within approximation of external electromagnetic and gravitational fields – several Feynman-type rules for diagrams in the momentum space. Some of the exposed subjects are intended to generalize the results already known in the specific literature.

The notations and index conventions are similar to those agreed in Chap. II, the possible exceptions being specified at the right time.

The problem of interaction between the gravitational field and the other fields appears in Gupta's pioneers papers [181, 182(1954), 183]. Using the minimal coupling principle (see Chapter IV), he obtained the first-order interaction Lagrangians in terms of the coupling constant $\kappa$ between the gravitational and matter fields, considering the linear approximation for the gravitational field (the weak gravitational field approximation – see Chapters I and II). Similar procedures were later developed by some other authors.

The interaction between the gravitational and scalar fields was analyzed by Gupta and collab. [25–27], Kimura [230], Feynman [142], De Witt [105] and some other researchers (*e.g.* [38(1969), 162, 412]), using either the functional formalism, or the "usual" procedure met in QFT. The interaction between gravitational and electromagnetic quantized fields was studied by Gupta and collab. [27, 181(p.608)], Piir [306], De Sabata and collab. [38], Weinberg [400], Tourrenc [353, 354] and some other authors [390–403], while the interaction between the gravitational and vectorial fields (within the framework offered by the Stueckelberg formalism) was succinctly investigated by Vladimirov [389(1964)]. The more complicated case of interaction between the gravitational and spinorial quantized fields was studied by Brill and Wheeler [49], Lias [246], Vladimirov [388, 389(1963)], Mitskevich [272, 275] and some others [11, 293, 296], in terms of the *vierbein* formalism, according to the ideas of Fock and Ivanenko.

The studies of the above mentioned authors mainly concern the scattering processes involving the elementary particles and the gravitons but – with some exceptions – they do not consider systematically the problem of finding the Feynman rules for diagrams in momentum space, or, if this subject is still approached and Feynman rules for some free particles, propagators and vertices (*e.g.* De Witt [105], De Sabbata and collab. [38], Kobzarev and Okun [234], or Papini and Valluri [301]) are obtained, this is done in a too general (or even incomplete) manner, being difficult to apply to concrete processes.

In the first part of this chapter, a general and unitary approach of the first and second-order processes related to the interaction between the gravitational and other fields is presented. By means of the $S$-matrix formalism applied within the electromagnetic and gravitational external fields approximation are obtained useful Feynman rules in the momentum space, necessary to write the $S$-matrix element corresponding to a given scattering process. These subjects were only slightly treated in the specific literature, the authors' results generalizing and completing the subjects approached in the above mentioned papers.

A problem that for a significant period of time caught the specialists attention is that concerning the hypotheses on the so-called *strong-gravity* (Salam, [320]), or the formal analogy between the weak interaction and that associated with the torsional forces (Cartan-type forces) (Hayashi [199], De Sabbata and Gasperini [97, 98]). An important role in the physicists endeavor to create a quantum unitary theory, that embraces the gravity together with the other fundamental interactions, is played by the supergravity theory – based on the concept of supersymmetry. This theory was initiated by Nieuwenhuizen and collab. [151] and later developed by other authors [114, 294, 379].

In connection with these subjects, at the end of this chapter, a new hypothesis regarding the influence of the gravitational field on the electromagnetic, weak and strong interactions between particles is presented. Besides, another hypothesis is issued on how the gravitational field can interact – independently of supersymmetry and supergravity reasons – with the Rarita–Schwinger field.

## V.1. Interaction with the scalar field

The interaction between the gravitational and scalar quantum fields is analyzed – by means of the $S$-matrix formalism – in the papers [26(1969), 105, 275]. The following presentation is dedicated to a general overview on this subject (concerning the first and second-order interacting processes, in terms of the powers of the coupling constant $\kappa$) and to the calculation of the $S$-matrix elements of the field operators, corresponding to the transition between well-determined states (the vacuum state and that containing a single particle possessing a given momentum and a fixed polarization state), this way obtaining Feynman-type rules for diagrams in the momentum space [358(1976)].

To fix the ideas, one shall consider the three states of the $\pi$ mesons. As it is known [252, 288], the $\pi$ meson triplet to which there corresponds the field operator $\vec{\varphi}\,(\varphi_1, \varphi_2, \varphi_3)$, with

$$\begin{cases} \varphi_1 = \dfrac{1}{\sqrt{2}}\Big(\varphi + \varphi^*\Big), \\ \varphi_2 = \dfrac{i}{\sqrt{2}}\Big(\varphi - \varphi^*\Big), \\ \varphi_3 = \varphi_0, \end{cases} \tag{5.1}$$

transforms according to the adjoint representation of the isospin group $SU(2)$ ($\vec{\varphi}$ transforms as a vector in the three-dimensional space of the isospin). Here $\varphi_0 = \varphi_{\pi^0}$ is the Hermitian operator of the scalar neutral field that describes the neutral meson $\pi^0$ (neutral pion), while $\varphi = \varphi_{\pi^+}$ and $\varphi^* = \varphi_{\pi^-}$ correspond to the scalar field[1] of the two charged mesons $\pi^+$ and $\pi^-$ (the two charged pions).

In this case, the Lagrangian density of the field

$$\mathcal{L} = -\frac{1}{2}\Big(\vec{\varphi}_{,\mu} \cdot \vec{\varphi}_{,\mu} + m^2 \vec{\varphi} \cdot \vec{\varphi}\Big) \tag{5.2}$$

is invariant under the $SU(2)$-group transformations and can be written as

$$\mathcal{L} = \mathcal{L}_{\pi^\pm} + \mathcal{L}_{\pi^0}. \tag{5.3}$$

In order to describe the interaction between the gravitational and pion's matter field, to Einstein's Lagrangian density $\mathcal{L}_g$ (given by (1.295)), the term representing the Lagrangian density of the pion field written in the curved space has to be added. This can be rigorously done in each case, in an appropriate manner (see subchapter IV.1) by means of the minimal coupling principle of quantum gravitation [97, 105], starting with the action functional in the absence of the gravitational field and replacing the usual derivatives by the covariant ones, the Minkowski metric by $g_{\mu\nu}$, and the volume element by $\sqrt{-g}\, d^4x$, that is, all the three substitutions presented in subchapter IV.1 (while describing the correct and consistent way of using the principle of minimal coupling in quantum gravity) must be performed. In this case the total Lagrangian density writes

$$\mathcal{L}_t = \mathcal{L}_{sc} + \mathcal{L}_{grav}, \tag{5.4}$$

where

$$\mathcal{L}_{sc} = \sqrt{-g}\Big(g^{\mu\nu} \varphi^*_{,\mu} \varphi_{,\nu} - m^2 \varphi^* \varphi\Big), \tag{5.5}$$

for the complex scalar field (charged pions), and

$$\mathcal{L}_{sc} = \sqrt{-g}\Big(g^{\mu\nu} \varphi_{0,\mu} \varphi_{0,\nu} - m^2 \varphi_0^2\Big), \tag{5.5$'$}$$

for the real scalar field (neutral pion).

[1] This choice is conventional.

Besides, the gravitational field must obey the De Donder–Fock harmonic coordinates condition mentioned in Chapter I (the subchapter regarding the quantization of the free weak gravitational field) as a subsidiary condition.

Following Gupta's ideas [181(p.608), 182(1954), 183] developed in the case of a weak gravitational field, the field quantities involved in (5.4) shall be expanded in power series in powers of the coupling constant $\kappa$, starting with the linearization (2.2). Considering the case of the charged scalar field in the flat space ($x_j$, $x_4 = it$) and taking into account (1.295), (1.297), (1.299), and (1.300), one obtains

$$\mathcal{L}_{int} = \mathcal{L}^{(1)}_{sc-g} + \mathcal{L}^{(1)}_{g} + \mathcal{L}^{(2)}_{sc-g} + \cdots , \tag{5.6}$$

where

$$\begin{cases} \mathcal{L}^{(1)}_{sc-g} = -\kappa\left(\varphi^*_{,\mu}\varphi_{,\nu}y_{\mu\nu} + \dfrac{1}{2}m^2\varphi^*\varphi\, y\right), \\ \mathcal{L}^{(2)}_{sc-g} = \dfrac{1}{4}\kappa^2 m^2\varphi^*\varphi\, y_{\mu\nu}h_{\mu\nu} \end{cases} \tag{5.7}$$

are, respectively, the first and second-order interaction Lagrangian densities between the complex scalar and gravitational fields[2], and

$$\mathcal{L}^{(1)}_{g} = \frac{1}{2}\kappa\delta_{\mu\nu,\lambda\rho,\gamma\sigma,\alpha\beta}\,y_{\mu\nu}y_{\lambda\rho,\alpha}y_{\gamma\sigma,\beta} \tag{5.8}$$

is the first-order self-coupling Lagrangian density of the gravitational field, where

$$\begin{aligned} \delta_{\mu\nu,\lambda\rho,\gamma\sigma,\alpha\beta} = {} & \delta_{\mu\lambda}\delta_{\nu\gamma}\delta_{\rho\sigma}\delta_{\alpha\beta} - \delta_{\mu\lambda}\delta_{\nu\gamma}\delta_{\rho\beta}\delta_{\sigma\alpha} - \frac{1}{2}\delta_{\mu\alpha}\delta_{\nu\beta}\delta_{\lambda\gamma}\delta_{\rho\sigma} \\ & - \frac{1}{2}\delta_{\mu\lambda}\delta_{\nu\rho}\delta_{\gamma\sigma}\delta_{\alpha\beta} + \frac{1}{4}\delta_{\mu\alpha}\delta_{\nu\beta}\delta_{\lambda\rho}\delta_{\gamma\sigma}, \end{aligned} \tag{5.9}$$

and $h_{\mu\nu}$ is given by (1.294).

The operator $S$ is provided by the QFT [2, 40, 356, 387],

$$S = 1 + \sum_n \frac{1}{n!}\int S_n(x_1, ..., x_n)d^4x_1 \cdots d^4x_n, \tag{5.10}$$

[2]In the cases of the real scalar field, the multiplicative factor 1/2 appears in front of the Lagrangian densities (5.7).

with

$$S_n(x_1, ..., x_n) = i^n T\Big\{N\Big[\mathcal{L}_{int}(x_1)\cdots\mathcal{L}_{int}(x_n)\Big]\Big\}, \tag{5.11}$$

where the symbols $T$ and $N$ denote the chronological (time-ordering)) and normal products of the field operators, respectively.

### V.1.1. First-order processes

It is known that the first-order processes can be produced only in the external field approximation, but in order to derive explicit Feynman-type rules for diagrams one must first study the free processes. In the case of interaction between the gravitational and scalar quantum fields, studied in the first-order approximation, the scattering matrix is given by

$$S_1 = i\int d^4x\ N\Big[\mathcal{L}^{(1)}_{sc-g}(x)\Big]. \tag{5.12}$$

Decomposing the field operators into positive and negative frequency parts, corresponding to annihilation and creation, respectively, of the particles (antiparticles), one obtains

$$N\Big[\mathcal{L}^{(1)}_{sc-g}(x)\Big] \to$$

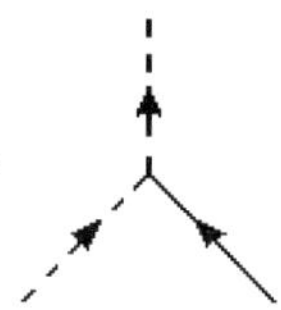

$$+\cdots, \tag{5.13}$$

to which one adds the diagrams obtained by changing the interaction channel (with the corresponding orientation of the time axis), and replacing the particles by their antiparticles (and vice versa). Thus, the above graph represents the absorption/emission of the graviton by the scalar particle/antiparticle, as well as creation of a pair $\pi^+ - \pi^-$ by a graviton, or its annihilation into a graviton.

In order to obtain the Feynman-type rules in the momentum space, one must calculate the matrix elements of the normal products of operators, corresponding to the transitions between states with well-determined numbers of free particles. In this case, following the well-known procedure used in QFT [2, 252], it is necessary to decompose the normal product of field operators into a sum of normal products of creation and annihilation operators, then commute the annihilation operators to the right until reaching the vacuum state vector $|0\rangle$, and creation operators to the left, until reaching the adjoint vector $\langle 0|$. According to definition of the vacuum state, the terms obtained this way cancel, and the only contributions to the matrix element are given by the commutation functions.

The operators $\varphi(x)$ and $\varphi^*(x)$, corresponding to the complex scalar field, admit the following Fourier expansions [252, 387][3],

$$\begin{aligned}\varphi(x) &= \frac{1}{(2\pi)^{3/2}} \int \frac{d^3k}{\sqrt{2k_0}} \left[a^{(+)}(\vec{k})e^{ikx} + a^{(-)}(\vec{k})e^{-ikx}\right] \\ &= \varphi^{(+)}(x) + \varphi^{(-)}(x),\end{aligned} \tag{5.14a}$$

$$\begin{aligned}\varphi^*(x) &= \frac{1}{(2\pi)^{3/2}} \int \frac{d^3k}{\sqrt{2k_0}} \left[a^{*(+)}(\vec{k})e^{ikx} + a^{*(-)}(\vec{k})e^{-ikx}\right] \\ &= \varphi^{*(+)}(x) + \varphi^{*(-)}(x),\end{aligned} \tag{5.14b}$$

with $k_0 = \sqrt{\vec{k}^2 + m^2}$. The annihilation and creation operators – $a^{(+)}(\vec{k})$, $a^{*(+)}(\vec{k})$ and $a^{(-)}(\vec{k})$, $a^{*(-)}(\vec{k})$, respectively – verify the commutation relations

$$\left[a^{(+)}(\vec{k}), a^{*(-)}(\vec{k}\,')\right] = \left[a^{*(+)}(\vec{k}), a^{(-)}(\vec{k}\,')\right] = \delta\left(\vec{k} - \vec{k}\,'\right), \tag{5.15}$$

all the other commutators being null. Since the vacuum state verifies the conditions

$$\begin{cases} a^{(+)}(\vec{k})|0\rangle = a^{*(+)}(\vec{k})|0\rangle = 0, \\ \langle 0|a^{(-)}(\vec{k}) = \langle 0|a^{*(-)}(\vec{k}) = 0, \\ \langle 0|0\rangle = 1, \end{cases} \tag{5.16}$$

the matrix elements of the operators $\varphi^{(\pm)}(x)$ and $\varphi^{*(\pm)}(x)$ are non-null only for the transitions between states whose numbers of particles differ by one, for example the vacuum state and that with a single particle of momentum $\vec{k}$. In this case, the evaluation of expressions

$$\langle 0|a^{(+)}(\vec{k}\,')\varphi^{*(-)}(x)|0\rangle, \quad \langle 0|\varphi^{(+)}(x)a^{*(-)}(\vec{k}\,')|0\rangle,$$

[3] $\varphi^{(+)}(x)$ and $\varphi^{*(-)}(x)$ are the annihilation and creation operators of the scalar particles (for example, the mesons $\pi^+$), respectively, while $\varphi^{*(+)}(x)$ and $\varphi^{(-)}(x)$ are the annihilation and creation operators of the corresponding antiparticles (the mesons $\pi^-$, in our case), respectively.

and

$$\langle 0|a^{*(+)}(\vec{k}\,')\varphi^{(-)}(x)|0\rangle, \quad \langle 0|\varphi^{*(+)}(x)a^{(-)}(\vec{k}\,')|0\rangle$$

lead – also taking into account (5.14), (5.15) and (5.16) – to the following correspondence between the operators of the charged scalar field and their matrix elements, corresponding to the transition between the state with one meson of momentum $\vec{k}$ and vacuum state[4]:

$$\varphi^{(\pm)}(x) = \varphi^{*(\pm)}(x) \rightarrow \frac{1}{(2\pi)^{3/2}\sqrt{2k_0}} e^{\pm ikx}. \tag{5.17}$$

As it has been shown in Chapter I, only the transversal gravitons of the type $y'_{11} = \frac{1}{2}(y_{11} - y_{22})$ and $y_{12}$ contribute to the expressions of momentum and energy. Following Mitskevich's ideas [275], and denoting $y'_{11} \equiv b_1$, $y_{12} \equiv b_2$, one shall consider the Fourier transform of the potential of the free gravitational field in a coordinate system with $z$-axis oriented along $\vec{p}$, that is

$$(y_{\mu\nu}(\vec{p})) = \begin{pmatrix} b_1 & b_2 & 0 & 0 \\ b_2 & -b_1 & 0 & 0 \\ 0 & 0 & 0 & 0 \\ 0 & 0 & 0 & 0 \end{pmatrix}. \tag{5.18}$$

The transition to an arbitrary coordinate system can be done taking into account the fact that the rank of polarization tensor has to be two, *i.e.*,

$$y_{\mu\nu} = \left(e_\mu^{(1)} e_\nu^{(1)} - e_\mu^{(2)} e_\nu^{(2)}\right) b_1 + \left(e_\mu^{(1)} e_\nu^{(2)} + e_\mu^{(2)} e_\nu^{(1)}\right) b_2. \tag{5.19}$$

Here $e_\mu^{(a)}$, $(a = 1, 2)$ are the orthogonal polarization vectors, analogous to those corresponding to photons in Quantum Electrodynamics and, taking the coefficients of the operators $b_1$ and $b_2$ as polarization tensors corresponding to the two types of real (observable) gravitons, it follows that

$$y_{\mu\nu}(\vec{p}) = e_{\mu\nu}^{(a)}(\vec{p}) b_a(\vec{p}), \tag{5.20}$$

where the index $a$ takes the values 1 and 2, corresponding to the two orientations of the graviton spin: parallel or anti-parallel with respect to direction of motion of the gravitons.

[4]The matrix elements obtained for the real scalar field are the same.

Starting from these reasons, the non-null matrix elements corresponding to the weak gravitational field shall be calculated in a different manner as against that in Ref. [275]. In this respect, taking into account (5.20), the Fourier series expansion of $y_{\mu\nu}(x)$ – given by (1.307) – becomes

$$\begin{aligned} y_{\mu\nu}(x) &= \frac{1}{(2\pi)^{3/2}} \int \frac{d^3p}{\sqrt{2p_0}} e^{(a)}_{\mu\nu}(\vec{p}) \left[ b_a^{(+)}(\vec{p}) e^{ipx} + b_a^{(-)}(\vec{p}) e^{-ipx} \right] \\ &= y^{(+)}_{\mu\nu}(x) + y^{(-)}_{\mu\nu}(x), \end{aligned} \tag{5.21}$$

where the annihilation and creation operators – $b_a^{(+)}(\vec{p})$ and $b_a^{(-)}(\vec{p})$, respectively – satisfy the commutation relation

$$\left[ b_a^{(+)}(\vec{p}), b_{a'}^{(-)}(\vec{p}\,') \right] = \delta_{aa'} \delta(\vec{p} - \vec{p}\,'), \tag{5.22}$$

all the other commutators being null. As an illustration of the method, let us calculate the matrix element for the emission operator $y^{(-)}_{\mu\nu}(x)$, corresponding to the transition between the vacuum state and that with a single graviton of momentum $\vec{p}\,'$ and polarization $a'$. Since the vacuum state satisfies the conditions

$$\begin{cases} b_a^{(+)}(\vec{p})|0\rangle = \langle 0|b_a^{(-)}(\vec{p}) = 0, \\ \langle 0|0\rangle = 1, \end{cases} \tag{5.23}$$

and taking into account (5.21)–(5.23), one has

$$\begin{aligned} &\langle 0|b_{a'}^{(+)}(\vec{p}\,') y^{(-)}_{\mu\nu}(x)|0\rangle \\ &= \langle 0| \left\{ y^{(-)}_{\mu\nu}(x) b_{a'}^{(+)}(\vec{p}\,') + \left[ b_{a'}^{(+)}(\vec{p}\,'), y^{(-)}_{\mu\nu}(x) \right] \right\} |0\rangle \\ &= \left[ b_{a'}^{(+)}(\vec{p}\,'), y^{(-)}_{\mu\nu}(x) \right] = \frac{1}{(2\pi)^{3/2}} \int \frac{d^3p}{\sqrt{2p_0}} e^{(a)}_{\mu\nu}(\vec{p}) \\ &\quad \times \left[ b_{a'}^{(+)}(\vec{p}\,'), b_a^{(-)}(\vec{p}) \right] e^{-ipx} = \frac{1}{(2\pi)^{3/2}\sqrt{2p_0'}} e^{(a')}_{\mu\nu}(\vec{p}\,') e^{-ip'x}. \end{aligned} \tag{5.24}$$

In a similar way, one obtains

$$\langle 0|y^{(+)}_{\mu\nu}(x) b_{a'}^{(-)}(\vec{p}\,')|0\rangle = \frac{1}{(2\pi)^{3/2}\sqrt{2p_0'}} e^{(a')}_{\mu\nu}(\vec{p}\,') e^{ip'x}, \tag{5.25}$$

leading to the following correspondence between the gravitational field operators and their matrix elements, corresponding to the transition between the state with one graviton (of momentum $\vec{p}$ and polarization $a$) and the vacuum state:

$$y_{\mu\nu}^{(\pm)}(x) \to \frac{1}{(2\pi)^{3/2}\sqrt{2p_0}} e_{\mu\nu}^{(a)}(\vec{p})e^{\pm ipx}. \tag{5.26}$$

The Feynman-type rules in momentum space are obtained by means of (5.7), (5.12), (5.17), and (5.26), also observing that the integration over exponentials leads to the Dirac delta function, this way resulting various types of first-order vertices, propagators and free particles; those corresponding to the first-order processes that are studied in the external field approximation – required by the applications presented in Chapters VI–VIII – are given in the Appendix F.

### V.1.2. Second-order processes

Since the applications contained in the next three chapters are not concerned with second-order processes, they are only briefly analyzed in this paragraph. In this case, applying the Wick's theorem, the $S$-operator leads to the following diagrams[5]:

$$S_2 = i\int d^4x\ N\left[\mathcal{L}_{sc-g}^{(2)}(x)\right]$$

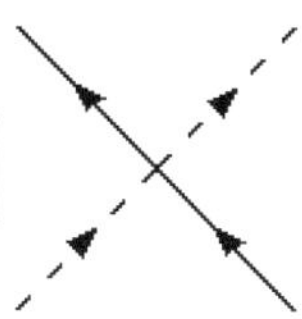

$(1) + \cdots$

$$+\frac{i^2}{2!}\iint d^4x\ d^4x'\ T\Big\{N\left[\mathcal{L}_{sc-g}^{(1)}(x)\mathcal{L}_{sc-g}^{(1)}(x')\right]$$

$(A)$

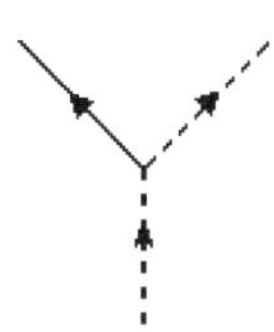

$+ \cdots +$

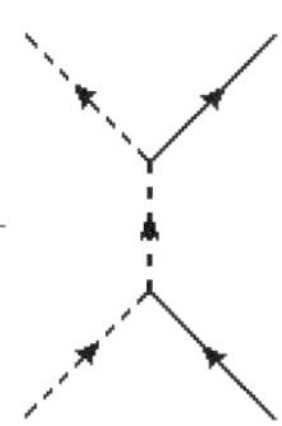

$(1) + \cdots$

[5]For the sake of simplicity, only the diagrams corresponding to the second term of the $S$-matrix expression are given in detail. In the other cases, only a representative diagram is provided.

$+$ 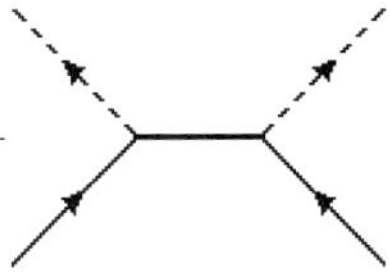 $(2)\cdots+$ 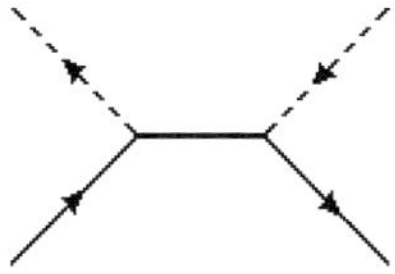 $(3)$

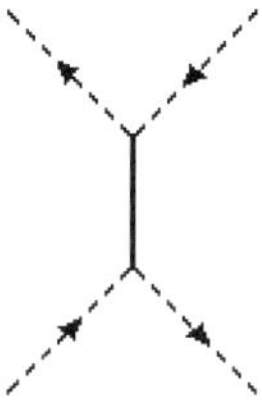 $\cdots+$ 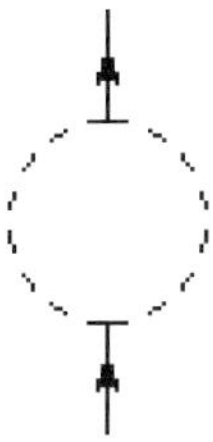 $(a)\quad+$ 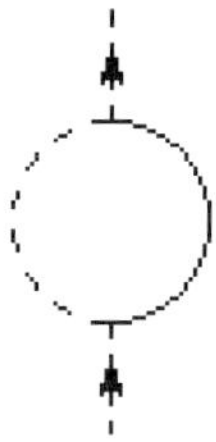 $(b)\cdots$

$+$ 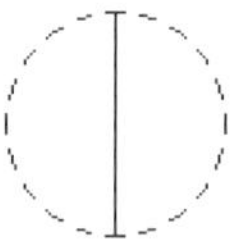 $(c)+N\Big[\mathcal{L}^{(1)}_{sc-g}(x')\mathcal{L}^{(1)}_{sc-g}(x)\Big]\cdots+$

$+$ 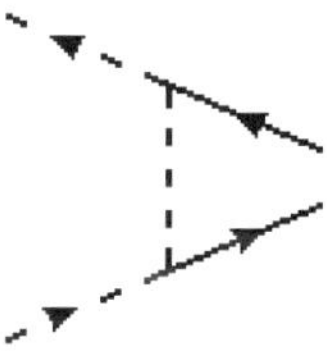 $(1)\ +\ \cdots+N\Big[\mathcal{L}^{(1)}_{sc-g}(x)\mathcal{L}^{(1)}_{g}(x')\Big]$

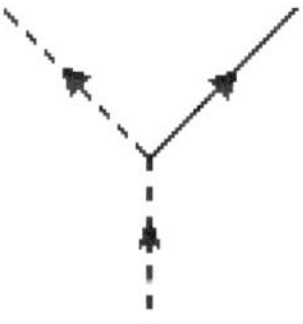 $(B)$  $\cdots+$ 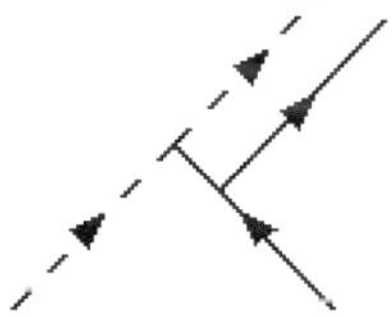 $(1)\cdots$

$+\cdots+N\Big[\mathcal{L}^{(1)}_{g}(x)\mathcal{L}^{(1)}_{sc-g}(x')\Big]\cdots+$ 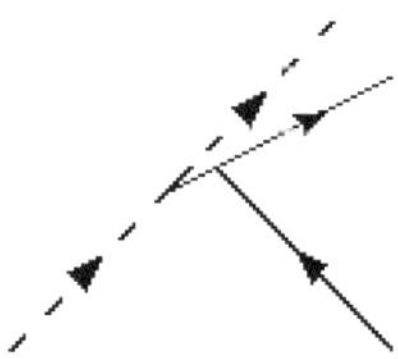 $(1)\cdots+\cdots\Big\},$

(5.27)

to which, as previously, one adds the diagrams obtained by changing the interaction channel (with the corresponding orientation of the time axis), and replacing the particles by their antiparticles (and vice versa)[6]. Thus, the graphs (1) represent the graviton diffusion on the scalar particle (antiparticle) (a Compton-type effect of quantum electrodynamics), as well as the transformation of a pair $\pi^+ - \pi^-$ into two gravitons, or conversely. The graphs (2) and (3) describe the diffusion of meson $\pi^+$ on $\pi^+$ or $\pi^-$ on $\pi^-$ (a Möller- type effect), and $\pi^+$ on $\pi^-$ (a Bhabha-type effect), respectively. The graphs $(a)$ and $(b)$ stand for processes of self-energy of the graviton and scalar particle (antiparticle), respectively, while the graph $(c)$ represents a vacuum-to-vacuum transition. Finally, the graphs $(A)$ and $(B)$ describe two first-order processes that take place simultaneously.

For all these Feynman diagrams there can be deduced Feynman-type rules in the momentum space, resulting various types of free particles, propagators and second-order vertices. These results were obtained by the authors [358(1976)] but, since they are not related to the next chapters, there shall be no further extension on the subject.

## V.2. Interaction with the vectorial field

### V.2.1. Interaction with the electromagnetic field

The interaction between the gravitational and electromagnetic quantum fields is studied – by means of the $S$-matrix formalism – in papers like [38(1967), 275, 306, 390, 392]. Using the method developed in the previous subchapter, the first and second-order processes of this type shall be investigated in a general and unitary manner, obtaining Feynman-type rules for diagrams in the momentum space [358(1976)]. At the same time, the subject shall be approached under a new variant, starting with an interaction Lagrangian density presented in Ref. [362].

Using the minimal coupling principle in quantum gravity and Gupta's ideas as well [181(p.608), 182(1954), 183], to the Lagrangian density of the gravitational field (1.295) one must add the Lagrangian density of the electromagnetic field written in the curved space,

$$\mathcal{L}_{tot} = \mathcal{L}_{em} + \mathcal{L}_g, \tag{5.28}$$

[6]In the case of real scalar field, the last type diagrams and those denoted by (3) vanish. This is also true for the electromagnetic and real vectorial mesonic fields.

where

$$\mathcal{L}_{em} = -\frac{1}{4}\sqrt{-g}\, g^{\mu\alpha} g^{\nu\beta} F_{\mu\nu} F_{\alpha\beta}, \tag{5.29}$$

with $F_{\mu\nu} = A_{\nu,\mu} - A_{\mu,\nu}$, $A_\mu(x)$ being the four-potential of the electromagnetic field.

As the subsidiary condition for the electromagnetic field one takes the general-covariant form of the Lorenz condition

$$A^\mu{}_{;\mu} = 0, \tag{5.30}$$

or, if the De Donder–Fock harmonic coordinates condition is considered, then the subsidiary condition writes

$$\sqrt{-g}\, g^{\mu\nu} A_{\mu,\nu} = 0. \tag{5.30$'$}$$

Going to the flat space, in view of (2.2) and (1.299), and expanding in power series in powers of $\kappa$, one has

$$\mathcal{L}_{int} = \mathcal{L}^{(1)}_{em-g} + \mathcal{L}^{(1)}_{g} + \mathcal{L}^{(2)}_{em-g} + \cdots, \tag{5.31}$$

where

$$\mathcal{L}^{(1)}_{em-g} = -\frac{1}{2}\kappa F_{\mu\nu} F_{\mu\alpha} u_{\nu\alpha} \tag{5.32$a$}$$

and

$$\mathcal{L}^{(2)}_{em-g} = -\frac{1}{4}\kappa^2 \left( F_{\mu\nu} F_{\alpha\beta} h_{\mu\alpha} h_{\nu\beta} + \frac{1}{4} F_{\mu\nu} F_{\mu\nu} y_{\alpha\beta} h_{\alpha\beta} \right) \tag{5.32$b$}$$

are the first and second-order interaction Lagrangian densities between the electromagnetic and gravitational fields, with

$$u_{\mu\nu} = y_{\mu\nu} - \frac{1}{4}\delta_{\mu\nu} y, \tag{5.33}$$

and $\mathcal{L}^{(1)}_g$ given by (5.8).

Alternatively, starting with one of Schweber's ideas [327], the electromagnetic field Lagrangian density in curved space can be written as

$$\mathcal{L}'_{em} = -\frac{1}{2}\sqrt{-g}\, g^{\mu\alpha} g^{\nu\beta} A_{\mu;\nu} A_{\alpha;\beta}, \tag{5.34}$$

where the covariant derivative of the potential four-vector writes

$$A_{\mu;\nu} = A_{\mu,\nu} - \Gamma^{\alpha}_{\mu\nu} A_{\alpha}, \tag{5.35}$$

and the Christoffel's symbols are given by

$$\Gamma^{\alpha}_{\mu\nu} = \frac{1}{2} g^{\alpha\lambda} \Big( g_{\mu\lambda,\nu} + g_{\nu\lambda,\mu} - g_{\mu\nu,\lambda} \Big).$$

Proceeding as in the previous case and taking advantage of (2.2), (1.299) and (1.300), for the first-order interaction Lagrangian density one obtains the following expression:

$$\begin{aligned} \mathcal{L}'^{(1)}_{em-g} = -\frac{1}{2}\kappa \Big[ & A_\mu A_{\nu,\alpha} \Big( h_{\mu\nu,\alpha} + h_{\mu\alpha,\nu} - h_{\nu\alpha,\mu} \Big) \\ & + A_{\mu,\nu} A_{\mu,\alpha} h_{\nu\alpha} + A_{\mu,\nu} A_{\alpha,\nu} y_{\mu\alpha} \Big]. \end{aligned} \tag{5.36}$$

Since the applications provided by the next chapters do not contain second-order processes, while the form of $\mathcal{L}'^{(2)}_{em-g}$ is very complicated (see [359, 361]), this part of the subject shall be skipped.

In order to obtain the matrix elements corresponding to the electromagnetic field operators, one uses the same procedure as in the previous subchapter. Thus, the Fourier series expansion corresponding to the real (observable) photons is given by

$$\begin{aligned} A_\mu(x) &= \frac{1}{(2\pi)^{3/2}} \int \frac{d^3k}{\sqrt{2k_o}} e^{(a)}_\mu(\vec{k}) \Big[ c^{(+)}_a(\vec{k}) e^{ikx} + c^{(-)}_a(\vec{k}) e^{-ikx} \Big] \\ &= A^{(+)}_\mu(x) + A^{(-)}_\mu(x), \end{aligned} \tag{5.37}$$

where $e^{(a)}_\mu(\vec{k})$ are the polarization orthogonal versors (the index $a$ taking the values 1 and 2, corresponding to the orientation of the photon spin, parallel or anti-parallel with respect to direction of motion of the photons), $k_0 = |\vec{k}|$, while the annihilation and creation operators – $c^{(+)}_a(\vec{k})$ and $c^{(-)}_a(\vec{k})$, respectively – satisfy the commutation relation

$$\Big[ c^{(+)}_a(\vec{k}), c^{(-)}_{a'}(\vec{k}') \Big] = \delta_{aa'} \delta\Big( \vec{k} - \vec{k}' \Big), \tag{5.38}$$

all the other commutators being null. Since the vacuum state satisfies the conditions

$$\begin{cases} c_a^{(+)}(\vec{k})|0\rangle = \langle 0|c_a^{(-)}(\vec{k}) = 0, \\ \langle 0|0\rangle = 1, \end{cases} \tag{5.39}$$

if the expressions $\langle 0|c_{a'}^{(+)}(\vec{k}\,')A_\mu^{(-)}(x)|0\rangle$ and $\langle 0|A_\mu^{(+)}(x)c_{a'}^{(-)}(\vec{k}\,')|0\rangle$ are calculated, and (5.37)–(5.39) are taken into account, then one obtains the following correspondence between the electromagnetic field operators and their matrix elements, corresponding to the transition between the one-photon state (of momentum $\vec{k}$ and polarization $a$) and the vacuum state:

$$A_\mu^{(\pm)}(x) \rightarrow \frac{1}{(2\pi)^{3/2}\sqrt{2k_0}} e_\mu^{(a)}(\vec{k}) e^{\pm ikx}. \tag{5.40}$$

The study of the first and second-order processes is performed following the same procedure as in the subchapter one and can be found in the papers [358(1976), 359, 361]. The obtained Feynman diagrams are similar to those given by (5.13) and (1.27) (where the scalar field is substituted by the electromagnetic field), suitable to be associated – by means of (5.26), (5.32), (5.36), and (5.40) – with the corresponding Feynman-type rules in momentum space. The operation shall also lead to various types of propagators, first and second-order vertices and free particles; those corresponding to the first-order processes that are studied in the external field approximation are given in Appendix F.

### V.2.2. Interaction with the vectorial mesonic field

The interaction between gravitational and vectorial mesonic quantized fields (in the Stueckelberg variant) is summarily analyzed in the paper [389(1964)]. Following the same procedure as in the previous cases, the subject shall be approached in a unitary manner, first considering the interaction with a generalized Yang–Mills field, and then – particularizing – with the vectorial field in the frame of Proca's and Stueckelberg's formalisms.

The problem of coupling a vectorial/isovectorial field with a conservative isotopic spin (isospin) current was first analyzed by Yang and Mills [409] in the framework of interaction between the $\pi$ and $\rho$ mesons, and the nucleons. As it is known, the $\rho$ meson triplet associated with the vectorial operator $\vec{B}_\mu\,(B_{1\mu}, B_{2\mu}, B_{3\mu})$ with

$$\begin{cases} B_{1\mu} = \dfrac{1}{2}\left(B_\mu + B_\mu^*\right), \\[2ex] B_{2\mu} = \dfrac{i}{\sqrt{2}}\left(B_\mu - B_\mu^*\right), \\[2ex] B_{3\mu} = B_{0\mu}, \end{cases} \tag{5.41}$$

transforms in accordance with the adjoint representation of the group $SU(2)$. Here $B_{0\mu} = B_\mu^{\rho^0}$ is the Hermitian operator of the neutral vectorial mesonic field, while $B_\mu = B_\mu^{\rho^+}$ and $B_\mu^* = B_\mu^{\rho^-}$ correspond to the vectorial field of charged mesons[7]. In this case, considering only the isospin current of the vectorial mesons, and adding a mass-term in agreement with Feynman's ideas [141, 142], the so-obtained Lagrangian density of the generalized Yang–Mills field,

$$\mathcal{L}_{YM} = -\frac{1}{4}\vec{E}_{\mu\nu}\cdot\vec{E}_{\mu\nu} - \frac{1}{2}\mu^2\vec{B}_\nu\cdot\vec{B}_\nu, \tag{5.42}$$

where

$$\begin{cases} \vec{E}_{\mu\nu} = \vec{G}_{\mu\nu} + 2f\vec{B}_\mu\times\vec{B}_\nu, \\ \vec{G}_{\mu\nu} = \vec{B}_{\nu,\mu} - \vec{B}_{\mu,\nu}, \end{cases} \tag{5.43}$$

is invariant under the $SU(2)$-group transformations and can be written as

$$\mathcal{L}_{YM} = \mathcal{L}_{\rho^\pm} + \mathcal{L}_{\rho^0} + \mathcal{L}_3 + \mathcal{L}_4. \tag{5.44}$$

The quantity $f$ appearing in (5.43) is a coupling constant, while the terms $\mathcal{L}_3$ and $\mathcal{L}_4$ in (5.44) are the third and fourth-order self-coupling Lagrangian densities of the vectorial mesonic field (representing the self-interaction of the $\rho$-triplet), their form being given in Refs. [361, 364].

The field corresponding to the vectorial mesons, charged or neutral, described by the Lagrangian densities $\mathcal{L}_{\rho^\pm}$ and $\mathcal{L}_{\rho^0}$ is called the

[7]This choice is conventional.

*Proca field*, and the corresponding operators satisfy a Lorenz-type condition [240, 252]:

$$B_{\mu,\nu} = 0. \tag{5.45}$$

By introducing a fifth component, with no effect upon the observable quantities, Stueckelberg brought the Lagrangian formalism of the vectorial mesons to a form similar to that found in electrodynamics [387]. He replaced the operator $B_\nu(x)$ with two fields $U_\nu(x)$ and $C(x)$,

$$B_\nu(x) = U_\nu(x) + \frac{1}{\mu} C_{,\nu}(x), \tag{5.46}$$

that verify the subsidiary condition

$$U_{\nu,\nu} + \mu C = 0, \tag{5.45$'$}$$

where $C(x)$ satisfies the wave equation corresponding to the scalar mesons.

Considering the case of the charged mesons[8] and using the minimal coupling principle met in theory of quantum gravity, the Lagrangian density of the gravitational field (1.295) has to be supplemented with the Lagrangian density of the complex vectorial field, written in the curved space,

$$\mathcal{L}_{tot} = \mathcal{L}_v + \mathcal{L}_g, \tag{5.47}$$

where, in agreement with Proca's formalism,

$$\mathcal{L}_v = \mathcal{L}_{Proca} \equiv \mathcal{L}_P = \sqrt{-g}\left(-\frac{1}{2} g^{\mu\alpha} g^{\nu\beta} G^*_{\mu\nu} G_{\alpha\beta} + \mu^2 g^{\nu\alpha} B^*_\nu B_\alpha\right), \tag{5.48}$$

while in the framework of Stueckelberg's formalism,

$$\mathcal{L}_v = \mathcal{L}_{Stueckelberg} + \mathcal{L}' \equiv \mathcal{L}_{St} + \mathcal{L}', \tag{5.49}$$

where

$$\begin{cases} \mathcal{L}_{St} = \sqrt{-g}\Big( - g^{\mu\alpha} g^{\nu\beta} U^*_{\mu;\nu} U_{\alpha;\beta} + \mu^2 g^{\nu\alpha} U^*_\nu U_\alpha \Big), \\ \mathcal{L}' = \sqrt{-g}\Big( g^{\mu\nu} C^*_{,\mu} C_{,\nu} - \mu^2 C^* C \Big). \end{cases} \tag{5.50}$$

[8]In the case of real vectorial field (neutral mesons) $B^*_\mu \to B_\mu \equiv B_{0\mu}$, $U^*_\mu \to U_\mu \equiv U_{0\mu}$ and $C^* \to C \equiv C_0$, while in the front of the r.h.s of Eqs. (5.48), (5.50), (5.52) and (5.53) one must appear the multiplicative factor 1/2.

As a subsidiary condition for the vectorial mesonic field one takes – by analogy with the electromagnetic field – the Lorenz-type condition (5.45) written in the general-covariant form.

Going to the flat space, taking into account (2.2), (1.297), (1.299), (1.300) and expanding in power series in powers of $\kappa$, one has

$$\mathcal{L}_{int} = \mathcal{L}^{(1)}_{v-g} + \mathcal{L}^{(1)}_{g} + \mathcal{L}^{(2)}_{v-g} + \cdots , \tag{5.51}$$

where the Lagrangian densities corresponding to the above two cases are

$$\begin{cases} \mathcal{L}^{(1)}_{P-g} = -\kappa\Big(G^*_{\mu\nu}G_{\mu\alpha}U_{\nu\alpha} + \mu^2 B^*_\nu B_\alpha y_{\nu\alpha}\Big), \\ \mathcal{L}^{(2)}_{P-g} = -\dfrac{1}{2}\kappa^2\left(G^*_{\mu\nu}G_{\alpha\beta}h_{\mu\alpha}h_{\nu\beta} + \dfrac{1}{4}G^*_{\mu\nu}G_{\mu\nu}y_{\alpha\beta}h_{\alpha\beta}\right), \end{cases} \tag{5.52}$$

and

$$\begin{aligned} \mathcal{L}^{(1)}_{St-g} = -\kappa\Big[&U^*_\mu U_{\nu,\alpha}\left(h_{\mu\nu,\alpha} + h_{\mu\alpha,\nu} - h_{\nu\alpha,\mu}\right) + U^*_{\mu,\nu}U_{\mu,\alpha}h_{\nu\alpha} \\ &+ U^*_{\mu,\nu}U_{\alpha,\nu}y_{\mu\alpha} + \mu^2 U^*_\nu U_\alpha y_{\nu\alpha}\Big], \end{aligned} \tag{5.53}$$

respectively.

The expressions of Lagrangian densities $\mathcal{L}'^{(1)}$ and $\mathcal{L}'^{(2)}$ are analogous to that given by (5.17), while the expression of $\mathcal{L}^{(2)}_{St-g}$ is very complicated (see [364]) and its presentation shall be skipped.

As we shall see in Chap. VI, in the present case, instead of the covariant derivative can be used the usual one, and so, the Lagrangian density given by (5.53) becomes

$$\mathcal{L}'^{(1)}_{St-g} \cong -\kappa\Big(U^*_{\mu,\nu}U_{\mu,\alpha}h_{\nu\alpha} + U^*_{\mu,\nu}U_{\alpha,\nu}y_{\mu\alpha} + \mu^2 U^*_\nu U_\alpha y_{\nu\alpha}\Big). \tag{5.53$'$}$$

The matrix elements of the vectorial field operators are obtained following the same procedure as that used in the previous subchapter. Thus, in the framework of Proca's formalism, the charged mesons are described by the following field functions [252][9],

$$B_\mu(x) = \frac{1}{(2\pi)^{3/2}} \int \frac{d^3k}{\sqrt{2k_0}} e_\mu^{(m)}(\vec{k}) \left[ b_m^{(+)}(\vec{k}) e^{ikx} + b_m^{(-)}(\vec{k}) e^{-ikx} \right]$$

$$= B_\mu^{(+)}(x) + B_\mu^{(-)}(x), \qquad (5.54a)$$

and

$$B_\mu^*(x) = \frac{1}{(2\pi)^{3/2}} \int \frac{d^3k}{\sqrt{2k_0}} e_\mu^{(m)}(\vec{k}) \left[ b_m^{*(+)}(\vec{k}) e^{ikx} + b_m^{*(-)}(\vec{k}) e^{-ikx} \right]$$

$$= B_\mu^{*(+)}(x) + B_\mu^{*(-)}(x), \qquad (5.54b)$$

where $e_\mu^{(m)}(\vec{k})$, $(m = 1, 2, 3)$ are orthogonal polarization versors, $k_0 = \sqrt{\vec{k}^2 + \mu^2}$, and the annihilation and creation operators – $b_m^{(+)}(\vec{k})$, $b_m^{*(+)}(\vec{k})$ and $b_m^{(-)}(\vec{k})$, $b_m^{*(-)}(\vec{k})$, respectively – satisfy the commutation relations

$$\left[ b_m^{(+)}(\vec{k}), b_{m'}^{*(-)}(\vec{k}\,') \right] = \left[ b_m^{*(+)}(\vec{k}), b_{m'}^{(-)}(\vec{k}\,') \right] = \delta_{mm'} \delta\left( \vec{k} - \vec{k}\,' \right), \quad (5.55)$$

all the other commutators being null. Since the vacuum state verifies the conditions

$$\begin{cases} b_m^{(+)}(\vec{k})|0\rangle = b_m^{*(+)}(\vec{k})|0\rangle = 0, \\ \langle 0|b_m^{(-)}(\vec{k}) = \langle 0|b_m^{*(-)}(\vec{k}) = 0, [10pt] \langle 0|0\rangle = 1, \end{cases} \qquad (5.56)$$

the calculation of the expressions

$$\langle 0|b_{m'}^{(+)}(\vec{k}\,') B_\mu^{*(-)}(x)|0\rangle, \ \langle 0|B_\mu^{(+)}(x) b_{m'}^{*(-)}(\vec{k}\,')|0\rangle,$$

$$\langle 0|b_{m'}^{*(+)}(\vec{k}\,') B_\mu^{(-)}(x)|0\rangle, \ \langle 0|B_\mu^{*(+)}(x) b_{m'}^{(-)}(\vec{k}\,')|0\rangle$$

leads, by means of (5.53)–(5.56), to the following correspondence between the operators of the charged vectorial field and their matrix elements, corresponding to the transition between the state with a

[9] $B_\mu^{(+)}(x)$ and $B_\mu^{*(-)}(x)$ are the annihilation and creation operators of the vectorial particles (for example, the mesons $\rho^+$), respectively, while $B_\mu^{*(+)}(x)$ and $B_\mu^{(-)}(x)$ are the annihilation and creation operators of the corresponding antiparticles (for the chosen example, the mesons $\rho^-$), respectively.

single meson (of momentum $\vec{k}$ and polarization $m$) and the vacuum state[10]:

$$\left(B_\mu^{\pm}(x);\ B_\mu^{*\pm}(x)\right) \to \frac{1}{(2\pi)^{3/2}\sqrt{2k_0}} e_\mu^{(m)}(\vec{k})e^{\pm ikx}. \tag{5.57}$$

In the frame of Stueckelberg formalism, the matrix elements corresponding to the operators $U_\mu^{(\pm)}(x)$ and $U_\mu^{*(\pm)}(x)$ – which admit power series expansions analogous to (5.54) – are similar to those given by (5.57), while the matrix elements of the operators $C^{*(\pm)}(x)$ and $C^{(\pm)}(x)$ are given by (5.17).

The study of the first and second-order processes follows the same procedure as that exposed in the first subchapter, being presented in detail in Refs. [361, 378]. The approaches corresponding to the two formalisms are equivalent (the resulting graphs are the same), because the auxiliary scalar field describes free particles which provide no contribution to the observed results[11]. This way are obtained diagrams similar to those expressed by (5.13) and (5.27) (the scalar field being replaced by the vectorial mesonic field), suitable for getting – by means of (5.26), (5.52), (5.53′), and (5.57) – the corresponding Feynman-type rules in momentum space. One obtains various types of propagators, first and second-order vertices and free particles; those corresponding to the first-order processes that are studied in the external field approximation are given in Appendix F.

## V.3. Interaction with the spinorial field

The interaction between the gravitational and spinorial quantized fields is studied – by means of the $S$-matrix formalism – in the papers [11, 246, 275, 388(1963), 393]. Using the same procedure, the problem shall be discussed in a consistent manner, first considering the interaction with the electron-positron field, and then with the massless spinorial field, this way obtaining Feynman-type rules for diagrams in the momentum space.

---

[10] In the case of real vectorial field, one obtains the same matrix elements.

[11] This field interferes only in the expressions of the propagators (see [23, 378, 387]).

### V.3.1. Interaction with the electron-positron field

Using the minimal coupling principle in quantum gravity, to the Lagrangian density of the gravitational field (1.295) one must add the Lagrangian density of the spinorial field written in the curved space,

$$\mathcal{L}_{tot} = \mathcal{L}_{sp} + \mathcal{L}_{g}, \tag{5.58}$$

where $\mathcal{L}_{sp}$ is obtained by means of the *vierbein* formalism. Thus, following the Lias' ideas in Ref. [246], but in a new and more natural variant, one has

$$\mathcal{L}_{sp} = \sqrt{-g}\left[\frac{i}{2}\left(\overline{\psi}\tilde{\gamma}^{\mu}\psi_{;\mu} - \overline{\psi}_{;\mu}\tilde{\gamma}^{\mu}\psi\right) - M\overline{\psi}\psi\right], \tag{5.59}$$

where $\tilde{\gamma}^{\mu}$ are the *Dirac's generalized matrices* (which are not constant, but functions of the coordinates $x^{\mu}$), $\overline{\psi} = \psi^{+}\gamma^{0}$, and the spinor covariant derivatives are given by

$$\begin{cases} \psi_{;\mu} = \psi_{,\mu} - \Gamma_{\mu}\psi, \\ \overline{\psi}_{;\mu} = \overline{\psi}_{,\mu} + \overline{\psi}\Gamma_{\mu}. \end{cases} \tag{5.60}$$

Here, $\Gamma_{\mu}$ are the *Fock–Ivanenko spin coefficients of the affine connection* (see Refs. [147, 275] for details),

$$\Gamma_{\mu} = \frac{1}{4}\tilde{\gamma}_{\nu;\mu}\tilde{\gamma}^{\nu}, \tag{5.61}$$

where

$$\tilde{\gamma}_{\mu;\nu} = \tilde{\gamma}_{\mu,\nu} - \tilde{\gamma}_{\lambda}\Gamma^{\lambda}_{\mu\nu}, \tag{5.62}$$

are the covariant derivatives of $\tilde{\gamma}^{\mu}$ with respect to coordinates $x^{\nu}$, while $\Gamma^{\lambda}_{\mu\nu}$ are the Christoffel's symbols of the second kind, given by the well-known formula

$$\Gamma^{\lambda}_{\mu\nu} = \frac{1}{2}g^{\lambda\alpha}\left(g_{\mu\alpha,\nu} + g_{\nu\alpha,\mu} - g_{\mu\nu,\alpha}\right).$$

Considering the case of a weak gravitational field and expressing the Dirac's generalized matrices by means of the *vierbein* coefficients $L^{\mu}(\alpha)$ (see Refs. [109, 316] for details),

$$\begin{cases} \tilde{\gamma}^{\mu} = L^{\mu}(\alpha)\gamma(\alpha), \\ \tilde{\gamma}_{\mu} = L_{\mu}(\alpha)\gamma(\alpha), \end{cases} \tag{5.63}$$

with

$$\begin{cases} L^{\mu}(\alpha)L^{\nu}(\alpha) = g^{\mu\nu}, \\ L_{\mu}(\alpha)L_{\nu}(\alpha) = g_{\mu\nu}, \end{cases} \tag{5.64}$$

one can expand $\tilde{\gamma}^{\mu}$ and $\tilde{\gamma}_{\mu}$ in power series in powers of both the usual Dirac matrices and quantities corresponding to the gravitational field. Thus, using (1.299), (1.300), (5.63), and (5.64) (see also Ref. [275]), one obtains

$$\begin{aligned} \tilde{\gamma}^{\mu} = \gamma^{\mu} &+ \frac{1}{2}\kappa\left(\frac{1}{2}\eta_{\alpha\beta}\gamma^{\mu} - \delta^{\mu}_{\beta}\gamma_{\alpha}\right)y^{\alpha\beta} \\ &+ \frac{1}{8}\kappa^2\Bigg(\frac{1}{4}\eta_{\alpha\beta}\eta_{\lambda\rho}\gamma^{\mu} + \eta_{\alpha\lambda}\eta_{\beta\rho}\gamma^{\mu} \\ &- \eta_{\alpha\beta}\delta^{\mu}_{\lambda}\gamma_{\rho} - \eta_{\alpha\lambda}\delta^{\mu}_{\beta}\gamma_{\rho}\Bigg)y^{\alpha\beta}y^{\lambda\rho} + \cdots \end{aligned} \tag{5.65}$$

and

$$\begin{aligned} \tilde{\gamma}_{\mu} = \gamma_{\mu} &- \frac{1}{2}\kappa\left(\frac{1}{2}\eta_{\alpha\beta}\gamma_{\mu} - \eta_{\mu\beta}\gamma_{\alpha}\right)y^{\alpha\beta} \\ &+ \frac{1}{8}\kappa^2\Bigg(\frac{1}{4}\eta_{\alpha\beta}\eta_{\lambda\rho}\gamma_{\mu} - \eta_{\alpha\lambda}\eta_{\beta\rho}\gamma_{\mu} \\ &- \eta_{\mu\beta}\eta_{\lambda\rho}\gamma_{\alpha} + 3\eta_{\mu\lambda}\eta_{\beta\rho}\gamma_{\alpha}\Bigg)y^{\alpha\beta}y^{\lambda\rho} - \cdots \end{aligned} \tag{5.66}$$

Going to the flat space[12], next using Eqs. (1.297), (5.58)–(5.62), (5.65), (5.66), and expanding in power series in powers of $\kappa$, one can write

$$\mathcal{L}_{int} = \mathcal{L}^{(1)}_{sp-g} + \mathcal{L}^{(1)}_{g} + \mathcal{L}^{(2)}_{sp-g} + \cdots, \tag{5.67}$$

where

$$\begin{cases} \mathcal{L}^{(1)}_{sp-g} = -\frac{1}{2}\kappa T_{\mu\nu}s_{\mu\nu} - \frac{1}{2}\kappa M\overline{\psi}\psi\, y, \\ \mathcal{L}^{(2)}_{sp-g} = \frac{1}{8}\kappa^2 T_{\alpha\alpha}y_{\mu\nu}h_{\mu\nu} + \frac{1}{8}\kappa^2 T_{\mu\nu}h_{\mu\alpha}h_{\nu\alpha} \\ \qquad + \frac{1}{4}\kappa^2 M\overline{\psi}\psi\, y_{\mu\nu}h_{\mu\nu} \\ \qquad + \frac{1}{16}\kappa^2\overline{\psi}\Big(\gamma_{\mu}y_{\nu\alpha} - \gamma_{\nu}y_{\mu\alpha} - \gamma_{\mu}\gamma_{\nu}\gamma_{\beta}y_{\alpha\beta}\Big)y_{\nu\alpha,\mu}\psi \end{cases} \tag{5.68}$$

[12]In which case $\overline{\psi} = \psi^{+}\gamma_4$ and $\gamma^{\mu} \to i\gamma_{\mu}$ (see Refs. [2, 240]).

are the first and second-order interaction Lagrangian densities between the spinorial and gravitational fields, with

$$\begin{cases} T_{\mu\nu} = \dfrac{1}{2}\left(\overline{\psi}\gamma_\mu\psi_{,\nu} - \overline{\psi}_{,\nu}\gamma_\mu\psi\right), \\ s_{\mu\nu} = y_{\mu\nu} + \dfrac{1}{2}\delta_{\mu\nu}y. \end{cases} \tag{5.69}$$

The Lagrangian densities $\mathcal{L}^{(1)}_{sp-g}$ and $\mathcal{L}^{(2)}_{sp-g}$ shown in Eq. (5.68) are more general than those found by Gupta[13] (in Ref. [181(p.608)]) and Vladimirov (in Ref. [388(1963)]), having a simpler form and thus, being more convenient to be used in applications, as compared with those used by Mitskevich in Ref. [275].

In order to obtain the matrix elements corresponding to the spinorial field operators, one starts from their Fourier expansions[14] (see Refs. [252, 387]):

$$\begin{cases} \psi(x) = \dfrac{1}{(2\pi)^{3/2}}\displaystyle\int d^3k\sqrt{\frac{M}{E_k}}\sum_{r=1}^{2}\Big[u^{(r)}(\vec{k})d_r^{(+)}(\vec{k})e^{ikx} \\ \qquad + v^{(r)}(\vec{k})d_r^{(-)}(\vec{k})e^{-ikx}\Big] = \psi^{(+)}(x) + \psi^{(-)}(x), \\ \overline{\psi}(x) = \dfrac{1}{(2\pi)^{3/2}}\displaystyle\int d^3k\sqrt{\frac{M}{E_k}}\sum_{r=1}^{2}\Big[\overline{v}^{(r)}(\vec{k})d_r^{*(+)}(\vec{k})e^{ikx} \\ \qquad + \overline{u}^{(r)}(\vec{k})d_r^{*(-)}(\vec{k})e^{-ikx}\Big] = \overline{\psi}^{(+)}(x) + \overline{\psi}^{(-)}(x). \end{cases} \tag{5.70}$$

Here $u^{(r)}(\vec{k})$, $\overline{u}^{(r)}(\vec{k})$ and $v^{(r)}(\vec{k})$, $\overline{v}^{(r)}(\vec{k})$ – where the index $s$ takes the values[15] 1 and 2 – are the spinors of the positive and negative energy, respectively, $E_k^2 = \vec{k}^2 + M^2$, and the annihilation and creation

[13]According to Ref. [181(p.608)], the first-order interaction Lagrangian density between the gravitational and matter fields is of the form $\mathcal{L}^{(1)}_{int} = -\frac{1}{2}\kappa T_{\mu\nu}h_{\mu\nu}$, where $T_{\mu\nu}$ is the energy-momentum tensor of the matter field, which in our case is the spinorial field.

[14]$\psi^{(+)}(x)$ and $\overline{\psi}^{(-)}(x)$ are the annihilation and creation operators of the spinorial particles (electrons $e^-$), respectively, while $\overline{\psi}^{(+)}(x)$ and $\psi^{(-)}(x)$ are the annihilation and creation operators of the corresponding antiparticles (positrons $e^+$), respectively.

[15]These values correspond to the two eigenvalues $\pm 1/2$ of the spin matrix of the Dirac's field particles (see Refs. [240, 252]).

operators – $d_r^{(+)}(\vec{k})$, $d_r^{*(+)}(\vec{k})$ and $d_r^{(-)}(\vec{k})$, $d_r^{*(-)}(\vec{k})$, respectively – satisfy the anticommutation relations

$$\left\{d_r^{(+)}(\vec{k}), d_{r'}^{*(-)}(\vec{k}\,')\right\} = \left\{d_r^{*(+)}(\vec{k}), d_{r'}^{(-)}(\vec{k}\,')\right\} = \delta_{rr'}\delta\left(\vec{k}-\vec{k}\,'\right), \quad (5.71)$$

all the other (anti)commutators being null. As usual, the vacuum state verifies the conditions

$$\begin{cases} d_r^{(+)}(\vec{k})|0\rangle = d_r^{*(+)}(\vec{k})|0\rangle = 0, \\ \langle 0|d_r^{*(-)}(\vec{k}) = \langle 0|d_r^{(-)}(\vec{k}) = 0, \\ \langle 0|0\rangle = 1. \end{cases} \quad (5.72)$$

Calculating the expressions

$$\langle 0|d_{r'}^{(+)}(\vec{k}\,')\overline{\psi}^{(-)}(x)|0\rangle, \quad \langle 0|\psi^{(+)}(x)d_{r'}^{*(-)}(\vec{k}\,')|0\rangle,$$

and

$$\langle 0|d_{r'}^{*(+)}(\vec{k}\,')\psi^{(-)}(x)|0\rangle, \quad \langle 0|\overline{\psi}^{(+)}(x)d_{r'}^{(-)}(\vec{k}\,')|0\rangle$$

this leads, by means of (5.70)–(5.72), to the following correspondence between the spinorial field operators and their matrix elements, corresponding to the transition between the one-electron (positron) state – of momentum $\vec{k}$, energy $E_k$ and spin $r$ – and the vacuum state:

$$\begin{cases} \psi^{(+)}(x) \to \dfrac{1}{(2\pi)^{3/2}}\sqrt{\dfrac{M}{E_k}}u^{(r)}(\vec{k})e^{ikx}, \\ \psi^{(-)}(x) \to \dfrac{1}{(2\pi)^{3/2}}\sqrt{\dfrac{M}{E_k}}v^{(r)}(\vec{k})e^{-ikx}, \\ \overline{\psi}^{(+)}(x) \to \dfrac{1}{(2\pi)^{3/2}}\sqrt{\dfrac{M}{E_k}}\overline{v}^{(r)}(\vec{k})e^{ikx}, \\ \overline{\psi}^{(-)}(x) \to \dfrac{1}{(2\pi)^{3/2}}\sqrt{\dfrac{M}{E_k}}\overline{u}^{(r)}(\vec{k})e^{-ikx}. \end{cases} \quad (5.73)$$

The procedure used to study the first and second-order processes is similar to that described in the first subchapter, being exposed in detail in Refs. [358(1977), 361]. In this case, like in the previous ones, there are also obtained Feynman diagrams similar to (5.13) and (5.27) (the scalar field being replaced by the spinorial field) and, in view of (5.26), (5.68) and (5.73), corresponding Feynman-type rules in momentum space can be obtained. Thus, several types of propagators, first and second-order vertices and free particles are resulting, and those corresponding to the first-order processes that are studied in the external field approximation are presented in Appendix F.

### V.3.2. Interaction with the massless spinorial field

The study of interaction between the gravitational and massless spinor fields, if the massless spinor particles are described by a four-component wave function, is usually performed by taking $M = 0$ in the Lagrangian densities (5.59) and (5.68). The matrix elements corresponding to the massless spinor field operators are obtained by taking out the factors $\sqrt{M/E_k}$ in (5.73), and the analysis of the first and second-order processes leads to the results already exposed.

The approach is somewhat more complicated if the massless spinor particle is described by a two-component spinor, the interaction associated with this case being studied in Refs. [38(1972), 124]. The following approach is based on a new variant, in the light of Landau and Lifshitz's ideas presented in Ref. [240].

Thus, one defines the matrix $\gamma^5 = -i\gamma^0\gamma^1\gamma^2\gamma^3$, whose form does not change in the curved space (see Ref. [98] for more details) and performs the following convenient well-known substitutions,

$$\begin{cases} \psi \to \dfrac{1}{2}\left(1+\gamma^5\right)\psi, \\ \psi \to \dfrac{1}{2}\overline{\psi}\left(1-\gamma^5\right), \end{cases} \tag{5.74}$$

in the Lagrangian density (5.59). Taking into account the properties of Dirac matrices, going to the flat space[16] and using the same procedure as that utilized in the case of electron-positron field, the

[16]In this case, $\gamma^5 \to \gamma_5 = \gamma_1\gamma_2\gamma_3\gamma_4$.

following expressions for $\mathcal{L}'^{(1)}_{sp-g}$ and $\mathcal{L}'^{(2)}_{sp-g}$ are obtained,

$$\begin{cases} \mathcal{L}'^{(1)}_{sp-g} = -\frac{1}{4}\kappa T_{5\mu\nu}s_{\mu\nu}, \\ \mathcal{L}'^{(2)}_{sp-g} = \frac{1}{16}\kappa^2 T_{5\alpha\alpha}y_{\mu\nu}h_{\mu\nu} + \frac{1}{16}\kappa^2 T_{5\mu\nu}h_{\mu\alpha}h_{\nu\alpha} \\ \qquad + \frac{1}{32}\kappa^2\Big[(\overline{\psi}\psi)_\mu y_{\nu\alpha} - (\overline{\psi}\psi)_\nu y_{\mu\alpha} - (\overline{\psi}\psi)_{\mu\nu\beta}y_{\alpha\beta}\Big]y_{\nu\alpha,\mu}, \end{cases} \tag{5.75}$$

where

$$T_{5\mu\nu} \equiv \frac{1}{2}\Big[\Big(\overline{\psi}\psi_{,\nu}\Big)_\mu - \Big(\overline{\psi}_{,\nu}\psi\Big)_\mu\Big], \tag{5.76}$$

with

$$\Big(\overline{\psi}\psi\Big)_{\alpha_1\ldots\alpha_n} \equiv \overline{\psi}\gamma_{\alpha_1}\ldots\gamma_{\alpha_n}\Big(1+\gamma_5\Big)\psi, \tag{5.77}$$

all the other notations being obvious.

The form of the interaction Lagrangian density $\mathcal{L}'^{(1)}_{sp-g}$ is more general than that obtained by Kobzarev and Okun in Ref. [234], or De Sabbata and collab. in Ref. [38(1972)] (who started with previously mentioned Gupta's ideas presented in Ref. [181(p.608)]), while $\mathcal{L}'^{(2)}_{sp-g}$ does not appear in the specific literature.

If the massless spinorial particles are described by a two-component wave function, then the matrix elements of the massless spinor field operators are given by

$$\begin{cases} \psi^{(+)})x) \to \frac{1}{(2\pi)^{3/2}}u^{(r)}(\vec{k})e^{ikx}, \\ \psi^{(-)}(x) \to \frac{1}{(2\pi)^{3/2}}v^{(r)}(\vec{k})e^{-ikx}, \\ \overline{\psi}^{(+)}(x) \to \frac{1}{(2\pi)^{3/2}}\overline{v}^{(r)}(\vec{k})e^{ikx}, \\ \overline{\psi}^{(-)}(x) \to \frac{1}{(2\pi)^{3/2}}\overline{u}^{(r)}(\vec{k})e^{-ikx}, \end{cases} \tag{5.78}$$

where, this time the matrix elements were calculated for the transition between the state containing only one massless spinor particle of energy $k_0$, momentum $\vec{k}$ (with $|\vec{k}| = k_0$) and polarization $r$, and the vacuum state. As is well known, the massless particles do not have spin states (like the massive particles), but instead they have polarization states. Here, the index $r$ indicating the polarization states of the massless spinorial particles can take only two values, namely 1 and 2, corresponding to the two eigenvalues $\pm 1/2$ of the polarization matrix. Thus, in this case the spin matrix (which is defined for the massive spinorial particles) must be replaced by the polarization matrix, because in the case of massless particles, instead of spin, one speaks about helicity (the spin projection along the direction of motion).

The study of the first and second-order processes is performed by analogy with that corresponding to the electron-positron field, being exposed in detail in Refs. [358(1977), 361]. Like in the previous cases, here also are obtained Feynman-type rules in the momentum space, which lead to various types of propagators, first and second-order vertices and free particles. The results corresponding to the first-order processes that are studied in the external field approximation, are given in Appendix F.

## V.4. The external field problem

As it was shown in Ref. [358], the interaction processes between the elementary particles and gravitons are most probable in the presence of external gravitational or electromagnetic fields. Following the $S$-matrix formalism, this subchapter is concerned with the first-order processes of this type (namely, particles diffusion in a static gravitational field described by Schwarzschild's metric [359, 362] and transformation of photons into gravitons in an external electromagnetic field in several concrete cases [361, 362]), and some appropriate Feynman-type rules for diagrams in the momentum space shall be deduced. The obtained results complete and generalize those offered by the specific literature.

### V.4.1. External gravitational field

Consider, for the beginning, the scattering process in an external gravitational field of the particles possessing various spin values, as follows:

*i*) In the case of a charged scalar field, following the standard procedure used in QFT [217, 252], the term of the Lagrangian density (5.7-1) – brought to the normal form – describing the interaction of scalar particles (*e.g.* the $\pi^+$ mesons) with the gravitational field is[17]

$$N\left[\mathcal{L}^{(1)}_{sc-g}(x)\right] = -\kappa\left[\varphi^{*(-)}_{,\mu}(x)\varphi^{(+)}_{,\nu}(x)y^{ext}_{\mu\nu}(x) + \frac{1}{2}m^2\varphi^{*(-)}(x)\varphi^{(+)}(x)y^{ext}(x)\right]. \tag{5.79}$$

*ii*) Taken under the normal product, the interaction Lagrangian density (5.32*a*) describing the interaction between the electromagnetic and gravitational fields – having in view that the electromagnetic field is a real field – becomes

$$N\left[\mathcal{L}^{(1)}_{em-g}(x)\right] = -\kappa F^{(-)}_{\mu\nu}(x)F^{(+)}_{\mu\alpha}(x)u^{ext}_{\nu\alpha}(x). \tag{5.80}$$

Since the static gravitational field with spherical symmetry described by Schwarzschild's metric is considered here as an external field (in which case, the only non-zero component is $y_{44}$), and the gauge used for photons in these applications is $e^{(a)}_4(\vec{k}) = e^{(b)}_4(\vec{k}') = 0$, the last term of the equivalent Lagrangian density (5.36) vanishes, and the only part presenting interest to our purpose gets the form

$$N\left[\mathcal{L}'^{(1)}_{em-g}(x)\right] = -\kappa\left\{\frac{1}{2}\left[A^{(-)}_{\mu}(x)A^{(+)}_{\nu,\alpha}(x) + A^{(-)}_{\nu,\alpha}(x)A^{(+)}_{\mu}(x)\right] \times\left[h^{ext}_{\mu\nu,\alpha}(x) + h^{ext}_{\mu\alpha,\nu}(x) - h^{ext}_{\nu\alpha,\mu}(x)\right] + A^{(-)}_{\mu,\nu}(x)A^{(+)}_{\mu,\alpha}(x)h^{ext}_{\nu\alpha}(x)\right\}. \tag{5.80'}$$

[17] In the case of corresponding antiparticles (for example, the $\pi^-$ mesons), the positive frequency parts appearing in (5.79) must be replaced by the negative parts and vice versa, also changing the operator order. If the scalar field is real, there appears a multiplicative factor 2 (because two equivalent normal products correspond to the same Feynman diagram) that simplifies with that existent in front of the interaction Lagrangian density, this fact leading to the same matrix element.

*iii*) In the case of charged vectorial mesonic field, the Lagrangian densities (5.52-1) and (5.53′) describing the interaction with gravitation [18] of the vectorial particles (*e.g.* the $\rho^+$ mesons) – in the framework of both the Proca's and Stueckelberg's formalisms – become

$$N\left[\mathcal{L}^{(1)}_{P-g}(x)\right] = -\kappa\Big[G^{*(-)}_{\mu\nu}(x)G^{(+)}_{\mu\alpha}(x)u^{ext}_{\nu\alpha}(x) + \mu^2 B^{*(-)}_{\nu}(x)B^{(+)}_{\alpha}(x)y^{ext}_{\nu\alpha}(x)\Big], \tag{5.81}$$

and

$$N\left[\mathcal{L}'^{(1)}_{St-g}(x)\right] = -\kappa\Big[U^{*(-)}_{\mu,\nu}(x)U^{(+)}_{\mu,\alpha}(x)h^{ext}_{\nu\alpha}(x) + U^{*(-)}_{\mu,\nu}(x)U^{(+)}_{\alpha,\nu}(x)y^{ext}_{\mu\alpha}(x) + \mu^2 U^{*(-)}_{\nu}(x)U^{(+)}_{\alpha}(x)y^{ext}_{\nu\alpha}(x)\Big], \tag{5.81′}$$

respectively, the last expression being useful (even indispensable) for the study of the diffusion process at small scattering angles.

*iv*) In the case of spinorial field studied in the framework of the *vierbein* formalism, the part of the Lagrangian density (5.68-1) describing the interaction between the spinorial particles (*e.g.* the electrons) and gravitation is given by

$$N\left[\mathcal{L}^{(1)}_{sp-g}(x)\right] = -\frac{1}{2}\kappa\left\{\frac{1}{2}\Big[\overline{\psi}^{(-)}(x)\gamma_\mu\psi^{(+)}_{,\nu}(x) - \overline{\psi}^{(-)}_{,\nu}(x)\gamma_\mu\psi^{(+)}(x)\Big]s^{ext}_{\mu\nu}(x) + M\overline{\psi}^{(-)}(x)\psi^{(+)}(x)y^{ext}(x)\right\}, \tag{5.82}$$

while in Mitskevich's approach (see Ref. [275]) – who used the Gupta's idea according to which the interaction Lagrangian density is obtained by coupling the energy-momentum tensor of the matter,

[18] With the same observation as previously evidentiated.

$T_{\mu\nu}$, with the gravitational field described by the symmetric tensor $h_{\mu\nu}$ – one obtains

$$N\Big[\mathcal{L}'^{(1)}_{sp-g}(x)\Big] = -\frac{1}{4}\kappa\Big[\overline{\psi}^{(-)}(x)\gamma_\mu\psi^{(+)}_{,\nu}(x) - \overline{\psi}^{(-)}_{,\nu}(x)\gamma_\mu\psi^{(+)}(x)\Big]h^{ext}_{\mu\nu}(x). \tag{5.82'}$$

*v*) For the massless spinorial field, if the corresponding particles are described by a two-component wave function, one has

$$N\Big[\mathcal{L}^{(1)}_{sp_0-g}(x)\Big] = -\frac{1}{8}\kappa\Big[\overline{\psi}^{(-)}(x)\gamma_\mu(1+\gamma_5)\psi^{(+)}_{,\nu}(x) - \overline{\psi}^{(-)}_{,\nu}(x)\gamma_\mu(1+\gamma_5)\psi^{(+)}(x)\Big]s^{ext}_{\mu\nu}(x). \tag{5.83}$$

*vi*) Finally, to study the scattering of gravitons in an external gravitational field, one starts with the self-coupling Lagrangian density (5.8). In this case, the terms must be grouped in such a way so as to be possible to separate the terms describing the real gravitons (the quantized field) from those describing the external (classical) scattering field. Such a separation is not completely univocal. The authors' solution is to symmetrically decompose both the free gravitons (for which $y = 0$) and the classical field (in our case, the gravitational field described by the Schwarzschild metric, for which the only non-null component is $y_{44}$), and to choose for gravitons the gauge

$$e^{(a)}_{4\alpha}(\vec{p}) = e^{(b)}_{4\beta}(\vec{p}\,') = 0.$$

The corresponding part of the Lagrangian density (5.8) then is

$$N\Big[\mathcal{L}^{(1)}_{g}(x)\Big] = -\frac{1}{2}\kappa\Big\{y^{(-)}_{\alpha\beta,\mu}(x)y^{(+)}_{\alpha\beta,\nu}(x)y^{ext}_{\mu\nu}(x) + \frac{1}{2}\Big[y^{(-)}_{\mu\nu}(x)y^{(+)}_{\mu\nu,\alpha}(x) + y^{(-)}_{\nu\beta}(x)y^{(+)}_{\alpha\beta,\nu}(x)\Big]y^{ext}_{,\alpha}(x)\Big\}, \tag{5.84}$$

the gravitational field being real.

To deduce Feynman-type rules for diagrams, necessary to study the diffusion processes in the Schwarzschild gravitational field, its metric shall be written in isotropic coordinates. Using the weak gravitational field approximation, one has [11, 275],

$$ds^2 \simeq \left(1 - 4\frac{C}{r}\right) dx_0^2 - \left(1 + 4\frac{C}{r}\right) \left(dx_1^2 + dx_2^2 + dx_3^2\right), \qquad (5.85)$$

where

$$C = \frac{\kappa^2 M_\odot}{32\pi}$$

is the gravitational radius of the source that generates the gravitational field, $M_\odot$ is the mass of the central body (the source of the gravitational field), and

$$r = (x_1^2 + x_2^2 + x_3^2)^{1/2}$$

is the distance determined to this centre.

It then follows that

$$g_{\mu\nu} \simeq \eta_{\mu\nu} - \frac{\kappa^2 M_\odot}{8\pi r}\delta_{\mu\nu}. \qquad (5.86)$$

Since, in the first-order approximation in terms of the powers of $\kappa$, the linearization (2.2) is equivalent to (1.292), and the gravitational field is static, its only non-zero component being

$$h_{00} = \frac{1}{2}y_{00}, \qquad (5.87)$$

according to (1.292), (5.86) and (5.87) one obtains

$$y_{00}^{ext}(\vec{x}) = -\frac{\kappa M_\odot}{4\pi|\vec{x}|}, \qquad (5.88\text{-}1)$$

$$y_{ij}^{ext}(\vec{x}) = 0, \quad r = |\vec{x}|. \qquad (5.88\text{-}2)$$

If the notations corresponding to the flat space are used,

$$x_\mu, \ (\mu = 0, 1, 2, 3) \ \rightarrow \ x_\mu, \ (\mu = 1, 2, 3, 4),$$

with

$$x_j, \ (j = 1, 2, 3) \text{ and } x_4 = ix_0 = it,$$

the above relations become

$$y_{\mu\nu}^{ext}(\vec{x}) = y^{ext}(\vec{x})\delta_{\mu 4}\delta_{\nu 4}, \ (\mu, \nu = 1, 2, 3, 4), \qquad (5.89\text{-}1)$$

$$y^{ext}(\vec{x}) = \frac{\kappa M_\odot}{4\pi|\vec{x}|}, \tag{5.89-2}$$

in agreement with [306].

In the matrix elements corresponding to the above-mentioned processes interferes the Fourier transform of the external gravitational field[19],

$$y^{ext}_{\mu\nu}(q) = \frac{1}{(2\pi)^{3/2}} \int e^{-iqx} y^{ext}_{\mu\nu}(x)\; d^4x, \tag{5.90}$$

with the following physical significance of the four-vector $q$: its spatial part $\vec{q}$ represents the momentum, and its temporal part $q_0$ stands for the energy, both being taken up by the external field. If the external gravitational field is static, being described by the Schwarzschild metric, then $y^{ext}_{\mu\nu}(q)$ has a single non-zero component and one can write

$$y^{ext}_{44}(\vec{x}) \equiv y^{ext}(\vec{x}) = \frac{1}{(2\pi)^{3/2}} \int e^{i\vec{q}\cdot\vec{x}} y_{44}(\vec{q})\; d^3\vec{q}, \tag{5.91}$$

that gives

$$y^{ext}_{44}(q) \equiv y^{ext}(q) = 2\pi\delta(q_0) y^{ext}(\vec{q}), \quad q_0 = 0, \tag{5.92}$$

with

$$y^{ext}(\vec{q}) = \frac{1}{(2\pi)^{3/2}} \int e^{-i\vec{q}\cdot\vec{x}} y^{ext}(\vec{x})\; d^3x = \frac{1}{(2\pi)^{3/2}} \frac{\kappa M_\odot}{|\vec{q}|^2}. \tag{5.93}$$

Here the Eq. (5.88-2) has been taken into account, and the integral (see Ref. [275]),

$$\int \frac{1}{|\vec{x}|} e^{-iqx}\; d^4x = 2(2\pi)^2 \frac{\delta(q_0)}{|\vec{q}|^2} \tag{5.94}$$

has been also considered. Besides, in the delta Dirac function the quantity $q_0$ must be replaced by the sum of the energies resulting from the conservation law[20]

$$q_0 = k'_0 - k_0 = 0.$$

---

[19] The factor $1/(2\pi)^{3/2}$ has been conveniently chosen, in agreement with the Fourier expansions of the quantized fields (see Refs. [40, 217]).

[20] In this case the momentum does not conserve: $\vec{k}\,' - \vec{k} = \vec{q}$.

The Feynman diagram illustrating the scattering process of various particles in external gravitational field is given in Fig. 5.1.

In order to establish the Feynman-type rules in the momentum space, besides the expressions for $y^{ext}_{\mu\nu}$ and $y^{ext}$ or their combinations, the Fourier transforms of the derivatives of the quantities corresponding to the gravitational field are also necessary; for instance,

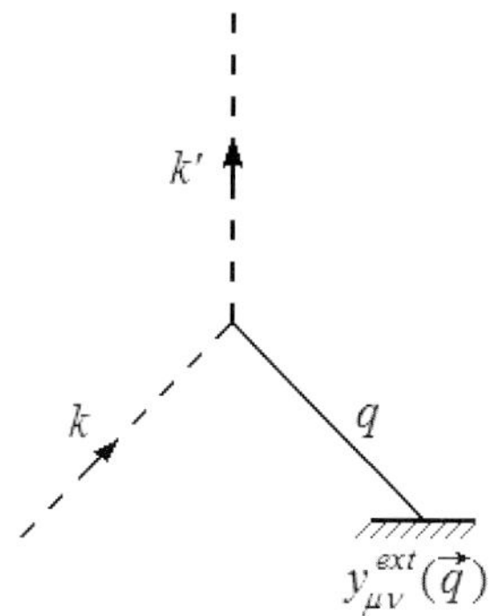

Fig. 5.1

$$y^{ext}_{\mu\nu,\alpha}(\vec{x}) = \frac{1}{(2\pi)^{3/2}} \int e^{i\vec{q}\cdot\vec{x}} y^{ext}_{\mu\nu,\alpha}(\vec{q}) d^3 q. \tag{5.95}$$

If the external gravitational field is static, one has

$$\begin{cases} y^{ext}_{\mu\nu,\alpha}(\vec{x}) = y^{ext}_{,\alpha}(\vec{x})\delta_{\mu 4}\delta_{\nu 4}, \\ y^{ext}_{,\alpha}(\vec{x}) = \delta_{\alpha j} y^{ext}_{,j}(\vec{x}), \end{cases} \tag{5.96}$$

while (5.91), (5.93), (5.95), and (5.96) yield

$$y^{ext}_{,j}(\vec{q}) = i q_j y^{ext}(\vec{q}). \tag{5.97}$$

By using the expressions of the interaction Lagrangian densities and the Fourier transforms established above, one obtains different kinds of external gravitational fields and vertices, given in the Tables 2 and 3 of Appendix F. The vertices are obtained by integrating over exponentials and taking into account that[21] $\delta(k' - k - q) = \delta(\vec{k}' - \vec{k} - \vec{q})\delta(k'_0 - k_0)$.

## V.4.2. External electromagnetic field

Consider now the process of conversion of photons into gravitons (or conversely) in an external electromagnetic field. In this case[22],

[21] Even if the Fourier expansions corresponding to the external gravitational field contain the exponential $\exp(i\vec{q}\cdot\vec{x})$, unlike the matrix elements of the free particles, which contain the exponential $\exp(ikx)$, when integrating over exponentials – also taking into account (5.92) – one obtains the same multiplicative factor $(2\pi)^4$ for all types of vertices.

[22] In the case of inverse process, in Eq. (5.98) the parts of positive frequency must be replaced with those of negative frequency and vice versa, also changing the order of the corresponding operators.

since for the real gravitons one has $y = 0$, the interaction Lagrangian density (5.32$a$) – brought to the normal form – becomes

$$N\Big[\mathcal{L}^{(1)}_{em-g}(x)\Big] = -\frac{1}{2}\kappa F^{ext}_{\mu\nu}(x) y^{(-)}_{\nu\alpha}(x) F^{(+)}_{\mu\alpha}(x). \qquad (5.98)$$

If the electromagnetic field is static (*e.g.* the nucleus Coulombian field, the electrostatic field, or that of the electric dipole), the only non-zero component of the four-potential $A_\mu$ of the external electromagnetic field is $A_4$, so that, considering for photons and gravitons the gauge relations $e^{(a)}_4(\vec{k}) = 0$ and $e^{(b)}_{4\alpha}(\vec{p}) = 0$, respectively, the suitable part of the equivalent Lagrangian density (5.36) – brought to a symmetric form – is given by

$$\begin{aligned} N\Big[\mathcal{L}'^{(1)}_{em-g}(x)\Big] = \frac{1}{4}\kappa \Big[ & A^{ext}_\mu(x) y^{(-)}_{\nu\alpha,\mu}(x) A^{(+)}_{\nu,\alpha}(x) \\ & - A^{ext}_{\nu,\alpha}(x) y^{(-)}_{\mu\alpha,\nu}(x) A^{(+)}_\mu(x)\Big], \end{aligned} \qquad (5.98')$$

and thus, the external electromagnetic field can be chosen in two different but equivalent ways.

In the matrix elements corresponding to the studied process, for various concrete configurations of the external electromagnetic field, there always appear their Fourier transforms; for instance,

$$A^{ext}_\mu(q) = \frac{1}{(2\pi)^{3/2}} \int e^{-iqx} A^{ext}_\mu(x)\, d^4x, \qquad (5.99)$$

or/and those of the form $F^{ext}_{\mu\nu}(q)$ and $A^{ext}_{\nu,\alpha}(q)$. The field being static, one also has [40, 217],

$$A^{ext}_\mu(\vec{x}) = \frac{1}{(2\pi)^{3/2}} \int e^{i\vec{q}\cdot\vec{x}} A^{ext}_\mu(\vec{q})\, d^3q, \qquad (5.100)$$

and

$$A^{ext}_\mu(q) = 2\pi\delta(q_0) A^{ext}_\mu(\vec{q}), \quad q_0 = 0, \qquad (5.101)$$

or/and similar series expansions for $F^{ext}_{\mu\nu}(\vec{x})$ and $A^{ext}_{\nu,\alpha}(\vec{x})$. As before, in the Dirac delta function $q_0$ must be replaced by the energy sum resulting from the conservation law

$$q_0 = p_0 - k_0 = 0.$$

Besides, there is no momentum conservation:

$$\vec{p} - \vec{k} = \vec{q}.$$

The Feynman diagram illustrating the process of transformation of a photon into a graviton, in an external electromagnetic field, is given in Fig. 5.2.

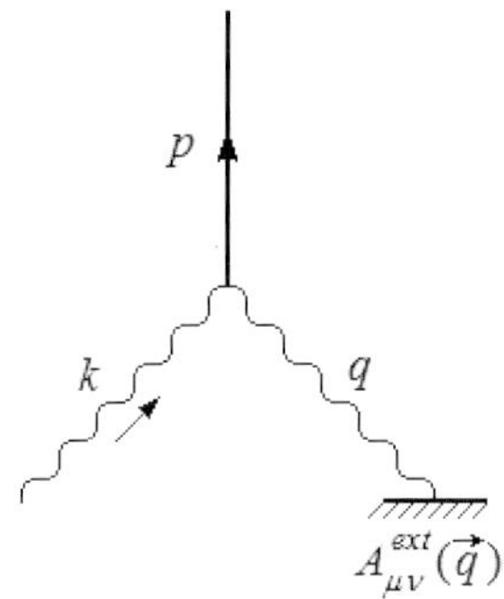


Fig. 5.2

*i*) In the case of the spherically symmetric Coulombian field $\vec{E}(\vec{x})$, of a nucleus of charge $Ze$, only one component of $A_\mu$ is different from zero and one can write (see Refs. [225, 387]),

$$A_\mu^{ext}(\vec{x}) = i\delta_{\mu 4} V(\vec{x}), \tag{5.102}$$

where $V(\vec{x})$ is the scalar potential of the external electromagnetic field. The electromagnetic field tensor is given by

$$F_{\mu\nu}^{ext}(\vec{x}) = i\left(\delta_{\mu 4}\delta_{\nu j} - \delta_{\mu j}\delta_{\nu 4}\right) E_j(\vec{x}), \tag{5.102'}$$

and the derivative of the electromagnetic field four-potential writes

$$A_{\nu,\alpha}^{ext}(\vec{x}) = -i\delta_{\nu 4}\delta_{\alpha j} E_j(\vec{x}), \tag{5.102''}$$

with

$$\begin{cases} V(\vec{x}) = \dfrac{Ze}{4\pi|\vec{x}|}, \\ E_j(\vec{x}) = -\dfrac{\partial V(\vec{x})}{\partial x_j} = \dfrac{Ze}{4\pi|\vec{x}|^3} x_j. \end{cases} \tag{5.103}$$

Since, as previously mentioned, the expansion of $1/|\vec{x}|$ is of the form

$$\frac{1}{|\vec{x}|} = \frac{2}{(2\pi)^2}\int e^{i\vec{q}\cdot\vec{x}} \frac{1}{|\vec{q}|^2}\, d^3q, \tag{5.104}$$

it follows that

$$\frac{x_j}{|\vec{x}|^3} = -\frac{\partial}{\partial x_j}\left(\frac{1}{|\vec{x}|}\right) = -\frac{2i}{(2\pi)^2}\int e^{i\vec{q}\cdot\vec{x}} \frac{q_j}{|\vec{q}|^2}\, d^3q, \tag{5.105}$$

in agreement with [275]. In view of Eqs. (5.99)–(5.102″), one also obtains

$$V(\vec{q}) = \frac{1}{(2\pi)^{3/2}} \frac{Ze}{|\vec{q}|^2}, \tag{5.106}$$

as well as

$$E_j(\vec{q}) = -iq_j V(\vec{q}). \tag{5.107}$$

*ii*) The Eqs. (5.102)–(5.192″) remain also valid for an homogeneous electrostatic field (*e.g.* that of a capacitor), the Fourier transform of $\vec{E}(\vec{x})$ having the same form as that given by Eq. (5.107). In the matrix element corresponding to the process, everything has to be expressed in terms of $E_j(\vec{q})$, its concrete form varying from one case to another.

*iii*) In the case of homogeneous magnetostatic field $\vec{H}(\vec{x})$, one obtains (see Ref. [307]),

$$F^{ext}_{\mu\nu}(\vec{x}) = \delta_{\mu j}\delta_{\nu l}\varepsilon_{jlm}H_m(\vec{x}), \tag{5.108}$$

where $\varepsilon_{jlm}$ is the Levi-Civita symbol. The Fourier transform $H_m(\vec{q})$ depends on the concrete analyzed process.

*iv*) The potential and the field intensity of an electric dipole of moment $\vec{d}$ are

$$\begin{cases} V(\vec{x}) = \dfrac{d_i x_i}{4\pi|\vec{x}|^3}, \\[2ex] E_j(\vec{x}) = -\dfrac{\partial V(\vec{x})}{\partial x_j}. \end{cases} \tag{5.109}$$

Here are also valid the Eqs. (5.102)–(5.102″), as well as the relation (5.107) between Fourier transforms of the field and potential. In this case, taking into account the Eq. (5.109) and the expansions (5.100) and (5.105), it follows that

$$V(\vec{q}) = \frac{1}{(2\pi)^{3/2}} \frac{d_j q_j}{|\vec{q}|^2}. \tag{5.110}$$

*v*) Finally, in the case of magnetic dipole of moment $\vec{M}$, the relationship (5.108) is also valid, while the corresponding potential and field intensity are given by[23]

$$\begin{cases} A_i(\vec{x}) = \varepsilon_{ijk} \dfrac{M_j x_k}{4\pi |\vec{x}|^3}, \\ H_j(\vec{x}) = \varepsilon_{jni} \dfrac{\partial}{\partial x_n} A_i(\vec{x}). \end{cases} \tag{5.111}$$

Taking into account the obvious equality

$$\varepsilon_{ijk}\varepsilon_{iln} = \delta_{jl}\delta_{kn} - \delta_{jn}\delta_{kl}, \tag{5.112}$$

from the Eq. (5.111) and expansions given by Eqs. (5.100) and (5.195), one obtains

$$H_l(\vec{q}) = \frac{4\pi}{(2\pi)^{3/2}} \left( M_l - q_l \frac{M_n q_n}{|\vec{q}|^2} \right), \tag{5.113}$$

or

$$\begin{cases} \vec{H}(\vec{q}) = i\vec{q} \times \vec{A}(\vec{q}), \\ \vec{A}(\vec{q}) = -\dfrac{i}{(2\pi)^{3/2}} \dfrac{\vec{M} \times \vec{q}}{|\vec{q}|^2}, \end{cases} \tag{5.114}$$

in agreement with [301].

Using the interaction Lagrangian densities (5.98) and/or (5.98′) and the above mentioned Fourier transforms, one obtains various types of external electromagnetic fields and vertices, which are presented in the Tables 2 and 3 of Appendix F. The vertices are obtained integrating over exponentials and using the relation

$$\delta(p - k - q) = \delta(\vec{p} - \vec{k} - \vec{q})\delta(p_0 - k_0).$$

We conclude this section by mentioning that in the case of diffusion of spinorial particles in an external gravitational field, in the expression of the corresponding matrix element, the spinorial factors and the matrices $\gamma_\mu$ and $\gamma_5$ are written in accordance with their order of appearance in the diagram, following the fermionic line in the opposite sense to the arrow. In the rest of cases, the bosonic factors

---

[23]Unlike the considerations presented in Ref. [307], where the Fourier transform of the dipole magnetic field $\vec{H}(\vec{x})$ is obtained in a laborious way, in the authors' approach one starts from the vector potential $\vec{A}(\vec{x})$, as a simpler and more natural procedure.

can be written anywhere. Some illustrations of these rules can be found in the next chapters.

## V.5. Interactions between particles in the presence of a gravitational field

The combined interaction of the spinorial (or scalar charged) field with the electromagnetic and gravitational fields is analyzed in Refs. [27, 107, 388(1963), 391]. On the other hand, after 1975 were formulated new hypotheses about the so-called *strong gravitation* [98, 102, 320] and later started to be exploited the formal analogy between the weak interactions and those of torsion (Cartan forces [97]). At the same time, the supergravitation theory [113, 114, 151, 294] started its vertiginous development.

In this subchapter, using the minimal coupling principle and the $S$-matrix formalism, the authors expose their hypothesis regarding the influence of the gravitational field on the electromagnetic, weak and strong interactions, as well as the way of its possible interaction – independently of supergravitation and supersymmetry reasons – with the Rarita–Schwinger field. The utilized method is consequent, consistent and has a general character. This way are obtained appropriate Feynman-type rules for diagrams in the momentum space, the results being presented in Chap. III (since the Rarita–Schwinger field is considered a high-spin field).

The next subchapters are dedicated to a short presentation of the main problems connected to the above mentioned interactions [359, 360(p.277), 378] and the most relevant conclusions shall be discussed.

### V.5.1. Electromagnetic interactions

The interaction Lagrangian density for the interaction process between the electromagnetic field and a charged matter field can be easily obtained by means of minimal coupling principle. As was shown in Chap. IV, this principle is based on the minimal substitution, which in this case writes

$$\partial_\mu \to \nabla_\mu = \partial_\mu - ieA_\mu$$

and must be performed in the Lagrangian density of the free matter field. Here $e$ is the electric charge of an electron, also playing the role of the electromagnetic coupling constant (see Chap. IV). In this case, the corresponding Lagrangian density

$$\mathcal{L} = ej^{\mu}A_{\mu} \tag{5.115}$$

contains the conservative electromagnetic current, composed by two terms, the leptonic and hadronic currents (see Ref. [240] for details),

$$j^{\mu} = j^{\mu}_{lept} + j^{\mu}_{hadr}. \tag{5.116}$$

If the interaction is considered as being punctiform and neglecting the renormalization effects, the leptonic electromagnetic current is given by [288],

$$j^{\mu}_{lept} = \overline{\psi}_k \gamma^{\mu} \psi_k, \tag{5.117}$$

where $k = e$, $\mu$ and $\tau$ corresponds to the electrons, muons, tauons, and their antiparticles; the massless spinor particles and their antiparticles do not interfere, since the corresponding current is axial.

As it is known, to the expression of hadronic electromagnetic current contribute the nucleons, the hyperons and the pseudoscalar mesons [252, 288]. Starting with the isospin group $SU(2)$ and extending the concept of electromagnetic current over the vectorial mesons, one can obtain[24] an interaction Lagrangian density between these fields, in agreement with that presented in some other specific papers (see, *e.g.* [356]).

Neglecting the strong interactions, the contribution of the pseudoscalar mesons $\pi$ and vectorial mesons $\rho$ to the generalized hadronic electromagnetic current obtained by the authors is

$$j^{\mu}_{hadr} = \left(\frac{\partial \vec{\varphi}}{\partial x_{\mu}} \times \vec{\varphi}\right)_3 + \left(\vec{G}^{\mu\nu} \times \vec{B}_{\nu}\right)_3 + f\left(\vec{B}^{\mu} \times \vec{B}^{\nu}\right)_3 \frac{\partial}{\partial x^{\nu}}, \tag{5.118}$$

where $\vec{\varphi}$ and $\vec{G}^{\mu\nu}$ are given by (5.1) and (5.43), respectively, $f$ being a coupling constant. The expression corresponding to the contribution of nucleons, hyperons, pseudoscalar mesons $K$ and vectorial mesons $\tilde{K}_{\mu}$ are given in Ref. [378].

[24] See, in this respect, the article [378].

Next, starting with (5.115)–(5.118)[25], the main types of electromagnetic interactions in the presence of a weak gravitational field shall be analyzed.

*i*) Consider the interaction between the electromagnetic and electron-positron fields. Using the *vierbein*[26] formalism as well as the linearization (2.2), going to the flat space and expanding in power series in powers of $\kappa$, one obtains

$$\mathcal{L}_{int} = \mathcal{L}^{(1)}_{em-sp} + \mathcal{L}^{(2)}_{em-sp-g} + \cdots , \tag{5.119}$$

with

$$\begin{cases} \mathcal{L}^{(1)}_{em-sp} = ie\overline{\psi}\gamma_\mu\psi A_\mu, \\ \mathcal{L}^{(2)}_{em-sp-g} = \dfrac{1}{2}ie\kappa A_\mu\overline{\psi}\gamma_\nu\psi s_{\mu\nu}. \end{cases} \tag{5.120}$$

The operator $S$ "leads" in this case to the following diagrams:

$$S_2 = i\int d^4x\ N\Big[\mathcal{L}^{(2)}_{em-sp-g}(x)\Big]\cdots$$

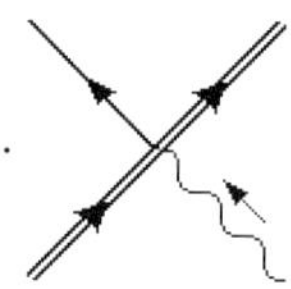

$$+\ \cdots$$

$$+\frac{i^2}{2!}\iint d^4x\,d^4x'\ T\Big\{N\Big[\mathcal{L}^{(1)}_{em-sp}(x)\mathcal{L}^{(1)}_{em-g}(x')\Big]\cdots$$

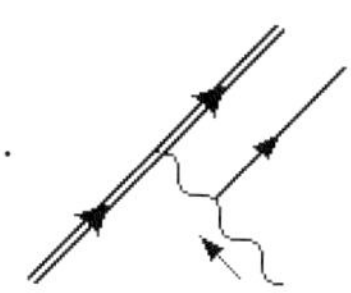

$$+\ \cdots + N\Big[\mathcal{L}^{(1)}_{em-sp}(x)\mathcal{L}^{(1)}_{sp-g}(x')\Big]\ \cdots$$

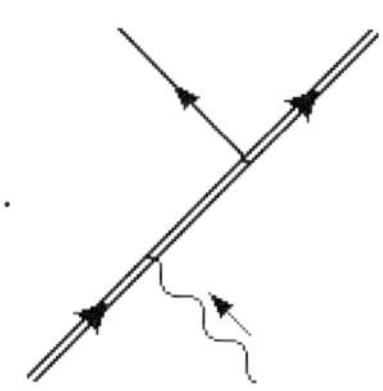

$$+\ \cdots$$

$$+N\Big[\mathcal{L}^{(1)}_{em-g}(x)\mathcal{L}^{(1)}_{sp-g}(x')\Big]\ \cdots$$

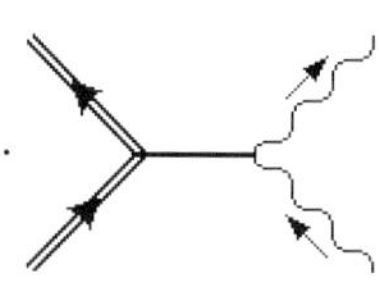

$$+\ \cdots\Big\} \tag{5.121}$$

[25]One considers only the second-order processes, in terms of $e\kappa$ and $\kappa^2$.

[26]This formalism implies the substitution of Dirac matrices $\gamma^\mu$ by the generalized Dirac matrices $\tilde{\gamma}^\mu$, and replacement of the Lagrangian by its density, corresponding to the curved space.

with $\mathcal{L}^{(1)}_{em-g}$ and $\mathcal{L}^{(1)}_{sp-g}$ given by (5.32) and (5.68), respectively, together with the diagrams obtained by changing the chronological order and the interaction channel (with the corresponding orientation of the time axis), and by replacing the particles by their antiparticles (or conversely).

In order to calculate a diffusion/scattering process, one considers the sum of contributing diagrams. For example, the Compton effect of a photon scattered by an electron with emission of a graviton is described by the first three diagrams of (5.121). The last diagram describes a special process (here, the electron-photon elastic scattering).

*ii*) Consider now the interaction between the electromagnetic and charged scalar ($\pi^{\pm}$ mesons) fields. By virtue of the "usual"[27] formalism, one obtains

$$\mathcal{L}_{int} = \mathcal{L}^{(1)}_{em-sc} + \mathcal{L}^{(2)}_{em-sc-g} + \cdots , \tag{5.122}$$

where

$$\mathcal{L}^{(1)}_{em-sc} = ie\Big(\varphi^*_{,\mu}\varphi - \varphi^*\varphi_{,\mu}\Big)A_\mu, \tag{5.123a}$$

and

$$\mathcal{L}^{(2)}_{em-sc-g} = ie\kappa A_\mu\Big(\varphi^*_{,\nu}\varphi - \varphi^*\varphi_{,\nu}\Big)y_{\mu\nu}. \tag{5.123b}$$

The $S$-operator and the corresponding diagrams are similar to those displayed at the point *i*), the spinorial field being replaced by the scalar one. The first three types of diagrams describe effects involving non-conservation of the spin (and its projection along the direction of motion, for the massless particles, respectively) and they can be considered only in the presence of the external gravitational field. The only possible free processes are given by the last type of diagrams (for example, the meson-photon elastic scattering, by the exchange of a virtual graviton).

*iii*) Consider, finally, the interaction between electromagnetic and charged vectorial ($\rho^{\pm}$ mesons) fields. In this case, by means of the

[27] Here, by "usual" one should understand the method of obtaining the interaction Lagrangian densities used in subchapters 1 and 2.

"usual" formalism, one can write

$$\mathcal{L}_{int} = \mathcal{L}^{(1)}_{em-v} + \mathcal{L}^{(2)}_{em-v-g} + \cdots , \tag{5.124}$$

with

$$\begin{cases} \mathcal{L}^{(1)}_{em-v} = ieA_\mu\Big(G^*_{\mu\nu}B_\nu - B^*_\nu G_{\mu\nu}\Big) - iefF_{\mu\nu}B^*_\mu B_\nu, \\ \mathcal{L}^{(2)}_{em-v-g} = -ie\kappa\Big[A_\mu\Big(B^*_\nu G_{\lambda\rho} - G^*_{\lambda\rho}B_\nu\Big) + fF_{\mu\nu}B^*_\lambda B_\rho\Big] \\ \qquad\qquad \times \Big(\delta_{\mu\lambda}u_{\nu\rho} + \delta_{\nu\rho}u_{\mu\lambda}\Big). \end{cases} \tag{5.125}$$

The corresponding diagrams are also analogous to those displayed at the point $i$), this time the spinorial field being replaced by the mesonic vectorial field. The first three types of diagrams describe effects that could be possible only in the presence of external electromagnetic or gravitational fields, among the free processes being real (possible) only those of the last type.

It therefore follows, in the author's opinion, that the electromagnetic interactions are influenced by the gravitational field, most of the mentioned processes taking place effectively, and those excluded due to the spin/helicity non-conservation becoming utmost probable in the presence of external electromagnetic or gravitational fields.

### V.5.2. Strong interactions

As it is known [240, 252], the unitary symmetry group $SU(3)$, that is specific to strong interactions, leads to the unitary spin octet[28] $J^\mu_i$, the only conserved quantities being the isospin current and the hypercharge. The third and eighth components of $J^\mu_i$ appear in the electromagnetic interactions.

As shown in Ref. [360(p.277)], using the ideas of Mani and Nearing (exposed in Ref. [255]), the unitary spin vectorial current, obtained by a gauge transformation[29],

$$J^\mu_i = f_{ijk}\left(\overline{B}^j\gamma^\mu B^k - P^j\frac{\partial}{\partial x_\mu}P^k\right), \tag{5.126}$$

[28] Here $i$, $j$, $k$ take the values 1,2,...,8.

[29] Only the currents corresponding to the barions and pseudoscalar mesons are considered.

is supplemented with a term that does not come from a gauge transformation[30],

$$J_i'^{\mu} = g_1 \varepsilon^{\mu\nu\lambda\rho} d_{ijk} \frac{\partial}{\partial x^\nu} V_\lambda^j \frac{\partial}{\partial x^\rho} P^k, \qquad (5.126')$$

this way being obtained a *generalized current*. Here $f_{ijk}$ are the $SU(3)$-group structure constants, $d_{ijk}$ are constants related to $f_{ijk}$ (see Ref. [240] for details), $\varepsilon^{\mu\nu\lambda\rho}$ is the four-dimensional Levi-Civita symbol, while $B^i$, $P^i$ and $V_\mu^i$ are vectorial operators in the eight-dimensional space, corresponding to the following field octets: barionic, mesonic pseudoscalar and vectorial, respectively, their form being given in [255, 288].

On the other hand, the Yukawa coupling between the barions and pseudoscalar mesons is described by the Lagrangian density [288]:

$$\mathcal{L} = \Big( i g_D d_{ijk} - g_F f_{ijk} \Big) \overline{B}^i \gamma_5 B^j P^k. \qquad (5.127)$$

Next, by means of the above definitions, the main types of strong coupling in the presence of a weak gravitational field shall be analyzed.

*i*) Consider the interaction between barions and vectorial mesons. The corresponding Lagrangian density is obtained by coupling the octet $V_\mu^i$ to the barionic current – corresponding to the first term of (5.126). As a concrete example, one shall analyze the interaction between the neutral mesons $\rho^0$ (described by the operator $B_{0\mu}$) and the nucleons (described by the operators $\overline{\psi}$ and $\psi$). Using the *vierbein* formalism and the linearization (2.2), then going to the flat space and expanding in power series in powers of $\kappa$, one has

$$\mathcal{L}_{int} = \mathcal{L}^{(1)}_{v-sp} + \mathcal{L}^{(2)}_{v-sp-g} + ..., \qquad (5.128)$$

with

$$\begin{cases} \mathcal{L}^{(1)}_{v-sp} = i g_2 B_{0\mu} \overline{\psi} \gamma_\mu \psi, \\ \mathcal{L}^{(2)}_{v-sp-g} = \dfrac{1}{2} i g_2 \kappa B_{0\mu} \overline{\psi} \gamma_\nu \psi s_{\mu\nu}. \end{cases} \qquad (5.129)$$

---

[30]Here and hereafter $g_1, ...$ are coupling constants.

The $S$-operator and the corresponding Feynman diagrams are similar to those given by (5.121), the electromagnetic field being replaced by the vectorial mesonic field.

*ii*) Consider now the interaction between barions and the pseudoscalar mesons, described by the Lagrangian density (5.127). As a concrete example, here is the pseudoscalar-type coupling of neutral mesons $\pi^0$ (described by the operator $\varphi_0$) and the nucleons (described by the operators $\overline{\psi}$ and $\psi$). By means of the "usual" formalism, for the corresponding interaction Lagrangian density one obtains

$$\mathcal{L}_{int} = \mathcal{L}^{(1)}_{sc-sp} + \mathcal{L}^{(2)}_{sc-sp-g} + \cdots \tag{5.130}$$

where

$$\begin{cases} \mathcal{L}^{(1)}_{sc-sp} = ig_3\overline{\psi}\gamma_5\psi\varphi_0, \\ \mathcal{L}^{(2)}_{sc-sp-g} = \dfrac{1}{2}ig_3\kappa\overline{\psi}\gamma_5\psi\varphi_0\, y. \end{cases} \tag{5.131}$$

The $S$-operator and the corresponding Feynman diagrams are analogous to those of (5.121), the electromagnetic field being replaced by the scalar field. The first three types of diagrams describe effects which are possible only in the presence of the external gravitational field[31].

Of all free processes, only those of the last type – for example, the nucleon-meson elastic scattering, through the exchange of a virtual graviton – are real/possible.

*iii*) Consider, finally, the interaction between the vectorial and pseudoscalar mesons. The appropriate Lagrangian density is obtained by coupling the octet $V^i_\mu$ with the pseudoscalar mesons' current – corresponding to the second term of (5.126). It is worth analyzing, as a concrete example, the interaction between the neutral mesons $\rho^0$ and $\pi^\pm$ mesons (described by the operators $\varphi$ and $\varphi^*$ of the associated complex scalar field). Proceeding as above, one obtains

$$\mathcal{L}_{int} = \mathcal{L}^{(1)}_{v-sc} + \mathcal{L}^{(2)}_{v-sc-g} + \cdots, \tag{5.132}$$

[31]The first type of diagrams is also "forbidden" due to the non-existence of the free gravitons, $y(x)$.

with

$$\begin{cases} \mathcal{L}^{(1)}_{v-sc} = ig_4 B_{0\mu}\Big(\varphi^*_{,\mu}\varphi - \varphi^*\varphi_{,\mu}\Big), \\ \mathcal{L}^{(2)}_{v-sc-g} = ig_4\kappa B_{0\mu}\Big(\varphi^*_{,\nu}\varphi - \varphi^*\varphi_{,\nu}\Big)y_{\mu\nu}. \end{cases} \tag{5.133}$$

The $S$-operator, together with the corresponding Feynman diagrams, are similar to those given by (5.121), the places of the electromagnetic and spinorial fields being taken by the mesonic vectorial and scalar fields, respectively. Here, again, the first three types of diagrams are considered only in the presence of an external gravitational field.

An interesting case is obtained by coupling $V^i_\mu$ to the conservative phenomenological current given by (5.126′). Consider, for example, the interaction between the charged vectorial mesons $\tilde{K}^{\pm}_{\mu}$ (described by the operators $B_\mu$, $B^*_\mu$) and the neutral pseudoscalar mesons $\pi^0$. Proceeding as before, one gets an expansion identical with that of (5.132), where

$$\begin{cases} \mathcal{L}^{(1)}_{v-sc} = -ig_5\varepsilon_{\mu\nu\lambda\rho}\Big(B^*_{\mu,\nu}B_\lambda - B^*_\mu B_{\lambda,\nu}\Big)\varphi_{0,\rho}, \\ \mathcal{L}^{(2)}_{v-sc-g} = -ig_5\kappa\varepsilon_{\alpha\nu\lambda\rho}y_{\alpha\mu}\Big(B^*_{\mu,\nu}B_\lambda - B^*_\mu B_{\lambda,\nu}\Big)\varphi_{0,\rho}. \end{cases} \tag{5.134}$$

The $S$-operator, together with the corresponding Feynman diagrams, are similar to those given by (5.121), the places of the electromagnetic and spinorial fields being taken, this time, by the scalar and mesonic vectorial fields, respectively. All four types of diagrams describe real/possible effects, as a result of spin conservation.

As already mentioned, according to the Salam's hypothesis (see Ref. [320]) regarding the so-called *strong gravitation*, one considers that only the leptons are able to directly interact with the gravitational field, while the hadrons interact indirectly, through the medium of the "mixture" $f - g$. More precisely, the hadrons interact with the tensorial mesonic field $f_{\mu\nu}$ (for example, the field corresponding to $f^0$ neutral mesons), the quanta of the field $f$ transform into $g$-quanta, corresponding to the gravitational field tensor $g_{\mu\nu}$, and this one interact with the leptons.

The authors consider, nevertheless, that in the first approximation the interactions hadron-lepton, or hadron-hadron, can also take place through the exchange of a virtual graviton, the processes excluded due to the spin non-conservation becoming considerably more probable in the presence of an external gravitational field.

### V.5.3. Weak interactions

According to the Fermi's hypothesis regarding the weak interactions between elementary particles, they can be phenomenologically described by means of a Lagrangian density of the type current × current (see Refs. [240, 252, 288]),

$$\mathcal{L} = \frac{G_w}{\sqrt{2}} O^*_\alpha O^\alpha, \tag{5.135}$$

where $G_w$ is a universal coupling constant[32], while the current $O^\alpha$ is composed by two parts – the leptonic $l^\alpha$ and weak hadronic $h^\alpha$ currents,

$$O^\alpha = l^\alpha + h^\alpha. \tag{5.136}$$

According to the Feynman–Gell-Mann [138, 163] and Sudarshan–Marshak [346] theories, the leptonic current corresponding to a coupling of the form $V - A$ (only the vectorial and axial-vectorial couplings are dominant) can be written as

$$l^\alpha = \left(\overline{\psi}_e \psi_{\nu_e}\right)^\alpha + \left(\overline{\psi}_\mu \psi_{\nu_\mu}\right)^\alpha, \tag{5.137}$$

where $\psi_i$, with $i = e, \mu, \nu_e, \nu_\mu$, are the wave functions of the electron, muon, electron neutrino and muon neutrino (considered as being massless particles), respectively.

Extension of the unitary symmetry $SU(3)$ over the weak interactions can be expressed as Cabibbo's hypothesis [55], according to which the weak hadronic current separates in a part $s^\alpha$, that conserves the strangeness, and a part $g^\alpha$, that doesn't, each term being decomposed into a vectorial part and an axial-vectorial part,

$$h^\alpha = s^\alpha + g^\alpha, \tag{5.138}$$

[32]Assumption that the weak interactions have a universal character consists in the supposition that the form of the currents $O^\alpha$ can be the same for all the particles, the strong interactions being suppressed (see Ref. [240] for details).

with

$$\begin{cases} s^\alpha = \cos\theta\Big(J_1^\alpha - iJ_2^\alpha + K_1^\alpha - iK_2^\alpha\Big), \\ g^\alpha = \sin\theta\Big(J_4^\alpha - iJ_5^\alpha + K_4^\alpha - iK_5^\alpha\Big). \end{cases} \tag{5.139}$$

Here $\theta = 0.26$ is the *Cabibbo's angle*, while $J_i^\alpha$ and $K_i^\alpha$ are the unitary vectorial currents – given by (5.126) – and axial-vectorial currents (both being members of the corresponding $SU(3)$ octets), respectively, the last one being given by (see Ref. [255]),

$$K_i^\alpha = f_{ijk}\overline{B}^j\gamma^\alpha\gamma^5 B^k + \lambda\frac{\partial}{\partial x_\alpha}P_i, \tag{5.140}$$

where $\lambda$ is a coupling constant.

In the paper [359] are studied the main types of weak disintegration (leptonic, semileptonic – with conservation and change of the strangeness – and non-leptonic) in the presence of a weak gravitational field. It is also considered the possibility that these interactions be mediated by gravitons and virtual vectorial bosons (see Ref. [245]), or even by virtual gravitons only. Since the last two cases are mostly improbable, hereinafter the first variant shall be briefly discussed.

In order to study the above-mentioned types of disintegration in the presence of gravity, the suitable Lagrangian densities must be considered in the curved space. Supposing that the gravitational field is weak, the linearization (2.2) shall be used. Besides, for the spinorial field we shall use the *vierbein* formalism, and the scalar field shall be described in the framework of the usual formalism. Expanding in power series in powers of $\kappa$, the Lagrangian densities describing the weak interactions write

$$\mathcal{L}_j = \mathcal{L}^0\Big(G_j\Big) + \mathcal{L}^{(1)}\Big(\kappa G_j\Big) + \mathcal{L}^{(2)}\Big(\kappa^2 G_j\Big) + \cdots, \tag{5.141}$$

where $G_j$ (with $j = \mu, n, K, \Lambda, ...$) are, respectively, the coupling constants corresponding to the disintegration of the muon (*e.g.* $\mu^- \to e^- + \overline{\nu}_e + \nu_\mu$), neutron ($n \to p + e^- + \overline{\nu}_e$), kaon (*e.g.* $K^- \to e^- + \overline{\nu}_e$), $\Lambda$-hyperon (*e.g.* $\Lambda^0 \to p + \pi^-$), etc.

To fix ideas one shall consider the purely leptonic disintegration of the muon. In this case, going to the flat space, the Lagrangian density writes

$$\begin{aligned}\mathcal{L} &= G_{\mu^-}\left(\overline{\psi}_{\nu_\mu}\psi_{\mu^-}\right)_\alpha\left(\overline{\psi}_{e^-}\psi_{\nu_e}\right)_\alpha \\ &+ \frac{1}{2}\kappa G_{\mu^-}\left(\overline{\psi}_{\nu_\mu}\psi_{\mu^-}\right)_\alpha\left(\overline{\psi}_{e^-}\psi_{\nu_e}\right)_\alpha y \\ &- \frac{1}{4}\kappa^2 G_{\mu^-}\left(\overline{\psi}_{\nu_\mu}\psi_{\mu^-}\right)_\alpha\left(\overline{\psi}_{e^-}\psi_{\nu_e}\right)_\alpha y_{\lambda\rho}h_{\lambda\rho} + \cdots\end{aligned} \tag{5.142}$$

In the zeroth-order approximation with respect to gravitational interaction, the $S$-operator and the corresponding Feynman diagram are of the form

$$S_1 = i\int d^4x\ N\left[\mathcal{L}^{(0)}(x)\right]$$

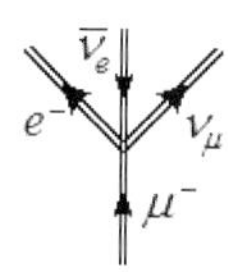


, (5.143)

this way being obtained the result known in the weak interactions theory.

To the first-order approximation in terms of the powers of $\kappa$ corresponds a set of more complex diagrams, which can be obtained from the previous one by attaching an emergent supplementary gravitonic line[33] to each fermionic line or in the vertex, which would correspond to a process of internal gravitational bremsstrahlung. But these diagrams describe free processes characterized by the spin/helicity non-conservation, which processes, in our opinion, do not have to be taken into consideration[34].

[33] The attachment of an incident gravitonic line has – in the author' opinion – a concrete physical significance only in the external field approximation.

[34] The disintegration in this order of approximation is also forbidden by the fact that the "y"-type free gravitons do not exist.

In the second-order approximation with respect to gravitational interaction, the $S$-operator writes

$$S_3 = \int d^4x \; N\Big[\mathcal{L}^{(2)}(x)\Big] + \frac{(i)^2}{2!} \iint d^4x\, d^4x' \; T\Big\{N\Big[\mathcal{L}^{(1)}(x)\mathcal{L}^{(1)}_{sp-g}(x')\Big] + \Big[\mathcal{L}^{(0)}(x)\mathcal{L}^{(2)}_{sp-g}(x')\Big] + \cdots\Big\} + \frac{(i)^3}{3!} \iiint d^4x\, d^4x'\, d^4x'' \times T\Big\{N\Big[\mathcal{L}^{(0)}(x)\mathcal{L}^{(1)}_{sp-g}(x')\mathcal{L}^{(1)}_{sp-g}(x'')\Big] + \cdots\Big\}, \tag{5.144}$$

where $\mathcal{L}^{(1)}_{sp-g}$ and $\mathcal{L}^{(2)}_{sp-g}$ are given by (5.68) and (5.75), respectively, while the dots signify the normal products obtained from the internal ones by the mutual permutation of the Lagrangian densities. In this case, the corresponding diagrams are very complicated, describing weak disintegration processes of the muon, with simultaneous emission of two gravitons, supplemented with gravitational radiative corrections. Here are some graphs containing radiative corrections in the process of muon disintegration:

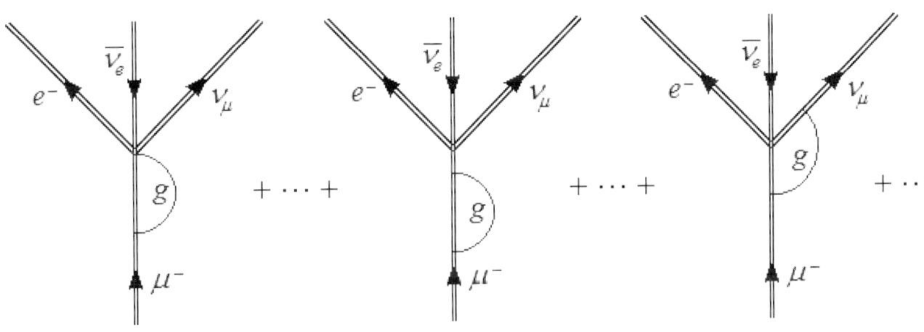


Fig. 5.3

In Fig. 5.3 the dots signify analogous diagrams, in which the gravitonic loop is attached to the other fermionic lines.

To conclude, in the authors' opinion the weak interactions are not influenced by the gravitational field in the first-order approximation (in terms of the powers of coupling constant $\kappa$) of the perturbation theory, due to the spin (helicity) non-conservation. There can be "real" (*i.e.* possible) only those processes of weak disintegration of the second-order (including the gravitational radiative corrections), but their occurrence probability is very low – at least as compared to the weak disintegration with photon emission, or those concerning the electromagnetic radiative corrections. This is so, because the gravitational coupling constant is much smaller than the coupling constants of the electromagnetic and weak interactions.

### V.5.4. Interactions involving Rarita–Schwinger field

According to the hadron dynamical symmetry [240] connected to the group $SU(3)$, there exists a spin-3/2 barion decuplet, composed by nine unstable particles (nuclear resonances) and one metastable (the isosinglet $\Omega^-$). These particles are described by the Rarita–Schwinger equation[35] [252, 314] (together with two subsidiary conditions), similar to Dirac's equation, but with a spin-vector $\psi_\mu$ as the wave function, that transforms according to the reducible representations $[(1/2,0)\otimes\ (0,1/2)]\otimes(1/2,1/2)$ of the homogeneous Lorentz group [352].

Starting with the Cohen's investigation [80] and using the $S$-matrix formalism, the possibility of interaction of the Rarita–Schwinger field (mostly of the $\Omega^-$ barions) with the gravitational field, without involving any supergravitation and supersymmetry considerations is briefly approached. The following presentation is concerned with the main results of this investigation.

Using the minimal coupling principle and the *vierbein* formalism, the Lagrangian density of the gravitational field (1.295) must be supplemented with a term representing the Rarita–Schwinger Lagrangian density written in the curved space,

$$\mathcal{L}_{tot} = \mathcal{L}_{RS} + \mathcal{L}_g. \tag{5.145}$$

Following the procedure used in Quantum Electrodynamics [2], one can obtain a symmetrical form of this Lagrangian density, even more general than that found in the specific literature, which, in the

[35]See Chap. I, paragraph I.3.8.3.

presence of gravitation, becomes (see also [208]),

$$\mathcal{L}_{RS} = \sqrt{-g}g^{\mu\nu}\left[\frac{i}{2}\left(\overline{\psi}_{\mu;\lambda}\tilde{\gamma}^{\lambda}\psi_{\nu} - \overline{\psi}_{\mu}\tilde{\gamma}^{\lambda}\psi_{\nu;\lambda}\right) + M\overline{\psi}_{\mu}\psi_{\nu}\right], \quad (5.146)$$

together with the following two subsidiary conditions,

$$\begin{aligned} \tilde{\gamma}^{\mu}\psi_{\mu} &= 0, \\ \tilde{g}^{\mu\nu}\psi_{\mu;\nu} &= 0, \end{aligned} \quad (5.146')$$

where $\tilde{g}^{\mu\nu} = \sqrt{-g}g^{\mu\nu}$, $M$ is the mass of the spin-3/2 particles, while the covarint derivatives of the spin-vectors are given by

$$\begin{aligned} \psi_{\mu;\nu} &= \psi_{\mu,\nu} - \Gamma_{\nu}\psi_{\mu}, \\ \overline{\psi}_{\mu;\nu} &= \overline{\psi}_{\mu,\nu} + \overline{\psi}_{\mu}\Gamma_{\nu}, \end{aligned} \quad (5.147)$$

where the Fock–Ivanenko spin coefficients of the affine connection, $\Gamma_{\mu}$, are given by (5.61).

In the weak gravitational field approximation, taking into account (1.297), (5.62), (5.65), (5.145)–(5.147), expanding in power series in powers of $\kappa$ and going to the flat space, one obtains

$$\mathcal{L}_{int} = \mathcal{L}^{(1)}_{RS-g} + \mathcal{L}^{(1)}_{g} + \mathcal{L}^{(2)}_{RS-g} + \cdots, \quad (5.148)$$

where

$$\begin{aligned} \mathcal{L}^{(1)}_{RS-g} &= \frac{1}{2}\kappa\left(\overline{\psi}_{\mu,\lambda}\gamma_{\lambda}\psi_{\nu} - \overline{\psi}_{\mu}\gamma_{\lambda}\psi_{\nu,\lambda}\right)y_{\mu\nu} \\ &\quad + \frac{1}{4}\kappa\left(\overline{\psi}_{\lambda,\mu}\gamma_{\nu}\psi_{\lambda} - \overline{\psi}_{\lambda}\gamma_{\nu}\psi_{\lambda,\mu}\right)h_{\mu\nu} \\ &\quad - \kappa M\overline{\psi}_{\mu}\psi_{\nu}y_{\mu\nu}, \end{aligned} \quad (5.149)$$

and $\mathcal{L}^{(2)}_{RS-g}$ are the first and second-order interaction Lagrangian densities, respectively, corresponding to the interaction process between the Rarita–Schwinger and gravitational fields. The expression of $\mathcal{L}^{(2)}_{RS-g}$ – which is very complicated – is given in Ref. [359].

The matrix elements of the operators of the Rarita–Schwinger field are analogous to those corresponding to the Dirac field (given by (5.73)), the positive and negative energy spinors being replaced by the positive and negative energy spin-vectors $u^{(r)}_{\mu}(\vec{k})$ and $v^{(r)}_{\mu}(\vec{k})$, and the mass of the spin-1/2 particles by the mass of the spin-3/2 particles.

For all the above mentioned Feynman diagrams – involving electromagnetic, strong and weak processes in the presence of the gravitational field, or those with participation of the Rarita–Schwinger field – one can obtain Feynman-type rules in the momentum space, resulting in various types of free particles, propagators and vertices [359, 360(p.277), 378].

Chapter VI

# Interaction of Scalar, Spinorial, Spin-Vectorial and Tensorial Particles, and the Gravitational Field Described by the Schwarzschild Metric

## VI.1. Gravitational scattering of spin 0, 1/2, 1, 3/2, and 2 massive particles in the centrally-symmetric gravitational field described by the Schwarzschild metric

### VI.1.1. General considerations

The aim of this subchapter is to point out the main aspects regarding the gravitational scattering process of scalar, spinorial, vectorial, spin-vectorial and tensorial massive particles in the framework of the theory of quantum gravity[1]. A part of the processes that will be presented below (those concerning the spin-0, 1/2, and 1 particles, both massive and massless, as well as spin-2 massless particles) have been studied by several authors before 1977, but only for the particular case of small scattering angles approximation (see, *e.g.*, [38(1972), 271, 275, 389(1964), 391]) and another part (those concerning the spin-3/2 particles, both massive and massless, as well as spin-2 massive particles) was the subject of relatively recent research [310, 312, 313], the results of these investigations being also presented further below.

The theme of this chapter is motivated, among other reasons, by the increasing interest regarding the spin-2 massive field (see, for

[1]Even if a complete and consequent theory of quantum gravity was not elaborated so far, the specific terminology is used as such, for both the linear approximation (formulated ever since 1952 [184]), as in our case, and the most general (non-linear) case, in the context of continuous efforts of nowadays investigations in this respect.

example, Refs. [84, 194, 195]), its approach in this context being intended to bring to light some aspects connected to the interaction with the gravitational field, within the framework of the external field formalism.

A similar subject, but regarding only the massless fields of spin-0, 1, and 2 was approached by Walter K. De Logi and Sandor J. Kovacs, Jr. [91], by using a Feynman-type perturbative method (based on Feynman diagrams). The Feynman diagram technique was initially utilized in quantum electrodynamics, but its efficiency as a general tool of investigation was widely proved in the study of quantum interactions, including those of quantum gravity.

The results obtained by De Logi and Kovacs agree – in the limit of small scattering angles – to those found by Peters [305], whose procedure was based on the Green functions formalism. The two authors showed that the method proposed in [91] can be also successfully utilized in the study of gravitational scattering processes of both massive and half-integer spin particles.

Unlike the method used by De Logi and Kovacs (and, also, by some other authors), in what follows, the interaction Lagrangians between the gravitational and massive considered fields have been determined by using the minimal coupling principle in quantum gravity [66]. This leads, as shall be seen, to obtaining of some corrective terms in the final expressions of the differential cross-sections. From the physical point of view, this fact leads to widening the range of values that can be taken by the scattering angle $\theta$. This procedure also allows the approach of the important particular case of backscattering (for the value $\theta = \pi$ of the scattering angle).

The use of minimal coupling principle in quantum gravity requires – besides the other two "substitutions" regarding the metric tensor and the four-volume element, presented in Chap. IV – writing the matter field Lagrangians in the curved space, which implies replacement of the usual derivatives by the covariant ones. As a result, in the interaction Lagrangians appear certain corrective terms, even in the first-order approximation. The following analysis shall concern only the interaction processes proportional to $\kappa^2$ (*i.e.* those processes that imply the exchange of a single graviton); in other words, it will be shown how the scattering cross-sections corresponding to the first-order Born approximation of the quantum perturbation theory can be determined.

Usually, when approaching this subject, most of the authors use the Gupta-type interaction Lagrangian densities (within the framework of the so-called *Gupta-type coupling* (see Ref. [366])):

$$\mathcal{L}_{int}^{Gupta} = -\frac{1}{2}\kappa h_{\mu\nu}T^{\mu\nu}.$$

But the Lagrangian density can be also obtained in a more general form (using the above described method), by taking

$$\mathcal{L}_{int} = \mathcal{L}_{int}^{Gupta} + \mathcal{L}_{int}^{(1)corr}(\kappa),$$

where $\mathcal{L}_{int}^{(1)corr}(\kappa)$ is the first-order correction in the coupling constant $\kappa$. For example, in the case of the massive spin-2 field, this term is of the following form:

$$\mathcal{L}_{int}^{(1)corr}(\kappa) = \kappa\Big[H^{*}_{\mu\nu\lambda}\Phi_{\rho\mu}\Big(h_{\lambda\rho,\nu} + h_{\nu\rho,\lambda} - h_{\lambda\nu,\rho}\Big) + \Phi^{*}_{\rho\lambda}\Big(h_{\mu\rho,\nu} + h_{\nu\rho,\mu} - h_{\mu\nu,\rho}\Big)H_{\lambda\nu\mu}\Big].$$

Nevertheless, due to the particular form of the Lagrangian densities of spin-0, 1, and 3/2 fields, the corrective term $\mathcal{L}_{int}^{(1)corr}(\kappa)$ is identically null. This fact can be easily observed, for example, in the case of the spin-1 field, where the property $B_{\nu;\mu} - B_{\mu;\nu} \equiv B_{\nu,\mu} - B_{\mu,\nu}$ signifies that $\tilde{G}_{\mu\nu} = G_{\mu\nu}$ (differently speaking, the symmetric second-rank tensor of the massive vectorial field is the same in both formalisms, covariant and non-covariant/ordinary). A similar observation can also be made about the spin-3/2 massive field; using the Rarita–Schwinger equation and its adjoint, following a relatively simple calculation, one can easily realize that the first-order interaction Lagrangian density obtained in the framework of the minimal coupling formalism applied in quantum gravity (that is, using the covariant derivative) is the same with that obtained on the basis of Gupta-type formalism. As far as the scalar field is concerned, here this property is obvious.

Another aspect that deserves to be mentioned concerns the fact that, unlike most papers about the massive tensorial field, in which this field is described by the ordinary Lagrangian density,

$$\mathcal{L}_{tens} = \sqrt{-g}\, g^{\mu\lambda}\, g^{\nu\rho}\Big(g^{\alpha\beta}\Phi^*_{\mu\nu;\alpha}\Phi_{\lambda\rho;\beta} - m^2\Phi^*_{\mu\nu}\Phi_{\lambda\rho}\Big),$$

in this subchapter, as a Lagrangian density for the massive tensorial field, that one proposed by J. Schwinger (see Ref. [331]) is considered, namely,

$$\mathcal{L}_{tens} = -\sqrt{-g}\, g^{\mu\rho}\, g^{\nu\beta}\Big(g^{\lambda\alpha}\tilde{H}^*_{\mu\nu\lambda}\tilde{H}_{\alpha\beta\rho} + m^2\Phi^*_{\mu\nu}\Phi_{\rho\beta}\Big).$$

The use of this Lagrangian density is more justified in order to describe the behavior of the massive tensorial particles, taking into account the correspondence and analogy principles, entitled by the relevant resemblance between this theory and that corresponding to the non-massive case (the reader is invited to compare the above Lagrangian density of the massive tensorial field used by Schwinger – where $\tilde{H}_{\mu\nu\lambda} = \Phi_{\nu\lambda;\mu} + \Phi_{\mu\lambda;\nu} - \Phi_{\mu\nu;\lambda}$ is the third-rank tensor of the massive tensorial field with the property $\tilde{H}_{\mu\nu\lambda} = \tilde{H}_{\nu\mu\lambda}$, and $\Phi_{\mu\nu;\lambda}$ is the covariant derivative of the massive tensorial field function – with that corresponding to the gravitational field (Einstein):

$$\mathcal{L}_g = \frac{1}{\kappa^2}\sqrt{-g}\, g^{\mu\nu}\Big(\Gamma^\alpha_{\mu\nu}\Gamma^\beta_{\alpha\beta} - \Gamma^\alpha_{\mu\beta}\Gamma^\beta_{\nu\alpha}\Big),$$

where, as it is well-known,

$$\Gamma^\lambda_{\mu\nu} = \frac{1}{2}g^{\lambda\alpha}\Big(g_{\mu\alpha,\nu} + g_{\nu\alpha,\mu} - g_{\mu\nu,\alpha}\Big)$$

are the Christoffel symbols of the second kind.

Considering, for the moment, only those geometries of the gravitational background generated by classical momentum-energy distributions (which are not considerably affected by the scattering process), the external field approximation can be conveniently used. As a particular case, in this subchapter as background shall be considered the static field with central symmetry described by the Schwarzschild metric.

## VI.1.2. Gravitational scattering of massive scalar, spinorial, vectorial, spin-vectorial and tensorial particles

As it is known, the massive field theories on curved backgrounds are non-renormalizable. It is not very clear – at least so far – to what extent the semiclassical approximation (*i.e.* a quantization theory of the massive fields on a background described by a classical geometry) is able to furnish reliable results. Nevertheless, the processes approached in the following study prove to be finite; in other words, the concrete calculations can be carried through (can be done to the end), without obtaining divergent terms, at least in the first-order Born approximation.

Following Feynman's ideas exposed in Ref. [142] and, since the approximation $|y^{\mu\nu}| \ll 1$ is required to be valid everywhere, the gravitational field can be expanded about the Minkowskian flat background [184],

$$\sqrt{-g}\, g^{\mu\nu} = \eta^{\mu\nu} - \kappa y^{\mu\nu}, \tag{6.1}$$

where $g^{\mu\nu}$, $\eta^{\mu\nu}$, and $y^{\mu\nu}$ are the metric tensor, the Minkowski tensor – diag $(+1, -1, -1, -1)$, and the weak gravitational field tensor, respectively, $g = \det(g_{\mu\nu})$, and $\kappa = \sqrt{16\pi G}$ (in the natural unit system, $G$ being the Newtonian constant).

In order to obtain the interaction Lagrangian densities between the gravitational field and the massive scalar, spinorial, vectorial, spin-vectorial (Rarita–Schwinger), and tensorial fields, one uses the minimal coupling principle in quantum gravity. According to this principle, to get the covariant form of the total Lagrangian densities of spin-0, 1 and 2 massive fields, besides the other two "substitutions" regarding the metric tensor and the four-volume element (which substitutions are presented in Chap. IV), the Lagrangian density of the gravitational field must be supplemented with the corresponding Lagrangian densities of the massive complex scalar, vectorial and tensorial fields, written in the curved space (*i.e.*, using the covariant derivative of the field functions instead of the ordinary one), namely [364, 366]:

$$\mathcal{L}_{scalar} = \sqrt{-g}\left[g^{\mu\nu}\varphi^*_{;\mu}\varphi_{;\nu} - \left(m^2 + \frac{1}{4}R\right)\varphi^*\varphi\right], \tag{6.2}$$

$$\mathcal{L}^{Proca}_{vectorial} = \sqrt{-g}\left(-\frac{1}{2}g^{\mu\alpha}g^{\nu\beta}\tilde{G}^*_{\mu\nu}\tilde{G}_{\alpha\beta} + m^2 g^{\nu\alpha}B^*_\nu B_\alpha\right), \tag{6.3}$$

$$\begin{aligned}\mathcal{L}^{Stueckelberg}_{vectorial} &= \sqrt{-g}\Big(-g^{\mu\alpha}g^{\nu\beta}U^*_{\mu;\nu}U_{\alpha;\beta} + m^2 g^{\nu\alpha}U^*_\nu U_\alpha\Big)\\ &\quad + \sqrt{-g}\Big(g^{\mu\nu}C^*_{,\mu}C_{,\nu} - m^2 C^* C\Big),\end{aligned} \tag{6.3$'$}$$

$$\mathcal{L}_{tensorial} = -\sqrt{-g}g^{\mu\rho}g^{\nu\beta}\left(g^{\lambda\alpha}\tilde{H}^*_{\mu\nu\lambda}\tilde{H}_{\alpha\beta\rho} + m^2\Phi^*_{\mu\nu}\Phi_{\rho\beta}\right). \tag{6.4}$$

As it can easily be observed, the Klein–Gordon formalism has been used for the massive scalar field and, in addition to the minimal coupling introduced by the minimal substitution, a non-minimal coupling – by means of the scalar curvature $R$ – has also been used.

As far as the vectorial field is concerned, two formalisms have been considered, namely the Proca and Stueckelberg's formalisms. The quantities that appear in (6.3) have the following significance:

- $\tilde{G}_{\mu\nu} = B_{\nu;\mu} - B_{\mu;\nu}$ ($B_{\mu;\nu}$ being the covariant derivative of the massive vectorial field function) is the second-rank symmetric tensor of the massive vectorial field in the Proca formalism;
- $U_\mu(x)$ and $C(x)$ appearing in (6.3′) are the two fields that describe the massive vectorial particles in the Stueckelberg formalism; they are equivalent to $B_\mu(x)$ of the Proca formalism, the connection between them being given by (5.46).

Due to the reasons exposed in the beginning of this subchapter, the spin-2 massive tensorial field is supposed to be described – independently from the Fierz–Pauli theory – by the Lagrangian density (6.4) proposed by Schwinger in Ref. [331].

In Eq. (6.4), $\tilde{H}_{\mu\nu\lambda} = \Phi_{\nu\lambda;\mu} + \Phi_{\mu\lambda;\nu} - \Phi_{\mu\nu;\lambda}$ is the third-rank tensor (symmetric in the first two indices, $\tilde{H}_{\mu\nu\lambda} = \tilde{H}_{\nu\mu\lambda}$) of the spin-2 massive field, and $\Phi_{\mu\nu;\lambda}$ is the covariant derivative of the corresponding field function.

The massive fermionic fields (spinorial and spin-3/2 fields) present several approaching particularities.

First, the massive spinorial particles are considered as being described by the Lagrangian density given by (5.59), namely

$$\mathcal{L}_{spinorial} = \sqrt{-g}\left[\frac{i}{2}\left(\overline{\psi}\tilde{\gamma}^\mu\psi_{;\mu} - \overline{\psi}_{;\mu}\tilde{\gamma}^\mu\psi\right) - m\overline{\psi}\psi\right], \tag{6.5}$$

where $\tilde{\gamma}^{\mu}$ are the Dirac's generalized matrices (in the curved space), and $\psi_{;\mu}$ denotes the covariant derivative of the spinor that describes the spin-1/2 massive particles. In order to describe the Rarita–Schwinger field, the Lagrangian density (5.146) has been considered,

$$\begin{aligned}\mathcal{L}_{\text{Rarita-Schwinger}} \equiv \mathcal{L}_{RS} &= \sqrt{-g}g^{\mu\nu} \\ &\quad\times \left[\frac{i}{2}\left(\overline{\psi}_{\mu}\overleftarrow{\nabla}_{\lambda}\tilde{\gamma}^{\lambda}\psi_{\nu} - \overline{\psi}_{\mu}\tilde{\gamma}^{\lambda}\overrightarrow{\nabla}_{\lambda}\psi_{\nu}\right) + m\overline{\psi}_{\mu}\psi_{\nu}\right] \\ &= \mathcal{L}_1 + \mathcal{L}_2,\end{aligned} \tag{6.6}$$

where $\overrightarrow{\nabla}_{\lambda}\psi_{\nu}$ is the covariant derivative of the Rarita–Schwinger field function $\psi_{\nu}$ with respect to coordinate $x^{\lambda}$, the arrow above the nabla symbol shows that the covariant derivative operator $\nabla$ acts towards the right (in the sense of the arrow),

$$\mathcal{L}_1 = \sqrt{-g}g^{\mu\nu}\left[\frac{i}{2}\left(\overline{\psi}_{\mu,\lambda}\tilde{\gamma}^{\lambda}\psi_{\nu} - \overline{\psi}_{\mu}\tilde{\gamma}^{\lambda}\psi_{\nu,\lambda}\right) + m\overline{\psi}_{\mu}\psi_{\nu}\right], \tag{6.7}$$

and

$$\mathcal{L}_2 = \frac{i}{8}\sqrt{-g}g^{\mu\nu}\overline{\psi}_{\mu}\left(\tilde{\gamma}_{\rho;\lambda}\tilde{\gamma}^{\rho}\tilde{\gamma}^{\lambda} + \tilde{\gamma}^{\lambda}\tilde{\gamma}_{\rho;\lambda}\tilde{\gamma}^{\rho}\right)\psi_{\nu}, \tag{6.8}$$

where $\psi_{\mu,\nu}$ is the ordinary derivative of the Rarita–Schwinger field function.

Second – taking into account the specific character of the wave functions of the fermionic fields – the minimal coupling principle has to be utilized within the framework of the *vierbein* formalism.

There also has to be considered the fact that there is a major difference between the study of massless particles interaction, on the one side, and that corresponding to massive particles, on the other, or, differently speaking, between a Lagrangian density containing mass terms and the one that does not contain such terms [365]. The presence of a mass term in the Lagrangian density seriously affects the physical content of the theory. This problem is solved/clear for the massive spinorial field, but in the case of massive Rarita–Schwinger field the situation is different. In this respect, for the massive spin-vectorial field there shall be utilized the mass term proposed by S. Deser, J.H. Kay and K.S. Stelle [113], which is $im\overline{\psi}_{\mu}\sigma^{\mu\nu}\psi_{\nu}$. To facilitate the calculations, this term can be written as $mg^{\mu\nu}\overline{\psi}_\mu\psi_\nu$, which is possible due to the constraints valid for this field: $\tilde{\gamma}^\mu\psi_\mu = 0$, $\partial_\mu\psi^\mu = 0$ [113]. Indeed, a simple calculation shows that

$$\begin{aligned} im\overline{\psi}_\mu\sigma^{\mu\nu}\psi_\nu &= im\overline{\psi}_\mu\frac{i}{2}\Big(\tilde{\gamma}^\mu\tilde{\gamma}^\nu - \tilde{\gamma}^\nu\tilde{\gamma}^\mu\Big)\psi_\nu = m\overline{\psi}_\mu\Big(g^{\mu\nu} - \tilde{\gamma}^\mu\tilde{\gamma}^\nu\Big)\psi_\nu \\ &= mg^{\mu\nu}\overline{\psi}_\mu\psi_\nu - m\overline{\psi}_\mu\tilde{\gamma}^\mu\tilde{\gamma}^\nu\psi_\nu = mg^{\mu\nu}\overline{\psi}_\mu\psi_\nu = m\overline{\psi}_\mu\psi^\mu. \end{aligned} \tag{6.9}$$

In regard with the terms $\mathcal{L}_1$ and $\mathcal{L}_2$ given by (6.7) and (6.8), their expressions have been obtained by using the covariant derivative of the field spin-vectors,

$$\overrightarrow{\nabla}_\mu\psi_\nu = \psi_{\nu,\mu} - \Gamma_\mu\psi_\nu, \quad \overline{\psi}_\nu\overleftarrow{\nabla}_\mu = \overline{\psi}_{\nu,\mu} + \overline{\psi}_\nu\Gamma_\mu, \tag{6.10}$$

where $\Gamma_\mu$ are the Fock–Ivanenko spin coefficients of the affine connection. According to the vierbein formalism, these coefficients are given by

$$\Gamma_\mu = \frac{1}{4}\tilde{\gamma}_{\lambda;\mu}\tilde{\gamma}^\lambda, \tag{6.11}$$

where the generalized Dirac matrices $\tilde{\gamma}^\mu$ are defined as [246],

$$\tilde{\gamma}^\mu = L^\mu(\alpha)\gamma(\alpha); \quad \tilde{\gamma}_\mu = L_\mu(\alpha)\gamma(\alpha). \tag{6.12}$$

Here $\gamma(\alpha)$ are the ordinary Dirac matrices, and $\tilde{\gamma}_{\lambda;\mu}$ appearing in (6.11) are given by

$$\tilde{\gamma}_{\lambda;\mu} = \tilde{\gamma}_{\lambda,\mu} - \tilde{\gamma}_\rho\Gamma^\rho_{\lambda\mu}. \tag{6.13}$$

The *vierbein coefficients* $L^\mu(\alpha)$ and $L_\mu(\alpha)$ occurring in (6.12) satisfy the following constraints:

$$L^\mu(\alpha)L^\nu(\alpha) = g^{\mu\nu}\ ; \quad L_\mu(\alpha)L_\nu(\alpha) = g_{\mu\nu}. \tag{6.14}$$

Since the following considerations concern only the first-order approximation in terms of the coupling constant $\kappa$, here are the linearized expressions for the quantities used in the calculations,

$$g^{\mu\nu} = \eta^{\mu\nu} - \kappa h^{\mu\nu} + \mathcal{O}(\kappa^2), \tag{6.15}$$

$$g_{\mu\nu} = \eta_{\mu\nu} + \kappa h_{\mu\nu} + \mathcal{O}(\kappa^2), \tag{6.16}$$

$$\Gamma^\rho_{\lambda\mu} = \frac{1}{2}\kappa\Big(h^\rho_{\lambda,\mu} + h^\rho_{\mu,\lambda} - h_{\lambda\mu,}{}^\rho\Big) + \mathcal{O}(\kappa^2), \tag{6.17}$$

$$\tilde{\gamma}^\lambda = \gamma^\lambda - \frac{1}{2}\kappa\gamma_\beta h^{\lambda\beta} + \mathcal{O}(\kappa^2), \tag{6.18}$$

$$\tilde{\gamma}_\lambda = \gamma_\lambda + \frac{1}{2}\kappa\gamma_\beta h^\beta_\lambda + \mathcal{O}(\kappa^2), \tag{6.19}$$

where

$$h^{\mu\nu} = y^{\mu\nu} - \frac{1}{2}\eta^{\mu\nu}y\ ; \quad h^\beta_\lambda = y^\beta_\lambda - \frac{1}{2}\delta^\beta_\lambda y\ ; \quad y = y^\alpha_\alpha. \tag{6.20}$$

It is easy to show that – in the first-order approximation – the term $\mathcal{L}_2$ met in (6.6) brings no contribution. Indeed, one has

$$\begin{aligned}\mathcal{L}_2^{(1)}(\kappa) &= \frac{i}{16}\kappa\overline{\psi}_\mu\gamma^\lambda\gamma_\alpha\gamma^\rho\psi^\mu\left(h^\alpha_{\rho,\lambda} - h^\alpha_{\lambda\rho}\right)\\ &= \frac{i}{16}\kappa\overline{\psi}_\mu\gamma^\lambda\left(\gamma^\alpha\gamma^\rho - \gamma^\rho\gamma^\alpha\right)\psi^\mu h_{\alpha\rho,\lambda} = 0,\end{aligned} \tag{6.21}$$

where the well-known anticommutation relations satisfied by the Dirac matrices

$$\left\{\gamma^\mu, \gamma^\nu\right\} = \gamma^\mu\gamma^\nu + \gamma^\nu\gamma^\mu = 2\eta^{\mu\nu} \tag{6.22}$$

have been used.

Taking into account the previous considerations, the first-order interaction Lagrangian density between the weak gravitational and Rarita–Schwinger fields becomes

$$\begin{aligned}\mathcal{L}_{int}^{(1)}(\kappa) &= -\frac{i}{4}\kappa\left(\overline{\psi}_{\lambda,\mu}\gamma_\nu\psi^\lambda - \overline{\psi}_\lambda\gamma_\nu\psi^\lambda{}_{,\mu}\right)h^{\mu\nu}\\ &\quad - \frac{i}{2}\kappa\left(\overline{\psi}_{\mu,\lambda}\gamma^\lambda\psi_\nu - \overline{\psi}_\mu\gamma^\lambda\psi_{\nu,\lambda}\right)y^{\mu\nu} - \kappa m\overline{\psi}_\mu\psi_\nu y^{\mu\nu}.\end{aligned} \tag{6.23}$$

Using the De Donder Fock gauge, $\left(\sqrt{-g}g^{\mu\nu}\right)_{,\nu} = 0$, and passing to the flat space through the following replacements,

$$\begin{aligned}&x^0 = t, x^j\ (j = 1,2,3)\ \rightarrow\ x_j\ (j - 1,2,3),\ x_4 = it;\\ &\partial_\mu\gamma^\mu \rightarrow i\partial_\mu\gamma_\mu;\ \eta^{\mu\nu} \rightarrow -\delta_{\mu\nu};\ \partial_\mu \equiv \tfrac{\partial}{\partial x^\mu} \rightarrow \tfrac{\partial}{\partial x_\mu},\end{aligned} \tag{6.24}$$

the first-order interaction Lagrangian densities between the gravitational field and the spin-0, 1/2, 1, 3/2, and 2 massive fields write as follows:

$$\mathcal{L}_{int}^{(1),0}(\kappa) = -\kappa\left[\varphi^*_{,\mu}\varphi_{,\nu}y_{\mu\nu} + \frac{1}{2}\varphi^*\varphi\left(m^2 y - \frac{1}{4}y_{,\mu\mu}\right)\right], \tag{6.25}$$

$$\mathcal{L}_{int}^{(1),1/2}(\kappa) = -\frac{1}{4}\kappa\left(\overline{\psi}\gamma_\mu\psi_{,\nu} - \overline{\psi}_{,\nu}\gamma_\mu\psi\right)s_{\mu\nu} - \frac{1}{2}\kappa\, m\,\overline{\psi}\psi\, y, \qquad (6.26)$$

$$\mathcal{L}_{int}^{(1),1,Proca}(\kappa) = -\kappa\left(G^*_{\mu\nu}G_{\mu\alpha}u_{\nu\alpha} + m^2 B^*_\nu B_\alpha y_{\nu\alpha}\right), \qquad (6.27)$$

$$\mathcal{L}_{int}^{(1),1,Stueckelberg}(\kappa) = -\kappa\Big[U^*_\mu U_{\nu,\alpha}\left(h_{\mu\nu,\alpha} + h_{\mu\alpha,\nu} - h_{\nu\alpha,\mu}\right) + U^*_{\mu,\nu}U_{\mu,\alpha}h_{\nu\alpha} + U^*_{\mu,\nu}U_{\alpha,\nu}y_{\mu\alpha} + m^2U^*_\nu U_\alpha y_{\nu\alpha}\Big], \qquad (6.27')$$

$$\mathcal{L}_{int}^{(1),3/2}(\kappa) = \frac{1}{4}\kappa\left(\overline{\psi}_{\alpha,\nu}\gamma_\mu\psi_\alpha - \overline{\psi}_\alpha\gamma_\mu\psi_{\alpha,\nu}\right)y_{\mu\nu} - \frac{1}{4}\kappa\, m\,\overline{\psi}_\alpha\psi_\alpha y, \qquad (6.28)$$

$$\begin{aligned}\mathcal{L}_{int}^{(1),2}(\kappa) = {}& \kappa\Big[H^*_{\mu\nu\alpha}\left(H_{\alpha\nu\lambda} + H_{\alpha\lambda\nu}\right) + H^*_{\alpha\nu\mu}H_{\nu\lambda\alpha} - 2m^2\Phi^*_{\mu\nu}\Phi_{\lambda\nu}\Big]y_{\mu\lambda} \\ & - \kappa\left(H^*_{\mu\nu\lambda}H_{\lambda\nu\mu} - \tfrac{1}{2}m^2\Phi^*_{\mu\nu}\Phi_{\mu\nu}\right)y \\ & + \kappa\Big[H^*_{\mu\nu\lambda}\Phi_{\rho\mu}\left(h_{\lambda\rho,\nu} + h_{\nu\rho,\lambda} - h_{\lambda\nu,\rho}\right) \\ & + \Phi^*_{\rho\lambda}\left(h_{\mu\rho,\nu} + h_{\nu\rho,\mu} - h_{\mu\nu,\rho}\right)H_{\lambda\nu\mu}\Big], \end{aligned} \qquad (6.29)$$

where

$$s_{\mu\nu} = y_{\mu\nu} + \frac{1}{2}\delta_{\mu\nu}y, \qquad (6.30)$$

$$u_{\mu\nu} = y_{\mu\nu} - \frac{1}{4}\delta_{\mu\nu}y_{\alpha\alpha}, \qquad (6.31)$$

$$G_{\mu\nu} = B_{\nu,\mu} - B_{\mu,\nu}\ ; \quad H_{\mu\nu\lambda} = \Phi_{\nu\lambda,\mu} + \Phi_{\mu\lambda,\nu} - \Phi_{\mu\nu,\lambda}, \qquad (6.32)$$

$B_{\mu,\nu}$ and $\Phi_{\mu\nu,\lambda}$ being the ordinary derivatives of the vectorial and tensorial field functions, respectively. It has also been used the Rarita–Schwinger equation for the massive spin-3/2 field,

$$\gamma_\mu\psi_{\alpha,\mu} = -m\psi_\alpha, \qquad (6.33)$$

and its adjoint as well.

According to the standard theory of quantum fields, the parts of the Lagrangian densities (6.25)–(6.29) – brought to the normal form – that describe the interaction between gravitation and the scalar, spinorial, vectorial (in Proca and Stueckelberg formalisms), spin-vectorial (Rarita–Schwinger) and tensorial massive particles are [366],

$$N\Big[\mathcal{L}^{(1)}_{sc-g}(x)\Big] = -\kappa\Bigg\{\varphi^{*(-)}_{,\mu}(x)\varphi^{(+)}_{,\nu}(x)y^{ext}_{\mu\nu}(x) + \frac{1}{2}\varphi^{*(-)}(x)\varphi^{(+)}(x) \times \Big[m^2 y^{ext}(x) - \frac{1}{4}y^{ext}_{,\mu\mu}(x)\Big]\Bigg\}, \tag{6.34}$$

$$N\Big[\mathcal{L}^{(1)}_{sp-g}(x)\Big] = -\frac{1}{4}\kappa\Big(\overline{\psi}^{(-)}(x)\gamma_\mu\psi^{(+)}_{,\nu}(x) - \overline{\psi}^{(-)}_{,\nu}(x)\gamma_\mu\psi^{(+)}(x)\Big)s^{ext}_{\mu\nu}(x) - \frac{1}{2}\kappa\, m\overline{\psi}^{(-)}(x)\psi^{(+)}(x)y^{ext}(x), \tag{6.35}$$

$$N\Big[\mathcal{L}^{(1)Proca}_{vect-g}(x)\Big] = -\kappa\Big[G^{*(-)}_{\mu\nu}(x)G^{(+)}_{\mu\alpha}(x)u^{ext}_{\nu\alpha}(x) + m^2 B^{*(-)}_{\nu}(x)B^{(+)}_{\alpha}(x)y^{ext}_{\nu\alpha}(x)\Big], \tag{6.36}$$

$$N\Big[\mathcal{L}^{(1)Stueck}_{vect-g}(x)\Big] = -\kappa\Big\{U^{*(-)}_{\mu}(x)U^{(+)}_{\nu,\alpha}(x) \times \Big[h^{ext}_{\mu\nu,\alpha}(x) + h^{ext}_{\mu\alpha,\nu}(x) - h^{ext}_{\nu\alpha,\mu}(x)\Big] + U^{*(-)}_{\mu,\nu}(x)U^{(+)}_{\mu,\alpha}(x)h^{ext}_{\nu\alpha}(x) + U^{*(-)}_{\mu,\nu}(x)U^{(+)}_{\alpha,\nu}(x) \times y^{ext}_{\mu\alpha}(x) + m^2 U^{*(-)}_{\nu}(x)U^{(+)}_{\alpha}(x)y^{ext}_{\nu\alpha}(x)\Big\}, \tag{6.36$'$}$$

$$N\Big[\mathcal{L}^{(1)}_{RS-g}(x)\Big] = \frac{1}{4}\kappa\Big[\overline{\psi}^{(-)}_{\alpha,\nu}\gamma_\mu\psi^{(+)}_{\alpha}(x) - \overline{\psi}^{(-)}_{\alpha}(x)\gamma_\mu\psi^{(+)}_{\alpha,\nu}(x)\Big]y^{ext}_{\mu\nu}(x) - \frac{1}{4}\kappa\, m\overline{\psi}^{(-)}_{\alpha}(x)\psi^{(+)}_{\alpha}(x)y^{ext}(x), \tag{6.37}$$

and

$$N\Big[\mathcal{L}^{(1)}_{tens-g}(x)\Big] = \kappa\Big\{\Phi^{*(-)}_{\alpha\nu,\mu}(x)\Big[3\Phi^{(+)}_{\lambda\nu,\alpha}(x) - \Phi^{(+)}_{\lambda\alpha,\nu}(x) - \Phi^{(+)}_{\nu\alpha,\lambda}(x)\Big]$$

$$+ \Phi^{*(-)}_{\alpha\mu,\nu}(x)\Big[\Phi^{(+)}_{\lambda\nu,\alpha}(x) + \Phi^{(+)}_{\lambda\alpha,\nu}(x) + \Phi^{(+)}_{\alpha\nu,\lambda}(x)\Big] + \Phi^{*(-)}_{\mu\nu,\alpha}(x)\Big[\Phi^{(+)}_{\alpha\lambda,\nu}(x)$$

$$+ \Phi^{(+)}_{\alpha\nu,\lambda}(x) - 3\Phi^{(+)}_{\lambda\nu,\alpha}(x)\Big] - 2m^2\Phi^{*(-)}_{\mu\nu}(x)\Phi^{(+)}_{\lambda\nu}(x)\Big\}y^{ext}_{\mu\lambda}(x)$$
$$- \kappa\Big\{\Big[\Phi^{*(-)}_{\lambda\nu,\mu}(x) + \Phi^{*(-)}_{\lambda\mu,\nu}(x) - \Phi^{*(-)}_{\mu\nu,\lambda}(x)\Big]\Big[\Phi^{(+)}_{\mu\nu,\lambda}(x) + \Phi^{(+)}_{\mu\lambda,\nu}(x) - \Phi^{(+)}_{\lambda\nu,\mu}(x)\Big]$$
$$- \frac{1}{2}m^2\Phi^{*(-)}_{\mu\nu}(x)\Phi^{(+)}_{\mu\nu}(x)\Big\}y^{ext}(x) + \kappa\Big\{\Big[\Phi^{*(-)}_{\lambda\nu,\mu}(x) + \Phi^{*(-)}_{\lambda\mu,\nu}(x) - \Phi^{*(-)}_{\mu\nu,\lambda}(x)\Big]$$
$$\times \Phi^{(+)}_{\mu\sigma}(x)\Big[h^{ext}_{\lambda\sigma,\nu}(x) + h^{ext}_{\nu\sigma,\lambda}(x) - h^{ext}_{\lambda\nu,\sigma}(x)\Big] + \Phi^{*(-)}_{\sigma\lambda}(x)\Big[\Phi^{(+)}_{\mu\nu,\lambda}(x)$$
$$+ \Phi^{(+)}_{\mu\lambda,\nu}(x) - \Phi^{(+)}_{\lambda\nu,\mu}(x)\Big]\Big[h^{ext}_{\mu\sigma,\nu}(x) + h^{ext}_{\nu\sigma,\mu}(x) - h^{ext}_{\mu\nu,\sigma}(x)\Big]\Big\}, \tag{6.38}$$

respectively.

The studied processes are described by the Feynman diagram depicted in Fig. 6.1, where $p$ and $(r)$ on the one side, and $p'$ and $(s)$ on the other are the four-momenta and spin-state indices $(r, s = \overline{1, 2s+1}$, where $s$ is the particle spin) of the initial and final particles, respectively, and $q$ is the four-momentum of the virtual graviton.

Using the $S$-matrix formalism, one can deduce Feynman-type rules for diagrams in the external gravitational field (described by the Schwarzschild metric) approximation, allowing to calculate the matrix elements $\langle p'|S|p\rangle$ in this approximation.

Thus, as a preliminary result, one finds [366]:

$$s^{ext}_{\mu\nu}(\vec{q}) = \left(\delta_{\mu 4}\delta_{\nu 4} + \frac{1}{2}\delta_{\mu\nu}\right) y^{ext}(\vec{q}), \tag{6.39}$$

$$u^{ext}_{\mu\nu}(\vec{q}) = \left(\delta_{\mu 4}\delta_{\nu 4} - \frac{1}{4}\delta_{\mu\nu}\right) y^{ext}(\vec{q}), \tag{6.40}$$

$$y^{ext}_{\mu\nu}(\vec{q}) = \delta_{\mu 4}\delta_{\nu 4} y^{ext}(\vec{q}), \tag{6.41}$$

$$h^{ext}_{\mu\nu,\alpha}(\vec{q}) = \left(\delta_{\mu 4}\delta_{\nu 4} - \frac{1}{2}\delta_{\mu\nu}\right) y^{ext}_{,\alpha}(\vec{q}), \tag{6.42}$$

$$y^{ext}_{,\alpha}(\vec{q}) = i\delta_{\alpha j}q_j y^{ext}(\vec{q})\ ; \quad \left(i = \sqrt{-1}\right). \tag{6.43}$$

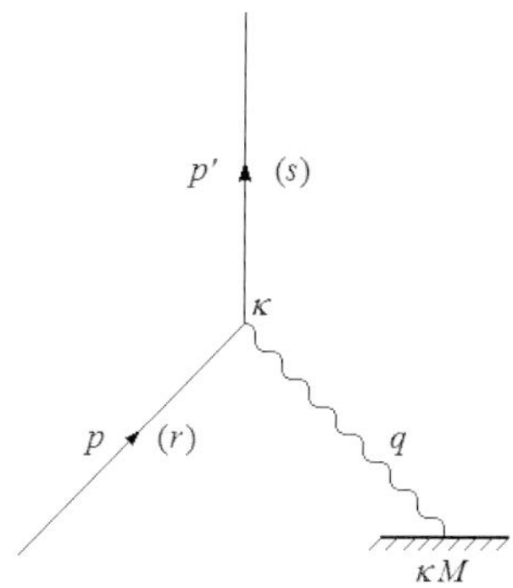


Fig. 6.1. The vertex (massive field)–graviton–(massive field). The curly line represents the graviton, and the straight lines describe the scalar, spinorial, vectorial, spin-vectorial and tensorial field quanta.

Taking into account the Fourier transform of the external gravitational field potential,

$$y^{ext}(\vec{q}) = \frac{1}{(2\pi)^{3/2}} \int e^{-i\vec{q}\cdot\vec{x}} \frac{\kappa M}{4\pi|\vec{x}|} \, d^3x = \frac{\kappa M}{(2\pi)^{3/2}|\vec{q}|^2}, \tag{6.44}$$

where $M$ is the mass of the central body that generates the gravitational field, and $|\vec{x}|$ is the distance from this centre, the matrix elements – considered in the external field approximation – corresponding to the Feynman diagram depicted in Fig. 6.1 are, respectively,

$$\begin{aligned} S^{(0)}_{p'p} = \langle p'|S^{(0)}|p\rangle = &-\frac{i\kappa^2 M}{2(2\pi)^2\sqrt{p'_0 p_0}} \int \frac{\delta(q_0)}{|\vec{q}|^2} \Big[ p'_\mu p_\nu \delta_{\mu 4}\delta_{\nu 4} \\ &+ \frac{1}{2}\left(m^2 + \frac{1}{4}\delta_{\mu\alpha}\delta_{\mu\beta} q_\alpha q_\beta\right)\Big] \delta(\vec{p} - \vec{k} - \vec{q}) \, d^3q \\ = &\, F_{sc}(p', p)\delta(p'_0 - p_0), \end{aligned} \tag{6.45}$$

$$\begin{aligned} S^{(1/2)}_{p'p} = \langle p'|S^{(1/2)}|p\rangle = &-\frac{i\kappa^2 m M}{4(2\pi)^2\sqrt{p'_0 p_0}} \int \frac{\delta(q_0)}{|\vec{q}|^2} \\ &\times \left\{ \left[ i p_\nu \overline{u}^{(s)}(\vec{p}')\gamma_\mu u^{(r)}(\vec{p}) + i p'_\nu \overline{u}^{(s)}(\vec{p}')\gamma_\mu u^{(r)}(\vec{p}) \right] \right. \\ &\times \left. \left( \delta_{\mu 4}\delta_{\nu 4} + \frac{1}{2}\delta_{\mu\nu} \right) + 2m\overline{u}^{(s)}(\vec{p}')u^{(r)}(\vec{p}) \right\} \\ &\times \delta(\vec{p}' - \vec{p} - \vec{q}) \, d^3q = F_{sp}(p', p)\delta(p'_0 - p_0), \end{aligned} \tag{6.46}$$

$$S_{p'p}^{(1)Proca} = \langle p'|S^{(1)Proca}|p\rangle = -\frac{i\kappa^2 M}{2(2\pi)^2\sqrt{p'_0 p_0}}\int \frac{\delta(q_0)}{|\vec{q}|^2}$$
$$\times \left\{\left[e_\nu^{(s)}(\vec{p}\,')p'_\mu - e_\mu^{(s)}(\vec{p}\,')p'_\nu\right]\left[e_\alpha^{(r)}(\vec{p})p_\mu - e_\mu^{(r)}(\vec{p})p_\alpha\right]\right.$$
$$\left.\times \left(\delta_{\nu 4}\delta_{\alpha 4} - \frac{1}{4}\delta_{\nu\alpha}\right) + m^2 e_\nu^{(s)}(\vec{p}\,')e_\alpha^{(r)}(\vec{p})\delta_{\nu 4}\delta_{\alpha 4}\right\}$$
$$\times\, \delta(\vec{p}\,' - \vec{p} - \vec{q})\; d^3q = F_{vect}^{Proca}(p', p)\delta(p'_0 - p_0), \quad (6.47)$$

$$S_{p'p}^{(1)Stueck} = \langle p'|S^{(1)Stueck}|p\rangle = -\frac{i\kappa^2 M}{2(2\pi)^2\sqrt{p'_0 p_0}}\int \frac{\delta(q_0)}{|\vec{q}|^2}$$
$$\times \left\{e_\mu^{(s)}(\vec{p}\,')e_\nu^{(r)}(\vec{p})p_\alpha\left[\left(\delta_{\mu 4}\delta_{\nu 4} - \frac{1}{2}\delta_{\mu\nu}\right)\delta_{\alpha\beta}q_\beta\right.\right.$$
$$\left. + \left(\delta_{\mu 4}\delta_{\alpha 4} - \frac{1}{2}\delta_{\mu\alpha}\right)\delta_{\nu\beta}q_\beta - \left(\delta_{\nu 4}\delta_{\alpha 4} - \frac{1}{2}\delta_{\nu\alpha}\right)\delta_{\mu\beta}q_\beta\right]$$
$$+ e_\mu^{(s)}(\vec{p}\,')e_\mu^{(r)}(\vec{p})p'_\nu p_\alpha\left(\delta_{\nu 4}\delta_{\alpha 4} - \frac{1}{2}\delta_{\nu\alpha}\right) + e_\mu^{(s)}(\vec{p}\,')e_\alpha^{(r)}(\vec{p})p'_\nu$$
$$\left.\times\, p_\nu\delta_{\mu 4}\delta_{\alpha 4} + m^2 e_\nu^{(s)}(\vec{p}\,')e_\alpha^{(r)}(\vec{p})\delta_{\nu 4}\delta_{\alpha 4}\right\}\delta(\vec{p}\,' - \vec{p} - \vec{q})\; d^3q$$
$$= F_{vect}^{Stueck}(p', p)\delta(p'_0 - p_0), \quad (6.47')$$

$$S_{p'p}^{(3/2)} = \langle p'|S^{(3/2)}|p\rangle = -\frac{i\kappa^2 mM}{2(2\pi)^2\sqrt{p'_0 p_0}}\int \frac{\delta(q_0)}{|\vec{q}|^2}$$
$$\times \left\{\left[ip'_\nu \overline{u}_\alpha^{(s)}(\vec{p}\,')\gamma_\mu u_\alpha^{(r)}(\vec{p})\delta_{\mu 4}\delta_{\nu 4} + i\overline{u}_\alpha^{(s)}(\vec{p}\,')\gamma_\mu p_\nu u_\alpha^{(r)}(\vec{p})\delta_{\mu 4}\delta_{\nu 4}\right.\right.$$
$$\left. + m\overline{u}_\alpha^{(s)}(\vec{p}\,')u_\alpha^{(r)}(\vec{p})\right]\delta(\vec{p}\,' - \vec{p} - \vec{q})\; d^3q = F_{RS}(p', p)\delta(p'_0 - p_0), \quad (6.48)$$

and

$$S_{p'p}^{(2)} = \langle p'|S^{(2)}|p\rangle = -\frac{i\kappa^2 M}{2(2\pi)^2\sqrt{p'_0 p_0}}\int \frac{\delta(q_0)}{|\vec{q}|^2}$$
$$\times \left\{\left[e_{\alpha\nu}^{(s)}(\vec{p}\,')p'_\mu\left(3e_{\lambda\nu}^{(r)}(\vec{p})p_\alpha - e_{\lambda\alpha}^{(r)}(\vec{p})p_\nu - e_{\nu\alpha}^{(r)}(\vec{p})p_\lambda\right)\right.\right.$$
$$+ e_{\alpha\mu}^{(s)}(\vec{p}\,')p'_\nu\left(e_{\lambda\nu}^{(r)}(\vec{p})p_\alpha + e_{\lambda\alpha}^{(r)}(\vec{p})p_\nu + e_{\alpha\nu}^{(r)}(\vec{p})p_\lambda\right)$$

$$+ e^{(s)}_{\mu\nu}(\vec{p}\,')p'_\alpha\Big(e^{(r)}_{\alpha\lambda}(\vec{p})p_\nu + e^{(r)}_{\alpha\nu}(\vec{p})p_\lambda - 3e^{(r)}_{\lambda\nu}(\vec{p})p_\alpha\Big) - 2m^2 e^{(s)}_{\mu\nu}(\vec{p}\,')$$

$$\times\, e^{(r)}_{\lambda\nu}(\vec{p})\Big]\delta_{\mu 4}\delta_{\lambda 4} - \Big[\Big(e^{(s)}_{\lambda\nu}(\vec{p}\,')p'_\mu + e^{(s)}_{\lambda\mu}(\vec{p}\,')p'_\nu - e^{(s)}_{\mu\nu}(\vec{p}\,')p'_\lambda\Big)$$

$$\times\,\Big(e^{(r)}_{\mu\nu}(\vec{p})p_\lambda + e^{(r)}_{\mu\lambda}(\vec{p})p_\nu - e^{(r)}_{\lambda\nu}(\vec{p})p_\mu\Big) - \frac{1}{2}m^2 e^{(s)}_{\mu\nu}(\vec{p}\,')e^{(r)}_{\mu\nu}(\vec{p})\Big]$$

$$+ \Big[\Big(e^{(s)}_{\lambda\nu}(\vec{p}\,')p'_\mu + e^{(s)}_{\lambda\mu}(\vec{p}\,')p'_\nu - e^{(s)}_{\mu\nu}(\vec{p}\,')p'_\lambda\Big)e^{(r)}_{\mu\sigma}(\vec{p})$$

$$\times\,\Big[\Big(\delta_{\lambda 4}\delta_{\sigma 4} - \frac{1}{2}\delta_{\lambda\sigma}\Big)\delta_{\nu j} + \Big(\delta_{\nu 4}\delta_{\sigma 4} - \frac{1}{2}\delta_{\nu\sigma}\Big)\delta_{\lambda j} - \Big(\delta_{\lambda 4}\delta_{\nu 4} - \frac{1}{2}\delta_{\lambda\nu}\Big)$$

$$\times\,\delta_{\sigma j}\Big] - e^{(s)}_{\sigma\lambda}(\vec{p}\,')\Big(e^{(r)}_{\mu\nu}(\vec{p})p_\lambda + e^{(r)}_{\mu\lambda}(\vec{p})p_\nu - e^{(r)}_{\lambda\nu}(\vec{p})p_\mu\Big)\Big[\Big(\delta_{\mu 4}\delta_{\sigma 4}$$

$$- \frac{1}{2}\delta_{\mu\sigma}\Big)\delta_{\nu j} + \Big(\delta_{\nu 4}\delta_{\sigma 4} - \frac{1}{2}\delta_{\nu\sigma}\Big)\delta_{\mu j} - \Big(\delta_{\mu 4}\delta_{\nu 4} - \frac{1}{2}\delta_{\mu\nu}\Big)\delta_{\sigma j}\Big]\Big] q_j\Big\}$$

$$\times\,\delta(\vec{p}\,' - \vec{p} - \vec{q})\; d^3q = F_{tens}(p', p)\delta(q_0), \tag{6.49}$$

where

$$u^{(r)}(\vec{p}),\ \Big[\overline{u}^{(r)}(\vec{p}) = (u^{(r)})^+(\vec{p})\gamma_4\Big],\ e^{(r)}_\mu(\vec{p}),\ u^{(r)}_\mu(\vec{p}),$$

$$\Big[\overline{u}^{(r)}_\mu(\vec{p}) = (u^{(r)}_\mu)^+(\vec{p})\gamma_4\Big],\ e^{(r)}_{\mu\nu}(\vec{p}),$$

on the one side, and

$$u^{(s)}(\vec{p}\,'),\ \Big[\overline{u}^{(s)}(\vec{p}\,') = (u^{(s)})^+(\vec{p}\,')\gamma_4\Big],\ e^{(s)}_\mu(\vec{p}\,'),\ u^{(s)}_\mu(\vec{p}\,'),$$

$$\Big[\overline{u}^{(s)}_\mu(\vec{p}\,') = (u^{(s)}_\mu)^+(\vec{p}\,')\gamma_4\Big],\ e^{(s)}_{\mu\nu}(\vec{p}\,'),$$

on the other side, are the spinors, vectors, spin-vectors and tensors of the corresponding spin-states of the initial and final massive particles, respectively. As one can see, the same notations ($m$ – mass, $p$ – four-momentum, and $p_0$ – energy) have been used for all field types, $q_0 = p'_0 - p_0 = 0$ expresses the energy conservation law, while the superscript + signifies the Hermitian conjugation.

The differential scattering cross-section is given by the well-known expression

$$d\sigma = (2\pi)^2 \left\langle \sum_{fin\,sp} \left|F(p',p)\right|^2 \right\rangle_{in\,sp} p_0^2\, d\Omega, \tag{6.50}$$

where $d\Omega = 2\pi \sin\theta\, d\theta$, $\theta$ being the scattering angle. In order to determine the scattering cross-sections one must first calculate the quantities $\left\langle \sum_{fin\,sp} |F(p',p)|^2 \right\rangle_{in\,sp}$. To this end, using (6.45)–(6.49) one obtains the following expressions for the quantity $|F(p',p)|^2$, corresponding to the scalar, spinorial, vectorial, spin-vectorial and second-rank tensorial massive fields:

$$\left|F_{sc}(p',p)\right|^2 = \left( \frac{\kappa^2 M}{8(2\pi)^2 p_0 \vec{p}^2 \sin^2\dfrac{\theta}{2}} \right)^2 \left|Q_{sc}(p',p)\right|^2, \tag{6.51}$$

$$\left|F_{sp}(p',p)\right|^2 = \left( \frac{\kappa^2 m M}{16(2\pi)^2 p_0 (p_0^2 - m^2) \sin^2\dfrac{\theta}{2}} \right)^2 \left|Q_{sp}(p',p)\right|^2, \tag{6.52}$$

$$\left|F_{vect}^{Proca}(p',p)\right|^2 = \left( \frac{\kappa^2 M}{8(2\pi)^2 p_0 \vec{p}^2 \sin^2\dfrac{\theta}{2}} \right)^2 \left|Q_{vect}^{Proca}(p',p)\right|^2, \tag{6.53}$$

$$\left|F_{vect}^{Stueck}(p',p)\right|^2 = \left( \frac{\kappa^2 M}{8(2\pi)^2 p_0 \vec{p}^2 \sin^2\dfrac{\theta}{2}} \right)^2 \left|Q_{vect}^{Stueck}(p',p)\right|^2, \tag{6.53'}$$

$$\left|F_{RS}(p',p)\right|^2 = \left( \frac{\kappa^2 m M}{16(2\pi)^2 p_0 (p_0^2 - m^2) \sin^2\dfrac{\theta}{2}} \right)^2 \left|Q_{RS}(p',p)\right|^2, \tag{6.54}$$

and

$$\left|F_{tens}(p',p)\right|^2 = \left(\frac{\kappa^2 M}{8(2\pi)^2 p_0 \vec{p}^2 \sin^2\dfrac{\theta}{2}}\right)^2 \left|Q_{tens}(p',p)\right|^2, \qquad (6.55)$$

respectively.

The quantities $\left\langle \sum_{fin\,sp} |F(p',p)|^2 \right\rangle_{in\,sp}$ corresponding to the scalar, spinorial, vectorial, spin-vectorial and tensorial fields are given by

$$\left\langle \sum_{fin\,sp} \left|F_{sc}(p',p)\right|^2 \right\rangle_{in\,sp} \equiv |F_{sc}(p',p)|^2$$
$$= \left(\frac{\kappa^2 M}{8(2\pi)^2 p_0 \vec{p}^2 \sin^2\dfrac{\theta}{2}}\right)^2 |Q_{sc}(p',p)|^2, \qquad (6.56)$$

$$\left\langle \sum_{fin\,sp} \left|F_{sp}(p',p)\right|^2 \right\rangle_{in\,sp} = \left(\frac{\kappa^2 m M}{16(2\pi)^2 p_0 (p_0^2 - m^2)\sin^2\dfrac{\theta}{2}}\right)^2$$
$$\times \left(\frac{1}{2}\sum_{spins} Q_{sp}^2\right)$$
$$= \left(\frac{\kappa^2 m M}{16(2\pi)^2 p_0 (p_0^2 - m^2)\sin^2\dfrac{\theta}{2}}\right)^2 \frac{1}{2}\sum_{r,s=1}^{2} \left|\overline{u}^{(s)}(\vec{p}')\, t\, u^{(r)}(\vec{p})\right|^2;$$

$$t = 2i\gamma_4 p_0 + \frac{1}{2}\gamma_\alpha(p'_\alpha + p_a) - 2im, \qquad (6.57)$$

$$\left\langle \sum_{fin\,sp} \left|F_{vect}^{Proca}(p',p)\right|^2 \right\rangle_{in\,sp} = \left(\frac{\kappa^2 M}{8(2\pi)^2 p_0 \vec{p}^2 \sin^2\dfrac{\theta}{2}}\right)^2 \qquad (6.58)$$
$$\times \left(\frac{1}{3}\sum_{pol} \left(Q_{vect}^{Proca}\right)^2\right),$$

$$\left\langle \sum_{fin\,sp} \left| F_{vect}^{Stueck}(p',p) \right|^2 \right\rangle_{in\,sp} = \left( \frac{\kappa^2 M}{8(2\pi)^2 p_0 \vec{p}^2 \sin^2 \frac{\theta}{2}} \right)^2 \times \left[ \frac{1}{3} \sum_{pol} \left( Q_{vect}^{Stueck} \right)^2 \right], \tag{6.58'}$$

$$\left\langle \sum_{fin\,sp} \left| F_{RS}(p',p) \right|^2 \right\rangle_{in\,sp} = \left( \frac{\kappa^2 mM}{16(2\pi)^2 p_0 (p_0^2 - m^2) \sin^2 \frac{\theta}{2}} \right)^2$$

$$\times \left( \frac{1}{4} \sum_{sp-pol} Q_{RS}^2 \right) = \left( \frac{\kappa^2 mM}{16(2\pi)^2 p_0 (p_0^2 - m^2) \sin^2 \frac{\theta}{2}} \right)^2 \times \frac{1}{4} \sum_{r,s=1}^{4} \left| \overline{u}_{\alpha}^{(s)}(\vec{p}')\, t\, u_{\alpha}^{(r)}(\vec{p}) \right|^2 ; \quad t = 2i\gamma_4 p_0 - im, \tag{6.59}$$

and

$$\left\langle \sum_{fin\,sp} \left| F_{tens}(p',p) \right|^2 \right\rangle_{in\,sp} = \left( \frac{\kappa^2 M}{8(2\pi)^2 p_0 \vec{p}^2 \sin^2 \frac{\theta}{2}} \right)^2 \left( \frac{1}{5} \sum_{pol} Q_{tens}^2 \right), \tag{6.60}$$

respectively, where

$$\sum_{spins} Q_{sp}^2, \sum_{pol} \left( Q_{vect}^{Proca} \right)^2, \sum_{pol} \left( Q_{vect}^{Stueck} \right)^2, \sum_{sp-pol} Q_{RS}^2 \text{ and } \sum_{pol} Q_{tens}^2$$

are the sums over spins of the spinorial, vectorial, spin-vectorial and tensorial particles, respectively; since these expressions are very long, they are not given explicitly here.

To evaluate the above sums one must consider the following relations satisfied by the spinors, vectors, spin-vectors and tensors of the corresponding spin-states of massive particles (see Refs. [252, 331]),

$$\sum_{r=1}^{2} u^{(r)}(\vec{p}) \overline{u}^{(r)}(\vec{p}) = \frac{\gamma_\alpha p_\alpha + im}{2im}, \tag{6.61}$$

$$\sum_{\lambda=1}^{3} e_{\mu}^{(\lambda)}(\vec{p})e_{\nu}^{(\lambda)}(\vec{p}) = d_{\mu\nu}, \quad \left(\mu,\nu=\overline{1,4}\right), \tag{6.62}$$

$$\begin{aligned}\sum_{r=1}^{4} u_{\mu}^{(r)}(\vec{p})\overline{u}_{\nu}^{(r)}(\vec{p}) = \frac{\gamma_\alpha p_\alpha + im}{2im}\bigg[\delta_{\mu\nu} - \frac{1}{3}\gamma_\mu\gamma_\nu \\ + \frac{i}{3m}\left(\gamma_\mu p_\nu - \gamma_\nu p_\mu\right) + \frac{2}{3m^2}p_\mu p_\nu\bigg],\end{aligned} \tag{6.63}$$

and

$$\Pi_{\mu\nu,\lambda\rho} \equiv \sum_{r=1}^{5} e_{\mu\nu}^{(r)}(\vec{p})e_{\lambda\rho}^{(r)}(\vec{p}) = \frac{1}{2}\left(d_{\mu\lambda}d_{\nu\rho} + d_{\mu\rho}d_{\nu\lambda}\right) - \frac{1}{3}d_{\mu\nu}d_{\lambda\rho}, \tag{6.64}$$

respectively, where

$$d_{\mu\nu} = \delta_{\mu\nu} + \frac{p_\mu p_\nu}{m^2}. \tag{6.65}$$

After long, laborious calculations, the following expressions for the differential scattering cross-sections of scalar, spinorial, vectorial, spin-vectorial and tensorial massive particles have been obtained,

$$d\sigma_{sc} = \left(\frac{\kappa^2 M}{16\pi}\right)^2 \frac{d\Omega}{\sin^4\frac{\theta}{2}}\left(\frac{1+v^2\cos^2\frac{\theta}{2}}{2v^2}\right)^2, \tag{6.66}$$

$$d\sigma_{sp} = \left(\frac{\kappa^2 M}{16\pi}\right)^2 \frac{d\Omega}{\sin^4\frac{\theta}{2}}\left[\left(\frac{1+v^2}{2v^2}\right)^2 - \frac{3+v^2}{4v^2}\sin^2\frac{\theta}{2}\right], \tag{6.67}$$

$$d\sigma_{vect}^{Proca} = \left(\frac{\kappa^2 M}{16\pi}\right)^2 \frac{d\Omega}{\sin^4\frac{\theta}{2}}\left[\left(\frac{1+v^2}{2v^2}\right)^2 - \frac{2}{3}\sin^2\frac{\theta}{2}\left(\frac{2}{v^2} - \sin^2\frac{\theta}{2}\right)\right], \tag{6.68}$$

$$d\sigma_{vect}^{Stueck} = \left(\frac{\kappa^2 M}{16\pi}\right)^2 \frac{d\Omega}{\sin^4\frac{\theta}{2}}\left\{\left(\frac{1+v^2}{2v^2}\right)^2 - \frac{1}{3}\sin^2\frac{\theta}{2}\left\{\frac{1-6v^2+v^4}{v^2(1-v^2)}\right.\right.$$

$$+\frac{4}{(1-v^2)^2}\sin^2\frac{\theta}{2}\left[\frac{5-10v^2+v^4}{4}+v^2\sin^2\frac{\theta}{2}\left(1+v^2-v^2\sin^2\frac{\theta}{2}\right)\right]\Bigg\}\Bigg\}, \tag{6.68'}$$

$$d\sigma_{RS} = \left(\frac{\kappa^2 M}{16\pi}\right)^2 \frac{d\Omega}{\sin^4\frac{\theta}{2}}\left\{\left(\frac{1+v^2}{2v^2}\right)^2 - \frac{1}{36v^4(1-v^2)^2}\right.$$
$$\times\left[v^2(15-41v^2+5v^4+21v^6)\sin^2\frac{\theta}{2}+4v^4(3-6v^2\right.$$
$$\left.\left.-5v^4)\sin^4\frac{\theta}{2}+8v^6(3+v^2)\sin^6\frac{\theta}{2}\right]\right\}, \tag{6.69}$$

and

$$d\sigma_{tens} = \left(\frac{\kappa^2 M}{16\pi}\right)^2 \frac{d\Omega}{\sin^4\frac{\theta}{2}}\left\{\left(\frac{1+v^2}{2v^2}\right)^2 - \frac{4}{v^2}\sin^2\frac{\theta}{2}+\frac{2}{45(1-v^2)^4}\right.$$
$$\times\sin^4\frac{\theta}{2}\left[9\left(31-108v^2+146v^4-92v^6+23v^8\right)\right.$$
$$+48v^2\left(7-18v^2+19v^4-8v^6\right)\sin^2\frac{\theta}{2}$$
$$-24v^4\left(5-6v^2-11v^4\right)\sin^4\frac{\theta}{2}$$
$$\left.\left.-192v^6\left(1+v^2\right)\sin^6\frac{\theta}{2}+128v^8\sin^8\frac{\theta}{2}\right]\right\}, \tag{6.70}$$

respectively, where $v$ denotes the ratio $|\vec{p}|/p_0$.

In the limit of small scattering angles, the corresponding sums over spins become

$$|Q_{sc}|^2 = \vec{p}^4\left(\frac{1+v^2}{2v^2}\right)^2, \tag{6.71}$$

$$\sum_{spins} Q_{sp}^2 = 2p_0^2\frac{(1+v^2)^2}{1-v^2}, \tag{6.72}$$

$$\sum_{pol}\left(Q_{vect}^{Proca}\right)^2 = 3\vec{p}^4\left(\frac{1+v^2}{2v^2}\right)^2, \tag{6.73}$$

$$\sum_{pol}\left(Q_{vect}^{Stueck}\right)^2 = 3\vec{p}^{\,4}\left(\frac{1+v^2}{2v^2}\right)^2, \tag{6.74}$$

$$\sum_{sp-pol} Q_{RS}^2 = 4p_0^2\frac{(1+v^2)^2}{1-v^2}, \tag{6.75}$$

and

$$\sum_{pol} Q_{tens}^2 = 5\vec{p}^{\,4}\left(\frac{1+v^2}{2v^2}\right)^2, \tag{6.76}$$

respectively.

According to the above considerations, in the small scattering angle approximation, all the differential scattering cross-sections shall have the same form,

$$d\sigma_{\theta\ll 1} = \left(\frac{\kappa^2 M}{16\pi}\right)^2 \frac{d\Omega}{\sin^4\dfrac{\theta}{2}}\left(\frac{1+v^2}{2v^2}\right)^2 = d\sigma_{Ruth}, \tag{6.77}$$

therefore, the differential scattering cross-sections in all six cases are of Rutherford-type. As one observes, in Eqs. (6.67)–(6.70) the expression $d\sigma_{Ruth}$ is contained as a first term. Since this term (*i.e.* $d\sigma_{Ruth}$) is precisely the differential scattering cross-section of the massive scalar particles (for example, the massive scalar mesons) in the case of minimal coupling (that is, without the non-minimal coupling term introduced through the medium of the scalar curvature $R$), the second term of Eqs. (6.67)–(6.70) can be interpreted as the contribution of the non-null spin of the spinorial, vectorial, Rarita–Schwinger and tensorial massive particles, respectively.

A special interest concerns the limiting case of backscattering. In such a case, the differential scattering cross-sections of the spin-0, 1/2, 1, 3/2, and 2 massive particles are given by

$$d\sigma_{sc}^* = \left(\frac{\kappa^2 M}{16\pi}\right)^2 \frac{d\Omega^*}{4v^4}, \tag{6.78}$$

$$d\sigma_{sp}^* = \left(\frac{\kappa^2 M}{16\pi}\right)^2 \frac{\left(1-v^2\right)}{4v^4} d\Omega^*, \tag{6.79}$$

$$\left(d\sigma_{vect}^{Proca}\right)^* = \left(\frac{\kappa^2 M}{16\pi}\right)^2 \left[\left(\frac{1+v^2}{2v^2}\right)^2 - \frac{2}{3}\frac{2-v^2}{v^2}\right] d\Omega^*, \tag{6.80}$$

$$\left(d\sigma_{vect}^{Stueck}\right)^* = \left(\frac{\kappa^2 M}{16\pi}\right)^2 \left[\left(\frac{1+v^2}{2v^2}\right)^2 - \frac{1}{3v^2}\right] d\Omega^*, \tag{6.80$'$}$$

$$d\sigma_{RS}^* = \left(\frac{\kappa^2 M}{16\pi}\right)^2 \frac{9-6v^2+5v^4}{36v^4(1-v^2)} d\Omega^*, \tag{6.81}$$

and

$$\begin{aligned} d\sigma_{tens}^* = \left(\frac{\kappa^2 M}{16\pi}\right)^2 \frac{d\Omega^*}{180v^4(1-v^2)^4}\Big(45 - 810v^2 \\ + 5067v^4 - 9228v^6 + 5475v^8 - 522v^{10} + 229v^{12}\Big), \end{aligned} \tag{6.82}$$

respectively, where $d\sigma^* = d\sigma|_{\theta=\pi}$ and $d\Omega^* = 2\pi\, d\theta$.

In the ultrarelativistic case ($v \to 1$), the following results are obtained from Eqs. (6.78)–(6.80$'$):

$$\frac{d\sigma_{sc}^*}{d\Omega^*} = \frac{1}{4}(GM)^2, \quad \text{(U.R.)} \tag{6.83}$$

$$\frac{d\sigma_{sp}^*}{d\Omega^*} = 0, \quad \text{(U.R.)} \tag{6.84}$$

$$\frac{\left(d\sigma_{vect}^{Proca}\right)^*}{d\Omega^*} = \frac{1}{3}(GM)^2, \quad \text{(U.R.)} \tag{6.85}$$

$$\frac{\left(d\sigma_{vect}^{Stueck}\right)^*}{d\Omega^*} = \frac{2}{3}(GM)^2. \quad \text{(U.R.)} \tag{6.85$'$}$$

Thus, in the backscattering case and, also, in the ultrarelativistic limit, the differential cross-section $d\sigma^*/d\Omega^*$ for the massive vectorial field is a constant (and different from zero), meaning that in this special limit case the helicity of the vectorial particles – quanta of the corresponding field – is not conserved; this result is in concordance with those obtained in Refs. [91, 367, 368].

We draw the reader's attention upon the fact that, even if our discussion concerns massive particles, the notion of helicity still makes sense because of the particular analyzed situation, *i.e.*, the special case of ultrarelativistic motion of massive particles. For these massive particles, when taking the limit $v \to c$ one must obtain such results that are similar (even identical) to those corresponding to massless particles. Indeed – a little forcing the phrase – in the limit $v \to c$ the massive particles "turn" into non-massive ones (in agreement with TSR (Theory of Special Relativity), only massless particles can have velocity c).

Regarding the spin-3/2 and 2 fields, as one can easily observe from (6.81) and (6.82), in the backscattering case and ultrarelativistic limit the differential scattering cross-sections are divergent. This fact is graphically illustrated in Figs. 6.10 and 6.12, where the integral/total scattering cross-sections are represented in terms of the ratio $v \equiv |\vec{p}|/p_0$. As one can observe, in the ultrarelativistic limit the corresponding graphs asymptotically approach the straight line $v = 1$.

The value of the backscattering cross-section in the ultrarelativistic limit for the massive scalar field is also a non-null constant, while the spinorial field gives a singular value (namely, zero) for this quantity, meaning that in this case the particle helicity is conserved in the scattering process.

Since in the limit $\theta \to 0$ all the differential scattering cross-sections are of Rutherford-type, some problems occur when calculating the integral cross-sections. More precisely, some divergent terms are obtained in the limit $\theta \to 0$ due to the fact that the differential cross-sections depend on $\sin^{-4}\frac{\theta}{2}$, so that the corresponding integral cross sections tend to infinity. The cause of this divergence is the same with that common to every scattering process which leads to a Rutherford-type cross-section: the infinite range combined with the $r^{-2}$ dependence of the field's force – in our case, the gravitational force.

In order to surpass this difficulty, the primitives of the functions $f(v,\theta) = \frac{d\sigma}{d\Omega}(v,\theta)$ have been first determined; in other words, first the total scattering cross-sections have been calculated as indefinite integrals of the corresponding differential cross-sections,

$$\sigma(v,\theta) = \int \frac{d\sigma}{d\Omega}(v,\theta)\, d\Omega = \int f(v,\theta)\, d\Omega, \tag{6.86}$$

and then the integral cross-sections for the five studied interaction processes have been determined by using the Leibniz–Newton formula

$$\sigma_{tot}(v) = \sigma(v,\theta)|_{\theta=\pi} - \sigma(v,\theta)|_{\theta=\varepsilon\ll 1}\,, \tag{6.87}$$

where $\varepsilon$ is a very small positive real number. As a matter of fact, this is an example for the so-called *cut-off procedure*. In the concrete estimations, the value of $\varepsilon$ was taken as $\varepsilon = 10^{-6}$. The graphs obtained under these circumstances, rigorously speaking, furnish only qualitative dependencies of the represented quantities. Obviously, for $\varepsilon \to 0$ one obtains divergent expressions for the total scattering cross-sections $\sigma_{tot}(v)$ in all six cases (the reason being already explained). The concrete calculations lead to the following (exact) expressions,

$$\sigma_{sc}(v,\theta) = -\frac{1}{2v^2}\left[2(1+v^2)\csc^2\frac{\theta}{2} + 4v^2\ln\left(\sin\frac{\theta}{2}\right)\right], \tag{6.88}$$

$$\sigma_{sp}(v,\theta) = \frac{1}{4v^2}\left[2(1+v^2)^2\csc\frac{\theta}{2} + 4v^2(3+v^2)\ln\left(\sin\frac{\theta}{2}\right)\right], \tag{6.89}$$

$$\sigma_{vect}^{Proca}(v,\theta) = -\frac{2}{3}\cos\theta - \frac{(1+v^2)^2}{2v^4}\cos^2\frac{\theta}{2} - \frac{16}{3v^2}\ln\left(\sin\frac{\theta}{2}\right), \tag{6.90}$$

$$\begin{aligned}\sigma_{vect}^{Stueck}(v,\theta) = \frac{T_1(v,\theta)}{3T_2(v,\theta)}\Bigg[&24v^4(20-32v^2+7v^4)\cos\theta - 48v^6\cos 2\theta \\ &- 8v^8\cos 3\theta - 144(1-v^4)^2\csc^2\frac{\theta}{2} \\ &- 384v^2(1-7v^2+7v^4-v^6)\ln\left(\sin\frac{\theta}{2}\right)\Bigg],\end{aligned} \tag{6.90'}$$

$$\sigma_{RS}(v,\theta) = \frac{3-3v^2-4v^4}{9(v^2-1)^2}\cos\theta - \frac{v^2(3+v^2)}{36(v^2-1)^2}\cos 2\theta - \frac{(1+v^2)^2}{2v^4}\csc^2\frac{\theta}{2} + \frac{15-26v^2-21v^4}{9v^2(v^2-1)}\ln\left(\sin\frac{\theta}{2}\right), \tag{6.91}$$

$$\begin{aligned}\sigma_{tens}(v,\theta) = &-\frac{186-536v^2+563v^4-246v^6+51v^8}{15(v^2-1)^4}\cos\theta \\ &+\frac{2v^2(14-41v^2+37v^4-8v^6)}{15(v^2-1)^4}\cos 2\theta \\ &+\frac{v^4(5+6v^2-8v^4)}{45(v^2-1)^4}\cos 3\theta - \frac{v^6(3-v^2)}{90(v^2-1)^4} \\ &\times\cos 4\theta - \frac{v^8}{225(v^2-1)^4}\cos 5\theta \\ &-\frac{(1+v^2)^2}{2v^4}\csc^2\frac{\theta}{2} - \frac{16}{v^2}\ln\left(\sin\frac{\theta}{2}\right),\end{aligned} \tag{6.92}$$

where

$$T_1(v,\theta) = \left(\frac{1+v^2}{2v^2}\right)^2 - \frac{1}{3}T_3(v,\theta)\sin^2\frac{\theta}{2},$$

$$\begin{aligned}T_2(v,\theta) = &\,24-16v^2+4v^4-32v^6+23v^8+4v^2(4-8v^2+3v^4+4v^6) \\ &\times\cos\theta - 4v^4(5-4v^2)\cos 2\theta + 4v^6(1-v^2)\cos 3\theta + v^8\cos 4\theta,\end{aligned}$$

and

$$T_3(v,\theta) = \frac{1-6v^2+v^4}{v^2(1-v^2)} + \frac{(5-10v^2+v^4)\sin^2\frac{\theta}{2} + 4v^2\left(1+v^2-v^2\sin^2\frac{\theta}{2}\right)\sin^4\frac{\theta}{2}}{(1-v^2)^2}.$$

The graphs for $\sigma_{tot}(v) = \sigma(v,\theta=\pi) - \sigma(v,\theta=\varepsilon=10^{-6})$ are represented in Figs. (6.2), (6.4), (6.8), (6.10) and (6.12). Since in the domain of very small values of the ratio $v$ the integral scattering cross-sections present a very rapid variation, it is not possible to clearly observe their dependence on $v$, in the above mentioned domain. An easier observation of the variation of $\sigma_{tot}(v)$ with $v$, for small and very small values of this ratio, is possible if one uses a logarithmic scale, as it has been shown in Figs. (6.3), (6.5), (6.7), (6.9), (6.11) and (6.13).

In this paragraph, beginning with relation (6.88), the factor $\frac{\kappa^4 M^2}{128\pi}$ $= 2\pi\left(\frac{\kappa^2 M}{16\pi}\right)^2$ has been omitted. Being a constant, this quantity affects in no way the shape of the curves.

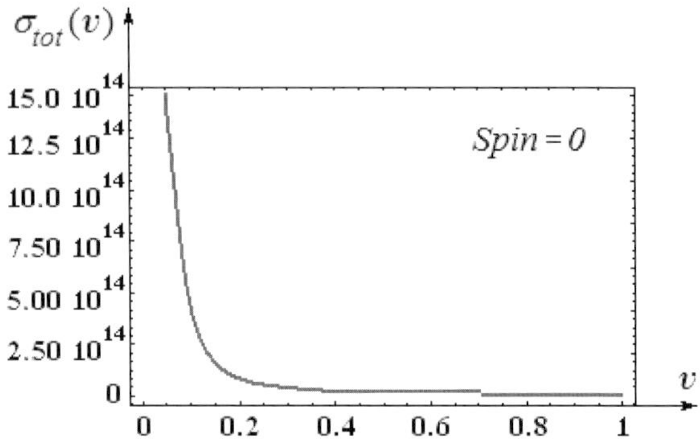


Fig. 6.2. Variation of the integral scattering cross-section $\sigma_{tot}(v)$ with the ratio $\frac{|\vec{p}|}{p_0}$ for the massive scalar field.

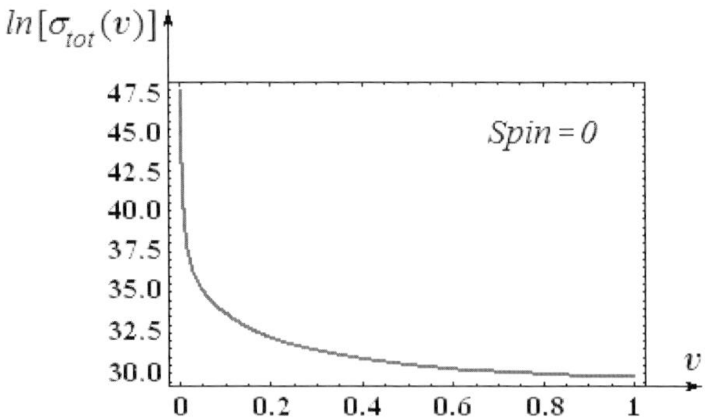


Fig. 6.3. Variation of $\ln[\sigma_{tot}(v)]$ with the ratio $\frac{|\vec{p}|}{p_0}$ for the massive scalar field.

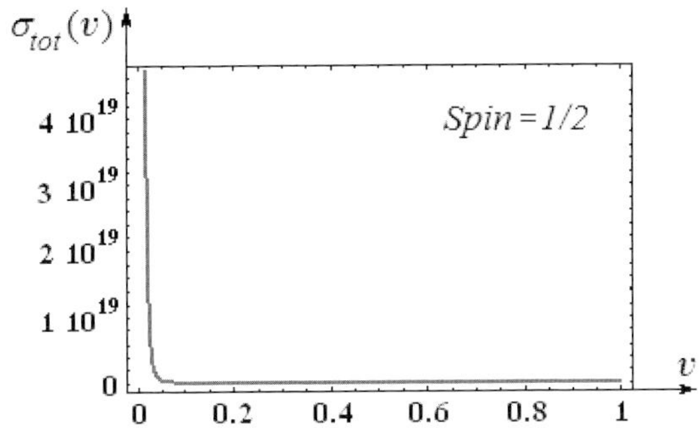


Fig. 6.4. Variation of the integral scattering cross-section $\sigma_{tot}(v)$ with the ratio $\frac{|\vec{p}|}{p_0}$ for the massive spinorial field.

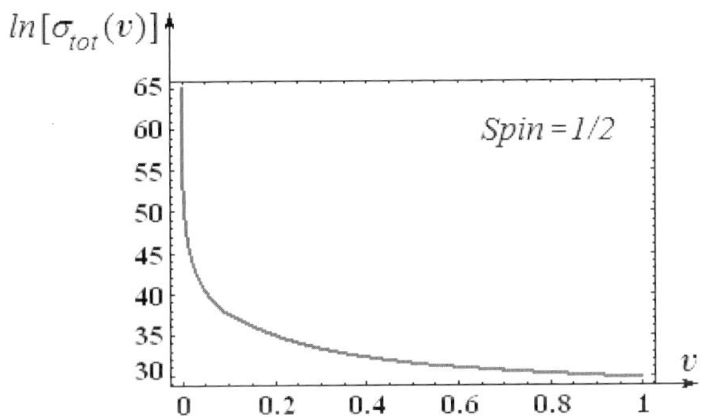


Fig. 6.5. Variation of $\ln[\sigma_{tot}(v)]$ with the ratio $\frac{|\vec{p}|}{p_0}$ for the massive spinorial field.

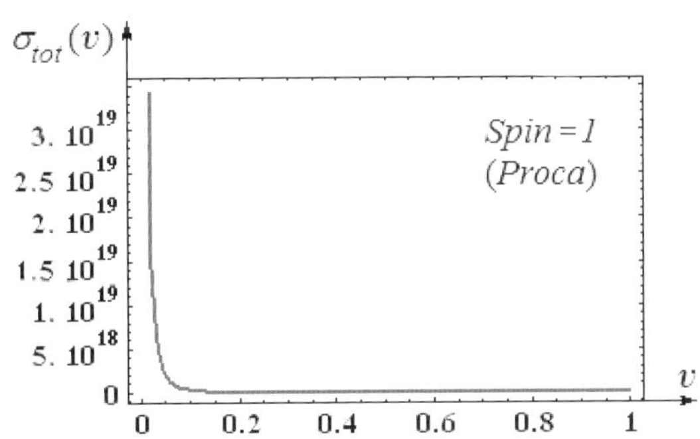


Fig. 6.6. Variation of the integral scattering cross-section $\sigma_{tot}(v)$ with the ratio $\frac{|\vec{p}|}{p_0}$ for the massive vectorial field (Proca formalism).

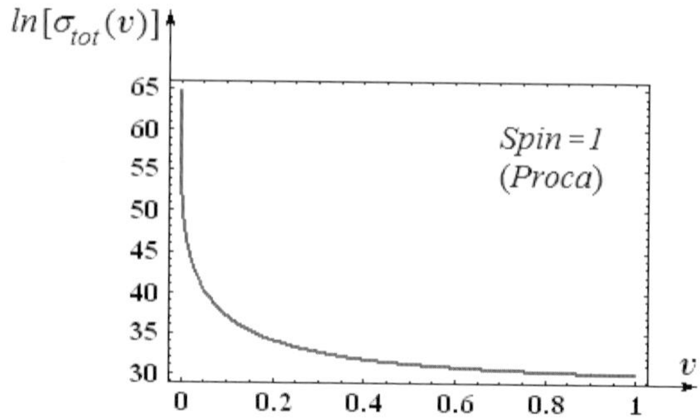


Fig. 6.7. Variation of $\ln[\sigma_{tot}(v)]$ with the ratio $\frac{|\vec{p}|}{p_0}$ for the massive spinorial field.

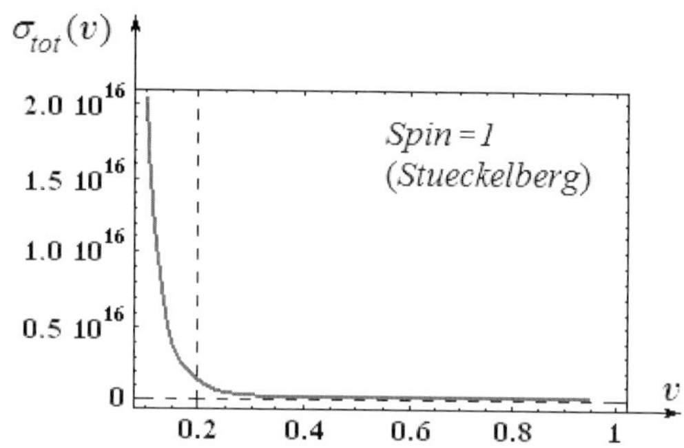


Fig. 6.8. Variation of the integral scattering cross-section $\sigma_{tot}(v)$ with the ratio $\frac{|\vec{p}|}{p_0}$ for the massive vectorial field (Stueckelberg formalism).

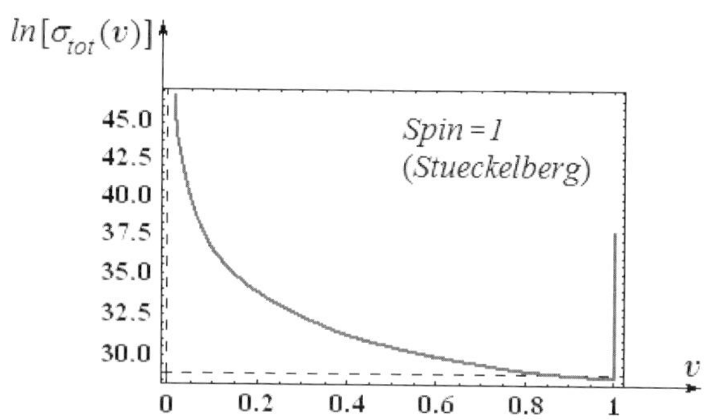


Fig. 6.9. Variation of $\ln[\sigma_{tot}(v)]$ with the ratio $\frac{|\vec{p}|}{p_0}$ for the massive vectorial field (Stueckelberg formalism).

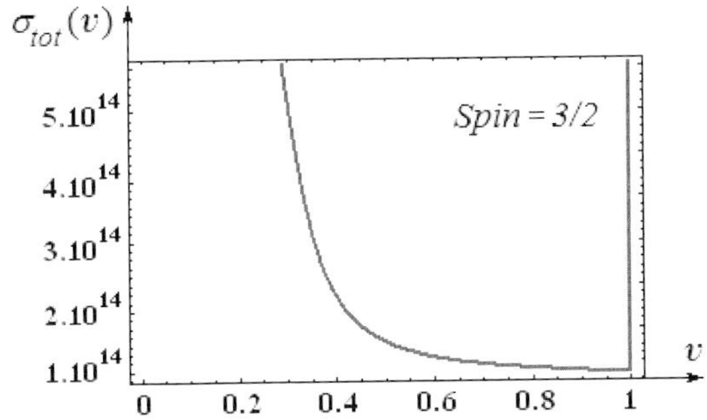


Fig. 6.10. Variation of the integral scattering cross-section $\sigma_{tot}(v)$ with the ratio $\frac{|\vec{p}|}{p_0}$ for the massive spin-3/2 field (Rarita–Schwinger).

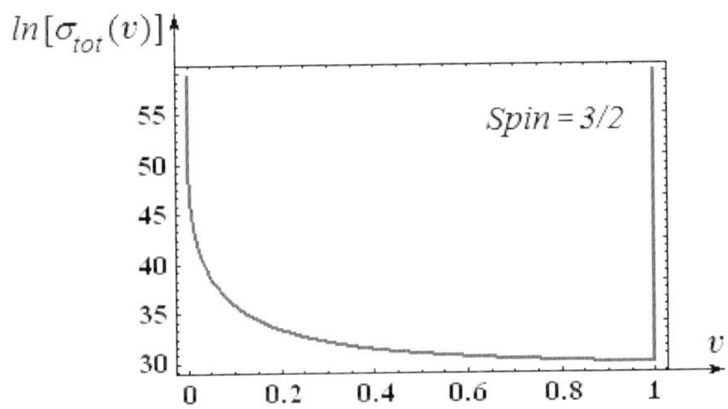


Fig. 6.11. Variation of $\ln[\sigma_{tot}(v)]$ with the ratio $\frac{|\vec{p}|}{p_0}$ for the massive spin-3/2 field (Rarita–Schwinger).

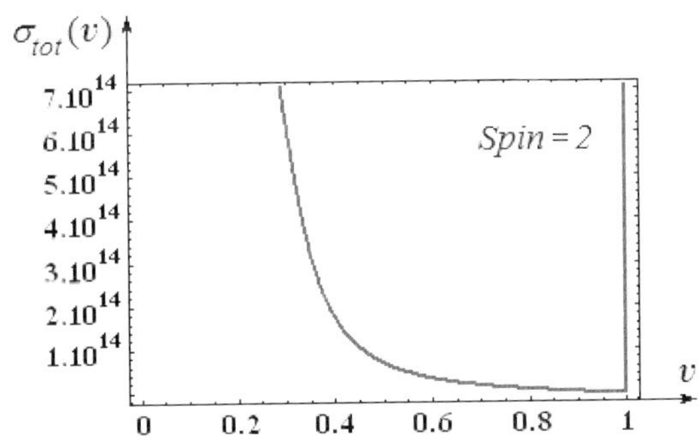


Fig. 6.12. Variation of the integral scattering cross-section $\sigma_{tot}(v)$ with the ratio $\frac{|\vec{p}|}{p_0}$ for the massive tensorial field.

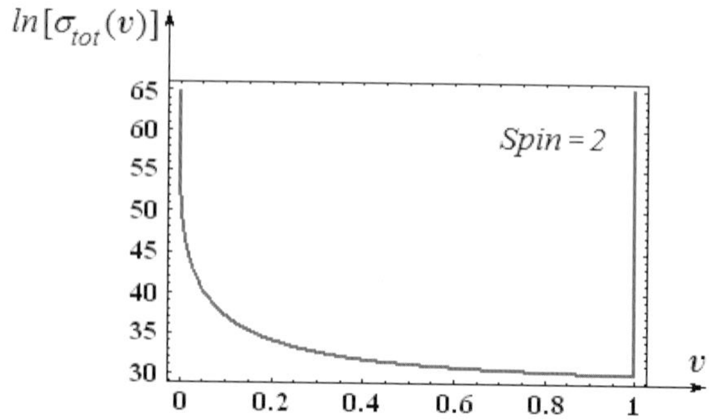


Fig. 6.13. Variation of $\ln[\sigma_{tot}(v)]$ with the ratio $\frac{|\vec{p}|}{p_0}$ for the massive tensorial field.

## VI.2. Gravitational scattering of spin 0, 1/2, 1, 3/2, and 2 massless particles in the centrally-symmetric gravitational field described by the Schwarzschild metric

### VI.2.1. General considerations

We specify from the start that, regarding the massless bosonic fields (of spin-0, 1, and 2), in this subchapter both terminologies, namely, that based on the wave character and that based on the corpuscular character of the substance, are going to be used.

Thus, it is talking about both the scattering of the scalar, electromagnetic (the mostly studied – and also with most applications – non-massive vectorial field) and gravitational (the most interesting and mostly investigated massless tensorial field) waves, and, respectively, the scattering of the non-massive scalar mesons[2], photons and gravitons.

Due to the specific character of the massless fermionic fields, the terminology is based on the corpuscular nature of the matter; that is why, the further investigations shall refer to the processes of neutrino[3] (quanta of the massless spinorial field) and gravitino (quanta of the massless spin-vectorial field) scattering.

---

[2]Hypothetical particles – at least so far – and the associated waves as well.

[3]Even if – as it is known – the rest mass of all types of neutrinos (electron-neutrino, muon-neutrino, tau-neutrino) is different from zero, in the following analysis the same term shall be used for the spin-1/2 non-massive particles.

The investigation procedure being – broadly speaking – the same as that utilized in the first subchapter, the conventions and general working frame shall not be mentioned anymore, but the study of the diffusion/scattering of the spin-0, 1/2, 1, 3/2, and 2 non-massive "waves"/particles shall be straightly approached. Nevertheless, there is an important aspect connected to the massless fields, namely the concrete way of calculation of the quantities $\left\langle \sum_{fin\,sp} |F(p',p)|^2 \right\rangle_{in\,sp}$, interfering in the differential scattering cross-sections. Since the particles with zero rest mass have only two polarization states, the usage of the same names/denominations like in the case of the massive particles, for the quantity $\left\langle \sum_{fin\,sp} |F_{sc}(p',p)|^2 \right\rangle_{in\,sp}$ – such as *summation* over the particle spins in the final state (after scattering), or *average* over the particle spins in the initial state (before scattering) – are somewhat inappropriate. More precisely, using the wave terminology, if the incident waves are polarized (polarized light – in the case of electromagnetic field), then there is no sense to perform an averaging over the polarization states of the incident photons. Otherwise (if the light is not polarized), summation and averaging over the photon polarizations in the final and initial states, respectively, have to be performed. However, but keeping in mind the above mentioned aspect, for the sake of convenience, the same notation $\left\langle \sum_{fin\,sp} |F(p',p)|^2 \right\rangle_{in\,sp}$ shall be maintained for the quantity that determines the matrix elements corresponding to the scattering processes.

## VI.2.2. Gravitational scattering of massless scalar, spinorial, vectorial, spin-vectorial and tensorial particles

This investigation is performed within the approximation of the external gravitational field, and the first-order Born approximation as well.

The fields specified in the title shall not be approached in a general manner, but the reader's attention shall be specifically drawn upon certain massless fields – which are of special interest – namely the following fields: *neutrinic* (as an example of non-massive spinorial field), *electromagnetic* (as an example of massless vectorial field), *non-massive Rarita–Schwinger* (as an example of massless spin-vectorial field) and *gravitational* (as an example of non-massive second-rank tensorial field).

Here are the starting Lagrangian densities:

1) for the massless complex scalar field, considering the Klein–Fock formalism and a non-minimal coupling introduced through the medium of the scalar curvature $R$,

$$\mathcal{L}_{scalar} \equiv \mathcal{L}_{0,0} = \sqrt{-g}\Big[g^{\mu\nu}\varphi^*_{,\mu}\varphi_{,\nu} - bR\varphi^*\varphi\Big], \tag{6.93}$$

where $b$ is a non-null constant, whose values depend on the procedure of obtaining the general equation of the massless complex scalar field;

2) for the massless spinorial field, using the two-component neutrino theory (which starts with the Lagrangian density of the non-massive Dirac field, also using the Landau and Lifshitz' ideas that imply the substitutions $\left\{\begin{array}{l}\psi \to \frac{1}{2}(1+\gamma^5)\psi,\\ \overline{\psi} \to \frac{1}{2}\overline{\psi}(1-\gamma^5),\end{array}\right.$),

$$\mathcal{L}_{spinor} \equiv \mathcal{L}_{0,1/2} = \frac{i}{4}\sqrt{-g}\Big[\overline{\psi}\gamma^\mu(1+\gamma^5)\psi_{;\mu} - \overline{\psi}_{;\mu}\gamma^\mu(1+\gamma^5)\psi\Big], \tag{6.94}$$

where $\overline{\psi} = \psi^+\tilde{\gamma}^0$;

3) for the electromagnetic field,

$$\mathcal{L}_{electromagnetic} \equiv \mathcal{L}_{0,1} = -\tfrac{1}{2}\sqrt{-g}g^{\mu\lambda}g^{\nu\rho} \times \Big(A_{\mu;\nu}A_{\lambda;\rho} - R_{\mu\nu}A_\lambda A_\rho\Big), \tag{6.95}$$

which is obtained from the standard Lagrangian density of the electromagnetic field, $\mathcal{L}_{em} = -\frac{1}{4}\sqrt{-g}g^{\mu\lambda}g^{\nu\rho}F_{\mu\nu}F_{\lambda\rho}$, using the Mitskevich's idea to explicitly take into account the covariant derivative of the electromagnetic field function[4];

[4]For more details, see the corresponding subchapter of Chap. VIII.

4) for the massless spin-vectorial (Rarita–Schwinger) field,

$$4.a)\ \mathcal{L}_{spin-vector} \equiv \mathcal{L}_{0,3/2}^{Rarita-Schwinger} = \frac{i}{4}\sqrt{-g}g^{\mu\nu} \times \left[\overline{\psi}_{\mu}\overleftarrow{D}_{\lambda}\tilde{\gamma}^{\lambda}\left(1+\tilde{\gamma}^{5}\right)\psi_{\nu} - \overline{\psi}_{\mu}\tilde{\gamma}^{\lambda}\left(1+\tilde{\gamma}^{5}\right)\overrightarrow{D}_{\lambda}\psi_{\nu}\right], \tag{6.96}$$

as the variant analogous to the non-massive spinorial field, in which case one starts with the Lagrangian density of the non-massive Rarita–Schwinger field written in curved space, where the following substitutions have been performed:

$$\begin{cases} \psi_{\mu} \to \frac{1}{2}\left(1+\tilde{\gamma}^{5}\right)\psi_{\mu}, \\ \overline{\psi} \to \frac{1}{2}\overline{\psi}_{\mu}\left(1-\tilde{\gamma}^{5}\right); \end{cases}$$

$$4.b)\qquad \mathcal{L}_{spin-vector} \equiv \mathcal{L}_{0,3/2}^{SG} = \frac{i}{2}\sqrt{-g}\,\varepsilon^{\alpha\beta\mu\nu} \times \left(\overline{\psi}_{\alpha}\overleftarrow{D}_{\nu}\tilde{\gamma}_{5}\tilde{\gamma}_{\mu}\psi_{\beta} - \overline{\psi}_{\alpha}\tilde{\gamma}_{5}\tilde{\gamma}_{\mu}\overrightarrow{D}_{\nu}\psi_{\beta}\right), \tag{6.96'}$$

as the variant of usage of the Lagrangian density of the *simple supergravitation theory*, where $\overline{\psi}_{\mu} = \psi_{\mu}^{+}\tilde{\gamma}^{0}$;

5) for the gravitational field

$$5.a)\qquad \mathcal{L}_{gravitational}^{(1)} \equiv \mathcal{L}_{0,2}^{(1)} = -\frac{1}{2}\kappa h_{\mu\nu}T_{\mu\nu}, \tag{6.97}$$

where the self-coupling Lagrangian density of the weak gravitational field has been written (directly) in Gupta's variant [181]. Here,

$$T_{\mu\nu} = \frac{1}{2}\left[y_{\alpha\beta,\mu}y_{\alpha\beta,\nu} - \frac{1}{2}y_{,\mu}y_{,\nu} - \frac{1}{2}\delta_{\mu\nu}\left(y_{\alpha\beta,\lambda}y_{\alpha\beta,\lambda} - \frac{1}{2}y_{,\lambda}y_{,\lambda}\right)\right]$$

is the energy-momentum pseudotensor of the weak gravitational field;

$$5.b)\quad \mathcal{L}_{gravitational} \equiv \mathcal{L}_{0,2} = \frac{\sqrt{-g}}{\kappa^{2}}g^{\mu\nu}\left(\Gamma_{\mu\nu}^{\alpha}\Gamma_{\alpha\beta}^{\beta} - \Gamma_{\beta\mu}^{\alpha}\Gamma_{\alpha\nu}^{\beta}\right) + \frac{\partial}{\partial x^{\alpha}}\left[\frac{\sqrt{-g}}{\kappa^{2}}\left(\Gamma_{\beta\nu}^{\nu}g^{\alpha\beta} - \Gamma_{\beta\nu}^{\alpha}g^{\beta\nu}\right)\right], \tag{6.97'}$$

if the variant of Einstein's Lagrangian density, written in a more convenient (from the calculation point of view) form by Mitskevich, is considered.

In the case of spinorial and spin-vectorial fields, in order to write the covariant derivatives (whose necessity is required by the minimal coupling principle in quantum gravity), the customary *vierbein* formalism shall be used. According to this procedure, one has

$$\begin{aligned} &\psi_{;\nu} = \psi_{,\nu} - \Gamma_\nu \psi, \quad \overline{\psi}_{;\nu} = \overline{\psi}_{,\nu} + \overline{\psi}\Gamma_\nu \\ &\overrightarrow{D}_\nu \psi_\mu = \psi_{\mu,\nu} - \Gamma_\nu \psi_\mu, \quad \overline{\psi}_\mu \overleftarrow{D}_\nu = \overline{\psi}_{\mu,\nu} + \overline{\psi}_\mu \Gamma_\nu, \end{aligned} \tag{6.98}$$

where

$$\Gamma_\mu = \frac{1}{4}\tilde{\gamma}_{\lambda;\mu}\tilde{\gamma}^\lambda, \tag{6.99}$$

(with

$$\tilde{\gamma}_{\lambda;\mu} = \tilde{\gamma}_{\lambda,\mu} - \tilde{\gamma}_\rho \Gamma^\rho_{\lambda\mu}\,) \tag{6.100}$$

are the Fock–Ivanenko spin coefficients of the affine connection, $\Gamma^\alpha_{\mu\nu}$ are the Christoffel symbols of the second kind, the semicolon (for the spinorial field) and the operators $\overrightarrow{D}_\nu$ and $\overleftarrow{D}_\nu$ (for the spin-vectorial field) stand for the covariant derivatives of the corresponding field functions, while the comma designates the ordinary partial derivative of the same functions. In the above relations $\tilde{\gamma}^\mu = L^\mu(\alpha)\hat{\gamma}(\alpha)$ and $\tilde{\gamma}_\mu = L_\mu(\alpha)\hat{\gamma}(\alpha)$, $(\alpha = \overline{1,4})$ represent the generalized Dirac matrices (whose elements depend on coordinates), $L^\mu(\alpha)$ and $L_\mu(\alpha)$ are the *vierbein* coefficients satisfying the constraints $L^\mu(\alpha)L^\nu(\alpha) = g^{\mu\nu}$, $L_\mu(\alpha)L_\nu(\alpha) = g_{\mu\nu}$, and $\hat{\gamma}(\alpha)$ are the ordinary Dirac matrices in the hyperbolic representation of signature $-2$ of the Minkowski space.

Making use of: *i*) the De Donder–Fock gauge $\left(\sqrt{-g}\, g^{\mu\nu}\right)_{,\nu} = 0$; *ii*) the Rarita–Schwinger massless field equations (if the variant 4.*a* is considered),

$$\begin{cases} i\hat{\gamma}^\lambda \left(1 + \hat{\gamma}^5\right) \psi_{\nu,\lambda} = 0, \\ i\overline{\psi}_{\nu,\lambda}\hat{\gamma}^\lambda \left(1 + \hat{\gamma}^5\right) = 0 \end{cases} \tag{6.101}$$

and, respectively, the equations

$$\begin{cases} i\varepsilon^{\alpha\beta\mu\nu}\hat{\gamma}_5\hat{\gamma}_\mu\psi_{\alpha,\beta} = 0, \\ i\varepsilon^{\alpha\beta\mu\nu}\overline{\psi}_{\alpha,\beta}\hat{\gamma}_5\hat{\gamma}_\mu = 0, \end{cases} \tag{6.102}$$

(for the variant 4.*b*); *iii*) of the anticommutation relations satisfied by the Dirac matrices,

$$\left\{\hat{\gamma}^\mu, \hat{\gamma}^\nu\right\} = \hat{\gamma}^\mu\hat{\gamma}^\nu + \hat{\gamma}^\nu\hat{\gamma}^\mu = 2\eta^{\mu\nu}, \tag{6.103}$$

then expanding in power series in powers of $\kappa$ all the necessary quantities and neglecting the terms higher than the second-order $\big(\mathcal{O}(\kappa^2) = 0\big)$, namely,

$$\sqrt{-g} = 1 - \frac{1}{2}\kappa y, \quad y = y^\alpha_\alpha, \tag{6.104}$$

$$g^{\mu\nu} = \eta^{\mu\nu} - \kappa h^{\mu\nu}, \tag{6.105}$$

$$g_{\mu\nu} = \eta_{\mu\nu} + \kappa h_{\mu\nu}, \tag{6.105'}$$

$$\Gamma^\rho_{\lambda\mu} = \frac{1}{2}\kappa\left(h^\rho_{\lambda,\mu} + h^\rho_{\mu.\lambda} - h_{\lambda\mu,}{}^\rho\right), \tag{6.106}$$

$$\tilde{\gamma}^\lambda = \hat{\gamma}^\lambda - \frac{1}{2}\kappa\hat{\gamma}_\beta h^{\lambda\beta}, \tag{6.107}$$

$$\tilde{\gamma}_\lambda = \hat{\gamma}_\lambda + \frac{1}{2}\kappa\hat{\gamma}_\beta h^\beta_\lambda, \tag{6.107'}$$

$$h^{\mu\nu} = y^{\mu\nu} - \frac{1}{2}\eta^{\mu\nu}y, \tag{6.108}$$

$$h^\beta_\lambda = y^\beta_\lambda - \frac{1}{2}\delta^\beta_\lambda y, \tag{6.108'}$$

$$R_{\mu\nu} = \frac{1}{2}\kappa h_{\mu\nu,\lambda}{}^\lambda, \quad R = -\frac{1}{2}\kappa y_{,\lambda}{}^\lambda \tag{6.109}$$

and, finally, passing to the complex flat Minkowski space-time,

$$x^0 = t,\ x^j \to x_j,\ x_4 = it; \quad \hat{\gamma}^\mu\partial_\mu \to i\gamma_\mu\partial_\mu;\ \eta^{\mu\nu} \to -\delta_{\mu\nu};$$

$$\partial_\mu \equiv \frac{\partial}{\partial x^\mu} \to \frac{\partial}{\partial x_\mu};\ \gamma^5 = -i\gamma^0\gamma^1\gamma^2\gamma^3 \to \gamma_5 = \gamma_1\gamma_2\gamma_3\gamma_4,$$

the interaction Lagrangian densities between the weak gravitational field and the scalar, spinor, vector (electromagnetic), spin-vector (Rarita–Schwinger) and second-rank tensor (gravitational) massless fields are

$$\mathcal{L}^{(1)}_{(int)0,0}(\kappa) = -\kappa\left[\varphi^*_{,\mu}\varphi_{,\nu}y_{\mu\nu} - \frac{1}{2}b\varphi^*\varphi y_{,\mu\mu}\right], \tag{6.110}$$

$$\mathcal{L}^{(1)}_{(int)0,1/2}(\kappa) = -\frac{1}{8}\kappa\left[\overline{\psi}\gamma_\mu(1+\gamma_5)\psi_{,\nu} - \overline{\psi}_{,\nu}\gamma_\mu(1+\gamma_5)\psi\right]s_{\mu\nu}, \tag{6.111}$$

$$\begin{aligned}\mathcal{L}^{(1)}_{(int)0,1}(\kappa) \equiv \mathcal{L}^{(1)}_{em}(\kappa) = -\frac{1}{2}\kappa\Big[&A_\mu A_{\nu,\lambda}\Big(h_{\mu\nu,\lambda} + h_{\mu\lambda,\nu} - h_{\nu\lambda,\mu}\Big)\\ &+ A_{\mu,\nu}A_{\mu,\lambda}h_{\nu\lambda} + A_{\mu,\nu}A_{\lambda,\nu}y_{\mu\lambda}\\ &+ \frac{1}{2}A_\mu A_\nu h_{\mu\nu,\lambda\lambda}\Big],\end{aligned} \tag{6.112}$$

$$\begin{aligned}\mathcal{L}^{(1)RS}_{(int)0,3/2}(\kappa) = \frac{1}{8}\kappa&\left[\overline{\psi}_{\mu,\lambda}\gamma_\nu(1+\gamma_5)\psi_\mu - \overline{\psi}_\mu\gamma_\nu(1+\gamma_5)\psi_{\mu,\lambda}\right]y_{\lambda\nu}\\ &- \frac{1}{32}\kappa\overline{\psi}_\mu\left[\gamma_\rho\gamma_\nu\gamma_\lambda(1+\gamma_5) + \gamma_\lambda(1+\gamma_5)\gamma_\rho\gamma_\nu\right]\\ &\times \psi_\mu\Big(h_{\nu\lambda,\rho} - h_{\rho\lambda,\nu}\Big),\end{aligned} \tag{6.113}$$

$$\begin{aligned}\mathcal{L}^{(1)SG}_{(int)0,3/2}(\kappa) = -\frac{1}{4}&\kappa\varepsilon_{\alpha\beta\mu\nu}\Big(\overline{\psi}_{\alpha,\nu}\gamma_5\gamma_\rho\psi_\beta - \overline{\psi}_\alpha\gamma_5\gamma_\rho\psi_{\beta,\nu}\Big)y_{\mu\rho}\\ &+ \frac{1}{16}\kappa\varepsilon_{\alpha\beta\mu\nu}\overline{\psi}_\alpha\Big(\gamma_\rho\gamma_\lambda\gamma_5\gamma_\mu + \gamma_5\gamma_\mu\gamma_\rho\gamma_\lambda\Big)\\ &\times \psi_\beta\Big(h_{\rho\nu,\lambda} - h_{\lambda\nu,\rho}\Big),\end{aligned} \tag{6.113$'$}$$

$$\begin{aligned}\mathcal{L}^{(1)Gupta}_{(int)0,2}(\kappa) = -\frac{1}{4}\kappa h_{\mu\nu}\Big[&y_{\alpha\beta,\mu}y_{\alpha\beta,\nu}\\ &- \frac{1}{2}y_{,\mu}y_{,\nu} - \frac{1}{2}\delta_{\mu\nu}\Big(y_{\alpha\beta,\lambda}y_{\alpha\beta,\lambda} - \frac{1}{2}y_{,\lambda}y_{,\lambda}\Big)\Big],\end{aligned} \tag{6.114}$$

and

$$\mathcal{L}^{(1)Einstein}_{(int)0,2}(\kappa) = \frac{1}{2}\kappa\Big[y_{\alpha\beta}y_{\alpha\beta,\mu\nu} + \frac{1}{2}\Big(y_{\alpha\beta,\mu}y_{\alpha\beta,\nu} - y_{\mu\nu,\alpha}y_{,\alpha}\Big)\Big]y_{\mu\nu}, \tag{6.114'}$$

respectively, where

$$s_{\mu\nu} = y_{\mu\nu} + \frac{1}{2}\delta_{\mu\nu}y, \tag{6.115}$$

and

$$h_{\mu\nu} = y_{\mu\nu} - \frac{1}{2}\delta_{\mu\nu}y. \tag{6.116}$$

In the case of Schwarzschild field (considered here as a background) one has

$$y^{ext}_{\mu\nu}(\vec{x}) = \delta_{\mu 4}\delta_{\nu 4}y(\vec{x}), \tag{6.117}$$

where

$$y(\vec{x}) = \frac{\kappa M}{4\pi|\vec{x}|}, \tag{6.118}$$

$M$ being the mass of the central body (the source of the gravitational/background field) and $|\vec{x}|$ – the distance from this source.

In agreement with the standard theory of quantum fields, the parts of the first-order interaction Lagrangian densities corresponding to the interaction process between the matter fields and the weak gravitational field – brought to the normal form – are:

$$\begin{aligned} N\Big[\mathcal{L}^{(1)}_{(int)0,0}(x)\Big] = -\kappa\Big[&\varphi^{*(-)}_{,\mu}(x)\varphi^{(+)}_{,\nu}(x)y^{ext}_{\mu\nu}(x) \\ &- \frac{1}{2}b\varphi^{*(-)}(x)\varphi^{(+)}(x)y^{ext}_{\mu\mu}(x)\Big], \end{aligned} \tag{6.119}$$

$$\begin{aligned} N\Big[\mathcal{L}^{(1)}_{(int)0,1/2}(x)\Big] = -\frac{1}{8}\kappa\Big[&\overline{\psi}^{(-)}(x)\gamma_\mu(1+\gamma_5)\psi^{(+)}_{,\nu}(x) \\ &- \overline{\psi}^{(-)}_{,\nu}(x)\gamma_\mu(1+\gamma_5)\psi^{(+)}(x)\Big]s^{ext}_{\mu\nu}(x), \end{aligned} \tag{6.120}$$

$$N\Big[\mathcal{L}^{(1)}_{(int)0,1}(x)\Big] \equiv N\Big[\mathcal{L}^{(1)}_{em}(x)\Big] = -\kappa\Bigg\{\frac{1}{2}\Big[A^{(-)}_{\mu}(x)A^{(+)}_{\nu,\lambda}(x)$$

$$+ A^{(-)}_{\nu,\lambda}(x)A^{(+)}_{\mu}(x)\Big]\Big[h^{ext}_{\mu\nu,\lambda}(x) + h^{ext}_{\mu\lambda,\nu}(x) - h^{ext}_{\nu\lambda,\mu}(x)\Big]$$

$$+ A^{ext}_{\mu,\nu}(x)A^{ext}_{\mu,\lambda}(x)h^{ext}_{\nu\lambda}(x) + A^{ext}_{\mu,\nu}(x)A^{ext}_{\lambda,\nu}(x)y^{ext}_{\mu\lambda}(x)$$

$$+ \frac{1}{2}A^{ext}_{\mu}(x)A^{ext}_{\nu}(x)h^{ext}_{\mu\nu,\lambda\lambda}(x)\Bigg\}, \tag{6.121}$$

$$N\Big[\mathcal{L}^{(1)RS}_{(int)0,3/2}(x)\Big] = \frac{\kappa}{8}\Big[\overline{\psi}^{(-)}_{\mu,\lambda}(x)\gamma_{\nu}(1+\gamma_5)\psi^{(+)}_{\mu}(x)$$

$$- \overline{\psi}^{(-)}_{\mu}(x)\gamma_{\nu}(1+\gamma_5)\psi^{(+)}_{\mu,\lambda}(x)\Big]y^{ext}_{\lambda\nu}(x)$$

$$- \frac{\kappa}{32}\overline{\psi}^{(-)}_{\mu}(x)\Big[\gamma_{\rho}\gamma_{\nu}\gamma_{\lambda}(1+\gamma_5) + \gamma_{\lambda}(1+\gamma_5)\gamma_{\rho}\gamma_{\nu}\Big]$$

$$\times \psi^{(+)}_{\mu}(x)\Big[h^{ext}_{\nu\lambda,\rho}(x) - h^{ext}_{\rho\lambda,\nu}(x)\Big], \tag{6.122}$$

$$N\Big[\mathcal{L}^{(1)SG}_{(int)0,3/2}(x)\Big] = -\frac{\kappa}{4}\varepsilon_{\alpha\beta\mu\nu}\Big[\overline{\psi}^{(-)}_{\alpha,\nu}(x)\gamma_5\gamma_{\rho}\psi^{(+)}_{\beta}(x)$$

$$- \psi^{(-)}_{\alpha}(x)\gamma_5\gamma_{\rho}\psi^{(+)}_{\beta,\nu}(x)\Big]y^{ext}_{\mu\rho}(x)$$

$$+ \frac{\kappa}{16}\varepsilon_{\alpha\beta\mu\nu}\overline{\psi}^{(-)}_{\alpha}(x)\Big(\gamma_{\rho}\gamma_{\lambda}\gamma_5\gamma_{\mu} + \gamma_5\gamma_{\mu}\gamma_{\rho}\gamma_{\lambda}\Big)$$

$$\times \psi^{(+)}_{\beta}(x)\Big[h^{ext}_{\rho\nu,\lambda}(x) - h^{ext}_{\lambda\nu,\rho}(x)\Big], \tag{6.122$'$}$$

$$N\Big[\mathcal{L}^{(1)Gupta}_{(int)0,2}(x)\Big] = -\frac{1}{2}\kappa y^{(-)}_{\lambda\rho,\mu}(x)y^{(+)}_{\lambda\rho,\nu}(x)y^{ext}_{\mu\nu}(x), \tag{6.123}$$

$$N\Big[\mathcal{L}^{(1)Einstein}_{(int)0,2}(x)\Big] = \frac{\kappa}{2}\Bigg\{\Big[y^{(-)}_{\alpha\beta}(x)y^{(+)}_{\alpha\beta,\mu\nu}(x) + y^{(-)}_{\alpha\beta,\mu\nu}(x)y^{(+)}_{\alpha\beta}(x)$$

$$+ y^{(-)}_{\alpha\beta,\mu}(x)y^{(+)}_{\alpha\beta,\nu}(x)\Big]y^{ext}_{\mu\nu}(x) - \frac{1}{2}\Big[y^{(-)}_{\mu\nu}(x)y^{(+)}_{\mu\nu,\alpha}(x)$$

$$+ y^{(-)}_{\mu\nu,\alpha}(x)y^{(+)}_{\mu\nu}(x)\Big]y^{ext}_{,\alpha}(x)\Bigg\}, \tag{6.123$'$}$$

where $(+)$ and $(-)$ designate the positive and negative frequency parts of the wave functions, corresponding to annihilation and creation of a particle at the point of coordinate $x$, respectively.

The process is described by the Feynman diagram shown in Fig. 6.14, where $k$, $e_\mu^{(a)}(\vec{k})$ and $e_{\mu\nu}^{(a)}(\vec{k})$, on the one side, and $p$, $e_\nu^{(b)}(\vec{p})$ and $e_{\mu\nu}^{(b)}(\vec{p})$, on the other side, are the four-momenta and polarization vectors and tensors of the initial and final particles, respectively, while $a, b = 1, 2$ for the real photons and gravitons (they have only two helicity-states). Besides,

$$u^{(r)}(\vec{k}), \quad \left[\overline{u}^{(r)}(\vec{k}) = \left(u^{(r)}\right)^+ (\vec{k})\gamma_4\right]$$

and

$$u_\mu^{(r)}(\vec{k}), \quad \left[\overline{u}_\mu^{(r)}(\vec{k}) = \left(u_\mu^{(r)}\right)^+ (\vec{k})\gamma_4\right]$$

on the one side, and

$$u^{(s)}(\vec{p}), \quad \left[\overline{u}^{(s)}(\vec{p}) = \left(u^{(s)}\right)^+ (\vec{p})\gamma_4\right]$$

and

$$u_\mu^{(s)}(\vec{p}), \quad \left[\overline{u}_\mu^{(s)}(\vec{p}) = \left(u_\mu^{(s)}\right)^+ (\vec{p})\gamma_4\right]$$

on the other side, are the polarization spinors and spin-vectors (spectors) of the massless spin-1/2 and spin-3/2 particles, respectively, for initial $(r = 1, 2)$ and final $(s = 1, 2)$ polarization states, and $q$ is the virtual graviton four-momentum ($\vec{q} = \vec{p} - \vec{k}$ and $q_0 = p_0 - k_0 = 0$ – the energy conservation law).

Using the Fourier transform of the external static gravitational potential

$$y(\vec{q}) = \frac{1}{(2\pi)^{3/2}} \int e^{-i\vec{q}\cdot\vec{x}}\, y(\vec{x})\, d^3x = \frac{\kappa M}{(2\pi)^{3/2}|\vec{q}|^2}, \tag{6.124}$$

considering $y = 0$ for the real gravitons, and choosing the gauge in which

$$e_4^{(a)}(\vec{k}) = 0, \quad e_4^{(b)}(\vec{p}) = 0,$$

$$\vec{e}^{(a)}(\vec{k}) \cdot \vec{k} = 0, \quad \vec{e}^{(b)}(\vec{p}) \cdot \vec{p} = 0$$

for photons, and the $TT$-gauge characterized by

$$e^{(a)}_{4\mu}(\vec{k}) = 0, \quad e^{(b)}_{4\nu}(\vec{p}) = 0,$$

$$e^{(a)}_{\alpha\beta}(\vec{k})k_\beta = e^{(b)}_{\alpha\beta}(\vec{p})p_\beta = 0,$$

for gravitons, the $S$-matrix elements – in the external field approximation – corresponding to the Feynman diagram shown in Fig. 6.14 are

$$S^{(*)}_{pk} = \left\langle p \left| S^{(*)} \right| k \right\rangle = F^{(*)}(p,k)\delta(q_0), \tag{6.125}$$

where the symbol $(*)$ stands for each of the values 0, 1/2, 1, 3/2 and 2, depending on the considered massless field (scalar, spinorial, vectorial, spin-vectorial and tensorial, respectively). The quantities $F^{(*)}(p,k)$ appearing in (6.125) are given in each case as follows,

$$F^{(0)}(p,k) = \frac{i\kappa^2 M}{8k_0(2\pi)^2 k_0^2 \sin^2 \dfrac{\theta}{2}} Q^{(0)}(p,k), \tag{6.126}$$

$$F^{(1/2)}(p,k) = \frac{\kappa^2 M}{32(2\pi)^2 k_0^2 \sin^2 \dfrac{\theta}{2}} Q^{(1/2)}(p,k), \tag{6.127}$$

$$F^{(1)}(p,k) = \frac{i\kappa^2 M}{8k_0(2\pi)^2 k_0^2 \sin^2 \dfrac{\theta}{2}} Q^{(1)}(p,k), \tag{6.128}$$

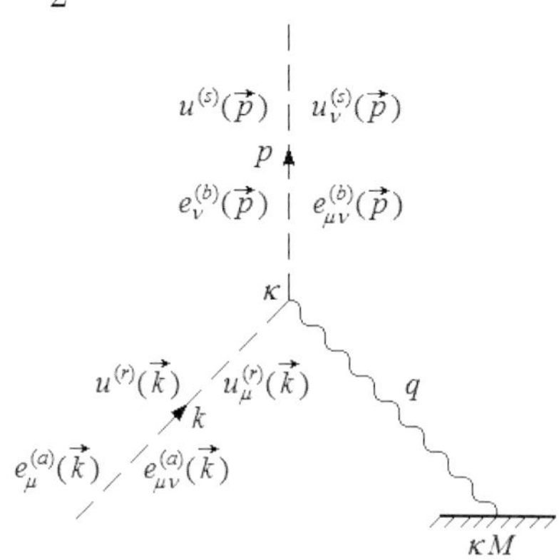


Fig. 6.14. The first-order Feynman vertex. The curly line represents the virtual/exchange graviton (external static gravitational field). The broken lines stand for the quanta of the scalar, spinorial, electromagnetic, spin-vectorial and gravitational fields, respectively.

$$F^{(3/2)}_{RS}(p,k) = \frac{-\kappa^2 M}{32(2\pi)^2 4k_0^2 \sin^2\dfrac{\theta}{2}} Q^{(3/2)}_{RS}(p,k), \tag{6.129}$$

$$F^{(3/2)}_{SG}(p,k) = \frac{\kappa^2 M}{16(2\pi)^2 4k_0^2 \sin^2\dfrac{\theta}{2}} Q^{(3/2)}_{SG}(p,k), \tag{6.129'}$$

$$F^{(2)}_{Gupta}(p,k) = \frac{\kappa^2 M}{16(2\pi)^2 k_0 \sin^2\dfrac{\theta}{2}} Q^{(2)}_{Gupta}(p,k), \tag{6.130}$$

$$F^{(2)}_{Einstein}(p,k) = \frac{i\kappa^2 M}{16k_0(2\pi)^2 k_0^2 \sin^2\dfrac{\theta}{2}} Q^{(2)}_{Einstein}(p,k), \tag{6.130'}$$

where $k_0$ denotes the energy of the initial particles, and the quantities $Q^{(*)}(p,k)$ are given by the following expressions (see also Appendix F),

$$Q^{(0)}(p,k) = k_\mu p_\nu \tilde{y}^{ext}_{\mu\nu}(\vec{q}) - \frac{b}{2}\tilde{y}^{ext}_{,\mu\mu}(\vec{q}), \tag{6.131}$$

$$Q^{(1/2)}(p,k) = \overline{u}^{(s)}(\vec{p}) t^{(1/2)}(p,k) u^{(r)}(\vec{k}), \tag{6.132}$$

$$\begin{aligned} Q^{(1)}(p,k) = \frac{1}{2}&\left[e^{(r)}_\nu(\vec{k}) e^{(s)}_\mu(\vec{p}) k_\lambda - e^{(r)}_\mu(\vec{k}) e^{(s)}_\nu(\vec{p}) p_\lambda\right] \\ &\times \left[q_\lambda \tilde{h}^{ext}_{\mu\nu}(\vec{q}) + q_\nu \tilde{h}^{ext}_{\mu\lambda}(\vec{q}) - q_\mu \tilde{h}^{ext}_{\nu\lambda}(\vec{q})\right] - e^{(r)}_\mu(\vec{k}) e^{(s)}_\mu(\vec{p}) k_\lambda p_\nu \\ &- e^{(r)}_\lambda(\vec{k}) e^{(s)}_\mu(\vec{p}) k_\nu p_\nu \tilde{y}^{ext}_{\mu\lambda}(\vec{q}) + \tfrac{1}{2} e^{(r)}_\nu(\vec{k}) e^{(s)}_\mu(\vec{p}) q_\lambda q_\lambda \tilde{h}^{ext}_{\mu\nu}(\vec{q}), \end{aligned} \tag{6.133}$$

$$Q^{(3/2)}_{RS}(p,k) = \overline{u}^{(s)}_\mu(\vec{p}) t^{(3/2)}_{RS}(p,k) u^{(r)}_\mu(\vec{k}), \tag{6.134}$$

$$Q^{(3/2)}_{SG}(p,k) = \overline{u}^{(s)}_\alpha(\vec{p}) \left[t_{\alpha\beta}(p,k)\right]^{(3/2)}_{SG} u^{(r)}_\beta(\vec{k}), \tag{6.134'}$$

$$Q^{(2)}_{Gupta}(p,k) = e^{(a)}_{\alpha\beta}(\vec{k}) e^{(b)}_{\alpha\beta}(\vec{p}), \tag{6.135}$$

$$Q^{(2)}_{Einstein}(p,k) = (p_\mu k_\nu - p_\mu p_\nu - k_\mu k_\nu)\, e^{(r)}_{\alpha\beta}(\vec{k})e^{(s)}_{\alpha\beta}(\vec{p})\tilde{y}^{ext}_{\mu\nu}(\vec{q})$$
$$- \frac{1}{2}\delta_{\alpha\beta}q_\beta\,(p_\alpha - k_\alpha)\, e^{(r)}_{\mu\nu}(\vec{k})e^{(s)}_{\mu\nu}(\vec{p})\tilde{y}^{ext}(\vec{q}), \tag{6.135'}$$

where

$$t^{(1/2)}(p,k) = \gamma_\mu(1+\gamma_5)(k_\nu + p_\nu)\tilde{s}^{ext}_{\mu\nu}(\vec{q}), \tag{6.136}$$

$$t^{(3/2)}_{RS}(p,k) = 8ik_0\gamma_4(1+\gamma_5) + (\gamma_\rho\gamma_\lambda\gamma_5\gamma_\nu + \gamma_5\gamma_\nu\gamma_\rho\gamma_\lambda)$$
$$\times \left[q_\rho\left(\delta_{\nu4}\delta_{\lambda4} - \frac{1}{2}\delta_{\nu\lambda}\right) - q_\nu\left(\delta_{\rho4}\delta_{\lambda4} - \frac{1}{2}\delta_{\rho\lambda}\right)\right], \tag{6.137}$$

$$[t_{\alpha\beta}(p,k)]^{(3/2)}_{SG} = \varepsilon_{\alpha\beta\mu\nu}\Big\{4(p_\nu + k_\nu)\gamma_5\gamma_4\delta_{\mu4}$$
$$+ (\gamma_\rho\gamma_\lambda\gamma_5\gamma_\mu + \gamma_5\gamma_\mu\gamma_\rho\gamma_\lambda)\Big[q_\lambda\left(\delta_{\rho4}\delta_{\nu4} - \frac{1}{2}\delta_{\rho\nu}\right)$$
$$- q_\rho\left(\delta_{\lambda4}\delta_{\nu4} - \frac{1}{2}\delta_{\lambda\nu}\right)\Big]\Big\}, \tag{6.138}$$

$$\tilde{y}^{ext}(\vec{q}) = (2\pi)^{3/2}\frac{|\vec{q}|^2}{\kappa M}y^{ext}(\vec{q}) = 1, \tag{6.139}$$

$$\tilde{y}^{ext}_{\mu\nu}(\vec{q}) = (2\pi)^{3/2}\frac{|\vec{q}|^2}{\kappa M}y^{ext}_{\mu\nu}(\vec{q}), \tag{6.140}$$

$$\tilde{s}^{ext}_{\mu\nu}(\vec{q}) = (2\pi)^{3/2}\frac{|\vec{q}|^2}{\kappa M}s^{ext}_{\mu\nu}(\vec{q}), \tag{6.141}$$

$$\tilde{h}^{ext}_{\mu\nu}(\vec{q}) = (2\pi)^{3/2}\frac{|\vec{q}|^2}{\kappa M}h^{ext}_{\mu\nu}(\vec{q}). \tag{6.142}$$

The differential scattering cross-section is given by the well-known expression

$$d\sigma = (2\pi)^2\left\langle \sum_{fin\,sp} |F(p,k)|^2 \right\rangle_{in\,sp} k_0^2\, d\Omega, \tag{6.143}$$

where $d\Omega = \sin\theta\, d\theta\, d\varphi$, $\theta$ being the scattering angle. In order to determine the differential scattering cross-sections, one must calculate the quantities $\left\langle \sum_{fin\, sp} \left|F^{(*)}(p,k)\right|^2 \right\rangle_{in\, sp}$ corresponding to each field,

$$\begin{aligned}\left\langle \sum_{fin\, sp} \left|F^{(*)}(p,k)\right|^2 \right\rangle_{in\, sp} &= \left\langle \sum_{fin\, sp} C^{(*)} \left|Q^{(*)}(p,k)\right|^2 \right\rangle_{in\, sp} \\ &= \frac{1}{2} C^{(*)} \sum_{pol} \left|Q^{(*)}(p,k)\right|^2 ,\end{aligned} \tag{6.144}$$

where the coefficients $C^{(*)}$ can be easily identified, and $\sum_{pol} \left|Q^{(*)}(p,k)\right|^2$ represent the corresponding sums over polarizations; since their expressions are very long, this step shall be skipped. In the case of spinorial and spin-vectorial massless fields, these sums are in fact two traces over the polarization states indices, namely

$$\begin{aligned}\sum_{pol} \left|Q^{(1/2)}(p,k)\right|^2 &= \sum_{r,s=1}^{2} \left|\overline{u}^{(s)}(\vec{p}) t^{(1/2)}(p,k) u^{(r)}(\vec{k})\right|^2 \\ &= \mathrm{Tr}\left\{\gamma_4 \left[t^{(1/2)}(p,k)\right]^{+} \gamma_4 P^{(1/2)}_{(+)}(\vec{p}) t^{(1/2)}(p,k) P^{(1/2)}_{(+)}(\vec{k})\right\},\end{aligned} \tag{6.145}$$

for the Dirac field, and

$$\begin{aligned}\sum_{pol} \left|Q^{(3/2)}_{SG}(p,k)\right|^2 &= \sum_{r,s=1}^{2} \left|\overline{u}^{(s)}_{\alpha}(\vec{p}) \left(t^{(3/2)}_{SG}(p,k)\right)_{\alpha\beta} u^{(r)}_{\beta}(\vec{k})\right|^2 \\ &= \mathrm{Tr}\left[\gamma_4 \left(t^{(3/2)}_{SG}(p,k)\right)^{*}_{\beta\alpha} \gamma_4 \left(P^{(3/2)}_{(+)}(\vec{p})\right)_{\alpha\gamma} \right. \\ &\quad \left. \times \left(t^{(3/2)}_{SG}(p,k)\right)_{\gamma\lambda} \left(P^{(3/2)}_{(+)}(\vec{k})\right)_{\lambda\beta}\right],\end{aligned} \tag{6.146}$$

for the spin-vectorial field in the simple supergravitation model. Here $P^{(1/2)}_{(+)}$ and $P^{(3/2)}_{(+)}$ are the projection covariant operators (projectors) for the spinors and spin-vectors of positive energy, respectively.

Unlike the massive particles, in the case of massless particles one must make distinction between the *integral* and *total* (more exactly, *total differential*) scattering cross-sections. Usually, a *total scattering cross-section* is understood as an *integral scattering cross-section.* But, due to the property of massless fields of having two linearly independent eigenvectors for the helicity operator, in the case of massless fields one can also calculate *partial differential scattering cross-sections*, the total differential scattering cross-sections being obtained by summing the partial differential scattering cross-sections (which is equivalent to perform a summation over the final polarization states of the particles).

For example, to directly determine the total differential scattering cross-sections, one must use the following equations satisfied by the polarization spinors, vectors, spin-vectors and tensors, respectively, corresponding to the massless particles [252, 331],

$$P^{(1/2)}_{(+)}(\vec{p}) = \sum_{r=1}^{2} u^{(r)}(\vec{p})\overline{u}^{(r)}(\vec{p}) = \frac{\gamma_\mu p_\mu}{2ip_0}, \tag{6.147}$$

$$\Pi_{\mu\nu}(\vec{p}) = \sum_{r=1}^{2} e^{(r)}_\mu(\vec{p})e^{(r)}_\nu(\vec{p}) = d_{\mu\nu}(\vec{p}), \tag{6.148}$$

$$\begin{aligned}\left(P^{(3/2)}_{(+)}(\vec{p})\right)_{\mu\nu} &= \sum_{r=1}^{2} u^{(r)}_\mu(\vec{p})\overline{u}^{(r)}_\nu(\vec{p})\\ &= \frac{1}{2ip_0}\left(\delta_{\mu\nu}\gamma_\alpha p_\alpha + \frac{1}{2}\gamma_\mu\gamma_\alpha\gamma_\nu p_\alpha - \gamma_\mu p_\nu - \gamma_\nu p_\mu\right)\\ &= -\frac{\gamma_\nu\gamma_\alpha\gamma_\mu p_\alpha}{4ip_0},\end{aligned} \tag{6.149}$$

and

$$\begin{aligned}\Pi_{\mu\nu,\lambda\rho}(\vec{p}) = \sum_{r=1}^{2} e^{(r)}_{\mu\nu}(\vec{p})e^{(r)}_{\lambda\rho}(\vec{p}) &= d_{\mu\lambda}(\vec{p})d_{\nu\rho}(\vec{p})\\ &+ d_{\mu\rho}(\vec{p})d_{\nu\lambda}(\vec{p}) - d_{\mu\nu}(\vec{p})d_{\lambda\rho}(\vec{p}),\end{aligned} \tag{6.150}$$

where

$$d_{\mu\nu}(\vec{p}) = \delta_{\mu\nu} - \frac{p_\mu p_\nu}{\vec{p}^2}. \tag{6.151}$$

After relatively easy, but laborious calculations, the following expressions are obtained for the *total differential scattering cross-sections* corresponding to scalar, spinorial, vectorial, spin-vectorial and tensorial massless[5] particles:

$$d\sigma^{(0)} = (GM)^2 \frac{d\Omega}{\sin^4 \dfrac{\theta}{2}} \left(1 - 2b\sin^2 \frac{\theta}{2}\right)^2, \tag{6.152}$$

$$d\sigma^{(1/2)} = (GM)^2 \frac{d\Omega}{\sin^4 \dfrac{\theta}{2}} \cos^2 \frac{\theta}{2}, \tag{6.153}$$

$$d\sigma^{(1)} = (GM)^2 \frac{d\Omega}{\sin^4 \dfrac{\theta}{2}} \cos^4 \frac{\theta}{2}, \tag{6.154}$$

$$d\sigma^{(3/2)}_{RS} = d\sigma^{(3/2)}_{SG} \equiv d\sigma^{(3/2)} = (GM)^2 \frac{d\Omega}{\sin^4 \dfrac{\theta}{2}} \cos^2 \frac{\theta}{2}, \tag{6.155}$$

and

$$d\sigma^{(2)}_{Gupta} = (GM)^2 \frac{d\Omega}{\sin^4 \dfrac{\theta}{2}} \left(\cos^2 \frac{\theta}{2} + \frac{1}{8}\sin^4 \theta\right), \tag{6.156}$$

$$d\sigma^{(2)}_{Einstein} = (GM)^2 \frac{d\Omega}{\sin^4 \dfrac{\theta}{2}} \cos^2 \theta \left(\cos^2 \theta + \frac{1}{8}\sin^4 \theta\right), \tag{6.156$'$}$$

respectively.

If, in addition, the angle $\theta$ takes only small and/or very small values, then all[6] the differential scattering cross-sections are of the same form

$$\left(d\sigma^{(*)}_{Schw}\right)_{\theta \ll 1} = (GM)^2 \frac{d\Omega}{\sin^4 \dfrac{\theta}{2}} \equiv d\sigma_{Ruth}, \tag{6.157}$$

[5]For gravitons, the $TT$-gauge (transversality and null trace) has been considered.

[6]Even more than that, in the case of massless scalar field this result remains valid for any value of the scattering angle, if the scalar field is minimally coupled to gravity ($b = 0$).

therefore they all are *Rutherford-type differential cross-sections*. This result coincides with that obtained for the corresponding massive particles – in the same approximation ($\theta \ll 1$) and supplementary in the ultrarelativistic limit [310, 362, 365, 367, 369, 370, 379, 380] – showing that in this special limiting case the particle gravitational scattering is spin-independent.

As it can be observed, in the case of Schwarzschild geometry the total scattering cross-sections of the spinorial and spin-vectorial particles are given by the same expression, namely

$$d\sigma^{(1/2)}_{Schw} = d\sigma^{(3/2)}_{Schw} = (GM)^2 \frac{d\Omega}{\sin^4 \dfrac{\theta}{2}} \cos^2 \frac{\theta}{2}, \tag{6.158}$$

meaning that both massless spinor particles and gravitino are identically scattered by a Schwarzschild background with small curvature.

As far as the massless vectorial and tensorial particles backscattering case is concerned (in our case, the photons and gravitons), it is worth specifying that – as it can easily be observed – the photon differential scattering cross-section vanishes in this limit ($\theta \to \pi$), meaning that in this case the photon helicity conserves in the scattering process, while in the case of gravitons this result is not valid. Here one must consider the fact that the analysis of this particular limit case ($\theta \to \pi$) can not be done within Gupta's formalism, because in this case the graviton differential scattering cross-section is not desirable. Indeed, the starting Lagrangian density obtained by means of the Gupta-type coupling between the energy-momentum pseudotensor $T_{\mu\nu}$ and the tensor $h_{\mu\nu}$ of the weak gravitational field is able to correctly describe only the small angle scatterings. It is easy to observe that in the limit ($\theta \to \pi$) the graviton differential cross-section obtained in Gupta's formalism vanishes, but this result leads to the wrong conclusion according to which the graviton helicity also conserves in the backscattering process, which is not true. Thus, in order to investigate this particular limit case ($\theta \to \pi$) for gravitons, the Einstein's formalism has to be used instead, according to which the graviton differential scattering cross-section is obtained starting with the *general* interaction Lagrangian density – formulated by Einstein and written in a more convenient form by Mitskevich – valid for any value of the scattering angle (including $\theta \to \pi$).

The interested reader is asked to determine the integral scattering cross-sections and – eventually – to study the diffusion of the electromagnetic/gravitational polarized waves.

Chapter VII

# Scattering of Electrons and Photogeneration of Gravitons in External Gravitational Field

## VII.1. General considerations

This chapter is concerned with the electron diffusion/scattering process and graviton photogeneration (photon to graviton conversion) within the framework of the external field formalism, for the following four particular cases of electromagnetic field:

a) the *homogeneous electrostatic field* $\vec{E}$ of a capacitor,
b) the *electric dipole field* of moment $\vec{D}$,
c) the *homogeneous magnetostatic field* $\vec{H}$, and
d) the *magnetic dipole field* of moment $\vec{M}$.

The problem of studying the scattering of the massive Dirac particles that "propagate" in a domain occupied by the electromagnetic field can be approached by quantization of both the electromagnetic background and the scattered field. In such a scenario the two fields are coupled in agreement with Feynman's rules, the interaction Lagrangian density being obtained through the medium of the minimal coupling principle applied in quantum electrodynamics. Nevertheless, since this investigation concerns only those configurations of the electromagnetic background – generated by classical momentum-energy distributions – which are not considerably affected by the scattering process, the virtual/exchange photon has been replaced by an appropriate external field. In particular, there were considered only the four static electromagnetic fields listed above.

Regarding the second type of process that is being studied in this chapter (the graviton photogeneration in external electromagnetic field), it is worth mentioning that, from the classical point of view, to the particles which are quanta of the electromagnetic and gravitational fields correspond two types of waves, namely the electromagnetic and gravitational waves. As far as the gravitational waves are concerned, their existence was put into evidence (first, indirectly, in 1974 by Russell A. Hulse and Joseph H. Taylor Jr. from Princeton University, and then in a direct way, in September 2015 – the LIGO experiment).

The study of graviton photogeneration in external electromagnetic fields possessing various symmetries has been also approached by some other authors (see, for example, Refs. [247, 248, 270, 300]), but only for some particular cases and by means of different approaching ways. The following investigation generalizes the previously obtained results in the field, being in agreement with those obtained by Mitskevich [270], Papini and Valluri [300], De Logi and Mickelson [92], and some other authors in several particular cases.

The method used in the study of the above mentioned processes is offered by the minimal coupling principle applied in quantum electrodynamics and quantum gravity, in strong connection with the Feynman diagram technique which is applied to a generic problem of covariant perturbations theory within the quantum fields theory formalism [66].

The obtained results allow one to observe a remarkable similarity between the differential scattering cross-sections corresponding to the diffusion process of Dirac particles, on the one side, and those corresponding to the process of graviton photogeneration (considered under the same conditions, regarding the nature and symmetry of the applied external electromagnetic fields), on the other side.

All the processes studied in this chapter are investigated in the first-order approximation (they are linear in the coupling constants). This is a very plausible hypothesis for the scattering process of the Dirac particles in external electromagnetic fields, but especially for the process of graviton photogeneration in the same field configurations. In this last case, due to the smallness of the gravitational constant, the scattering cross-sections are too small for being of practical interest in laboratory experiments. But in astrophysics the situation is different. Thus, as it was shown by D. Boccaletti, in astrophysics there are certain possibilities for the gravitons generation rate in quantum processes to be at least as large as in classical processes. Due to this fact, a special paragraph dedicated to several astrophysical considerations has been inserted in this chapter.

Connected to the problem of detectability of the graviton photogeneration process in external electromagnetic fields, it is worth mentioning that several recent results in this respect show that the scattering cross-sections can have measurable (detectable) values even in laboratory experiments. So, according to the scenario proposed by Long, Soa and Tuan [247, 248], the scattering cross-section corresponding to the process of graviton photogeneration in the electrostatic field of a capacitor and in an axially-symmetric magnetic field of a solenoid can be made "as large as possible" (the limitations being imposed by the nowadays technology), on account of increasing the (electric or magnetic) field intensities, or on account of enlargement of the 3D spatial domain where the interaction takes place. In addition, the same authors showed that the inverse process is also possible; according to their considerations, the conversion of gravitons into photons can take place both in static and periodic external electromagnetic fields. In the latter case their evaluations concerning the differential scattering cross-sections were performed in the quasi-static limit.

## VII.2. Notations and writing conventions

Denoting by $g^{\mu\nu}$, $\eta^{\mu\nu}$, and $y^{\mu\nu}$ the metric tensor, the Minkowski tensor – diag $(+1, -1, -1, -1)$ – and the weak gravitational field tensor, respectively, the Gupta linear approximation writes

$$\sqrt{-g}\, g^{\mu\nu} = \eta^{\mu\nu} - \kappa y^{\mu\nu},$$

where $g = \det(g_{\mu\nu})$. The electrostatic and magnetostatic field intensities are denoted by $\vec{E}$ and $\vec{H}$ respectively, while $\vec{D}$ and $\vec{M}$ designate the moments of the electric and magnetic dipoles, respectively. For the first type of classical field, the potential of the static Maxwell field is given by

$$A_\mu^{ext}(\vec{x}) = i\delta_{\mu 4}V(\vec{x}), \quad V(\vec{x}) = \frac{\vec{D}\cdot\vec{x}}{4\pi|\vec{x}|^3}, \quad \vec{E}(\vec{x}) = -\nabla V(\vec{x}), \quad (7.1)$$

while for the second case (the magnetostatic field) by

$$\begin{aligned} A_\mu^{ext}(\vec{x}) &= \delta_{\mu i}A_i(\vec{x}), \quad \vec{A}(\vec{x}) = \frac{\vec{M}\times\vec{x}}{4\pi|\vec{x}|^3}, \\ \vec{H}(\vec{x}) &= -\nabla\times\vec{A}(\vec{x}), \quad \nabla\cdot\vec{A}(\vec{x}) = 0. \end{aligned} \quad (7.2)$$

In the first case, the tensor of the classical Maxwell field is

$$F_{\mu\nu}^{ext}(\vec{x}) = i\Big(\delta_{\mu 4}\delta_{\nu j} - \delta_{\mu j}\delta_{\nu 4}\Big)E_j(\vec{x}), \quad (7.3)$$

and in the second case,

$$F_{\mu\nu}^{ext}(\vec{x}) = i\delta_{\mu i}\delta_{\nu j}\varepsilon_{ijk}H_k(\vec{x}). \quad (7.4)$$

In the above equations $\varepsilon_{ijk}$ is the totally antisymmetric unit third-rank tensor (the Levi-Civita symbol)[1], the components of the three-dimensional vectors are denoted by Roman indices, and those of four-vectors by Greek indices.

## VII.3. Electron scattering in external static electromagnetic fields

Using the minimal coupling principle applied in quantum electrodynamics [341], one obtains the well-known interaction Lagrangian density between the Dirac and electromagnetic fields,

$$\mathcal{L}_{M-D} = ie\overline{\psi}\gamma_\mu\psi A_\mu. \quad (7.5)$$

Here $\psi$ and $\overline{\psi}(=\psi^+\gamma_4)$ are the operators–field functions corresponding to the massive spinorial field, $\gamma_\mu$ represent the ordinary Dirac matrices, the cross stands for Hermitian conjugation, and $e$ is the electromagnetic coupling constant (the electron electric

[1]In this subchapter, the notations corresponding to the Euclidean complex space-time have been used: $x_\mu$ $(x_j, x_4 = it)$.

charge). Using the $S$-matrix formalism, one can deduce Feynman-type rules for diagrams in the external electromagnetic field approximation, used to obtain the corresponding differential scattering cross-sections.

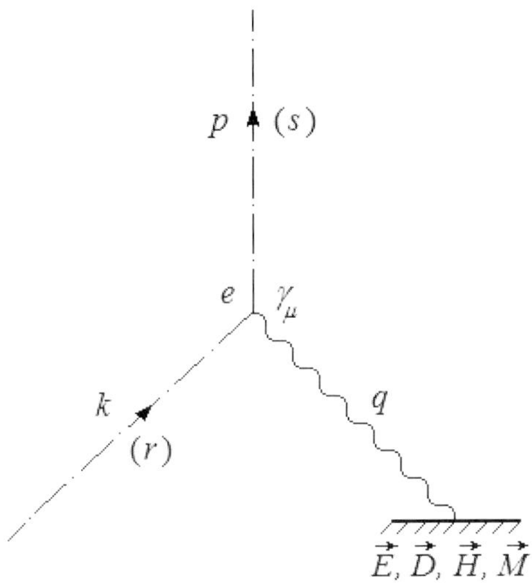


Fig. 7.1. Feynman diagram for the electron–photon–electron interaction process. The photon is represented by the curly line and the Dirac particles are represented by broken lines.

The process is described by the Feynman diagram shown in Fig. 7.1, where $k$ and $(r)$, and also $p$ and $(s)$ represent the four-momenta and the spin-states of the initial and final particles, respectively – with $r, s = 1, 2$ – and $q$ is the four-momentum of the virtual photon ($\vec{q} = \vec{p} - \vec{k}$ and $q_0 = p_0 - k_0 = 0$ – the momentum and energy conservation laws).

According to the standard quantum field theory, the part of the Lagrangian density – brought to the normal form – that describes the interaction between the massive Dirac particles and the Maxwell external field is

$$N\left[\mathcal{L}_{M-D}\right] = ie\overline{\psi}^{(-)}(x)\gamma_\mu\psi^{(+)}(x)A_\mu^{ext}(x), \tag{7.6}$$

where $(+)$ and $(-)$ denote the positive and negative frequency parts of the field function, corresponding to annihilation and creation of the spinorial particles at the point $x$, respectively. The matrix element – in the external electromagnetic field approximation – corresponding to the process described by the Feynman diagram shown in Fig. 7.1 writes

$$\begin{aligned}\langle p|S|k\rangle = -\frac{me}{(2\pi k_0 p_0)^{1/2}} \int d^3q\, \overline{u}^{(s)}(\vec{p})\gamma_\mu u^{(r)}(\vec{k}) A_\mu^{ext}(\vec{q}) \\ \times\, \delta(\vec{p}-\vec{k}-\vec{q})\delta(p_0-k_0) = F(k,p)\delta(p_0-k_0),\end{aligned} \tag{7.7}$$

where $u$ and $\overline{u}$ are Dirac spinors of positive energy, and $m$ is the mass of the spinorial particles. If the integration is performed, then one obtains

$$\begin{aligned}F(k,p) &= -\frac{me}{(2\pi)^{1/2}k_0} Q(k,p), \\ Q(k,p) &= \overline{u}^{(s)}(\vec{p})\gamma_\mu u^{(r)}(\vec{k}) A_\mu^{ext}(\vec{q}), \quad \vec{q}=\vec{p}-\vec{k},\end{aligned} \tag{7.8}$$

where the Fourier transform of the classical electromagnetic field potential

$$A_\mu^{ext}(\vec{q}) = \frac{1}{(2\pi)^{3/2}} \int d^3x\, e^{-i\vec{q}\cdot\vec{x}} A_\mu^{ext}(\vec{x}) \tag{7.9}$$

has to be introduced.

The differential scattering cross-section of the process is obtained as a result of averaging over the spin-states of the incident Dirac particles, and summation over the spin-states of the emergent particles [317],

$$d\sigma = (2\pi)^2 k_0^2 \left\langle \sum_{fin\,sp} |F(k,p)|^2 \right\rangle_{in\,sp} d\Omega, \tag{7.10}$$

where $d\Omega = \sin\theta\, d\theta\, d\varphi$ is the elementary solid angle corresponding to the emergent particles, and $\theta$ is the scattering angle (between the vectors $\vec{k}$ and $\vec{p}$). One then finds

$$\left\langle \sum_{fin\,sp} |F(k,p)|^2 \right\rangle_{in\,sp} = \frac{m^2e^2}{2\pi k_0^2}\frac{1}{2}\sum_{sp} |Q(k,p)|^2 . \tag{7.11}$$

Using the projection operator – written in the covariant form – for the massive spinorial particles,

$$P_+(\vec{k}) = \sum_{r=1}^{2} u^{(r)}(\vec{k})\overline{u}^{(r)}(\vec{k}) = \frac{\gamma_\mu k_\mu + im}{2im}, \tag{7.12}$$

and taking into account the real character of the electromagnetic field potential four-vector $A_\mu^{ext}(x)$, one obtains

$$\begin{aligned}\frac{1}{2}\sum_{sp}|Q(k,p)|^2 &= -\frac{1}{2}A_\mu^{ext}(\vec{q})A_\nu^{ext}(-\vec{q})\,\mathrm{Tr}\left[\gamma_\mu P_+(\vec{k})\gamma_\nu P_+(\vec{p})\right]\\ &= \frac{1}{2m}\left(k_\mu p_\nu + p_\mu k_\nu + 2\delta_{\mu\nu}\vec{k}^{\,2}\sin^2\frac{\theta}{2}\right)A_\mu^{ext}(\vec{q})A_\nu^{ext}(-\vec{q}),\end{aligned} \tag{7.13}$$

so that, the differential scattering cross-section becomes

$$\frac{d\sigma}{d\Omega} = \pi e^2\left(k_\mu p_\nu + p_\mu k_\nu + 2\delta_{\mu\nu}\vec{k}^{\,2}\sin^2\frac{\theta}{2}\right)A_\mu^{ext}(\vec{q})A_\nu^{ext}(-\vec{q}),\ \vec{q} = \vec{p} - \vec{k}. \tag{7.14}$$

This general result can be particularized for various types of external Maxwell fields. To determine the Fourier transforms of the classical electromagnetic fields $\vec{F} \equiv \vec{E}, \vec{D}, \vec{H}, \vec{M}$, suppose that $\vec{F}$ lies in the $xOz$-plane, and denote by $\alpha$ the angle between $\vec{F}$ and the direction of motion of the incident electron (taken along $z$-axis), that is,

$$\begin{cases} \vec{F} \equiv (F\sin\alpha,\ 0,\ F\cos\alpha), \quad \vec{k} = \left(0, 0, |\vec{k}|\right), \\ \vec{p} \equiv \left(|\vec{k}|\sin\theta\cos\varphi,\ |\vec{k}|\sin\theta\sin\varphi,\ |\vec{k}|\cos\theta\right), \\ \vec{q} \equiv \left(|\vec{k}|\sin\theta\cos\varphi,\ |\vec{k}|\sin\theta\sin\varphi,\ -|\vec{k}|(1-\cos\theta)\right), \end{cases} \tag{7.15}$$

where $\varphi$ designates the angle between the $(\vec{F}, \vec{k})$ and $(\vec{p}, \vec{k})$ planes, as shown in Fig. 7.2.

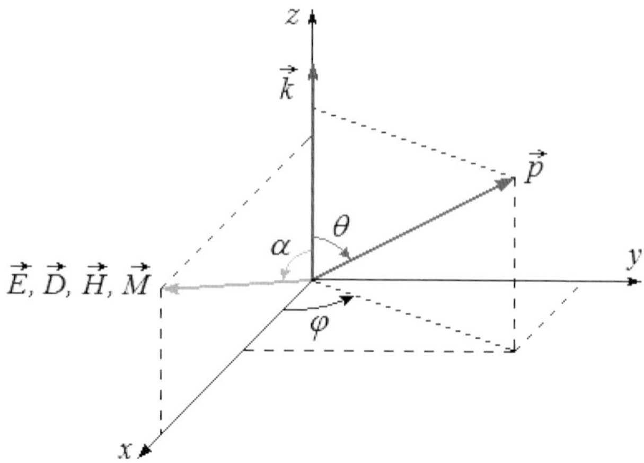


Fig. 7.2. Spatial orientation of moving directions of incident ($\vec{k}$) and emergent ($\vec{p}$) particles with respect to direction of the electromagnetic fields $\vec{E}, \vec{D}, \vec{H}, \vec{M}$.

### VII.3.1. The electric field case

Using the Maxwell field potential (7.1) and the Fourier transform (7.9), one has

$$A^{ext}_{\mu}(\vec{q}) = i\delta_{\mu 4} V^{ext}(\vec{q}), \quad V^{ext}(\vec{q}) = iV(\vec{q}), \tag{7.16}$$

leading to the following expressions for the homogeneous electrostatic and electric dipole fields:

$$V(\vec{q}) = \frac{\vec{q} \cdot \vec{E}(\vec{q})}{\vec{q}^{\,2}}, \qquad V(\vec{q}) = -\frac{\vec{D} \cdot \vec{q}}{(2\pi)^{3/2} \vec{q}^{\,2}}. \tag{7.17}$$

The differential scattering cross-section then becomes

$$\frac{d\sigma}{d\Omega} = 2\pi e^2 \left( k_0^2 - \vec{k}^{\,2} \sin^2 \frac{\theta}{2} \right) [V(\vec{q})]^2, \quad \vec{q} = \vec{p} - \vec{k}. \tag{7.18}$$

Thus, if the electron energy is $k_0$ and the scattering process takes place in the electrostatic homogeneous field $\vec{E}$ acting in the volume $A \times B \times C$ of a capacitor then, utilizing the Fourier transform of the field,

$$\vec{E}(\vec{q}) = \frac{\vec{E}}{(2\pi)^{3/2}} \int_{-A/2}^{A/2} dx\ e^{-ix|\vec{k}|\sin\theta\cos\varphi} \times \int_{-B/2}^{B/2} dy\ e^{-iy|\vec{k}|\sin\theta\sin\varphi} \int_{-C/2}^{C/2} dz\ e^{-iz|\vec{k}|(1-\cos\theta)}, \tag{7.19}$$

one obtains

$$\frac{d\sigma}{d\Omega} = \frac{e^2E^2\left(k_0^2 - \vec{k}^{\,2}\sin^2\frac{\theta}{2}\right)}{\pi^2\vec{k}^{\,8}\sin^4\theta\sin^6\frac{\theta}{2}\sin^2\varphi\cos^2\varphi}\left(\sin\alpha\cos\frac{\theta}{2}\cos\varphi - \cos\alpha\sin\frac{\theta}{2}\right)^2 \times \sin^2\left(\frac{1}{2}A|\vec{k}|\sin\theta\cos\varphi\right)\sin^2\left(\tfrac{1}{2}B|\vec{k}|\sin\theta\sin\varphi\right) \times \sin^2\left(\frac{1}{2}C|\vec{k}|(1-\cos\theta)\right), \tag{7.20}$$

while for the electron scattering in the field of an electric dipole of moment $\vec{D}$ the result is

$$\frac{d\sigma}{d\Omega} = \frac{e^2D^2\left(k_0^2 - \vec{k}^{\,2}\sin^2\frac{\theta}{2}\right)}{(4\pi)^2\vec{k}^{\,2}\sin^2\frac{\theta}{2}}\left(\sin\alpha\cos\frac{\theta}{2}\cos\varphi - \cos\alpha\sin\frac{\theta}{2}\right)^2. \tag{7.21}$$

In the limit case $\varphi = \alpha = 2\pi$, the differential scattering cross-section given by (7.20) vanishes, while in the ultraralativistic limit ($|\vec{k}| \to k_0$), if $\theta = \alpha = \pi/2$ and $\varphi = 0$, then the differential scattering cross-section becomes

$$\frac{d\sigma}{d\Omega} = \frac{e^2E^2B^2}{2\pi^2k_0^4}\sin^2\left(\frac{1}{2}Ak_0\right)\sin^2\left(\frac{1}{2}Ck_0\right). \tag{7.22}$$

As one can easily observe, in both cases (the electrostatic and electric dipole fields) the differential scattering cross-section depends on the squared field intensity.

### VII.3.2. The magnetic field case

In this case, using again (7.1) and the Fourier transform (7.9) as well, one finds

$$A_\mu^{ext}(\vec{q}) = \delta_{\mu i} A_i^{ext}(\vec{q}), \quad (i = 1, 2, 3), \quad \vec{A}^{ext}(\vec{q}) = i\vec{A}(\vec{q}), \tag{7.23}$$

leading to the following expressions for the homogeneous magnetostatic and magnetic dipole fields,

$$\vec{A}(\vec{q}) = \frac{\vec{q} \times \vec{H}(\vec{q})}{\vec{q}^2}, \quad \vec{q} \cdot \vec{A}(\vec{q}) = 0, \quad \vec{A}(\vec{q}) = -\frac{\vec{M} \times \vec{q}}{(2\pi)^{3/2} \vec{q}^2}, \tag{7.24}$$

and the differential scattering cross-section (7.14) writes in the following form:

$$\frac{d\sigma}{d\Omega} = 2\pi e^2 \left\{ \left[\vec{k} \cdot \vec{A}(\vec{q})\right] \left[\vec{p} \cdot \vec{A}(\vec{q})\right] + \vec{k}^2 \sin^2 \frac{\theta}{2} \left[\vec{A}(\vec{q})\right]^2 \right\}, \quad \vec{q} = \vec{p} - \vec{k}. \tag{7.25}$$

Thus, for the scattering process of electrons with energy $k_0$ in the homogeneous[2] magnetostatic field $\vec{H}(\vec{x})$, within a domain of dimensions $A \times B \times C$, using the Fourier transform for the magnetostatic field similar to that given by (7.19), one obtains

$$\begin{aligned} \frac{d\sigma}{d\Omega} &= \frac{e^2 H^2 \left[\sin^2 \alpha \sin^2 \varphi + \sin^2 \frac{\theta}{2} \left(\sin \alpha \sin \frac{\theta}{2} \cos \varphi + \cos \alpha \cos \frac{\theta}{2}\right)^2\right]}{\pi^2 \vec{k}^6 \sin^4 \theta \sin^6 \frac{\theta}{2} \sin^2 \varphi \cos^2 \varphi} \\ &\quad \times \sin^2 \left(\frac{1}{2} A|\vec{k}| \sin \theta \cos \varphi\right) \sin^2 \left(\frac{1}{2} B|\vec{k}| \sin \theta \sin \varphi\right) \\ &\quad \times \sin^2 \left(\frac{1}{2} C|\vec{k}|(1 - \cos \theta)\right), \end{aligned} \tag{7.26}$$

and for the same process, but considered for the external field of a magnetic dipole of moment $\vec{M}$,

$$\frac{d\sigma}{d\Omega} = \frac{e^2 M^2 \left[\sin^2 \alpha \sin^2 \varphi + \sin^2 \frac{\theta}{2} \left(\sin \alpha \sin \frac{\theta}{2} \cos \varphi + \cos \alpha \cos \frac{\theta}{2}\right)^2\right]}{(4\pi)^2 \sin^2 \frac{\theta}{2}}. \tag{7.27}$$

[2] The process studied in this approach does not concern the synchrotronic radiation produced by the relativistic electrons moving in a magnetic field (for example, in synchrotrons, or in the storage rings), which is a second-order process.

In the limit case $\theta = \varphi = \alpha = \pi/2$, the differential scattering cross-section (7.26) associated to the ultrarelativistic electrons becomes

$$\frac{d\sigma}{d\Omega} = \frac{2e^2H^2A^2}{\pi^2 k_0^4} \sin^2\left(\frac{1}{2}Bk_0\right)\sin^2\left(\frac{1}{2}Ck_0\right), \tag{7.28}$$

while for $\theta = \alpha = \pi/2$ and $\varphi = 0$, the differential scattering cross-section is given by

$$\frac{d\sigma}{d\Omega} = \frac{e^2H^2B^2}{2\pi^2 k_0^4} \sin^2\left(\frac{1}{2}Ak_0\right)\sin^2\left(\frac{1}{2}Ck_0\right). \tag{7.29}$$

In accordance with (7.26) and (7.27), the magnetic field intensity-squared dependence of the differential scattering cross-section is similar to that found in the case of electric field.

## VII.4. Graviton photogeneration in external static electromagnetic fields

Following the standard prescription based on the minimal coupling principle in quantum gravity [33], in order to determine the interaction Lagrangian density between the gravitational and electromagnetic fields, the Lagrangian density of the gravitational field must be supplemented by a term representing the Maxwell field Lagrangian density written in the curved space-time,

$$\mathcal{L}_{Maxwell} \equiv \mathcal{L}_M = -\frac{1}{4}\sqrt{-g}g^{\mu\alpha}g^{\nu\beta}F_{\mu\nu}F_{\alpha\beta}. \tag{7.30}$$

Expanding all necessary quantities in power series in powers of the gravitational coupling constant $\kappa$,

$$\begin{cases} \sqrt{-g} = 1 - \dfrac{1}{2}\kappa y + \mathcal{O}(\kappa^2), \quad y = y_\alpha^\alpha, \\ g^{\mu\nu} = \eta^{\mu\nu} - \kappa h^{\mu\nu} + \mathcal{O}(\kappa^2), \quad h^{\mu\nu} = y^{\mu\nu} - \dfrac{1}{2}\eta^{\mu\nu}y, \end{cases} \tag{7.31}$$

and passing to the flat Minkowski space-time, the following expression is obtained for the first-order interaction Lagrangian density between the gravitational and Maxwell fields,

$$\mathcal{L}_{M-G}^{(1)}(\kappa) = -\frac{1}{2}\kappa F_{\mu\nu}F_{\mu\alpha}u_{\nu\alpha}, \tag{7.32}$$

where

$$u_{\mu\nu} = y_{\mu\nu} - \frac{1}{4}\delta_{\mu\nu}y. \tag{7.33}$$

Making allowance for the $S$-matrix formalism, one can deduce Feynman-type rules for diagrams in the external field approximation, used to determine the differential scattering cross-sections corresponding to the investigated processes.

The analyzed interaction process in this subchapter is described by the Feynman diagram shown in Fig. 7.3, where $k$ and $e_\mu^{(a)}(\vec{k})$, and also $p$ and $e_{\mu\nu}^{(b)}(\vec{p})$ are the four-momenta and polarization vectors (respectively, polarization tensors) of the incident and emergent particles, respectively ($a, b = 1, 2$ for the real photons and gravitons), and $q$ is the virtual photon four-momentum ($\vec{q} = \vec{p} - \vec{k}$ and $q_0 = p_0 - k_0$, expressing the momentum and energy conservation laws).

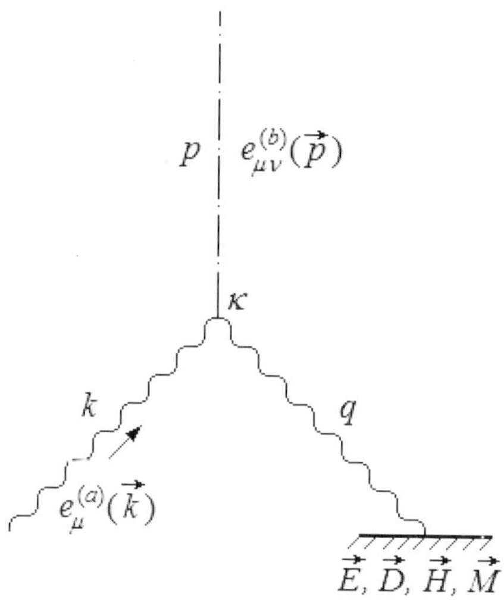


Fig. 7.3. Photon–photon–graviton Feynman diagram. The graviton is represented by a dotted line, and the photons by wavy lines.

Since the real gravitons are characterized by $y = 0$, the part of the Lagrangian density (7.32) that describes the analyzed interaction process – brought to the normal form – is

$$N\left[\mathcal{L}_{M-G}^{(1)}(x)\right] = -\frac{1}{2}\kappa F_{\mu\nu}^{ext}(x) y_{\nu\alpha}^{(-)}(x) F_{\mu\alpha}^{(+)}(x). \tag{7.34}$$

In order to determine the Fourier transform of the external electromagnetic fields $\vec{F} \equiv \vec{E}, \vec{D}, \vec{H}, \vec{M}$,

$$F_{\mu\nu}^{ext}(\vec{q}) = \frac{1}{(2\pi)^{3/2}} \int d^3x \; e^{-i\vec{q}\cdot\vec{x}} F_{\mu\nu}^{ext}(\vec{x}) \tag{7.35}$$

suppose, again, that $\vec{F}$ lies in the $xOz$-plane, and denote by $\alpha$ the angle between $\vec{F}$ and the direction of motion of the incident photon (taken along $z$-axis), and by $\varphi$ the angle between the $(\vec{F}, \vec{k})$ and $(\vec{p}, \vec{k})$ planes. In this case,

$$\begin{cases} \vec{F} \equiv (F \sin\alpha, \, 0, \, F \cos\alpha), \quad \vec{k} = (0, \, 0, \, k_0), \\ \vec{p} \equiv (k_0 \sin\theta \cos\varphi, \; k_0 \sin\theta \sin\varphi, \; k_0 \cos\theta), \\ \vec{q} \equiv \Big( k_0 \sin\theta \cos\varphi, \; k_0 \sin\theta \sin\varphi, \; -k_0(1 - \cos\theta) \Big). \end{cases} \tag{7.36}$$

### VII.4.1. The electric field case

By virtue of Eq. (7.3) and the Fourier transform of the electromagnetic field tensor (7.35), one obtains

$$F_{\mu\nu}^{ext}(\vec{q}) = \frac{i}{(2\pi)^{3/2}} \Big( \delta_{\mu 4}\delta_{\nu j} - \delta_{\mu j}\delta_{\nu 4} \Big) E_j(\vec{q}), \quad E_j(\vec{q}) = -\frac{\vec{D}\cdot\vec{q}}{(2\pi)^{3/2}} \frac{q_j}{\vec{q}^{\,2}}, \tag{7.37}$$

while the matrix element in the external field approximation is

$$\begin{aligned} \langle p|S|k\rangle &= \frac{i\kappa}{4(2\pi k_0 p_0)^{1/2}} \int d^3q \, \Big( \delta_{\mu 4}\delta_{\nu j} - \delta_{\mu j}\delta_{\nu 4} \Big) E_j(\vec{q}) e_{\nu\alpha}^{(b)}(\vec{p}) \\ &\quad \times \Big[ e_\alpha^{(a)}(\vec{k}) k_\mu - e_\mu^{(a)}(\vec{k}) k_\alpha \Big] \delta(\vec{p} - \vec{k} - \vec{q}) \delta(p_0 - k_0) \\ &= F(k,p)\delta(p_0 - k_0). \end{aligned} \tag{7.38}$$

Thus, for the interacting process of photons of energy $k_0$ in the homogeneous electrostatic field $\vec{E}(\vec{x})$ of a capacitor, within a 3D domain of spatial dimensions $A \times B \times C$, considering for the photons and gravitons the gauge $e_4^{(a)}(\vec{k}) = 0$ and $e_{4\alpha}^{(b)}(\vec{p}) = 0$, respectively, one obtains

$$F(k,p) = -\frac{\kappa}{4(2\pi)^{1/2}} Q(k,p), \quad Q(k,p) = e_i^{(a)}(\vec{k}) e_{ij}^{(b)}(\vec{p}) E_j(\vec{q}), \tag{7.39}$$

where

$$\vec{E}(\vec{q}) = \frac{\vec{E}}{(2\pi)^{3/2}} \int_{-A/2}^{A/2} dx\, e^{-ixk_0 \sin\theta\cos\varphi} \int_{-B/2}^{B/2} dy\, e^{-iyk_0 \sin\theta\sin\varphi} \times \int_{-C/2}^{C/2} dz\, e^{-izk_0(1-\cos\theta)}, \qquad \vec{q} = \vec{p} - \vec{k}. \tag{7.40}$$

Performing averaging over the polarization states of the incident photons and summation over the polarization states of the emergent gravitons, that is

$$\left\langle \sum_{fin\, sp} |F(k,p)|^2 \right\rangle_{in\, sp} = \frac{\kappa^2}{32\pi}\frac{1}{2}\sum_{pol}|Q(k,p)|^2 = \frac{\kappa^2}{64\pi}\left[1 + \frac{\left(\vec{k}\cdot\vec{p}\right)^2}{\vec{k}^2\vec{p}^2}\right] \times \left\{\left[\vec{E}(\vec{q})\right]^2 - \frac{\left[\vec{p}\cdot\vec{E}(\vec{q})\right]^2}{\vec{p}^2}\right\}, \tag{7.41}$$

and using for photons and gravitons the following completeness relations,

$$\begin{cases} \displaystyle\sum_{a=1}^{2} e_i^{(a)}(\vec{k})e_j^{(a)}(\vec{k}) = d_{ij}(\vec{k}) = \delta_{ij} - \frac{k_i k_j}{\vec{k}^2}, \quad (i,j = 1,2,3), \\ \displaystyle\sum_{b=1}^{2} e_{ij}^{(b)}(\vec{p})e_{kl}^{(b)}(\vec{p}) = d_{ik}(\vec{p})d_{jl}(\vec{p}) + d_{il}(\vec{p})d_{jk}(\vec{p}) \\ \qquad\qquad - d_{ij}(\vec{p})d_{kl}(\vec{p}), \quad (i,j,k,l = 1,2,3), \end{cases} \tag{7.42}$$

the differential scattering cross-section corresponding to the investigated process writes

$$\begin{aligned} \frac{d\sigma}{d\Omega} = \frac{1}{2}&\left(\frac{\kappa E}{2\pi k_0^2 \sin^2\theta \sin^2\dfrac{\theta}{2}\sin\varphi\cos\varphi}\right)^2 (1+\cos^2\theta) \\ &\times \left[1 - (\sin\alpha\sin\theta\cos\varphi + \cos\alpha\cos\theta)^2\right] \\ &\times \sin^2\left(\frac{1}{2}Ak_0\sin\theta\cos\varphi\right)\sin^2\left(\frac{1}{2}Bk_0\sin\theta\sin\varphi\right) \\ &\times \sin^2\left(\frac{1}{2}Ck_0(1-\cos\theta)\right), \end{aligned} \tag{7.43}$$

which generalizes the Mitskevich results presented in Ref. [270].

In the limit case $\theta = \alpha = \pi/2$, $\varphi = 0$, the differential scattering cross-section corresponding to the process of photon-graviton transformation in a homogeneous electrostatic field – considered as an external field – vanishes. If one chooses $\theta = \varphi = \alpha = \pi/2$, then the differential scattering cross-section becomes

$$\frac{d\sigma}{d\Omega} = \frac{\kappa^2 E^2 A^2}{2(2\pi)^2 k_0^2} \sin^2\left(\frac{1}{2} B k_0\right) \sin^2\left(\frac{1}{2} C k_0\right), \tag{7.44}$$

in agreement with Mitskevich's result, obtained in a different manner.

In the case of an electric dipole of moment $\vec{D}$, taking into account the transversality condition $p_i e_{ij}^{(b)}(\vec{p}) = 0$ satisfied by the gravitons and using the same gauge $e_4^{(a)}(\vec{k}) = 0$ for photons and $e_{4\alpha}^{(b)}(\vec{p}) = 0$ for gravitons, the matrix element corresponding to the analyzed interaction process – in the external field approximation – writes

$$\begin{aligned} \langle p|S|k\rangle &= \frac{-i\kappa}{(4\pi)^2 (k_0 p_0)^{1/2}} \int d^3q \left(\delta_{\mu 4}\delta_{\nu j} - \delta_{\mu j}\delta_{\nu 4}\right) \frac{\vec{D}\cdot\vec{q}}{\vec{q}^2} q_j e_{\nu\alpha}^{(b)}(\vec{p}) \\ &\quad \times \left[e_\alpha^{(a)}(\vec{k}) k_\mu - e_\mu^{(a)}(\vec{k}) k_\alpha\right] \delta(\vec{p} - \vec{k} - \vec{q}) \delta(p_0 - k_0) \\ &= F(k,p)\delta(p_0 - k_0), \end{aligned} \tag{7.45}$$

where

$$\begin{aligned} F(k,p) &= \frac{\kappa D}{2(4\pi)^2} \left(\cos\alpha - \cot\frac{\theta}{2}\cos\varphi \sin\alpha\right) Q(k,p), \\ Q(k,p) &= e_i^{(a)}(\vec{k}) e_{i3}^{(b)}(\vec{p}). \end{aligned} \tag{7.46}$$

Performing averaging over the polarization states of photons and summation over the polarization states of gravitons, and using the completeness relations (7.42), the following expression is obtained for the differential scattering cross-section corresponding to the photon-graviton conversion, in the external field of an electric dipole of moment $\vec{D}$:

$$\begin{aligned} \frac{d\sigma}{d\Omega} &= \frac{\kappa^2 D^2 k_0^2}{2(8\pi)^2} \left(1 + \cos^2\theta\right) \\ &\quad \times \left(\sin\alpha \cos\frac{\theta}{2}\cos\varphi - \cos\alpha \sin\frac{\theta}{2}\right)^2 \cos^2\frac{\theta}{2}. \end{aligned} \tag{7.47}$$

Since this kind of process has not been studied in such a context so far, the subject shall be a little bit more deeply investigated. Performing integration of (7.47) over the scattering angle $\theta$, one obtains the following $\varphi$-differential scattering cross-section,

$$\begin{aligned} f_1(\alpha,\varphi) &= \frac{1}{K}]\frac{d\sigma}{d\varphi}(\alpha,\varphi) = \frac{1}{K}\int_0^\pi \frac{d\sigma}{d\Omega}(\alpha,\theta,\varphi)\sin\theta\, d\theta \\ &= \frac{1}{30}\bigg\{13 - \cos 2\alpha + 7\cos 2\varphi - \frac{7}{2}\Big[\cos(2\alpha - 2\varphi) \\ &\quad + \cos(2\alpha + 2\varphi)\Big] - \frac{75\pi}{16}\sin 2\alpha\cos\varphi\bigg\}, \end{aligned} \tag{7.48}$$

and, if the same relation is integrated over the angle $\varphi$, one gets the following $\theta$-differential scattering cross-section:

$$\begin{aligned} f_2(\alpha,\theta) &= \frac{1}{K}\frac{d\sigma}{d\omega}(\alpha,\theta) = \frac{1}{K}\int_0^{2\pi} \frac{d\sigma}{d\Omega}(\alpha,\theta,\varphi)\, d\varphi = \frac{\pi}{16}\cos^2\frac{\theta}{2} \\ &\quad \times \Big[6 + 2\cos 2\alpha - 2\cos\theta - 3\cos(2\alpha - \theta) - 3\cos(2\alpha + \theta)\Big] \\ &\quad \times (3 + \cos 2\theta), \quad d\omega = \sin\theta\, d\theta. \end{aligned} \tag{7.49}$$

The two differential scattering cross-sections found above – corresponding to the process of graviton photogeneration in the external field of an electric dipole – are graphically represented in Fig. 7.4 and Fig. 7.5, respectively.

It is not difficult to integrate (7.47) over both variables $\theta$ and $\varphi$; the result is

$$\begin{aligned} \sigma^*_{tot}(\alpha) &= \frac{1}{K}\sigma_{tot}(\alpha) = \frac{1}{K}\int_0^{2\pi}\left(\int_0^\pi \frac{d\sigma}{d\Omega}(\alpha,\theta,\varphi)\sin\theta\, d\theta\right) d\varphi \\ &= \frac{1}{15}(13 - \cos 2\alpha), \end{aligned} \tag{7.50}$$

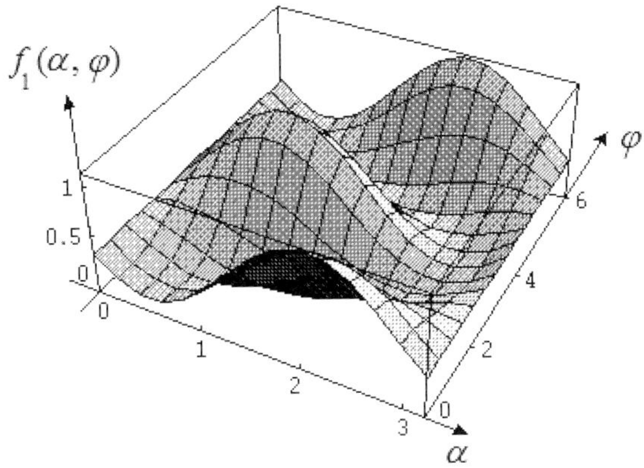


Fig. 7.4. Variation of $\varphi$-differential scattering cross-section $f_1(\alpha,\varphi) = \frac{1}{K}\frac{d\sigma}{d\varphi}(\alpha,\varphi)$ with respect to angles $\alpha$ and $\varphi$.

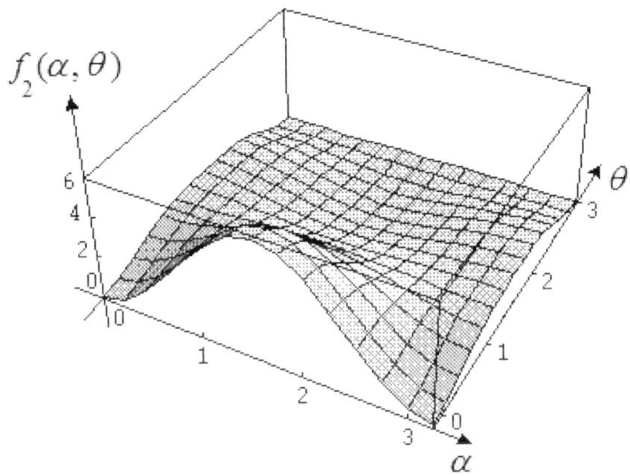


Fig. 7.5. Variation of $\theta$-differential scattering cross-section $f_2(\alpha,\theta) = \frac{1}{K}\frac{d\sigma}{d\omega}(\alpha,\theta)$ with respect to angles $\alpha$ and $\theta$.

and represents, obviously, the integral scattering cross-section corresponding to the graviton photogeneration process in the external field of an electric dipole. The quantity $K$ appearing in Eqs. (7.48), (7.49) and (7.50) signifies $K = \dfrac{\kappa^2 D^2 k_0^2}{2(8\pi)^2}$. The variation of the integral scattering cross-section $\sigma^*_{tot}(\alpha)$ with respect to angle $\alpha$ is graphically shown in Fig. 7.6. According to Eq. (7.47), the differential scattering cross-section of the graviton photogeneration process in the external field of an electric dipole of moment $\vec{D}$ shows a quadratic dependence on both the dipole moment amplitude $D = |\vec{D}|$ and the photon energy $k_0$.

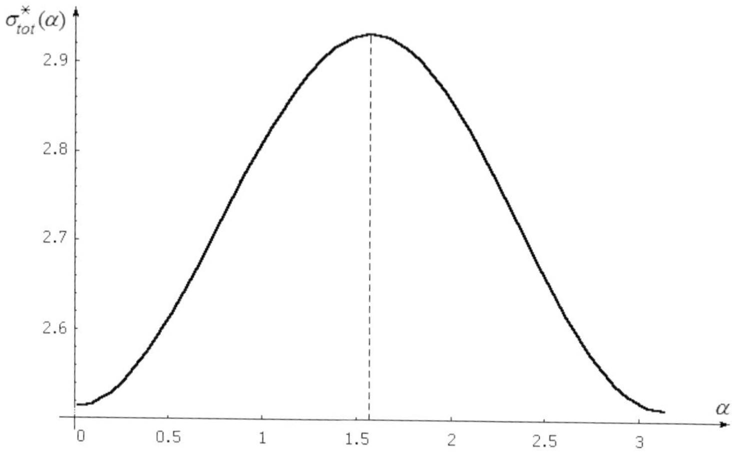


Fig. 7.6. Variation of the integral scattering cross-section $\sigma^*_{tot}(\alpha)$ with respect to angle $\alpha$, for a graviton photogeneration process in the external field of an electric dipole.

### VII.4.2. The magnetic field case

By means of Eqs. (7.4) and (7.9), the homogeneous magnetostatic and magnetic dipole fields respectively, are expressed by

$$F^{ext}_{\mu\nu}(\vec{q}) = \frac{1}{(2\pi)^{3/2}}\delta_{\mu i}\delta_{\nu j}\varepsilon_{ijk}H_k(\vec{q}),$$
$$H_k(\vec{q}) = \frac{1}{(2\pi)^{3/2}}\left(M_k - \frac{\vec{M}\cdot\vec{q}}{\vec{q}^{\,2}}q_k\right), \tag{7.51}$$

and the corresponding matrix element – in the external electromagnetic field approximation – writes

$$\begin{aligned}\langle p|S|k\rangle &= \frac{\kappa}{4\left(2\pi k_0 p_0\right)^{1/2}} \int d^3q\ \delta_{\mu i}\delta_{\nu j}\varepsilon_{ijk} H_k(\vec{q}) e^{(b)}_{\nu\alpha}(\vec{p}) \\ &\quad\times \left[e^{(a)}_{\alpha}(\vec{k})k_\mu - e^{(a)}_{\mu}(\vec{k})k_\alpha\right] \delta(\vec{p}-\vec{k}-\vec{q})\delta(p_0-k_0) \\ &= F(k,p)\delta(p_0-k_0).\end{aligned} \tag{7.52}$$

In this way, for the process of conversion of photons (possessing the energy $k_0$) into gravitons in the homogeneous magnetostatic field $\vec{H}(\vec{x})$, in a 3D domain of dimensions $A \times B \times C$, considering for the photons and gravitons the gauge specified by $e^{(a)}_4(\vec{k}) = 0$ and $e^{(b)}_{4\alpha}(\vec{k}) = 0$, respectively, one obtains

$$\begin{aligned}F(k,p) &= -\frac{\kappa}{4(2\pi)^{1/2}k_0} Q(k,p), \\ Q(k,p) &= \varepsilon_{ijk} H_k(\vec{q}) \left[e^{(a)}_l(\vec{k})k_i - e^{(a)}_i(\vec{k})k_l\right] e^{(b)}_{jl}(\vec{p}),\end{aligned} \tag{7.53}$$

where the Fourier transform of the homogeneous magnetostatic field $\vec{H}(\vec{q})$ is similar to that given by expression (7.40). Performing averaging over the polarization states of the initial photons, and summation over the polarization states of the emergent gravitons,

$$\begin{aligned}\left\langle \sum_{fin\,sp} |F(k,p)|^2 \right\rangle_{in\,sp} &= \frac{\kappa^2}{32\pi k_0^2}\frac{1}{2}\sum_{pol}|Q(k,p)|^2 \\ &= \frac{\kappa^2}{64\pi k_0^2}\left[\vec{k}^2 + \frac{\left(\vec{k}\cdot\vec{p}\right)^2}{\vec{p}^2}\right] \\ &\quad\times \left\{\left[\vec{H}(\vec{q})\right]^2 - \frac{\left[\vec{p}\cdot\vec{H}(\vec{q})\right]^2}{\vec{p}^2}\right\},\end{aligned} \tag{7.54}$$

and using the completeness relations (7.42), one obtains the following expression for the differential scattering cross-section of the analyzed process,

$$\frac{d\sigma}{d\Omega} = \frac{1}{2}\left(\frac{\kappa H}{2\pi k_0^2 \sin^2\theta \sin^2\frac{\theta}{2}\sin\varphi\cos\varphi}\right)^2 \left(1+\cos^2\theta\right)$$
$$\times\left[1-\left(\sin\alpha\sin\theta\cos\varphi+\cos\alpha\cos\theta\right)^2\right]\sin^2\left(\frac{1}{2}Ak_0\sin\theta\cos\varphi\right)$$
$$\times\sin^2\left(\frac{1}{2}Bk_0\sin\theta\sin\varphi\right)\sin^2\left(\frac{1}{2}Ck_0(1-\cos\theta)\right), \qquad (7.55)$$

this way generalizing the Mitskevich's results presented in Ref. [270]. In particular, for $\theta = \alpha = \pi/2$, $\varphi = 0$, the differential scattering cross-section for the process of photon-graviton conversion in a magnetic dipole field – considered as an external field – vanishes, while for $\theta = \varphi = \alpha = \pi/2$, the differential scattering cross-section becomes

$$\frac{d\sigma}{d\Omega} = \frac{\kappa^2 H^2 A^2}{2(2\pi)^2 k_0^2}\sin^2\left(\frac{1}{2}Bk_0\right)\sin^2\left(\frac{1}{2}Ck_0\right), \qquad (7.56)$$

in agreement with Mitskevich's results, obtained by a different procedure.

Finally, the matrix element corresponding to the process of photon-graviton conversion in the field of a magnetic dipole of moment $\vec{M}$ – in the external field approximation – writes

$$\langle p|S|k\rangle = \frac{\kappa}{(4\pi)^2(k_0p_0)^{1/2}}\int d^3q\;\delta_{\mu i}\delta_{\nu j}\varepsilon_{ijk}\left(M_k - q_k\frac{\vec{M}\cdot\vec{q}}{\vec{q}^2}\right)$$
$$\times\, e^{(b)}_{\nu\alpha}(\vec{p})\left[e^{(a)}_{\alpha}(\vec{k})k_\mu - e^{(a)}_{\mu}(\vec{k})k_\alpha\right]\delta(\vec{p}-\vec{k}-\vec{q})\delta(p_0-k_0)$$
$$= F(k,p)\delta(p_0-k_0). \qquad (7.57)$$

Considering for photons and gravitons the same gauge conditions as before, one obtains

$$
\begin{aligned}
F(k,p) &= \frac{\kappa}{(4\pi)^2 k_0} Q(k,p), \\
Q(k,p) &= \left[e_l^{(a)}(\vec{k})k_i - e_i^{(a)}(\vec{k})k_l\right]\varepsilon_{ijk}e_{jl}^{(b)}(\vec{p}) \\
&\quad \times \left[M_k - (p_k - k_k)\frac{\vec{M}\cdot(\vec{p}-\vec{k})}{4k_0^2\sin^2\dfrac{\theta}{2}}\right].
\end{aligned} \tag{7.58}
$$

Performing averaging and summation operations over the polarization states corresponding to incident photons and emergent gravitons, respectively, the differential scattering cross-section corresponding to graviton photogeneration process in the external field of a magnetic dipole of moment $\vec{M}$ is obtained in the form

$$
\begin{aligned}
\frac{d\sigma}{d\Omega} &= \frac{\kappa^2 M^2 k_0^2}{2(16\pi)^2}\left(1+\cos^2\theta\right) \\
&\quad \times\left[f_1^2(\alpha,\theta,\varphi)\sin^2\alpha + f_2^2(\alpha,\theta,\varphi) - f_3^2(\alpha,\theta,\varphi)\right],
\end{aligned} \tag{7.59}
$$

where

$$
\begin{cases}
f_1(\alpha,\theta,\varphi) = 2 + g(\alpha,\theta,\varphi)\cot\alpha\sin\theta\cos\varphi, \\
f_2(\alpha,\theta,\varphi) = \left[g^2(\alpha,\theta,\varphi)\sin^2\theta\sin\varphi \right. \\
\qquad\qquad\qquad \left. + 4\left(1 - g(\alpha,\theta,\varphi)\sin^2\dfrac{\theta}{2}\right)^2\right]^{1/2}, \\
f_3(\alpha,\theta,\varphi) = \sin\alpha\sin\theta\cos\varphi + \cos\alpha(1+\cos\theta), \\
g(\alpha,\theta,\varphi) = 1 - \tan\alpha\cot\dfrac{\theta}{2}\cos\varphi.
\end{cases} \tag{7.60}
$$

This result has the most general form and contains all the particular cases studied in the specific literature [92, 247, 248, 270]. Thus, if the photon direction of motion is oriented along the dipole moment $\vec{M}$ (*i.e.* $\alpha = 0$), the differential scattering cross-section becomes

$$
\frac{d\sigma_{||}}{d\Omega} = \frac{\kappa^2 M^2 k_0^2}{2(16\pi)^2}\sin^2\theta(1+\cos^2\theta), \tag{7.61}
$$

while if the same directions are orthogonal (*i.e.* $\alpha = \pi/2$), then one obtains

$$\frac{d\sigma_{\perp}}{d\Omega} = \frac{\kappa^2 M^2 k_0^2}{2(8\pi)^2}(1+\cos^2\theta)\left(\sin^2\varphi + \cos^2\varphi \sin^4\frac{\theta}{2}\right), \qquad (7.62)$$

in accordance with the results obtained by Papini and Valluri [300] for a particular case and using a different procedure. Following Eqs. (7.61) and (7.62), the differential scattering cross-section corresponding to the process of photon-graviton conversion in a magnetic dipole field – considered as an external field – is quadratically dependent on both the amplitude $M$ of the magnetic dipole moment $\vec{M}$, and the photon energy $k_0$.

It is worth mentioning the remarkable analogy between the expressions of the differential scattering cross-sections for the electron scattering process in external electromagnetic fields – given by Eqs. (7.20), (7.21), (7.26), and (7.27) – and the corresponding expressions of the differential scattering cross-sections for the graviton photogeneration process in the same external electromagnetic fields – expressed by Eqs. (7.43), (7.47), (7.55), and (7.59). One must specify, nevertheless, that in various particular cases, the spatial distribution of the emergent particles is mutually distinct.

More interesting and more dedicated to practical applications proves to be the process in which, as soon as the photon-graviton conversion has taken place, the resultant gravitational waves are involved in a reconversion process into electromagnetic waves. From the quantum field theory point of view, a somewhat similar process – that implies interaction with a double external static electromagnetic field, through the medium of a graviton propagator – can be described by the Feynman diagram shown in Fig. 7.7.

Since this is a second-order process, an exact result in this order of approximation is difficult to obtain and demands laborious calculations. Nevertheless, there is a simpler method to approach such a subject, namely the formalism offered by the ordinary quantum mechanics. In such a context there exists the possibility of obtaining correct – even if less rigorous – results.

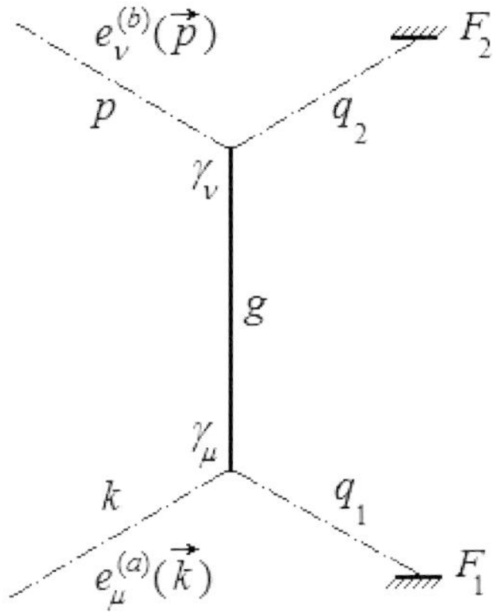


Fig. 7.7. The solid straight line represents the graviton, and the dash-dotted lines describe the photons.

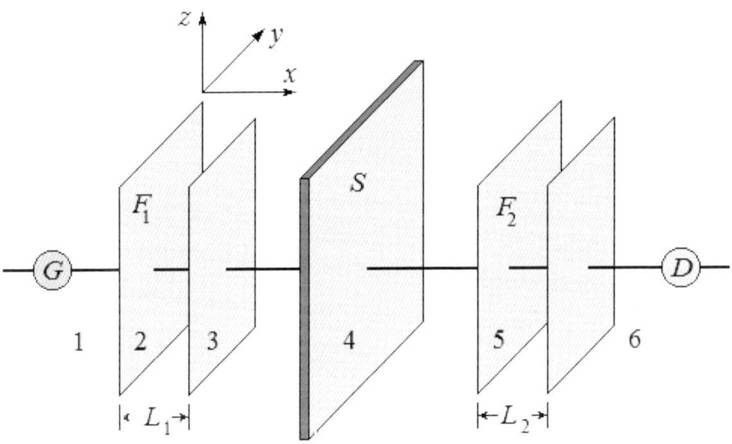


Fig. 7.8.

An experiment able to reproduce in reality such a process, proposed for the first time by Boccaletti and colab. [38(1970)], is schematically represented in Fig. 7.8. Here $G$ is a generator of linearly polarized electromagnetic waves, $D$ is an electromagnetic wave detector, and $S$ is a screen opaque to electromagnetic radiation. $F_1$ and $F_2$ represent the static fields. In region 1 are present the electromagnetic waves and, possibly, gravitational waves going to the left. In region 3 are existent the electromagnetic waves that have passed the region 2 and, possibly, gravitational waves going to the right.

The screen $S$ obstructs the electromagnetic waves to pass further towards the region 4, but it has no screening effect upon the gravitational waves. Consequently, in the region 4 are present only the "incident" gravitational waves and, possibly, the electromagnetic waves produced in region 5.

The energy density of the incident waves is (see Ref. [38(1970)] for details),

$$W_{inc} = \frac{H_0}{4\pi}, \tag{7.63}$$

where $H_0$ is the magnitude of the magnetic field intensity. If the detector registers only the magnetic field in which it is present (the contribution of the electric component being negligible), the density of the energy produced in region 6 is given by

$$\begin{aligned} W_{prod} = W_{inc}\frac{G^2 L_1^2 L_2^2}{c^8}\Big[\Big(H_{1z}H_{2z} + H_{2y}H_{1y}\Big)^2 \\ + \Big(H_{2z}H_{1y} - H_{2y}H_{1z}\Big)^2\Big]. \end{aligned} \tag{7.64}$$

In the above relation, the factor $G^2/c^8$ (which is extremely small) is due to the double gravitational interaction. There is an important characteristic specific to this type of experiment, resulting from Eq. (7.64), namely the independence of the produced power on the frequency of the electromagnetic waves. This fact allows one a convenient choice of the electromagnetic waves frequency, in agreement with the detector characteristics. One can also observe that, choosing $L_1 = L_2$ and $F_1 = F_2$, it follows that $W_{prod}$ depends on the fourth power of both the linear dimensions of the domain of field existence, and the field intensity itself. It then remains only the problem of determination of the best compromise between the dimensions and intensity.

## VII.5. Observations. Some astrophysical applications

As it has been already specified, due to the smallness of the gravitational constant, only the first-order processes in terms of the coupling constant are usually investigated, involving very interesting applications in both the laboratory experiments and in the astrophysics.

Thus, in the case of transformation of photons of energy $k_0$ into gravitons, in the electrostatic field $\vec{E}$ of a capacitor, or in a homogeneous magnetostatic field $\vec{H}$ (both fields acting in a 3D domain of dimensions $A \times B \times C$), the corresponding differential scattering cross-section is given by

$$\frac{d\sigma}{d\Omega} = \frac{\kappa^2 F_\perp^2 A^2}{2(2\pi)^2 k_0^2} \sin^2\left(\frac{1}{2}Bk_0\right) \sin^2\left(\frac{1}{2}Ck_0\right), \tag{7.65}$$

where $F_\perp = E, H$ is the field intensity oriented orthogonal to the photon direction of motion, which in this case is $A$, while $\kappa$ is directly connected to the gravitational coupling constant. In the case of diffusing bodies with linear dimensions very large as compared to the wavelength of the incident photon $\left(\lambda \ll V^{1/3}\right)$, the total scattering cross-section is

$$\sigma = \frac{1}{8}\kappa^2 L F_\perp^2 V,$$

where $L$ stands for the distance traveled by the photon in the field that occupies a 3D domain of volume $V$.

According to (7.65), the differential scattering cross-section vanishes if the photon direction of motion coincides with that of the magnetic field. An estimating calculation (see Ref. [92]) concludes that even in the laboratory one can create conditions for the integral scattering cross-section to reach values of the order of $\sigma \simeq 10^{-30}\,\mathrm{cm}^2$; in this case, the field intensity must be of about $E \simeq 10^4/A\lambda$ (respectively, $H \simeq 10^{13}/4\pi A\lambda$), where $\lambda$ is the wavelength of the photon passing through the field. This effect can also take place in the cosmic space, involving the photons that are crossing the galactic magnetic fields. In this case, for the same value of the total cross-section $\sigma$, it is necessary a lower intensity of the magnetic field, namely $H \simeq 10^{-3}\mathrm{A/m}$. In this case it is also possible the reverse process, namely the transformation of the gravitational into electromagnetic waves, when entering a spatial zone where there is a magnetic field. This quantum process lies at the basis of explanation of the phenomenon of intensification of radiation of the spiral galaxy arms with relatively intense magnetic field, within the domain of large wavelengths of the visible spectrum.

A similar process can also occur in the field of a magnetic dipole. In this case, if the photons of energy $k_0$ move along the direction of magnetic dipole moment $\vec{m}$, the total scattering cross-section is

$$\sigma \simeq (3/5)\kappa^2 m^2 k_0^2,$$

while if the same two directions are orthogonal, one obtains

$$\sigma \simeq \frac{1}{4}\kappa^2 m^2 k_0^2.$$

Estimated calculations show that the total scattering cross-section for the diffusion process of photons of energy $k_0 \simeq 10^9$ eV in the magnetic dipole field of an electron, is of about $\sigma \simeq 10^{-60}\,\mathrm{cm}^2$. Supposing one can produce in laboratory conditions a magnetic dipole field of moment $m \simeq 10^7\,\mathrm{Am}^2$, then for the visible light ($k_0 \simeq 1\,\mathrm{eV}$) one obtains $\sigma \simeq 10^{-32}\,\mathrm{cm}^2$.

In agreement with De Sabbata's ideas [99], the reverse process has many applications, mostly in astrophysics, being used to successfully explain the phenomenon of emission of electromagnetic waves by some cosmic objects, such as neutron stars and black holes. The explanation lies in the fact that, when traveling through the intense magnetic field of such an exotic cosmic object, due to the quantum effect of graviton-photon conversion, the gravitational field can produce an emission of electromagnetic waves. If these considerations are applied to a model of galactic centre formed by a neutron star cluster, there exists the possibility of interaction between the gravitational waves produced by the cluster itself with the magnetic dipole field of the neutron stars. During the process of neutron star collisions – which can take place between the two stars of any neutron star pair – the gravitational waves formed inside the cluster shall produce, through the medium of the above mentioned process, an emission of electromagnetic waves having a power of about $P \simeq 10^{24} - 10^{25}\,\mathrm{J/s}$ at the frequency $f = 1.6 \times 10^3\,\mathrm{Hz}$.

Finally, if these results are applied to the Earth's magnetic dipole field, under the hypothesis that a flux of gravitational waves coming from the centre of our galaxy arrives on the Earth, one obtains an electromagnetic waves emission having a power of about $P \simeq 10^{-11}$–$10^{-12}$ J/s, at the same frequency.

Chapter VIII

# Interaction of Scalar, Spinorial, Vectorial, Spin-Vectorial and Tensorial Particles with the Axially-Symmetric Gravitational Field Described by the Kerr Metric

## VIII.1. Gravitational scattering of spin-0, 1/2, 1, 3/2, and 2 massive particles in the axially-symmetric gravitational field described by the Kerr metric

### VIII.1.1. General considerations

The problem of gravitational scattering of particles possessing various spin values by an axially-symmetric gravitational field created by a massive body has already been considered in many scientific papers (see, *e.g.*, [91], [157], [158], [166], [305], [377]), but only in some particular cases. In this chapter, using the formalism offered by the minimal coupling principle applied in quantum gravity, a general and consistent approach of scattering process of scalar, spinorial, vectorial, spin-vectorial (Rarita–Schwinger) and second-rank tensorial massive particles in the external gravitational field described by the Kerr metric is developed. The obtained concrete results concerning the differential and integral scattering cross-sections, as well as the basic characteristics of the interaction processes between massive particles of spin-0, 1/2, and 1, and a stationary gravitational field described by a Kerr-type geometry generalize all the results previously obtained in papers like [166], [310], [369], [370], within the limits of a weak field and long wavelengths (which means that the wavelength corresponding to incident particles $\gg$ radius of the body which creates the gravitational background $\gg$ gravitational radius/mass of that body). These results also refer to the interaction process that implies the spin-3/2 massive particles and the above-mentioned gravitational field, which process has not yet been studied in such a context. In addition to the already mentioned investigations, the massive second-rank tensorial field has also been considered, in order to clarify several aspects regarding the interaction between this field and gravitation. In fact, there exists an increasing worldwide interest – reflected by the large number of scientific papers published by authors all over the world (see, *e.g.*, [79], [112], [136], [194–195]) – regarding the place and role of this field in contemporary theoretical physics.

The working technique used in this section is, roughly speaking, the same as that which has been applied in chapters VI and VII; the main differences appear, obviously, in the final results, and they are essentially due to the supplementary coupling between the spin of the interacting particles and the angular momentum of the body that creates the axially-symmetric gravitational background described by the Kerr metric.

As in the previous chapter, in order to determine the interaction Lagrangians [1] between the gravitational field and the considered matter fields, and unlike most of the authors, here one utilizes the minimal coupling principle in quantum gravity, which is equivalent to considering the matter field Lagrangians studied on curved spacetime. First of all, this procedure implies the necessity of replacement of the usual derivatives appearing in the matter field Lagrangians by their corresponding covariant derivatives [66]. The simple fact of doing the corresponding *minimal substitution* leads to appearance of some corrective terms in the interaction Lagrangians, even in the first-order approximation. One must specify the fact that, when considering the massive scalar field, a non-minimal coupling – introduced through the scalar curvature $R$ – has also been used. One of the main effects of the minimal substitution on the scattering cross-section is the full enlargement of the value range taken by the scattering angle, which is not forced to take only small values anymore. This

[1] For convenience, from now on, we shall use the appellative "Lagrangian" instead of "Lagrangian density"; in other words, in this chapter, by "Lagrangian" one must understood "Lagrangian density".

fact allows one to approach the special limit case of backwards scattering (backscattering).

As it has already been specified, in contrast to the majority of the authors that usually utilize a Gupta-type interaction Lagrangian (defined within the so-called *Gupta coupling formalism*),

$$L_{int}^{Gupta} = -\frac{1}{2}\kappa h_{\mu\nu} T^{\mu\nu},$$

in this chapter the interaction Lagrangians have been obtained by means of the already mentioned procedure as

$$L_{int} = L_{int}^{Gupta} + L_{int}^{(1)corr}(\kappa),$$

where $L_{int}^{(1)corr}(\kappa)$ is the first-order correction in the coupling constant $\kappa$. This corrective term is composed of the additional terms introduced by covariant derivative of the corresponding field functions; in the first-order approximation this corrective term is proportional to the first-order derivatives of the weak gravitational field, expressed through the second-rank tensorial function $h_{\mu\nu}$. For example, in the case of spin-2 massive particles this corrective term is of the form

$$L_{int}^{(1)corr}(\kappa) = -\kappa(h_{\mu\sigma,\alpha} + h_{\alpha\sigma,\mu} - h_{\mu\alpha,\sigma})(\overset{*}{\varphi}_{\mu\nu,\alpha}\varphi_{\sigma\nu} + \overset{*}{\varphi}_{\sigma\nu}\varphi_{\mu\nu,\alpha}),$$

where $\varphi_{\mu\nu,\alpha}$ is the usual derivative of the second-rank tensorial function which describes the matter field.

For supplementary details regarding the way of approaching the problems in this subchapter, the reader is advised to consult Chap. XIV of Ref. [323].

### VIII.1.2. Notations and writing conventions

Let $g^{\mu\nu}$, $\eta^{\mu\nu}$, and $y^{\mu\nu}$ denote the fundamental metric tensor, the Minkowski tensor, diag $(+1, -1, -1, -1)$, and the weak gravitational field tensor, respectively. Following Feynman's ideas [142], and asking that inequality $|y^{\mu\nu}| \ll 1$ to be valid everywhere, the gravitational field can be expanded around Minkowski's flat background [184],

$$\sqrt{-g}\, g^{\mu\nu} = \eta^{\mu\nu} - \kappa y^{\mu\nu},$$

where $g = \det(g_{\mu\nu})$ and $\kappa = \sqrt{16\pi G}$. Here $G$ is Newton's constant, and the natural unit system $(c = 1, \hbar = 1)$ has been used.

In Boyer–Lindquist coordinates, the Kerr geometry is described by the line element [269]:

$$ds^2 = \Sigma\left(\frac{dr^2}{\Delta} + d\theta^2\right) + \left(r^2 + a^2\right)\sin^2\theta\, d\varphi^2 - dt^2 + \frac{2MGr}{\Sigma}\left(a\sin^2\theta d\varphi - dt\right)^2, \tag{8.1}$$

$$\Sigma \equiv r^2 + a^2\cos^2\theta\ , \quad \Delta \equiv r^2 - 2MGr + a^2.$$

For large and very large distances $r$ from the source of the gravitational field, the line element (8.1) reduces to that corresponding to the flat space-time. So, if in the expression of $ds^2$ one performs the transition from spherical to Cartesian coordinates, $x = r\sin\theta\cos\varphi$, $y = r\sin\theta\sin\varphi$, $z = r\cos\theta$, then for the far-zone (*i.e.* far from the source of the gravitational field), one obtains

$$ds^2 = \left[1 - \frac{2MG}{r} + \mathcal{O}\left(\frac{1}{r^3}\right)\right]dt^2 + \left[4\varepsilon_{jkl}G\frac{S^k x^l}{r^3} + \mathcal{O}\left(\frac{1}{r^3}\right)\right]dtdx^j - \left[\left(1 + \frac{2MG}{r}\right)\delta_{jk} + \left(\begin{array}{c}\text{radiative gravitational terms}\\ \text{that vanish as } \mathcal{O}\left(\frac{1}{r}\right)\end{array}\right)\right]dx^j dx^k, \tag{8.2}$$

$$\left(j,k,l = \overline{1,3}\right).$$

In this relation $\vec{S} = M\vec{a}$ is the angular momentum of the gravitational source, and $M$ is its mass. According to Eq. (8.2), the components of the fundamental metric tensor $g_{\mu\nu}$ therefore are

$$g_{00} = 1 - \frac{2MG}{r}\ , \quad g_{0j} = g_{j0} = \frac{2MG}{r^3}\varepsilon_{jkl}a_k x_l\ , \quad g_{jk} = -\left(1 + \frac{2MG}{r}\right)\delta_{jk}\ , \quad \left(j,k,l = \overline{1,3}\right). \tag{8.3}$$

By using the relations

$$g_{\mu\nu} = \eta_{\mu\nu} + \kappa h_{\mu\nu}\ , \quad h_{\mu\nu} = y_{\mu\nu} - \frac{1}{2}y\,\eta_{\mu\nu}\ , \quad y = y^{\alpha}_{\alpha}, \tag{8.4}$$

one finds

$$h_{00} = -\frac{2MG}{r} , \quad h_{0j} = h_{j0} = 2\varepsilon_{jkl}\frac{MG}{r^3}a_k x_l ,$$
$$h_{jk} = -\frac{2MG}{r}\delta_{jk} , \quad \left(j,k,l=\overline{1,3}\right) , \tag{8.5}$$

while the Fourier transforms of the components of the weak gravitational field tensor $y_{\mu\nu}$ are

$$y_{0j}(\vec{q}) = y_{j0}(\vec{q}) = -iy_{4j}(\vec{q}) = -iy_{j4}(\vec{q}) = \frac{1}{(2\pi)^{3/2}}\frac{\kappa M}{2|\vec{q}|^2}\varepsilon_{jkl}a_k q_l ,$$

$$y_{00}(\vec{q}) = y_{44}(\vec{q}) = \frac{1}{(2\pi)^{3/2}}\frac{\kappa M}{|\vec{q}|^2} , \quad y_{jk}(\vec{q}) = 0 , \quad \left(j,k,l=\overline{1,3}\right) . \tag{8.6}$$

If the angular momentum per unit mass $\vec{a}$ vanishes, then obviously, the Schwarzschild linearized geometry is retrieved.

From now on, the components of the three-dimensional vectors shall be denoted by Latin indices, and those of the four-vectors by Greek indices. Under this agreement, the set of values taken by (significant or summation) vector and/or tensor indices shall not be specified anymore. In addition, comma shall stand for the usual derivative and semicolon for covariant derivative, both being considered in terms of space-time coordinates.

### VIII.1.3. Gravitational scattering of scalar, spinorial, vectorial, spin-vectorial and tensorial massive particles

The problem of scattering of massive particles, possessing various spin values and moving in a space-time of small curvature, can be approached – from the quantum point of view – in a consequent manner by quantifying both the gravitational background and the scattered matter field. In such a scenario, the two fields couple to each other according to the Feynman-type rules. Nevertheless, since our interest is restricted in this chapter to a gravitational background geometry generated by energy-momentum distributions that are not appreciably affected by the scattering process, the virtual graviton may successfully be replaced by an external classical gravitational field; in other words, the external field approximation for a stationary gravitational field described by the Kerr geometry [92] shall be used.

The following discussion shall refer to those interactions which are proportional to $\kappa^2$, *i.e.* to processes implying a single exchange graviton, the ultimate goal being to compute the differential and integral scattering cross-sections in the first-order Born approximation.

As one knows, the matter field theories in a fixed curved background are in general nonrenormalizable; it is not even clear, so far, to what extent the semiclassical approximation, *i.e.* quantized matter fields in a classical background geometry can provide reliable results. However, the processes studied in this chapter prove to be finite, that is, the concrete calculations can be done to the end without obtaining divergent terms, at least in the first-order of the Born approximation.

To determine the first-order interaction Lagrangians, describing the coupling between the gravitational and the massive studied fields (*i.e.* scalar, spinorial, vectorial, spin-vectorial and second-rank tensorial fields) one uses – as shown in the beginning of this subchapter – the minimal coupling principle in quantum gravity, while for the massive scalar field is also considered a non-minimal coupling, introduced via the scalar curvature $R$. According to this principle, to the Lagrangian of the gravitational field have to be added the Lagrangians corresponding to the massive scalar, spinorial, vectorial and tensorial fields, written in curved space ([364], [366]),

$$L_{scalar} \equiv L_{m,0} = \sqrt{-g}\left[g^{\mu\nu}\varphi^*_{;\mu}\varphi_{;\nu} - (m^2 + bR)\varphi^*\varphi\right], \tag{8.7}$$

$$L_{spinorial} \equiv L_{m,1/2} = \frac{i}{2}\sqrt{-g}\left(\overline{\psi}\tilde{\gamma}^{\mu}\psi_{;\mu} - \overline{\psi}_{;\mu}\tilde{\gamma}^{\mu}\psi\right) - \sqrt{-g}m\overline{\psi}\psi, \tag{8.8}$$

$$L_{vectorial} \equiv L_{m,1} = \sqrt{-g}\left(-\frac{1}{2}g^{\mu\lambda}g^{\nu\rho}\tilde{G}^*_{\mu\nu}\tilde{G}_{\lambda\rho} + m^2 g^{\mu\nu}B^*_{\mu}B_{\nu}\right) \tag{8.9}$$

and

$$L_{tensorial} \equiv L_{m,2} = \sqrt{-g}g^{\mu\lambda}g^{\nu\rho}\left(g^{\sigma\tau}\varphi^*_{\mu\nu;\sigma}\varphi_{\lambda\rho;\tau} - m^2\varphi^*_{\mu\nu}\varphi_{\lambda\rho}\right), \tag{8.10}$$

respectively.

One can easily observe that for massive scalar and vectorial fields the Klein–Fock and Proca formalisms respectively, have been considered. In Eq. (8.7) $R$ is the scalar curvature and $b$ is an arbitrary constant, whose particular values depend on the procedure used to obtain the general equation of the scalar field (or some other specific reasons). For example, $b = -1/4$ if the Klein–Fock equation is obtained by squaring the Dirac equation in the curved space-time, $b = -1/3$ if the same equation is found by using Feynman's functional formalism within the frame of special relativity, and $b = -1/6$ in the case of conformally invariant theory of the massless scalar field. In the above relations $\tilde{G}_{\mu\nu} = B_{\nu;\mu} - B_{\mu;\nu}$ stands for the massive vectorial field tensor, $B_{\mu;\nu}$ is the covariant derivative of the vector field function, and $\varphi_{\mu\nu;\lambda}$ is the covariant derivative of the second-rank tensor field function, which function satisfies the constraints $(\sqrt{-g}\,\varphi^{\mu\nu})_{;\nu} = 0$, and $\varphi_\alpha{}^\alpha = 0$.

In the case of massive Rarita–Schwinger field, in addition to the minimal coupling principle one also uses the *vierbein* formalism, so that the Lagrangian of this system can be written as (see Ref. [80] for details),

$$L_{spin\text{-}vector} \equiv L_{m,3/2} = \sqrt{-g}g^{\mu\nu}\left[\frac{i}{2}\left(\overline{\psi}_\mu \overleftarrow{D}_\lambda \tilde{\gamma}^\lambda \psi_\nu - \overline{\psi}_\mu \tilde{\gamma}^\lambda \overrightarrow{D}_\lambda \psi_\nu\right) + m\overline{\psi}_\mu \psi_\nu\right] = {}^1L_{m,3/2} + {}^2L_{m,3/2}, \tag{8.11}$$

where

$${}^1L_{m,3/2} = \sqrt{-g}g^{\mu\nu}\left[\frac{i}{2}\left(\overline{\psi}_{\mu,\lambda}\tilde{\gamma}^\lambda\psi_\nu - \overline{\psi}_\mu\tilde{\gamma}^\lambda\psi_{\nu,\lambda}\right) + m\overline{\psi}_\mu\psi_\nu\right]$$

and

$${}^2L_{m,3/2} = \frac{i}{8}\sqrt{-g}g^{\mu\nu}\overline{\psi}_\mu\left(\tilde{\gamma}_{\rho;\lambda}\tilde{\gamma}^\rho\tilde{\gamma}^\lambda + \tilde{\gamma}^\lambda\tilde{\gamma}_{\rho;\lambda}\tilde{\gamma}^\rho\right)\psi_\nu.$$

Here $\psi_{\mu,\lambda}$ denotes the usual derivative of the Rarita–Schwinger massive field function.

Obviously, the study of interaction processes involving massive particles is much different from that implying non-massive particles. More precisely, there is a great difference between two Lagrangians describing some interaction process, only one of them containing a mass term [365]. The presence of such a term in the Lagrangian seriously affects the physical content of the problem.

For example, even if nowadays are used (on a large scale) theories in which the gravitino is considered to be a non-massive particle, into a theory involving the cosmological constant, the gravitino is able to receive a non-zero rest mass (the role of the source for the gravitino rest-mass being played by the cosmological constant), as proved by Deser and Zumino (see Ref. [115] for details). Regarding the Rarita–Schwinger field, the mass term proposed by S. Deser, J.H. Kay, and K.S. Stelle in Ref. [113] has been used, namely $im\overline{\psi}_\mu\sigma^{\mu\nu}\psi_\nu$. For the sake of simplification, this mass term has been written under the more convenient form $mg^{\mu\nu}\overline{\psi}_\mu\psi_\nu$; this fact is possible, since the spin-vectorial field function obeys the constraints $\tilde{\gamma}^\mu\psi_\mu = 0$ and $\partial_\mu\psi^\mu = 0$, as shown in Ref. [113].

The terms ${}^1L_{m,3/2}$ and ${}^2L_{m,3/2}$ in the above relations have been obtained by using the covariant derivative of the Rarita–Schwinger massive field function,

$$\overrightarrow{D}_\mu\psi_\nu = \psi_{\nu,\mu} - \Gamma_\mu\psi_\nu \ ; \quad \overline{\psi}_\nu\overleftarrow{D}_\mu = \overline{\psi}_{\nu,\mu} - \overline{\psi}_\nu\Gamma_\mu, \tag{8.12}$$

where $\Gamma_\mu$ are the Fock–Ivanenko spin coefficients of the affine connection. As shown by the *vierbein* formalism, these coefficients have the following form,

$$\Gamma_\mu = \frac{1}{4}\tilde{\gamma}_{\lambda;\mu}\tilde{\gamma}^\lambda, \tag{8.13}$$

where $\tilde{\gamma}^\mu$ are the generalized Dirac matrices [246],

$$\tilde{\gamma}^\mu = L^\mu(\alpha)\hat{\gamma}(\alpha) \ , \quad \hat{\gamma}_\mu = L_\mu(\alpha)\hat{\gamma}(\alpha), \tag{8.14}$$

$\hat{\gamma}(\alpha)$ being the usual Dirac matrices in the hyperbolic representation of signature $-2$ of the Minkowski space-time. The covariant derivative of the generalized Dirac matrices $\tilde{\gamma}_\lambda$ is given by

$$\tilde{\gamma}_{\lambda;\mu} = \tilde{\gamma}_{\lambda,\mu} - \tilde{\gamma}_\rho\Gamma^\rho_{\lambda\mu}. \tag{8.15}$$

The *vierbein* coefficients $L^\mu(\alpha)$ and $L_\mu(\alpha)$ appearing in relation (8.14) satisfy the following constraints:

$$L^\mu(\alpha)L^\nu(\alpha) = g^{\mu\nu} \ ; \quad L_\mu(\alpha)L_\nu(\alpha) = g_{\mu\nu}. \tag{8.16}$$

Since only the first-order approximation in the gravitational coupling constant $\kappa$ is going to interfere in this subchapter, here are all the linearized quantities involved in this analysis ([270], [366]),

$$g^{\mu\nu} = \eta^{\mu\nu} - \kappa h^{\mu\nu} + \mathcal{O}(\kappa^2), \tag{8.17}$$

$$g_{\mu\nu} = \eta_{\mu\nu} + \kappa h_{\mu\nu} + \mathcal{O}(\kappa^2), \tag{8.18}$$

$$\Gamma^{\rho}_{\lambda\mu} = \frac{1}{2}\kappa(h^{\rho}_{\lambda,\mu} + h^{\rho}_{\mu,\lambda} - h_{\lambda\mu,}{}^{\rho}) + \mathcal{O}(\kappa^2), \tag{8.19}$$

$$\tilde{\gamma}^{\lambda} = \hat{\gamma}^{\lambda} - \frac{1}{2}\kappa\hat{\gamma}_{\beta}h^{\lambda\beta} + \mathcal{O}(\kappa^2), \tag{8.20}$$

$$\tilde{\gamma}_{\lambda} = \hat{\gamma}_{\lambda} + \frac{1}{2}\kappa\hat{\gamma}_{\beta}h^{\beta}_{\lambda} + \mathcal{O}(\kappa^2), \tag{8.21}$$

where

$$h^{\mu\nu} = y^{\mu\nu} - \frac{1}{2}\eta^{\mu\nu}y\ , \quad h^{\beta}_{\lambda} = y^{\beta}_{\lambda} - \frac{1}{2}\delta^{\beta}_{\lambda}y\ , \quad y = y^{\alpha}_{\alpha}. \tag{8.22}$$

It can be easily shown that the term ${}^{2}L_{m,3/2}$ appearing in equation (8.11) brings no contribution to the $L_{m,3/2}$ in the first-order approximation. Indeed, using the well-known anticommutation relations satisfied by the usual Dirac matrices,

$$\{\hat{\gamma}^{\mu}, \hat{\gamma}^{\nu}\} = \hat{\gamma}^{\mu}\hat{\gamma}^{\nu} + \hat{\gamma}^{\nu}\hat{\gamma}^{\mu} = 2\eta^{\mu\nu}, \tag{8.23}$$

one finds

$$\begin{aligned} {}^{2}L^{(1)}_{m,3/2}(\kappa) &= \frac{i}{16}\kappa\overline{\psi}_{\mu}\hat{\gamma}^{\lambda}\hat{\gamma}_{\alpha}\hat{\gamma}^{\rho}\psi^{\mu}\left(h^{\alpha}_{\rho,\lambda} - h^{\alpha}_{\lambda,\rho}\right) \\ &= \frac{i}{16}\kappa\overline{\psi}_{\mu}\hat{\gamma}^{\lambda}(\hat{\gamma}^{\alpha}\hat{\gamma}^{\rho} - \hat{\gamma}^{\rho}\hat{\gamma}^{\alpha})\psi^{\mu}h_{\alpha\rho,\lambda} = 0. \end{aligned}$$

In view of the above considerations, the first-order interaction Lagrangian between the gravitational and massive Rarita–Schwinger fields writes

$$\begin{aligned} L^{(1)}_{(int)m,3/2}(\kappa) = &-\frac{i}{4}\kappa\Big(\overline{\psi}_{\lambda,\mu}\hat{\gamma}_{\nu}\psi^{\lambda} - \overline{\psi}_{\lambda}\hat{\gamma}_{\nu}\psi^{\lambda}_{,\mu}\Big)h^{\mu\nu} \\ &- \frac{i}{2}\kappa\Big(\overline{\psi}_{\mu,\lambda}\hat{\gamma}^{\lambda}\psi_{\nu} - \overline{\psi}_{\mu}\hat{\gamma}^{\lambda}\psi_{\nu,\lambda}\Big)y^{\mu\nu} \\ &- \kappa\, m\,\overline{\psi}_{\mu}\psi_{\nu}y^{\mu\nu}. \end{aligned} \tag{8.24}$$

As is easy to see from formula (8.8), for the massive spinorial field, the usual Dirac Lagrangian has been used, while for the massive second-rank tensorial field, the simplest Lagrangian proposed by J. Schwinger in Ref. [331] has been considered.

Passing now to the flat complex Minkowski space-time,

$$x^0 = t,\, x^j \;\to\; x_j,\, x_4 = it\;; \quad \hat{\gamma}^\mu \partial_\mu \to i\gamma_\mu \partial_\mu\;;$$

$$\eta^{\mu\nu} \to -\delta_{\mu\nu}\,; \quad \partial_\mu \equiv \frac{\partial}{\partial x^\mu} \to \frac{\partial}{\partial x_\mu}, \tag{8.25}$$

the first-order interaction Lagrangians for the interaction processes between the gravitational and massive scalar, spinorial, vectorial, spin-vectorial and tensorial fields are given by

$$L^{(1)}_{(int)m,0}(\kappa) = -\kappa\left[\varphi^*_{,\mu}\varphi_{,\nu}y_{\mu\nu} + \frac{1}{2}\varphi^*\varphi\left(m^2 y - b y_{,\mu\mu}\right)\right], \tag{8.26-1}$$

$$L^{(1)}_{(int)m,1/2}(\kappa) = -\frac{1}{4}\kappa\left(\overline{\psi}\gamma_\mu\psi_{,\nu} - \overline{\psi}_{,\nu}\gamma_\mu\psi\right)s_{\mu\nu} - \frac{1}{2}\kappa\, m\overline{\psi}\psi y, \tag{8.26-2}$$

$$L^{(1)}_{(int)m,1}(\kappa) = -\kappa\left(G^*_{\mu\nu}G_{\mu\alpha}u_{\nu\alpha} + m^2 B^*_\nu B_\alpha y_{\nu\alpha}\right), \tag{8.26-3}$$

$$L^{(1)}_{(int)m,3/2}(\kappa) = +\frac{1}{4}\kappa\left(\overline{\psi}_{\alpha,\nu}\gamma_\mu\psi_\alpha - \overline{\psi}_\alpha\gamma_\mu\psi_{\alpha,\nu}\right)y_{\mu\nu} - \frac{1}{4}\kappa\, m\overline{\psi}_\alpha\psi_\alpha y, \tag{8.26-4}$$

and

$$\begin{aligned} L^{(1)}_{(int)m,2}(\kappa) = &-\kappa\varphi^*_{\mu\nu,\alpha}\varphi_{\mu\nu,\beta}y_{\alpha\beta} - 2\kappa\left(\varphi^*_{\mu\nu,\alpha}\varphi_{\lambda\nu,\alpha} + m^2\varphi^*_{\mu\nu}\varphi_{\lambda\nu}\right)y_{\mu\lambda} \\ &+ \kappa\left(\varphi^*_{\mu\nu,\alpha}\varphi_{\mu\nu,\alpha} + \frac{1}{2}m^2\varphi^*_{\mu\nu}\varphi_{\mu\nu}\right)y \\ &- \kappa\left(h_{\mu\sigma,\alpha} + h_{\alpha\sigma,\mu} - h_{\mu\alpha,\sigma}\right)\left(\varphi^*_{\mu\nu,\alpha}\varphi_{\sigma\nu} + \varphi^*_{\sigma\nu}\varphi_{\mu\nu,\alpha}\right), \end{aligned} \tag{8.26-5}$$

respectively, where

$$s_{\mu\nu} = y_{\mu\nu} + \frac{1}{2}\delta_{\mu\nu}y, \tag{8.27}$$

$$u_{\mu\nu} = y_{\mu\nu} - \frac{1}{4}\delta_{\mu\nu} y, \tag{8.28}$$

$$G_{\mu\nu} = B_{\nu,\mu} - B_{\mu,\nu}. \tag{8.29}$$

There have also been used the massive Rarita–Schwinger field equation $\gamma_\mu \psi_{\alpha,\mu} = -m\psi_\alpha$, and its adjoint as well. According to the standard quantum fields theory, the parts of the first-order interaction Lagrangians expressed through relations (8.26-1)–(8.26-5) – brought to the normal form – describing the interaction processes between gravitation and the above considered massive fields are ([365, 366]),

$$N\Big[L^{(1)}_{(int)m,0}(x)\Big] = -k\Big[\varphi^{*(-)}_{,\mu}(x)\varphi^{(+)}_{,\nu}(x)y^{(ext)}_{\mu\nu}(x) + \frac{1}{2}\varphi^{*(-)}(x)\varphi^{(+)}(x)\Big(m^2 y^{(ext)}(x) - b y^{(ext)}_{,\mu\mu}(x)\Big)\Big], \tag{8.30-1}$$

$$N\left[L^{(1)}_{(int)m,1/2}(x)\right] = -\frac{1}{4}\kappa\Big(\overline{\psi}^{(-)}(x)\gamma_\mu\psi^{(+)}_{,\nu}(x) - \overline{\psi}^{(-)}_{,\nu}(x)\gamma_\mu\psi^{(+)}(x)\Big) \times s_{\mu\nu}(x) - \frac{1}{2}\kappa\, m\overline{\psi}^{(-)}(x)\psi^{(+)}(x)y^{ext}(x), \tag{8.30-2}$$

$$N\Big[L^{(1)}_{(int)m,1}(x)\Big] = -\kappa\Big(G^{*(-)}_{\mu\nu}(x)G^{(+)}_{\mu\alpha}(x)u^{ext}_{\nu\alpha}(x) + m^2 B^{*(-)}_{\nu}(x)B^{(+)}_{\alpha}(x)y^{ext}_{\nu\alpha}(x)\Big), \tag{8.30-3}$$

$$N\left[L^{(1)}_{(int)m,3/2}(x)\right] = +\frac{1}{4}\kappa\left(\overline{\psi}^{(-)}_{\alpha,\nu}(x)\gamma_\mu\psi^{(+)}_{\alpha}(x) - \overline{\psi}^{(-)}_{\alpha}(x)\gamma_\mu\psi^{(+)}_{\alpha,\nu}(x)\right) \times y^{ext}_{\mu\nu}(x) - \frac{1}{4}\kappa\, m\overline{\psi}^{(-)}_{\alpha}(x)\psi^{(+)}_{\alpha}(x)y^{ext}(x), \tag{8.30-4}$$

and

$$
\begin{aligned}
N\left[L^{(1)}_{(int)m,2}(x)\right] &= -\kappa\varphi^{*(-)}_{\mu\nu,\alpha}(x)\varphi^{(+)}_{\mu\nu,\beta}(x)y^{ext}_{\alpha\beta}(x) \\
&- 2\kappa\left(\varphi^{*(-)}_{\mu\nu,\alpha}(x)\varphi^{(+)}_{\lambda\nu,\alpha}(x) + m^2\varphi^{*(-)}_{\mu\nu}(x)\varphi^{(+)}_{\lambda\nu}(x)\right)y^{ext}_{\mu\lambda}(x) \\
&+ \kappa\left(\varphi^{*(-)}_{\mu\nu,\alpha}(x)\varphi^{(+)}_{\mu\nu,\alpha}(x) + \frac{1}{2}m^2\varphi^{*(-)}_{\mu\nu}(x)\varphi^{(+)}_{\mu\nu}(x)\right)y^{ext}(x) \qquad (8.30\text{-}5) \\
&- \kappa\Big(h^{ext}_{\mu\sigma,\alpha}(x) + h^{ext}_{\alpha\sigma,\mu}(x) - h^{ext}_{\mu\alpha,\sigma}(x)\Big) \\
&\times\left(\varphi^{*(-)}_{\mu\nu,\alpha}(x)\varphi^{(+)}_{\sigma\nu}(x) + \varphi^{*(-)}_{\sigma\nu}(x)\varphi^{(+)}_{\mu\nu,\alpha}(x)\right),
\end{aligned}
$$

respectively.

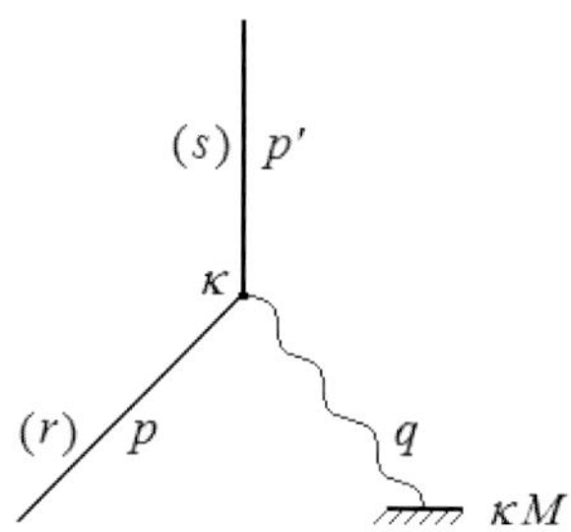


Fig. 8.1. The vertex (massive field)–graviton–(massive field). The curly line represents the graviton, while the solid straight lines represent the quanta of the scalar, spinorial, vectorial, spin-vectorial and tensorial fields, respectively.

In Fig. 8.1 is depicted the representative Feynman diagram for the investigated processes. Here $p$ and $(r)$, as well as $p'$ and $(s)$, are the four-momenta and spin-states indices ($r, s = \overline{1, 2s+1}$, where $s$ is the spin of the interacting particles) of the initial and final particles, respectively, and $q$ is the four-momentum of the virtual graviton.

In Fig. 8.2 are represented the spatial orientations of the angular momentum $\vec{a}$ and the "scattered direction" $\vec{p}\,'$, relative to the direction of incident particles' momentum $\vec{p}$.

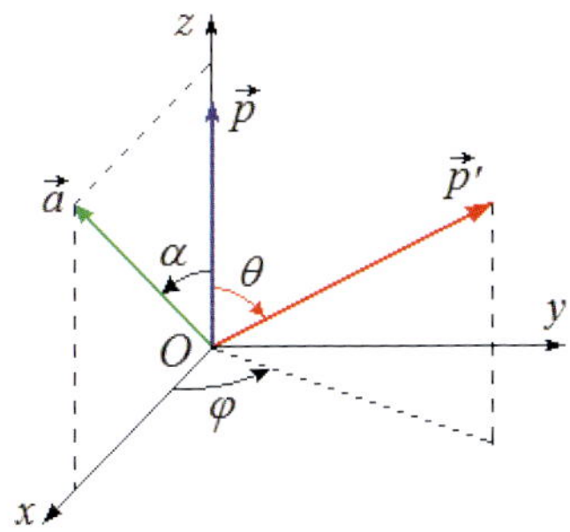


Fig. 8.2. Spatial orientation of the angular momentum $\vec{a}$ and of the scattered particle momentum $\vec{p}'$ relative to the momentum $\vec{p}$ of the incident particle.

In order to determine the $S$-matrix elements, $\langle p'|S|p\rangle$, which are necessary to calculate the differential scattering cross-sections, one must deduce the corresponding Feynman-type rules within the external field approximation. As usual, all these considerations are performed in the framework of the general $S$-matrix formalism.

Thus, if the external gravitational field is described by the Kerr metric, then the following expressions are obtained for the quantities $y_{\mu\nu}^{ext}(\vec{q})$, $s_{\mu\nu}^{ext}(\vec{q})$, $u_{\mu\nu}^{ext}(\vec{q})$, and $h_{\mu\nu}^{ext}(\vec{q})$ :

$$\begin{cases} y_{jk}^{ext}(\vec{q}) = 0, \\ y_{j4}^{ext}(\vec{q}) = y_{4j}^{ext}(\vec{q}) = \dfrac{1}{2}\dfrac{1}{(2\pi)^{3/2}}\dfrac{\kappa M}{|\vec{q}|^2}(\vec{a}\times\vec{q})_j, \\ y_{44}^{ext}(\vec{q}) = y^{ext}(\vec{q}) = \dfrac{1}{(2\pi)^{3/2}}\dfrac{\kappa M}{|\vec{q}|^2}. \end{cases} \tag{8.31}$$

$$\begin{cases} s_{jk}^{ext}(\vec{q}) = \dfrac{1}{2}\dfrac{1}{(2\pi)^{3/2}}\dfrac{\kappa M}{|\vec{q}|^2}\delta_{jk}, \\ s_{j4}^{ext}(\vec{q}) = s_{4j}^{ext}(\vec{q}) = \dfrac{1}{2}\dfrac{1}{(2\pi)^{3/2}}\dfrac{\kappa M}{|\vec{q}|^2}(\vec{a}\times\vec{q})_j, \\ s_{44}^{ext}(\vec{q}) = \dfrac{3}{2}\dfrac{1}{(2\pi)^{3/2}}\dfrac{\kappa M}{|\vec{q}|^2}. \end{cases} \tag{8.32}$$

$$\begin{cases} u_{jk}^{ext}(\vec{q}) = \frac{1}{4}\frac{1}{(2\pi)^{3/2}}\frac{\kappa M}{|\vec{q}|^2}\delta_{jk}, \\ u_{j4}^{ext}(\vec{q}) = u_{4j}^{ext}(\vec{q}) = \frac{1}{2}\frac{1}{(2\pi)^{3/2}}\frac{\kappa M}{|\vec{q}|^2}(\vec{a}\times\vec{q})_j, \\ u_{44}^{ext}(\vec{q}) = \frac{3}{4}\frac{1}{(2\pi)^{3/2}}\frac{\kappa M}{|\vec{q}|^2}, \end{cases} \tag{8.33}$$

and

$$\begin{cases} h_{jk}^{ext}(\vec{q}) = -\frac{1}{2}\frac{1}{(2\pi)^{3/2}}\frac{\kappa M}{|\vec{q}|^2}\delta_{jk}, \\ h_{j4}^{ext}(\vec{q}) = h_{4j}^{ext}(\vec{q}) = \frac{1}{2}\frac{1}{(2\pi)^{3/2}}\frac{\kappa M}{|\vec{q}|^2}(\vec{a}\times\vec{q})_j, \\ h_{44}^{ext}(\vec{q}) = \frac{1}{2}\frac{1}{(2\pi)^{3/2}}\frac{\kappa M}{|\vec{q}|^2}, \end{cases} \tag{8.34}$$

respectively.

It should be mentioned that any derivation with respect to coordinate $x_\alpha$ of the quantities $y_{\mu\nu}^{ext}(\vec{q})$, $s_{\mu\nu}^{ext}(\vec{q})$, $u_{\mu\nu}^{ext}(\vec{q})$, and $h_{\mu\nu}^{ext}(\vec{q})$ is equivalent to appearance of a supplementary factor $iq_\alpha$.

Taking into account the previous relations, the $S$-matrix elements in the external field approximation, corresponding to the Feynman diagram shown in Fig. 8.1 can be expressed as follows,

$$S_{p'p}^{(*)} \equiv \left\langle p' \middle| S^{(*)} \middle| p \right\rangle = F^{(*)}(p',p)\delta(q_0), \tag{8.35}$$

where the superscript symbol $(*)$ denotes one (every) of all possible values 0, 1/2, 1, 3/2, and 2, depending on the considered field (scalar, spinorial, vectorial, spin-vectorial, or tensorial). The equality $q_0 = p_0' - p_0 = 0$ appearing in Eq. (8.35) expresses the energy conservation law, while the quantities $F^{(*)}(p',p)$ are given by

$$F^{(0)}(p',p) = \frac{i\kappa^2 M}{2p_0(2\pi)^2|\vec{p}\,' - \vec{p}|^2}Q^{(0)}(p',p), \tag{8.36-1}$$

$$F^{(1/2)}(p',p) = \frac{\kappa^2 mM}{4p_0(2\pi)^2|\vec{p}\,' - \vec{p}|^2}Q^{(1/2)}(p',p), \tag{8.36-2}$$

$$F^{(1)}(p',p) = \frac{-i\kappa^2 M}{2p_0(2\pi)^2|\vec{p}\,' - \vec{p}|^2} Q^{(1)}(p',p), \tag{8.36-3}$$

$$F^{(3/2)}(p',p) = \frac{\kappa^2 mM}{p_0(2\pi)^2|\vec{p}\,' - \vec{p}|^2} Q^{(3/2)}(p',p), \tag{8.36-4}$$

and

$$F^{(2)}(p',p) = \frac{-i\kappa^2 M}{2p_0(2\pi)^2|\vec{p}\,' - \vec{p}|^2} Q^{(2)}(p',p), \tag{8.36-5}$$

respectively, where $p_0$ and $p_0'$ denote the energies of the incident and emergent particles, respectively, whereas the quantities $Q^{(*)}(p',p)$ are given by (see, also, Appendix F):

$$Q^{(0)}(p',p) = p_\mu p'_\nu \tilde{y}^{ext}_{\mu\nu}(\vec{q}) - \frac{b}{2}\tilde{y}^{ext}_{,\mu\mu}(\vec{q}) + \frac{m^2}{2}, \tag{8.37-1}$$

$$Q^{(1/2)}(p',p) = \overline{u}^{(s)}(\vec{p}\,')t^{(1/2)}u^{(r)}(\vec{p}), \tag{8.37-2}$$

$$\begin{aligned} Q^{(1)}(p',p) = &\Big[e^{(r)}_\nu(\vec{p})e^{(s)}_\lambda(\vec{p}\,')p'_\mu p_\mu - e^{(r)}_\nu(\vec{p})e^{(s)}_\mu(\vec{p}\,')p'_\mu p_\lambda \\ &- e^{(r)}_\mu(\vec{p})e^{(s)}_\lambda(\vec{p}\,')p'_\nu p_\mu + e^{(r)}_\mu(\vec{p})e^{(s)}_\mu(\vec{p}\,')p'_\nu p_\lambda\Big]\tilde{u}^{ext}_{\nu\lambda}(\vec{q}) \\ &+ m^2 e^{(r)}_\nu(\vec{p})e^{(s)}_\mu(\vec{p}\,')\tilde{y}^{ext}_{\mu\nu}(\vec{q}), \end{aligned} \tag{8.37-3}$$

$$Q^{(3/2)}(p',p) = \overline{u}^{(s)}_\alpha(\vec{p}\,')t^{3/2}u^{(r)}_\alpha(\vec{p}), \tag{8.37-4}$$

$$\begin{aligned} Q^{(2)}(p',p) = &\, e^{(s)}_{\mu\nu}(\vec{p}\,')e^{(r)}_{\mu\nu}(\vec{p})\left[p'_\alpha p_\beta \tilde{y}^{ext}_{\alpha\beta}(\vec{q}) - \left(p'_\alpha p_\alpha + \frac{1}{2}m^2\right)\tilde{y}^{ext}(\vec{q})\right] \\ &+ 2e^{(s)}_{\mu\nu}(\vec{p}\,')e^{(r)}_{\lambda\nu}(\vec{p})\left(p'_\alpha p_\alpha + m^2\right)\tilde{y}^{ext}_{\mu\lambda}(\vec{q}) + \Big[q_\alpha \tilde{h}^{ext}_{\mu\sigma}(\vec{q}) + q_\mu \tilde{h}^{ext}_{\alpha\sigma}(\vec{q}) \\ &- q_\sigma \tilde{h}^{ext}_{\mu\alpha}(\vec{q})\Big]\Big[p'_\alpha e^{(s)}_{\mu\nu}(\vec{p}\,')e^{(r)}_{\sigma\nu}(\vec{p}) - p_\alpha e^{(s)}_{\sigma\nu}(\vec{p}\,')e^{(r)}_{\mu\nu}(\vec{p})\Big], \end{aligned} \tag{8.37-5}$$

where

$$t^{(1/2)} = \gamma_\mu\left(p_\nu + p'_\nu\right)\tilde{s}^{ext}_{\mu\nu}(\vec{q}) - 2im\tilde{y}^{ext}(\vec{q}), \tag{8.38}$$

$$t^{(3/2)} = \frac{1}{4}\left[\left(p_\nu + p'_\nu\right)\gamma_\mu \tilde{y}^{ext}_{\mu\nu}(\vec{q}) - im\tilde{y}^{ext}(\vec{q})\right], \tag{8.39}$$

$$\tilde{y}^{ext}(\vec{q}) = (2\pi)^{3/2}\frac{|\vec{q}|^2}{\kappa M} y^{ext}(\vec{q}) = 1, \tag{8.40}$$

$$\tilde{y}^{ext}_{\mu\nu}(\vec{q}) = (2\pi)^{3/2}\frac{|\vec{q}|^2}{\kappa M} y^{ext}_{\mu\nu}(\vec{q}), \tag{8.41}$$

$$\tilde{s}^{ext}_{\mu\nu}(\vec{q}) = (2\pi)^{3/2}\frac{|\vec{q}|^2}{\kappa M} s^{ext}_{\mu\nu}(\vec{q}), \tag{8.42}$$

$$\tilde{u}^{ext}_{\mu\nu}(\vec{q}) = (2\pi)^{3/2}\frac{|\vec{q}|^2}{\kappa M} u^{ext}_{\mu\nu}(\vec{q}). \tag{8.43}$$

In the previous relations,

$$u^{(r)}(\vec{p}),\ \left[\overline{u}^{(r)}(\vec{p}) = (u^{(r)})^+(\vec{p})\gamma_4\right],\ e^{(r)}_\mu(\vec{p}),\ u^{(r)}_\mu(\vec{p}),$$

$$\left[\overline{u}^{(r)}_\mu(\vec{p}) = (u^{(r)}_\mu)^+(\vec{p})\gamma_4\right],\ e^{(r)}_{\mu\nu}(\vec{p}),$$

on the one side, and

$$u^{(s)}(\vec{p}\,'),\ \left[\overline{u}^{(s)}(\vec{p}\,') = (u^{(s)})^+(\vec{p}\,')\gamma_4\right],\ e^{(s)}_\mu(\vec{p}\,'),\ u^{(s)}_\mu(\vec{p}\,'),$$

$$\left[\overline{u}^{(s)}_\mu(\vec{p}\,') = (u^{(s)}_\mu)^+(\vec{p}\,')\gamma_4\right],\ e^{(s)}_{\mu\nu}(\vec{p}\,'),$$

on the other side, are the spinors, vectors, spin-vectors and second-rank tensors of the corresponding spin-states of the initial and final massive particles, respectively. We mention that the same notations, namely $m, \partial$, and $p_0$, have been used for the characteristic quantities (the mass, four-momentum and energy, respectively) of all five field types, $q_0 = p'_0 - p_0 = 0$ expresses the energy conservation law, while the superscript $+$ signifies the Hermitian conjugation.

The differential scattering cross-section is given by the well-known expression

$$d\sigma = (2\pi)^2 \left\langle \sum_{fin\,sp} |F(p', p)|^2 \right\rangle_{in\,sp} p_0^2\, d\Omega, \tag{8.44}$$

where $d\Omega = \sin\theta\, d\theta\, d\varphi$, $\theta$ being the scattering angle. To evaluate the differential scattering cross-sections, we first need to determine the quantities $\left\langle \sum_{fin\,sp} |F(p',p)|^2 \right\rangle_{in\,sp}$. In this respect, one can write

$$\begin{aligned} \left\langle \sum_{fin\,sp} \left|F^{(*)}(p',p)\right|^2 \right\rangle_{in\,sp} &= \left\langle \sum_{fin\,sp} C^{(*)} \left|Q^{(*)}(p',p)\right|^2 \right\rangle_{in\,sp} \\ &= \frac{C^{(*)}}{2s+1} \sum_{pol} \left|Q^{(*)}(p',p)\right|^2, \end{aligned} \tag{8.45}$$

where the coefficients $C^{(*)}$ can easily be identified, and $\sum_{pol} \left|Q^{(*)}(p',p)\right|^2$ are the sums over the corresponding spin-states; since these formulas are very long, they shall not be given here. In the case of massive spinorial and spin-vectorial fields, these sums are in fact two corresponding traces over the spinorial indices, that is

$$\begin{aligned} \sum_{pol} \left|Q^{(1/2)}(p',p)\right|^2 &= \sum_{r,s=1}^{2} \left|\overline{u}^{(s)}(\vec{p}\,')t^{(1/2)}u^{(r)}(\vec{p})\right|^2 \\ &= \mathrm{Tr}\left[\gamma_4 \left(t^{(1/2)}\right)^+ \gamma_4 P^{(1/2)}_{(+)}(\vec{p}\,')t^{(1/2)} P^{(1/2)}_{(+)}(\vec{p})\right], \end{aligned} \tag{8.46}$$

$$\begin{aligned} \sum_{pol} \left|Q^{(3/2)}(p',p)\right|^2 &= \sum_{r,s=1}^{4} \left|\overline{u}^{(s)}_{\alpha}(\vec{p}\,')t^{(3/2)}u^{(r)}_{\alpha}(\vec{p})\right|^2 \\ &= \mathrm{Tr}\left[\gamma_4 \left(t^{(3/2)}\right)^+ \gamma_4 P^{(3/2)}_{(+)}(\vec{p}\,')t^{(3/2)} P^{(3/2)}_{(+)}(\vec{p})\right], \end{aligned} \tag{8.47}$$

where $P^{(1/2)}_{(+)}$ and $P^{(3/2)}_{(+)}$ are the covariant projection operators (projectors) for the positive energy spinors and spin-vectors, respectively. To evaluate the sums over the spin-states, one must take into account the fact that the spinors, vectors, spin-vectors and second-rank tensors corresponding to the spin-states of the massive particles satisfy the following relations (see Refs. [252], [331]),

$$P^{(1/2)}_{(+)}(\vec{p}) = \sum_{r=1}^{2} u^{(r)}(\vec{p})\overline{u}^{(r)}(\vec{p}) = \frac{\gamma_\mu p_\mu + im}{2im}, \tag{8.48}$$

$$\Pi_{\mu\nu}(\vec{p}) = \sum_{k=1}^{3} e_{\mu}^{(k)}(\vec{p}) e_{\nu}^{(k)}(\vec{p}) = d_{\mu\nu}, \tag{8.49}$$

$$P_{(+)\mu\nu}^{(3/2)}(\vec{p}) = \sum_{r=1}^{4} u_{\mu}^{(r)}(\vec{p}) \overline{u}_{\nu}^{(r)}(\vec{p}) = \frac{\gamma_\alpha p_\alpha + im}{2im}$$
$$\times \left[ \delta_{\mu\nu} - \frac{1}{3}\gamma_\mu\gamma_\nu + \frac{i}{3m}\left(\gamma_\mu p_\nu - \gamma_\nu p_\mu\right) + \frac{2}{3m^2} p_\mu p_\nu \right], \tag{8.50}$$

$$\Pi_{\mu\nu,\lambda\rho}(\vec{p}) = \sum_{k=1}^{5} e_{\mu\nu}^{(k)}(\vec{p}) e_{\lambda\rho}^{(k)}(\vec{p}) = \frac{1}{2}\left(d_{\mu\lambda} d_{\nu\rho} + d_{\mu\rho} d_{\nu\lambda}\right)$$
$$- \frac{1}{3} d_{\mu\nu} d_{\lambda\rho}, \tag{8.51}$$

where $d_{\mu\nu} = \delta_{\mu\nu} + \frac{p_\mu p_\nu}{m^2}$.

After laborious calculations, for the differential scattering cross-sections of the interaction processes between the massive scalar, spinorial, vectorial, spin-vectorial and second-rank tensorial particles, and the external axially-symmetric gravitational field described by the Kerr metric one obtains the following expressions,

$$d\sigma^{(0)} = (GM)^2 \frac{d\Omega}{\sin^4 \dfrac{\theta}{2}} \left[ \left( \frac{1+v^2}{2v^2} \right)^2 - 2b \frac{1+v^2}{v^2} \sin^2 \frac{\theta}{2} + 4b^2 \sin^4 \frac{\theta}{2} \right]$$
$$+ (GM)^2 \frac{d\Omega}{\sin^4 \dfrac{\theta}{2}} a^2 p_0^2 \sin^2 \alpha \sin^2 \theta \sin^2 \varphi, \tag{8.52-1}$$

$$d\sigma^{(1/2)} = (GM)^2 \frac{d\Omega}{\sin^4 \dfrac{\theta}{2}} \left[ \left( \frac{1+v^2}{2v^2} \right)^2 - \frac{3+v^2}{4v^2} \sin^2 \frac{\theta}{2} \right]$$
$$+ (GM)^2 a^2 p_0^2 \Big[ f_1(\theta) + f_2(v,\theta) \sin^2 \alpha + f_3(\theta) \sin 2\alpha \cos \varphi$$
$$+ f_4(\theta) \cos^2 \varphi + f_5(v,\theta) \sin^2 \alpha \cos^2 \varphi \Big] d\Omega, \tag{8.52-2}$$

$$d\sigma^{(1)} = (GM)^2 \frac{d\Omega}{\sin^4\dfrac{\theta}{2}} \left[\left(\frac{1+v^2}{2v^2}\right)^2 - \frac{2}{3}\sin^2\frac{\theta}{2}\left(\frac{2}{v^2} - \sin^2\frac{\theta}{2}\right)\right]$$

$$+ (GM)^2 a^2 p_0^2 \Big[f_1(v,\theta) + f_2(v,\theta)\sin^2\alpha + f_3(v,\theta)\sin 2\alpha\cos\varphi$$

$$+ f_4(\theta)\sin^2\alpha\cos^2\varphi\Big] d\Omega, \qquad (8.52\text{-}3)$$

$$d\sigma^{(3/2)} = (GM)^2 \frac{d\Omega}{\sin^4\dfrac{\theta}{2}} \Bigg\{ \left(\frac{1+v^2}{2v^2}\right)^2 - \frac{1}{36v^4(1-v^2)^2}\Big[(15 - 41v^2$$

$$+ 5v^4 + 21v^6)v^2\sin^2\frac{\theta}{2} + 4\left(3 - 6v^2 - 5v^4\right)v^4\sin^4\frac{\theta}{2} + 8(3+v^2)$$

$$\times v^6\sin^6\frac{\theta}{2}\Big]\Bigg\} + (GM)^2 a^2 p_0^2\Big[f_1(v,\theta) + f_2(v,\theta)\sin^2\alpha$$

$$+ f_3(v,\theta)\sin 2\alpha\cos\varphi + f_4(v,\theta)\cos^2\frac{\alpha}{2}\cos^2\varphi\Big] d\Omega, \qquad (8.52\text{-}4)$$

and

$$d\sigma^{(2)} = (GM)^2 \frac{d\Omega}{\sin^4\dfrac{\theta}{2}} \Bigg\{ \left(\frac{1+v^2}{2v^2}\right)^2 + \frac{2}{5}\left[\frac{1}{3v(1-v^2)^2}\right]\Big[128v^{10}$$

$$\times \sin^{10}\frac{\theta}{2} - 64v^2(1+5v^2)\sin^8\frac{\theta}{2} + 4v^6(75v^4 + 50v^2 - 53)\sin^6\frac{\theta}{2}$$

$$+ 4v^4(23v^6 - 121v^4 + 255v^2 - 71)\sin^4\frac{\theta}{2} - v^2(31v^8 - 528v^6$$

$$+ 1114v^4 - 768v^2 + 151)\sin^2\frac{\theta}{2} + 15(v^{10} - 13v^8 + 34v^6 - 34v^4$$

$$+ 13v^2 - 1)\Big]\sin^2\frac{\theta}{2}\Bigg\} + (GM)^2 a^2 p_0^2\Big[f_1(v,\theta) + f_2(v,\theta)\cos^2\alpha$$

$$+ f_3(v,\theta)\sin 2\alpha\cos\varphi + f_4(v,\theta)\sin^2\alpha\cos^2\varphi\Big] d\Omega, \qquad (8.52\text{-}5)$$

respectively, where $v$ designates the ratio between the modulus of the momentum of incident particles and their energy, $v = \frac{|\vec{p}|}{p_0}$. The functions $f_i(v,\theta)$, $(i = \overline{1,5})$ appearing in the above relations are given in Appendix G.

According to Eqs. (8.52), all the expressions giving the differential scattering cross-sections are of the form

$$\left(\frac{d\sigma}{d\Omega}\right)_{Kerr} = \left(\frac{d\sigma}{d\Omega}\right)_{Schw} + \text{ terms proportional to } \vec{a}^2 p_0^2 \,.$$

If the angular momentum per unit mass of the gravitational field source $\vec{a}$ vanishes, then the corresponding results obtained for the Schwarzschild spherically symmetric geometry ([367, 368], [380]) are retrieved, namely,

$$d\sigma^{(0)}_{Schw} = (GM)^2 \frac{d\Omega}{\sin^4\dfrac{\theta}{2}} \left[\left(\frac{1+v^2}{2v^2}\right)^2 - 2b\frac{1+v^2}{v^2}\sin^2\frac{\theta}{2} + 4b^2\sin^4\frac{\theta}{2}\right], \tag{8.53-1}$$

$$d\sigma^{(1/2)}_{Schw} = (GM)^2 \frac{d\Omega}{\sin^4\dfrac{\theta}{2}} \left[\left(\frac{1+v^2}{2v^2}\right)^2 - \frac{3+v^2}{4v^2}\sin^2\frac{\theta}{2}\right], \tag{8.53-2}$$

$$d\sigma^{(1)}_{Schw} = (GM)^2 \frac{d\Omega}{\sin^4\dfrac{\theta}{2}} \left[\left(\frac{1+v^2}{2v^2}\right)^2 - \frac{2}{3}\sin^2\frac{\theta}{2}\left(\frac{2}{v^2} - \sin^2\frac{\theta}{2}\right)\right], \tag{8.53-3}$$

$$\begin{aligned} d\sigma^{(3/2)}_{Schw} = (GM)^2 \frac{d\Omega}{\sin^4\dfrac{\theta}{2}} \Bigg\{ \left(\frac{1+v^2}{2v^2}\right)^2 - \frac{1}{36v^4(1-v^2)^2}\Bigg[(15 - 41v^2 \\ + 5v^4 + 21v^6)v^2\sin^2\frac{\theta}{2} + 4(3 - 6v^2 - 5v^4)v^4\sin^4\frac{\theta}{2} \\ + 8(3+v^2)v^6\sin^6\frac{\theta}{2}\Bigg]\Bigg\}, \end{aligned} \tag{8.53-4}$$

and

$$d\sigma^{(2)}_{Schw} = \frac{(GM)^2}{\sin^4\dfrac{\theta}{2}}\left\{\left(\frac{1+v^2}{2v^2}\right)^2 + \frac{2}{5}\left[\frac{1}{3v\left(1-v^2\right)^2}\right]\left[128v^{10}\sin^{10}\frac{\theta}{2}\right.\right.$$
$$-64v^2\left(1+5v^2\right)\sin^8\frac{\theta}{2} + 4v^6\left(75v^4+50v^2-53\right)\sin^6\frac{\theta}{2}$$
$$+4v^4\left(23v^6-121v^4+255v^2-71\right)\sin^4\frac{\theta}{2} - v^2\left(31v^8\right.$$
$$\left.-528v^6+1114v^4-768v^2+151\right)\sin^2\frac{\theta}{2} + 15\left(v^{10}-13v^8\right.$$
$$\left.\left.\left.+34v^6-34v^4+13v^2-1\right)\right]\sin^2\frac{\theta}{2}\right\}d\Omega, \tag{8.53-5}$$

respectively.

If the scattering angle $\theta$ takes only small and very small values, then all[2] the differential scattering cross-sections become

$$d\sigma^{(*)}_{Schw} = (GM)^2\frac{d\Omega}{\sin^4\frac{\theta}{2}}\left(\frac{1+v^2}{2v^2}\right)^2 \equiv d\sigma_{Ruth}, \tag{8.54}$$

which is a Rutherford-type differential scattering cross-sections. As one observes, Eqs. (8.53-1)–(8.53-5) contain $d\sigma_{Ruth}$ as a first term. Since this term (*i.e.* $d\sigma_{Ruth}$) represents precisely the differential scattering cross-section of the massive scalar particles (for example, the scalar mesons) – which particles are minimally coupled to gravity – the second term of Eqs. (8.53-2)–(8.53-5) can be interpreted as being due to the active contribution of the non-null spin of spinorial, vectorial, spin-vectorial, and tensorial massive particles.

It is worth specifying that, under the same conditions as above (that is $\vec{a} = 0$), in the case of massive scalar field under the gravitational minimal coupling ($b = 0$), due to dependence in $r^{-1}$ of the gravitational field potential, the Eq. (8.53-1) reduces to a Rutherford-type differential scattering cross-section (*i.e.* the differential scattering cross-section is proportional to $\sin^{-4}\frac{\theta}{2}$) for any value of the

[2]Even more than that, in the case of massive scalar field minimally coupled to gravity ($b = 0$), this result is valid for any value of the scattering angle.

scattering angle. If the coupling is non-minimal ($b \neq 0$), then the differential scattering cross-section shows a Rutherford-type dependence only for small scattering angles ($\theta \ll 1$). This feature of the differential scattering cross-section for the massive scalar particles also appears in the case of massless scalar field. This fact should not be surprising, since the presence of the terms containing the $b$-constant in the differential scattering cross-section is due to the scalar curvature $R$; if $R$ is different from zero along the universe line of the scattering source, one observes that for large collision parameters (corresponding to small values of the scattering angle), the scalar curvature cannot significantly contribute to the value of differential scattering cross-section. In addition, Eq. (8.52-1) shows that only for the massive scalar field the angular momentum effect upon the scattering process yields the appearance of a positive semidefinite supplementary term in the differential scattering cross-section $d\sigma^{(0)}$ which, for small and very small scattering angles, is negligible with respect to $d\sigma^{(0)}_{Schw}$. This fact can be easily understood by taking into account that for large collision parameters, the $r^{-1}$ dependence of the gravitational potential $h_{00}$ dominates the $r^{-2}$ dependence of the magnetic-type term $h_{0j}$ of the gravitational field (which is, in fact, the origin of the angular momentum of the scattering source).

At the end of this section it should be noted that, even if a gravitational background given by an exact solution with $R = 0$ has been considered, from the quantum point of view, the first-order contributions of gravitons to the interaction process produce a purely quantum fluctuation of $R$ of the form

$$\sqrt{-g}R = \frac{1}{2}\kappa y_{,\mu\mu},$$

which generates a non-trivial coupling term in constant $b$, namely

$$\frac{1}{2}b\kappa y_{,\mu\mu}(x)\varphi^*(x)\varphi(x)$$

in the Lagrangian of the investigated model (see, in this respect, the Eq. (8.26-1)).

In regard to the invariance of the obtained results, one should specify that, if the particles' mass is set to be zero in Eq. (8.9), then the final result is invariant under the transformation

$$B_\mu(x) \to B'_\mu(x) = B_\mu(x) + \partial_\mu f(x),$$

which is, in fact, a gauge transformation, where $f$ is an arbitrary scalar function. This can be easily understood if one notices that, using the relation (8.29) and setting $m = 0$ in Eq. (8.9), one retrieves – up to a factor 1/2 – the Lagrangian of the electromagnetic field. Unlike this situation, if one takes $m = 0$ in Eq. (8.10) corresponding to the massive second-rank tensorial field, the differential scattering cross-section is no longer invariant under a similar gauge transformation of the form

$$\varphi_{\mu\nu}(x) \to \varphi'_{\mu\nu}(x) = \varphi_{\mu\nu}(x) + \partial_\mu f_\nu(x) + \partial_\nu f_\mu(x),$$

where $f_\mu(x)$ is an arbitrary vector function. This fact, in its turn, can be understood by observing that the Feynman diagrams formalism preserves the gauge invariance only if *all* the diagrams of the same order in coupling constant are considered. Therefore, the exterior field approximation simplifies fairly enough the algebra, but there is a price to be paid for the omitted diagrams, namely the loss of gauge invariance of the scattering amplitude.

Finally, it is worth mentioning that, in the case of $\vec{a} = 0$ and small scattering angles, the differential scattering cross-sections have the same form for all types of scattered massive particles. In addition, if the ultrarelativistic limit is also considered, then the so-obtained results coincide with those corresponding to the non-massive scalar particles, massless spinor particles, photons, gravitinos and gravitons. In other words, in this special limit case the gravitational scattering of particles is spin-independent ([362], [379]), this result being in agreement with that obtained in some particular cases by other authors through different procedures. Besides, this result represents nowadays a veritable test of correctitude/validity of various methods of study of the gravitational scattering of particles – quanta of various fields.

### VIII.1.4. The backscattering case

Since the previous results concerning the differential scattering cross-sections $d\sigma/d\Omega$ are valid for any value of the scattering angle, the method used to determine the differential scattering cross-sections also allows one to study the limit case of backscattering ($\theta \to \pi$). The expression obtained for the differential backscattering cross-sections are substantially different from one case to another, showing a strong dependence of these differential cross-sections on the spin of particles.

For example, the differential backscattering cross-section obtained for the massive scalar particles is finite and does not depend on the angular momentum per unit mass $\vec{a}$ of the gravitational source,

$$\left.\frac{d\sigma^{(0)}}{d\Omega}\right|_{\theta=\pi} = (GM)^2\left[4b^2 + \left(\frac{1+v^2}{2v^2}\right)^2 - 2b\frac{1+v^2}{v^2}\right]. \tag{8.55}$$

Besides, in the ultrarelativistic limit this differential backscattering cross-section coincides with the formula given by De Logi and Kovacs in Ref. [91], for the massless scalar particles,

$$\left.\frac{d\sigma^{(0)}}{d\Omega}\right|_{\substack{\theta=\pi\\ v=1}} = (GM)^2(1-2b)^2. \tag{8.56}$$

The massive spinorial and vectorial particles also show finite values for the differential backscattering cross-sections but, unlike the case of massive scalar particles, they depend on both the squared angular momentum per unit mass $\vec{a}^2$, of the gravitational source, and the angles $\alpha$ and $\varphi$. Besides, it can be observed that this angular dependence also maintains in the ultrarelativistic limit, in which case

$$\left.\frac{d\sigma^{(1/2)}}{d\Omega}\right|_{\substack{\theta=\pi\\ v=1}} = (GM)^2\vec{a}^2p_0^2\left(\frac{9}{2}\cos^2\varphi + \sin^2\alpha\right), \tag{8.57}$$

and

$$\left.\frac{d\sigma^{(1)}}{d\Omega}\right|_{\substack{\theta=\pi\\ v=1}} = (GM)^2\left(\frac{1}{3} + 4\vec{a}^2p_0^2\cos^2\alpha\right). \tag{8.58}$$

As it can easily be checked, the spin-3/2 and 2 massive particles display differential backscattering cross-sections dependent on the angle $\alpha$, both being divergent in the ultrarelativistic limit. This last property is valid not only in the backscattering limit case, but also for any value of the scattering angle $\theta$, and even for total scattering cross-sections, as it can easily be observed from Figs. 8.14 and 8.17, which show the dependence of the two integral scattering cross-sections on both the ratio $|\vec{p}|/p_0 \equiv v$ and angle $\alpha$. As it can be observed from these figures, in the ultrarelativistic limit the plotted surfaces tend asymptotically to infinity as the values of $v$ tend to 1. Just for comparison, the dependence of the integral scattering cross-sections on the same variables has also been graphically shown for the rest of particles (*i.e.* the scalar, spinorial, and vectorial particles).

## VIII.1.5. Integral scattering cross-sections

Since in the limit $\theta \to 0$ the differential scattering cross-sections show a Rutherford-type dependence, some difficulties appear when calculating the integral scattering cross-sections. More precisely, due to the dependence on $\sin^{-4}\frac{\theta}{2}$ of the differential scattering cross-sections, one obtains some divergent terms; the integral/total scattering cross-sections tend to infinity, and the physical reason for this behavior is the infinite range of the gravitational force. This difficulty has been overcome in a manner similar to that used in Chap. VI, namely, by means of a cut-off procedure based on the Leibniz–Newton formula.

In this respect, the primitives of the functions

$$f(v,\alpha,\theta,\varphi) \equiv \frac{d\sigma}{d\Omega}(v,\alpha,\theta,\varphi)$$

have been firstly determined; the new resulting functions were denoted by $\sigma(v,\alpha,\theta)$. They have as arguments only the variables $v, \alpha$ and $\theta$ because they were calculated as indefinite integrals over both angular variables $\theta$ and $\alpha$, and at the same time as definite integrals over the angular variable $\varphi$, of the corresponding differential scattering cross-sections,

$$\begin{aligned}\sigma(v,\alpha,\theta) &= \int_{(\Omega)} \frac{d\sigma}{d\Omega}(v,\alpha,\theta,\varphi)\, d\Omega \\ &= \int_\theta d\theta \sin\theta \int_0^{2\pi} f(v,\alpha,\theta,\varphi)\, d\varphi.\end{aligned} \tag{8.59}$$

Then, using the Leibniz–Newton formula, the integral scattering cross-sections of all the five processes have been obtained as

$$\sigma_{tot}(v,\alpha) = \sigma(v,\alpha,\theta)|_{\theta=\pi} - \sigma(v,\alpha,\theta)|_{\theta=\varepsilon\ll 1}\,, \tag{8.60}$$

where $\varepsilon$ is a very small real number (in the concrete estimations, the value $\varepsilon = 10^{-6}$ was considered). Of course, rigorously speaking, the graphs drawn this way show only a qualitative dependence on variables $v$ and $\alpha$ of the so-obtained functions. It is clear that for $\varepsilon \to 0$ in all five cases one obtains divergent expressions for the integral scattering cross-sections $\sigma_{tot}(v,\alpha)$, the reason being explained above. The calculations lead to the following expressions for the integral scattering cross-sections:

$$\begin{aligned}\sigma^{(0)}(v,\alpha,\theta) &= 4\pi\left(\vec{a}^2 p_0^2 \sin^2\alpha - 2b^2\right)\cos\theta - \frac{\pi(1+v^2)^2}{v^4}\csc^2\frac{\theta}{2}\\ &\quad + 16\pi\frac{v^2\vec{a}^2p_0^2\sin^2\alpha - b\left(1+v^2\right)}{v^2}\ln\left(\sin\frac{\theta}{2}\right),\end{aligned} \tag{8.61-1}$$

$$\begin{aligned}\sigma^{(1/2)}(v,\alpha,\theta) &= \frac{\pi}{4}\vec{a}^2p_0^2\left(3+2v^2\sin^2\alpha - 7\cos 2\alpha\right)\cos\theta\\ &- \frac{\pi}{16}\vec{a}^2p_0^2(1+2v^2\sin^2\alpha - 3\cos 2\alpha)\cos 2\theta - \frac{\pi}{v^4}\left(1+v^2\right)^2\csc^2\frac{\theta}{2}\\ &- \frac{2\pi}{v^2}\left[3+v^2(1-13\vec{a}^2p_0^2+4\vec{a}^2p_0^2\cos 2\alpha)\right]\ln\left(\sin\frac{\theta}{2}\right),\end{aligned} \tag{8.61-2}$$

$$\begin{aligned}\sigma^{(1)}(v,\alpha,\theta) &= -\frac{\pi}{6}\left[8+\vec{a}^2p_0^2(4-13v^2+12\cos 2\alpha + 19v^2\cos 2\alpha)\right]\\ &\times\cos\theta - \frac{2\pi}{3}\vec{a}^2p_0^2\cos^2\alpha\cos 2\theta + \frac{\pi}{9}v^2\vec{a}^2p_0^2\cos^2\alpha\cos 3\theta - \frac{\pi}{v^4}\\ &\times\left(1+v^2\right)^2\csc^2\frac{\theta}{2} - \frac{8\pi}{3v^2}\left[4 - v^2\vec{a}^2p_0^2\Big(11+\cos 2\alpha\Big)\right]\ln\left(\sin\frac{\theta}{2}\right),\end{aligned} \tag{8.61-3}$$

$$\sigma^{(3/2)}(v,\alpha,\theta) = \frac{\pi}{432(1-v^2)^2}\Big[6f_1(v,\alpha)\cos\theta + 6f_2(v,\alpha)\cos 2\theta$$
$$+ \vec{a}^2 p_0^2\left(f_3(v,\alpha)\cos 3\theta + \frac{3}{2}f_4(v,\alpha)\cos 4\theta\right)\Big]$$
$$- \frac{\pi v^4}{360(1-v^2)}\vec{a}^2 p_0^2 \cos^2\frac{\alpha}{2}\cos 5\theta - \frac{\pi(1-v^2)^2}{v^4}$$
$$\times \csc^2\frac{\theta}{2} - \frac{2\pi}{9v^2(1-v^2)} f_5(v,\alpha)\ln\left(\sin\frac{\theta}{2}\right), \tag{8.61-4}$$

$$\sigma^{(2)}(v,\alpha,\theta) = \frac{\pi}{900(1-v^2)^4}\Big[5f_1(v,\alpha)\cos\theta + 10 f_2(v,\alpha)\cos 2\theta$$
$$+ \frac{5}{3}v^2 f_3(v,\alpha)\cos 3\theta + \frac{5}{2}v^4 f_4(v,\alpha)\cos 4\theta + v^6 f_5(v,\alpha)\cos 5\theta\Big]$$
$$- \frac{\pi(1+v^2)^2}{v^4}\csc^2\frac{\theta}{2} - \frac{8\pi}{3v^2(1-v^2)} f_6(v,\alpha)\ln\left(\sin\frac{\theta}{2}\right). \tag{8.61-5}$$

The functions $f_i(v,\alpha)$, $i = \overline{1,6}$ are given in Appendix $G$, while the graphical representations of these integral cross-sections, namely

$$\sigma_{tot}(v,\alpha) = \sigma(v,\alpha,\theta=\pi) - \sigma(v,\alpha,\theta=\varepsilon=10^{-6}),$$

are shown in Figs. 8.3, 8.5, 8.8, 8.11, 8.14, and 8.17.

Since for small and very small values of the ratio $|\vec{p}|/p_0 \equiv v$ the integral scattering cross-sections $\sigma_{tot}(v,\alpha)$ vary very rapidly, a logarithmic scale has been used to draw the Figs. 8.4, 8.6, 8.9, 8.12, 8.15, and 8.18. This procedure allows one a more detailed observation of dependence of total scattering cross-sections $\sigma_{tot}(v,\alpha)$ on the small and very small values of the ratio $|\vec{p}|/p_0 \equiv v$. Besides, in the Figs. 8.7, 8.10, 8.13, 8.16, and 8.19 it has been separately represented (for the most interesting case, namely, in the ultrarelativistic limit, for the massive particles of spin-0, 1/2 and 1, and in the quasi-ultrarelativistic limit, for the massive particles of spin-3/2 and 2) the dependence of the integral scattering cross-sections on angular variable $\alpha$, because in the Figs. 8.3, 8.5, 8.8, 8.11, 8.14, and 8.17 this dependency can not clearly be observed. This is due to the fact that

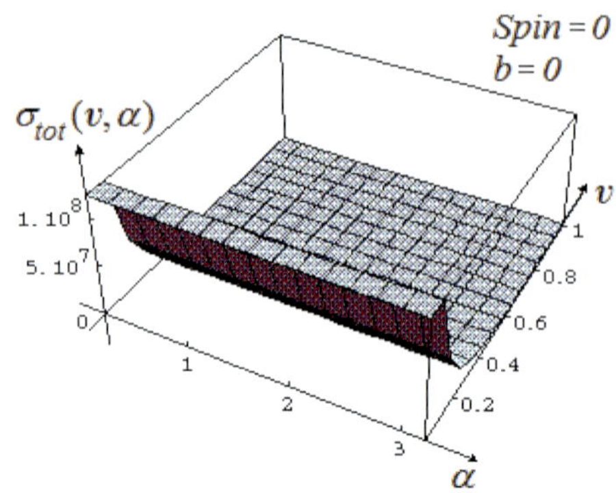


Fig. 8.3. Variation of the integral scattering cross-section $\sigma_{tot}(v, \alpha)$ with respect to both the ratio $v = \frac{|\vec{p}|}{p_0}$ and angle $\alpha$, for massive scalar particles, in the minimal coupling case ($b = 0$).

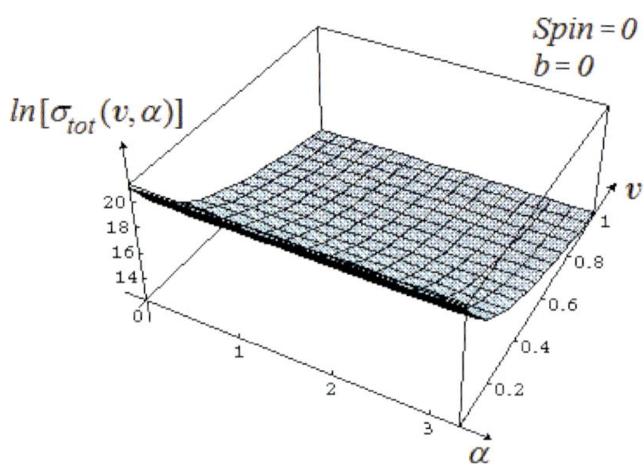


Fig. 8.4. Variation of $\ln[\sigma_{tot}(v, \alpha)]$ with respect to both the ratio $v = \frac{|\vec{p}|}{p_0}$ and angle $\alpha$, for massive scalar particles, in the minimal coupling case ($b = 0$).

the variation domain with respect to angular variable $\alpha$ of the integral scattering cross-sections is much larger than that with respect to variable $v$, and so the variation of $\sigma_{tot}(v, \alpha)$ with respect to $\alpha$ is masked by the variation with respect to $v$. We mention that in the present paragraph, beginning with relation (8.61-1), the constant factor $(GM)^2$ has been omitted. This fact in no respect affects the shape of the graphically represented curves and surfaces. Moreover, for convenience, in exact numerical calculations it has been considered that $\vec{a}^2 p_0^2 = 1$. Under these circumstances, obviously, the numerical values displayed in the graphs do not have a quantitative significance, but only a purely qualitative one.

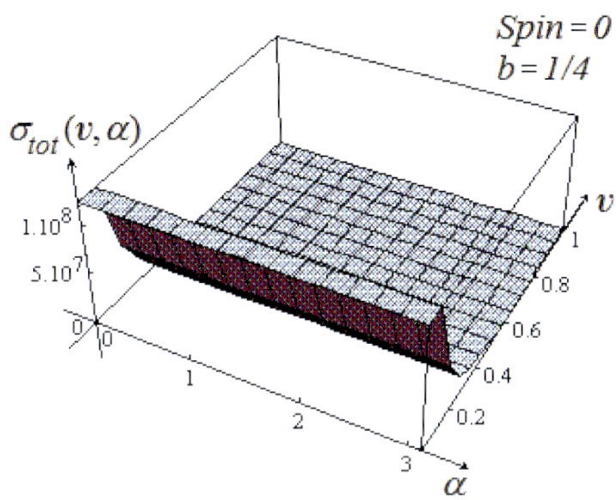


Fig. 8.5. Variation of the integral scattering cross-section $\sigma_{tot}(v, \alpha)$ with respect to both the ratio $v = \frac{|\vec{p}|}{p_0}$ and angle $\alpha$, for scalar massive particles, in the non-minimal coupling case ($b = 1/4$).

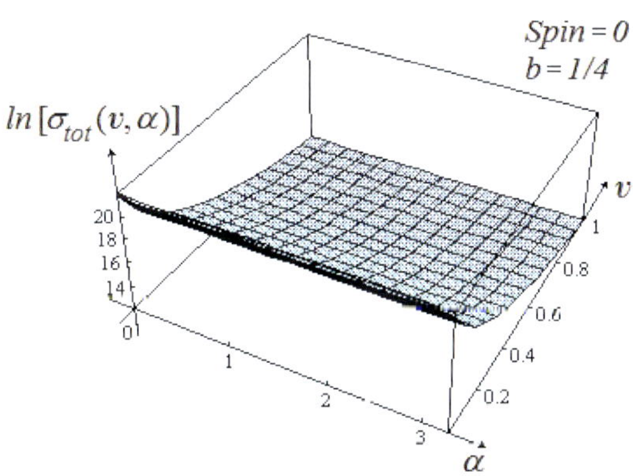


Fig. 8.6. Variation of $\ln[\sigma_{tot}(v, \alpha)]$ with respect to both the ratio $v = \frac{|\vec{p}|}{p_0}$ and angle $\alpha$, for massive scalar particles, in the non-minimal coupling case ($b = 1/4$).

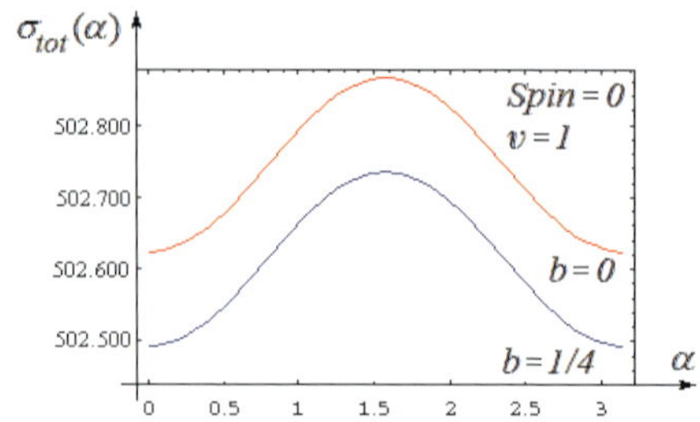


Fig. 8.7. Variation of the integral scattering cross-section $\sigma_{tot}$ $(\alpha)$ with respect to angle $\alpha$, for massive scalar particles, in the ultra-relativistic limit $(v = 1)$, for both minimal $(b = 0)$ and non-minimal $(b = 1/4)$ couplings.

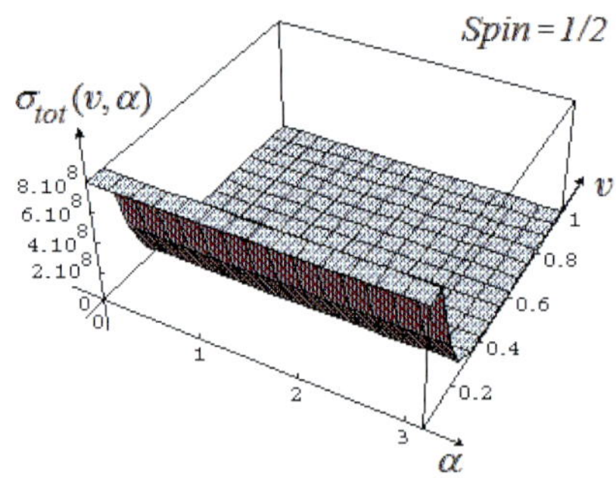


Fig. 8.8. Variation of the integral scattering cross-section $\sigma_{tot}$ $(v, \alpha)$ with respect to both the ratio $v = \frac{|\vec{p}|}{p_0}$ and angle $\alpha$, for massive spinorial particles.

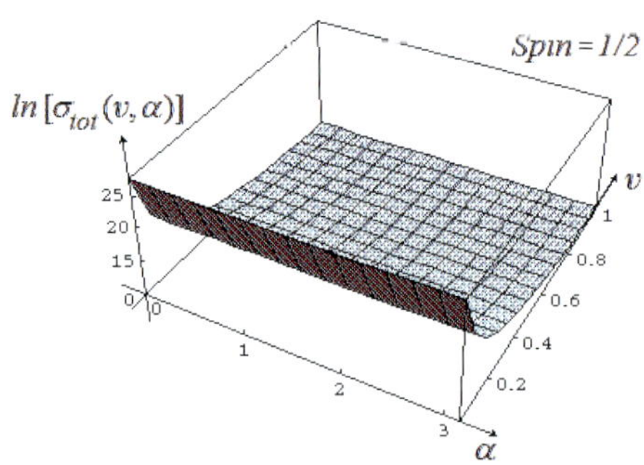


Fig. 8.9. Variation of $\ln[\sigma_{tot}(v, \alpha)]$ with respect to both the ratio $v = \frac{|\vec{p}|}{p_0}$ and angle $\alpha$, for massive spinorial particles.

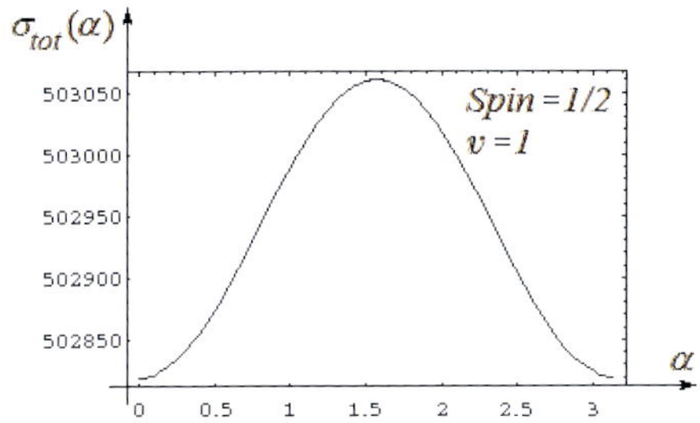


Fig. 8.10. Variation of the integral scattering cross-section $\sigma_{tot}(\alpha)$ with respect to angle $\alpha$ for massive spinorial particles, in the ultra-relativistic limit ($v = 1$).

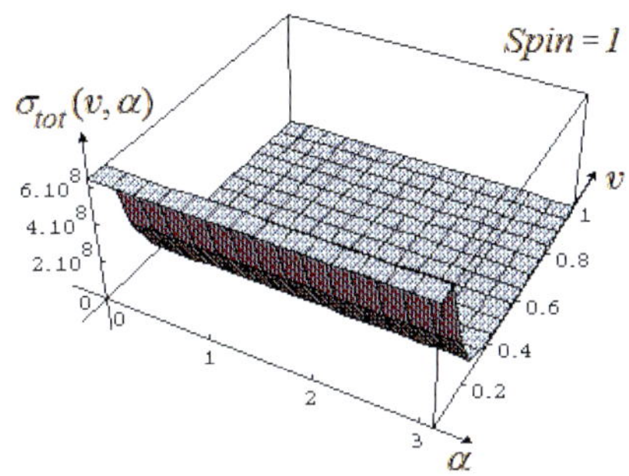


Fig. 8.11. Variation of the integral scattering cross-section $\sigma_{tot}(v, \alpha)$ with respect to both the ratio $v = \frac{|\vec{p}|}{p_0}$ and angle $\alpha$, for massive vectorial particles.

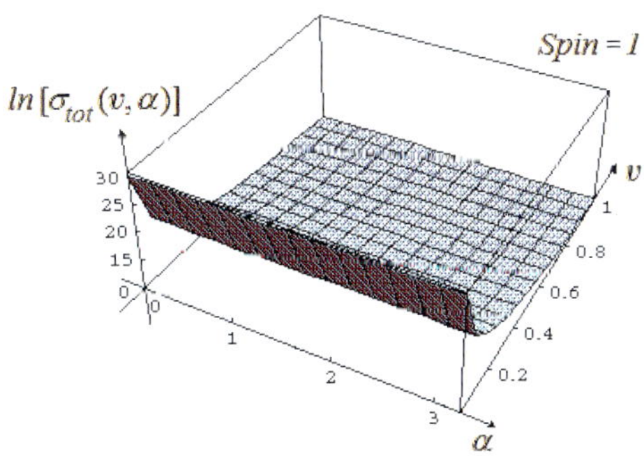


Fig. 8.12. Variation of $\ln[\sigma_{tot}(v, \alpha)]$ with respect to both the ratio $v = \frac{|\vec{p}|}{p_0}$ and angle $\alpha$, for massive vectorial particles.

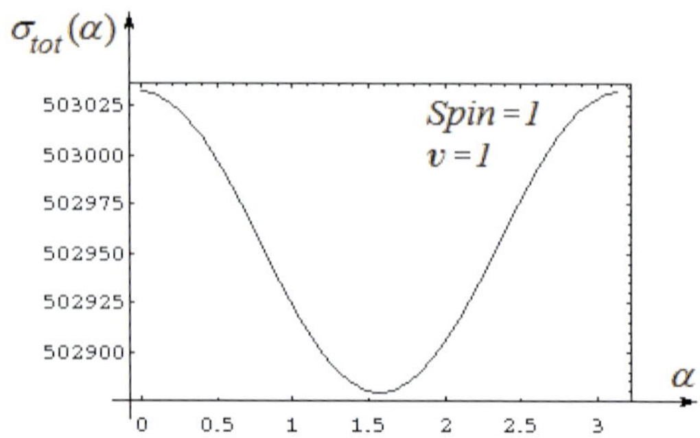


Fig. 8.13. Variation of the integral scattering cross-section $\sigma_{tot}(\alpha)$ with respect to angle $\alpha$, for massive vectorial particles, in the ultra-relativistic limit ($v = 1$).

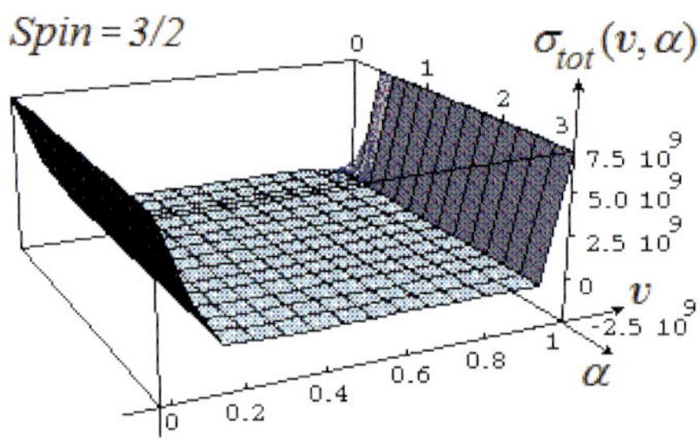


Fig. 8.14. Variation of the integral scattering cross-section $\sigma_{tot}(v, \alpha)$ with respect to both the ratio $v = \frac{|\vec{p}|}{p_0}$ and angle $\alpha$, for massive spin-vectorial particles.

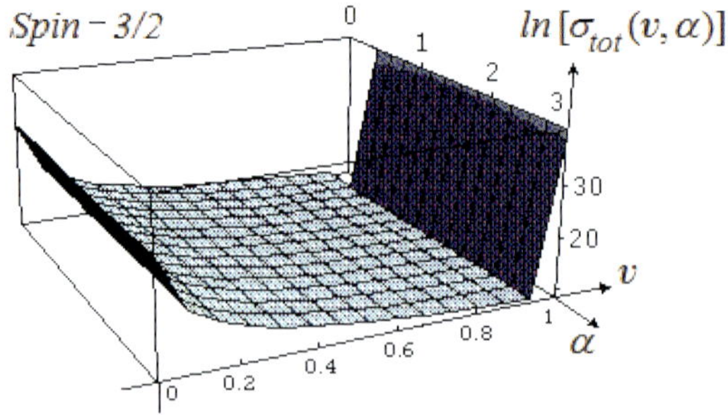


Fig. 8.15. Variation of $\ln[\sigma_{tot}(v, \alpha)]$ with respect to both the ratio $v = \frac{|\vec{p}|}{p_0}$ and angle $\alpha$, for massive spin-vectorial particles.

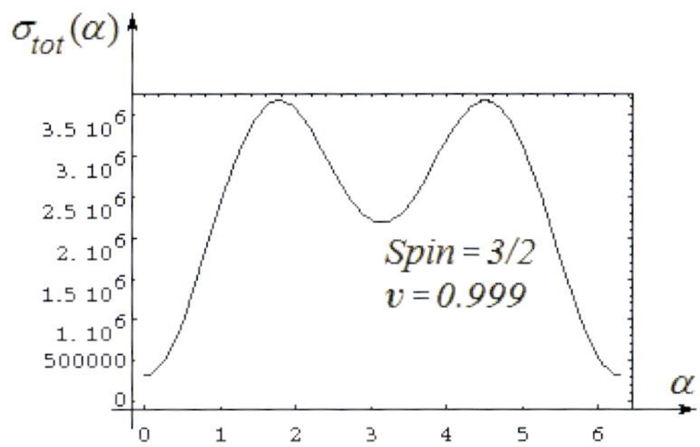


Fig. 8.16. Variation of the integral scattering cross-section $\sigma_{tot}(\alpha)$ with respect to angle $\alpha$ for massive spin-vectorial particles, in the quasi-ultrarelativistic limit ($v = 0.999$).

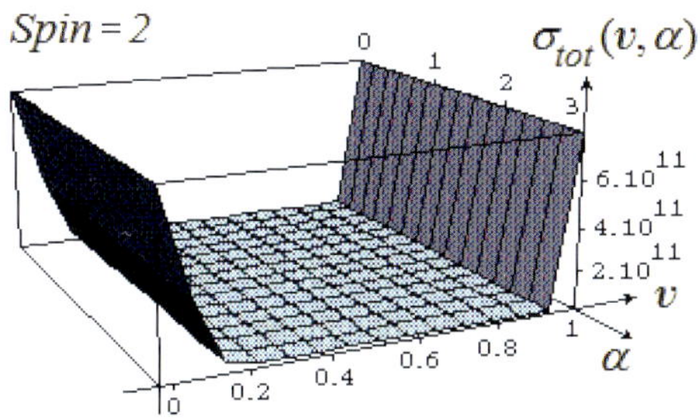


Fig. 8.17. Variation of the integral scattering cross-section $\sigma_{tot}(v, \alpha)$ with respect to both the ratio $v = \frac{|\vec{p}|}{p_0}$ and angle $\alpha$, for massive tensorial particles.

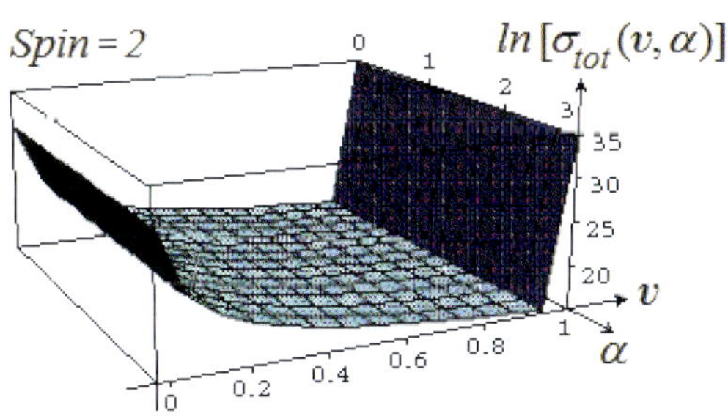


Fig. 8.18. Variation of $\ln[\sigma_{tot}(v, \alpha)]$ with respect to both the ratio $v = \frac{|\vec{p}|}{p_0}$ and angle $\alpha$, for massive tensorial particles.

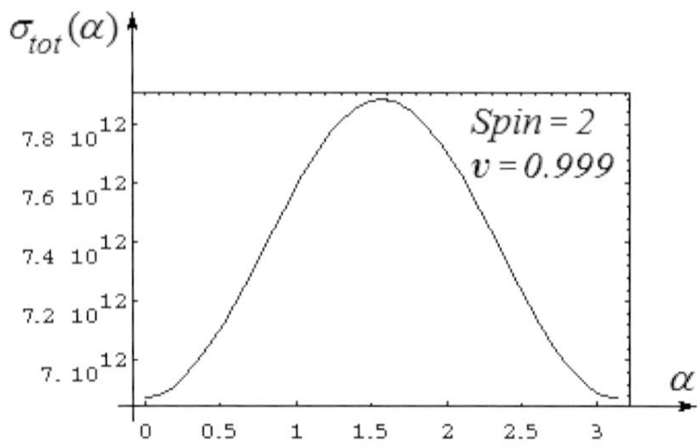


Fig. 8.19. Variation of the integral scattering cross-section $\sigma_{tot}(\alpha)$ with respect to angle $\alpha$ for massive tensorial particles, in the quasi-ultrarelativistic limit ($v = 0.999$).

## VIII.2. Gravitational scattering of massless, spin 0, 1/2, 1, 3/2, and 2 particles in an axially-symmetric field described by the Kerr metric

### VIII.2.1. General considerations

There are two main motivating reasons for a quantum investigation of the interaction process between non-massive particles[3] of spin 1 and 2, and the axially-symmetric gravitational field, namely,

1) A clear evidentiation of the gravitational waves (both indirectly, in 1974, by Russell A. Hulse and Joseph H. Taylor Jr., Princeton University, New Jersey, USA, and directly in September 2015 – the LIGO experiment): Rainer Weiss (professor at MIT and LSU, USA), Barry C. Barish (professor at Caltech, USA), and Kip S. Thorne (professor at Caltech, USA) were awarded the Nobel Prize in 2017);

2) Several papers published in the 90's ([242], [373, 374]), reconsidering the electromagnetism–gravitation coupling.

Regarding the direct discovery of the gravitational waves it's worth mentioning that the detection of gravitational waves is a highly complex experimental undertaking, due to the very weak signals which may reach the Earth even from very powerful cosmic sources.

[3]More precisely, these particles correspond, from the classical point of view, to electromagnetic and, respectively, gravitational waves.

The first indirect evidence for gravitational energy radiation, which is understood as a wave phenomenon, came in 1974 from the so-called *Hulse–Taylor binary* – a pair of stars, one of which is a pulsar (a radiating neutron star). Each of them has a mass around 1.4 $M_\odot$ and the distance between them is around $2 \times 10^6$ km, of the order of Sun's diameter. They are expected to radiate $10^{22}$ times the gravitational energy radiated by the Earth–Sun system. This causes the stars to gradually move closer together, in what is known as an *inspiral*, and this has an effect on the observed pulsar's signals. Russell A. Hulse (b. 1950) and Joseph H. Taylor, Jr. (b. 1941) were awarded the Nobel Prize in 1993 for their measurements which led to the discovery of the first binary pulsar, and allowed them to show that the gravitational radiation predicted by General Relativity matched the results of these observations with a precision within 0.2%. The search for direct evidence involves mainly detectors based on laser interferometry, like LIGO on Earth ground (Laser Interferometer Gravitational Wave Observatory) with two sites, in Livingston, Louisiana, and Hanford, Washington, or the planned eLISA (Evolved Laser Interferometer Space Antenna) orbiting in space. The principle of the laser interferometry in gravitational wave detection is conceptually rather simple: if a polarized gravitational wave propagates in the $z$-direction, the proper distance will dilate in one of the transverse directions and will decrease in the other. If in one of these directions we orient one arm of an interferometer, set initially to show destructive interference, the effect of the gravitational waves propagating in the arm will, for example, increase the travel time of one laser beam, leading to an interference signal. The basic experimental set-up is illustrated in Fig. 8.20.

The technical complexity of the laser interferometers is however staggering, since they have to detect amplitudes (or *strains*, as they are termed in the gravitational literature) of about $\mathcal{A} \approx 10^{-20}$ or less. The strongest source of gravitational waves is the coalescence of a black-hole/black-hole binary. A calculation which is beyond the scope of this book shows that the amplitude of the gravitational waves emitted by this system while the black holes circle each other is

$$\mathcal{A} \sim \frac{r_S^2}{rR},$$

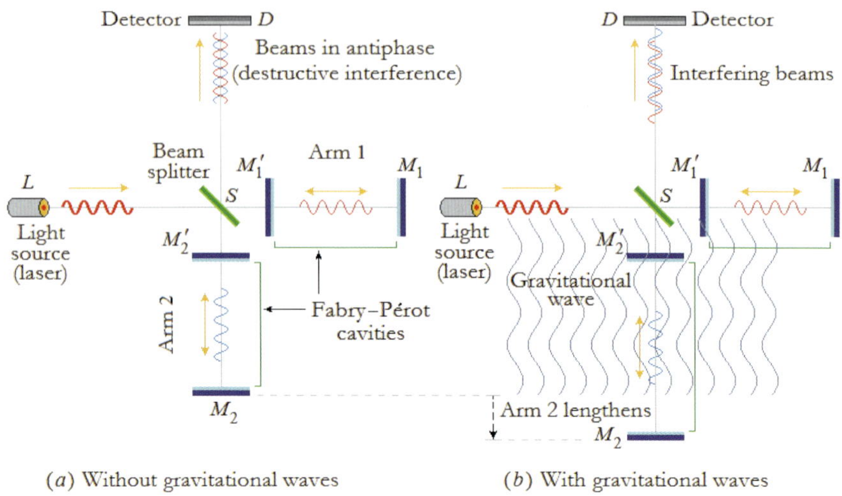


Fig. 8.20. Schematic representation of a laser-interferometer observatory: (*a*) the half-beams produced by the beam splitter $S$ follow optical paths calibrate so that in the absence of gravitational waves, the interference at the detector $D$ is destructive and no signal is observed; (*b*) when a gravitational wave arrives, it disturbs space-time, changing (in our example, lengthening) the light's path along arm 2; when the beams recombine and arrive at the detector $D$, an interference signal is registered.

where $r_S$ is the Schwarzschild radius (the black holes are assumed to have the same mass $M$), $R$ is the orbital radius, and $r$ is the distance between the pair of black holes and Earth. Typically, black holes are about 10 times more massive than the Sun and we can consider that the orbital radius is 10 times the Schwarzschild radius, while the distance to the observation point on Earth is of the order of 300 Mpc. This means $r_S \sim 3$ km, $R \sim 30$ km, and $r \sim 10^{22}$ km, leading to the order of magnitude for the amplitude quoted above.

On 14 September 2015, the two detectors of LIGO registered simultaneously a gravitational wave signal – the first direct detection ever achieved (see Fig. 8.21). The source was a pair of black holes merging together. The event happened at a distance of 410 Mpc, which means that it was observed by LIGO 1.3 billion years later.

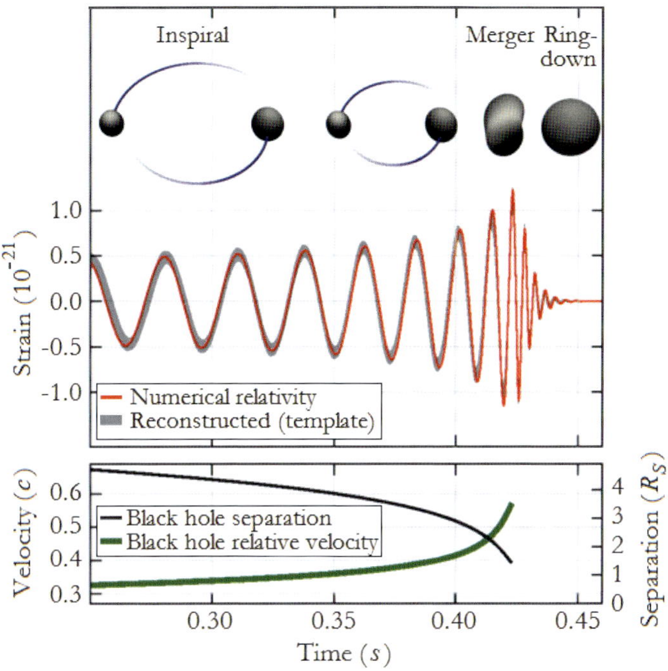


Fig. 8.21. *Top*: Estimated gravitational wave strain amplitude from the event observed by the LIGO Collaboration on 14 September 2015 (Hansford site). The inset images show numerical relativity models of the black hole horizons as the black holes coalesce. *Bottom*: The Keplerian effective black hole separation in units of Schwarzschild radii and the effective relative velocity given by the post-Newtonian parameter $v/c = (GM\pi f/c^3)^{1/3}$, where $f$ is the gravitational-wave frequency calculated with numerical relativity and M is the total mass. *Source*: B.P. Abbott *et al.* (LIGO Scientific Collaboration and Virgo Collaboration), *Phys. Rev. Lett.* 116, 061102 (2016).

The masses of the two initial black holes were $36\,M_\odot$ and $29\,M_\odot$, while the mass of the final one was $62\,M_\odot$, meaning that an energy of $3\,M_\odot c^2$ was radiated in gravitational waves. The observed signal took less then one second: during the first 0.2 s, the signal increased in frequency from 35 to 150 Hz, which is consistent with the theoretical prediction for the inspiral of two orbiting masses. The orbital frequency is half of the gravitational wave frequency, *i.e.* maximum 75 Hz in this case. This is a very high orbital frequency, that can be achieved without merging only if the orbiting masses are black holes. The signal peaks in amplitude during the merger and then the waveform decays in a manner consistent with the damped oscillations of a (final) black hole to a stationary Kerr configuration.

The LIGO interferometer is basically a Michelson interferometer, but with arms 4 km long. It is the largest interferometer ever built and the most precise, capable of measuring a change in the arm length 10,000 times smaller than a proton. The mirrors placed near the beam splitter cause multiple reflections of the laser beam, increasing the distance traveled in each arm to 1120 km. This system of mirrors forms an optical resonator known as the Fabry–Pérot cavity. Another ingenious stratagem for increasing the sensitivity are the power-boosting mirror, necessary for increasing the power of the laser beams from 200 W to 750 kW. This one-way mirror is placed before the beam splitter. The interferometer is aligned in such a way that almost all the laser light reflected by the arms is directed to the recycling mirror, and through it back to the source (instead of the detector), which greatly increases the power of the beam and consequently the sensitivity of the instrument.

The gravitational wave interferometers are not directional, they survey the whole sky. It is therefore of great importance to have simultaneous observations of gravitational waves at different locations, in order to confirm the signals and increase the accuracy in determining their origin. At the moment, there are three more observatories: GEO600 near Sarstedt in Germany is already operational, while VIRGO near Pisa in Italy and KAGRA (Kamioka Gravitational Wave Detector) in the Kamioka mines in Japan are under construction. The recent success of LIGO came together with the approval of a third LIGO Observatory in India, which is expected to start operation in 2023.

Coming back to our subject, this subchapter is intended to examine the problem of gravitational interaction of non-massive vector and second-rank tensor particles, and the external axially-symmetric gravitational field described by the Kerr metric.

To complete the picture of the considerations to follow and due to implications in various cosmological problems, this subchapter also investigates the problem of gravitational interaction of non-massive scalar, spinorial and spin-vectorial particles, that is the problem of interaction between scalar waves, massless spinor particles and gravitinos, and the external axially-symmetric gravitational field described by the Kerr metric.

As well-known, the study of massless fields is relatively difficult to motivate from the physical point of view (as a matter of fact, there is no known scalar field with zero rest-mass, even if various models of the modern theoretical physics frequently make use of it).

Nevertheless, on the one side, the massless fields can be utilized as an approximation of the corresponding massive fields when the energy scale interfering in the studied problem is much larger than that "set" by the rest-masses of the particles and, on the other side, these fields can be utilized in solving of some astrophysical problems, such that concerning the possibility to distinguish between various types of cosmic objects (for example, to distinguish between a black hole and a big compact object), by examining the behavior of the differential scattering cross-sections of non-massive particles at big distances relative to the scattering center, corresponding to the two cases. Last but not least, the investigation to follow is also intended to determine the effect of the scattering process upon various helicity states of the scattered particles or, differently speaking, which is the effect of coupling between the angular momentum of the gravitational source and the spin of the scattered particles, on their helicity states.

The main difference between the subject approached in this subchapter and the study performed in Chap. VI concerning the same type of particles consists in consideration of a different type of gravitational background, namely an axially-symmetric one, described by the Kerr metric. This last case is more interesting in regard with effects that can appear as a result of supplementary interaction between the spin of the scattered particles and that of the field source.

Since the working procedure follows step-by-step the method provided by the previous subchapter, the present investigation shall not be focused upon the general working frame, but shall rather evidentiate the main differences as compared to the massive particles case, studied in the first subchapter.

Thus, the main difference between non-massive and massive particles is that, since there is no possibility for massless particles to be referred to a rest frame, one must consider a particular direction with respect to which, the projection of particle's momentum can have only two orientations, corresponding to the two values of the non-massive particles' helicity. As a result, one becomes interesting to determine, for example, the *partial* differential scattering cross-sections (between different helicity states), and to study this way the effect of the scattering process on particles' helicity (that is, to find if the scattered particles change or don't change their helicity as a result of interaction with the scattering source field or, in other words, if the scattered particles conserve or don't conserve their helicity, for both linear and circular polarizations, as examples of pure helicity states of the particles).

Also, unlike the previous subchapter, the investigation to follow shall consider the non-minimal interaction of photons with axially-symmetric gravitational field. In other words, in this subchapter not one, but two types of non-minimal couplings are considered: one for the massless scalar field (introduced, as in the case of massive scalar field, via the scalar curvature $R$), and the other for the massless vector field (electromagnetic field); in this last case, the two fields shall be non-minimally coupled – as we shall see – through the medium of the Ricci tensor $R_{ik}$.

## VIII.2.2. Gravitational scattering of the scalar, spinorial, vectorial, spin-vectorial, and tensorial massless particles

The scattering problem of non-massive particles with various spin values, "propagating"[4] in a space-time with small curvature, can be solved by quantizing both the gravitational field, and the scattered

[4]This time, the use of term "propagating" is more suitable than in the case of massive particles, taking into account that a non-massive field has par excellence, from the phenomenological point of view, a wave-like character.

field. In such a scenario, the two fields will couple according to Feynman rules. Since the present investigation is limited to a gravitational background geometry generated by energy-momentum distributions which are not considerably affected by the scattering process, one can replace the virtual graviton by an external gravitational field. Differently speaking, one can use the external field approximation for the stationary gravitational field described by the Kerr geometry.

As in the previous subchapter, the investigation shall be limited to the interaction processes proportional to $\kappa^2$ (which processes involve a single exchange graviton), therefore the differential scattering cross-sections shall be calculated in the first-order Born approximation; in this case, the calculations can be carried on up to the end, without any interference of divergent terms.

In order to determine the interaction Lagrangians (up to the first-order in the gravitational coupling constant $\kappa$) between the gravitational field and the matter fields (that is the massless fields of spin-0, 1/2, 1, 3/2, and 2), the minimal coupling principle in quantum gravity shall be utilized. For the massless fields of spin-0 and 1 the non-minimal couplings mentioned at the beginning of this subchapter shall also be used.

According to minimal coupling principle applied in quantum gravity, to the gravitational field Lagrangian have to be added the matter field Lagrangians, written in curved space ([366], [375]), namely,

$$L_{scalar} \equiv L_{0,0} = \sqrt{-g}\left[g^{\mu\nu}\varphi^*_{,\mu}\varphi_{,\nu} - bR\varphi^*\varphi\right], \tag{8.62}$$

$$L_{spinor} \equiv L_{0,1/2} = \frac{i}{4}\sqrt{-g}\Big[\overline{\psi}\gamma^\mu(1+\gamma^5)\psi_{;\mu} - \overline{\psi}_{;\mu}\gamma^\mu(1+\gamma^5)\psi\Big], \tag{8.63}$$

$$L_{vector} \equiv L_{0,1} = -\frac{1}{2}\sqrt{-g}g^{\mu\lambda}g^{\nu\rho}\left(A_{\mu;\nu}A_{\lambda;\rho} - R_{\mu\nu}A_\lambda A_\rho\right), \tag{8.64}$$

$$L_{tensor} \equiv L_{0,2} = \frac{\sqrt{-g}}{\kappa^2}g^{\mu\nu}\left(\Gamma^\alpha_{\mu\nu}\Gamma^\beta_{\alpha\beta} - \Gamma^\alpha_{\beta\mu}\Gamma^\beta_{\alpha\nu}\right) + \frac{\partial}{\partial x^\alpha}\left[\frac{\sqrt{-g}}{\kappa^2}\left(\Gamma^\nu_{\beta\nu}g^{\alpha\beta} - \Gamma^\alpha_{\beta\nu}g^{\beta\nu}\right)\right]. \tag{8.65}$$

One can easily notice that, in the case of non-massive scalar field, the Klein–Fock formalism and a non-minimal coupling introduced through the medium of the scalar curvature have been considered. In Eq. (8.62) $R$ is the scalar curvature, and $b$ is a real, non-null arbitrary constant; the particular values of this constant are chosen as in the case of the massive scalar field.

Regarding the massless spinorial field, it is convenient to use the two-component non-massive neutrino theory, in which case, following Landau and Lifshitz [240], one appeals to the classical Dirac's Lagrangian in which the following substitutions have been previously performed,

$$\begin{cases} \psi \to \frac{1}{2}(1+\gamma^5)\psi, \\ \overline{\psi} \to \frac{1}{2}\overline{\psi}(1-\gamma^5), \end{cases}$$

where $\gamma^5 = -i\gamma^0\gamma^1\gamma^2\gamma^3 (= \gamma_5 = \gamma_1\gamma_2\gamma_3\gamma_4$, when passing to the flat space).

In order to write down the Lagrangian of the electromagnetic field we started with the Mitskevich idea (presented in Ref. [270]) to use the Lagrangian proposed by Schweber; besides, a non-minimal coupling introduced by means of the Ricci tensor $R_{\mu\nu}$, has been used. Unlike other authors, which use the standard Lagrangian of the electromagnetic field (see, *e.g.*, [91]),

$$L_{em} = -\frac{1}{4}\sqrt{-g}\, g^{\mu\lambda} g^{\nu\rho} F_{\mu\nu} F_{\lambda\rho}, \tag{8.66}$$

where $F_{\mu\nu}$ is the second-rank antisymmetric tensor of the electromagnetic field, it was preferred to employ the Lagrangian given by (8.64), which explicitly contains the covariant derivatives required by the minimal coupling principle, and this seems to be more natural. Following Ref. [270], it can be shown the equivalence of (8.64) and (8.66). Indeed, using the gauge $A^\mu_{;\mu} = 0$, the Maxwell field Lagrangian in the curved space writes

$$L_{em} = -\frac{1}{4}\sqrt{-g}\, F^{\mu\nu} F_{\mu\nu} - \frac{1}{2}\sqrt{-g}\, A^\mu_{;\mu} A^\nu_{;\nu},$$

and omitting the four-divergence

$$\left[\sqrt{-g}\,(A^{\mu}_{;\mu}A^{\nu} - A^{\nu}_{;\mu}A^{\mu})\right]_{;\nu}$$

one obtains

$$L_{em} = -\frac{1}{2}\sqrt{-g}\,g^{\mu\nu}g^{\rho\lambda}(A_{\mu;\rho}A_{\nu;\lambda} - R_{\mu\rho}A_{\nu}A_{\lambda}),$$

which is precisely the Lagrangian given by (8.64).

In the case of non-massive spin-3/2 field (massless Rarita–Schwinger field), by analogy with the case of massless spinorial field, the classical Lagrangian of the massless Rarita–Schwinger theory has been considered, in which the following substitutions (analogous to those utilized in the case of non-massive spinorial field),

$$\begin{cases} \psi_{\mu} \to \dfrac{1}{2}(1+\tilde{\gamma}^{5})\psi_{\mu}, \\ \overline{\psi}_{\mu} \to \dfrac{1}{2}\overline{\psi}_{\mu}(1-\tilde{\gamma}^{5}), \end{cases}$$

have been performed.

Taking into account the particularities of the non-massive Rarita–Schwinger field, the minimal coupling principle has been used within the framework of the *vierbein* formalism, so that the Lagrangian of this system can be written as

$$L_{spin\text{-}vector} \equiv L_{0,3/2} = \frac{i}{4}\sqrt{-g}\,g^{\mu\nu} \times \left[\overline{\psi}_{\mu}\overleftarrow{D}_{\lambda}\tilde{\gamma}^{\lambda}(1+\tilde{\gamma}^{5})\psi_{\nu} - \overline{\psi}_{\mu}\tilde{\gamma}^{\lambda}(1+\tilde{\gamma}^{5})\overrightarrow{D}_{\lambda}\psi_{\nu}\right], \tag{8.67}$$

where $\tilde{\gamma}^{\mu}$ are the generalized Dirac matrices,

$$\tilde{\gamma}^{\mu} = L^{\mu}(\alpha)\hat{\gamma}(\alpha)\ ; \quad \tilde{\gamma}_{\mu} = L_{\mu}(\alpha)\hat{\gamma}(\alpha),$$

and the *vierbein* coefficients $L^{\mu}(\alpha)$ satisfy the constraints

$$L^{\mu}(\alpha)L^{\nu}(\alpha) = g^{\mu\nu}\ ; \quad L_{\mu}(\alpha)L_{\nu}(\alpha) = g_{\mu\nu}.$$

The generalized Dirac matrices $\tilde{\gamma}^{\mu}$ are expressed in terms of the usual Dirac matrices $\hat{\gamma}(\alpha)$ written in the hyperbolic representation of signature $-2$ of the Minkowski space-time, and $D_{\mu}\psi_{\nu}$ is the covariant derivative of the spin-vector $\psi_{\nu}$ (the wave function of the Rarita–

Schwinger field),

$$\overrightarrow{D}_\mu \psi_\nu = \psi_{\nu,\mu} - \Gamma_\mu \psi_\nu \ ; \quad \overline{\psi}_\nu \overleftarrow{D}_\mu = \overline{\psi}_{\nu,\mu} - \overline{\psi}_\nu \Gamma_\mu .$$

Here $\Gamma_\mu$ are the Fock–Ivanenko spin coefficients of the affine connection,

$$\Gamma_\mu = \frac{1}{4}\tilde{\gamma}_{\lambda;\mu}\tilde{\gamma}^\lambda ,$$

where

$$\tilde{\gamma}_{\lambda;\mu} = \tilde{\gamma}_{\lambda,\mu} - \tilde{\gamma}_\rho \Gamma^\rho_{\lambda\mu} .$$

It is worth specifying that the final results are the same if, for the massless Rarita–Schwinger field, instead of the Lagrangian written above, one uses the corresponding non-massive spin-3/2 part of the Lagrangian of simple supergravity theory,

$$L^{SG}_{0,3/2} = \frac{i}{2}\sqrt{-g}\,\varepsilon^{\alpha\beta\mu\nu}\left(\overline{\psi}_\alpha \overleftarrow{D}_\nu \tilde{\gamma}_5 \tilde{\gamma}_\mu \psi_\beta - \overline{\psi}_\alpha \tilde{\gamma}_5 \tilde{\gamma}_\mu \overrightarrow{D}_\nu \psi_\beta\right). \quad (8.68)$$

This statement was proved in the papers [375, 376] and [383] for the Schwarzschild geometry, the choice of this alternative being intended to facilitate the calculations. In addition, it should be mentioned that the interaction process between the massless spin-vectorial field/particles and the gravitational field is quite rarely studied in concrete applications, even if there exists nowadays a rigorous general approach of this problem in the framework of simple and/or extended supergravitation theories (see, *e.g.*, [13], [137], [174], [287]).

Finally, in the case of massless second-rank tensor field, the Einstein Lagrangian written in a more convenient form by Mitskevich (see Ref. [270] for details), has been used; it is given by (8.65). Since the following considerations concern only the first-order approximation in the coupling constant $\kappa$ (weak coupling), below are given the relations expressing the linearized quantities interfering in calculations [366]. Thus, taking into account the De Donder–Fock gauge $(\sqrt{-g}\,g^{\mu\nu})_{,\nu} = 0$, and expanding all quantities in power series in powers of the coupling constant $\kappa$, one obtains

$$g^{\mu\nu} = \eta^{\mu\nu} - \kappa h^{\mu\nu} + \mathcal{O}(\kappa^2),$$

$$g_{\mu\nu} = \eta_{\mu\nu} + \kappa h_{\mu\nu} + \mathcal{O}(\kappa^2),$$

$$\Gamma^{\rho}_{\lambda\mu} = \frac{1}{2}\kappa(h^{\rho}_{\lambda,\mu} + h^{\rho}_{\mu,\lambda} - h_{\lambda\mu,}{}^{\rho}) + \mathcal{O}(\kappa^2),$$

$$\tilde{\gamma}^{\lambda} = \hat{\gamma}^{\lambda} - \frac{1}{2}\kappa\hat{\gamma}_{\beta}h^{\lambda\beta} + \mathcal{O}(\kappa^2),$$

$$\tilde{\gamma}_{\lambda} = \hat{\gamma}_{\lambda} + \frac{1}{2}\kappa\hat{\gamma}_{\beta}h^{\beta}_{\lambda} + \mathcal{O}(\kappa^2),$$

where

$$h^{\mu\nu} = y^{\mu\nu} - \frac{1}{2}\eta^{\mu\nu}y \; ; \quad h^{\beta}_{\lambda} = y^{\beta}_{\lambda} - \frac{1}{2}\delta^{\beta}_{\lambda}y \; ; \quad y = y^{\alpha}_{\alpha}.$$

In view of the above considerations and passing to flat complex Minkowski space-time,

$$x^0 = t\,,\, x^j \;\rightarrow\; x_j\,,\, x_4 = it \; ; \quad \hat{\gamma}^{\mu}\partial_{\mu} \rightarrow i\gamma_{\mu}\partial_{\mu} \; ; \quad \eta^{\mu\nu} \rightarrow -\delta_{\mu\nu};$$

$$\partial_{\mu} \equiv \frac{\partial}{\partial x^{\mu}} \rightarrow \frac{\partial}{\partial x_{\mu}},$$

the interaction Lagrangians corresponding to the interaction process between the weak gravitational field and the scalar, spinorial, vectorial, spin-vectorial and second-rank tensorial massless fields are expressed as

$$L^{(1)}_{(int)0,0}(\kappa) = -\kappa\left[\varphi^{*}_{,\mu}\varphi_{,\nu}y_{\mu\nu} - \frac{1}{2}b\varphi^{*}\varphi y_{,\mu\mu}\right], \tag{8.69-1}$$

$$L^{(1)}_{(int)0,1/2}(\kappa) = -\frac{1}{8}\kappa\left[\overline{\psi}\gamma_{\mu}(1+\gamma_5)\psi_{,\nu} - \overline{\psi}_{,\nu}\gamma_{\mu}(1+\gamma_5)\psi\right]s_{\mu\nu}, \tag{8.69-2}$$

$$L^{(1)}_{(int)0,1}(\kappa) \equiv L^{(1)}_{em}(\kappa) = -\frac{1}{2}\kappa\Big[A_{\mu}A_{\nu,\lambda}(h_{\mu\nu,\lambda} + h_{\mu\lambda,\nu} - h_{\nu\lambda,\mu})$$
$$+ A_{\mu,\nu}A_{\mu,\lambda}h_{\nu\lambda} + A_{\mu,\nu}A_{\lambda,\nu}y_{\mu\lambda} + \frac{1}{2}A_{\mu}A_{\nu}h_{\mu\nu,\lambda\lambda}\Big], \tag{8.69-3}$$

$$L^{(1)}_{(int)0,3/2}(\kappa) = \frac{1}{8}\kappa\left[\overline{\psi}_{\mu,\lambda}\gamma_\nu(1+\gamma_5)\psi_\mu - \overline{\psi}_\mu\gamma_\nu(1+\gamma_5)\psi_{\mu,\lambda}\right]y_{\lambda\nu}$$
$$-\frac{1}{32}\kappa\overline{\psi}_\mu\left[\gamma_\rho\gamma_\nu\gamma_\lambda(1+\gamma_5) + \gamma_\lambda(1+\gamma_5)\gamma_\rho\gamma_\nu\right]\psi_\mu\left(h_{\nu\lambda,\rho} - h_{\rho\lambda,\nu}\right), \tag{8.69-4}$$

and

$$L^{(1)}_{(int)0,2}(\kappa) = \frac{1}{2}\kappa\left[y_{\alpha\beta}y_{\alpha\beta,\mu\nu} + \frac{1}{2}(y_{\alpha\beta,\mu}y_{\alpha\beta,\nu} - y_{\mu\nu,\alpha}y_\alpha)\right]y_{\mu\nu}, \tag{8.69-5}$$

respectively, where $s_{\mu\nu} = y_{\mu\nu} + \frac{1}{2}\delta_{\mu\nu}y$. We mention that the first-order interaction Lagrangian for the massless second-rank tensor field has been obtained under the $TT$-gauge[5] (transversality and null-trace) [269]. The equations of the massless Rarita–Schwinger field

$$\begin{cases} i\hat{\gamma}^\lambda(1+\hat{\gamma}^5)\psi_{\nu,\lambda} = 0, \\ i\overline{\psi}_{\nu,\lambda}\hat{\gamma}^\lambda(1+\hat{\gamma}^5) = 0, \end{cases}$$

together with the well-known anti-commutation relations satisfied by the usual Dirac matrices

$$\{\hat{\gamma}^\mu, \hat{\gamma}^\nu\} = \hat{\gamma}^\mu\hat{\gamma}^\nu + \hat{\gamma}^\nu\hat{\gamma}^\mu = 2\eta^{\mu\nu}$$

have also been used.

In agreement with the standard theory of quantum fields, the parts of the Lagrangians (8.69-1)–(8.69-5) – brought to the normal form – expressing the interaction process between gravitation and scalar, spinorial, vectorial, spin-vectorial, and second-rank tensorial massless particles are

$$N[L^{(1)}_{(int)0,0}(x)] = -\kappa\left[\varphi^{*(-)}_{,\mu}(x)\varphi^{(+)}_{,\nu}(x)y^{ext}_{\mu\nu}(x) - \frac{1}{2}b\varphi^{*(-)}(x)\varphi^{(+)}(x)y^{(ext)}_{,\mu\mu}(x)\right], \tag{8.70-1}$$

[5] This gauge is always used in the study of gravitational waves. It supposes the following conditions on the tensor $h_{\mu\nu}$: 1) $h_{\mu 0} = 0$ (only spatial components are non-null), 2) $h_{ij,j}$ $(= h_{ij}k_j)$ $= 0$ (the spatial part is divergence-free, that is the wave is transversal), and 3) $h_{jj} = 0$ (the spatial components satisfy the condition of null trace). It is worth specifying that any spatial tensor $S_{ij}$ can be written as $S_{ij} = S^T_{ij} + S^L_{ij} + S^{TT}_{ij}$, where $S^L_{ij} = S^L_{i,j} + S^L_{j,i}$ and $S^T_{ij} = \frac{1}{2}(\delta_{ij}f_{,kk} - f_{,ij})$, with $S^T_{ij,j} = 0$ and $S^T_{kk} = \nabla^2 f = \Delta f$, where $f$ is an arbitrary $C^2$-class scalar field.

$$N[L^{(1)}_{(int)0,1/2}(x)] = -\frac{1}{8}\kappa\Big[\overline{\psi}^{(-)}(x)\gamma_\mu(1+\gamma_5)\psi^{(+)}_{,\nu}(x) - \overline{\psi}^{(-)}_{,\nu}(x)\gamma_\mu(1+\gamma_5)\psi^{(+)}(x)\Big]s^{ext}_{\mu\nu}(x), \tag{8.70-2}$$

$$\begin{aligned} N[L^{(1)}_{(int)0,1}(x)] &\equiv N[L^{(1)}_{em}(x)] \\ &= -\frac{\kappa}{2}\bigg\{\frac{1}{2}\Big[A^{(-)}_\mu(x)A^{(+)}_{\nu,\lambda}(x) + A^{(-)}_{\nu,\lambda}(x)A^{(+)}_\mu(x)\Big] \\ &\quad\times\Big[h^{ext}_{\mu\nu,\lambda}(x) + h^{ext}_{\mu\lambda,\nu}(x) - h^{ext}_{\nu\lambda,\mu}(x)\Big] \\ &\quad + A^{(-)}_{\mu,\nu}(x)A^{(+)}_{\mu,\lambda}(x)h^{ext}_{\nu\lambda}(x) + A^{(-)}_{\mu,\nu}(x)A^{(+)}_{\lambda,\nu}(x)y^{ext}_{\mu\lambda}(x) \\ &\quad + \frac{1}{2}A^{(-)}_\mu(x)A^{(+)}_\nu(x)h^{ext}_{\mu\nu,\lambda\lambda}(x)\bigg\}, \end{aligned} \tag{8.70-3}$$

$$\begin{aligned} N[L^{(1)}_{(int)0,3/2}(x)] &= \frac{1}{8}\kappa\Big[\overline{\psi}^{(-)}_{\mu,\lambda}(x)\gamma_\nu(1+\gamma_5)\psi^{(+)}_\mu(x) - \overline{\psi}^{(-)}_\mu(x) \\ &\quad\times\gamma_\nu(1+\gamma_5)\psi^{(+)}_{\mu,\lambda}(x)\Big]y^{ext}_{\lambda\nu}(x) - \frac{1}{32}\kappa\overline{\psi}^{(-)}_\mu(x) \\ &\quad\times\Big[\gamma_\rho\gamma_\nu\gamma_\lambda(1+\gamma_5) + \gamma_\lambda(1+\gamma_5)\gamma_\rho\gamma_\nu\Big] \\ &\quad\times\psi^{(+)}_\mu(x)\Big[h^{ext}_{\nu\lambda,\rho}(x) - h^{ext}_{\rho\lambda,\nu}(x)\Big], \end{aligned} \tag{8.70-4}$$

and

$$\begin{aligned} N[L^{(1)}_{(int)0,2}(x)] &= \frac{1}{2}\kappa\bigg\{\Big[y^{(-)}_{\alpha\beta}(x)y^{(+)}_{\alpha\beta,\mu\nu}(x) \\ &\quad + y^{(-)}_{\alpha\beta,\mu\nu}(x)y^{(+)}_{\alpha\beta}(x) + y^{(-)}_{\alpha\beta,\mu}(x)y^{(+)}_{\alpha\beta,\nu}(x)\Big]y^{ext}_{\mu\nu}(x) \\ &\quad - \frac{1}{2}\Big[y^{(-)}_{\mu\nu}(x)y^{(+)}_{\mu\nu,\alpha}(x) + y^{(-)}_{\mu\nu,\alpha}(x)y^{(+)}_{\mu\nu}(x)\Big]y^{ext}_{,\alpha}(x)\bigg\}, \end{aligned} \tag{8.70-5}$$

respectively.

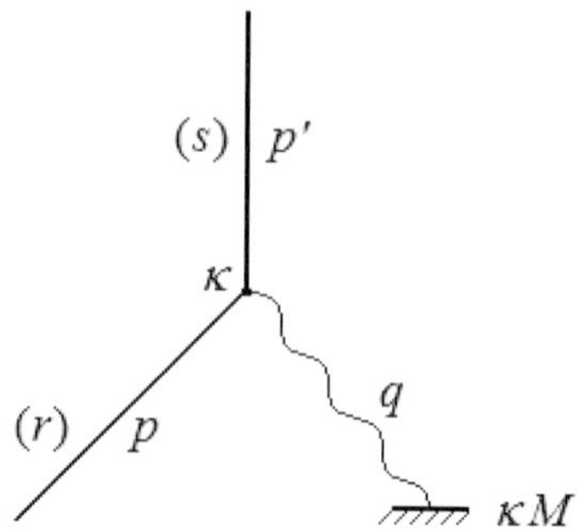


Fig. 8.22. The vertex (non-massive field)–graviton–(non-massive field). The solid straight lines represent the quanta of the scalar, spinorial, vectorial, spin-vectorial and second-rank tensorial fields, respectively, while the curly line represents the virtual graviton.

The processes are described by the Feynman diagram shown in Fig. 8.22, where $p$ and $(r)$, as well as $p'$ and $(s)$, are the four-momenta and indices of the polarization states for the initial and final particles, respectively, and $q$ is the four-momentum of the virtual graviton.

The spatial orientation of the angular momentum per unit mass $\vec{a}$ of the gravitational field source, together with the direction of motion of the scattered particles, $\vec{p}\,'$, relative to that of the incident particles, $\vec{p}$, are shown in Fig. 8.2.

Using the $S$-matrix formalism, one can deduce Feynman-type rules in the framework of external gravitational field approximation (the gravitational background being described by the Kerr metric), which allow one to calculate the corresponding $S$-matrix elements $\langle p'|S|p\rangle$.

Regarding the above-mentioned Feyman rules, the results are similar to those obtained in the first subchapter and are given by

$$\begin{cases} y_{jk}^{ext}(\vec{q}) = 0, \\ y_{j4}^{ext}(\vec{q}) = y_{4j}^{ext}(\vec{q}) = \dfrac{1}{2}\dfrac{1}{(2\pi)^{3/2}}\dfrac{\kappa M}{|\vec{q}|^2}(\vec{a}\times\vec{q})_j, \\ y_{44}^{ext}(\vec{q}) = y^{ext}(\vec{q}) = \dfrac{1}{(2\pi)^{3/2}}\dfrac{\kappa M}{|\vec{q}|^2}. \end{cases}$$

$$\begin{cases} s_{jk}^{ext}(\vec{q}) = \dfrac{1}{2}\dfrac{1}{(2\pi)^{3/2}}\dfrac{\kappa M}{|\vec{q}|^2}\delta_{jk}, \\ s_{j4}^{ext}(\vec{q}) = s_{4j}^{ext}(\vec{q}) = \dfrac{1}{2}\dfrac{1}{(2\pi)^{3/2}}\dfrac{\kappa M}{|\vec{q}|^2}(\vec{a}\times\vec{q})_j, \\ s_{44}^{ext}(\vec{q}) = \dfrac{3}{2}\dfrac{1}{(2\pi)^{3/2}}\dfrac{\kappa M}{|\vec{q}|^2}. \end{cases}$$

$$\begin{cases} u_{jk}^{ext}(\vec{q}) = -\dfrac{1}{4}\dfrac{1}{(2\pi)^{3/2}}\dfrac{\kappa M}{|\vec{q}|^2}\delta_{jk}, \\ u_{j4}^{ext}(\vec{q}) = u_{4j}^{ext}(\vec{q}) = \dfrac{1}{2}\dfrac{1}{(2\pi)^{3/2}}\dfrac{\kappa M}{|\vec{q}|^2}(\vec{a}\times\vec{q})_j, \\ u_{44}^{ext}(\vec{q}) = \dfrac{3}{4}\dfrac{1}{(2\pi)^{3/2}}\dfrac{\kappa M}{|\vec{q}|^2}. \end{cases}$$

$$\begin{cases} h_{jk}^{ext}(\vec{q}) = -\dfrac{1}{2}\dfrac{1}{(2\pi)^{3/2}}\dfrac{\kappa M}{|\vec{q}|^2}\delta_{jk}, \\ h_{j4}^{ext}(\vec{q}) = h_{4j}^{ext}(\vec{q}) = \dfrac{1}{2}\dfrac{1}{(2\pi)^{3/2}}\dfrac{\kappa M}{|\vec{q}|^2}(\vec{a}\times\vec{q})_j, \\ h_{44}^{ext}(\vec{q}) = \dfrac{1}{2}\dfrac{1}{(2\pi)^{3/2}}\dfrac{\kappa M}{|\vec{q}|^2}. \end{cases}$$

Here, as in the previous subchapter, any derivative with respect to coordinate $x_\alpha$ is equivalent to appearance of a supplementary factor $iq_\alpha$.

Taking into account the previous relations, the $S$-matrix elements corresponding to the Feynman diagram shown in Fig. 8.22 – within the external field approximation – can be written as

$$S_{p'p}^{(*)} = \langle p'|S^{(*)}|p\rangle = F^{(*)}(p',p)\delta(q_0),$$

where the superscript symbol $(*)$ stands for each of the values 0, 1/2, 1, 3/2, and 2, depending on the considered non-massive field (scalar, spinorial, vectorial, spin-vectorial, or tensorial). In the previous equation $q_0 = p_0' - p_0 = 0$ expresses the energy conservation law, while the quantities $F^{(*)}(p',p)$ are given by

$$F^{(0)}(p',p) = \frac{i\kappa^2 M}{8p_0(2\pi)^2 p_0^2 \sin^2\dfrac{\theta}{2}}Q^{(0)}(p',p), \tag{8.71-1}$$

$$F^{(1/2)}(p',p) = \frac{\kappa^2 mM}{32(2\pi)^2 p_0^2 \sin^2\frac{\theta}{2}} Q^{(1/2)}(p',p), \qquad (8.71\text{-}2)$$

$$F^{(1)}(p',p) = \frac{i\kappa^2 M}{8p_0(2\pi)^2 p_0^2 \sin^2\frac{\theta}{2}} Q^{(1)}(p',p), \qquad (8.71\text{-}3)$$

$$F^{(3/2)}(p',p) = \frac{\kappa^2 M}{32(2\pi)^2 p_0^2 \sin^2\frac{\theta}{2}} Q^{(3/2)}(p',p), \qquad (8.71\text{-}4)$$

$$F^{(2)}(p',p) = \frac{i\kappa^2 M}{16p_0(2\pi)^2 p_0^2 \sin^2\frac{\theta}{2}} Q^{(2)}(p',p). \qquad (8.71\text{-}5)$$

Here $p_0$ and $p'_0$ stand for the energies of the initial and final particles, respectively, and the quantities $Q^{(*)}(p',p)$ are given by (see also Appendix F),

$$Q^{(0)}(p',p) = p_\mu p'_\nu \tilde{y}^{ext}_{\mu\nu}(\vec{q}) - \frac{b}{2}\tilde{y}^{ext}_{,\mu\mu}(\vec{q}), \qquad (8.72\text{-}1)$$

$$Q^{(1/2)}(p',p) = \overline{u}^{(s)}(\vec{p}\,')t^{(1/2)}u^{(r)}(\vec{p}), \qquad (8.72\text{-}2)$$

$$\begin{aligned} Q^{(1)}(p',p) &= \frac{1}{2}\Big[e^{(r)}_\nu(\vec{p})e^{(s)}_\mu(\vec{p}\,')p_\lambda - e^{(r)}_\mu(\vec{p})e^{(s)}_\nu(\vec{p}\,')p'_\lambda\Big] \\ &\times\Big[q_\lambda\tilde{h}^{ext}_{\mu\nu}(\vec{q}) + q_\nu\tilde{h}^{ext}_{\mu\lambda}(\vec{q}) - q_\mu\tilde{h}^{ext}_{\nu\lambda}(\vec{q})\Big] - e^{(r)}_\mu(\vec{p})e^{(s)}_\mu(\vec{p}\,')p_\lambda p'_\nu \\ &- e^{(r)}_\lambda(\vec{p})e^{(s)}_\mu(\vec{p}\,')p_\nu p'_\nu\tilde{y}^{ext}_{\mu\lambda}(\vec{q}) + \frac{1}{2}e^{(r)}_\nu(\vec{p})e^{(s)}_\mu(\vec{p}\,')q_\lambda q_\lambda\tilde{h}^{ext}_{\mu\nu}(\vec{q}), \end{aligned} \qquad (8.72\text{-}3)$$

$$Q^{(3/2)}(p',p) = \overline{u}^{(s)}_\mu(\vec{p}\,')t^{(3/2)}u^{(r)}_\mu(\vec{p}), \qquad (8.72\text{-}4)$$

$$\begin{aligned} Q^{(2)}(p',p) &= (p'_\mu p_\nu - p'_\mu p'_\nu - p_\mu p_\nu)e^{(r)}_{\alpha\beta}(\vec{p})e^{(s)}_{\alpha\beta}(\vec{p}\,')\tilde{y}^{ext}_{\mu\nu}(\vec{q}) \\ &- \frac{1}{2}\delta_{\alpha\beta}q_\beta(p'_\alpha - p_\alpha)e^{(r)}_{\mu\nu}(\vec{p})e^{(s)}_{\mu\nu}(\vec{p}\,')\tilde{y}^{ext}(\vec{q}), \end{aligned} \qquad (8.72\text{-}5)$$

where

$$t^{(1/2)} = \gamma_\mu(1+\gamma_5)(p_\nu + p'_\nu)\tilde{s}^{ext}_{\mu\nu}(\vec{q}), \tag{8.73}$$

$$t^{(3/2)} = \gamma_\nu(1+\gamma_5)(p_\lambda + p'_\lambda)\tilde{y}^{ext}_{\lambda\nu}(\vec{q}) + \frac{1}{4}\Big[\gamma_\rho\gamma_\nu\gamma_\lambda$$
$$\times (1+\gamma_5) + \gamma_\lambda(1+\gamma_5)\gamma_\rho\gamma_\nu\Big]\Big[q_\rho\tilde{h}^{ext}_{\nu\lambda}(\vec{q}) - q_\nu\tilde{h}^{ext}_{\rho\lambda}(\vec{q})\Big], \tag{8.74}$$

$$\tilde{y}^{ext}(\vec{q}) = (2\pi)^{3/2}\frac{|\vec{q}|^2}{\kappa M}y^{ext}(\vec{q}) = 1,$$

$$\tilde{y}^{ext}_{\mu\nu}(\vec{q}) = (2\pi)^{3/2}\frac{|\vec{q}|^2}{\kappa M}y^{ext}_{\mu\nu}(\vec{q}),$$

$$\tilde{s}^{ext}_{\mu\nu}(\vec{q}) = (2\pi)^{3/2}\frac{|\vec{q}|^2}{\kappa M}s^{ext}_{\mu\nu}(\vec{q}),$$

$$\tilde{h}^{ext}_{\mu\nu}(\vec{q}) = (2\pi)^{3/2}\frac{|\vec{q}|^2}{\kappa M}h^{ext}_{\mu\nu}(\vec{q}).$$

In the above relations, the quantities

$$u^{(r)}(\vec{p}),\ \Big[\overline{u}^{(r)}(\vec{p}) = (u^{(r)})^+(\vec{p})\gamma_4\Big],\ e^{(r)}_\mu(\vec{p}),\ u^{(r)}_\mu(\vec{p}),$$

$$\Big[\overline{u}^{(r)}_\mu(\vec{p}) = (u^{(r)}_\mu)^+(\vec{p})\gamma_4\Big],\ e^{(r)}_{\mu\nu}(\vec{p}),$$

on the one side, and

$$u^{(s)}(\vec{p}\,'),\ \Big[\overline{u}^{(s)}(\vec{p}\,') = (u^{(s)})^+(\vec{p}\,')\gamma_4\Big],\ e^{(s)}_\mu(\vec{p}\,'),\ u^{(s)}_\mu(\vec{p}\,'),$$

$$\Big[\overline{u}^{(s)}_\mu(\vec{p}\,') = (u^{(s)}_\mu)^+(\vec{p}\,')\gamma_4\Big],\ e^{(s)}_{\mu\nu}(\vec{p}\,'),$$

on the other side, represent the polarization spinors, vectors, spin-vectors and second-rank tensors of initial and final particles, respectively. It is worthwhile to mention that the same notations have been used, namely $p$ and $p_0$ for the characteristic quantities (the four-momentum and energy) of all the five fields, while the superscript $+$ stands for the Hermitian conjugate.

The differential scattering cross-section is given by the well-known definition-formula

$$d\sigma = (2\pi)^2 \left\langle \sum_{fin\,sp} |F(p',p)|^2 \right\rangle_{in\,sp} p_0^2\, d\Omega,$$

where $d\Omega = \sin\theta\, d\theta\, d\varphi$, $\theta$ being the scattering angle. To determine the differential scattering cross-sections, one must first calculate the quantities $\left\langle \sum_{fin\,sp} |F(p',p)|^2 \right\rangle_{in\,sp}$. Thus, it follows that

$$\begin{aligned} \left\langle \sum_{fin\,sp} |F^{(*)}(p',p)|^2 \right\rangle_{in\,sp} &= \left\langle \sum_{fin\,sp} C^{(*)} |Q^{(*)}(p',p)|^2 \right\rangle_{in\,sp} \\ &= \frac{1}{2} C^{(*)} \sum_{pol} |Q^{(*)}(p',p)|^2, \end{aligned} \tag{8.75}$$

where the coefficients $C^{(*)}$ are easily determinable, and the terms $\sum_{pol} |Q^{(*)}(p',p)|^2$ are the corresponding sums over polarizations; since these expressions are very long, they shall not be given here. For example, in the case of massless spinorial and spin-vectorial fields, these sums are in fact two traces over the corresponding polarization states indices, namely,

$$\begin{aligned} \sum_{pol} \left| Q^{(1/2)}(p',p) \right|^2 &= \sum_{r,s=1}^{2} \left| \overline{u}^{(s)}(\vec{p}\,') t^{(1/2)} u^{(r)}(\vec{p}) \right|^2 \\ &= \mathrm{Tr}\left[ \gamma_4 \left( t^{(1/2)} \right)^+ \gamma_4 P_{(+)}^{(1/2)}(\vec{p}\,') t^{(1/2)} P_{(+)}^{(1/2)}(\vec{p}) \right], \end{aligned} \tag{8.76}$$

and

$$\begin{aligned} \sum_{pol} \left| Q^{(3/2)}(p',p) \right|^2 &= \sum_{r,s=1}^{2} \left| \overline{u}_\alpha^{(s)}(\vec{p}\,') t^{(3/2)} u_\alpha^{(r)}(\vec{p}) \right|^2 \\ &= \mathrm{Tr}\left[ \gamma_4 \left( t^{(3/2)} \right)^+ \gamma_4 P_{(+)}^{(3/2)}(\vec{p}\,') t^{(3/2)} P_{(+)}^{(3/2)}(\vec{p}) \right], \end{aligned} \tag{8.77}$$

where $P_{(+)}^{(1/2)}$ and $P_{(+)}^{(3/2)}$ are the covariant projection operators (projectors) for the spinors and spin-vectors of positive energy, respectively.

Unlike the case of massive particles, there is an obvious difference between *integral* and *total*[6] scattering cross-sections (more precisely, *total differential scattering cross-sections*) in the case of massless particles. Usually, by *total scattering cross-section* is understood an *integral* scattering cross-section. Nevertheless, due to particularity of massless fields to present two linearly-independent eigenvectors for the helicity operator (two independent states with opposite helicities), in this case one can also calculate *partial differential scattering cross-sections*. Under these circumstances, the total differential scattering cross-sections yield as a sum of the partial differential scattering cross-sections, which is equivalent to summation over the indices of the polarization final states of the particles. In order to directly determine the total differential scattering cross-sections, one must use the following relations satisfied by the polarization spinors, vectors, spin-vectors and second-rank tensors, corresponding to massless particles ([252], [331]),

$$P_{(+)}^{(1/2)}(\vec{p}) = \sum_{r=1}^{2} u^{(r)}(\vec{p})\overline{u}^{(r)}(\vec{p}) = \frac{\gamma_\mu p_\mu}{2ip_0}, \tag{8.78}$$

$$\Pi_{\mu\nu}(\vec{p}) = \sum_{r=1}^{2} e_\mu^{(r)}(\vec{p})e_\nu^{(r)}(\vec{p}) = d_{\mu\nu}(\vec{p}), \tag{8.79}$$

$$P_{(+)\mu\nu}^{(3/2)}(\vec{p}) = \sum_{r=1}^{2} u_\mu^{(r)}(\vec{p})\overline{u}_\nu^{(r)}(\vec{p}) = -\frac{\gamma_\nu\gamma_\alpha\gamma_\mu p_\alpha}{4ip_0}, \tag{8.80}$$

$$\begin{aligned}\Pi_{\mu\nu,\lambda\rho}(\vec{p}) &= \sum_{r=1}^{2} e_{\mu\nu}^{(r)}(\vec{p})e_{\lambda\rho}^{(r)}(\vec{p}) \\ &= d_{\mu\lambda}(\vec{p})d_{\nu\rho}(\vec{p}) + d_{\mu\rho}(\vec{p})d_{\nu\lambda}(\vec{p}) - d_{\mu\nu}(\vec{p})d_{\lambda\rho}(\vec{p}),\end{aligned} \tag{8.81}$$

[6]Here the term "total" must by no means be confounded with the term "integral" when referring to scattering cross-sections. It is allowed to talk about the *total differential scattering cross-section*, because there also exists the *partial differential scattering cross-section*, both of them being *differential* scattering cross-sections. Therefore, the syntagma "total differential scattering cross-section" does not represent – only in this context – any contradiction in terms. Therefore, in this paragraph the two designations/names "total" and "integral" are not equivalent anymore, as before.

where

$$d_{\mu\nu}(\vec{p}) = \delta_{\mu\nu} - \frac{p_\mu p_\nu}{\vec{p}^2}.$$

After laborious calculations, the following expressions are obtained for the *total differential scattering cross-sections*, corresponding to the massless scalar, spinorial, vectorial, spin-vectorial and second-rank tensorial[7] particles:

$$\begin{aligned} d\sigma^{(0)} &= (GM)^2 \frac{d\Omega}{\sin^4 \frac{\theta}{2}} \left(1 - 2b \sin^2 \frac{\theta}{2}\right)^2 \\ &\quad + (GM)^2 \frac{d\Omega}{\sin^4 \frac{\theta}{2}} a^2 p_0^2 \sin^2 \alpha \sin^2 \theta \sin^2 \varphi, \end{aligned} \tag{8.82-1}$$

$$\begin{aligned} d\sigma^{(1/2)} &= (GM)^2 \frac{d\Omega}{\sin^4 \frac{\theta}{2}} \cos^2 \frac{\theta}{2} - (GM)^2 a p_0 \Big[ f_1(\theta) \cos \alpha \\ &+ f_2(\theta) \sin \alpha \cos \varphi \Big] + (GM)^2 a^2 p_0^2 \Big[ f_3(\theta) + f_4(\theta) \sin^2 \alpha \\ &+ f_5(\theta) \sin 2\alpha \cos \varphi + f_6(\theta) \sin^2 \alpha \cos^2 \varphi \Big] d\Omega, \end{aligned} \tag{8.82-2}$$

$$\begin{aligned} d\sigma^{(1)} &= (GM)^2 \frac{d\Omega}{\sin^4 \frac{\theta}{2}} \cos^4 \frac{\theta}{2} + 4(GM)^2 a^2 p_0^2 \Big[ f_1(\theta) + f_2(\theta) \sin^2 \alpha \\ &\quad + f_3(\theta) \sin^2 \alpha \sin^2 \varphi + f_4(\theta) \sin 2\alpha \cos \varphi \Big] d\Omega, \end{aligned} \tag{8.82-3}$$

$$\begin{aligned} d\sigma^{(3/2)} &= (GM)^2 \frac{d\Omega}{\sin^4 \frac{\theta}{2}} \cos^2 \frac{\theta}{2} + (GM)^2 a p_0 f_1(\alpha, \theta, \varphi) \\ &\quad + (GM)^2 a^2 p_0^2 \Big[ f_2(\theta) + f_3(\theta) \sin^2 \alpha \\ &\quad + f_4(\theta) \sin 2\alpha \cos \varphi + f_5(\theta) \sin^2 \alpha \cos^2 \varphi \Big] d\Omega, \end{aligned} \tag{8.82-4}$$

[7]For gravitons, the *TT*-gauge has been considered.

and

$$d\sigma^{(2)} = (GM)^2 \frac{d\Omega}{\sin^4 \dfrac{\theta}{2}} \cos^2\theta \left( \cos^2\theta + \frac{1}{8}\sin^4\theta \right) + (GM)^2 a^2 p_0^2 f_1(\theta) \sin^2\alpha \sin^2\varphi, \tag{8.82-5}$$

respectively.

The corresponding expressions of the functions $f_i(\theta)$, $(i = \overline{1,6})$ appearing in the above relations are given in Appendix G.

One can easily observe that, according to Eqs. (8.82), the differential scattering cross-sections for bosonic particles (that is, in the case of scalar, electromagnetic and gravitational waves) have the following form,

$$\left(\frac{d\sigma}{d\Omega}\right)_{Kerr} = \left(\frac{d\sigma}{d\Omega}\right)_{Schw} + \text{ terms proportional to } \vec{a}^2 p_0^2\ ,$$

while in the case of fermionic particles, as a result of contribution of the angular momentum of the body that generates the gravitational background, these sections also contain a term proportional to $ap_0$. This particularity also appears in the case of electromagnetic field, but in a different sense: the contribution of the angular momentum of the gravitational field source (that creates the gravitational background) to the differential scattering cross-section is proportional to $\vec{a}^2 p_0^2$ in the case of linear polarizations, while for circular polarizations the same contribution includes a term in $ap_0$.

Obviously, for $\vec{a} = 0$ (Schwarzschild geometry), the results obtained in the previous chapter, as well as in the papers [366], [370], [376, 377], are retrieved, namely,

$$d\sigma^{(0)}_{Schw} = (GM)^2 \frac{d\Omega}{\sin^4 \dfrac{\theta}{2}} \left(1 - 2b\sin^2\frac{\theta}{2}\right)^2, \tag{8.83-1}$$

$$d\sigma^{(1/2)}_{Schw} = (GM)^2 \frac{d\Omega}{\sin^4 \dfrac{\theta}{2}} \cos^2\frac{\theta}{2}, \tag{8.83-2}$$

$$d\sigma^{(1)}_{Schw} = (GM)^2 \frac{d\Omega}{\sin^4 \dfrac{\theta}{2}} \cos^4\frac{\theta}{2}, \tag{8.83-3}$$

$$d\sigma^{(3/2)}_{Schw} = (GM)^2 \frac{d\Omega}{\sin^4 \dfrac{\theta}{2}} \cos^2 \frac{\theta}{2}, \tag{8.83-4}$$

$$d\sigma^{(2)}_{Schw} = (GM)^2 \frac{d\Omega}{\sin^4 \dfrac{\theta}{2}} \cos^2 \theta \left( \cos^2 \theta + \frac{1}{8} \sin^4 \theta \right). \tag{8.83-5}$$

If, in addition, it is required that the scattering angle $\theta$ takes only small and very small values, then all[8] the differential scattering cross-sections have the same form, namely,

$$d\sigma^{(*)}_{Schw} = (GM)^2 \frac{d\Omega}{\sin^4 \dfrac{\theta}{2}} \equiv d\sigma_{Ruth}, \tag{8.84}$$

which is a *Rutherford-type differential scattering cross-section.*

This result coincides with that obtained in the case of corresponding massive particles – under the same approximation ($\theta \ll 1$) and, in addition, in the ultrarelativistic limit ([310], [362], [365], [367], [369, 370], [379, 380]) – which shows that in this special limit case the gravitational particles scattering is spin-independent.

Regarding the massless scalar particles, it is worth specifying that all the results obtained while studying the analogous process for scalar massive particles remain valid, because the term expressing the contribution of the angular momentum of the gravitational field source is the same. Besides, the same observation as in the previous subchapter, concerning the non-null contribution brought by the curvature invariant in the Lagrangian of the scalar field remains also valid here (*i.e.* in the case of the massless scalar field).

The next analysis concerns massless particles with half-integer spin.

The most interesting property of the total differential scattering cross-sections associated with massless spinorial and spin-vectorial particles consists in appearance of a supplementary term proportional to $ap_0$, in addition to the term $a^2p_0^2$ appearing in all other

[8]More than that, in the case of massless scalar field this result remains valid for any value of the scattering angle, if the massless scalar field is minimally coupled to the gravitational field ($b = 0$).

cases for the massless particles (where *only* this term appears). As shall be seen, such a dependence (in $ap_0$) also appears in the case of electromagnetic field (photon) scattering, even if in a slightly different way. More precisely, this dependence appears only in the *partial differential photon scattering cross-sections*, the photon polarization being purely space-like, with complex polarization vectors.

Another interesting feature of the differential scattering cross-sections corresponding to massless, half-integer spin particles, relies on the fact that, in the case of Schwarzschild geometry (that is for $\vec{a}=0$), these sections have the same expression, namely,

$$d\sigma^{(1/2)}_{Schw} = d\sigma^{(3/2)}_{Schw} = (GM)^2 \frac{d\Omega}{\sin^4 \dfrac{\theta}{2}} \cos^2 \frac{\theta}{2}, \tag{8.85}$$

meaning that both massless spinor particle and gravitino are similarly scattered by the Schwarzschild background of small curvature.

In the following investigation, the reader's attention shall be focused on the gravitational scattering of massless vectorial and tensorial particles (*i.e.* the gravitational scattering process of electromagnetic and gravitational waves). These processes are of particular importance, not only because of their theoretical significance, but also because of their experimental implications.

In order to determine the partial differential scattering cross-sections of the electromagnetic waves, one chooses purely space-like polarizations for photons, corresponding to the gauge in which $e^{(r)}_4(\vec{p}) = 0$. Under this gauge, one can write $e^{(r)}_\mu(\vec{p}) = (\vec{e}_1, 0)$, $e^{(s)}_\mu(\vec{p}\,') = (\vec{e}_2, 0)$, and the general expression (8.72-3) for $Q^{(1)}(p', p)$ becomes

$$Q^{(1)}(p', p) = \xi + i\omega, \tag{8.86}$$

where

$$\begin{aligned} \xi = \frac{1}{2}\Bigg\{ & \Big[p_0^2 + (\vec{p}\cdot\vec{p}\,')\Big](\vec{e}_1\cdot\vec{e}^{\,*}_2) + \frac{1}{2}(\vec{q}\cdot\vec{e}^{\,*}_2) \\ & \times \Big[(\vec{p}+\vec{p}\,')\cdot\vec{e}_1\Big] - \frac{1}{2}(\vec{q}\cdot\vec{e}_1)\Big[(\vec{p}+\vec{p}\,')\cdot\vec{e}^{\,*}_2\Big]\Bigg\}, \end{aligned} \tag{8.87}$$

$$\omega = \frac{1}{2} p_0 \varepsilon_{ijk} a_j q_k \Big[ (\vec{e}^{\,*}_2)_i (\vec{q} \cdot \vec{e}_1) - (\vec{e}_1)_i (\vec{q} \cdot \vec{e}^{\,*}_2) \quad - (p_i + p'_i)(\vec{e}_1 \cdot \vec{e}^{\,*}_2) \Big]. \tag{8.88}$$

In the case of linear polarizations $\vec{e}_1$ and $\vec{e}_2$ are real, while for circular polarizations the photon polarization vectors $\vec{e}_1$ and $\vec{e}_2$ are complex. The second case is, obviously, the most interesting, and for the polarization states of the photons one chooses the basis in which

$$\begin{cases} \vec{e}^{\,R}_1 = \dfrac{1}{\sqrt{2}} (\hat{e}_x + i\hat{e}_y), \quad \vec{e}^{\,L}_1 = \dfrac{1}{\sqrt{2}} (\hat{e}_x - i\hat{e}_y), \\ \vec{e}^{\,R}_2 = \dfrac{1}{\sqrt{2}} (\hat{e}_\theta + i\hat{e}_\varphi), \quad \vec{e}^{\,L}_2 = \dfrac{1}{\sqrt{2}} (\hat{e}_\theta - i\hat{e}_\varphi), \end{cases} \tag{8.89}$$

where $\hat{e}_x$, $\hat{e}_y$, $\hat{e}_\theta$, and $\hat{e}_\varphi$ are unit vectors in the directions of $x$, $y$, $\theta$ and $\varphi$, respectively. For concrete calculations, it was considered the configuration depicted in Fig. 8.2, namely,

$$\begin{cases} \hat{e}_x = (1, 0, 0); \\ \hat{e}_y = (0, 1, 0); \\ \hat{e}_\theta = (\cos\theta \cos\varphi,\ \cos\theta \sin\varphi,\ -\sin\theta); \\ \hat{e}_\varphi = (-\sin\varphi,\ \cos\varphi,\ 0); \end{cases}$$
$$\begin{cases} \vec{a} = (a \sin\alpha,\ 0,\ a\cos\alpha); \\ \vec{p} = p_0 (0,\ 0,\ 1); \\ \vec{p}\,' = p_0 (\sin\theta\cos\varphi,\ \sin\theta\sin\varphi,\ \cos\theta); \\ \vec{q} = \vec{p}\,' - \vec{p}. \end{cases} \tag{8.90}$$

After some algebraic manipulations, the following expressions for the partial differential scattering cross-sections are obtained,

$$\left( \frac{d\sigma}{d\Omega}^{(1)} \right)_{RL} = \left( \frac{d\sigma}{d\Omega}^{(1)} \right)_{LR} = 0, \tag{8.91}$$

$$\left(\frac{d\sigma}{d\Omega}^{(1)}\right)_{LL} = (GM)^2\left\{\left[\cot^2\frac{\theta}{2} - 2ap_0\cos\frac{\theta}{2}\left(\cos\alpha\cos\frac{\theta}{2} + \sin\alpha\sin\frac{\theta}{2}\cos\varphi\right)\right]^2 + 4a^2p_0^2\sin^2\alpha\cot^2\frac{\theta}{2}\sin^2\varphi\right\}, \tag{8.92}$$

$$\left(\frac{d\sigma}{d\Omega}^{(1)}\right)_{RR} = (GM)^2\left\{\left[\cot^2\frac{\theta}{2} + 2ap_0\cos\frac{\theta}{2}\left(\cos\alpha\cos\frac{\theta}{2} + \sin\alpha\sin\frac{\theta}{2}\cos\varphi\right)\right]^2 + 4a^2p_0^2\sin^2\alpha\cot^2\frac{\theta}{2}\sin^2\varphi\right\}, \tag{8.93}$$

which are in agreement with the results obtained by other authors (see, *e.g.*, [91]). If $\vec{e}_1$ and $\vec{e}_2$ are real (linear polarizations), then the contribution of angular momentum of the gravitational source to the partial differential scattering cross-section is proportional to $a^2p_0^2$, while if $\vec{e}_1$ and $\vec{e}_2$ are complex (circular polarizations) the mentioned contribution involves an extra term proportional to $ap_0$, as it has been already specified.

The *total differential scattering cross-section* is obtained by summing over the photon final polarization states, which is equivalent to the summation of the partial differential scattering cross-sections (that is the differential scattering cross-sections for the pure helicity states). Here is the result:

$$\frac{d\sigma}{d\Omega}^{(1)} = (GM)^2\left\{\cot^4\frac{\theta}{2} + 4a^2p_0^2\left[\cos^2\frac{\theta}{2}\left(\cos\frac{\theta}{2}\cos\alpha + \sin\frac{\theta}{2}\sin\alpha\cos\varphi\right)^2 + \left(\cot\frac{\theta}{2}\sin\alpha\sin\varphi\right)^2\right]\right\}. \tag{8.94}$$

This way, one recovers the result given by (8.82-3), but in a more explicit manner. According to relations (8.91)–(8.93), the following *scattering matrix* is obtained (for the employed polarization basis),

$$\begin{pmatrix} \left(\dfrac{d\sigma}{d\Omega}^{(1)}\right)_{RR} & \left(\dfrac{d\sigma}{d\Omega}^{(1)}\right)_{RL} \\ \left(\dfrac{d\sigma}{d\Omega}^{(1)}\right)_{LR} & \left(\dfrac{d\sigma}{d\Omega}^{(1)}\right)_{LL} \end{pmatrix} = \begin{pmatrix} \left(\dfrac{d\sigma}{d\Omega}^{(1)}\right)_{RR} & 0 \\ 0 & \left(\dfrac{d\sigma}{d\Omega}^{(1)}\right)_{LL} \end{pmatrix}$$

$$= \begin{pmatrix} \Xi + \Theta & 0 \\ 0 & \Xi - \Theta \end{pmatrix},$$

where by $\Xi$ and $\Theta$ we have denoted the quantities

$$\Xi = (GM)^2\left[\cot^4\frac{\theta}{2} + 4a^2p_0^2\cos^2\frac{\theta}{2}\left(\cos\alpha\cos\frac{\theta}{2} + \sin\alpha\sin\frac{\theta}{2}\cos\varphi\right)^2 + 4a^2p_0^2\sin^2\alpha\cot^2\frac{\theta}{2}\sin^2\varphi\right],$$

and

$$\Theta = 4(GM)^2ap_0\cos\frac{\theta}{2}\cot^2\frac{\theta}{2}\left(\cos\alpha\cos\frac{\theta}{2} + \sin\alpha\sin\frac{\theta}{2}\cos\varphi\right),$$

respectively. The scattering matrix is diagonal and, consequently, the helicity conserves in the scattering process. The diagonal elements are not equal, meaning that the scattering process induces a partial polarization to the scattered waves. The *polarization degree* of these waves (coming from an initial totally non-polarized wave fascicle) is by definition,

$$P = \frac{\left(\dfrac{d\sigma}{d\Omega}^{(1)}\right)_{RR} - \left(\dfrac{d\sigma}{d\Omega}^{(1)}\right)_{LL}}{\left(\dfrac{d\sigma}{d\Omega}^{(1)}\right)_{RR} + \left(\dfrac{d\sigma}{d\Omega}^{(1)}\right)_{LL}} = \frac{\Theta}{\Xi}$$

$$= 4ap_0\cos\frac{\theta}{2}\left(\cos\alpha\cos\frac{\theta}{2} + \sin\alpha\sin\frac{\theta}{2}\cos\varphi\right)\left[\cot^2\frac{\theta}{2} + 4a^2p_0^2\sin^2\frac{\theta}{2}\right.$$

$$\times\left(\cos\alpha\cos\frac{\theta}{2}+\sin\alpha\sin\frac{\theta}{2}\cos\varphi\right)^2+4a^2p_0^2\sin^2\alpha\sin^2\varphi\Bigg]^{-1}$$

$$=\Bigg[ap_0\Gamma(\alpha,\theta,\varphi)\sin\frac{\theta}{2}\tan\frac{\theta}{2}+ap_0\sin^2\alpha\sin^2\varphi\Gamma^{-1}(\alpha,\theta,\varphi)\sec\frac{\theta}{2}$$

$$+\Big(4ap_0\Gamma(\alpha,\theta,\varphi)\Big)^{-1}\cot\frac{\theta}{2}\csc\frac{\theta}{2}\Bigg]^{-1},$$

where $\Gamma(\alpha,\theta,\varphi)$ denotes the following quantity:

$$\Gamma(\alpha,\theta,\varphi)=\cos\alpha\cos\frac{\theta}{2}+\sin\alpha\sin\frac{\theta}{2}\cos\varphi.$$

As one can easily be observed, the polarization degree of the scattered electromagnetic waves is a function of three variables, namely the angles $\alpha\in[0,\pi]$, $\theta\in[0,\pi]$, and $\varphi\in[0,2\pi]$. Figures 8.23–8.27 show the variation of the *polarization degree function* $P(\alpha,\theta,\varphi)$ for several values of the angle $\alpha$ (the angle between direction of the angular momentum of the gravitational field source and that of the incident photons), namely, $\alpha=0$, $\alpha=\frac{\pi}{4}$, $\alpha=\frac{\pi}{2}$, $\alpha=\frac{3\pi}{4}$, and $\alpha=\pi$. Since this investigation concerns the qualitative aspect of the dependence $P=P(\theta,\varphi)$ only, one can take for convenience $ap_0=1$. Because of this, the numerical values on the ordinate axis are proportional and not equal to the possible real values. Figures 8.28–8.30 also show the qualitative dependence (here also $ap_0=1$) of the polarization degree $P(\alpha,\varphi)$ for the non-deviated particles ($\theta=0$), those scattered at $\theta=\frac{\pi}{2}$, and those backscattered ($\theta-\pi$).

Turning to the case of gravitational waves, as it is well-known from the linear theory, for any kind of such a wave, the metric perturbation $h_{\mu\nu}$ can be put in the $TT$-form [269]. Thus, in order to obtain the partial differential scattering cross-sections corresponding to the diffusion process of gravitational waves, one chooses the $TT$-gauge, so that $e^{(r)}_{\mu\nu}(\vec{p})\to\overset{\leftrightarrow}{e}{}^{(r)}(\vec{p})$, and $e^{(s)}_{\mu\nu}(\vec{p}')\to\overset{\leftrightarrow}{e}{}^{(s)}(\vec{p}')$. In the case of circular polarizations, $\overset{\leftrightarrow}{e}{}^{(r)}(\vec{p})\equiv\overset{\leftrightarrow}{e}_1$, and $\overset{\leftrightarrow}{e}{}^{(s)}(\vec{p}')\equiv\overset{\leftrightarrow}{e}_2$ are complex and, by analogy with the photons' case, the following polarization basis for gravitons is considered:

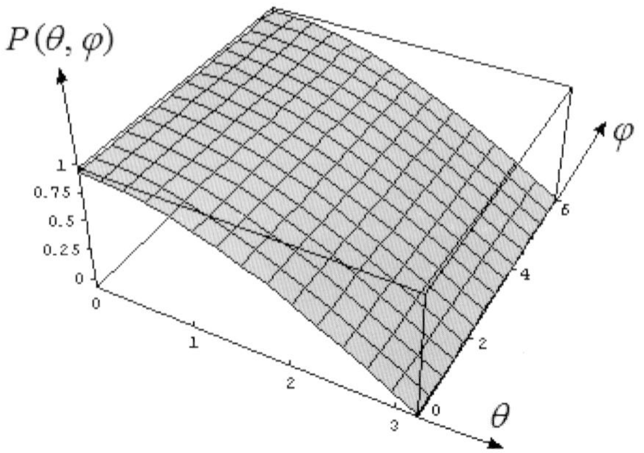


Fig. 8.23. Variation of the polarization degree of the scattered electromagnetic waves for $\alpha = 0$ in terms of both angles $\theta$ (between directions of the incident and emergent photons), and $\varphi$ (between the $Ox$-axis and projection in $xOy$-plane of the direction of the emergent photons).

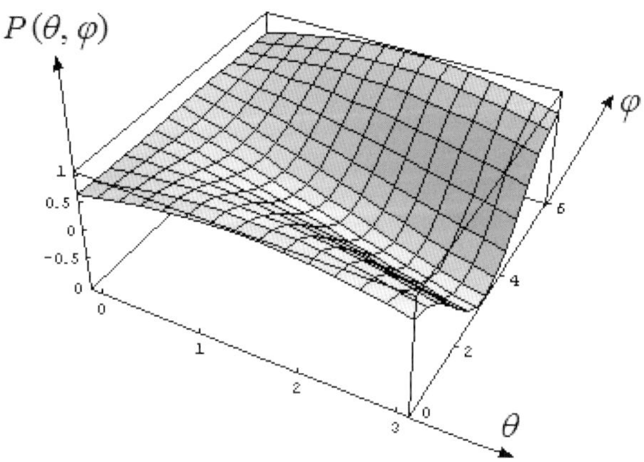


Fig. 8.24. Variation of the polarization degree of the scattered electromagnetic waves for $\alpha = \frac{\pi}{4}$ in terms of both angles $\theta$ (between directions of the incident and emergent photons), and $\varphi$ (between the $Ox$-axis and projection in $xOy$-plane of the direction of the emergent photons).

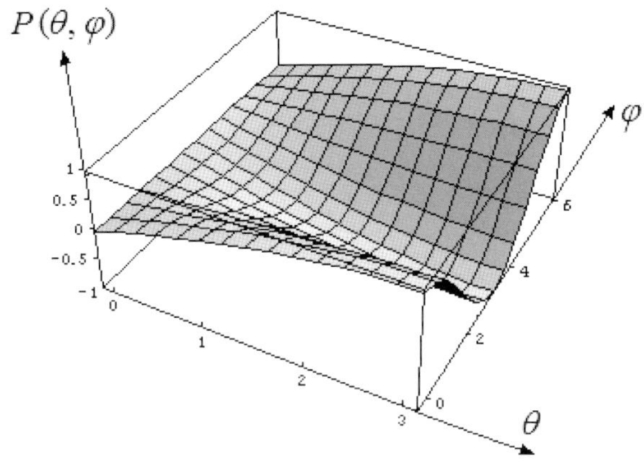


Fig. 8.25. Variation of the polarization degree of the scattered electromagnetic waves for $\alpha = \frac{\pi}{2}$ in terms of both angles $\theta$ (between directions of the incident and emergent photons), and $\varphi$ (between the $Ox$-axis and projection in $xOy$-plane of the direction of the emergent photons).

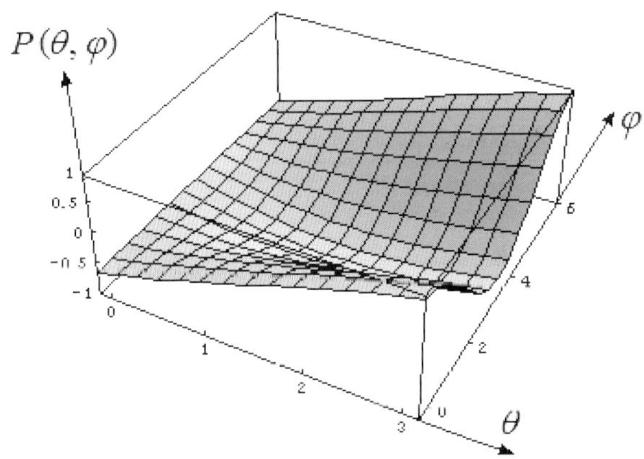


Fig. 8.26. Variation of the polarization degree of the scattered electromagnetic waves for $\alpha = \frac{3\pi}{4}$ in terms of both angles $\theta$ (between directions of the incident and emergent photons), and $\varphi$ (between the $Ox$-axis and projection in $xOy$-plane of the direction of the emergent photons).

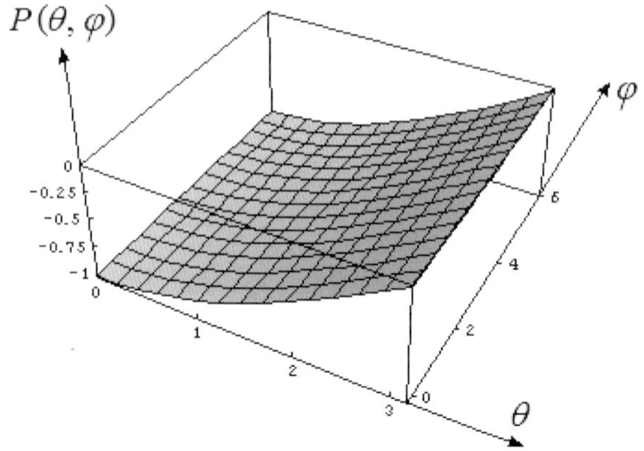


Fig. 8.27. Variation of the polarization degree of the scattered electromagnetic waves for $\alpha = \pi$ in terms of both angles $\theta$ (between directions of the incident and emergent photons), and $\varphi$ (between the $Ox$-axis and projection in $xOy$-plane of the direction of the emergent photons).

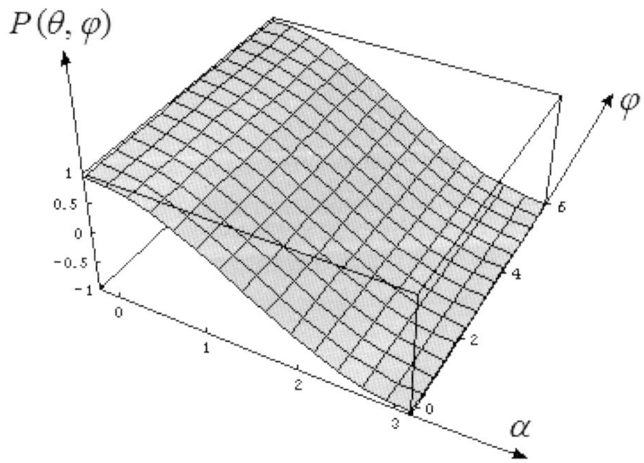


Fig. 8.28. Variation of the polarization degree of the scattered electromagnetic waves for $\theta = 0$ (non-deviated fascicle) in terms of both angles $\alpha$ (between direction of the incident photons and that of the angular momentum of the gravitational background source) and $\varphi$ (between the $Ox$-axis and projection in $xOy$-plane of the direction of the emergent photons).

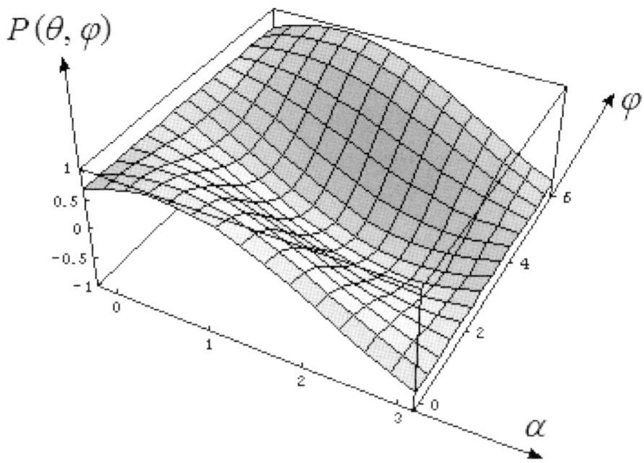


Fig. 8.29. Variation of the polarization degree of the scattered electromagnetic waves for $\theta = \frac{\pi}{2}$ in terms of both angles $\alpha$ (between direction of the incident photons and that of the angular momentum of the gravitational background source) and $\varphi$ (between the $Ox$-axis and projection in $xOy$-plane of the direction of the emergent photons).

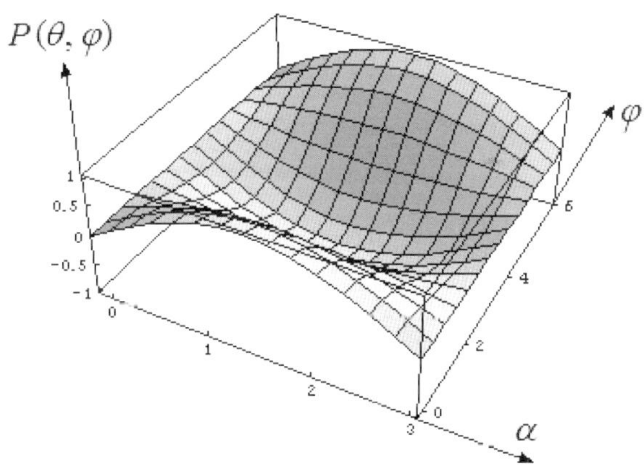


Fig. 8.30. Variation of the polarization degree of the scattered electromagnetic waves for $\theta = \pi$ (backscattering case) in terms of both angles $\alpha$ (between direction of the incident photons and that of the angular momentum of the gravitational background source) and $\varphi$ (between the $Ox$-axis and projection in $xOy$-plane of the direction of the emergent photons).

$$\begin{cases} \overset{\leftrightarrow R}{e}_1 = \dfrac{1}{2}\Big[\hat{e}_x\hat{e}_x - \hat{e}_y\hat{e}_y + i(\hat{e}_x\hat{e}_y + \hat{e}_y\hat{e}_x)\Big], \\ \overset{\leftrightarrow L}{e}_1 = \dfrac{1}{2}\Big[\hat{e}_x\hat{e}_x - \hat{e}_y\hat{e}_y - i(\hat{e}_x\hat{e}_y + \hat{e}_y\hat{e}_x)\Big], \end{cases} \tag{8.95-1}$$

$$\begin{cases} \overset{\leftrightarrow R}{e}_2 = \dfrac{1}{2}\Big[\hat{e}_\theta\hat{e}_\theta - \hat{e}_\varphi\hat{e}_\varphi + i(\hat{e}_\theta\hat{e}_\varphi + \hat{e}_\varphi\hat{e}_\theta)\Big], \\ \overset{\leftrightarrow L}{e}_2 = \dfrac{1}{2}\Big[\hat{e}_\theta\hat{e}_\theta - \hat{e}_\varphi\hat{e}_\varphi - i(\hat{e}_\theta\hat{e}_\varphi + \hat{e}_\varphi\hat{e}_\theta)\Big]. \end{cases} \tag{8.95-2}$$

Under these conditions, the general expression (8.72-5) for $Q^{(2)}(p', p)$ becomes

$$Q^{(2)}(p', p) = p_0\left(p_0\cos\theta + i\varepsilon_{jkl}a_j p_k p'_l\right)\left[\overset{\leftrightarrow}{e}_1(\vec{p}) \otimes \overset{\leftrightarrow *}{e}_2(\vec{p}\,')\right], \tag{8.96}$$

where $\overset{\leftrightarrow}{e}_1(\vec{p}) \otimes \overset{\leftrightarrow}{e}_2(\vec{p}\,')$ represents the inner tensor product. As a result of calculation, the following expressions for the *partial differential scattering cross-sections* of the gravitational waves, between helicity pure states, are obtained:

$$\begin{aligned} \left(\frac{d\sigma^{(2)}}{d\Omega}\right)_{RL} &= \left(\frac{d\sigma^{(2)}}{d\Omega}\right)_{LR} \\ &= (GM)^2(\cos^2\theta + a^2p_0^2\sin^2\theta\sin^2\alpha\sin^2\varphi)\sin^4\frac{\theta}{2}, \end{aligned} \tag{8.97}$$

$$\begin{aligned} \left(\frac{d\sigma^{(2)}}{d\Omega}\right)_{LL} &= \left(\frac{d\sigma^{(2)}}{d\Omega}\right)_{RR} \\ &= (GM)^2(\cos^2\theta + a^2p_0^2\sin^2\theta\sin^2\alpha\sin^2\varphi)\cos^4\frac{\theta}{2}\cot^4\frac{\theta}{2}. \end{aligned} \tag{8.98}$$

The *total differential cross-section* is

$$\begin{aligned} \left(\frac{d\sigma^{(2)}}{d\Omega}\right) &= \frac{(GM)^2}{\sin^4\dfrac{\theta}{2}}\left(\cos^2\theta + \frac{1}{8}\sin^4\theta\right) \\ &\quad\times\left(\cos^2\theta + a^2p_0^2\sin^2\theta\sin^2\alpha\sin^2\varphi\right), \end{aligned} \tag{8.99}$$

or, written in a more suggestive form,

$$\left(\frac{d\sigma}{d\Omega}^{(2)}\right)_{Kerr} = \left(\frac{d\sigma}{d\Omega}^{(2)}\right)_{Schw}$$
$$+ 4(GM)^2 a^2 p_0^2 \left(\cos^2\theta + \frac{1}{8}\sin^4\theta\right)\cot^2\frac{\theta}{2}\sin^2\alpha\sin^2\varphi, \tag{8.100}$$

this way retrieving the result (8.82-5), but in a more explicit manner.

Eq. (8.97) shows that, unlike the electromagnetic waves, in the case of scattering of gravitational waves on an axially-symmetric background, the *helicity is not conserved*. Also, according to Eq. (8.98) the partial differential scattering cross-section is the same for both the left and right circular polarization states. Therefore, in this situation there is neither different scattering of particles with opposite helicity states, nor partial polarization of the incident gravitational non-polarized waves. In other words, the gravitational radiation does not produce any of the effects induced by the angular momentum of the gravitational base-field source to the electromagnetic waves.

Even if both electromagnetic and gravitational fields are massless bosonic fields, the scattering process conserves the helicity in the case of electromagnetic waves, in contrast to gravitational waves, for which this result is not valid. The explanation of this fact must be searched in the lack of symmetry under the gauge-type transformations

$$e^{(i)}_{\alpha\beta} \to e^{(i)}_{\alpha\beta} + k^{(i)}_{\alpha}\chi_\beta + k^{(i)}_{\beta}\chi_\alpha, \quad (i - 1, 2),$$

(where $\chi_\alpha$ is an arbitrary four-vector) of the transition amplitude corresponding to massless spin-2 field as against the case of massless spin-1 field, characterized by the gauge transformation

$$e^{(i)}_{\alpha} \to e^{(i)}_{\alpha} + \gamma^{(i)}k_\alpha, \quad (i = 1, 2),$$

(where $\gamma^{(i)}$ is an arbitrary scalar) in which case such a symmetry (invariance) exists.

Also, in the case of photons' scattering process, for a Schwarzschild-type geometry, the differential scattering cross-section is independent of helicity, while in the case of a rotating body (source of an axially-symmetric gravitational background) possessing a non-zero angular momentum, this cross-section clearly depends on the particles helicity. On the other side, in the case of gravitons (gravitational waves), the scattering process does not depend on the particles helicity, since the angular momentum of the gravitational field source does not have any polarization effect on the incident non-polarized gravitons/gravitational waves (the reason being explained above).

### VIII.2.3. The backscattering case

The method utilized to determine the differential scattering cross-sections allows one to study the backscattering limit case, since the results obtained for the quantities $\frac{d\sigma}{d\Omega}$ are valid for any value of the scattering angle $\theta$.

Thus, for the scalar, vectorial and second-rank tensorial massless particles, the backscattering cross-sections are finite and independent of the angular momentum per unit mass of the gravitational source $\vec{a}$,

$$\left.\frac{d\sigma}{d\Omega}^{(0)}\right|_{\theta=\pi} = (GM)^2(1-2b)^2, \tag{8.101}$$

$$\left.\frac{d\sigma}{d\Omega}^{(1)}\right|_{\theta=\pi} = 0, \tag{8.102}$$

$$\left.\frac{d\sigma}{d\Omega}^{(2)}\right|_{\theta=\pi} = (GM)^2. \tag{8.103}$$

The constant, non-zero values of the differential backscattering cross-sections of the scalar and second-rank tensorial massless particles show that, if the incident particles have a pure helicity state, then the backscattered particles must have an helicity state opposite to the initial one. The massless vectorial and spin-vectorial particles also have finite values for the backscattering cross-sections, but unlike the other cases, these sections depend on both the squared angular momentum per unit mass of the source of gravitational field

$\vec{a}^2$, and the angle $\alpha$,

$$\left.\frac{d\sigma}{d\Omega}^{(1/2)}\right|_{\theta=\pi} = \left.\frac{d\sigma}{d\Omega}^{(3/2)}\right|_{\theta=\pi} = (GM)^2 \vec{a}^2 p_0^2 \sin^2\alpha. \qquad (8.104)$$

## VIII.2.4. Integral scattering cross-sections

Calculation of the integral scattering cross-sections of massless particles faces exactly the same difficulties as in the case of massive particles, which difficulties were overcome as in the previous subchapter. Consequently, here are the results obtained by means of the same algorithm as that used to determine the integral scattering cross-sections for the corresponding massive particles. The only (minor) difference between the two cases consists in the fact that the total differential scattering cross-sections functions for massless particles do not depend anymore on the ratio between the modulus of the momentum of incident particles and their energy, $v = \frac{|\vec{p}|}{p_0}$, as in the previous subchapter. Differently speaking, in order to determine the integral scattering cross-sections, indefinite integrations have to be performed over the functions

$$f(\alpha, \theta, \varphi) = \frac{d\sigma}{d\Omega}(\alpha, \theta, \varphi).$$

The calculations give the following expressions for the integral scattering cross-sections of massless particles scattered on an axially-symmetric gravitational background described by the Kerr metric:

$$\begin{aligned}\sigma^{(0)}(\alpha, \theta) = 4\pi\Bigg[&\left(\vec{a}^2 p_0^2 \sin^2\alpha - 2b^2\right)\cos\theta - \csc^2\frac{\theta}{2} \\ &+ 4\left(\vec{a}^2 p_0^2 \sin^2\alpha - 2b\right)\Bigg]\ln\left(\sin\frac{\theta}{2}\right),\end{aligned} \qquad (8.105\text{-}1)$$

$$\sigma^{(1/2)}(\alpha,\theta) = 4\pi\left\{\frac{1}{4}\left[\vec{a}^2p_0^2(1-2\cos 2\alpha) - 4ap_0\cos\alpha\right]\cos\theta\right.$$
$$-\frac{1}{16}\vec{a}^2p_0^2\cos 2\alpha\cos 2\theta - \csc^2\frac{\theta}{2} + 2\Big(2\vec{a}^2p_0^2\sin^2\alpha$$
$$\left.-\,2ap_0\cos\alpha - 1\Big)\ln\left(\sin\frac{\theta}{2}\right)\right\}, \qquad (8.105\text{-}2)$$

$$\sigma^{(1)}(\alpha,\theta) = 4\pi\left\{\frac{1}{32}\left[\vec{a}^2p_0^2(3-23\cos 2\alpha) - 16\right]\cos\theta\right.$$
$$-\frac{1}{4}\vec{a}^2p_0^2\cos^2\alpha\cos 2\theta - \frac{1}{96}(1+3\cos 2\alpha)\vec{a}^2p_0^2\cos 3\theta$$
$$\left.-\csc^2\frac{\theta}{2} - 4\left(1-\vec{a}^2p_0^2\sin^2\alpha\right)\ln\left(\sin\frac{\theta}{2}\right)\right\}, \qquad (8.105\text{-}3)$$

$$\sigma^{(3/2)}(\alpha,\theta) = 4\pi\left\{\frac{1}{4}\left[\vec{a}^2p_0^2(1-2\cos 2\alpha) + 4ap_0\cos\alpha\right]\cos\theta\right.$$
$$-\frac{1}{16}\vec{a}^2p_0^2\cos 2\alpha\cos 2\theta - \csc^2\frac{\theta}{2} - 2\Big(1-2ap_0 \qquad (8.105\text{-}4)$$
$$\left.\times\cos\alpha - 2\vec{a}^2p_0^2\sin^2\alpha\Big)\ln\left(\sin\frac{\theta}{2}\right)\right\},$$

$$\sigma^{(2)}(\alpha,\theta) = \frac{\pi}{320}\Big[10f_1(\alpha)\cos\theta + 20f_2(\alpha)\cos 2\theta + \frac{5}{3}f_3(\alpha)$$
$$\times\cos 3\theta + 5f_4(\alpha)\cos 4\theta + f_5(\alpha)\cos 5\theta\Big] \qquad (8.105\text{-}5)$$
$$-\,4\pi\csc^2\frac{\theta}{2} + f_6(\alpha)\ln\left(\sin\frac{\theta}{2}\right).$$

The functions $f_i$ ($i = \overline{1,6}$) are given in Appendix G, and the graphs for the integral scattering cross-sections

$$\sigma(\alpha) = \sigma(\alpha, \theta = \pi) - \sigma(\alpha, \theta = \varepsilon = 10^{-6})$$

are shown in Figs. (8.31)–(8.35). As one can easily see, for concrete calculation of the integral scattering cross-sections it has been used the same value $\varepsilon = 10^{-6}$ when applying the cut-off procedure, as in the previous subchapter.

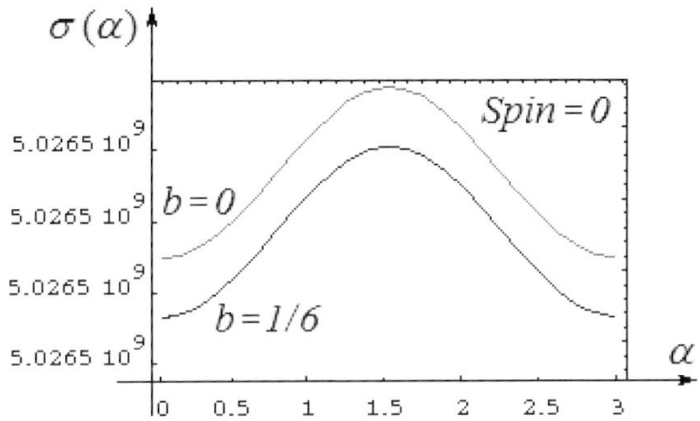


Fig. 8.31. Variation of the integral scattering cross-section $\sigma(\alpha)$ with respect to angle $\alpha$, for non-massive scalar particles, in the case of minimal ($b = 0$) and a non-minimal ($b = 1/6$) couplings.

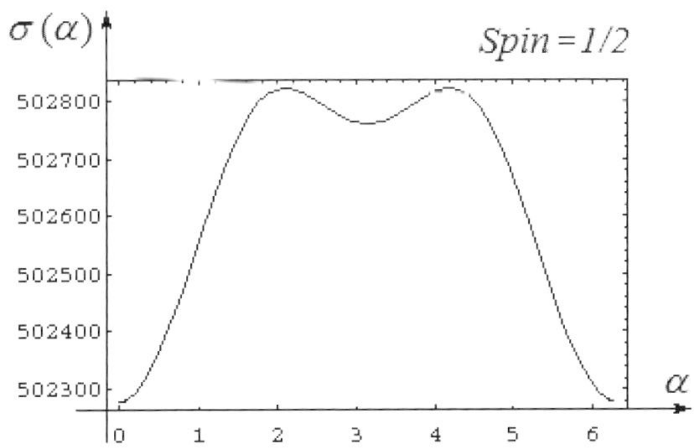


Fig. 8.32. Variation of the integral scattering cross-section $\sigma(\alpha)$ with respect to angle $\alpha$, for non-massive spinorial particles.

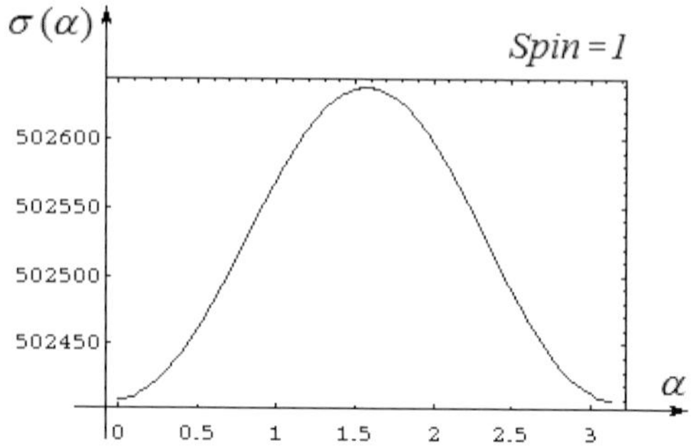


Fig. 8.33. Variation of the integral scattering cross-section $\sigma(\alpha)$ with respect to angle $\alpha$, for non-massive vectorial particles.

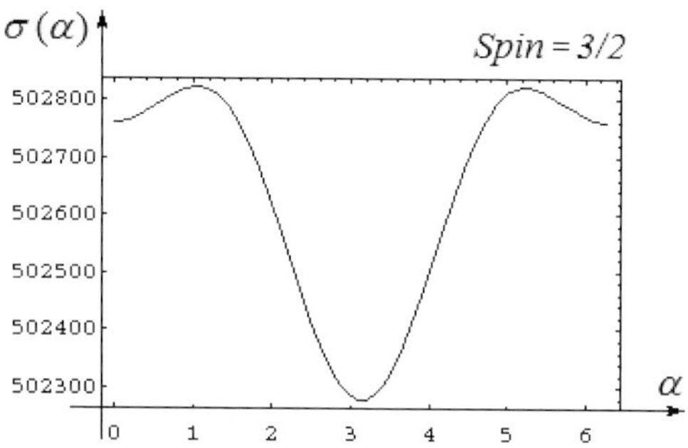


Fig. 8.34. Variation of the integral scattering cross-section $\sigma(\alpha)$ with respect to angle $\alpha$, for non-massive, spin-3/2 particles.

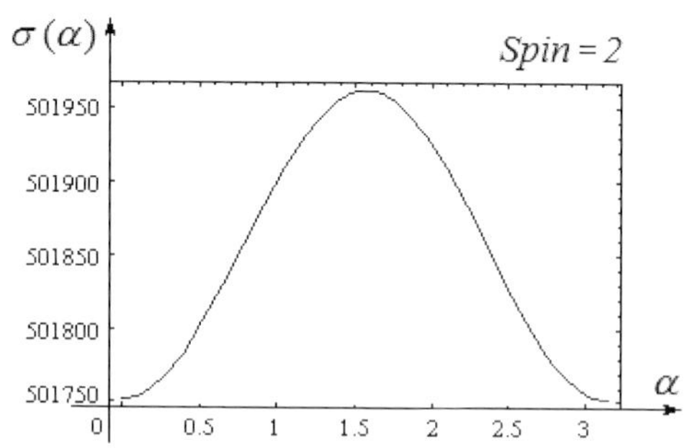


Fig. 8.35. Variation of the integral scattering cross-section $\sigma(\alpha)$ with respect to angle $\alpha$, for non-massive, spin-2 particles.

It should also be mentioned that in this paragraph (beginning with relation (8.105–1)), the constant factor $(GM)^2$, which in no way affects the allure of the graphically represented curves, has been omitted. To facilitate the concrete numerical calculations, $\vec{a}^2 p_0^2 = 1$ has also been used and, consequently, the numerical values appearing on the graphs do not have a quantitative significance, but only a qualitative one.

Chapter IX

# Software Package for Anaytical Calculation of Differential Cross-Sections for Gravitational Scattering of High-Spin Particles

The complexity of calculations implied in the study of interaction processes of high-spin particles led to creation and usage of some special analytical programs necessary to determine the differential and integral scattering cross-sections of these particles for the interaction processes which take place in both electromagnetic and gravitational external fields.

The specialized software employed to approach the investigations in this respect was *Mathematica*, the computing programs being run on a Dell XPS 8900 MT computer. In addition to the specific commands and instructions of the *Mathematica* software [46, 407, 408] and the already implemented functions of this software, in order to introduce the quantities specific to our problem, the following notations have been used:

- the Kronecker symbol $\delta_{ij}$ has been denoted by kr[i, j];
- the three-dimensional Levi-Civita symbol $\varepsilon_{ijk}$ has been denoted by eps[i, j, k];
- the four-dimensional Levi-Civita symbol $\varepsilon_{ijkl}$ has been denoted by eps[i, j, k, l];
- the four-momentum components of the incident particles are k[1], k[2], k[3], k[4], where k[4] = I k0; k0 is the energy of the incident particles, while the first three components have different expressions, denoted as appropriate;
- the four-momentum components of the emergent (scattered) particles are kp[1], kp[2], kp[3], kp[4], where kp[4] = I k0, k0 being

the energy of the emergent particles (the same with the energy of incident particles, because the energy conserves in the scattering process), while the first three components have different expressions, denoted as appropriate;

- the usual Dirac matrices have been denoted by gamma[1], gamma[2], gamma[3], gamma[4], and gamma[5];
- the contraction between the incident particles four-momentum and the usual Dirac matrices $p_k\gamma_k$, $(k = 1, 2, 3, 4)$, has been denoted by gammap;
- the contraction between the emergent particles four-momentum and the usual Dirac matrices $p'_k\gamma_k$, $(k = 1, 2, 3, 4)$, has been denoted by gammapp, while gammaqq represents the contraction between the usual Dirac matrices and the four-momentum transfer to the external field, $q = p' - p$ (which in program notation writes qq[i]=kp[i]-k[i], $i = 1, 2, 3, 4$).

## IX.1. Program for analytical calculation of the differential cross-section for the gravitational scattering of massive vector particles

```
Off[General::spell]

Remove["Global`*"];

Unprotect[In,Out];

Clear[In,Out];


a[1]=aa Sin[alfa];

a[2]=0;

a[3]=aa Cos[alfa];


k[1]=0;

k[2]=0;

k[3]=v k0;

k[4]=I k0;
```

```
kp[1]=v k0 Sin[th] Cos[fi];
kp[2]=v k0 Sin[th] Sin[fi];
kp[3]=v k0 Cos[th];
kp[4]=I k0;

q[1]=kp[1]-k[1];
q[2]=kp[2]-k[2];
q[3]=kp[3]-k[3];

eps[i_,j_,k_]: =Signature[{i,j,k}];
kr[1,1]=kr[2,2]=kr[3,3]=kr[4,4]=1;
kr[1,2]=kr[1,3]=kr[2,3]=kr[2,1]=kr[3,1]=kr[3,2]=0;
kr[1,4]=kr[4,1]=kr[2,4]=kr[4,2]=kr[3,4]=kr[4,3]=0;

Do[y[i,j]=0,{i,1,3},{j,1,3}]
Do[y[4,i]=y[i,4]=1/2 Sum[eps[i,j,k] a[j] q[k],{j,1,3},{k,1,3}],{i,1,3}]
   y[4,4]=1;
Do[u[i,j]=-1/4 kr[i,j],{i,1,3},{j,1,3}]
Do[u[4,i]=u[i,4]=y[4,i],{j,1,3},{k,1,3},{i,1,3}]
         u[4,4]=3/4;

e[a,m_] e[a,n_] ^: =kr[m,n]+kp[m] kp[n]/((1-v^2) k0^2)
e[b,m_] e[b,n_] ^: =kr[m,n]+k[m] k[n]/((1-v^2) k0^2)

e[a,1]^x_ ^: =e[a,1] e[a,1]
```

```
e[a,2]^x_ ^: =e[a,2] e[a,2]
e[a,3]^x_ ^: =e[a,3] e[a,3]
e[a,4]^x_ ^: =e[a,4] e[a,4]
e[b,1]^x_ ^: =e[b,1] e[b,1]
e[b,2]^x_ ^: =e[b,2] e[b,2]
e[b,3]^x_ ^: =e[b,3] e[b,3]
e[b,4]^x_ ^: =e[b,4] e[b,4]

suma=Expand[4(1-v^2)/k0^4 Expand[(Sum[e[a,4] e[b,4] kp[m] k[m]
u[4,4],{m,1,4}]+
        Sum[e[a,4] e[b,i] kp[m] k[m] u[4,i],{m,1,4},{i,1,3}]+
        Sum[e[a,i] e[b,4] kp[m] k[m] u[i,4],{m,1,4},{i,1,3}]+
        Sum[e[a,i] e[b,j] kp[m] k[m] u[i,j],{m,1,4},{i,1,3},{j,1,3}]-
        Sum[e[a,4] e[b,m] kp[m] k[4] u[4,4],{m,1,4}]-
        Sum[e[a,4] e[b,m] kp[m] k[i] u[4,i],{m,1,4},{i,1,3}]-
        Sum[e[a,i] e[b,m] kp[m] k[4] u[i,4],{m,1,4},{i,1,3}]-
        Sum[e[a,i] e[b,m] kp[m] k[j] u[i,j],{m,1,4},{i,1,3},{j,1,3}]-
        Sum[e[a,m] e[b,4] kp[4] k[m] u[4,4],{m,1,4}]-
        Sum[e[a,m] e[b,i] kp[4] k[m] u[4,i],{m,1,4},{i,1,3}]-
        Sum[e[a,m] e[b,4] kp[i] k[m] u[i,4],{m,1,4},{i,1,3}]-
        Sum[e[a,m] e[b,j] kp[i] k[m] u[i,j],{m,1,4},{i,1,3},{j,1,3}]+
        Sum[e[a,m] e[b,m] kp[4] k[4] u[4,4],{m,1,4}]+
        Sum[e[a,m] e[b,m] kp[4] k[i] u[4,i],{m,1,4},{i,1,3}]+
        Sum[e[a,m] e[b,m] kp[i] k[4] u[i,4],{m,1,4},{i,1,3}]+
        Sum[e[a,m] e[b,m] kp[i] k[j] u[i,j],{m,1,4},{i,1,3},{j,1,3}]+
```

```
    k0^2 (1-v^2) e[a,4] e[b,4] y[4,4]+
    k0^2 (1-v^2) Sum[e[a,4] e[b,i] y[4,i],{i,1,3}]+
    k0^2 (1-v^2) Sum[e[a,i] e[b,4] y[i,4],{i,1,3}]+
    k0^2 (1-v^2) Sum[e[a,i] e[b,j] y[i,j],{i,1,3},{j,1,3}])*

(Sum[e[a,4] e[b,4] kp[m] k[m] u[4,4],{m,1,4}]-
    Sum[e[a,4] e[b,i] kp[m] k[m] u[4,i],{m,1,4},{i,1,3}]-
    Sum[e[a,i] e[b,4] kp[m] k[m] u[i,4],{m,1,4},{i,1,3}]+
    Sum[e[a,i] e[b,j] kp[m] k[m] u[i,j],{m,1,4},{i,1,3},{j,1,3}]-
    Sum[e[a,4] e[b,m] kp[m] k[4] u[4,4],{m,1,4}]+
    Sum[e[a,4] e[b,m] kp[m] k[i] u[4,i],{m,1,4},{i,1,3}]+
    Sum[e[a,i] e[b,m] kp[m] k[4] u[i,4],{m,1,4},{i,1,3}]-
    Sum[e[a,i] e[b,m] kp[m] k[j] u[i,j],{m,1,4},{i,1,3},{j,1,3}]-
    Sum[e[a,m] e[b,4] kp[4] k[m] u[4,4],{m,1,4}]+
    Sum[e[a,m] e[b,i] kp[4] k[m] u[4,i],{m,1,4},{i,1,3}]+
    Sum[e[a,m] e[b,4] kp[i] k[m] u[i,4],{m,1,4},{i,1,3}]-
    Sum[e[a,m] e[b,j] kp[i] k[m] u[i,j],{m,1,4},{i,1,3},{j,1,3}]+
    Sum[e[a,m] e[b,m] kp[4] k[4] u[4,4],{m,1,4}]-
    Sum[e[a,m] e[b,m] kp[4] k[i] u[4,i],{m,1,4},{i,1,3}]-
    Sum[e[a,m] e[b,m] kp[i] k[4] u[i,4],{m,1,4},{i,1,3}]+
    Sum[e[a,m] e[b,m] kp[i] k[j] u[i,j],{m,1,4},{i,1,3},{j,1,3}]+
    k0^2 (1-v^2) e[a,4] e[b,4] y[4,4]-
    k0^2 (1-v^2) Sum[e[a,4] e[b,i] y[4,i],{i,1,3}]-
    k0^2 (1-v^2) Sum[e[a,i] e[b,4] y[i,4],{i,1,3}]+
    k0^2 (1-v^2) Sum[e[a,i] e[b,j] y[i,j],{i,1,3},{j,1,3}])
```

```
]];
suma1=Factor[k0^4/4/(1-v^2) Factor[suma]];
suma2=4/k0^4 suma1;
suma3=Expand[suma2/.{Sin[alfa]^2->1-Cos[alfa]^2,
    Sin[th]^2->1-Cos[th]^2,Sin[th]^3->Sin[th] (1-Cos[th]^2),
    Sin[th]^4->(1-Cos[th]^2)^2,
    Sin[fi]^2->1-Cos[fi]^2,Sin[fi]^3->Sin[fi] (1-Cos[fi]^2),
    Sin[fi]^4->(1-Cos[fi]^2)^2}];

suma3/.{aa->0}
    3 - 2*v^2 + 5*v^4 + 8*v^2*Cos[th] - 4*v^4*Cos[th] +
2*v^4*Cos[th]^2
suma4=Expand[suma3-(suma3/.{aa->0})];

x2=Factor[Coefficient[suma4,Cos[fi]]]
    4*aa^2*k0^2*v^4*Cos[alfa]*(-1 + Cos[th])^2*(2 - v^2 +
    v^2*Cos[th])*Sin[alfa]*Sin[th]

x3=Factor[Coefficient[suma4,Cos[fi]^2]]
2*aa^2*k0^2*v^4*(-1 + Cos[alfa])*(1 + Cos[alfa])*(-1 + Cos[th])*(1 +
Cos[th])*(-8 + 3*v^2 + 2*Cos[th] - 4*v^2*Cos[th] +v^2*Cos[th]^2)

x1=Factor[Expand[suma4-Cos[fi] x2-Cos[fi]^2 x3]]
2*aa^2*k0^2*v^4*(-1 + Cos[th])*(-10 + 4*v^2 + 8*Cos[alfa]^2 -
3*v^2*Cos[alfa]^2 - 2*Cos[th] - 4*v^2*Cos[th] +2*Cos[alfa]^2*Cos[th]
```

```
+3*v^2*Cos[alfa]^2*Cos[th] +2*Cos[alfa]^2*Cos[th]^2 -
v^2*Cos[alfa]^2*Cos[th]^2 +v^2*Cos[alfa]^2*Cos[th]^3)
```

## IX.2. Program for analytical calculation of the differential cross-section for gravitational scattering of massless vector particles

```
Off[General::spell]
Remove["Global`*"];
Unprotect[In,Out];
Clear[In,Out];

ex={1,0,0};
ey={0,1,0};
eth={Cos[th] Cos[fi],Cos[th] Sin[fi], -Sin[th]};
efi={-Sin[fi],Cos[fi],0};
av={a Sin[alfa],0,a Cos[alfa]};
kv=k0 {0,0,1};
kvp=k0 {Sin[th] Cos[fi],Sin[th] Sin[fi],Cos[th]};
qv=kvp-kv;
eR1=1/Sqrt[2] (ex+I ey);
eL1=1/Sqrt[2] (ex-I ey);
eR2=1/Sqrt[2] (eth-I efi);
eL2=1/Sqrt[2] (eth+I efi);
eps[l_,m_,n_]:=Signature[{l,m,n}]
```

```
S[e1_,e2_]: =1/2((k0^2+kv.kvp)(e1.e2)+1/2(qv.e2)((kv+kvp).e1)-
        1/2(qv.e1)((kv+kvp).e2))
Q[e1_,e2_]: =1/2 k0 Sum[eps[i,j,k] av[[j]] qv[[k]] (e2[[i]] (qv.e1)-
        e1[[i]] (qv.e2)-(kv[[i]]+kvp[[i]]) (e1.e2)),{i,1,3},{j,1,3},{k,1,3}]
F[e1_,e2_]: =S[e1,e2]+I Q[e1,e2]

suma1=Factor[F[eR1,eR2] 8/k0^2] /.{I->0,2I->0,4I->0,-I->0,-2I->0,
        -4I->0};
suma2=Factor[Expand[-I (F[eR1,eR2] 8/k0^2-suma1)]];
A1=Factor[Expand[suma1^2+suma2^2]];
suma1=Factor[F[eL1,eL2] 8/k0^2] /.{I->0,2I->0,4I->0,-I->0,-2I->0,
        -4I->0};
suma2=Factor[Expand[-I (F[eL1,eL2] 8/k0^2-suma1)]];
A2=Factor[Expand[suma1^2+suma2^2]];

suma1=Factor[F[eR1,eL2] 8/k0^2] /.{I->0,2I->0,4I->0,-I->0,-2I->0,
        -4I->0};
suma2=Factor[Expand[-I (F[eR1,eL2] 8/k0^2-suma1)]];
A3=Factor[Expand[suma1^2+suma2^2]];
suma1=Factor[F[eL1,eR2] 8/k0^2] /.{I->0,2I->0,4I->0,-I->0,-2I->0,
        -4I->0};
suma2=Factor[Expand[-I (F[eL1,eR2] 8/k0^2-suma1)]];
A4=Factor[Expand[suma1^2+suma2^2]];

Factor[Expand[A1/.{Cos[fi]^2->1-Sin[fi]^2,
```

```
Cos[fi]^4->(1-Sin[fi]^2)^2,Cos[fi]^3->Cos[fi] (1-Sin[fi]^2),
Cos[fi]^5->Cos[fi] (1-Sin[fi]^2)^2,Cos[fi]^6->(1-Sin[fi]^2)^3,
Cos[th]^2->1-Sin[th]^2,
Cos[th]^4->(1-Sin[th]^2)^2,Cos[th]^3->Cos[th] (1-Sin[th]^2),
Cos[th]^5->Cos[th] (1-Sin[th]^2)^2,Cos[th]^6->(1-Sin[th]^2)^3,
Cos[alfa]^2->1-Sin[alfa]^2}]]

16*(2 + 2*Cos[th] - Sin[th]^2 + 2*a*k0*Cos[alfa]*Sin[th]^2 +
2*a*k0*Cos[alfa]*Cos[th]*Sin[th]^2 +2*a^2*k0^2*Sin[alfa]^2*Sin[th]^2 -
2*a^2*k0^2*Cos[th]*Sin[alfa]^2*Sin[th]^2+2*a^2*k0^2*Sin[alfa]^2*
Sin[fi]^2*Sin[th]^2+2*a^2*k0^2*Cos[th]*Sin[alfa]^2*Sin[fi]^2*Sin[th]^2+
2*a*k0*Cos[fi]*Sin[alfa]*Sin[th]^3+2*a^2*k0^2*Cos[alfa]*Cos[fi]*
Sin[alfa]*Sin[th]^3 - 2*a^2*k0^2*Cos[alfa]*Cos[fi]*Cos[th]*Sin[alfa]*
Sin[th]^3 + a^2*k0^2*Sin[th]^4 - 2*a^2*k0^2*Sin[alfa]^2*Sin[th]^4
+a^2*k0^2*Sin[alfa]^2*Sin[fi]^2*Sin[th]^4)

Factor[Expand[A2/.{Cos[fi]^2->1-Sin[fi]^2,
Cos[fi]^4->(1-Sin[fi]^2)^2,Cos[fi]^3->Cos[fi] (1-Sin[fi]^2),
Cos[fi]^5->Cos[fi] (1-Sin[fi]^2)^2,Cos[fi]^6->(1-Sin[fi]^2)^3,
Cos[th]^2->1-Sin[th]^2,
Cos[th]^4->(1-Sin[th]^2)^2,Cos[th]^3->Cos[th] (1-Sin[th]^2),
Cos[th]^5->Cos[th] (1-Sin[th]^2)^2,Cos[th]^6->(1-Sin[th]^2)^3,
Cos[alfa]^2->1-Sin[alfa]^2}]]

16*(2 + 2*Cos[th] - Sin[th]^2 - 2*a*k0*Cos[alfa]*Sin[th]^2 -
```

```
2*a*k0*Cos[alfa]*Cos[th]*Sin[th]^2 + 2*a^2*k0^2*Sin[alfa]^2*Sin[th]^2 -
2*a^2*k0^2*Cos[th]*Sin[alfa]^2*Sin[th]^2+2*a^2*k0^2*Sin[alfa]^2*
Sin[fi]^2*Sin[th]^2+2*a^2*k0^2*Cos[th]*Sin[alfa]^2*Sin[fi]^2*Sin[th]^2-
2*a*k0*Cos[fi]*Sin[alfa]*Sin[th]^3+2*a^2*k0^2*Cos[alfa]*Cos[fi]*
Sin[alfa]*Sin[th]^3 - 2*a^2*k0^2*Cos[alfa]*Cos[fi]*Cos[th]*Sin[alfa]*
Sin[th]^3 + a^2*k0^2*Sin[th]^4 - 2*a^2*k0^2*Sin[alfa]^2*Sin[th]^4 +
a^2*k0^2*Sin[alfa]^2*Sin[fi]^2*Sin[th]^4)
```

```
Factor[Expand[((A1+A2)/2)/.{Cos[fi]^2->1-Sin[fi]^2,
Cos[fi]^4->(1-Sin[fi]^2)^2,Cos[fi]^3->Cos[fi] (1-Sin[fi]^2),
Cos[fi]^5->Cos[fi] (1-Sin[fi]^2)^2,Cos[fi]^6->(1-Sin[fi]^2)^3,
Cos[th]^2->1-Sin[th]^2,Cos[th]^4->(1-Sin[th]^2)^2,
Cos[th]^3->Cos[th] (1-Sin[th]^2),Cos[th]^5->Cos[th] (1-Sin[th]^2)^2,
Cos[th]^6->(1-Sin[th]^2)^3,Cos[alfa]^2->1-Sin[alfa]^2}]]
```

```
16*(2 + 2*Cos[th] - Sin[th]^2 + 2*a^2*k0^2*Sin[alfa]^2*Sin[th]^2 -
2*a^2*k0^2*Cos[th]*Sin[alfa]^2*Sin[th]^2 + 2*a^2*k0^2*Sin[alfa]^2*
Sin[fi]^2*Sin[th]^2+2*a^2*k0^2*Cos[th]*Sin[alfa]^2*Sin[fi]^2*
Sin[th]^2+2*a^2*k0^2*Cos[alfa]*Cos[fi]*Sin[alfa]*Sin[th]^3 -
2*a^2*k0^2*Cos[alfa]*Cos[fi]*Cos[th]*Sin[alfa]*Sin[th]^3 +
a^2*k0^2*Sin[th]^4 - 2*a^2*k0^2*Sin[alfa]^2*Sin[th]^4 +
a^2*k0^2*Sin[alfa]^2*Sin[fi]^2*Sin[th]^4)
```

## IX.3. Program for analytical calculation of the differential cross-section for gravitational scattering of massive spin-vector particles

```
Off[General::spell]
Remove["Global`*"];
Unprotect[In,Out];
Clear[In,Out];

a[1]=aa Sin[alfa];
a[2]=0;
a[3]=aa Cos[alfa];

k[1]=0;
k[2]=0;
k[3]=v k0;
k[4]=I k0;
kp[1]=v k0 Sin[th] Cos[fi];
kp[2]=v k0 Sin[th] Sin[fi];
kp[3]=v k0 Cos[th];
kp[4]=I k0;

q[1]=kp[1]-k[1];
q[2]=kp[2]-k[2];
q[3]=kp[3]-k[3];
q[4]=kp[4]-k[4];
```

```
eps[i_Integer,j_Integer,k_Integer]: =Signature[{i,j,k}];
eps4[i_Integer,j_Integer,k_Integer,l_Integer]: =Signature[{i,j,k,l}];
gamma[1]={{0, 0, 0, -I}, {0, 0, -I, 0}, {0, I, 0, 0}, {I, 0, 0, 0}};
gamma[2]={{0, 0, 0, -1}, {0, 0, 1, 0}, {0, 1, 0, 0}, {-1, 0, 0, 0}};
gamma[3]={{0, 0, -I, 0}, {0, 0, 0, I}, {I, 0, 0, 0}, {0, -I, 0, 0}};
gamma[4]={{1, 0, 0, 0}, {0, 1, 0, 0}, {0, 0, -1, 0}, {0, 0, 0, -1}};
gamma[5]=gamma[1].gamma[2].gamma[3].gamma[4];

hermitic[x_]: =ComplexExpand[Conjugate[Transpose[x]]]
trace[x_]: =Sum[x[[i,i]],{i,1,4}]

gammapp=Sum[gamma[j] kp[j],{j,1,4}]+I m IdentityMatrix[4];
gammap=Sum[gamma[j] k[j],{j,1,4}]+I m IdentityMatrix[4];

kr[1,1]=kr[2,2]=kr[3,3]=kr[4,4]=1;
kr[1,2]=kr[1,3]=kr[2,3]=kr[2,1]=kr[3,1]=kr[3,2]=0;
kr[1,4]=kr[4,1]=kr[2,4]=kr[4,2]=kr[3,4]=kr[4,3]=0;
Stx[al_,be_]:=Expand[
Sum[eps4[al,be,mi,4] gamma[5].((kp[mi]+k[mi]) (gamma[4]+
1/2 Sum[eps[i,j,l] a[i] (kp[j]-k[j]) gamma[l],{i,1,3},{j,1,3},{l,1,3}])+I k0
gamma[mi]), {mi,1,4}]+1/2 Sum[eps4[al,be,mi,i] gamma[5].((kp[i]+k[i])
gamma[mi]+Sum[eps[i,j,l] (kp[mi]+k[mi]) a[j] (kp[l]-k[l])
gamma[4],{i,1,3},{j,1,3},{l,1,3}]), {mi,1,4},{i,1,3}]];

Do[St[al,be]=Stx[al,be];Print[al,be],{al,1,4},{be,1,4}]
```

```
reg={Sin[alfa]^2->1-Cos[alfa]^2,
     Sin[th]^2->1-Cos[th]^2,Sin[th]^3->Sin[th] (1-Cos[th]^2),
     Sin[th]^4->(1-Cos[th]^2)^2,Sin[th]^5->Sin[th] (1-Cos[th]^2)^2,
     Sin[th]^6->(1-Cos[th]^2)^3,
     Sin[fi]^2->1-Cos[fi]^2,Sin[fi]^3->Sin[fi] (1-Cos[fi]^2),
     Sin[fi]^4->(1-Cos[fi]^2)^2,Sin[fi]^5->Sin[fi] (1-Cos[fi]^2)^2,
     Sin[fi]^6->(1-Cos[fi]^2)^3};
S=0;
Do[
S=S+Expand[Expand[trace[gamma[4].hermitic[St[al,be]].gamma[4].
gammapp.
   (kr [al, ga]
IdentityMatrix[4]-1/3gamma[al].gamma[ga]+I/(3m)(kp[ga]gamma[al]-
kp[al] gamma[ga]) +2/(3m^2) kp[al] kp[ga]
IdentityMatrix[4]).St[ga,la].gammap.(kr[la, be] IdentityMatrix[4]-1/3
gamma[la].gamma[be]+I/(3m)(k[be] gamma[la]-k[la] gamma[be])
+2/(3m^2) k[la] k[be] IdentityMatrix[4])]]/.reg
        ];

Print[al,be,ga,la," ",ByteCount[S]," ",MemoryInUse[],"
",MaxMemoryUsed[]];,
        {al,1,4},{be,1,4},{ga,1,4},{la,1,4}]

sumaA=Factor[S/.{aa->0}];
sumaB=Factor[Factor[Expand[S-sumaA]]/.{m^2->k0^2(1-v^2),
```

```
        m^4->k0^4(1-v^2)^2,
        m^6->(k0^2(1-v^2))^3,m^8->(k0^2(1-v^2))^4}];
sumaC=Factor[sumaB 9m^2/(4aa k0^7 v^4 (1+Cos[th]))];

x12=Factor[Coefficient[sumaC,Cos[fi]^2]/aa]

((-1 + Cos[th])*(-72*aa*k0 + 30*aa*k0*v^2 + 8*I*Cos[alfa] -
6*I*v^2*Cos[alfa] + 64*aa*k0*Cos[alfa]^2 - 14*aa*k0*v^2*Cos[alfa]^2 –
40*aa*k0*Cos[th] + 56*aa*k0*v^2*Cos[th] + 6*aa*k0*v^4*Cos[th] +
8*I*Cos[alfa]*Cos[th] - 8*I*v^2*Cos[alfa]*Cos[th] + 40*aa*k0*
Cos[alfa]^2*Cos[th] - 40*aa*k0*v^2*Cos[alfa]^2*Cos[th] –
6*aa*k0*v^4*Cos[alfa]^2*Cos[th] + 8*aa*k0*Cos[th]^2 + 10*aa*k0*
v^2*Cos[th]^2 + 12*aa*k0*v^4*Cos[th]^2 - 2*I*v^2*Cos[alfa]*
Cos[th]^2 - 26*aa*k0*v^2*Cos[alfa]^2*Cos[th]^2 -
12*aa*k0*v^4*Cos[alfa]^2*Cos[th]^2 - 16*aa*k0*v^2*Cos[th]^3 +
6*aa*k0*v^4*Cos[th]^3 - 6*aa*k0*v^4*Cos[alfa]^2*Cos[th]^3 -
4*I*Sin[alfa] + 3*I*v^2*Sin[alfa] + 100*aa*k0*Cos[alfa]*Sin[alfa] -
56*aa*k0*v^2*Cos[alfa]*Sin[alfa] +3*aa*k0*v^4*Cos[alfa]*Sin[alfa] -
4*I*Cos[th]*Sin[alfa] + 4*I*v^2*Cos[th]*Sin[alfa] +
76*aa*k0*Cos[alfa]*Cos[th]*Sin[alfa] - 100*aa*k0*v^2*Cos[alfa]*
Cos[th]*Sin[alfa] - 9*aa*k0*v^4*Cos[alfa]*Cos[th]*Sin[alfa] +
I*v^2*Cos[th]^2*Sin[alfa] - 24*aa*k0*Cos[alfa]*Cos[th]^2
*Sin[alfa] + 4*aa*k0*v^2*Cos[alfa]*Cos[th]^2*Sin[alfa] -
27*aa*k0*v^4*Cos[alfa]*Cos[th]^2*Sin[alfa] + 48*aa*k0*v^2*
Cos[alfa]*Cos[th]^3*Sin[alfa] - 5*aa*k0*v^4*Cos[alfa]*Cos[th]^3*
```

```
Sin[alfa] - 12*aa*k0*Sin[fi]*Sin[th] + 6*aa*k0*v^2*Sin[fi]*Sin[th] –
3*aa*k0*v^4*Sin[fi]*Sin[th] + 4*aa*k0*Cos[alfa]^2*Sin[fi]*Sin[th] -
2*aa*k0*v^2*Cos[alfa]^2*Sin[fi]*Sin[th] - 3*aa*k0*v^4*Cos[alfa]^2*
Sin[fi]*Sin[th] - 4*aa*k0*Cos[th]*Sin[fi]*Sin[th] + 14*aa*k0*v^2*Cos[th]*
Sin[fi]*Sin[th] - 6*aa*k0*v^4*Cos[th]*Sin[fi]*Sin[th] - 4*aa*k0*
Cos[alfa]^2*Cos[th]*Sin[fi]*Sin[th] + 6*aa*k0*v^2*Cos[alfa]^2*
Cos[th]*Sin[fi]*Sin[th] - 6*aa*k0*v^4*Cos[alfa]^2*Cos[th]*Sin[fi]*
Sin[th]+8*aa*k0*v^2*Cos[th]^2*Sin[fi]*Sin[th]- 3*aa*k0*v^4*Cos[th]^2*
Sin[fi]*Sin[th] + 8*aa*k0*v^2*Cos[alfa]^2*Cos[th]^2*
Sin[fi]*Sin[th] - 3*aa*k0*v^4*Cos[alfa]^2*Cos[th]^2*Sin[fi]*Sin[th] +
40*aa*k0*Cos[alfa]*Sin[alfa]*Sin[fi]*Sin[th] - 32*aa*k0*v^2*Cos[alfa]*
Sin[alfa]*Sin[fi]*Sin[th] + 12*aa*k0*v^4*Cos[alfa]*Sin[alfa]*
Sin[fi]*Sin[th] + 8*aa*k0*Cos[alfa]*Cos[th]*Sin[alfa]*Sin[fi]*Sin[th] -
48*aa*k0*v^2*Cos[alfa]*Cos[th]*Sin[alfa]*Sin[fi]*Sin[th] +
24*aa*k0*v^4*Cos[alfa]*Cos[th]*Sin[alfa]*Sin[fi]*Sin[th] –
16*aa*k0*v^2*Cos[alfa]*Cos[th]^2*Sin[alfa]*Sin[fi]*Sin[th] +
12*aa*k0*v^4*Cos[alfa]*Cos[th]^2*Sin[alfa]*Sin[fi]*Sin[th]))/aa

x13=Factor[Coefficient[sumaC,Cos[fi]]]
x11=Factor[Expand[sumaC-x12 Cos[fi]^2-x13 Cos[fi]]]
Length[S[[1]]]
2
Factor[sumaA/.{m^2->k0^2(1-v^2),m^4->k0^4(1-v^2)^2,
m^6->(k0^2(1-v^2))^3,m^8->(k0^2(1-v^2))^4}]
```

```
(8*k0^6*(2 + v^2 + v^2*Cos[th])*(-9 + 9*v^2 - 4*v^4 + 3*v^2*Cos[th]
+ v^4*Cos[th]))/(9*m^2)
```

*

* *

The above analytical computing program (and that in the following subchapter as well) was subsequently improved, according to the following "nucleus":

```
(* Mathematica *)
(* Variant no. 3 *)

Unprotect[{p,pp,lst,gamma,gammap,gammapp,qq,gammaqq,kr,eps,regula,
hermitic}];
Unprotect[{trace,tr1}];

trace[x_List]:=Sum[x[[i,i]],{i,1,Length[x]}];
tr1[matrix,elem_List]:=Block[{expression,indic,cond,lsti,lstj,lungimei,lungi
mej,jx,kk},
        expression=0;
        cond={};

        lsti=Permutations[Complement[{1,2,3,4},elem[[1]] ]];
        lungimei=Length[lsti];

        lstj=Permutations[Complement[{1,2,3,4},elem[[2]] ]];
```

```
        lungimej=Length[lstj];

        Do[ If[ NumberQ[elem[[i,j]]],,
                cond=Join[cond,{elem[[i,j]]->indic[i,j]}] ],{i,1,2},{j,1,4} ];

        Do[(Do[ ( kk=0;
                 Do[ If[ NumberQ[elem[[1,jx]]],, (kk=kk+1;
indic[1,jx]=lsti[[i,kk]];) ] ,
                        {jx,1,4}]; kk=0;
                        Do[ If[ NumberQ[elem[[2,jx]]],, (kk=kk+1;
indic[2,jx]=lstj[[j,kk]];) ] ,{jx,1,4}];
                expression=Expand[expression+trace[matrix/.cond] ] ),
                {i,1,lungimei}]; Print[lungimej-j-1] ),{j,1,lungimej}];

        Together[expression]/.regula ];

lst=Permutations[{1,2,3,4}];

hermitic[x_List]: =ComplexExpand[Conjugate[Transpose[x]]];

eps[x_Integer,y_Integer,z_Integer,w_Integer]:
=0/;x==y||x==z||x==w||y==z||y==w||z==w
eps[x_Integer,y_Integer,z_Integer,w_Integer]: =Signature[{x,y,z,w}]

kr[x_Integer,y_Integer]: =1/;x==y
```

```
kr[x_Integer,y_Integer]: =0

regula=AlgebraicRules[{p0^2-m^2==p[1]^2+p[2]^2+p[3]^2,\
              p0^2-m^2==pp[1]^2+pp[2]^2+pp[3]^2,\
              (p0^2-m^2) Cos[th] ==p[1] pp[1]+\
              p[2] pp[2]+p[3] pp[3]},\
              {p[1],pp[1],p[3],pp[3]}];
gamma[1]={  { 0, 0, 0,-I}, { 0, 0,-I, 0}, { 0, I, 0, 0}, { I, 0, 0, 0}    };
gamma[2]={  { 0, 0, 0,-1}, { 0, 0, 1, 0}, { 0, 1, 0, 0}, {-1, 0, 0, 0}    };
gamma[3]={  { 0, 0,-I, 0}, { 0, 0, 0, I}, { I, 0, 0, 0}, { 0,-I, 0, 0}    };
gamma[4]={  { 1, 0, 0, 0}, { 0, 1, 0, 0}, { 0, 0,-1, 0}, { 0, 0, 0,-1}    };
gamma[5]=gamma[1].gamma[2].gamma[3].gamma[4];

p[4]=I p0;
pp[4]=p[4];

gammap=Sum[p[i] gamma[i],{i,1,4}];
gammapp=Sum[pp[i] gamma[i],{i,1,4}];
qq[x_]:=pp[x]-p[x];
gammaqq=Sum[qq[i] gamma[i],{i,1,4}];

Protect[{p,pp,lst,gamma,gammap,gammapp,qq,gammaqq,kr,eps,regula,her
mitic}];

Protect[{trace,tr1}];
```

## IX.4. Program for analytical calculation of the differential cross-section for gravitational scattering of massless spin-vector particles

```
Off[General::spell]

Remove["Global`*"];

Unprotect[In,Out];

Clear[In,Out];

a[1]=aa Sin[alfa];

a[2]=0;

a[3]=aa Cos[alfa];

k[1]=0;

k[2]=0;

k[3]=k0;

k[4]=I k0;

kp[1]=k0 Sin[th] Cos[fi];

kp[2]=k0 Sin[th] Sin[fi];

kp[3]=k0 Cos[th];

kp[4]=I k0;


q[1]=kp[1]-k[1];

q[2]–kp[2]-k[2];

q[3]=kp[3]-k[3];

q[4]=kp[4]-k[4];


eps[i_Integer,j_Integer,k_Integer]:=Signature[{i,j,k}];
```

```
eps4[i_Integer,j_Integer,k_Integer,l_Integer]:=Signature[{i,j,k,l}];

gamma[1]={{0, 0, 0, -I}, {0, 0, -I, 0}, {0, I, 0, 0}, {I, 0, 0, 0}};
gamma[2]={{0, 0, 0, -1}, {0, 0, 1, 0}, {0, 1, 0, 0}, {-1, 0, 0, 0}};
gamma[3]={{0, 0, -I, 0}, {0, 0, 0, I}, {I, 0, 0, 0}, {0, -I, 0, 0}};
gamma[4]={{1, 0, 0, 0}, {0, 1, 0, 0}, {0, 0, -1, 0}, {0, 0, 0, -1}};
gamma[5]=gamma[1].gamma[2].gamma[3].gamma[4];

hermitic[x_]:=ComplexExpand[Conjugate[Transpose[x]]]
trace[x_]: =Sum[x[[i,i]],{i,1,4}]
gammap=Sum[gamma[j] k[j],{j,1,4}];
gammapp=Sum[gamma[j] kp[j],{j,1,4}];

Stx[al_,be_]:=Expand[Sum[
        eps4[al,be,mi,4] (k[mi]+kp[mi]) gamma[5].(gamma[4]+1/2
Sum[eps[i,j,l] a[i]
        (kp[j]-k[j])gamma[l],{i,1,3},{j,1,3},{l,1,3}]),{mi,1,4}]+1/2
Sum[eps4[al,be,mi,i]
        eps[i,j,l](kp[mi]+k[mi])a[j](kp[l]-
k[l])gamma[5].gamma[4],{mi,1,4},{i,1,3},
{j,1,3}, {l,1,3}] ]
Do[St[al,be]=Stx[al,be];Print[al,be],{al,1,4},{be,1,4}]
11
12
13
```

14

21

22

23

24

31

32

33

34

41

42

43

44

```
reg={Sin[alfa]^2->1-Cos[alfa]^2,
     Sin[th]^2->1-Cos[th]^2,Sin[th]^3->Sin[th] (1-Cos[th]^2),
     Sin[th]^4->(1-Cos[th]^2)^2,Sin[th]^5->Sin[th](1-Cos[th]^2)^2,
     Sin[th]^6->(1-Cos[th]^2)^3,Sin[fi]^2->1-Cos[fi]^2,
     Sin[fi]^3->Sin[fi] (1-Cos[fi]^2),
     Sin[fi]^4->(1-Cos[fi]^2)^2,Sin[fi]^5->Sin[fi] (1-Cos[fi]^2)^2,
     Sin[fi]^6->(1-Cos[fi]^2)^3};

S=Factor[Expand[Expand[Sum[trace[gamma[4].hermitic[St[al,be]].gamma
[4].gamma[ga].gammapp.gamma[al].St[ga,la].gamma[be].gammap.gamma[l
a]],
{al,1,4},{be,1,4},{ga,1,4},{la,1,4}]]/.reg]];
```

```
sumaA=Factor[S/.{aa->0}]
(-64*k0^4*(1 + Cos[th]))

sumaB=Factor[S-sumaA]

16*aa^2*k0^6*(-1 + Cos[th])*(3 + Cos[th])*(3 - 2*Cos[alfa]^2 -
2*Cos[fi]^2 + 2*Cos[alfa]^2*Cos[fi]^2 + Cos[th] –
2*Cos[fi]^2*Cos[th] + 2*Cos[alfa]^2*Cos[fi]^2*
Cos[th] + 2*Cos[alfa]*Cos[fi]*Sin[alfa]*Sin[th])
```

## IX.5. Program for analytical calculation of the differential cross-section for gravitational scattering of massive tensor particles

```
Off[General::spell]
Remove["Global`*"];
Unprotect[In,Out];
Clear[In,Out];

a[1]=aa Sin[alfa];
a[2]=0;
a[3]=aa Cos[alfa];

k[1]=0;
k[2]=0;
k[3]=v k0;
```

```
k[4]=I k0;

kp[1]=v k0 Sin[th] Cos[fi];
kp[2]=v k0 Sin[th] Sin[fi];
kp[3]=v k0 Cos[th];
kp[4]=I k0;

q[1]=kp[1]-k[1];
q[2]=kp[2]-k[2];
q[3]=kp[3]-k[3];
q[4]=kp[4]-k[4];

eps[i_,j_,k_]:=Signature[{i,j,k}];
kr[1,1]=kr[2,2]=kr[3,3]=kr[4,4]=1;
kr[1,2]=kr[1,3]=kr[2,3]=kr[2,1]=kr[3,1]=kr[3,2]=0;
kr[1,4]=kr[4,1]=kr[2,4]=kr[4,2]=kr[3,4]=kr[4,3]=0;

dp[m_,n_]: =kr[m,n]+kp[m] kp[n]/(k0^2 (1-v^2))
d[m_,n_]: =kr[m,n]+k[m] k[n]/(k0^2 (1-v^2))
e[s,m_,n_] e[s,l_,ro_] ^: =1/2(dp[m,l] dp[n,ro]+dp[m,ro] dp[n,l])-1/3
dp[m,n] dp[l,ro]
e[r,m_,n_] e[r,l_,ro_] ^: =1/2(d[m,l] d[n,ro]+d[m,ro] d[n,l])-1/3 d[m,n]
d[l,ro]
e[s,1,1]^x_ ^: =e[s,1,1] e[s,1,1]
e[s,1,2]^x_ ^: =e[s,1,2] e[s,1,2]
```

```
e[s,1,3]^x_ ^: =e[s,1,3] e[s,1,3]
e[s,1,4]^x_ ^: =e[s,1,4] e[s,1,4]
e[s,2,1]^x_ ^: =e[s,2,1] e[s,2,1]
e[s,2,2]^x_ ^: =e[s,2,2] e[s,2,2]
e[s,2,3]^x_ ^: =e[s,2,3] e[s,2,3]
e[s,2,4]^x_ ^: =e[s,2,4] e[s,2,4]
e[s,3,1]^x_ ^: =e[s,3,1] e[s,3,1]
e[s,3,2]^x_ ^: =e[s,3,2] e[s,3,2]
e[s,3,3]^x_ ^: =e[s,3,3] e[s,3,3]
e[s,3,4]^x_ ^: =e[s,3,4] e[s,3,4]
e[s,4,1]^x_ ^: =e[s,4,1] e[s,4,1]
e[s,4,2]^x_ ^: =e[s,4,2] e[s,4,2]
e[s,4,3]^x_ ^: =e[s,4,3] e[s,4,3]
e[s,4,4]^x_ ^: =e[s,4,4] e[s,4,4]

e[r,1,1]^x_ ^: =e[r,1,1] e[r,1,1]
e[r,1,2]^x_ ^: =e[r,1,2] e[r,1,2]
e[r,1,3]^x_ ^: =e[r,1,3] e[r,1,3]
e[r,1,4]^x_ ^: =e[r,1,4] e[r,1,4]
e[r,2,1]^x_ ^: =e[r,2,1] e[r,2,1]
e[r,2,2]^x_ ^: =e[r,2,2] e[r,2,2]
e[r,2,3]^x_ ^: =e[r,2,3] e[r,2,3]
e[r,2,4]^x_ ^: =e[r,2,4] e[r,2,4]
e[r,3,1]^x_ ^: =e[r,3,1] e[r,3,1]
e[r,3,2]^x_ ^: =e[r,3,2] e[r,3,2]
```

```
e[r,3,3]^x_ ^: =e[r,3,3] e[r,3,3]
e[r,3,4]^x_ ^: =e[r,3,4] e[r,3,4]
e[r,4,1]^x_ ^: =e[r,4,1] e[r,4,1]
e[r,4,2]^x_ ^: =e[r,4,2] e[r,4,2]
e[r,4,3]^x_ ^: =e[r,4,3] e[r,4,3]
e[r,4,4]^x_ ^: =e[r,4,4] e[r,4,4]

i1=Factor[Expand[I/2 k0 Sum[eps[j,l,m] (k[j]+kp[j]) a[l]
                        q[m],{j,1,3},{l,1,3},{m,1,3}]]];
i2=Factor[Expand[(k[1] q[1]+k[2] q[2]+k[3] q[3])
        Sum[eps[1,l,m] a[l] q[m],{l,1,3},{m,1,3}]]];
i3=Factor[Expand[(k[1] q[1]+k[2] q[2]+k[3] q[3])
        Sum[eps[2,l,m] a[l] q[m],{l,1,3},{m,1,3}]]];
i4=Factor[Expand[(k[1] q[1]+k[2] q[2]+k[3] q[3])
        Sum[eps[3,l,m] a[l] q[m],{l,1,3},{m,1,3}]]];
i5=Factor[Expand[2(k[1] q[1]+k[2] q[2]+k[3] q[3]) ]];
i6=Factor[Expand[1/2 k0^2 (1-v^2) +k[1] kp[1]+k[2] kp[2]+k[3] kp[3]]]];
i7=Factor[Expand[1/2(q[1]^2+q[2]^2+q[3]^2)]/.{Sin[alfa]^2->1-
Cos[alfa]^2, Sin[th]^2->1-Cos[th]^2,Sin[th]^3->Sin[th] (1-Cos[th]^2),
        Sin[th]^4->(1-Cos[th]^2)^2, Sin[th]^5->Sin[th] (1-Cos[th]^2)^2,
        Sin[th]^6->(1-Cos[th]^2)^3, Sin[fi]^2->1-Cos[fi]^2,
        Sin[fi]^3->Sin[fi] (1-Cos[fi]^2), Sin[fi]^4->(1-Cos[fi]^2)^2,
        Sin[fi]^5->Sin[fi] (1-Cos[fi]^2)^2,Sin[fi]^6->(1-Cos[fi]^2)^3}];
i8=Factor[Expand[I k0]];
i9=+1/2;
```

```
i10=Factor[Expand[i7 Sum[eps[1,l,m] a[l] q[m],{l,1,3},{m,1,3} ]]];
i11=Factor[Expand[i7 Sum[eps[2,l,m] a[l] q[m],{l,1,3},{m,1,3} ]]];
i12=Factor[Expand[i7 Sum[eps[3,l,m] a[l] q[m],{l,1,3},{m,1,3} ]]];
i13=Factor[Expand[1/2 Sum[eps[j,l,m] (k[j]+kp[j]) a[l]
                q[m],{j,1,3},{l,1,3},{m,1,3}]]];
i14=Factor[Expand[I k0 Sum[eps[1,l,m] a[l] q[m],{l,1,3},{m,1,3}]]];
i15=Factor[Expand[I k0 Sum[eps[2,l,m] a[l] q[m],{l,1,3},{m,1,3}]]];
i16=Factor[Expand[I k0 Sum[eps[3,l,m] a[l] q[m],{l,1,3},{m,1,3}]]];

S1=Sum[e[s,m,n] e[r,m,n],{m,1,4},{n,1,4}];
S2=Sum[(e[s,4,n] e[r,1,n]+e[s,1,n] e[r,4,n]),{n,1,4}];
S3=Sum[(e[s,4,n] e[r,2,n]+e[s,2,n] e[r,4,n]),{n,1,4}];
S4=Sum[(e[s,4,n] e[r,3,n]+e[s,3,n] e[r,4,n]),{n,1,4}];
S5=Sum[e[s,4,n] e[r,4,n],{n,1,4}];
S7=Sum[e[s,4,n] e[r,4,n],{n,1,4}]-Sum[e[s,j,n] e[r,j,n],{j,1,3},{n,1,4}];
S8=Sum[q[m] (e[s,m,n] e[r,4,n]-e[s,4,n] e[r,m,n]),{m,1,4},{n,1,4}];
S9=Sum[q[m] (k[j]+kp[j]) (e[s,m,n] e[r,j,n]-e[s,j,n]
      e[r,m,n]),{j,1,3},{n,1,4},{m,1,4}];
S14=Sum[q[m] (e[s,m,n] e[r,1,n]-e[s,1,n] e[r,m,n]),{n,1,4},{m,1,4}];
S15=Sum[q[m] (e[s,m,n] e[r,2,n]-e[s,2,n] e[r,m,n]),{n,1,4},{m,1,4}];
S16=Sum[q[m] (e[s,m,n] e[r,3,n]-e[s,3,n] e[r,m,n]),{n,1,4},{m,1,4}];

s1s1=Factor[Expand[Factor[Expand[S1*S1]]/.{Sin[alfa]^2->1-
Cos[alfa]^2,
        Sin[th]^2->1-Cos[th]^2, Sin[th]^3->Sin[th] (1-Cos[th]^2),
```

```
        Sin[th]^4->(1-Cos[th]^2)^2, Sin[th]^5->Sin[th](1-Cos[th]^2)^2,
        Sin[th]^6->(1-Cos[th]^2)^3, Sin[fi]^2->1-Cos[fi]^2,
        Sin[fi]^3->Sin[fi] (1-Cos[fi]^2), Sin[fi]^4->(1-Cos[fi]^2)^2,
        Sin[fi]^5->Sin[fi] (1-Cos[fi]^2)^2, Sin[fi]^6->(1-Cos[fi]^2)^3}]] ;

s2s2=Factor[Expand[Factor[Expand[S2*S2]]/.{Sin[alfa]^2->1-
Cos[alfa]^2,
        Sin[th]^2->1-Cos[th]^2, Sin[th]^3->Sin[th] (1-Cos[th]^2),
        Sin[th]^4->(1-Cos[th]^2)^2, Sin[th]^5->Sin[th](1-Cos[th]^2)^2,
        Sin[th]^6->(1-Cos[th]^2)^3, Sin[fi]^2->1-Cos[fi]^2,
        Sin[fi]^3->Sin[fi] (1-Cos[fi]^2), Sin[fi]^4->(1-Cos[fi]^2)^2,
        Sin[fi]^5->Sin[fi] (1-Cos[fi]^2)^2, Sin[fi]^6->(1-Cos[fi]^2)^3}]];

s3s3=Factor[Expand[Factor[Expand[S3*S3]]/.{Sin[alfa]^2->1-
Cos[alfa]^2,
        Sin[th]^2->1-Cos[th]^2, Sin[th]^3->Sin[th] (1-Cos[th]^2),
        Sin[th]^4->(1-Cos[th]^2)^2, Sin[th]^5->Sin[th] (1-Cos[th]^2)^2,
        Sin[th]^6->(1-Cos[th]^2)^3, Sin[fi]^2->1-Cos[fi]^2,
        Sin[fi]^3->Sin[fi] (1-Cos[fi]^2), Sin[fi]^4->(1-Cos[fi]^2)^2,
        Sin[fi]^5->Sin[fi] (1-Cos[fi]^2)^2, Sin[fi]^6->(1-Cos[fi]^2)^3}]];

s4s4=Factor[Expand[Factor[Expand[S4*S4]]/.{Sin[alfa]^2->1-
Cos[alfa]^2,
        Sin[th]^2->1-Cos[th]^2, Sin[th]^3->Sin[th] (1-Cos[th]^2),
        Sin[th]^4->(1-Cos[th]^2)^2, Sin[th]^5->Sin[th] (1-Cos[th]^2)^2,
```

```
        Sin[th]^6->(1-Cos[th]^2)^3, Sin[fi]^2->1-Cos[fi]^2,
        Sin[fi]^3->Sin[fi] (1-Cos[fi]^2), Sin[fi]^4->(1-Cos[fi]^2)^2,
        Sin[fi]^5->Sin[fi] (1-Cos[fi]^2)^2, Sin[fi]^6->(1-Cos[fi]^2)^3}]];

s8s8=Factor[Expand[Factor[Expand[S8*S8]]/.{Sin[alfa]^2->1-
Cos[alfa]^2,
        Sin[th]^2->1-Cos[th]^2, Sin[th]^3->Sin[th] (1-Cos[th]^2),
        Sin[th]^4->(1-Cos[th]^2)^2, Sin[th]^5->Sin[th] (1-Cos[th]^2)^2,
        Sin[th]^6->(1-Cos[th]^2)^3, Sin[fi]^2->1-Cos[fi]^2,
        Sin[fi]^3->Sin[fi] (1-Cos[fi]^2), Sin[fi]^4->(1-Cos[fi]^2)^2,
        Sin[fi]^5->Sin[fi] (1-Cos[fi]^2)^2, Sin[fi]^6->(1-Cos[fi]^2)^3}]];

s14s14=Factor[Expand[Factor[Expand[S14*S14]]/.{Sin[alfa]^2->1-
Cos[alfa]^2,
        Sin[th]^2->1-Cos[th]^2, Sin[th]^3->Sin[th] (1-Cos[th]^2),
        Sin[th]^4->(1-Cos[th]^2)^2, Sin[th]^5->Sin[th] (1-Cos[th]^2)^2,
        Sin[th]^6->(1-Cos[th]^2)^3, Sin[fi]^2->1-Cos[fi]^2,
        Sin[fi]^3->Sin[fi] (1-Cos[fi]^2), Sin[fi]^4->(1-Cos[fi]^2)^2,
        Sin[fi]^5->Sin[fi] (1-Cos[fi]^2)^2, Sin[fi]^6->(1-Cos[fi]^2)^3}]];

s15s15=Factor[Expand[Factor[Expand[S15*S15]]/.{Sin[alfa]^2->1-
Cos[alfa]^2,
        Sin[th]^2->1-Cos[th]^2, Sin[th]^3->Sin[th] (1-Cos[th]^2),
        Sin[th]^4->(1-Cos[th]^2)^2, Sin[th]^5->Sin[th] (1-Cos[th]^2)^2,
        Sin[th]^6->(1-Cos[th]^2)^3, Sin[fi]^2->1-Cos[fi]^2,
```

```
        Sin[fi]^3->Sin[fi] (1-Cos[fi]^2), Sin[fi]^4->(1-Cos[fi]^2)^2,
        Sin[fi]^5->Sin[fi] (1-Cos[fi]^2)^2, Sin[fi]^6->(1-Cos[fi]^2)^3}]];

s16s16=Factor[Expand[Factor[Expand[S16*S16]]/.{Sin[alfa]^2->1-
Cos[alfa]^2,
        Sin[th]^2->1-Cos[th]^2,Sin[th]^3->Sin[th] (1-Cos[th]^2),
        Sin[th]^4->(1-Cos[th]^2)^2,Sin[th]^5->Sin[th] (1-Cos[th]^2)^2,
        Sin[th]^6->(1-Cos[th]^2)^3, Sin[fi]^2->1-Cos[fi]^2,
        Sin[fi]^3->Sin[fi] (1-Cos[fi]^2), Sin[fi]^4->(1-Cos[fi]^2)^2,
        Sin[fi]^5->Sin[fi] (1-Cos[fi]^2)^2, Sin[fi]^6->(1-Cos[fi]^2)^3}]];

s1s2=Factor[Expand[Factor[Expand[S1*S2]]/.{Sin[alfa]^2->1-
Cos[alfa]^2,
        Sin[th]^2->1-Cos[th]^2, Sin[th]^3->Sin[th] (1-Cos[th]^2),
        Sin[th]^4->(1-Cos[th]^2)^2, Sin[th]^5->Sin[th] (1-Cos[th]^2)^2,
        Sin[th]^6->(1-Cos[th]^2)^3, Sin[fi]^2->1-Cos[fi]^2,
        Sin[fi]^3->Sin[fi] (1-Cos[fi]^2), Sin[fi]^4->(1-Cos[fi]^2)^2,
        Sin[fi]^5->Sin[fi] (1-Cos[fi]^2)^2, Sin[fi]^6->(1-Cos[fi]^2)^3}]];

s1s3=Factor[Expand[Factor[Expand[S1*S3]]/.{Sin[alfa]^2->1-
Cos[alfa]^2,
        Sin[th]^2->1-Cos[th]^2, Sin[th]^3->Sin[th] (1-Cos[th]^2),
        Sin[th]^4->(1-Cos[th]^2)^2, Sin[th]^5->Sin[th] (1-Cos[th]^2)^2,
        Sin[th]^6->(1-Cos[th]^2)^3, Sin[fi]^2->1-Cos[fi]^2,
        Sin[fi]^3->Sin[fi] (1-Cos[fi]^2), Sin[fi]^4->(1-Cos[fi]^2)^2,
```

```
        Sin[fi]^5->Sin[fi] (1-Cos[fi]^2)^2,Sin[fi]^6->(1-Cos[fi]^2)^3}]];

s1s4=Factor[Expand[Factor[Expand[S1*S4]]/.{Sin[alfa]^2->1-
Cos[alfa]^2,
        Sin[th]^2->1-Cos[th]^2, Sin[th]^3->Sin[th] (1-Cos[th]^2),
        Sin[th]^4->(1-Cos[th]^2)^2, Sin[th]^5->Sin[th] (1-Cos[th]^2)^2,
        Sin[th]^6->(1-Cos[th]^2)^3, Sin[fi]^2->1-Cos[fi]^2,
        Sin[fi]^3->Sin[fi] (1-Cos[fi]^2), Sin[fi]^4->(1-Cos[fi]^2)^2,
        Sin[fi]^5->Sin[fi] (1-Cos[fi]^2)^2, Sin[fi]^6->(1-Cos[fi]^2)^3}]];

s1s8=Factor[Expand[Factor[Expand[S1*S8]]/.{Sin[alfa]^2->1-
Cos[alfa]^2,
        Sin[th]^2->1-Cos[th]^2, Sin[th]^3->Sin[th] (1-Cos[th]^2),
        Sin[th]^4->(1-Cos[th]^2)^2, Sin[th]^5->Sin[th] (1-Cos[th]^2)^2,
        Sin[th]^6->(1-Cos[th]^2)^3, Sin[fi]^2->1-Cos[fi]^2,
        Sin[fi]^3->Sin[fi] (1-Cos[fi]^2), Sin[fi]^4->(1-Cos[fi]^2)^2,
        Sin[fi]^5->Sin[fi] (1-Cos[fi]^2)^2,Sin[fi]^6->(1-Cos[fi]^2)^3}]];

 s1s14=Factor[Expand[Factor[Expand[S1*S14]]/.{Sin[alfa]^2->1-
Cos[alfa]^2,
        Sin[th]^2->1-Cos[th]^2, Sin[th]^3->Sin[th] (1-Cos[th]^2),
        Sin[th]^4->(1-Cos[th]^2)^2, Sin[th]^5->Sin[th] (1-Cos[th]^2)^2,
        Sin[th]^6->(1-Cos[th]^2)^3, Sin[fi]^2->1-Cos[fi]^2,
        Sin[fi]^3->Sin[fi] (1-Cos[fi]^2), Sin[fi]^4->(1-Cos[fi]^2)^2,
        Sin[fi]^5->Sin[fi] (1-Cos[fi]^2)^2, Sin[fi]^6->(1-Cos[fi]^2)^3}]];
```

```
s1s15=Factor[Expand[Factor[Expand[S1*S15]]/.{Sin[alfa]^2->1-
Cos[alfa]^2,
      Sin[th]^2->1-Cos[th]^2, Sin[th]^3->Sin[th] (1-Cos[th]^2),
      Sin[th]^4->(1-Cos[th]^2)^2, Sin[th]^5->Sin[th] (1-Cos[th]^2)^2,
      Sin[th]^6->(1-Cos[th]^2)^3, Sin[fi]^2->1-Cos[fi]^2,
      Sin[fi]^3->Sin[fi] (1-Cos[fi]^2), Sin[fi]^4->(1-Cos[fi]^2)^2,
      Sin[fi]^5->Sin[fi] (1-Cos[fi]^2)^2, Sin[fi]^6->(1-Cos[fi]^2)^3}]];

s1s16=Factor[Expand[Factor[Expand[S1*S16]]/.{Sin[alfa]^2->1-
Cos[alfa]^2,
      Sin[th]^2->1-Cos[th]^2,Sin[th]^3->Sin[th] (1-Cos[th]^2),
      Sin[th]^4->(1-Cos[th]^2)^2,Sin[th]^5->Sin[th] (1-Cos[th]^2)^2,
      Sin[th]^6->(1-Cos[th]^2)^3, Sin[fi]^2->1-Cos[fi]^2,
      Sin[fi]^3->Sin[fi] (1-Cos[fi]^2), Sin[fi]^4->(1-Cos[fi]^2)^2,
      Sin[fi]^5->Sin[fi] (1-Cos[fi]^2)^2, Sin[fi]^6->(1-Cos[fi]^2)^3}]];

s2s3=Factor[Expand[Factor[Expand[S2*S3]]/.{Sin[alfa]^2->1-
Cos[alfa]^2,
      Sin[th]^2->1-Cos[th]^2, Sin[th]^3->Sin[th] (1-Cos[th]^2),
      Sin[th]^4->(1-Cos[th]^2)^2, Sin[th]^5->Sin[th] (1-Cos[th]^2)^2,
      Sin[th]^6->(1-Cos[th]^2)^3, Sin[fi]^2->1-Cos[fi]^2,
      Sin[fi]^3->Sin[fi] (1-Cos[fi]^2), Sin[fi]^4->(1-Cos[fi]^2)^2,
      Sin[fi]^5->Sin[fi] (1-Cos[fi]^2)^2, Sin[fi]^6->(1-Cos[fi]^2)^3}]];
```

```
s2s4=Factor[Expand[Factor[Expand[S2*S4]]/.{Sin[alfa]^2->1-
Cos[alfa]^2,
      Sin[th]^2->1-Cos[th]^2, Sin[th]^3->Sin[th] (1-Cos[th]^2),
      Sin[th]^4->(1-Cos[th]^2)^2, Sin[th]^5->Sin[th] (1-Cos[th]^2)^2,
      Sin[th]^6->(1-Cos[th]^2)^3, Sin[fi]^2->1-Cos[fi]^2,
      Sin[fi]^3->Sin[fi] (1-Cos[fi]^2), Sin[fi]^4->(1-Cos[fi]^2)^2,
      Sin[fi]^5->Sin[fi] (1-Cos[fi]^2)^2, Sin[fi]^6->(1-Cos[fi]^2)^3}]];

s2s8=Factor[Expand[Factor[Expand[S2*S8]]/.{Sin[alfa]^2->1-
Cos[alfa]^2,
      Sin[th]^2->1-Cos[th]^2, Sin[th]^3->Sin[th] (1-Cos[th]^2),
      Sin[th]^4->(1-Cos[th]^2)^2, Sin[th]^5->Sin[th] (1-Cos[th]^2)^2,
      Sin[th]^6->(1-Cos[th]^2)^3, Sin[fi]^2->1-Cos[fi]^2,
      Sin[fi]^3->Sin[fi] (1-Cos[fi]^2), Sin[fi]^4->(1-Cos[fi]^2)^2,
      Sin[fi]^5->Sin[fi] (1-Cos[fi]^2)^2, Sin[fi]^6->(1-Cos[fi]^2)^3}]];

s2s14=Factor[Expand[Factor[Expand[S2*S14]]/.{Sin[alfa]^2->1-
Cos[alfa]^2,
      Sin[th]^2->1-Cos[th]^2, Sin[th]^3->Sin[th] (1-Cos[th]^2),
      Sin[th]^4->(1-Cos[th]^2)^2, Sin[th]^5->Sin[th] (1-Cos[th]^2)^2,
      Sin[th]^6->(1-Cos[th]^2)^3, Sin[fi]^2->1-Cos[fi]^2,
      Sin[fi]^3->Sin[fi] (1-Cos[fi]^2), Sin[fi]^4->(1-Cos[fi]^2)^2,
      Sin[fi]^5->Sin[fi] (1-Cos[fi]^2)^2, Sin[fi]^6->(1-Cos[fi]^2)^3}]];
```

```
s2s15=Factor[Expand[Factor[Expand[S2*S15]]/.{Sin[alfa]^2->1-
Cos[alfa]^2,
        Sin[th]^2->1-Cos[th]^2, Sin[th]^3->Sin[th] (1-Cos[th]^2),
        Sin[th]^4->(1-Cos[th]^2)^2, Sin[th]^5->Sin[th] (1-Cos[th]^2)^2,
        Sin[th]^6->(1-Cos[th]^2)^3, Sin[fi]^2->1-Cos[fi]^2,
        Sin[fi]^3->Sin[fi] (1-Cos[fi]^2), Sin[fi]^4->(1-Cos[fi]^2)^2,
        Sin[fi]^5->Sin[fi] (1-Cos[fi]^2)^2, Sin[fi]^6->(1-Cos[fi]^2)^3}]];

s2s16=Factor[Expand[Factor[Expand[S2*S16]]/.{Sin[alfa]^2->1-
Cos[alfa]^2,
        Sin[th]^2->1-Cos[th]^2, Sin[th]^3->Sin[th] (1-Cos[th]^2),
        Sin[th]^4->(1-Cos[th]^2)^2, Sin[th]^5->Sin[th] (1-Cos[th]^2)^2,
        Sin[th]^6->(1-Cos[th]^2)^3, Sin[fi]^2->1-Cos[fi]^2,
        Sin[fi]^3->Sin[fi] (1-Cos[fi]^2), Sin[fi]^4->(1-Cos[fi]^2)^2,
        Sin[fi]^5->Sin[fi] (1-Cos[fi]^2)^2, Sin[fi]^6->(1-Cos[fi]^2)^3}]];

s3s4=Factor[Expand[Factor[Expand[S3*S4]]/.{Sin[alfa]^2->1-
Cos[alfa]^2,
        Sin[th]^2->1-Cos[th]^2, Sin[th]^3->Sin[th] (1-Cos[th]^2),
        Sin[th]^4->(1-Cos[th]^2)^2, Sin[th]^5->Sin[th] (1-Cos[th]^2)^2,
        Sin[th]^6->(1-Cos[th]^2)^3, Sin[fi]^2->1-Cos[fi]^2,
        Sin[fi]^3->Sin[fi] (1-Cos[fi]^2), Sin[fi]^4->(1-Cos[fi]^2)^2,
        Sin[fi]^5->Sin[fi] (1-Cos[fi]^2)^2, Sin[fi]^6->(1-Cos[fi]^2)^3}]];
```

```
s3s8=Factor[Expand[Factor[Expand[S3*S8]]/.{Sin[alfa]^2->1-
Cos[alfa]^2,
        Sin[th]^2->1-Cos[th]^2, Sin[th]^3->Sin[th] (1-Cos[th]^2),
        Sin[th]^4->(1-Cos[th]^2)^2, Sin[th]^5->Sin[th] (1-Cos[th]^2)^2,
        Sin[th]^6->(1-Cos[th]^2)^3, Sin[fi]^2->1-Cos[fi]^2,
        Sin[fi]^3->Sin[fi] (1-Cos[fi]^2), Sin[fi]^4->(1-Cos[fi]^2)^2,
        Sin[fi]^5->Sin[fi] (1-Cos[fi]^2)^2, Sin[fi]^6->(1-Cos[fi]^2)^3}]];

s3s14=Factor[Expand[Factor[Expand[S3*S14]]/.{Sin[alfa]^2->1-
Cos[alfa]^2,
        Sin[th]^2->1-Cos[th]^2, Sin[th]^3->Sin[th] (1-Cos[th]^2),
        Sin[th]^4->(1-Cos[th]^2)^2, Sin[th]^5->Sin[th] (1-Cos[th]^2)^2,
        Sin[th]^6->(1-Cos[th]^2)^3, Sin[fi]^2->1-Cos[fi]^2,
        Sin[fi]^3->Sin[fi] (1-Cos[fi]^2), Sin[fi]^4->(1-Cos[fi]^2)^2,
        Sin[fi]^5->Sin[fi] (1-Cos[fi]^2)^2, Sin[fi]^6->(1-Cos[fi]^2)^3}]];

s3s15=Factor[Expand[Factor[Expand[S3*S15]]/.{Sin[alfa]^2->1-
Cos[alfa]^2,
        Sin[th]^2->1-Cos[th]^2, Sin[th]^3->Sin[th] (1-Cos[th]^2),
        Sin[th]^4->(1-Cos[th]^2)^2, Sin[th]^5->Sin[th] (1-Cos[th]^2)^2,
        Sin[th]^6->(1-Cos[th]^2)^3, Sin[fi]^2->1-Cos[fi]^2,
        Sin[fi]^3->Sin[fi] (1-Cos[fi]^2), Sin[fi]^4->(1-Cos[fi]^2)^2,
        Sin[fi]^5->Sin[fi] (1-Cos[fi]^2)^2, Sin[fi]^6->(1-Cos[fi]^2)^3}]];
```

```
s3s16=Factor[Expand[Factor[Expand[S3*S16]]/.{Sin[alfa]^2->1-
Cos[alfa]^2,
        Sin[th]^2->1-Cos[th]^2, Sin[th]^3->Sin[th] (1-Cos[th]^2),
        Sin[th]^4->(1-Cos[th]^2)^2, Sin[th]^5->Sin[th] (1-Cos[th]^2)^2,
        Sin[th]^6->(1-Cos[th]^2)^3, Sin[fi]^2->1-Cos[fi]^2,
        Sin[fi]^3->Sin[fi] (1-Cos[fi]^2), Sin[fi]^4->(1-Cos[fi]^2)^2,
        Sin[fi]^5->Sin[fi] (1-Cos[fi]^2)^2, Sin[fi]^6->(1-Cos[fi]^2)^3}]];

s4s8=Factor[Expand[Factor[Expand[S4*S8]]/.{Sin[alfa]^2->1-
Cos[alfa]^2,
        Sin[th]^2->1-Cos[th]^2, Sin[th]^3->Sin[th] (1-Cos[th]^2),
        Sin[th]^4->(1-Cos[th]^2)^2, Sin[th]^5->Sin[th] (1-Cos[th]^2)^2,
        Sin[th]^6->(1-Cos[th]^2)^3, Sin[fi]^2->1-Cos[fi]^2,
        Sin[fi]^3->Sin[fi] (1-Cos[fi]^2), Sin[fi]^4->(1-Cos[fi]^2)^2,
        Sin[fi]^5->Sin[fi] (1-Cos[fi]^2)^2, Sin[fi]^6->(1-Cos[fi]^2)^3}]];

s4s14=Factor[Expand[Factor[Expand[S4*S14]]/.{Sin[alfa]^2->1-
Cos[alfa]^2,
        Sin[th]^2->1-Cos[th]^2, Sin[th]^3->Sin[th] (1-Cos[th]^2),
        Sin[th]^4->(1-Cos[th]^2)^2, Sin[th]^5->Sin[th] (1-Cos[th]^2)^2,
        Sin[th]^6->(1-Cos[th]^2)^3, Sin[fi]^2->1-Cos[fi]^2,
        Sin[fi]^3->Sin[fi] (1-Cos[fi]^2), Sin[fi]^4->(1-Cos[fi]^2)^2,
        Sin[fi]^5->Sin[fi] (1-Cos[fi]^2)^2, Sin[fi]^6->(1-Cos[fi]^2)^3}]];
```

```
s4s15=Factor[Expand[Factor[Expand[S4*S15]]/.{Sin[alfa]^2->1-
Cos[alfa]^2,
      Sin[th]^2->1-Cos[th]^2, Sin[th]^3->Sin[th] (1-Cos[th]^2),
      Sin[th]^4->(1-Cos[th]^2)^2, Sin[th]^5->Sin[th] (1-Cos[th]^2)^2,
      Sin[th]^6->(1-Cos[th]^2)^3, Sin[fi]^2->1-Cos[fi]^2,
      Sin[fi]^3->Sin[fi] (1-Cos[fi]^2), Sin[fi]^4->(1-Cos[fi]^2)^2,
      Sin[fi]^5->Sin[fi] (1-Cos[fi]^2)^2, Sin[fi]^6->(1-Cos[fi]^2)^3}]];
```

```
s4s16=Factor[Expand[Factor[Expand[S4*S16]]/.{Sin[alfa]^2->1-
Cos[alfa]^2,
      Sin[th]^2->1-Cos[th]^2, Sin[th]^3->Sin[th] (1-Cos[th]^2),
      Sin[th]^4->(1-Cos[th]^2)^2, Sin[th]^5->Sin[th] (1-Cos[th]^2)^2,
      Sin[th]^6->(1-Cos[th]^2)^3, Sin[fi]^2->1-Cos[fi]^2,
      Sin[fi]^3->Sin[fi] (1-Cos[fi]^2), Sin[fi]^4->(1-Cos[fi]^2)^2,
      Sin[fi]^5->Sin[fi] (1-Cos[fi]^2)^2, Sin[fi]^6->(1-Cos[fi]^2)^3}]];
```

```
s8s14=Factor[Expand[Factor[Expand[S8*S14]]/.{Sin[alfa]^2->1-
Cos[alfa]^2,
      Sin[th]^2->1-Cos[th]^2, Sin[th]^3->Sin[th] (1-Cos[th]^2),
      Sin[th]^4->(1-Cos[th]^2)^2, Sin[th]^5->Sin[th] (1-Cos[th]^2)^2,
      Sin[th]^6->(1-Cos[th]^2)^3, Sin[fi]^2->1-Cos[fi]^2,
      Sin[fi]^3->Sin[fi] (1-Cos[fi]^2), Sin[fi]^4->(1-Cos[fi]^2)^2,
      Sin[fi]^5->Sin[fi] (1-Cos[fi]^2)^2, Sin[fi]^6->(1-Cos[fi]^2)^3}]];
```

```
s8s15=Factor[Expand[Factor[Expand[S8*S15]]/.{Sin[alfa]^2->1-
Cos[alfa]^2,
        Sin[th]^2->1-Cos[th]^2, Sin[th]^3->Sin[th] (1-Cos[th]^2),
        Sin[th]^4->(1-Cos[th]^2)^2, Sin[th]^5->Sin[th] (1-Cos[th]^2)^2,
        Sin[th]^6->(1-Cos[th]^2)^3, Sin[fi]^2->1-Cos[fi]^2,
        Sin[fi]^3->Sin[fi] (1-Cos[fi]^2), Sin[fi]^4->(1-Cos[fi]^2)^2,
        Sin[fi]^5->Sin[fi] (1-Cos[fi]^2)^2, Sin[fi]^6->(1-Cos[fi]^2)^3}]];

s8s16=Factor[Expand[Factor[Expand[S8*S16]]/.{Sin[alfa]^2->1-
Cos[alfa]^2,
        Sin[th]^2->1-Cos[th]^2, Sin[th]^3->Sin[th] (1-Cos[th]^2),
        Sin[th]^4->(1-Cos[th]^2)^2, Sin[th]^5->Sin[th] (1-Cos[th]^2)^2,
        Sin[th]^6->(1-Cos[th]^2)^3, Sin[fi]^2->1-Cos[fi]^2,
        Sin[fi]^3->Sin[fi] (1-Cos[fi]^2), Sin[fi]^4->(1-Cos[fi]^2)^2,
        Sin[fi]^5->Sin[fi] (1-Cos[fi]^2)^2, Sin[fi]^6->(1-Cos[fi]^2)^3}]];

s14s15=Factor[Expand[Factor[Expand[S14*S15]]/.{Sin[alfa]^2->1-
Cos[alfa]^2,
        Sin[th]^2->1-Cos[th]^2, Sin[th]^3->Sin[th] (1-Cos[th]^2),
        Sin[th]^4->(1-Cos[th]^2)^2, Sin[th]^5->Sin[th] (1-Cos[th]^2)^2,
        Sin[th]^6->(1-Cos[th]^2)^3, Sin[fi]^2->1-Cos[fi]^2,
        Sin[fi]^3->Sin[fi] (1-Cos[fi]^2), Sin[fi]^4->(1-Cos[fi]^2)^2,
        Sin[fi]^5->Sin[fi] (1-Cos[fi]^2)^2, Sin[fi]^6->(1-Cos[fi]^2)^3}]];
```

```
s14s16=Factor[Expand[Factor[Expand[S14*S16]]/.{Sin[alfa]^2->1-
Cos[alfa]^2,
        Sin[th]^2->1-Cos[th]^2, Sin[th]^3->Sin[th] (1-Cos[th]^2),
        Sin[th]^4->(1-Cos[th]^2)^2, Sin[th]^5->Sin[th] (1-Cos[th]^2)^2,
        Sin[th]^6->(1-Cos[th]^2)^3, Sin[fi]^2->1-Cos[fi]^2,
        Sin[fi]^3->Sin[fi] (1-Cos[fi]^2), Sin[fi]^4->(1-Cos[fi]^2)^2,
        Sin[fi]^5->Sin[fi] (1-Cos[fi]^2)^2, Sin[fi]^6->(1-Cos[fi]^2)^3}]];

s15s16=Factor[Expand[Factor[Expand[S15*S16]]/.{Sin[alfa]^2->1-
Cos[alfa]^2,
        Sin[th]^2->1-Cos[th]^2, Sin[th]^3->Sin[th] (1-Cos[th]^2),
        Sin[th]^4->(1-Cos[th]^2)^2, Sin[th]^5->Sin[th] (1-Cos[th]^2)^2,
        Sin[th]^6->(1-Cos[th]^2)^3, Sin[fi]^2->1-Cos[fi]^2,
        Sin[fi]^3->Sin[fi] (1-Cos[fi]^2), Sin[fi]^4->(1-Cos[fi]^2)^2,
        Sin[fi]^5->Sin[fi] (1-Cos[fi]^2)^2, Sin[fi]^6->(1-Cos[fi]^2)^3}]];

sumaA=Factor[Expand[i5^2 suma55+i6^2 suma11+i7^2 suma77+i8^2
suma88+i9^2 suma99-2i5 i6 suma51+2i5 i7 suma57+2i5 i8 suma58-2i5 i9
suma59-2i6 i7 suma17-2i6 i8 suma18+2i6 i9 suma19+2i7 i8 suma78-2i7 i9
suma79-2i8 i9 suma89]/.{Sin[fi]^2->1-Cos[fi]^2}];

sumaB=Factor[Expand[Expand[i1^2s1s1+(i2+i10)^2s2s2+(i3+i11)^2s3s
3+(i4+i12)^2 s4s4+i13^2 s8s8+i14^2 s14s14+i15^2 s15s15+i16^2
s16s16+2i1(i2+i10) s1s2+2i1(i3+i11) s1s3+2i1(i4+i12) s1s4+2i1 i13
s1s8+2i1 i14 s1s14+2i1 i15 s1s15+2i1 i16 s1s16+2(i2+i10)(i3+i11)
```

```
s2s3+2(i2+i10)(i4+i12) s2s4+2(i2+i10)i13 s2s8+2(i2+i10) i14
s2s14+2(i2+i10) i15 s2s15+2(i2+i10)i16 s2s16+2(i3+i11)(i4+i12)
s3s4+2(i3+i11) i13 s3s8+2(i3+i11) i14 s3s14+2(i3+i11)i15
s3s15+2(i3+i11) i16 s3s16+2(i4+i12) i13 s4s8+2(i4+i12) i14 s4s14+
2(i4+i12) i15 s4s15+2(i4+i12) i16 s4s16+2i13 i14 s8s14+2i13 i15 s8s15+
2i13 i16 s8s16+2i14 i15 s14s15+2i14 i16 s14s16+2i15 i16 s15s16]/.{
    Sin[alfa]^2->1-Cos[alfa]^2, Sin[th]^2->1-Cos[th]^2,
    Sin[th]^3->Sin[th] (1-Cos[th]^2), Sin[th]^4->
    (1-Cos[th]^2)^2,Sin[th]^5-> Sin[th] (1-Cos[th]^2)^2,
    Sin[th]^6->(1-Cos[th]^2)^3, Sin[fi]^2->
    (1-Cos[fi]^2), Sin[fi]^3->Sin[fi] (1-Cos[fi]^2), Sin[fi]^4->
    (1-Cos[fi]^2)^2, Sin[fi]^5->Sin[fi](1-Cos[fi]^2)^2,Sin[fi]^6->
    (1-Cos[fi]^2)^3}]];

sumaC=Factor[sumaB/Product[sumaB[[i]],{i,1,7}]];
x2=Factor[Coefficient[sumaC,Cos[fi]]]

6*(-1 + v)*(1 + v)*Cos[alfa]*(-1 + Cos[th])*(-2 + v^2 + v^2*Cos[th])*(-
15+18*v^2-7*v^4+12*v^2*Cos[th]-4*v^4*Cos[th]-
4*v^4*Cos[th]^2)*Sin[alfa]*Sin[th]
x3=Factor[Coefficient[sumaC,Cos[fi]^2]]

(-1 + Cos[alfa])*(1 + Cos[alfa])*(1 + Cos[th])*(180 - 453*v^2 + 596*v^4 -
382*v^6 +152*v^8 - 21*v^10 - 90*Cos[th] - 24*v^2*Cos[th] +
8*v^4*Cos[th] - 172*v^6*Cos[th] + 2*v^8*Cos[th] - 12*v^10*Cos[th] +
```

```
117*v^2*Cos[th]^2 - 4*v^4*Cos[th]^2 +242*v^6*Cos[th]^2 +
68*v^8*Cos[th]^2 + 9*v^10*Cos[th]^2 – 60*v^4*Cos[th]^3 -
60*v^6*Cos[th]^3 - 180*v^8*Cos[th]^3 + 12*v^10*Cos[th]^3 +
12*v^6*Cos[th]^4 + 48*v^8*Cos[th]^4 + 12*v^10*Cos[th]^4)

x1=Factor[Expand[sumaC-x2 Cos[fi]-x3 Cos[fi]^2]];

x12=Factor[Coefficient[x1,Cos[alfa]^2]]

(-180 + 453*v^2 - 596*v^4 + 382*v^6 - 152*v^8 +  21*v^10 +
90*Cos[th] - 9*v^2*Cos[th] - 106*v^4*Cos[th] + 320*v^6*
Cos[th] - 112*v^8*Cos[th] + 33*v^10*Cos[th] - 90*Cos[th]^2 +
159*v^2*Cos[th]^2 - 40*v^4*Cos[th]^2 - 130*v^6*Cos[th]^2 -
46*v^8*Cos[th]^2 + 3*v^10*Cos[th]^2 + 117*v^2*Cos[th]^3 -
278*v^4*Cos[th]^3 - 56*v^6*Cos[th]^3 + 94*v^8*Cos[th]^3 -
21*v^10*Cos[th]^3 - 60*v^4*Cos[th]^4 + 192*v^6*Cos[th]^4 +
108*v^8*Cos[th]^4 - 24*v^10*Cos[th]^4 + 12*v^6*Cos[th]^5 -
72*v^8*Cos[th]^5 - 12*v^10*Cos[th]^5)

x11=Factor[Expand[x1-x12 Cos[alfa]^2]]

(270 - 696*v^2 + 845*v^4 - 499*v^6 + 173*v^8 - 21*v^10 - 90*Cos[th] -
108*v^2*Cos[th] + 337*v^4*Cos[th] - 467*v^6*Cos[th] +
145*v^8*Cos[th] - 33*v^10*Cos[th] + 84*v^2*Cos[th]^2 -
149*v^4*Cos[th]^2 + 163*v^6*Cos[th]^2 + 49*v^8*Cos[th]^2 -
```

```
3*v^10*Cos[th]^2 + 47*v^4*Cos[th]^3 + 191*v^6*Cos[th]^3 -
115*v^8*Cos[th]^3 + 21*v^10*Cos[th]^3 - 108*v^6*Cos[th]^4 -
132*v^8*Cos[th]^4 + 24*v^10*Cos[th]^4 + 60*v^8*Cos[th]^5 +
12*v^10*Cos[th]^5)
```

## Appendix A

# Isotopic Formalism

The *isotopic spin formalism* was introduced by Heisenberg for the purpose of describing the fact that the nuclear forces are independent of electric charge. In this formalism, the proton and neutron are considered as two different states of one and the same particle – the *nucleon*.

In relation to the rotations in a fictive three-dimensional space, called *isotopic spin space* (or, simpler, *isotopic space*), the nucleon wave function is endowed with the properties of a two-component spinor. Analogue properties are also attributed to the wave functions of some other charge multiplets. For example, the pionic triplet $\pi^+$, $\pi^0$, $\pi^-$ is described by three functions, which can be regarded as the components of a vector in the isotopic space, and so on.

From the physical point of view, the invariance to rotations in the isotopic space corresponds to the particle "equivalence" within the multiplets. As far as is known, this "equity" property is broken by the electromagnetic interactions, but in the case of nuclear interactions the isotopic invariance is always fulfilled.

Here are the main equations of the isotopic formalism for nucleons and pions.

### 1. The nucleon doublet

The first topic concerns the nucleon *isotopic doublet*. The spinor associated to this physical system in the three-dimensional space has two components, corresponding to proton and neutron field functions, $\psi_p$ and $\psi_n$:

$$\psi = \begin{pmatrix} \psi_p \\ \psi_n \end{pmatrix}, \qquad \psi^+ = \begin{pmatrix} \psi_p^*, \psi_n^* \end{pmatrix}. \tag{A.1}$$

The rotation operators in the isotopic space are constructed by second rank matrices, which are coincident with the Pauli matrices, but denoted by different symbols,

$$\tau_1 = \begin{pmatrix} 0 & 1 \\ 1 & 0 \end{pmatrix}, \qquad \tau_2 = \begin{pmatrix} 0 & -i \\ i & 0 \end{pmatrix}, \qquad \tau_3 = \begin{pmatrix} 1 & 0 \\ 0 & -1 \end{pmatrix}. \tag{A.2}$$

The $\tau$ matrices anticommute with each other,

$$\tau_a \tau_b + \tau_b \tau_a = 0, \quad (a \neq b), \quad \tau_a^2 = 1, \tag{A.3}$$

and satisfy the following relations:

$$\tau_1 \tau_2 = i\tau_3 \text{ (together with cyclic permutations of 1, 2, 3)}. \tag{A.4}$$

The isotopic rotation operators are connected to the $\tau$-matrices by

$$I_a = \frac{\tau_a}{2}. \tag{A.5}$$

According to (A.3) and (A.4), these operators satisfy the commutation relations for the rotation operators

$$[I_1, I_2] = iI_3 \text{ (together with cyclic permutations of 1, 2, 3)}, \tag{A.6}$$

as well as the relation

$$I^2 = I_1^2 + I_2^2 + I_3^2 = \frac{3}{4},$$

since, according to quantum mechanics, the eigenvalue $I$ of the total isotopic momentum of rotation $\vec{I}$ is determined by

$$\vec{I}^2 = I(I+1),$$

where it can be shown that $I = \frac{1}{2}$.

The following two linear combinations of the matrices $\tau_1$ and $\tau_2$,

$$\tau_+ = \tau_1 + i\tau_2 = \begin{pmatrix} 0 & 1 \\ 0 & 0 \end{pmatrix} \quad \text{and} \quad \tau_- = \tau_1 - i\tau_2 = \begin{pmatrix} 0 & 0 \\ 1 & 0 \end{pmatrix} \tag{A.7}$$

transform the neutronic into protonic states and vice versa,

$$\tau_+ \begin{pmatrix} 0 \\ \psi \end{pmatrix} = \begin{pmatrix} \psi \\ 0 \end{pmatrix}, \quad \tau_- \begin{pmatrix} \psi \\ 0 \end{pmatrix} = \begin{pmatrix} 0 \\ \psi \end{pmatrix},$$

where $\tau_1$ and $\tau_2$ are also called "raising" (increasing) and "lowering" (decreasing) electric charge operators.

In the case of a rotation of angle $\theta$ in the isotopic space about the $Ox_a$ axis, the transformation relations are

$$\psi' = \Lambda_a(\theta)\psi, \quad \overline{\psi}' = \Lambda_a^{-1}(\theta)\overline{\psi},$$

where

$$\Lambda_a(\theta) = e^{-iI_a\theta} = e^{-i\tau_a\theta/2} = \cos\frac{\theta}{2} - i\tau_a \sin\frac{\theta}{2}.$$

The bilinear expressions $\overline{\psi}\tau_a\psi$, $(a = 1, 2, 3)$ stand for the spatial density of the isotopic spin vector

$$\vec{I} = \int \frac{1}{2}\overline{\psi}(x)\vec{\tau}\psi(x)\, dx,$$

the third component being related to the electric charge $Q$ and hypercharge $Y$ by *Gell-Mann–Nishijima formula*:

$$Q = I_3 + \frac{Y}{2}. \tag{A.8}$$

## 2. The pion triplet

The three pseudoscalar pions $\pi^+, \pi^0, \pi^-$ form an *isotopic triplet*, written by means of the three pseudoscalar field functions that define the isotopic vector. One usually employ two representations of these functions. In the first representation the functions are real,

$$\vec{\pi} = \{\pi_1, \pi_2, \pi_3\}, \quad \pi_a^*(x) = \pi_a(x), \tag{A.9}$$

while in the second, the real functions $\pi_1$ and $\pi_2$ form two complex combinations,

$$\Pi_1 = \frac{\pi_1 - i\pi_2}{\sqrt{2}}, \quad \Pi_2 = \pi_3, \quad \Pi_3 = \frac{\pi_1 + i\pi_2}{\sqrt{2}}. \tag{A.10}$$

By means of (A.9), the triplet

$$\vec{\Pi} = \{\Pi_1, \Pi_2, \Pi_3\} \tag{A.11}$$

can be written under a matrix form

$$\Pi = U\pi, \quad U = \begin{pmatrix} \frac{1}{\sqrt{2}} & \frac{-i}{\sqrt{2}} & 0 \\ 0 & 0 & 1 \\ \frac{1}{\sqrt{2}} & \frac{i}{\sqrt{2}} & 0 \end{pmatrix}, \tag{A.12}$$

where $U$ is an unitary matrix,

$$U^+U = 1.$$

Under a rotation in the three-dimensional isotopic space, the isotopic vector (A.9) transforms according to

$$\pi \to \pi' = \Lambda(\beta)\pi, \quad \Lambda(\beta) = e^{-il_n\beta_n}$$

where $\beta_n$, $(n = 1, 2, 3)$ are the rotation angles, and $l_n$ are the isotopic spin matrices in representation (A.9),

$$l_1 = \begin{pmatrix} 0 & 0 & 0 \\ 0 & 0 & -i \\ 0 & i & 0 \end{pmatrix} \quad l_2 = \begin{pmatrix} 0 & 0 & i \\ 0 & 0 & 0 \\ -i & 0 & 0 \end{pmatrix} \quad l_3 = \begin{pmatrix} 0 & -i & 0 \\ i & 0 & 0 \\ 0 & 0 & 0 \end{pmatrix}. \tag{A.13}$$

The squared matrix-vector $\vec{l}$ is determined by means of

$$l(l+1) = l_1^2 + l_2^2 + l_3^3 = 2.$$

Due to this property, $\vec{l}$ can be interpreted as the angular momentum operator, with eigenvalues 1, 0, and $-1$.

The operator corresponding to the transformation of rotation of angle $\beta_3$, about the axis $Oz \equiv Ox_3$ can be represented as

$$\Lambda(\beta_3) = e^{-il_3\beta_3} = \begin{pmatrix} \cos\beta_3 & -\sin\beta_3 & 0 \\ \sin\beta_3 & +\cos\beta_3 & 0 \\ 0 & 0 & 1 \end{pmatrix}. \tag{A.14}$$

The components of the matrices (A.13) can be expressed by means of the unitary antisymmetric third rank tensor

$$(l_a)_{bc} = -i\varepsilon_{abc}.$$

The matrices $l_a$ satisfy the commutation relation

$$[l_1, l_2] = il_3 \text{ (together with cyclic permutations of 1, 2, 3).} \tag{A.15}$$

The transition to representation (A.11) is performed by means of the unitary transformation

$$T_n = U l_n U^+.$$

The matrices $T_n$ have the following form,

$$T_1 = \frac{1}{\sqrt{2}} \begin{pmatrix} 0 & -1 & 0 \\ -1 & 0 & 1 \\ 0 & 1 & 0 \end{pmatrix}, \quad T_2 = \frac{1}{\sqrt{2}} \begin{pmatrix} 0 & i & 0 \\ -i & 0 & -i \\ 0 & i & 0 \end{pmatrix},$$

$$T_3 = \begin{pmatrix} 1 & 0 & 0 \\ 0 & 0 & 0 \\ 0 & 0 & -1 \end{pmatrix}, \tag{A.16}$$

and satisfy the commutation relations (A.15). The matrix $T_3$ is, obviously, diagonal. This fact shows that in the representation (A.11), the components $\Pi_n$ correspond to certain values of the electric charge. On the basis of (A.8), one can write

$$\Pi_1 \to \pi^+, \quad \Pi_2 \to \pi^0, \quad \Pi_3 \to \pi^-. \tag{A.17}$$

# Appendix B

# The Dirac Matrices and the Dirac Equation

## 1. The Dirac matrices

The properties of the four "hypercomplex" numbers (which shall be identified afterwards with the usual[1] Dirac matrices),

$$\gamma_\nu \ (\nu = 0, 1, 2, 3)$$

are determined from the basic anticommutation relation

$$\gamma_\mu \gamma_\nu + \gamma_\nu \gamma_\mu = 2g_{\mu\nu}. \tag{B.1}$$

According to these relations, the following sixteen quantities,

$$\left.\begin{array}{c} \text{the unit matrix, } I = g^{\nu\nu}\gamma^\nu\gamma^\nu, \quad (\text{ no summation }), \\ \text{the four quantities } \gamma_\nu, \\ \text{the six quantities } \sigma_{\nu\mu} = (\gamma_\nu\gamma_\mu - \gamma_\mu\gamma_\nu)/2i, \ (\nu < \mu), \\ \text{the quantity } \gamma_5 = i\gamma_0\gamma_1\gamma_2\gamma_3, \\ \text{and the four quantities } D_\nu = \gamma_\nu\gamma_5, \end{array}\right\} \tag{B.2}$$

form the algebra of the complex number corpus/field. In other words, the set of elements determined by all possible linear combinations with complex coefficients of the quantities (B.2) and, also, various products of these quantities and their mutual combinations is closed with respect to summation of elements, their mutual multiplication and multiplication with complex numbers as well.

---

[1]The term "usual" is used here to distinguish the *usual Dirac matrices*, or simply, Dirac matrices as against the *generalized Dirac matrices*, used in Chapters VI and VIII. However, for convenience, from now on (in all appendices) we shall use the name "Dirac matrices" instead of "usual Dirac matrices".

It is worth mentioning that the forms $\sigma_{\mu\nu}$ are bilinear in $\gamma$-matrices, the forms $D_\nu$ are trilinear, and $\gamma_5$ is the only fourth-rank form. All the other forms of the fifth or higher rank are reducible by means of (B.1) to one of the quantities included in (B.2).

On the basis of certain algebra theorems, one can demonstrate that the rank $r$ of a system of matrix representations of hypercomplex numbers is connected to the number $n$ of linearly independent elements through the relation

$$r^2 = n,$$

which appears somewhat obvious, if the hypercomplex numbers are presented as square matrices.

It can be proved that the sixteen matrices are linearly independent. To this end, it is necessary to study the trace ("spur", in German) of all matrices and be sure that all traces – except that of the unit matrix ($\text{Tr } I = r$) – are null. Demonstration is based on the possibility of doing cyclic permutations of the matrices (appearing as factors) under the trace sign. Consider, as an example, the trace of the product $\gamma_\nu\gamma_\mu$ ($\nu \neq \mu$). Using (B.1) and the above mentioned property, one can write

$$\text{Tr } \gamma_\nu\gamma_\mu = \text{Tr } \gamma_\mu\gamma_\nu = \frac{1}{2}\text{Tr } (\gamma_\nu\gamma_\mu + \gamma_\mu\gamma_\nu) = 0.$$

In the second stage, it is necessary to consider the linear form

$$F = aI + b^\nu\gamma_\nu + c^{\nu\mu}\sigma_{\nu\mu} + d^\nu D_\nu + e\gamma_5,$$

and, by calculating the traces of $F$ and of the products $F\gamma_\nu, F\gamma_{\nu\mu}$, $FD_\nu, F\gamma_5$, one can conclude that the supposition $F = 0$ leads to the following sequence of relations:

$$a = 0, \quad b^\nu = 0, \quad c^{\nu\mu} = 0, \quad d^\nu = 0, \quad e = 0.$$

In the given case $n = 16$ and, according to the formula $n = r^2$, one can conclude that the hypercomplex numbers $\gamma^\nu$ determined by (B.1) can be represented as fourth-rank square matrices – the *Dirac matrices*.

According to (B.1), the four matrices $\gamma_\nu$ $(\nu = 0,1,2,3)$ can be chosen as unitary if the Hermitian conjugation operation is defined as

$$(\gamma^\nu)^+ = \gamma_\nu, \quad (\gamma^\nu)^+_{\alpha\beta} \equiv (\gamma^\nu)^*_{\beta\alpha}\,. \tag{B.3}$$

Since the matrix $\gamma_5$ anticommute with $\gamma_\nu$ $(\nu = 0,1,2,3)$, and its square equals 1, one can write

$$\gamma_m\gamma_n + \gamma_n\gamma_m = 2g_{mn}, \ (m,n = 0,1,2,3,5), \tag{B.4}$$

where, by definition

$$g_{55} = 1.$$

This way,

$$\gamma_5^+ = \gamma_5 = \gamma^5.$$

The basic relations (B.1) together with their consequences are invariant with respect to unitary transformation $\gamma_\nu \to U\gamma_\nu U^{-1}$, where $U$ is an arbitrary invertible matrix which, in view of (B.3), has to be a unitary matrix.

The Dirac matrices are therefore determined up to a unitary transformation. One usually chooses a representation of the type

$$\gamma^0 = \begin{pmatrix} 1 & . & . & . \\ . & 1 & . & . \\ . & . & -1 & . \\ . & . & . & -1 \end{pmatrix}, \quad \gamma^1 = \begin{pmatrix} . & . & . & 1 \\ . & . & 1 & . \\ . & -1 & . & . \\ -1 & . & . & . \end{pmatrix},$$

$$\gamma^2 = \begin{pmatrix} . & . & . & -i \\ . & . & i & . \\ . & i & . & . \\ -i & . & . & . \end{pmatrix}, \quad \gamma^3 = \begin{pmatrix} . & . & 1 & . \\ . & . & . & -1 \\ -1 & . & . & . \\ . & 1 & . & . \end{pmatrix}, \tag{B.5}$$

$$\gamma^5 = \begin{pmatrix} . & . & -1 & . \\ . & . & . & -1 \\ -1 & . & . & . \\ . & -1 & . & . \end{pmatrix},$$

where the dots signify 0 (zero). As one can see, the matrix $\gamma^0$ is diagonal. This representation is very convenient for the study of non-relativistic limit of Dirac equation (see further) and it shall be called the *standard representation of Dirac matrices.*

This representation of Dirac matrices is sometime connected to the matrices $\vec{\alpha}$, $\beta$, $\vec{\sigma}$ and $\vec{\rho}$ by means of the following relations:

$$\alpha_n = \gamma^0\gamma^n \ (n = 1, 2, 3), \quad \beta = \gamma^0;$$

$$\sigma_n = \gamma^5\gamma^n\gamma^0 \ (n = 1, 2, 3), \quad \sigma_1 = \sigma^{23}, \quad \sigma_2 = \sigma^{31}, \quad \sigma_3 = \sigma^{12}; \tag{B.6}$$

$$\rho_1 = -\gamma^5, \quad \rho_2 = i\gamma^0\gamma^5, \quad \rho_3 = \gamma^0.$$

The standard representation (B.5) can be written under the so-called *compact* form,

$$\gamma^0 = \begin{pmatrix} \mathbf{I} & \mathbf{0} \\ \mathbf{0} & \mathbf{I} \end{pmatrix}, \ \gamma^n = -\begin{pmatrix} \mathbf{0} & \sigma_n \\ -\sigma_n & \mathbf{0} \end{pmatrix}, \ \gamma^5 = \begin{pmatrix} \mathbf{0} & -\mathbf{I} \\ -\mathbf{I} & \mathbf{0} \end{pmatrix}, \tag{B.7}$$

by means of the Pauli matrices

$$\sigma_1 = \begin{pmatrix} 0 & 1 \\ 1 & 0 \end{pmatrix}, \quad \sigma_2 = \begin{pmatrix} 0 & -i \\ i & 0 \end{pmatrix}, \quad \sigma_3 = \begin{pmatrix} 1 & 0 \\ 0 & -1 \end{pmatrix}, \tag{B.8}$$

that satisfy the relations

$$\sigma_i\sigma_j + \sigma_j\sigma_i = 2\delta_{ij}, \quad \sigma_k\sigma_j = \delta_{kj} + i\varepsilon_{kjl}\sigma_l, \tag{B.9}$$

as well as of the unit and null matrices of the second-rank,

$$\mathbf{I} = \begin{pmatrix} 1 & 0 \\ 0 & 1 \end{pmatrix}, \quad \mathbf{0} = \begin{pmatrix} 0 & 0 \\ 0 & 0 \end{pmatrix}.$$

Return, now, to the traces of Dirac matrices and their products. The trace of the unit matrix coincides with the rank of representation, that is

$$\mathrm{Tr}\,\mathbf{I} = \mathrm{Tr}\,\gamma_\nu\gamma^\nu = 4\,, \ \text{(no summation over } \nu).$$

Since the traces of the matrices $\sigma_{\mu\nu}$ and $D_\nu$ are null, one obtains

$$\mathrm{Tr}\,\gamma_\nu = 0, \quad \mathrm{Tr}\,\gamma_\nu\gamma_\mu = 4g_{\nu\mu}, \quad \mathrm{Tr}\,\gamma_\nu\gamma_\mu\gamma_\rho = 0,$$

$$\frac{1}{4}\,\mathrm{Tr}\,\gamma_\nu\gamma_\mu\gamma_\rho\gamma_\sigma = g_{\nu\mu}g_{\rho\sigma} + g_{\nu\sigma}g_{\mu\rho} - g_{\nu\rho}g_{\mu\sigma}, \tag{B.10}$$

$$(\nu, \mu, \rho, \sigma = 0, 1, 2, 3).$$

As a general rule, one can state that the traces of products of an odd number of $\gamma$-matrices are always zero, while the traces of products of an even number of $\gamma$-matrices, up to a factor, are expressed by means of antisymmetrized products of factors $g_{\mu\nu}$, the sign of the term $4g_{\mu_1\mu_2}g_{\mu_3\mu_4}\cdots g_{\mu_{2n-1}\mu_{2n}}$ existent in $\mathrm{Tr}\,(\gamma_{\nu_1}\gamma_{\nu_2}\cdots\gamma_{\nu_{2n}})$ being determined by the parity of the permutation $P^{\mu_1,\mu_2,\ldots,\mu_{2n}}_{\nu_1,\nu_2,\ldots,\nu_{2n}}$.

Finally, here are some formulas for the traces of contracted products of $\gamma$-matrices with four-vectors ($\hat{a} = a^\nu\gamma_\nu$), following from (B.10),

$$\mathrm{Tr}\,\hat{a}\hat{b} = 4(ab), \tag{B.11$a$}$$

$$\mathrm{Tr}\,\hat{a}\hat{b}\hat{c}\hat{d} = 4[(ab)(cd) + (ad)(bc) - (ac)(bd)], \tag{B.11$b$}$$

$$\begin{aligned}&\frac{1}{4}\mathrm{Tr}\,[(\hat{k}_1 + m_1)(\hat{k}_2 + m_2)(\hat{k}_3 + m_3)(\hat{k}_4 + m_4)]\\ &\quad = (K_1K_2)(K_3K_4) + (K_1K_4)(K_2K_3) - (K_1K_3)(K_2K_4),\end{aligned} \tag{B.11$c$}$$

where $K_iK_j = k_ik_j + \eta_{ij}m_im_j$, $\eta_{ij} = (-1)^{i-j+1}$.

## 2. The Dirac equation

This is a first-order linear matricial equation in the partial derivatives of the field function $\psi(x)$:

$$\Big[(i\hat{\partial}_x - m)\psi(x)\Big]_\alpha \equiv \left(i\gamma_\nu\frac{\partial}{\partial x_\nu} - mI\right)_{\alpha\beta}\psi_\beta(x) = 0. \tag{B.12}$$

The field function $\psi(x)$ has four components $\psi_\alpha$, ($\alpha = 1, 2, 3, 4$). Acting with the operator $(i\hat{\partial} + m)$ on the left hand side of Eq. (B.12) and using (B.1), one obtains

$$(i\hat{\partial} + m)_{\sigma\alpha}(i\hat{\partial} - m)_{\alpha\beta}\psi_\beta = (-\partial^2 - m^2)_{\sigma\beta}\psi_\beta = (\Box - m^2)_{\sigma\beta}\psi_\beta = 0.$$

Thus, each component $\psi_\alpha$ ($\alpha = 1, 2, 3, 4$) satisfies the Klein–Gordon equation.

The conjugated equation is obtained from the Dirac equation in two stages. Using the Hermitian conjugate of the equation (B.12), and by means of (B.3), one obtains

$$i\frac{\partial \psi^+}{\partial x_\nu}\gamma^\nu + m\psi^+ = 0.$$

But this equation does not have a "correct" form from the point of view of the Lorentz transformation symmetry. So, multiplying on the right by the matrix $\gamma^0$ and taking into account that

$$\gamma_0\gamma^\nu\gamma_0 = \gamma_\nu = (\gamma^\nu)^+\,, \tag{B.13}$$

one obtains

$$i\frac{\partial \overline{\psi}(x)}{\partial x^\nu}\gamma_\nu + m\overline{\psi}(x) = 0, \tag{B.14}$$

called the *Hermitian conjugate of the Dirac equation*. The spinor

$$\overline{\psi} = \psi^+\gamma^0 \tag{B.15}$$

is called *conjugate* in Dirac's sense with respect to $\psi$. The operators $\psi$ and $\overline{\psi}$ are usually written as rows and columns with four components, but this kind of writing is not very comfortable. Most frequently, in practice is used the component-form (B.12).

## 3. Transformation properties

This paragraph is concerned with presentation – without demonstration – of the transformation formulas of the field spinorial functions.

According to the general considerations exposed in paragraph I.1.3, under a *translation* of the coordinate axes one has

$$x \to x' = x + a \quad \Rightarrow \quad \psi(x) \to \psi'(x') = \psi(x).$$

Under a spatial rotation of angle $\theta$ in the plane $x_m x_n$ expressed by

$$x'_m = x_m\cos\theta - x_n\sin\theta, \quad x'_n = x_n\cos\theta + x_m\sin\theta,$$

the matrix form of the corresponding transformation of the field function writes

$$\psi(x) \to \psi'(x') = \Lambda_{(mn)}(\theta)\psi(x), \tag{B.16}$$

where the rotation matrix $\Lambda$ has the following form,

$$\Lambda_{(mn)}(\theta) = \exp\left(-i\sigma_{mn}\frac{\theta}{2}\right).$$

Here the matrix $\sigma_{mn}$, introduced by (B.2), is the *spin-tensor matrix*. Then one can write

$$\psi'(x') = \cos\frac{\theta}{2}\psi(x) - i\sin\frac{\theta}{2}\sigma_{mn}\psi(x). \tag{B.17}$$

Under a Lorentz rotation in the plane $x^0x^n$, $(n = 1, 2, 3)$,

$$x'_0 = x_0\cosh\theta - x_n\sinh\theta, \quad x'_n = x_n\cosh\theta - x_0\sinh\theta$$
$$\Big(\tan\theta = v(= v/c)\Big),$$

one obtains

$$\Lambda_{(0n)}(\theta) = \exp\left(i\sigma_{0n}\frac{\theta}{2}\right) = \cosh\frac{\theta}{2} + i\sigma_{0n}\sinh\frac{\theta}{2}.$$

The expressions for $\Lambda$ in these two cases (spatial and Lorentz-type rotations) can be unified in a single relation

$$\Lambda_{(\mu\nu)}(\theta) = \exp\left(-i\sigma^{\mu\nu}\frac{\theta}{2}\right). \tag{B.18}$$

It can be shown that the above results are in accordance with the Dirac equation. In this respect, one must show that the matrices $\Lambda_{(\mu\nu)}$, corresponding to the transformation

$$x \to x' - Lx, \quad x'_\nu - L_{\nu\mu}x^\mu,$$

satisfy the following relations:

$$\Lambda^{-1}_{(\lambda\rho)}\gamma_\nu\Lambda_{(\lambda\rho)} = L_{\nu\mu}\gamma_\mu, \quad \Lambda\gamma_\mu\Lambda^{-1} = L^{-1}_{\mu\nu}\gamma_\nu. \tag{B.19}$$

Multiplying now the left hand side of the Dirac equation (B.12) by $\Lambda$, one obtains

$$\Lambda\left(i\gamma_\nu\partial^\nu - m\right)\psi = \left[i\left(\Lambda\gamma_\nu\Lambda^{-1}\right)\partial^\nu - m\right]\Lambda\psi$$
$$= \left(iL^{-1}_{\nu\mu}\gamma_\mu\partial^\nu - m\right)\psi'(x') = \left(i\hat{\partial}' - m\right)\psi'(x') = 0.$$

As it is known, an equation is called *covariant* with respect to certain transformations if after transformation (*i.e.* rewritten in terms of the transformed functions and variables) keeps its initial form.

Therefore, it has been shown that, by virtue of the matrix transformations (B.16), the Dirac equation is covariant with respect to Lorentz transformations.

The same conclusion can be obtained for the conjugate Dirac's equation, by means of the transformation formulas for the conjugated spinor,

$$\overline{\psi}(x') = \overline{\psi}(x)\Lambda^{-1}(\theta), \tag{B.20}$$

which follows from (B.16) by Hermitian conjugation and multiplication of the r.h.s by $\gamma_0$, and also taking into account that

$$\gamma_0\Lambda^+\gamma_0 = \gamma_0\exp\left(i\left(\sigma^{\mu\nu}\right)^+\frac{\theta}{2}\right)\gamma_0 = \exp\left(i\sigma^{\mu\nu}\frac{\theta}{2}\right) = \Lambda^{-1}.$$

According to Eq. (B.17) (which expresses a spatial rotation), the non-uniqueness of spinorial functions follows in a straight manner. Taking $\theta = 2\pi$, it follows that to a complete spatial rotation of the coordinate system corresponds the transformation matrix $\Lambda(2\pi) = -1$, meaning that under such a transformation the sign of the field function changes. Since under a $2\pi$-rotation the coordinate system comes to the initial position (identical transformation), one can conclude that the spinorial wave functions are always determined up to a sign.

Here is another form of the $\Lambda$-matrix for the transformations of reflection of different number of coordinate axes. Observing that the transformation formulas for reflections of an even number of different spatial axes, which reduce to rotations, follow from Eq. (B.17), one must consider the reflection transformations of all three spatial axes (that is a parity transformation, $P$):

$$x'^n = -x^n \quad (n = 1, 2, 3), \qquad x'^0 = x^0,$$
$$\psi'(x') = \eta(P)\Lambda_{123}\psi(x), \quad \Lambda_{123} = \gamma^0, \tag{B.21}$$

since the spinorial representation is two valued, and thus the phase factor $\eta$ obeys the condition $\eta^2(P) = \pm 1$.

The "Lorentz transformation" formulas for the matrices $\gamma_\mu$, together with transformation formulas for $\psi$ and $\overline{\psi}$ allow one to obtain the laws of transformation for the bilinear forms $\overline{\psi} M \psi$. Indeed, by virtue of (B.16) and (B.20), one has

$$\overline{\psi}'(x') M \psi'(x') = \overline{\psi}(x) \Lambda^{-1} M \Lambda \psi(x).$$

In view of (B.19), one obtains:

a) $M = 1, \quad \overline{\psi}'(x')\psi'(x') = \overline{\psi}(x)\psi(x),$

b) $M = \gamma_\nu, \quad \overline{\psi}'(x')\gamma_\nu\psi'(x') = L_{\nu\mu}\overline{\psi}(x)\gamma_\mu\psi(x),$

c) $M = \sigma_{\nu\mu}, \quad \overline{\psi}'(x')\sigma_{\nu\mu}\psi'(x') = L_{\nu\nu'}L_{\mu\mu'}\overline{\psi}(x)\sigma_{\nu'\mu'}\psi(x).$

This way, the analyzed quadratic forms transform according to the tensorial representations of the Lorentz group. Particularly interesting is the case when the matrix $\gamma_5$ appears as a factor in $M$. It can be easily shown that the rotation matrix (B.18) commutes with $\gamma_5$:

$$\Lambda^{-1}(\theta)\gamma_5\Lambda(\theta) = \gamma_5.$$

At the same time, the matrix (B.21) of reflection of an odd number of spatial axes anti-commutes with $\gamma_5$:

$$\Lambda_{123}^{-1}\gamma_5\Lambda_{123} = \gamma_0\gamma_5\gamma_0 = -\gamma_5.$$

Consequently, the following classification of bilinear forms is obtained:

$\overline{\psi}\psi$ – scalar;

$\overline{\psi}\gamma_\nu\psi$ – four-vector;

$\overline{\psi}\sigma_{\mu\nu}\psi$ – second rank tensor;

$\overline{\psi}\gamma_5\psi$ – pseudoscalar;

$\overline{\psi}\gamma_\nu\gamma_5\psi$ – pseudo four-vector.

## 4. Non-relativistic limit

The non-relativistic limit of the Dirac equation corresponds to the case when $|\vec{p}| \ll m$. In order to pass to the non-relativistic limit it is convenient to represent the four-component spinor in the four-dimensional space-time by two 2-component spinors, $\chi$ and $\varphi$, in the Euclidean 3-dimensional space

$$u(p) = \begin{pmatrix} \varphi(p) \\ \chi(p) \end{pmatrix}, \quad \varphi = \begin{pmatrix} \varphi_1 \\ \varphi_2 \end{pmatrix}, \quad \chi = \begin{pmatrix} \chi_1 \\ \chi_2 \end{pmatrix}. \tag{B.22}$$

Thus, choosing the standard representation (B.5) of Dirac matrices, one arrives to matrices $\vec{\alpha}$ and $\beta$:

$$\vec{\alpha} = \begin{pmatrix} 0 & \vec{\sigma} \\ \vec{\sigma} & 0 \end{pmatrix}, \quad \beta = \begin{pmatrix} I & 0 \\ 0 & -I \end{pmatrix}. \tag{B.23}$$

Here $\vec{\sigma}$ are the Pauli (second-rank and quadratic) matrices given by (B.8). From Eq. (B.12), written in momentum representation, it follows that

$$\left(p^0\gamma_0 - \vec{p}\cdot\vec{\gamma} - mI\right) u(\vec{p}) = 0, \tag{B.24}$$

where by $u(\vec{p})$ is denoted the negative frequency part $u^-(\vec{p})$ of the spinor. Multiplying (B.24) on the left by $\gamma_0 = \beta$ and using (B.22) and (B.23), the following two equations for the 2-component spinors $\varphi$ and $\chi$ are obtained:

$$\left(p^0 - m\right) \varphi(\vec{p}) + (\vec{p}\cdot\vec{\sigma})\,\chi(\vec{p}) = 0, \tag{B.25}$$

$$(\vec{p}\cdot\vec{\sigma})\,\varphi(\vec{p}) + \left(p^0 + m\right) \chi(\vec{p}) = 0. \tag{B.26}$$

In the non-relativistic approximation, the spinor $\chi$ is much smaller than the spinor $\varphi$. Indeed, from (B.26) it follows that

$$\chi(\vec{p}) = \frac{(\vec{\sigma}\cdot\vec{p})}{m+p^0}\varphi(\vec{p}) \simeq \frac{(\vec{\sigma}\cdot\vec{p})}{2m}\varphi(\vec{p}). \tag{B.27}$$

That is why, usually, the spinor $\varphi$ describes the electronic components (see (1.158)), and spinor $\chi$ – the positronic components.

Substituting (B.27) into (B.25) and (B.26), the following equation for $\varphi$ is obtained:

$$p^0\varphi(\vec{p}) = \left[m + \frac{(\vec{\sigma}\cdot\vec{p})\,(\vec{\sigma}\cdot\vec{p})}{2m}\right]\varphi(\vec{p}).$$

Using the matrix identity

$$(\vec{\sigma}\cdot\vec{a})\left(\vec{\sigma}\cdot\vec{b}\right) = \left(\vec{a}\cdot\vec{b}\right) - i\vec{\sigma}\cdot\left(\vec{a}\times\vec{b}\right), \tag{B.28}$$

one can observe that the operator corresponding to the r.h.s. of equation for $\varphi$ is diagonal in relation to spinorial indices. Switching to coordinate representation and putting into evidence the factor $e^{imt}$,

$$\varphi(\vec{x},t)e^{-imt} = \int e^{-ip^0t+i\vec{p}\cdot\vec{x}}\varphi(\vec{p})\,d\vec{p},$$

one obtains for $\varphi(\vec{x},t)$,

$$i\frac{\partial\varphi(\vec{x},t)}{\partial t} = -\frac{1}{2m}\nabla^2\varphi(\vec{x},t),$$

which is Schrödinger's equation.

In the presence of the external electromagnetic field, instead of (B.24) one has to write

$$\left[\left(p^0 - eA_0\right)\gamma_0 - \left(\vec{p} - e\vec{A}\right)\cdot\vec{\gamma} - mI\right]u(\vec{p}) = 0. \tag{B.29}$$

As a result of calculations similar to those used above for the 2-component spinor, neglecting terms in $eA_0$ as compared to $2m$, one obtains

$$i\frac{\partial\varphi}{\partial t} = \left[-\frac{1}{2m}\left(\nabla - ie\vec{A}\right)^2 + eA_0 - \frac{e}{2m}\left(\vec{\sigma}\cdot\vec{H}\right)\right]\varphi. \tag{B.30}$$

Here $\vec{H} = \operatorname{curl}\vec{A}$ is the magnetic field intensity vector. Equation (B.30) is called *Pauli's equation*. It is similar to Schrödinger's equation for the Hamiltonian

$$H = \frac{1}{2m}\left(\vec{p} - e\vec{A}\right)^2 + eA_0 - \left(\vec{\mu}\cdot\vec{H}\right), \tag{B.31}$$

where the last term represents the energy of the magnetic dipole of moment $\vec{\mu}$,

$$\vec{\mu} = \mu_0\vec{\sigma}, \quad \mu_0 = \frac{e}{2m}\left(= \frac{e\hbar}{2mc}\right), \tag{B.32}$$

in the external magnetic field $\vec{H}$. The quantity $\mu_0$ is the electron magnetic moment, called *Bohr's magneton*.

## Appendix C

# Operatorial Transformations

### 1. Continuous linear transformations

Consider a continuous linear transformation of the operators $a_i$, depending on the parameter $\varphi$,

$$a_i - a_i' = L_{ij}(\varphi)a_j, \quad a' = U^{-1}(\varphi)aU(\varphi)$$

and possessing the group property

$$L(\varphi_1)L(\varphi_2) = L(\varphi_1 + \varphi_2).$$

In this case, the concrete form of the operator $U$ which, due to the above mentioned properties, is necessarily written in the exponential form

$$U(\varphi) = \exp(i\varphi V), \quad V^+ = V,$$

can be determined by means of the following general example.

Consider the following infinitesimal ($\varphi \to \varepsilon \ll 1$) transformation:

$$a' \simeq a + \varepsilon A, \quad U = 1 + i\varepsilon V.$$

Equalizing

$$a + \varepsilon A \simeq (1 - i\varepsilon V)a(1 + i\varepsilon V),$$

one finds

$$A = i[a, V], \tag{C.1}$$

which is the basic relation used to determine the concrete form of the operator $V$. Here are several important examples.

**a) Translation transformation with quantity (number) c:**

$$a' = a + \varphi, \quad a^{+\prime} = a^+ + \varphi^+,$$

or

$$a' = a + \varphi + i\psi, \quad a^{+\prime} = a^+ + \varphi^+ - i\psi^+,$$

$$U = \exp(i\varphi u + i\psi v).$$

The basic equation gives

$$1 = i(au - ua), \quad 1 = i[a^+, u],$$

$$1 = [a, v], \qquad -1 = [a^+, v].$$

One observes that $u$ and $v$ must be linear in terms of the bosonic operators $a$ and $a^+$:

$$[a, a^+] = aa^+ - a^+a = 1, \quad [a, a] = [a^+, a^+] = 0.$$

One finds

$$u = i(a - a^+), \quad v = (a + a^+),$$

which gives

$$U = \exp(a^+\varphi - a\varphi^+). \tag{C.2}$$

It is worth mentioning that a similar solution corresponding to fermionic operators does not exist. In this case, the translation operation requires replacement of the quantities (numbers) $c$ of the objects $\varphi$, $\varphi^+$ by the generators of the corresponding Grassmann algebra.

**b) Rotation transformation $\{a = (a_1, a_2)\}$:**

$$a_1' = a_1 \cos\varphi - a_2 \sin\varphi, \quad \text{and similarly for } a_1^{+\prime};$$

$$a_2' = a_2 \cos\varphi + a_1 \sin\varphi, \quad \text{and similarly for } a_2^{+\prime}.$$

To such transformations reduce, for example, the rotation transformations in the isotopic space.

The basic equation gives

$$a_1 = i[a_2, V], \quad -a_2 = i[a_1, V], \tag{C.3}$$

and for the Hermitian conjugate quantities,

$$a_1^+ = i[a_2^+, V], \quad -a_2^+ = i[a_2^+, V]. \tag{C.3$'$}$$

Unlike the previous case, the operator $V$ must be bilinear with respect to $a_k$ and $a_k^+$, $(k = 1, 2)$:

$$V = ca_1^+ a_1 + da_2^+ a_2 + e(a_1^+ a_2 + a_2^+ a_1) + if(a_1^+ a_2 - a_2^+ a_1).$$

Substituting this general Hermitian expression into (C.3) and (C.3′) one finds $c = d = e = 0,\ f = 1$, so that

$$U = \exp[\varphi(a_2^+ a_1 - a_1^+ a_2)]. \tag{C.4}$$

This expression is valid bot both bosonic and fermionic operators.

**c) Bogoliubov transformation for bosonic operators**:

$$\begin{aligned} b_1 \to \beta_1 &= b_1 \cosh\varphi + b_2^+ \sinh\varphi, & \beta_1^+ &= b_1^+ \cosh\varphi + b_2 \sinh\varphi, \\ b_2 \to \beta_2 &= b_1^+ \sinh\varphi + b_2 \cosh\varphi, & \beta_2^+ &= b_1 \sinh\varphi + b_2^+ \cosh\varphi. \end{aligned} \tag{C.5}$$

The basic equation yields

$$b_2^+ = i[b_1, V], \quad b_1^+ = i[b_2, V]$$

and the corresponding Hermitian conjugate as well. The solution to this equations leads to

$$V = i\left(b_1 b_2 - b_2^+ b_1^+\right) = i\left(\beta_1 \beta_2 - \beta_2^+ \beta_1^+\right),$$

so that

$$U(\varphi) = \exp\left[\varphi\left(b_2^+ b_1^+ - b_1 b_2\right)\right]. \tag{C.6}$$

**d) Bogoliubov transformation for fermionic operators**:

$$\begin{aligned} \alpha_1 &= +a_1 \cos\varphi + a_2^+ \sin\varphi, & \alpha_1^+ &= +a_1^+ \cos\varphi + a_2 \sin\varphi, \\ \alpha_2 &= -a_1^+ \sin\varphi + a_2 \cos\varphi, & \alpha_2^+ &= -a_1 \sin\varphi + a_2^+ \cos\varphi. \end{aligned} \tag{C.7}$$

The basic equation yields

$$a_1^+ = -i[a_2, V], \quad a_2^+ = i[a_1, V],$$

and the corresponding Hermitian conjugate as well. The solution to this equation leads to

$$V = i(a_1 a_2 - a_2^+ a_1^+) = i\left(\alpha_1\alpha_2 - \alpha_2^+\alpha_1^+\right),$$

and, therefore,

$$U(\varphi) = \exp\left[\varphi\left(a_2^+ a_1^+ - a_1 a_2\right)\right]. \quad \text{(C.8)}$$

**e) Lorentz-type rotations of Dirac matrices**

Let $U(\varphi)$ be the operator of the Dirac matrices transformation

$$\gamma_\mu \to \tilde{\gamma}_\mu = L_{\mu\nu}(\varphi)\gamma_\nu,$$

where $L$ is the Lorentz-type 4-dimensional rotations matrix of the coordinate vector, $x' = L(\varphi)x$. For the spatial (Euclidean) rotation in the $x_1Ox_2$ plane, one has

$$\begin{aligned}
\tilde{\gamma}_1 &= \gamma_1 \cos\varphi - \gamma_2 \sin\varphi,\\
\tilde{\gamma}_2 &= \gamma_1 \sin\varphi + \gamma_2 \cos\varphi,\\
\tilde{\gamma}_0 &= \gamma_0,\\
\tilde{\gamma}_3 &= \gamma_3,
\end{aligned}$$

and for the Lorentz-type rotation in $x_0Ox_1$ plane,

$$\begin{aligned}
\tilde{\gamma}_0 &= \gamma_0 \cosh\varphi + \gamma_1 \sinh\varphi,\\
\tilde{\gamma}_1 &= \gamma_0 \sinh\varphi + \gamma_1 \cosh\varphi,\\
\tilde{\gamma}_2 &= \gamma_2,\\
\tilde{\gamma}_3 &= \gamma_3.
\end{aligned}$$

The basic equation yields

$$\begin{aligned}
-\gamma_2 = i[\gamma_1, V_{12}],\ \gamma_1 = i[\gamma_2, V_{12}],\ [\gamma_0, V_{12}] = [\gamma_3, V_{12}] = 0;\\
\gamma_1 = i[\gamma_0, V_{01}],\ \gamma_0 = i[\gamma_1, V_{01}],\ [\gamma_2, V_{01}] = [\gamma_3, V_{01}] = 0.
\end{aligned}$$

The solution to these equations is of the form

$$V_{12} = -\frac{i}{2}\gamma_1\gamma_2, \quad V_{01} = -\frac{i}{2}\gamma_0\gamma_1.$$

The obtained results can be unified in a single formula

$$U_{\mu\nu} = \exp\left(\frac{\varphi}{2}\gamma_\mu\gamma_\nu\right). \tag{C.9}$$

## 2. Exponential simplification

Consider

$$U = \exp(A + B),$$

where $A$ and $B$ do not commute with each other. The dependence on operators $A$ and $B$ has to be factored. If, in particular, $A$ contains only creation operators and $B$ only annihilation operators, such a problem appears when the operator $U$ is brought to normal form.

In general, the operator $U$ can be represented as

$$U = e^A O e^B,$$

where the factor $O$, generally speaking, contains the dependence on operators $A$ and $B$ as well as on their repeated commutators.

In order to determine the concrete form of $O$, the method of infinitesimal transformations shall be used. To this end, one introduces the auxiliary parameter $\tau$,

$$U \to U(\tau) = e^{\tau(A+B)} = e^{\tau A} O\left(\tau, A, B, [A, B], \ldots\right) e^{\tau B}$$

and, differentiating in terms of $\tau$, one obtains for $O$ the following operatorial differential equation:

$$O'(\tau) = e^{-\tau A} B e^{\tau A} O(\tau) - O(\tau)B, \quad O(\tau = 0) = 1.$$

The case becomes very simple if the commutators of the operators $A$ and $B$

$$[A, B]$$

is a $c$-number (see (C.2)). Therefore,

$$e^{-\tau A} B e^{\tau A} = B - \tau C, \quad O(\tau) = \exp\left(-\frac{\tau^2 C}{2}\right),$$

leading to the so-called *Baker–Hausdorff formula*,

$$e^{A+B} = e^A e^{-\frac{C}{2}} e^B, \quad (C = [A, B] \sim \text{a } c - \text{number}). \tag{C.10}$$

In order to simplify the formulas (C.6) and (C.8) appearing as a result of the Bogoliubov's canonical transformation, an important case arises if the commutator $C$ is an operator, and the operators $A, B$ and $C$ form a closed commutative algebra of a special form,

$$[A, B] = C, \quad [A, C] = -2\xi A, \quad [B, C] = 2\xi B. \tag{C.11}$$

This case can be also analyzed by a differential method. The reader has the opportunity to convince himself that the searched solutions are of the following form,

$$e^{\tau(A+B)} = e^{\alpha(\tau)A} e^{\gamma(\tau)C} e^{\alpha(\tau)B}, \tag{C.12}$$

where

$$\alpha(\tau, \xi) = \frac{1}{\sqrt{\xi}} \tanh\left(\tau\sqrt{\xi}\right), \quad \gamma(\tau, \xi) = \frac{1}{\xi} \ln \cosh\left(\tau\sqrt{\xi}\right). \tag{C.13}$$

In the case of bosonic operators

$$A = b_1^+ b_2^+, \quad B = -b_1 b_2, \quad C = 1 - b_1 b_1^+ - b_2 b_2^+, \quad \xi = 1,$$

and for the fermionic operators

$$A = a_1^+ a_2^+, \quad B = -a_1 a_2, \quad C = 1 - a_1 a_1^+ - a_2 a_2^+, \quad \xi = 1.$$

## 3. Commutators with particle density operators $n(\vec{k})$

The reader is reminded of the following useful formula for the commutators with the particle density operators:

$$n(\vec{k}) = a^+(\vec{k}) a^-(\vec{k}).$$

It is worth defining the following linear superposition of the operators $a^+$ and $a^-$:

$$A = A_+ + A_-, \quad A_\pm = \int \lambda_\pm(\vec{q}) a^\pm(\vec{q})\, d\vec{q}. \tag{C.14}$$

Then

$$\left[n(\vec{k}), A_\pm\right] = \pm\lambda_\pm(\vec{k}) a^\pm(\vec{k})$$

and

$$\left[n(\vec{k}), A\right] = K(\vec{k}) = \lambda_+(\vec{k}) a^+(\vec{k}) - \lambda_-(\vec{k}) a^-(\vec{k}),$$

which yields a $c$-number for the commutator of operators $A$ and $K$:

$$\left[A, K(\vec{k})\right] = -2\lambda^2(\vec{k}), \quad \lambda^2(\vec{k}) \equiv \lambda_+(\vec{k})\lambda_-(\vec{k}).$$

Therefore

$$\left[n(\vec{k}), A^\nu\right] = \nu K(\vec{k}) A^{\nu-1} - \nu(\nu-1)\lambda^2 A^{\nu-2}$$

and, consequently,

$$\left[n(\vec{k}), e^{\pm A}\right] = \left\{\pm K(\vec{k}) + \lambda^2(\vec{k})\right\} e^{\pm A}. \tag{C.15}$$

# Appendix D

# Singular Functions

## 1. Some useful singular functions

The four-dimensional Dirac function writes

$$\delta(x) = \frac{1}{(2\pi)^4} \int e^{ikx}\, dk, \tag{D.1}$$

$$\delta(x) = \delta(x^0)\delta(\vec{x}) = \delta(x^0)\delta(x^1)\delta(x^2)\delta(x^3). \tag{D.2}$$

The related functions $\theta(\alpha)$ and $\varepsilon(\alpha)$ are

$$\theta(\alpha) = \frac{1}{2\pi i} \int_{-\infty}^{+\infty} \frac{e^{i\alpha\tau}}{\tau - i\varepsilon}\, d\tau = \begin{cases} 1, & \alpha > 0, \\ 0, & \alpha < 0. \end{cases} \qquad (\varepsilon \to +0), \tag{D.3}$$

$$\varepsilon(\alpha) = \frac{1}{\pi i} \mathcal{P} \int_{-\infty}^{+\infty} e^{i\alpha\tau} \frac{d\tau}{\tau} = \begin{cases} +1, & \alpha > 0, \\ -1, & \alpha < 0, \end{cases} \tag{D.4}$$

where $\mathcal{P}$ designates the *principal part* of the quantity that follows.

The positive and negative frequency parts of the $\delta$-function are

$$\delta_{\pm}(\alpha) = \frac{1}{2\pi} \int_0^{\infty} e^{\pm i\alpha\tau}\, d\tau = \frac{1}{2} \left[ \delta(\alpha) \pm \frac{i}{\pi} \mathcal{P} \frac{1}{\alpha} \right]. \tag{D.5}$$

Some useful relations:

$$\frac{1}{\alpha + i\varepsilon} = \frac{2\pi}{i} \delta_{+}(\alpha) = \frac{\pi}{i} \delta(\alpha) + \mathcal{P} \frac{1}{\alpha}, \quad (\varepsilon \to +0), \tag{D.6}$$

$$\frac{1}{\alpha - i\varepsilon} = 2\pi i \delta_{-}(\alpha) = \pi i \delta(\alpha) + \mathcal{P} \frac{1}{\alpha}, \tag{D.7}$$

$$\mathcal{P}\frac{1}{\alpha} = \frac{1}{2i}\int_{-\infty}^{+\infty} \varepsilon(\tau)e^{i\alpha\tau}\, d\tau. \tag{D.8}$$

## 2. The scalar field functions

The Pauli–Dirac *commutation function* $D$ is introduced by means of

$$[\varphi(x), \varphi(y)] = \frac{1}{i}D(x-y), \tag{D.9}$$

being given by

$$\begin{aligned} D(x) &= \frac{i}{(2\pi)^3}\int e^{-ikx}\varepsilon(k^0)\delta(k^2-m^2)\, dk \\ &= \frac{1}{(2\pi)^3}\int \frac{d\vec{k}}{\sqrt{k^2+m^2}} e^{i\vec{k}\cdot\vec{x}} \sin\left(x^0\sqrt{\vec{k}^{\,2}+m^2}\right). \end{aligned} \tag{D.10}$$

One has

$$\left.\frac{\partial D(x^0,\vec{x})}{\partial x^0}\right|_{x^0=0} = \delta(\vec{x}), \tag{D.11}$$

$$\begin{aligned} D(x) &= \frac{1}{2\pi}\varepsilon(x^0)\delta(\lambda) - \frac{m}{4\pi\sqrt{\lambda}}\varepsilon(x^0)\theta(\lambda)J_1\left(m\sqrt{\lambda}\right), \\ \lambda &= x^2 = x_0^2 - \vec{x}^{\,2}. \end{aligned} \tag{D.12}$$

In the vicinity of the light cone, $D$ gets the form

$$D(x) \sim \frac{1}{2\pi}\varepsilon(x^0)\delta(\lambda) - \frac{m^2}{8\pi}\varepsilon(x^0)\theta(\lambda). \tag{D.13}$$

The positive and negative frequency parts $D^+$ and $D^-$ of the Pauli–Dirac function are introduced by

$$\left[\varphi^-(x), \varphi^+(y)\right] = \langle\varphi(x)\varphi(y)\rangle_0 = \frac{1}{i}D^-(x-y), \tag{D.14}$$

$$\left[\varphi^+(x), \varphi^-(y)\right] = \frac{1}{i}D^+(x-y) = iD^-(y-x), \tag{D.15}$$

and are given by

$$D^{\pm}(x) = \frac{\pm 1}{(2\pi)^3 i}\int e^{ikx}\theta(\pm k^0)\delta(k^2-m^2)\,dk$$
$$= \frac{\mp 1}{(2\pi)^3}\int \frac{d\vec{k}}{2\sqrt{\vec{k}^{\,2}+m^2}} e^{\pm i x^0\sqrt{\vec{k}^{\,2}+m^2}-\vec{x}\cdot\vec{k}}$$
$$= \frac{1}{8\pi}\varepsilon(x^0)\delta(\lambda) - \frac{m\theta(\lambda)}{8\pi\sqrt{\lambda}}\left[\varepsilon(x^0)J_1\left(m\sqrt{\lambda}\right) \mp iN_1\left(m\sqrt{\lambda}\right)\right]$$
$$\mp \frac{mi\theta(-\lambda)}{4\pi^2|\lambda|^{1/2}}K_1\left(m\sqrt{-\lambda}\right). \tag{D.16}$$

In the vicinity of the light cone, $D^+$ and $D^-$ become

$$D^{\pm}(x) = \frac{\varepsilon(x^0)\delta(\lambda)}{4\pi} \pm \frac{i}{4\pi^2\lambda} \mp \frac{im^2}{8\pi^2}\ln\frac{m|\lambda|^{1/2}}{2} - \frac{m^2}{16\pi^2}\varepsilon(x^0)\theta(\lambda). \tag{D.17}$$

The parity and "reality" properties are expressed by

$$D^{\pm}(-x) = -D^{\mp}(x) = -\left(D^{\pm}(x)\right)^*, \quad D(x) = -D(-x) = D^*(x). \tag{D.18}$$

The causal Green function $D^c$ is introduced by

$$\langle T\varphi(x)\varphi(y)\rangle_0 = \frac{1}{i}D^c(x-y), \tag{D.19}$$

$$\left(\Box - m^2\right)D^c(x) = -\delta(x), \tag{D.20}$$

being given by

$$D^c(x) = \frac{1}{(2\pi)^4}\int e^{-i\vec{k}\cdot\vec{x}}D^c(k)dk, \quad D^c(k) = \frac{1}{m^2-k^2-i\varepsilon}, \tag{D.21}$$

$$D^c(x) = \frac{mi}{4\pi^2}\frac{K_1\left(m\sqrt{-\lambda+i\varepsilon}\right)}{\sqrt{-\lambda+i\varepsilon}}. \tag{D.22}$$

One also has

$$D_0^c(x) = \frac{1}{4\pi}\left(\delta(\lambda) - \frac{i}{\pi\lambda}\right) = \frac{1}{4\pi^2 i(\lambda - i\varepsilon)}. \tag{D.23}$$

In the vicinity of the light cone,

$$D^c(x) \simeq \frac{1}{4\pi}\delta(\lambda) - \frac{i}{4\pi^2\lambda} + \frac{im^2}{8\pi^2}\ln|\lambda|^{1/2} - \frac{m^2}{16\pi}\theta(\lambda). \tag{D.24}$$

The retarded and advanced Green functions are given by

$$D^{ret}(x) = 0 \quad \text{for} \quad x^0 < 0 \quad \text{and} \quad D^{adv}(x) = 0 \quad \text{for} \quad x^0 > 0,$$

$$D^{ret}(x) = \frac{1}{(2\pi)^4}\int \frac{e^{-ikx}}{m^2 - k^2 - i\varepsilon k^0}\,dk = \theta\left(x^0\right)D(x), \tag{D.25}$$

$$D^{adv}(x) = \frac{1}{(2\pi)^4}\int \frac{e^{-ikx}}{m^2 - k^2 + i\varepsilon k^0}\,dk = -\theta\left(-x^0\right)D(x), \tag{D.26}$$

$$D^{ret}(x) = \frac{1}{(2\pi)}\theta\left(x^0\right)\left\{\delta(\lambda) - \theta(\lambda)\frac{m}{2\sqrt{\lambda}}J_1\left(m\sqrt{\lambda}\right)\right\}, \tag{D.27}$$

$$D^{adv}(x) = \frac{1}{(2\pi)}\theta\left(-x^0\right)\left\{\delta(\lambda) - \theta(\lambda)\frac{m}{2\sqrt{\lambda}}J_1\left(m\sqrt{\lambda}\right)\right\}. \tag{D.28}$$

The connections between functions $D$, $D^+$, $D^-$, $D^c$, $D^{ret}$, and $D^{adv}$ are

$$D(x) = D^+(x) + D^-(x), \tag{D.29}$$

$$D^c(x) = \theta\left(x^0\right)D^-(x) - \theta\left(-x^0\right)D^+(x), \tag{D.30}$$

$$D^{ret}(x) = \theta\left(x^0\right)D(x) = D^c(x) + D^+(x), \tag{D.31}$$

$$D^{adv}(x) = -\theta\left(-x^0\right)D(x) = D^c(x) - D^-(x), \tag{D.32}$$

$$D(x) = D^{ret}(x) - D^{adv}(x). \tag{D.33}$$

## 3. Functions of the electromagnetic, massive-vector and spinor fields

The electromagnetic field:

$$[A_\mu(x), A_\nu(y)] = ig_{\mu\nu}D_0(x-y), \tag{D.34}$$

$$\langle A_\mu(x)A_\nu(y)\rangle_0 = ig_{\mu\nu}D_0^-(x-y), \tag{D.35}$$

$$\langle TA_\mu(x)A_\nu(y)\rangle_0 = ig_{\mu\nu}D_0^c(x-y). \tag{D.36}$$

The functions $D_0$, $D_0^{\pm}$, $D_0^c$ and similar other ones are obtained from the corresponding functions of the scalar field $D$, $D^{\pm}$, $D^c$ by setting $m = 0$. For example,

$$D_0^c(x) = D^c(x)\Big|_{m=0} = -\frac{1}{(2\pi)^4}\int \frac{e^{-ikx}}{k^2 + i\varepsilon}\, dk. \tag{D.37}$$

For the massive vector field, one has

$$[U_\mu(x), U_\nu(y)] = iD_{\mu\nu}(x-y), \tag{D.38}$$

$$\langle U_\mu(x)U_\nu(y)\rangle_0 = iD^-_{\mu\nu}(x-y), \tag{D.39}$$

$$\langle TU_\mu(x)U_\nu(y)\rangle_0 = iD^c_{\mu\nu}(x-y). \tag{D.40}$$

The functions $D_{\mu\nu}$, $D^{\pm}_{\mu\nu}$, $D^c_{\mu\nu}$ and other similar ones can be obtained from the corresponding functions of the scalar field by means of the derivative operator $g_{\mu\nu} + \partial_\mu\partial_\nu/m^2$. For example,

$$D^c_{\mu\nu}(x) = \frac{1}{(2\pi)^4}\int \frac{g_{\mu\nu} - \dfrac{k_\mu k_\nu}{m^2}}{m^2 - k^2 - i\varepsilon} e^{-ikx}\, dk. \tag{D.41}$$

For the spinor field, one has

$$\left[\psi_\alpha(x), \overline{\psi}_\beta(y)\right] = -iS_{\alpha\beta}(x-y), \tag{D.42}$$

$$\left\langle \psi_\alpha(x)\overline{\psi}_\beta(y)\right\rangle_0 = -iS^-_{\alpha\beta}(x-y), \tag{D.43}$$

$$\left\langle T\psi_\alpha(x)\overline{\psi}_\beta(y)\right\rangle_0 = -iS^c_{\alpha\beta}(x-y). \tag{D.44}$$

The functions $S$, $S^{\pm}$, $S^c$ and other similar ones can be obtained from the corresponding functions of the scalar field by means of the operator

$$\left(i\hat{\partial} + m\right)_{\alpha\beta} = i\gamma^\nu_{\alpha\beta}\partial_\nu + mI_{\alpha\beta}.$$

For example,

$$S(x) = \frac{i}{(2\pi)^3}\int e^{-ikx}\varepsilon\left(k^0\right)(\hat{k} + m)\delta\left(k^2 - m^2\right)\, dk, \tag{D.45}$$

$$S(x)|_{x^0=0} = i\gamma^0\delta(x), \tag{D.46}$$

$$S^c(x) = \frac{i}{(2\pi)^4}\int \frac{m + \hat{p}}{m^2 - p^2 - i\varepsilon} e^{-ipx}\, dp. \tag{D.47}$$

## Appendix E

# Integration Formulas in Momentum Space

### 1. Alpha representation

Transition to *alpha representation* is given by

$$\frac{1}{m^2 - p^2 - i\varepsilon} = i \int_0^\infty e^{i\alpha(p^2 - m^2 + i\varepsilon)} \, d\alpha, \tag{E.1}$$

and its generalization is

$$\frac{1}{(D - i\varepsilon)^k} = \frac{i^k}{\Gamma(k)} \int_0^\infty e^{i\alpha(-D + i\varepsilon)} \alpha^{k-1} \, d\alpha, \quad k > 0. \tag{E.2}$$

The fundamental Gaussian integrals (with $a$ – a positive number, and $b$ – a four-vector) are

$$\frac{i}{\pi^2} \int dk \; e^{i(ak^2 + 2bk)} = \frac{1}{a^2} e^{-ib^2/a} \equiv I(a, b^2), \tag{E.3}$$

$$\frac{i}{\pi^2} \int dk \; e^{i(ak^2 + 2bk)} [k^\nu] = \left[ -\frac{b^\nu}{a} \right] I(a, b^2), \tag{E.4}$$

$$\frac{i}{\pi^2} \int dk \; e^{i(ak^2 + 2bk)} [k^\nu k^\mu] = \left[ \frac{2b^\nu b^\mu + iag^{\mu\nu}}{2a^2} \right] I(a, b^2), \tag{E.5}$$

$$\frac{i}{\pi^2} \int dk \; e^{i(ak^2 + 2bk)} [k^2] = \left[ \frac{b^2 + 2ia}{a^2} \right] I(a, b^2), \tag{E.6}$$

$$\frac{i}{\pi^2} \int dk \; e^{i(ak^2 + 2bk)} [k^\nu k^2] = \left[ -\frac{b^\nu (b^2 + 3ia)}{a^3} \right] I(a, b^2). \tag{E.7}$$

The integrals with infinity as superior limit, of the type $\int_0^\infty d\alpha_1 \cdots \int_0^\infty d\alpha_n$, are written as

$$\int_0^\infty d\alpha_1 \cdots \int_0^\infty d\alpha_n = \int_0^1 \{dx\}_n \int_9^\infty a^{n-1}\, da, \tag{E.8}$$

where

$$a = \alpha_1 + \alpha_2 + \cdots + \alpha_n, \quad x_\nu = \frac{\alpha_\nu}{a}, \tag{E.9}$$

$$\{dx\}_n = dx_1 dx_2 \cdots dx_n \delta(1 - x_1 - x_2 - \cdots - x_n). \tag{E.10}$$

Some typical integrals with respect to $a$:

$$\frac{1}{i}\int_0^\infty da\ e^{ia(A+i\varepsilon)} = \frac{1}{A+i\varepsilon}, \tag{E.11}$$

$$\int_0^\infty \frac{da}{a}\left(e^{iaA} - e^{iaB}\right)e^{-\varepsilon a} = \ln\frac{B+i\varepsilon}{A+i\varepsilon}, \tag{E.12}$$

$$\begin{aligned} &\frac{1}{i}\int_0^\infty \frac{da}{a^2}\left(e^{iaA} - e^{iaB}\right)\left(e^{iaC} - e^{iaD}\right)e^{-\varepsilon a} \\ &= (A+D)\ln(A+D+i\varepsilon) + (B+C)\ln(B+C+i\varepsilon) \\ &\quad - (A+C)\ln(A+C+i\varepsilon) - (B+D)\ln(B+D+i\varepsilon). \end{aligned} \tag{E.13}$$

If $B$ and $D$ contain the large parameter $\Lambda^2\left(B = b\Lambda^2,\ D = d\Lambda^2\right)$, the equation (E.13), in the limit $\Lambda^2 \to \infty$, leads to

$$\begin{aligned} &\Lambda^2[b\ln b + d\ln d - (b+d)\ln(b+d)] + (A+C)\ln\Lambda^2 \\ &\quad + (A+C)[1 - \ln(A+C)] + A\ln d + C\ln b. \end{aligned} \tag{E.14}$$

## 2. Feynman parametrization

The starting formula is

$$\frac{1}{a_1 a_2 \ldots a_n} = (n-1)! \int_0^1 \frac{\{dx\}_n}{\left(\sum\limits_{1\le\nu\le n} a_\nu x_\nu\right)^n}, \tag{E.15}$$

where the symbol $\{dx\}_n$ is given by (E.10).

For practical purposes, the integrals over momentum of the type

$$J_L(D) = \frac{i}{\pi^2}\int \frac{dq}{(g^2 - D + i\varepsilon)^L} = \frac{(-1)^{L+1}(L-3)!}{(D-i\varepsilon)^{L-2}(L-1)!}, \quad L \ge 3, \tag{E.16}$$

can be easily generalized, resulting shift integrals, such as

$$\begin{aligned} F_L(D,k) = J_L(k^2 - D) &= \frac{i}{\pi^2}\int \frac{dp}{\left[p^2 - 2kp + D\right]^L} \\ &= -\frac{(L-3)!}{(D-k^2+i\varepsilon)^{L-2}(L-1)!}, \end{aligned} \tag{E.17}$$

$$\begin{aligned} F_L^\mu(D,k) &= \frac{i}{\pi^2}\int \frac{p^\mu dp}{\left[p^2 - 2kp + D\right]^L} \\ &= -\frac{k^\mu (L-3)!}{(D-k^2)^{L-2}(L-1)!} = k^\mu F_L(D,k), \end{aligned} \tag{E.18}$$

$$\begin{aligned} F_L^{\mu\nu}(D,k) &= \frac{i}{\pi^2}\int \frac{p^\mu p^\nu dp}{\left[p^2 - 2kp + D\right]^L} \\ &= \frac{g^{\mu\nu}(D-k^2) + 2(L-3)k^\mu k^\nu}{2(L-3)} F_L(D,k), \end{aligned} \tag{E.19}$$

$$\begin{aligned} F_L^{(2)}(D,k) &= \frac{i}{\pi^2}\int \frac{p^2 dp}{\left[p^2 - 2kp + D\right]^L} \\ &= \frac{2D - (L-1)k^2}{(L-3)} F_L(D,k), \end{aligned} \tag{E.20}$$

and others of the same type.

## 3. Measure regularization

$$\int dp = \int d^4p \to \int d^n p = \mu^{2\varepsilon} \int\limits_{\Omega(n)} d\Omega \int\limits_0^\infty p^{n-1} dp, \quad n = 4 - 2\varepsilon. \tag{E.21}$$

Here are some modified Gaussian integrals:

$$\frac{i}{\pi^2} \int d^n k \; e^{i(ak^2+2bk)} = \left(\frac{ia\mu^2}{\pi}\right)^\varepsilon \frac{1}{a^2} e^{-ib^2/a} = I_\varepsilon(a, b^2), \tag{E.22}$$

$$\frac{i}{\pi^2} \int d^n k \; e^{i(ak^2+2bk)} \left[k^\nu\right] = \left[-\frac{b^\nu}{a}\right] I_\varepsilon(a, b^2), \tag{E.23}$$

$$\frac{i}{\pi^2} \int d^n k \; e^{i(ak^2+2bk)} \left[k^\nu k^\mu\right] = \left[\frac{2b^\nu b^\mu + iag^{\nu\mu}}{2a^2}\right] I_\varepsilon(a, b^2) \tag{E.24}$$

(these three formulas are different from (E.3), (E.4) and (E.5) only by the change $I(a, b^2) \to I_\varepsilon(a, b^2)$);

$$\frac{i}{\pi^2} \int d^n k \; e^{i(ak^2+2bk)} \left[k^2\right] = \left[\frac{b^2 + ia(2-\varepsilon)}{a^2}\right] I_\varepsilon(a, b^2), \tag{E.25}$$

$$\frac{i}{\pi^2} \int d^n k \; e^{i(ak^2+2bk)} \left[k^2 k^\nu\right] = \left[-\frac{b^\nu(b^2 + ia(3-\varepsilon)}{a^3}\right] I_\varepsilon(a, b^2), \tag{E.26}$$

and some others of the same type.

The formula (E.25) can be deduced from either (E.22), by taking the derivative with respect to parameter $a$, or (E.24), by summing over the Lorentzian indices. In the second case one can easily show that

$$g_{\mu\nu} g^{\nu\mu} = \operatorname{Tr} g_{\mu\nu} = n = 4 - 2\varepsilon. \tag{E.27}$$

Using this property and performing the corresponding operations with the Dirac matrices, one obtains the well-known useful relations

$$\gamma_\mu \gamma_\nu + \gamma_\nu \gamma_\mu = 2g_{\mu\nu}, \quad (\mu, \nu = 0, 1, 2, 3). \tag{E.28}$$

The Feynman-type modified integration formulas:

$$\frac{i}{\pi^2} \int \frac{d^n p}{[p^2 - 2kp + D + i\varepsilon]^L} = \left(-\frac{i^2\mu^2}{\pi}\right)^\varepsilon \frac{\Gamma(L + \varepsilon - 2)}{(D-k)^{L+\varepsilon-2}\Gamma(L)}$$

$$\equiv F_L^\varepsilon(D, k), \tag{E.29}$$

$$\frac{i}{\pi^2}\int \frac{p^\nu d^n p}{[p^2-2kp+D]^L} = k^\nu F_L^\varepsilon(D,k), \tag{E.30}$$

$$\frac{i}{\pi^2}\int \frac{p^\nu p^\mu d^n p}{[p^2-2kp+D]^L} = \left\{k^\nu k^\mu + \frac{g^{\nu\mu}(D-k^2)}{2(L+\varepsilon-3)}\right\} F_L^\varepsilon(D,k), \tag{E.31}$$

$$\frac{i}{\pi^2}\int \frac{p^2 d^n p}{[p^2-2kp+D]^L} = \left\{k^2 + (D-k^2)\frac{2-\varepsilon}{L+\varepsilon-3}\right\} F_L^\varepsilon(D,k^2), \tag{E.32}$$

$$\frac{i}{\pi^2}\int \frac{p_\nu p^2 d^n p}{[p^2-2kp+D]^L} = k_\nu \left\{k^2 + (D-k^2)\frac{3-\varepsilon}{L+\varepsilon-3}\right\} F_L^\varepsilon(D,k^2). \tag{E.33}$$

Using the above formulas, it can be shown that for $\varepsilon \to 0$ and $\Gamma(\varepsilon) \to \frac{1}{\varepsilon} - C$ (where $C = 0.5772...$ is the *Euler constant*) one can also get

$$\frac{i}{\pi^2}\int \frac{d^n p}{p^2-2kp+D} = \frac{D-k^2}{\varepsilon} + \left\{\ln\frac{\mu^2}{(k^2-D)\pi} + 1 - C\right\}(D-k^2), \tag{E.34}$$

$$\frac{i}{\pi^2}\int \frac{d^n p}{[p^2-2kp+D]^2} = -\frac{1}{\varepsilon} + \ln\frac{(k^2-D)\pi}{\mu^2} + C, \tag{E.35}$$

$$\frac{i}{\pi^2}\int \frac{p_\nu d^n p}{[p^2-2kp+D]^2} = k_\nu\left\{-\frac{1}{\varepsilon} + \ln\frac{(k^2-D)\pi}{\mu^2} + C\right\}, \tag{E.36}$$

$$\begin{aligned}\frac{i}{\pi^2}\int \frac{p_\nu p_\mu d^n p}{[p^2-2kp+D]^2} &= \left\{-\frac{1}{\varepsilon} + \ln\frac{(k^2-D)\pi}{\mu^2} + C\right\}\\ &\quad\times\left\{k_\mu k_\nu + g_{\mu\nu}\frac{k^2-D}{2}\right\},\end{aligned} \tag{E.37}$$

$$\begin{aligned}\frac{i}{\pi^2}\int \frac{p_\nu p_\mu d^n p}{[p^2-2kp+D]^3} &= -\frac{g_{\mu\nu}}{4\varepsilon} + \frac{k_\nu k_\mu}{2(k^2-D)}\\ &\quad + \frac{g_{\mu\nu}}{4}\left[\ln\frac{(k^2-D)\pi}{\mu^2} + C\right],\end{aligned} \tag{E.38}$$

and other similar results.

# Appendix F

# Matrix Elements of the Field Operators and First-Order Vertices for the Gravitational Interactions of Particles

In order to construct the $S$-matrix elements, $\langle p'|S^{(*)}|p\rangle$ for the transition $p \xrightarrow{\text{g}} p'$ considered in the external field approximation, one must know the matrix elements of the field operators corresponding to particle creation and annihilation between one-particle state (with given four-momentum and spin-state/helicity) and the vacuum state, on the one hand, and the first-order vertices corresponding to interaction processes of particles with the external gravitational field, on the other. The matrix elements for the spin-0, 1/2, 1, 3/2, and 2 particles, determined according to the algorithm presented in Chap. III and specifications exposed in Chap. IV are synthetically presented in the following tables.

**Table 1.** Free particles.

| Particle | Type | S-matrix factor |
|---|---|---|
| Scalar mesons | $\varphi^{*(\pm)}(k);\ \varphi^{(\pm)}(k);\ \varphi_0^{(\pm)}(k)$ | $\dfrac{1}{(2\pi)^{3/2}\sqrt{2k_0}}$ |
| | $\varphi^{*(\pm)}_{,\mu}(k);\ \varphi^{(\pm)}_{,\mu}(k);\ \varphi^{(\pm)}_{0,\mu}(k)$ | $\pm\dfrac{ik_\mu}{(2\pi)^{3/2}\sqrt{2k_0}}$ |
| Photons | $A^{(a)(\pm)}_\mu(k)$ | $\dfrac{e^{(a)}_\mu(\vec{k})}{(2\pi)^{3/2}\sqrt{2k_0}}$ |
| | $A^{(a)(\pm)}_{\mu,\nu}(k)$ | $\pm\dfrac{ie^{(a)}_\mu(\vec{k})k_\nu}{(2\pi)^{3/2}\sqrt{2k_0}}$ |
| | $F^{(a)(\pm)}_{\mu\nu}(k)$ | $\pm\dfrac{i[e^{(a)}_\nu(\vec{k})k_\nu - e^{(a)}_\mu(\vec{k})k_\nu]}{(2\pi)^{3/2}\sqrt{2k_0}}$ |

| | | |
|---|---|---|
| Vector mesons | $B_\mu^{*(m)(\pm)}(k);\ B_\mu^{(m)(\pm)}(k);\ B_{0\mu}^{(m)(\pm)}(k)$ | $\dfrac{e_\mu^{(m)}(\vec{k})}{(2\pi)^{3/2}\sqrt{2k_0}}$ |
| | $G_{\mu\nu}^{*(m)(\pm)}(k);\ G_{\mu\nu}^{(m)(\pm)}(k);\ G_{0\mu\nu}^{(m)(\pm)}(k)$ | $\pm\dfrac{i[e_\nu^{(m)}(\vec{k})k_\mu - e_\mu^{(m)}(\vec{k})k_\nu]}{(2\pi)^{3/2}\sqrt{2k_0}}$ |
| | $U_\mu^{*(m)(\pm)}(k);\ U_\mu^{(m)(\pm)}(k);\ U_{0\mu}^{(m)(\pm)}(k)$ | $\dfrac{e_\mu^{(m)}(\vec{k})}{(2\pi)^{3/2}\sqrt{2k_0}}$ |
| | $U_{\mu,\nu}^{*(m)(\pm)}(k);\ U_{\mu,\nu}^{(m)(\pm)}(k);\ U_{0\mu,\nu}^{(m)(\pm)}(k)$ | $\pm\dfrac{ie_\mu^{(m)}(\vec{k})k_\nu}{(2\pi)^{3/2}\sqrt{2k_0}}$ |
| Electrons (positrons) | $\psi^{(r)(+)}(k)$ | $\dfrac{1}{(2\pi)^{3/2}}\sqrt{M/E_k}\,u^{(r)}(\vec{k})$ |
| | $\psi^{(r)(-)}(k)$ | $\dfrac{1}{(2\pi)^{3/2}}\sqrt{M/E_k}\,v^{(r)}(\vec{k})$ |
| | $\overline{\psi}^{(r)(+)}(k)$ | $\dfrac{1}{(2\pi)^{3/2}}\sqrt{M/E_k}\,\overline{v}^{(r)}(\vec{k})$ |
| | $\overline{\psi}^{(r)(-)}(k)$ | $\dfrac{1}{(2\pi)^{3/2}}\sqrt{M/E_k}\,\overline{u}^{(r)}(\vec{k})$ |
| | $\psi_{,\nu}^{(r)(+)}(k)$ | $\dfrac{ik_\nu}{(2\pi)^{3/2}}\sqrt{M/E_k}\,u^{(r)}(\vec{k})$ |
| | $\psi_{,\nu}^{(r)(-)}(k)$ | $-\dfrac{ik_\nu}{(2\pi)^{3/2}}\sqrt{M/E_k}\,v^{(r)}(\vec{k})$ |
| | $\overline{\psi}_{,\nu}^{(r)(+)}(k)$ | $\dfrac{ik_\nu}{(2\pi)^{3/2}}\sqrt{M/E_k}\,\overline{v}^{(r)}(\vec{k})$ |
| | $\overline{\psi}_{,\nu}^{(r)(-)}(k)$ | $-\dfrac{ik_\nu}{(2\pi)^{3/2}}\sqrt{M/E_k}\,\overline{v}^{(r)}(\vec{k})$ |

| | | |
|---|---|---|
| Spinor zero rest mass particles (and the corresponding antiparticles) | $\psi_0^{(r)(+)}(k)$ | $\dfrac{u^{(r)}(\vec{k})}{(2\pi)^{3/2}}$ |
| | $\psi_0^{(r)(-)}(k)$ | $\dfrac{v^{(r)}(\vec{k})}{(2\pi)^{3/2}}$ |
| | $\overline{\psi}_0^{(r)(+)}(k)$ | $\dfrac{\overline{v}^{(r)}(\vec{k})}{(2\pi)^{3/2}}$ |
| | $\overline{\psi}_0^{(r)(-)}(k)$ | $\dfrac{\overline{u}^{(r)}(\vec{k})}{(2\pi)^{3/2}}$ |
| | $\psi_{0,\nu}^{(r)(+)}(k)$ | $\dfrac{ik_\nu u^{(r)}(\vec{k})}{(2\pi)^{3/2}}$ |
| | $\psi_{0,\nu}^{(r)(-)}(k)$ | $-\dfrac{ik_\nu v^{(r)}(\vec{k})}{(2\pi)^{3/2}}$ |
| | $\overline{\psi}_{0,\nu}^{(r)(+)}(k)$ | $\dfrac{ik_\nu \overline{v}^{(r)}(\vec{k})}{(2\pi)^{3/2}}$ |
| | $\overline{\psi}_{0,\nu}^{(r)(-)}(k)$ | $-\dfrac{ik_\nu \overline{u}^{(r)}(\vec{k})}{(2\pi)^{3/2}}$ |
| Gravitons | $y_{\mu,\nu}^{(a)(\pm)}(p)$ | $\dfrac{e_{\mu\nu}^{(a)}(\vec{p})}{(2\pi)^{3/2}\sqrt{2p_0}}$ |
| | $y_{\mu,\nu,\alpha}^{(a)(\pm)}(p)$ | $\pm\dfrac{ie_{\mu\nu}^{(a)}(\vec{p})p_\alpha}{(2\pi)^{3/2}\sqrt{2p_0}}$ |

In the above relations, for the $\begin{cases}\text{incident}\\ \text{emergent}\end{cases}$ particles, one corresponds

$$\begin{cases}(\pm)\to(+),\ \vec{p}\to\vec{p},\\ (\pm)\to(-),\ \vec{p}\to\vec{p}\,'.\end{cases}$$

**Table 2.** External fields.

| Field | Type | S-matrix factor |
|---|---|---|
| Schwarzschild | $y^{ext}(\vec{q})$ | $\frac{1}{(2\pi)^{3/2}} y^{ext}(\vec{q})$; $y^{ext}(\vec{q}) = \frac{\kappa M_\odot}{(2\pi)^{3/2}\vert\vec{q}\vert^2}$ |
| | $y^{ext}_{\mu\nu}(\vec{q})$ | $\frac{1}{(2\pi)^{3/2}} y^{ext}_{\mu\nu}(\vec{q})$; $y^{ext}_{\mu\nu}(\vec{q}) = \delta_{\mu 4}\delta_{\nu 4} y^{ext}(\vec{q})$ |
| | $u^{ext}_{\mu\nu}(\vec{q})$ | $\frac{1}{(2\pi)^{3/2}} u^{ext}_{\mu\nu}(\vec{q})$; $u^{ext}_{\mu\nu}(\vec{q}) = \left(\delta_{\mu 4}\delta_{\nu 4} - \frac{1}{4}\delta_{\mu\nu}\right) y^{ext}(\vec{q})$ |
| | $h^{ext}_{\mu\nu}(\vec{q})$ | $\frac{1}{(2\pi)^{3/2}} h^{ext}_{\mu\nu}(\vec{q})$; $h^{ext}_{\mu\nu}(\vec{q}) = \left(\delta_{\mu 4}\delta_{\nu 4} - \frac{1}{2}\delta_{\mu\nu}\right) y^{ext}(\vec{q})$ |
| | $s^{ext}_{\mu\nu}(\vec{q})$ | $\frac{1}{(2\pi)^{3/2}} s^{ext}_{\mu\nu}(\vec{q})$; $s^{ext}_{\mu\nu}(\vec{q}) = \left(\delta_{\mu 4}\delta_{\nu 4} + \frac{1}{2}\delta_{\mu\nu}\right) y^{ext}(\vec{q})$ |
| | $y^{ext}_{,\alpha}(\vec{q})$ | $\frac{1}{(2\pi)^{3/2}} y^{ext}_{,\alpha}(\vec{q})$; $y^{ext}_{,\alpha}(\vec{q}) = i\delta_{\alpha j} q_j y^{ext}(\vec{q})$ |
| | $h^{ext}_{\mu\nu,\alpha}(\vec{q})$ | $\frac{1}{(2\pi)^{3/2}} h^{ext}_{\mu\nu,\alpha}(\vec{q})$; $h^{ext}_{\mu\nu,\alpha}(\vec{q}) = \left(\delta_{\mu 4}\delta_{\nu 4} - \frac{1}{2}\delta_{\mu\nu}\right) y^{ext}_{,\alpha}(\vec{q})$ |

| | | |
|---|---|---|
| Kerr | $y^{ext}_{jk}(\vec{q});\ j,k=\overline{1,3}$ | $0$ |
| | $y^{ext}_{j4}(\vec{q})=y^{ext}_{4j}(\vec{q})$ | $\frac{1}{2}\frac{1}{(2\pi)^{3/2}}\frac{\kappa M_{\odot}}{\|\vec{q}\|^2}(\vec{a}\times\vec{q})_j$ |
| | $y^{ext}_{44}(\vec{q})=y^{ext}(\vec{q})$ | $\frac{1}{(2\pi)^{3/2}}\frac{\kappa M_{\odot}}{\|\vec{q}\|^2}$ |
| | $s^{ext}_{jk}(\vec{q})$ | $\frac{1}{2}\frac{1}{(2\pi)^{3/2}}\frac{\kappa M_{\odot}}{\|\vec{q}\|^2}\delta_{jk}$ |
| | $s^{ext}_{j4}(\vec{q})=s^{ext}_{4j}(\vec{q})$ | $\frac{1}{2}\frac{1}{(2\pi)^{3/2}}\frac{\kappa M_{\odot}}{\|\vec{q}\|^2}(\vec{a}\times\vec{q})_j$ |
| | $s^{ext}_{44}(\vec{q})$ | $\frac{3}{2}\frac{1}{(2\pi)^{3/2}}\frac{\kappa M_{\odot}}{\|\vec{q}\|^2}$ |
| | $h^{ext}_{jk}(\vec{q})$ | $-\frac{1}{2}\frac{1}{(2\pi)^{3/2}}\frac{\kappa M_{\odot}}{\|\vec{q}\|^2}\delta_{jk}$ |
| | $h^{ext}_{j4}(\vec{q})=h^{ext}_{4j}(\vec{q})$ | $\frac{1}{2}\frac{1}{(2\pi)^{3/2}}\frac{\kappa M_{\odot}}{\|\vec{q}\|^2}(\vec{a}\times\vec{q})_j$ |
| | $h^{ext}_{jk}(\vec{q})$ | $-\frac{1}{2}\frac{1}{(2\pi)^{3/2}}\frac{\kappa M_{\odot}}{\|\vec{q}\|^2}\delta_{jk}$ |
| | $h^{ext}_{44}(\vec{q})$ | $\frac{1}{2}\frac{1}{(2\pi)^{3/2}}\frac{\kappa M_{\odot}}{\|\vec{q}\|^2}$ |
| Coulombian | $A^{ext}_{\mu}(\vec{q})$ | $\frac{1}{(2\pi)^{3/2}}A^{ext}_{\mu}(\vec{q}),$ <br> $A^{ext}_{\mu}(\vec{q})=i\delta_{\mu4}V(\vec{q}),\quad V(\vec{q})=\frac{Ze}{(2\pi)^{3/2}\|\vec{q}\|^2}$ |
| | $A^{ext}_{\nu,\alpha}(\vec{q})$ | $\frac{1}{(2\pi)^{3/2}}A^{ext}_{\nu,\alpha}(\vec{q}),$ <br> $A^{ext}_{\nu,\alpha}(\vec{q})=-\delta_{\nu4}\delta_{\alpha j}q_jV(\vec{q})$ |
| | $F^{ext}_{\mu,\nu}(\vec{q})$ | $\frac{1}{(2\pi)^{3/2}}F^{ext}_{\mu\nu}(\vec{q}),$ <br> $F^{ext}_{\mu\nu}(\vec{q})=(\delta_{\mu4}\delta_{\nu j}-\delta_{\mu j}\delta_{\nu4})q_jV(\vec{q})$ |

| | | |
|---|---|---|
| Homogeneous electrostatic | $A_\mu^{ext}(\vec{q})$ | $\frac{1}{(2\pi)^{3/2}} A_\mu^{ext}(\vec{q})$, $A_\mu^{ext}(\vec{q}) = i\delta_{\mu 4} V(\vec{q}), \quad q_j V(\vec{q}) = iE_j(\vec{q})$ |
| | $A_{\nu,\alpha}^{ext}(\vec{q})$ | $\frac{1}{(2\pi)^{3/2}} A_{\nu,\alpha}^{ext}(\vec{q})$, $A_{\nu,\alpha}^{ext}(\vec{q}) = -i\delta_{\nu 4}\delta_{\alpha j} E_j(\vec{q})$ |
| | $F_{\mu\nu}^{ext}(\vec{q})$ | $\frac{1}{(2\pi)^{3/2}} F_{\mu\nu}^{ext}(\vec{q})$, $F_{\mu\nu}^{ext}(\vec{q}) = i(\delta_{\mu 4}\delta_{\nu j} - \delta_{\mu j}\delta_{\nu 4}) E_j(\vec{q})$ |
| Homogeneous magnetostatic | $F_{\mu\nu}^{ext}(\vec{q})$ | $\frac{1}{(2\pi)^{3/2}} F_{\mu\nu}^{ext}(\vec{q})$, $F_{\mu\nu}^{ext}(\vec{q}) = \delta_{\mu j}\delta_{\nu l}\varepsilon_{jlm} H_m(\vec{q})$ |
| Electric dipole | $A_\mu^{ext}(\vec{q})$ | $\frac{1}{(2\pi)^{3/2}} A_\mu^{ext}(\vec{q}), \quad A_\mu^{ext}(\vec{q}) = i\delta_{\mu 4} V(\vec{q})$, $V(\vec{q}) = -\frac{i}{(2\pi)^{3/2}} \frac{d_i q_i}{\lvert\vec{q}\rvert^2}$ |
| | $A_{\nu,\alpha}^{ext}(\vec{q})$ | $\frac{A_{\nu,\alpha}^{ext}(\vec{q})}{(2\pi)^{3/2}}, \quad A_{\nu,\alpha}^{ext}(\vec{q}) = i\delta_{\nu 4}\delta_{\alpha j} E_j(\vec{q})$, $E_j(\vec{q}) = -iq_j V(\vec{q})$ |
| | $F_{\mu\nu}^{ext}(\vec{q})$ | $\frac{1}{(2\pi)^{3/2}} F_{\mu\nu}^{ext}(\vec{q})$, $F_{\mu\nu}^{ext}(\vec{q}) = i(\delta_{\mu 4}\delta_{\nu j} - \delta_{\mu j}\delta_{\nu 4}) E_j(\vec{q})$ |
| Magnetic dipole | $F_{\mu\nu}^{ext}(\vec{q})$ | $\frac{F_{\mu\nu}^{ext}(\vec{q})}{(2\pi)^{3/2}}, \quad F_{\mu\nu}^{ext}(\vec{q}) = \delta_{\mu i}\delta_{\nu k}\varepsilon_{ikl} H_l(\vec{q})$, $H_l(\vec{q}) = \frac{1}{(2\pi)^{3/2}}\left(M_l - q_l \frac{M_n q_n}{\lvert\vec{q}\rvert^2}\right)$ |

**Table 3.** First-order vertices.

| Process | Type | S-matrix factor |
|---|---|---|
| Particle scattering in the exterior gravitational field | $V_{1sc-g}(k,k',q)$ | $-i\kappa(2\pi)^4\delta(\vec{k}'-\vec{k}-\vec{q})\delta(q_0)$ |
| | $V_{2sc-g}(k,k',q)$ | $-\frac{1}{2}i\kappa m^2(2\pi)^4\delta(\vec{k}'-\vec{k}-\vec{q})\delta(q_0)$ |
| | $V_{em-g}(k,k',q)$ | $-i\kappa(2\pi)^4\delta(\vec{k}'-\vec{k}-\vec{q})\delta(q_0)$ |
| | $V'_{1em-g}(k,k',q)$ $= -V'_{2em-g}(k,k',q)$ | $\frac{1}{2}i\kappa(2\pi)^4\delta(\vec{k}'-\vec{k}-\vec{q})\delta(q_0)$ |
| | $V'_{3em-g}(k,k',q)$ | $-i\kappa(2\pi)^4\delta(\vec{k}'-\vec{k}-\vec{q})\delta(q_0)$ |
| | $V_{1P-g}(k,k',q)$ | $-i\kappa(2\pi)^4\delta(\vec{k}'-\vec{k}-\vec{q})\delta(q_0)$ |
| | $V_{2P-g}(k,k',q)$ | $-i\kappa\mu^2(2\pi)^4\delta(\vec{k}'-\vec{k}-\vec{q})\delta(q_0)$ |
| | $V_{1St-g}(k,k',q)$ | $-i\kappa(2\pi)^4\delta(\vec{k}'-\vec{k}-\vec{q})\delta(q_0)$ |
| | $V_{2St-g}(k,k',q)$ | $-i\kappa\mu^2(2\pi)^4\delta(\vec{k}'-\vec{k}-\vec{q})\delta(q_0)$ |
| | $V_{1sp-g}(k,k',q)$ $= -V_{2sp-g}(k,k',q)$ | $\frac{1}{4}i\kappa\gamma_\mu(2\pi)^4\delta(\vec{k}'-\vec{k}-\vec{q})\delta(q_0)$ |
| | $V_{3sp-g}(k,k',q)$ | $-\frac{1}{2}i\kappa(2\pi)^4\delta(\vec{k}'-\vec{k}-\vec{q})\delta(q_0)$ |
| | $V'_{1sp-g}(k,k',q)$ $= -V'_{2sp-g}(k,k',q)$ | $\frac{1}{4}i\kappa\gamma_\mu(2\pi)^4\delta(\vec{k}'-\vec{k}-\vec{q})\delta(q_0)$ |
| | $V_{1n-g}(k,k',q)$ $= -V_{2n-g}(k,k',q)$ | $\frac{1}{8}i\kappa\gamma_\mu(1+\gamma_5)(2\pi)^4\delta(\vec{k}'-\vec{k}-\vec{q})\delta(q_0)$ |
| | $V_{1g}(p,p',q)$ | $-\frac{1}{2}\kappa(2\pi)^4\delta(\vec{k}'-\vec{k}-\vec{q})\delta(q_0)$ |
| | $V_{2g}(p,p',q)$ | $-\frac{1}{4}\kappa(2\pi)^4\delta(\vec{k}'-\vec{k}-\vec{q})\delta(q_0)$ |

| | | |
|---|---|---|
| Transformation of photons into gravitons in the external electromagnetic field | $V_{em-g}(k,p,q)$ | $-\frac{1}{2}i\kappa(2\pi)^4\delta(\vec{k}'-\vec{k}-\vec{q})\delta(q_0)$ |
| | $V'_{1em-g}(k,k',q)$ $= -V'_{2em-g}(k,k',q)$ | $\frac{1}{4}i\kappa(2\pi)^4\delta(\vec{k}'-\vec{k}-\vec{q})\delta(q_0)$ |
| | | with $q_0 = p_0 - k_0 = 0$ and $\int d^3q$ |

Appendix G

# Expressions of the Coefficients-Functions of the Differential and Integral Scattering Cross-Sections of Particles in the External Axially-Symmetric Gravitational Field Described by the Kerr Metric

This appendix presents the expressions of the coefficient-functions appearing in the differential and integral/total scattering cross-sections of massive and massless particles in an external axially-symmetric gravitational field, determined in Chap. VIII. These coefficient-functions are the following:

**1.** In Eq. (8.52-2):

$$f_1(\theta) = \frac{1}{4}\sin^2\theta\csc^2\frac{\theta}{2}.$$

$$f_2(v,\theta) - \frac{1}{4}\csc^2\frac{\theta}{2}\Big(9 + 6\cos\theta + \cos^2\theta - v^2\sin^2\theta\Big),$$

$$f_3(\theta) = \frac{1}{2}\sin\theta,$$

$$f_4(\theta) = \frac{9}{2}\csc^2\frac{\theta}{2},$$

$$f_5(v,\theta) = \frac{1}{4}\csc^2\frac{\theta}{2}\left(-9 - 8\cos\theta + \cos^2\theta + v^2\sin^2\theta\right),$$

**2.** In Eq. (8.52-3):

$$f_1(v,\theta) = \frac{4}{3}\left[3 + 2v^2 - 2\sin^2\frac{\theta}{2}\left(1 - v^2\cos^2\frac{\theta}{2}\right)\right] - \frac{8}{3}v^2 - \frac{4}{3},$$

$$f_2(v,\theta) = 4\csc^2\frac{\theta}{2} - \frac{8}{3}v^2 - \frac{4}{3} - f_1(v,\theta),$$

$$f_3(v,\theta) = \frac{4}{3}\sin\theta\left(1 - v^2\sin^2\frac{\theta}{2}\right),$$

$$f_4(\theta) = -\frac{4}{3}\cot^2\frac{\theta}{2}.$$

**3.** In Eq. (8.52-4):

$$\begin{aligned} f_1(v,\theta) = \frac{1}{36}\csc^2\frac{\theta}{2}\Big\{&82 - v^2\left(129 - 68v^2 + 3v^4\right) \\ &+ 2\cos\theta\left(31 - 70v^2 + 27v^4\right) - v^2\cos^2\theta\left(19 - 14v^2 - v^4\right) \\ &+ 2v^4\cos^3\theta\left(4 + v^2\cos\theta\right) - \Big[77 - v^2\left(123 - 65v^2 + 3v^4\right) \\ &+ 2\cos\theta\left(31 - 68v^2 + 27v^4\right) + \cos^2\theta\left(5 - 25v^2 + 15v^4 + v^6\right) \\ &- 4v^2\cos^3\theta\left(1 - 2v^2\right) + 2v^4\cos^4\theta\left(1 + v^2\right)\Big]\Big\}\left(1 - v^2\right)^{-2}, \end{aligned}$$

$$\begin{aligned} f_2(v,\theta) = \frac{1}{36}\csc^2\frac{\theta}{2}\Big[&82 - v^2\left(129 - 68v^2 + 3v^4\right) \\ &+ 2\cos\theta\left(31 - 70v^2 + 27v^4\right) - v^2\cos^2\theta\left(19 - 14v^2 - v^4\right) \\ &+ 2v^4\cos^3\theta\left(4 + v^4\cos\theta\right)\Big] \times \left(1 - v^2\right)^{-2} - f_1(v,\theta), \end{aligned}$$

$$f_3(v,\theta) = \frac{1}{18}\sin\theta\left(5 - 6v^2 + 3v^4 - 4v^2\cos\theta + 2v^4\cos^2\theta\right)\left(1 - v^2\right)^{-2},$$

$$f_4(v,\theta) = -\frac{2}{9}\cos^2\frac{\theta}{2}\Big[77 - v^2\Big(123 - 65v^2 + 3v^4\Big)$$
$$- \cos\theta\Big(5 + 25v^2 + 5v^4 - 3v^6\Big) + 2v^2\cos^2\theta\Big(2 + 7v^2 - v^4\Big)$$
$$- 2v^4\cos^3\theta\Big(1 - v^2\Big)\Big]\Big(1 - v^2\Big)^{-2}.$$

**4.** In Eq. (8.52-5):

$$f_1(v,\theta) = \frac{1}{45}\csc^2\frac{\theta}{2}\Big[270 - 696v^2 + 845v^4 - 499v^6 + 173v^8 - 21v^{10}$$
$$- \cos\theta\Big(90 + 108v^2 - 337v^4 + 467v^6 - 145v^8 + 33v^{10}\Big)$$
$$+ \cos^2\theta\Big(84v^2 - 149v^4 + 163v^6 + 49v^8 - 3v^{10}\Big)$$
$$+ \cos^3\theta\Big(47v^4 + 191v^6 - 115v^8 + 21v^{10}\Big)$$
$$- \cos^4\theta\Big(108v^6 + 132v^8 - 24v^{10}\Big)$$
$$+ 12v^8\Big(5 + v^2\Big)\cos^5\theta\Big]\Big(1 - v^2\Big)^{-4},$$

$$f_2(v,\theta) = \frac{1}{45}\csc^2\frac{\theta}{2}\Big[-180 + 453v^2 - 596v^4 + 382v^6 - 152v^8 + 21v^{10}$$
$$+ \cos\theta\Big(90 - 9v^2 - 106v^4 + 320v^6 - 112v^8 + 33v^{10}\Big)$$
$$- \cos^2\theta\Big(90 - 159v^2 + 40v^4 + 130v^6 + 46v^8 - 3v^{10}\Big)$$
$$+ \cos^3\theta\Big(117v^2 - 278v^4 - 56v^6 + 94v^8 - 21v^{10}\Big)$$
$$- \cos^4\theta\Big(60v^4 - 192v^6 - 108v^8 + 24v^{10}\Big)$$
$$+ 12v^6\Big(1 - 6v^2 - v^4\Big)\cos^5\theta\Big]\Big(1 - v^2\Big)^{-4},$$

$$f_3(v,\theta) = \frac{4}{15}\Big(1 - v^2\Big)^{-3}\Big(1 - v^2\cos^2\frac{\theta}{2}\Big)\Big[15\Big(1 - v^2\Big)^2$$
$$+ 24v^2\Big(1 - v^2\Big)\sin^2\frac{\theta}{2} + 16v^4\sin^4\frac{\theta}{2}\Big]\sin\theta,$$

$$f_4(v,\theta) = -\frac{2}{45}\cot^2\frac{\theta}{2}\Big[180 - 453v^2 + 596v^4 - 382v^6 + 152v^8 - 21v^{10}$$
$$- 2\cos\theta\Big(45 + 12v^2 - 4v^4 + 86v^6 - v^8 + 6v^{10}\Big)$$
$$+ \cos^2\theta\Big(117v^2 - 4v^4 + 242v^6 + 68v^8 + 9v^{10}\Big)$$
$$- 12\cos^3\theta\Big(5v^4 + 5v^6 + 15v^8 - v^{10}\Big)$$
$$+ 12v^6\cos^4\theta\Big(1 + 4v^2 + v^4\Big)\Big]\Big(1 - v^2\Big)^{-4}.$$

**5.** In Eq. (8.61-4):

$$f_1(v,\alpha) = 48\Big(1 - v^2\Big) - 64v^4 + \vec{a}^{\,2}p_0^2\Big[\Big(531 - 1121v^2 + 566v^4 + 4v^6\Big)$$
$$+ \Big(303 - 513v^2 + 268v^4 - 10v^6\Big)\cos\alpha$$
$$- \Big(268 - 664v^2 + 326v^4 + 14v^6\Big)\cos 2\alpha\Big].$$

$$f_2(v,\alpha) = -4v^2\Big(3 + v^2\Big) + \vec{a}^{\,2}p_0^2\Big[\Big(67 - 155v^2 + 84v^4 + 4v^6\Big)$$
$$+ \Big(72 - 146v^2 + 66v^4\Big)\cos\alpha$$
$$- \Big(5 - 29v^2 + 26v^4 + 4v^6\Big)\cos 2\alpha\Big],$$

$$f_3(v,\alpha) = -10 - 34v^2 + 27v^4 + 9v^6 - \Big(10 + 42v^2 - 15v^4 - 5v^6\Big)\cos\alpha$$
$$+ \Big(8v^2 - 20v^4 - 4v^6\Big)\cos 2\alpha,$$

$$f_4(v,\alpha) = 2 + 5v^2 + v^4 + \Big(2 + 6v^2\Big)\cos\alpha - v^2\Big(1 + v^2\Big)\cos 2\alpha,$$

$$f_5(v,\alpha) = 15 - 26v^2 - 21v^4 - 144v^2\vec{a}^{\,2}p_0^2\Big(1 - v^2\Big)\sin^2\alpha.$$

**6.** In Eq. (8.61-5):

$$f_1(v,\alpha) = 2416 - 10016v^2 + 10724v^4 - 5352v^6 - 956v^8 - \vec{a}^{\,2}p_0^2$$
$$\times \Big[\Big(1440 - 2937v^2 + 2412v^4 - 1008v^6 + 492v^8 - 147v^{10}\Big)$$
$$- v^2\Big(1419 - 2736v^2 + 1812v^4 - 96v^6 - 147v^8\Big)\cos 2\alpha\Big],$$

$$f_2(v,\alpha) = v^2\Big( - 284 + 808v^2 - 340v^4 + 208v^6\Big) - \vec{a}^{\,2}p_0^2$$
$$\times \Big[\Big(45 - 339v^2 + 537v^4 - 405v^6 + 162v^8 - 36v^{10}\Big)$$
$$+ \Big(135 - 381v^2 + 483v^4 - 219v^6 - 18v^8 + 36v^{10}\Big)\cos 2\alpha\Big],$$

$$f_3(v,\alpha) = v^2\Big(212 - 104v^2 - 36v^4\Big)$$
$$+ \vec{a}^{\,2}p_0^2\Big[\Big(117 - 364v^2 + 491v^4 - 256v^6 + 66v^8\Big)$$
$$+ \Big(351 - 800v^2 + 469v^4 - 8v^6 - 66v^8\Big)\cos 2\alpha\Big],$$

$$f_4(v,\alpha) = -8v^2\Big(1 + v^2\Big) - \vec{a}^{\,2}p_0^2\Big[\Big(15 - 3v^2 + 33v^4 - 9v^6\Big)$$
$$+ \Big(45 - 93v^2 + 3v^4 + 9v^6\Big)\cos 2\alpha\Big],$$

$$f_5(v,\alpha) = -8v^2 + \vec{a}^{\,2}p_0^2\Big[\Big(3 + 12v^2 + 3v^4\Big) + \Big(9 - 24v^2 - 3v^4\Big)\cos 2\alpha\Big].$$

**7.** In Eq. (8.82-2):

$$f_1(\theta) = 2\cot^2\frac{\theta}{2},$$

$$f_2(\theta) = \sin\theta\csc^2\frac{\theta}{2},$$

$$f_3(\theta) = \cos^2\frac{\theta}{2},$$

$$f_4(\theta) = \frac{1}{2}\csc^2\frac{\theta}{2}\Big(4 + 3\cos\theta + \cos^2\theta\Big),$$

$$f_5(\theta) = \frac{1}{2}\sin\theta,$$

$$f_6(\theta) = -4\cot^2\frac{\theta}{2}.$$

**8.** In Eq. (8.82-3):

$$f_1(\theta) = \cos^4\frac{\theta}{2},$$

$$f_2(\theta) = \cos^2\frac{\theta}{2}\left(1 - 2\cos^2\frac{\theta}{2}\right),$$

$$f_3(\theta) = \cot^2\frac{\theta}{2}\left(1 - \sin^4\frac{\theta}{2}\right),$$

$$f_4(\theta) = \frac{1}{2}\sin\theta\cos^2\frac{\theta}{2}.$$

**9.** In Eq. (8.82-4):

$$f_1(\alpha,\theta,\varphi) = 2\cot^2\frac{\theta}{2}\left(\sin\alpha\tan\frac{\theta}{2}\cos\varphi + \cos\alpha\right),$$

$$f_2(\theta) = \cos^2\frac{\theta}{2},$$

$$f_3(\theta) = \frac{1}{2}\csc^2\frac{\theta}{2}\left(4 + 3\cos\theta + \cos^2\theta\right),$$

$$f_4(\theta) = \frac{1}{2}\sin\theta,$$

$$f_5(\theta) = -4\cot^2\frac{\theta}{2}.$$

**10.** In Eq. (8.82-5):

$$f_1(\theta) = \cot^2\frac{\theta}{2}\left(4\cos^2\theta + \frac{1}{2}\sin^4\theta\right).$$

**11.** In Eq. (8.105-5):

$$f_1(\alpha) = -844 + 274\,\vec{a}^{\,2} p_0^2 \sin^2\alpha,$$

$$f_2(\alpha) = -68 + 30\,\vec{a}^{\,2} p_0^2 \sin^2\alpha,$$

$$f_3(\alpha) = -156 + 70\,\vec{a}^{\,2} p_0^2 \sin^2\alpha,$$

$$f_4(\alpha) = -2\Big(2 - \vec{a}^{\,2} p_0^2 \sin^2\alpha\Big),$$

$$f_5(\alpha) = -2\Big(2 - \vec{a}^{\,2} p_0^2 \sin^2\alpha\Big),$$

$$f_6(\alpha) = -16\Big(4 - \vec{a}^{\,2} p_0^2 \sin^2\alpha\Big).$$

Appendix H

# Values of Some Physical Quantities in CGS and Natural $\left(\hbar = 1,\ c = 1\right)$ Unit Systems. Equivalence Between Different Units of Measurement

## Notations

$c$ — speed of light in vacuum;

$\hbar = \dfrac{h}{2\pi}$ — reduced Planck constant;

$e$ — electron electric charge;

$\dfrac{e^2}{\hbar c}$ — fine-structure constant;

$G$ — Newtonian constant of gravitation;

$\kappa$ — gravitational coupling constant;

$g$ — strong interaction coupling constant;

$f$ — weak interaction coupling constant;

$m_e$ — electron mass;

$m_p$ — proton mass;

$m_{\pi^0}$ — neutral meson $\pi^0$ mass;

$m_{\nu_e}$ — electron neutrino mass;

$m_{\nu_\mu}$ — muon-neutrino mass;

Some constants and relations between different measurement units in CGS unit system

$$c = 3 \times 10^{10}\,\frac{\text{cm}}{\text{s}}; \quad \hbar = 10^{-27}\,\text{erg} \cdot \text{s} = 10^{-27}\,\frac{\text{g} \cdot \text{cm}^2}{\text{s}};$$

$$1\,\frac{\text{g}\cdot\text{cm}^2}{\text{s}} = \frac{\hbar}{10^{-27}}; \quad 1\,\frac{\text{cm}}{\text{s}} = \frac{c}{3\times 10^{10}} \quad 1\,\text{g}\cdot\text{cm} = \frac{\hbar}{c}\,\frac{3\times 10^{10}}{10^{-27}};$$

$$1\,\text{g} = \frac{3\times 10^{10}}{10^{-27}}\frac{\hbar}{c}\,\text{cm}^{-1} = 3\times 10^{37}\frac{\hbar}{c}\,\text{cm}^{-1}; \quad 1\,\text{s} = \frac{3\times 10^{10}}{c}\,\text{cm};$$

$$G = 6.67\times 10^{-8}\frac{\text{cm}^3}{\text{g}\cdot\text{s}^2}; \quad \kappa^2 = \frac{16\pi G}{c^4} = 4\times 10^{-48}\frac{\text{s}^2}{\text{g}\cdot\text{cm}};$$

$$m_e = 9.1\times 10^{-28}\,\text{g}; \quad m_{\pi^0} = 264\,m_e;$$

$$\frac{m_e c}{\hbar} = 2.7\times 10^{10}\,\text{cm}^{-1}; \quad \frac{m_{\pi^0} c}{\hbar} = 7\times 10^{12}\,\text{cm}^{-1}$$

$$\frac{e^2}{\hbar c} = \frac{1}{137}; \quad \frac{g^2}{\hbar c} \approx 1 - 20; \quad \frac{f^2}{\hbar c} \approx 10^{-12} - 10^{-14}; \quad \frac{G m_p^2}{\hbar c} \approx 5\times 10^{-39}.$$

In the natural unit system ($\hbar = 1$, $c = 1$), we have

$$1\,\text{g} = 3\times 10^{37}\text{cm}^{-1}; \quad 1\,\text{s} = 3\times 10^{10}\text{cm};$$

$$\kappa^2 = 16\pi G = 4\times 10^{-48}\frac{\text{s}^2}{\text{g}\cdot\text{cm}} = 4\times 10^{-48}\frac{(3\times 10^{10})^2\,\text{cm}^2}{3\times 10^{37}\,\text{cm}^{-1}\cdot\text{cm}}$$

$$= 1.2\times 10^{-64}\,\text{cm}^2;$$

$$\kappa = 1.1\times 10^{-32}\,\text{cm}; \; m_e = 9.1\times 10^{-28}\times 3\times 10^{37}\text{cm}^{-1} = 2.7\times 10^{10}\text{cm}^{-1};$$

$$m_{\pi^0} = 264\,m_e = 7\times 10^{12}\text{cm}^{-1}; \quad m_{\nu_e} < 4\times 10^{-4} m_e = 10^7\text{cm}^{-1};$$

$$m_{\nu_\mu} < 8\times 10^{-4} m_e = 2\times 10^7\text{cm}^{-1};$$

$$G = 6.67\times 10^{-8}\frac{\text{cm}^3}{3\times 10^{37}\text{cm}^{-1}(3\times 10^{10})^2\,\text{cm}^2} = 2.4\times 10^{-66}\text{cm}^2;$$

$$\frac{\kappa}{m_e^2} \approx 10^{-53}\text{cm}^3; \; \frac{\kappa}{m_{\pi^0}^2} \approx 10^{-58}\text{cm}^3; \quad \frac{\kappa}{m_\nu^2} > 10^{-46}\text{cm}^3;$$

$$\frac{\kappa}{m_e} \approx 10^{-42}\text{cm}^2; \quad \frac{\kappa}{m_{\pi^0}} \approx 10^{-45}\text{cm}^2; \; \frac{\kappa}{m_\nu} > 10^{-39}\text{cm}^2.$$

# Bibliography


[1] Aguilar, A.C. and Natale, A.A., *A dynamical gluon mass solution in a coupled system of the Schwinger–Dyson equations*, Journal of High Energy Physics (8): 057, arXiv: hep-ph/0408254, Bibcode: 2004 JHEP ...08..057A, doi:10.1088/1126-6708/2004/08/057 (2004).

[2] Akhiezer, A. and Berestetskii, V., *Kvantovaia electrodinamika*, Izd-vo Nauka, Moskva, 1969.

[3] Álvarez, E., González-Martín, S., Herrero-Valea, M. and Martín, C.P., *Quantum Corrections to Unimodular Gravity*, JHEP 1508 (2015) 078, doi:10.1007/JHEP08(2015)078, e-Print: arXiv:1505.01995 [hep-th].

[4] Ambjørn, J., Coumbe, D.N., Gizbert-Studnicki, J. and Jurkiewicz, J., *Signature Change of the Metric in CDT Quantum Gravity?*, JHEP 1508 (2015) 033, doi:10.1007/JHEP08 (2015) 033, e-Print: arXiv:1503.08580 [hep-th].

[5] Anderson, J., *Phys. Rev.*, 110, 1197 (1958); 111, 965 (1958); 114, 1182 (1959).

[6] Aragone, C. and Deser, S., *Higher Spin Vierbein Gauge Fermions and Hypergravities*, Nucl. Phys., B170, [FS1], 329 (1980).

[7] Aragone, C. and Deser, S., *Hamiltonian form for massless higher-spin fermions*, Phys. Rev., D21, 352, (1980).

[8] Argüelles, C., Grandi, N. and Park, Mu-In, *Complete classification of four-dimensional black hole and membrane solutions in IR-modified Hořava gravity*, *JHEP* 1510 (2015) 100, doi:10.1007/JHEP10 (2015) 100, e-Print: arXiv:1508.04380 [hep-th].

[9] Arnowitt, R. and Deser, S., *Phys. Rev.*, 113, 745 (1959).

[10] Arnowitt, R., Deser, S. and Misner, C., *Phys. Rev.*, 116, 1322 (1959); 117, 1595 (1960); 121, 1556 (1961); 122, 997 (1961).

[11] Aslanian, E., *Vestnik M.G.U., Fizika i astronomia*, 4, 438 (1972).

[12] *ATLAS Experiment*, ATLAS, 1 January 2006, retrieved 2012-07-05.

[13] Baaklini, N.S., *New Superalgebra and Goldstone Spin-3/2 Particle*, Phys. Lett., 67B, No. 3, 335 (1977).

[14] Baaklini, N.S. and Tuite, M., *Dirac quantization of massive spin-3/2 field*, J. Phys., A11, L 139 (1978).

[15] Baaklini, N.S. and Tuite, M., *Dirac quantization of spin-2 field*, J. Phys., A12, L 13 (1979).

[16] BaBar Collaboration, *Evidence for an excess of $B \to D^{(*)}\ \tau^- \nu_\tau$ decays*, Phys. Rev. Lett., 109 (10) (2012): 101802, arXiv:1205.5442, Bibcode: 2012PhRvL.109j1802L, doi:10.1103/PhysRevLett.109.101802.

[17] *BaBar data hint at cracks in the Standard Model*, Science News, 18 June 2012, retrieved 2013-08-06.

[18] *BABAR Data in Tension with the Standard Model*, SLAC, 31 May 2012, retrieved 2013-08-06.

[19] Badalyan, R.G. and Nazaryan, A.E., *Quark-gluon mixing and inclusive spectra of pseudoscalar and tensor mesons*, preprint, EFI 931-82-86; ERFI, 86-931 (25 p) (1986).

[20] Bae, J., Imbimbo, C., Rey, S.-J. and Rosa, D., *New Supersymmetric Localizations from Topological Gravity*, September 30, 2015, 45 pp., e-Print: arXiv:1510.00006 [hep-th].

[21] Bagan, E. and Narison, S., *Tensor mesons in QCD: revisited*, preprint, PM 88-24; MU PM 88-24, (15pp.) (1988).

[22] Bagdonaite, J. *et al.*, *A Stringent Limit on a Drifting Proton-to-Electron Mass Ratio from Alcohol in the Early Universe*, Science 339 (6115): 46 (2012), Bibcode: 2013Sci...339...46B, doi:10.1126/science.1224898.

[23] Bailin, D., *Phys. Rev.*, 135, 1B, 166 (1964); *Nuovo Cimento*, 33, 633 (1964).

[24] Barcelos-Neto, J. and Cabo, A., *Faddeev–Jackiw Quantization of Massive Tensor Fields*, preprint, ICTP, IC/95/246 (1995).

[25] Barker, B., Gupta, S. and Haracz, R., *Phys. Rev.*, 149, 1027 (1966).

[26] Barker, B., Bhatia, M. and Gupta, S., *Phys.Rev.*, 158, 1498 (1967); 162, 1750 (1967); 182, 1387 (1969).

[27] Barker, B., Gupta, S. and Kaskas, J., *Phys. Rev.*, 182, 1391 (1969).

[28] Basu, R., *Higher Spin de Sitter Quantum Gravity*, JHEP 1510 (2015) 151, doi:10.1007/JHEP10 (2015) 151, e-Print: arXiv:1507.04757 [hep-th].

[29] Berends, F.A., van Holten, J.W. and Nieuwenhuizen, P. van, *On Field Theory For Massive and Massless Spin-5/2 Particles*, Nucl. Phys., B154, 261 (1979).

[30] Berends, F.A., van Holten, J.W. and Nieuwenhuizen, P. van, *On Spin 5/2 Gauge Fields*, Phys. Lett., 83B, 188 (1979).

[31] Bergström, L., Hulth, G. and Snellman, H., *Relativistic calculation of the two photon coupling of scalar and tensor mesons*, preprint, CERN TH 3381. Publ. Ref.: Publ. in Z. Phys., C16 (1983), 263–269, (19 pp.), (1983).

[32] Bern, Z., Cheung, C., Chi, H.-H., Davies, S., Dixon, L. and Nohle, J., *Evanescent Effects Can Alter Ultraviolet Divergences in Quantum Gravity Without Physical Consequences*, July 22, 2015, 6 pp., SLAC-PUB-16332, CERN-PH-TH-2015-162, UCLA-15-TEP-101, CALT-2015-036, e-Print: arXiv:1507.06118 [hep-th].

[33] Birrell, N. and Davies, P., *Quantum Fields in Curved Space*, Cambridge Univ. Press, 1984.

[34] Biswas, S., Guha, J. and Sarkar, N.G., *Classical and Quantum Gravity*, 12, 1591 (1995).

[35] Blagojević, M., *Gravitation and Gauge Symmetries*, IOP Publishing, Bristol, 2002.

[36] Blagojević, M. and Hehl, F.W., *Gauge Theories of Gravitation: A Reader with Commentaries*, World Scientific, ISBN 978-184- 8167-26-1, 2013.

[37] Boccaletti, D., De Sabbata, V. and Gualdi, C., *Nuovo Cimento*, 36, 685 (1965).

[38] Boccaletti, D., De Sabbata, V., Gualdi, C. and Fortini, P., *Nuovo Cimento*, 48 A, 58 (1967); 54 B, 134 (1968); 60 B, 320 (1969); 70 B, 129 (1970); 11 B, Nr. 2, 289 (1972).

[39] Boccaletti, D., De Sabbata, V., Gualdi, C. and Fortini, P., *Nature*, 232, 53 (1971).

[40] Bogoliubov, N.N. and Chirkov, D.V., *Introduction to the Theory of Quantized Fields*, John Wiley and Sons Inc; 3rd edition, 1980.

[41] Bogoliubov, N.N. and Shirkov, D.V., *Kvantovie Polia*, Moskva, Izd-vo Nauka, 1980.

[42] Bogolubsky, I.L., Ilgenfritz, E.M., Müller-Preussker, M. and Sternbeck, A., *Landau-gauge gluon and ghost propagators in 4D SU(3) gluodynamics on large lattice volumes* (PDF), Lattice 2007, Trieste: Proceedings of Science, p. 290, arXiv:0710.1968, retrieved 2013-11-18 (2007).

[43] Bogolubsky, I.L., Ilgenfritz, E.-M., Müller-Preussker, M. and Sternbeck, A., *Lattice gluodynamics computation of Landau-gauge Green's functions in the deep infrared*, Physics Letters B 676 (1-3): 69–73 (2009), arXiv:0901.0736, Bibcode:2009 PhLB..676...69B, doi:10.1016/j.physletb.2009.04.076.

[44] Bond, J.R., Szalay, A.S. and Turner, M.S., *Formation of galaxies in a gravitino-dominated universe*, preprint, EFI 82-13, Publ. Ref.: Phys. Rev. Lett., 48 (1982), 1636, (8 p) (1982).

[45] Boucaud, Ph., Brüntjen, T., Leroy, J.P., Le Yaouanc, A., Lokhov, A., Micheli, J., Pène, O. and Rodriguez-Quintero, J., *Is the QCD ghost dressing function finite at zero momentum?*, Journal of High Energy Physics (6): 001. arXiv:hep-ph/0604056, Bibcode: 2006JHEP... 06..001B, doi:10.1088/1126-6708/2006/06/001 (2006).

[46] Boyland, P., Keiper, J. *et al.*, *Guide to Standard MATHEMATICA. Packages, Version 2.1*, Wolfram Research, 1992.

[47] Braibant, S., Giacomelli, G. and Spurio, M., *Particles and Fundamental Interactions: An Introduction to Particle Physics*, Springer, pp. 313–314, ISBN 978-94-007-2463-1, 2009.

[48] Bramon, A., Casas, R. and Casulleras, J., *Gluonium contaminations in tensor mesons*, preprint, UAB FT 120, (13 pp.) (1985).

[49] Brill, D. and Wheeler, J., *Rev. Mod. Phys.*, 29, 465 (1957).

[50] Bromley, D.A., *Gauge Theory of Weak Interactions*, Springer, ISBN 3-540-67672-4, 2000.

[51] Bronnikov, K.A. and Galiakhmetov, A.M., *Wormholes without exotic matter in Einstein–Cartan theory*, Aug 5, 2015, 7 pp., e-Print: arXiv:1508.01114 [gr-qc].

[52] Brown, E.G. and Louko, J., *Smooth and sharp creation of a Dirichlet wall in 1 + 1 quantum field theory: how singular is the sharp creation*

*limit?*, JHEP 1508 (2015) 061, doi:10.1007/JHEP08 (2015) 061, e-Print: arXiv:1504.05269 [hep-th].

[53] Brustein, R. and Medved, A.J.M., *Quantum state of the black hole interior*, JHEP 1508 (2015) 082, doi:10.1007/JHEP08(2015)082, e-Print: arXiv:1505.07131 [hep-th].

[54] Buijs, A., *Observation of the formation of the tensor mesons* $f_2'$ (1525) *and* $\chi_{c2}$ *in photon-photon collisions at LEP*, submitted to 10th International Workshop on Photon–Photon Collisions: Photon '95 Sheffield, UK; 8–13 April 1995. Publ. in Proceedings, Miller, D.J., Cartwright, S.L. and Khoze, V., World Scientific, Singapore, 1995, (232–237) (1996).

[55] Cabibbo, N., *Phys. Rev. Lett.*, 10, 531 (1963); 12, 62 (1964).

[56] Cabo, A., Chaichian, M. and Louis Martinez, D., *An Alternative Derivation of Faddeev–Popov Path Integral*, preprint, HU-TFT-90-24, (10 pp.) (1990).

[57] Callum, M., in *Quantum Gravity, An Oxford Symposium*, Clarendon Press, Oxford, 1975.

[58] Camanho, X.O., Edelstein, J.D., Gomberoff, A. and Sierra-Garcia, A., *On AdS to dS transitions in higher-curvature gravity*, JHEP 1510 (2015) 179, doi:10.1007/JHEP10(2015)179, e-Print: arXiv:1504.04496 [hep-th].

[59] Caprini, I., Colangelo, G. and Leutwyler, H., *Mass and width of the lowest resonance in QCD*, Physical Review Letters 96 (13): 132001, arXiv:hep-ph/0512364, Bibcode:2006PhRvL..96m2001C, doi:10.1103/PhysRevLett.96.132001 (2006).

[60] Carmeli, M., *Phys. Rev.*, 158, 1243 (1967).

[61] Carr, B.J., Mureika, J. and Nicolini, P., *Sub-Planckian black holes and the Generalized Uncertainty Principle*, JHEP 1507 (2015) 052, doi:10.1007/JHEP07 (2015) 052, e-Print: arXiv:1504.07637 [gr-qc].

[62] Carroll, S., *Dark Matter, Dark Energy: The Dark Side of the Universe*, Ph.D., Cal Tech, The Teaching Company, Guidebook Part 2, 2007.

[63] Caruso, F. and Predazzi, E., *Quark and gluon content of the neutral tensor mesons* $f, f', \theta$, preprint, DFTT 9-86; DFTT 86-9, (31 pp.) (1986).

[64] Castro, G.L. and Munoz, J.H., *Non-leptonic B decays involving tensor mesons*, preprint, hep-ph/9702238 (1997).

[65] *CERN experiments observe particle consistent with long-sought Higgs boson*, CERN, 4 July 2012, retrieved 2012-07-04.

[66] Chaichian, M. and Nelipa, N.F., *Introduction to Gauge Field Theories*, Springer-Verlag, 1984.

[67] Chaichian, M., Prešnajder, P. and Tureanu, A., *New concept of relativistic invariance in NC space-time: twisted Poincaré symmetry and its implications*, Phys. Rev. Lett., 94 (2005).

[68] Chaichian, M., Merches, I. and Tureanu, A., *Mechanics – An Intensive Course*, Springer (2012).

[69] Chang, S.J., *Quantization of Spin-2 Fields*, Phys. Rev., 148, 1259 (1966).

[70] Chang, S.J., *Lagrange Formulation for Systems with Higher Spin*, Phys. Rev., 161, 1308 (1967).

[71] Chang, S.J., *Quantization of Multispinor Fields*, Phys. Rev., 161, 1316 (1967).

[72] Chavleishvili, M.P., *Low-energy theorems for gravitino-scalar particle scattering*, preprint, E 2-87-69, JINR E 87–69 (1987).

[73] Chester, A., *Phys. Rev.*, 143, 1275 (1966).

[74] Chiu, H., *Phys. Rev.*, 123, 1040 (1961).

[75] Chow, C.K. and Rey S.J., *Quenched and Partially Quenched Chiral Perturbation Theory for Vector and Tensor Mesons*, preprint, hep-ph/9708432 (1997).

[76] Chow, C.K. and Rey S.J., *Chiral Perturbation Theory for Tensor Mesons*, preprint, hep-ph/9708355 (1997).

[77] Close, F.E. and Li, Z.P., *On the polarisation of tensor mesons in $\psi$ radiative decays and possible implications for the gluonic content of the $\theta$ (1720)*, preprint, UKT 90-091; RAL 90-091 (1990).

[78] Codello, A. and Jain, R.K., *Covariant Effective Field Theory of Gravity I: Formalism and Curvature Expansion*, July 22, 2015, 35 pp., CP3-ORIGINS-2015-28, DNRF90, DIAS-2015-28, CP3-ORIGINS-2015-28-DNRF 90-AND-DIAS-2015-28, e-Print: arXiv:1507.06308 [gr-qc].

[79] Coffman, D., *Decays of $J/\psi$ into a nonstrange vector and tensor mesons pair*, preprint, SLAC PUB 5162 (1990).

[80] Cohen, H., *Nuovo Cimento*, 52 A, 1242 (1967).

[81] *Confirmed: CERN discovers new particle likely to be the Higgs boson*, YouTube, Russia Today, 4 July 2012, retrieved 2013-08-06.

[82] Cordin, A., *Pseudoscalar and tensor mesons production in gamma–gamma reactions*, preprint, LAL 84-41, (29 pp.) (1984).

[83] Cruz, M., Gonzalez-Espinoza, M., Saavedra, J. and Vargas-Arancibia, D., *Scalar Perturbations of two-dimensional Hořava–Lifshitz Black Holes*, August 3, 2015, 16 pp., e-Print: arXiv:1508.00650 [gr-qc].

[84] Cucchieri, A., Porrati, M. and Deser, S., *Tree-Level Unitarity Constraints on the Gravitational Couplings of Higher-Spin Massive Fields*, preprint, hep-th/9408073, CERN-TH.7388/94 (1994).

[85] Cucchieri, A. and Mendes, T., *What's up with IR gluon and ghost propagators in Landau gauge? A puzzling answer from huge lattices*, Lattice 2007, Trieste: Proceedings of Science, arXiv:0710.0412, retrieved 2013-11-18, 2007.

[86] Curtright, T.L., *Massless Field Supermultiplets with Arbitrary Spin*, Phys. Lett., 85B, No. 2–3, 219 (1979).

[87] Dai, Y., Shen, Z. and Shi, Y., *Killing quantum entanglement by acceleration or a black hole*, JHEP 1509 (2015) 071, doi:10.1007/JHEP09 (2015) 071, e-Print: arXiv:1507.00612 [gr-qc].

[88] Daniels, R.D. and Shore, G.M., *"Faster than Light" Photons and Rotating Black Holes*, preprint, CERN-TH/95-229, gr-qc 9508048 (1995).

[89] Das, A. and Freedman, D.Z., *Gauge Quantization for Spin-3/2 Fields*, Nucl. Phys., B114, 271 (1976).

[90] Delepouve, T. and Gurau, R., *Phase Transition in Tensor Models*, JHEP 1506 (2015) 178, doi:10.1007/JHEP06(2015)178, e-Print: arXiv:1504.05745 [hep-th].

[91] De Logi, W.K. and Kovacs, S.J., Jr., *Gravitational scattering of zero-rest-mass plane waves*, Phys. Rev., D16, 237 (1977).
[92] De Logi, W.K. and Mickelson, A.R., *Electrogravitational conversion cross-sections in static electromagnetic fields*, Phys. Rev., D16, 2915 (1977).
[93] Demmel, M., Saueressig, F. and Zanusso, O., *A proper fixed functional for four-dimensional Quantum Einstein Gravity*, JHEP 1508 (2015) 113, doi:10.1007/JHEP08 (2015) 113, e-Print: arXiv:1504.07656 [hep-th].
[94] de Pádua Santos, A. and Bezerra de Mello, E.R., *Non-Abelian cosmic strings in de Sitter and anti-de Sitter space*, Sep 15, 2015, 12 pp., e-Print: arXiv:1509.04668 [gr-qc].
[95] De Rey Neto, E.C., De Araujo, J.C.N. and Aguiar, O.D., *Classical and Quantum Gravity*, 21, No.5, 541 (2004).
[96] De Sabbata, V., Boccaletti, D. and Gualdi, C., *Iad. fiz.*, 8, 924 (1968).
[97] De Sabbata, V. and Gasperini, M., *Lett. Nuovo Cimento*, 21, 328 (1978).
[98] De Sabbata, V. and Gasperini, M., *Strong Gravity and Weak Interaction*, preprint, Instituto di Fisica dell'Università, Bologna (1979).
[99] De Sabbata, V., Fortini, P., Fortini, L. and Gualdi, C., *Models of Galactic Nuclei*, Acta Phys. Pol., No. 2 B7, 99 (1976).
[100] De Sabbata, V., Fortini, P. and Gualdi, C., *Astrophys. Lett.*, 12, 87 (1972).
[101] De Sabbata, V. and Collab., *Acta Physica Polonica*, 115, 13 (1974); B7, 19 (1976).
[102] De Sabbata, V. and Rizzati, P., *Lett. Nuovo Cimento*, 20, 525 (1977).
[103] Detournay, S. and Zwikel, C., *Phase transitions in warped AdS3 gravity*, JHEP 1505 (2015) 074, doi:10.1007/JHEP05 (2015) 074, e-Print: arXiv: 1504.00827 [hep-th].
[104] De Witt, B., in *Relativity, Groups and Topology*, Gordon and Breach, New York (1964).
[105] De Witt, B., *Phys. Rev.*, 160, 1113 (1967); 162, 1195 (1967); 162, 1239 (1967).
[106] De Witt, B., in *Magic without Magic: J.A. Wheeler*, W.H. Freeman and Co., San Francisco (1972).
[107] De Witt, C. and Wesley, W., *General Relativity and Gravitation*, 2, 235 (1971).
[108] De Witt, B. and Freedman, D.Z., *Systematics of higher-spin gauge fields*, Phys. Rev., D21, 358 (1979).
[109] Deser, S., *Phys. Lett.*, 7, 42 (1963).
[110] Deser, S., in *Quantum Gravity*, An Oxford Symposium, Clarendon Press, Oxford, 1975.
[111] Deser, S., *On Consistency of Spin 3/2 Couplings*, J. Phys. A: Math. Gen., Vol. 12, No. 5, L 97 (1979).
[112] Deser, S. and Waldron, A., *Phys. Lett.*, B508, 347 (2001).
[113] Deser, S., Kay, J.H. and Stelle, K.S., *Hamiltonian formulation of supergravity*, Phys. Rev., D16, 2448 (1977).
[114] Deser, S. and Zumino B., *Phys. Lett.*, 62 B, 335 (1976).
[115] Deser, S. and Zumino, B., *Broken Supersymmetry and Supergravity*, Phys. Rev. Lett., 38, 1433 (1977); see also Cremmer, E., Ferrara, S., Girardello,

L. and Proeyen, A. van, Phys. Lett., B116, 231 (1982); Nucl. Phys., B212, 413 (1983).
[116] Dirac, P.M.A., *Proc. Roy. Soc.*, A 246, 326 (1958); A 246, 333 (1958).
[117] Dirac, P.M.A., *Phys. Rev.*, 114, 924 (1959).
[118] Dobrev, V., *Vestnik M.G.U., Physics and Astronomy*, 2, 184 (1972).
[119] D'Odorico, G., Goossens, J.-W. and Saueressig, F., *Covariant computation of effective actions in Hořava–Lifshitz gravity*, JHEP 1510 (2015) 126, doi:10.1007/JHEP10 (2015) 126, e-Print: arXiv:1508.00590 [hep-th].
[120] Donà, P., Giaccari, S., Modesto, L., Rachwal, L. and Zhu, Y., *Scattering amplitudes in super-renormalizable gravity*, JHEP 1508 (2015) 038, doi:10.1007/JHEP08 (2015) 038, e-Print: arXiv:1506.04589 [hep-th].
[121] Donoghue, J.F. and El-Menoufi, B.K., *Covariant non-local action for massless QED and the curvature expansion*, July 22, 2015, 24 pp., JHEP 1510 (2015) 044, doi:10.1007/JHEP10 (2015) 044, e-Print: arXiv:1507.06321 [hep-th].
[122] Douglas, M.R. and Nekrasov, N.A., *Noncommutative field theory*, Rev. Mod. Phys. 73, 977–1029 (2001).
[123] Drijard, D. *et al.*, *Production of vector and tensor mesons in proton–proton collisions at $\sqrt{s} = 52.5$ GeV*, preprint, CERN EP 81-12, Publ. Ref.: Publ. in Z. Phys., C9 (1981), 293–303, (38 pp.) (1981).
[124] Duff, M., in *Quantum Gravity*, An Oxford Symposium, Clarendon Press, Oxford (1975).
[125] Einstein, A., *Sitzber. Preuss. Akad. Wiss.*, Berlin, 154 (1918).
[126] Einstein, A., *The Meaning of Relativity*, Princeton University Press, 2014.
[127] Ellis, J., Linde, A.D. and Nanopoulos, D.V., *Inflation can save the gravitino*, preprint, CERN TH 3356, Publ. in: Phys. Lett., B118 (1982), 59–64, (12 pp.) (1982).
[128] Ellis, J. and Nanopoulos, D.V., *The gravitino and spin-zero boson masses*, preprint, CERN-TH 3319, Publ. in: Phys. Lett., B116 (1982), 133–136, (8 pp.) (1982).
[129] Ellis, J., Kim, J.E. and Nanopoulos, D.V., *Cosmological gravitino regeneration and decay*, preprint, CERN TH 3839, Publ. Ref.: Phys. Lett., B145 (1984), 181–186; *The early universe* – reprints, Kolb, E.W. and Turner, M.S., Addison-Wesley, Redwood City, 1988, Frontiers in physics; Vol. 70, 501–506, (12 pp.) (1984).
[130] Endo, R. and Kimura, T., *Axial anomaly for gravitino and unconventional FP ghosts*, preprint, RRK 80-1 (1980).
[131] Eremyan, Sh.S. and Nazaryan, A.E., *Quark-gluon mixing in pseudoscalar and tensor mesons*, preprint, EFI 903-54-86; ERFI 86–903 (1986).
[132] Faddeev, L. and Popov, V., *Usp. fiz. nauk*, 111, 451 (1973).
[133] Fan, W., Fathizadeh, F. and Marcolli, M., *Spectral Action for Bianchi Type-IX Cosmological Models*, JHEP 1510 (2015) 085, doi:10.1007/JHEP10 (2015) 085, e-Print: arXiv:1506.06779 [hep-th].
[134] Fang, J. and Fronsdal, C., *Massless fields with half-integer spin*, Phys. Rev., D18, No. 10, 3630 (1978).

[135] Fayet, P., *Effects of the spin-1 partner of the goldstino (gravitino) on neutral current phenomenology*, preprint, CERN TH 2895, Publ. in: Phys. Lett., B95 (1980), 285–289, (13 pp.) (1980).

[136] Feindt, M. and Harjes, J., *Two photon couplings of scalar and tensor mesons*, preprint, DESY 90–146, Subm. to: Rheinfels '90 Workshop on the Hadron Mass Spectrum Sankt Goar, Germany; 3–6 Sep. 1990. Publ. in: *The hadron mass spectrum*, Proceedings, Klempt E. and Peters, K., Nucl. Phys., B, Proc. Suppl., 21 61–74 (1991).

[137] Ferrara, S. and van Nieuwenhuizen, P., *Consistent Supergravity with Complex Spin-3/2 Gauge Fields*, Phys. Rev. Lett., 37, No. 25, 1669 (1976).

[138] Feynman, R. and Gell-Mann, M., *Phys. Rev.*, 109, 193 (1958).

[139] Feynman, R.P., *The Character of Physical Law*, MIT Press, ISBN 0-262-56003-8 [1964] (2001).

[140] Feynman, R.P. and Hibbs, A.R., *Quantum Mechanics and Path Integrals*, McGraw-Hill, New York, 1965.

[141] Feynman, R., in *Magic without Magic: J.A. Wheeler*, W.H. Freeman and Co., San Francisco (1972).

[142] Feynman, R.P., *Quantum Theory of Gravitation*, Acta Phys. Pol., 24, 697 (1963), presents in an easy-to-follow manner, the algorithm for gravitational interactions "calculation" by means of the Feynman diagrams technique; this is one of the most useful and clear papers dedicated to the presentation of Feynman diagrams application in gravity.

[143] Feynman, R.P., *QED: The Strange Theory of Light and Matter*, Princeton University Press, ISBN 0-691-12575-9 [1985] (2006).

[144] Field, J.H., *What have we learned from measurement of the 2 gamma radiative widths of the pseudoscalar and tensor mesons?*, preprint, LPNHE 83-05, Publ. Ref.: Proceedings of HEP83, Guy J. and Costain C., RAL, Chilton, 1983, 168–179, (12 pp.) (1983).

[145] Fierz, M., *Über die relativistische Theorie kräftefreier Teilchen mit beliebigem Spin*, Helvetica Physica Acta, 12, 3 (1939).

[146] Fierz, M. and Pauli, W., *On relativistic wave equations for particles of arbitrary spin in an electromagnetic field*, Proc. Roy. Soc., A173, 211 (1939).

[147] Fock, V., Ivanenko, D., *C. R. Acad. Sci.*, 188, 1470 (1929).

[148] Fock, V., *Theory of Space, Time and Gravitation*, 2nd Revised Edition, Pergamon Press, 1964.

[149] Fradkin, E., Tiutin, I., *Phys. Rev.*, D2, 2841 (1970).

[150] Frampton, P.H., *Gauge Field Theories, Frontiers in Physics* (2nd ed.), Wiley, 2000, *Gauge Field Theories* (3rd edn.), Wiley-VCH, ISBN 978-3-527-40835-1, 2008.

[151] Freedman, D., Nieuwenhuizen, P. van and Ferrara, S., *Phys. Rev.*, D13, 3214 (1976).

[152] Frolov, V.P., Zelnikov, A. and de Paula Netto, T., *Spherical collapse of small masses in the ghost-free gravity*, JHEP 1506 (2015) 107, ALBERTA-THY-6-15, doi:10.1007/JHEP06 (2015) 107, e-Print: arXiv:1504.00412 [hep-th].

[153] Fronsdal, C., *Massless fields with integer spin*, Phys. Rev., D18, No. 10, 3624 (1978).
[154] Fujisaki, H., Kumekawa, K., Yamaguchi, M. and Yoshimura, M., *Particle production and gravitino abundance after inflation*, preprint, hep-ph/9511381 (1995).
[155] Fukuma, M., Sugishita, S. and Umeda, N., *Random volumes from matrices*, JHEP 1507 (2015) 088, KUNS-2547, doi:10.1007/JHEP07 (2015) 088, e-Print: arXiv:1503.08812 [hep-th].
[156] Fung, M.K., van Nieuwenhuizen, P. and Jones, D.R.T., *Transversality of the gravitino self-energy, Ward identities and dimensional regularization*, preprint, ITP SB 80-5 (1980).
[157] Gaina, A.B. and Kochorbe, F.G., *S matrix and resonance states in the Kerr–Newman geometry*, Zh. Eksp. Teor. Fiz., 92, 369–378 (1987) [Sov. Phys. JETP 65 (2) (1987)].
[158] Gaina, A.B. and Chizhov, G.A., *The scattering and absorption of slow particles by a black hole* (*Rassejanie i pogloshenje medlenyh cheastits cheornyi dyry*) [English translation of this paper in bull point of the Moscow State University, ser. 3, Physics and Astronomy, No. 2, 1983]; Vestn. Mosk. Univ. Ser. 3 Fiz. Astronomiya, 24, 3 (1983).
[159] Gambini, R., *The Loop Representations in Gauge Theories and Quantum Gravity*, preprint, hep-th/9403006 (1994).
[160] Gandel'man, G. and Pinaev, V., *Jurn. Eksp. Teor. Fiz.*, 37, 1572 (1959).
[161] Garcia Maia, M.R. de and Lima, J.A.S., *Graviton production in elliptical and hyperbolic universes*, Phys. Rev., D54, No. 10, 6111 (1996).
[162] Garrison, J., *Phys. Rev.*, 129, 1424 (1963).
[163] Gell-Mann, M., *Phys. Rev.*, 125, 1067 (1962).
[164] Gerstenhaber, M., *On the Deformation of Rings and Algebras*, Ann. Math., 79, 59 (1964).
[165] Gertzenstein, M., *Jurn. Eksp. Teor. Fiz.*, 41, 113 (1961).
[166] Gibbons, G.W., *Vacuum polarization and the spontaneous loss of charge by black holes*, Commun. Math. Phys., 44, 245 (1975).
[167] Gielen, S., *Identifying cosmological perturbations in group field theory condensates*, JHEP 1508 (2015) 010, IMPERIAL-TP-2015-SG-1, doi:10.1007/JHEP08 (2015) 010, e-Print: arXiv:1505.07479 [gr-qc].
[168] Gottlieb, I., *Scientific studies and researches, Physics and technical sciences*, Iasi, XI, fasc. 2, 239 (1960).
[169] Gottlieb, I., *Scientific Annals of Alexandru Ioan Cuza University, Iasi*, (n.s.), Sect. Ib, fiz., XI, 1 (1965).
[170] Greiner, W. and Müller, B., *Gauge Theory of Weak Interactions*, Springer, ISBN 3-540-67672-4, 2000.
[171] Grensing, G., *Structural Aspects of Quantum Field Theory and Noncommutative Geometry*, World Scientific, ISBN 978-981-4472-69-2, 2013.
[172] Gribbin, J., *Q is for Quantum: Particle Physics from A to Z*, Weidenfeld and Nicolson, ISBN 0-297-81752-3, 1998.
[173] Grisaru, D. and Jackiw, R., *Phys. Rev.*, 166, 1287 (1968).

[174] Grisaru, M.T. and Pendleton, H.N., *Supergravity and the S matrix*, Phys. Rev., D15, 15 (1976).
[175] Grisaru, M.T. and Pendelton, H.N., *Soft Spin-3/2 Fermions Require Gravity and Supersymmetry*, Phys. Lett., 67B, 323 (1977).
[176] Grosche, C. and Steiner, F., *Classification of solvable Feynman path integrals*, preprint, DESY 92–189 (1993).
[177] Grosche, C., *Path Integrals, Hyperbolic Spaces and Selberg Trace Formulae*, DESY 95-021, Hamburg, February (1995).
[178] Grosche, C., *Boundary-Conditions in Path Integrals*, DESY 95-032, Hamburg, February (1995).
[179] Gross, D. and Jackiw, R., *Phys. Rev.*, 166, 1287 (1968).
[180] Grossiord, J., *C. R. Acad. Sci.*, 275, Nr. 11, B.365 (1972).
[181] Gupta, S., *Proc. Phys. Soc.*, A65, 161, 608 (1952).
[182] Gupta, S., *Phys. Rev.*, 96, 1683 (1954); 172, 1303 (1968).
[183] Gupta, S., *Rev. Mod. Phys.*, 29, 334 (1957).
[184] Gupta, S., *Kvantovanie gravitatzionnovo polia. Lineinoe priblijenie*, Noveishie problemy gravitatzii, Izdatelistvo Innostrannoi Literatury, Moskva, p. 325 (1961); *Kvantovanie gravitatzionnovo polia. Obshciaia teoria, Noveishie problemy gravitatzii*, Izdatelistvo Innostrannoi Literatury, Moskva, p. 341 (1961).
[185] Gupta, K.S., Harikumar, E., Jurić, T., Meljanac, S. and Samsarov, A., *Non-commutative scalar quasinormal modes and quantization of entropy of a BTZ black hole*, JHEP 1509 (2015) 025, doi:10.1007/JHEP09 (2015) 025, e-Print: arXiv:1505.04068 [hep-th].
[186] Guralnik, G. S., *The History of the Guralnik, Hagen and Kibble development of the Theory of Spontaneous Symmetry Breaking and Gauge Particles*, International Journal of Modern Physics A 24 (14): 2601–2627, arXiv:0907.3466. Bibcode: 2009IJMPA..24.2601G, doi:10.1142/S0217751X09045431 (2009).
[187] Gursoy, U., Jarvinen, M. and Policastro, G., *Late time behavior of non-conformal plasmas*, July 30, 2015, 32 pp., e-Print: arXiv:1507.08628 [hep-th].
[188] Habib Mazharimousavi, S. and Halilsoy, M., $2+1$-*dimensional wormhole from a doublet of scalar fields*, July 26, 2015, 6 pp., Phys. Rev. D92 (2015) 2, 024040, doi:10.1103/PhysRevD.92.024040, e-Print: arXiv:1507.07257 [gr-qc].
[189] Hagen, R. and Hurley, W.J., *Magnetic Moment of a Particle with Arbitrary Spin*, Phys. Rev. Lett., 24, 1381 (1970).
[190] Hagen, C.R., *Minimal Electromagnetic Coupling of Spin-Two Fields*, Phys. Rev., D6, 984 (1972).
[191] Haidt, D., *The discovery of the weak neutral currents*, CERN Courier, 4 October 2004, retrieved 8 May 2008.
[192] Halpern, L., *Nuovo Cimento*, 25, 1239 (1962).
[193] Halpern, L., Laurent, B., *Nuovo Cimento*, 33, 728 (1964).
[194] Hamamoto, S., *Gauge Theory of Massive Tensor Field*, preprint, Toyama-84, hep-th/9509144 (1995).

[195] Hamamoto, S., *Massless Limits of Massive Tensor Fields*, preprint, Toyama-90, hep-th/9605084 (1996).

[196] Hanson, A., Regge, T. and Teitelboim, C., *Constrained Hamiltonian Systems*, Academia Nazionale dei Lincei, Rome (1976).

[197] Hartong, J. and Obers, N.A., *Hořava–Lifshitz gravity from dynamical Newton–Cartan geometry*, JHEP 1507 (2015) 155, doi:10.1007/JHEP07 (2015) 155, e-Print: arXiv:1504.07461 [hep-th].

[198] Hawking, S., in *Quantum Gravity*, An Oxford Symposium, Clarendon Press, Oxford (1975).

[199] Hayashi, K., *Nuovo Cimento*, A16, 639 (1973).

[200] Hazewinkel, M., ed., *Quantum field theory, Yang–Mills field, Gauge transformation*, Encyclopedia of Mathematics, Springer, ISBN 978-1-55608-010-4, 2001.

[201] Hearn, A.C. (editor), *REDUCE – User's manual*, The Rand Corporation, Santa Monica, 1985.

[202] Heisenberg, W., *Physics and Philosophy*, Allen and Unwin, London, p. 96, 1962.

[203] Hendi, S.H., Sheykhi, A., Sepehri Rad, M. and Matsuno, K., *Slowly rotating dilatonic black holes with exponential form of nonlinear electrodynamics*, Gen. Rel. Grav. 47 (2015) 10, 117, doi:10.1007/s10714-015-1959-4.

[204] Huber, M., Maas, A. and von Smekal, L., *Two- and three-point functions in two-dimensional Landau-gauge Yang–Mills theory: Continuum results*, Journal of High Energy Physics, 2012: 035, arXiv:1207.0222, Bibcode: 2012JHEP...11..035H, doi:10.1007/JHEP11, 035 (2012).

[205] *Huge $10 billion collider resumes hunt for "God particle"*, CNN, 11 November 2009, retrieved 2010-05-04.

[206] Husain, V., Rahmati, S. and Ziprick, J., *Linearized 3D gravity with dust*, September 20, 2015, 7 pp., e-Print: arXiv:1509.06091 [gr-qc].

[207] Ionescu-Pallas, N., *General relativity and cosmology*, Scientific and Encyclopedic Publishing house, Bucharest, 1980.

[208] Isemberg, J., Nester, J. and Skinner, R., in *Abstracts of Contributed Papers, 8th International Conference on General Relativity and Gravitation*, University of Waterloo, Ontario, August 7–12, 1977.

[209] Isham, C., Salam, A. and Strathdee, J., *Phys. Rev.*, D3, 1805 (1971); D5, 2548 (1972).

[210] Isham, C., in *Quantum Gravity*, An Oxford Symposium, Clarendon Press, Oxford (1975).

[211] Ivanenko, D., *Usp. fiz. nauk*, 32, 149 (1947).

[212] Ivanenko, D. and Sokolov, A., *Vestnik M.G.U.*, Nr. 8, 103 (1947).

[213] Jackiw, R., *Phys. Rev.*, 168, 1623 (1968).

[214] Jackson, J.D., *From Lorenz to Coulomb and other explicit gauge transformations*, Am.J.Phys. 70(9): 917–928, arXiv:physics/0204034, Bibcode: 2002 Am. J. Phys. 70..917J, doi:10.1119/1.1491265 (2002).

[215] Jasinschi, R.S. and Smith, A.W., *Dynamical mass generation for the gravitino in simple $N = 1$ supergravity*, preprint, IFUSP P 444 (1983).

[216] Jauch, J.M. and Rohrlich, F., *The Theory of Photons and Electrons*, Addison-Wesley Publ. Comp. Inc., Cambridge, Mass., 1955.

[217] Jauch, J. and Rohrlich, F., *The Theory of Photos and Electrons*, Springer-Verlag, New York, 1976.

[218] Jennings, R.J., Tasson, J.D. and Yang, S., *Matter-Sector Lorentz Violation in Binary Pulsars*, October 13, 2015, 8 pp., e-Print: arXiv:1510.03798 [gr-qc].

[219] Just, K., *Nuovo Cimento*, 24, 567 (1964).

[220] Kabat, D. and Lifschytz, G., *Bulk equations of motion from CFT correlators*, JHEP 1509 (2015) 059, doi:10.1007/JHEP09 (2015) 059, e-Print: arXiv:1505.03755 [hep-th].

[221] Kalaydzhyan, T., *Gravitational mass of positron from LEP synchrotron losses*, July 13, 2015, 3 pp., e-Print: arXiv:1508.04377 [physics.gen-ph].

[222] Kashiwa, T., Sakoda, S. and Zenkin, S.V., *Ordering, Symbols and Finite-Dimensional Approximations of Path Integrals*, preprint, hep-th/9401063 (1994).

[223] Kawasaki, M. and Kobayashi, M., *Lagrange formulation of the symmetric theory of massive spin-3 fields*, Phys. Rev., D17, 446 (1978).

[224] Kawasaki, M. and Moroi, T., *Gravitino production in the inflationary universe and the effects on Big-Bang nucleosynthesis*, preprint, hep-ph/9403364, astro-ph/9403061 (1994).

[225] Källén, G., *Quantum Electrodynamics*, Springer-Verlag, Berlin, 1972.

[226] Khriplovich, I. B., *Iad. fiz.*, 3, 575 (1966).

[227] Kibble, T., *Usp. fiz. nauk*, 96, 497 (1968).

[228] Kim, J.E., *The photino, the axino and the gravitino in cosmology*, preprint, CERN TH 3904, Subm. to: Conference on Inner Space/Outer Space Batavia, IL, USA; 2–5 May 1984. Publ. in: *The interface between cosmology and particle physics*, Proceedings, Kolb, E.W. *et al.*, Univ. Chicago Press, Chicago, 1985 (543–549) (1985).

[229] Kim, J. and Porrati, M., *On a Canonical Quantization of 3D Anti de Sitter Pure Gravity*, JHEP 1510 (2015) 096, CERN-PH-TH-2015-188, doi:10.1007/JHEP10 (2015) 096, e-Print: arXiv:1508.03638 [hep-th].

[230] Kimura, T., *Progr. Theoret. Phys.*, 16, 157 (1956); 16, 555 (1956).

[231] Kirsch, I., *A Higgs mechanism for gravity*, Phys. Rev. D72, 024001, arXiv: hep-th/0503024 (2005).

[232] Kleinert, H. and Schulte-Frohlinde, V., *Critical Properties of $\varphi^4$-Theories*, World Scientific, ISBN 981-02-4658-7, 2001.

[233] Kleinert, H., Multivalued Fields in *Condensed Matter, Electrodynamics, and Gravitation* (PDF), World Scientific, ISBN 978-981-279-170-2, 2008.

[234] Kobzarev, I. and Okun, L.B., *Jurn. Eksp. Teor. Fiz.*, 43, 1904 (1962).

[235] Konkowski, D.A. and Helliwell, T.M., *Classical and Quantum Gravity*, 21, No. 1, 265 (2004).

[236] Korkina, M., *Ukr. fiz. jurnal*, 5, 763 (1960).

[237] Korotkin, D. and Nikolai, H., *An Integrable Model of Quantum Gravity*, preprint, DESY 95-058 (1995).

[238] Koyama, Y. and Chu, C.-S., *Dilaton, Screening of the Cosmological Constant and IR-Driven Inflation*, JHEP 1509 (2015) 024, NCTS-TH-1501, doi:10.1007/JHEP09 (2015) 024, e-Print: arXiv:1506.02848 [hep-th].

[239] Kramer, G. and Palmer, W.F., *The decay of strange bottom into vector mesons*, preprint, DESY 92-043 (1992).

[240] Landau, L.D. and Lifshitz, E.M., *Théorie quantique relativiste*, Éditions Mir, Moskou, 1972, (première partie), 1973 (deuxième partie).

[241] Landau, L.D. and Lifshitz, E.M., *The Classical Theory of Fields*, (Volume 2 of A Course of Theoretical Physics), Pergamon Press, 1971.

[242] Landi, G., Viet, N.A. and Wali, K.C., *Gravity and Electromagnetism in Noncommutative Geometry*, preprint, hep-th/9402046, SU-4240-566 (1994).

[243] Laurent, B., *Nuovo Cimento*, 4, 1445 (1956).

[244] Leclerk, M., *The Higgs sector of gravitational gauge theories*, Annals of Physics 321, 708, doi:10.1016/j.aop.2005.08.009 (2006).

[245] Lee, T. and Yang, C., *Phys. Rev.*, 119, 1410 (1960).

[246] Lias, R., *K kvantovaniiu nelineinyh uravneniy gravitatzionnovo polia*, Tr. Inst. Fiz. Astr., A.N.E.S.S.R., No. 5, 26 (1957).

[247] Long, H.N., Soa, D.V. and Tuan, A. Tran, *Electromagnetic-Gravitational Conversion Cross-Sections in External Electromagnetic Fields*, Mod. Phys. Lett., A9, 3619 (1994), [preprint, astro-ph/9410003, IC/94/285 (1994)].

[248] Long, H.N., Soa, D.V. and Tran, T.A., *Physics Letters*, A186, 382 (1994).

[249] Lopes, D.V., Mamiya, A. and Pinzul, A., *IR Hořava–Lifshitz gravity coupled to Lorentz violating matter: A spectral action approach*, Aug. 1, 2015, 21 pp., e-Print: arXiv:1508.00137 [hep-th].

[250] Luo, M.J., *Dark energy from quantum uncertainty of distant clock*, JHEP 1506 (2015) 063, doi:10.1007/JHEP06 (2015) 063, e-Print: arXiv:1401.2488 [physics.gen-ph].

[251] Lupanov, G., *Jurn. Eksp. Teor. Fiz.*, 52, 118 (1967).

[252] Lurié, D., *Particles and Fields*, J. Wiley and Sons, New York (1968).

[253] Mandelstam, S., Phys. Rev., 175, 1580 (1968); 175, 1604 (1968).

[254] Mangiarotti, L. and Sardanashvily, G., *Connections in Classical and Quantum Field Theory*, World Scientific, ISBN 981-02-2013-8, 2000.

[255] Mani, S. and Nearing, C., *Phys. Rev.*, 135, 1009 (1964).

[256] Mann R., *An Introduction to Particle Physics and the Standard Model*, CRC Press, ISBN 978-1-4200-8298-2, 2010.

[257] McMahon, D., *Quantum Field Theory*, McGraw-Hill, ISBN 978-0-07-154382-8, 2008.

[258] Merches , I., Tatomir, D. and Lupu, E.R., *Basics of Quantum Electrodynamics*, CRC Press (2013).

[259] Merches, I. and Agop, M., *Differentiability and Fractality in Dynamics of Physical Systems*, World Scientific (2015).

[260] Minami, S., *Radially excited tensor mesons with $I = 0$*, preprint, OCU 102 (1983).

[261] Minwalla, Sh., Van Raamsdonk, M. and Seiberg, N., *Noncommutative Perturbative Dynamics*, Journal of High Energy Physics (2000) and Matusis, A., Susskind, L. and Toumbas, N., *The IR/UV Connection in the Non-Commutative Gauge Theories*, Journal of High Energy Physics (2000).

[262] Mironovski, V., *Jurn. Eksp. Teor. Fiz.*, 48, 358 (1965).
[263] Mironovski, V., *Astronom. jurn.*, 42, 977 (1965).
[264] Misner, C., *Rev. Mod. Phys.*, 29, 497 (1957 ).
[265] Misner, C., *Phys. Rev.*, 186, 1319 (1969).
[266] Misner, C., in *Magic without Magic: J.A. Wheeler*, W.H. Freeman and Co., San Francisco (1972).
[267] Misner, C., *Phys. Rev. Lett*, 28, Nr.15, 994 (1972).
[268] Misner, C., Breuer, R., Brill, D., Chrzanowski, P., Hughes, H. and Pereira, C., *Phys. Rev. Lett*, 28, Nr.15, 998 (1972).
[269] Misner, C.W., Thorne, K.S. and Wheeler, J.A., *Gravitation*, Freeman, San Francisco, 1973.
[270] Mitskevich, N.V., *Physical Fields in General Relativity*, Nauka, Moskow (1969), (in Russian: Fiz. polia v O.T.O.).
[271] Mitskevich, N., *Jurn. Eksp. Teor. Fiz.*, 34, 1656 (1958).
[272] Mitskevich, N., *Jurn. Eksp. Teor. Fiz.*, 36, 1207 (1959).
[273] Mitskevich, N., *Wiss. Zs. Fr. Schiller – Univ. Jena*, Heft 4/5, 341 (1959).
[274] Mitskevich, N., *Izvestia V. U. Zov., Fizika*, Nr. 6, 64 (1964).
[275] Mitskevich, N., *Fiz. polia v obschcei teorii otnosit.*, Izd.-vo Nauka, Moskva, 1969.
[276] Moldauer, P.A. and Case, K.M., *Properties of Half-Integer Spin Dirac–Fierz–Pauli Particles*, Phys. Rev., 102, 279 (1956).
[277] Moore, G.W., *Computation of Some Zamolodchikov Volumes, with an Application*, Aug 23, 2015, 20 pp., e-Print: arXiv:1508.05612 [hep-th].
[278] Moretti V., *Aspects of noncommutative Lorentzian geometry for globally hyperbolic spacetimes*, Rev. Math. Phys. 15, 1171–1218 (2003).
[279] Moroi, T., *Effects of the gravitino on the inflationary universe*, preprint, hep-ph/9503210 (1995).
[280] Moskowitz, C., *Phew! Universe's Constant Has Stayed Constant*, Space.com., 13 December 2012, retrieved 2012-12-14.
[281] Nakayama, Y. and Nomura, Y., *Weak Gravity Conjecture in AdS/CFT*, September 4, 2015, 8 pp., CALT-TH-2015-045, IPMU15-0145, UCB-PTH-15-07, e-Print: arXiv:1509.01647 [hep-th].
[282] Namba, R., Peloso, M., Shiraishi, M., Sorbo, L. and Unal, C., *Scale-dependent gravitational waves from a rolling axion*, Sep 24, 2015, 41+1 pp., e-Print: arXiv:1509.07521 [astro-ph.CO].
[283] Narison, S., *Tensor mesons and gluonium of QCD*, preprint, PM 88-45; MU PM 88-45. Subm. to: 24th International Conference on High-energy Physics Munich, F.R. Germany; 4–10 Aug. 1988. Publ. in: Proceedings, Kotthaus R. and Kühn J.H., Springer, Heidelberg, 610–614 (1989).
[284] Ne'eman, Y., *Nucl. Phys.*, 26, 222 (1962).
[285] *New results indicate that new particle is a Higgs boson*, CERN, 14 March 2013, retrieved 2013-08-06.
[286] Nieto, J.A. and Obregon, O., *Classical supersymmetric spin-3/2 particle*, Phys. Lett., A175, 11 (1993).
[287] Nieuwenhuizen, P. van, *Physics Reports*, 68, 189 (1981).

[288] Novacu, V. and Vescan, T.T., *Theory of elementary particles*, R.S.R. Publishing House, Bucharest (1970).
[289] Nowakowski, M. and Rindani, S.D., *Astrophysical limits on gravitino mass*, preprint, hep-ph/9410262 (1994).
[290] *Observation of a New Particle with a Mass of 125 GeV*, CERN, 4 July 2012, retrieved 2012-07-05.
[291] Obukhov, Yu., *Poincaré gauge gravity: selected topics*, Int. J. Geom. Methods Mod. Phys. 3, 95–138, arXiv:gr-qc/0601090 (2006).
[292] Oerter R., in *The Theory of Almost Everything: The Standard Model, the Unsung Triumph of Modern Physics* (Kindle ed.), Penguin Group, p. 2, ISBN 0-13-236678-9, 2006.
[293] Ogievetski, V. and Polubarinov, I., *Jurn. Eksp. Teor. Fiz.*, 48, 1625 (1965).
[294] Ogievetski, V. and Mezincescu, L., *Usp. Fiz. Nauk*, 117, 637 (1975).
[295] Okada, N. and Raut, D., *Running Non-Minimal Inflation with Stabilized Inflaton Potential*, September 15, 2015, 20 pp., e-Print: arXiv:1509.04439 [hep-ph].
[296] Oliveira, C. and Tiomno, J., *Nuovo Cimento*, 23, nr. 4 (1962).
[297] Oliveira, O., Silva, P.J., Ilgenfritz, E.M. and Sternbeck, A., *The gluon propagator from large asymmetric lattices* (PDF), Lattice 2007, Trieste: Proceedings of Science, arXiv:0710.1424, retrieved 2013-11-18 (2007).
[298] Overbye D., *A New Particle Could Be Physics' Holy Grail*, New York Times, 4 July 2012, retrieved 2012-07-04.
[299] Pagels, H., *Galaxy structure in a gravitino dominated universe*, preprint, RU 82 B 17, RU B 82-17 (1982).
[300] Papini, G. and Valluri, S.-R., *Photoproduction of Gravitons in Static Electromagnetic Fields*, Can. J. Phys., 53, 2306 (1975).
[301] Papini, G. and Valluri, S., *Phys. Rep.*, 33, Nr.2, 51 (1977).
[302] Park, S., Prokopec, T. and Woodard, R.P., *Quantum Scalar Corrections to the Gravitational Potentials on de Sitter Background*, Oct 12, 2015, 21 pp., UFIFT-QG-15-08, e-Print: arXiv:1510.03352 [gr-qc].
[303] *Particle chameleon caught in the act of changing*, CERN, 31 May 2010, retrieved 2012-07-05.
[304] Peet, F., *Can. J. Phys.*, 49, Nr.7, 831 (1971).
[305] Peters, P.C., *Differential cross-sections for weak-field gravitational scattering*, Phys. Rev., D13, 775 (1976); In this paper are obtained the differential scattering cross-sections of the scalar, electromagnetic and gravitational waves in the long wavelength limit, on a weak Schwarzschild background. The Green functions on a low curvature space-time are used in this procedure.
[306] Piir, I., *Tr. inst. fiz. astr., A.N.E.S.S.R.*, Nr. 5, 41 (1957).
[307] Poznanin, P., *Izvestia V. U. Zov., Fizika*, Nr.10, 78 (1969).
[308] Privman, V. and Singer, P., *The width and decay modes of charmed tensor mesons*, preprint, PRE 23544 (1979).
[309] Pustovoit, V. and Gertzenstein, M., *Jurn. Eksp. Teor. Fiz.*, 42, 163 (1962).
[310] Radu, D. and Tatomir, D., *Gravitational scattering of the minimally coupled 1, 3/2 and 2-spin particles on Schwarzschild Background*, Il Nuovo Cimento, 111 B, No. 12, 1465–1480 (1996).

[311] Radu, D., Tatomir, D. and Mihalache, O., *The Electron Scattering and Transformation of Photons into Gravitons in Magnetic Fields*, Rom. Rep. Phys., 50, Nos. 3–4, 859 (1998).
[312] Radu, D. and Tatomir, D., *Gravitational Scattering of Massive Particles on a Background with Axial Symmetry*, Can. J. Phys., 76, Nr. 1, pp. 1–22 (1998).
[313] Radu, D., *Quantum Interaction of Fermions with the Far Field of a Kerr Black-Hole*, Acta Physica Slovaca, 54, No. 3, p. 307–320 (2004).
[314] Rarita, W. and Schwinger, J., *Phys. Rev.*, 59, 436 (1941).
[315] Reifler, N., *Conditions for exact equivalence of Kaluza–Klein and Yang–Mills theories*, arXiv:0707.3790 [gr-qc] (2007).
[316] Rodicev, V., in *Einsteinovskii sbornik*, Izd-vo Nauka, Moskva (1968).
[317] Ryder, L., *Quantum Field Theory*, Cambridge Univ. Press (1985), reprinted (1992).
[318] Ryttov, T. and Sannino, F., *Supersymmetry Inspired QCD Beta Function*, Physical Review D 78 (6): 065001, arXiv:0711.3745, Bibcode: 2008PhRvD..78f5001R. doi:10.1103/PhysRevD, 78, 065001 (2008).
[319] Saha, A.K. and Sil, A., *A Dynamic Modification to Sneutrino Chaotic Inflation*, Sep 1, 2015, 22 pp., e-Print: arXiv:1509.00218 [hep-ph].
[320] Salam, A., in *Quantum Gravity*, An Oxford Symposium, Clarendon Press, Oxford (1975).
[321] Sardanashvily, G., *Classical gauge gravitation theory*, Int. J. Geom. Methods Mod. Phys. 8, 1869–1895, arXiv:1110.1176 (2011).
[322] Sasakura, N. and Sato, Y., *Constraint algebra of general relativity from a formal continuum limit of canonical tensor model*, JHEP 1510 (2015) 109, YITP-15-52, WITS-MITP-014, doi:10.1007/JHEP10 (2015) 109, e-Print: arXiv:1506.04872 [hep-th].
[323] Scadron, M.D., *Advanced Quantum Theory and its Applications Through Feynman Diagrams*, Springer-Verlag, Berlin, Germany, 1991.
[324] Schuler, G.A. and Terron, J., *Elastic and diffractive photoproduction of* $J/\psi$ *mesons*, preprint, DESY92-017 (1992).
[325] Schumm, B., *Deep Down Things*, John Hopkins University Press, 2004.
[326] Schwartz, M.D., *Quantum Field Theory and the Standard Model*, Cambridge University Press, ISBN 978-1107034730, 2014.
[327] Schweber, S., *Vvedenie v rel. kvant. teor. polia*, Izd-vo Mir, Moskva (1963).
[328] Schwinger, J., *Phys. Rev.*, 130, 1253 (1963); 132, 1317 (1963).
[329] Schwinger, J., *The Theory of Quantized Fields. I*, Phys. Rev., 82, 914 (1951).
[330] Schwinger, J., *The Theory of Quantized Fields. II*, Phys. Rev., 91, 713 (1953).
[331] Schwinger, J., *Particles, Sources and Fields*, Addison-Wesley, Reading, Mass., 1970.
[332] Senjanovic, G., *Hamiltonian formulation and quantization of the spin-3/2 field*, Phys. Rev., D16, No. 2, 307 (1977).
[333] Shima, K., Kasuya, M., *Exact classical solution of gravitino equation of motion and topology of flat space in supergravity*, preprint, INS 363 SIT 79-09 (1979).

[334] Shklovski, I. and Kardashev, N., *Dokl. Akad. Nauk S.S.S.R.*, 155, 1039 (1964).
[335] Shklovski, I., *Astron. jurnal*, 46, 715 (1969).
[336] Shnir, Ya., *Gravitating sphalerons in the Skyrme model*, Aug 26, 2015, 10 pp., Phys. Rev. D 92 (2015) 8, 085039, doi:10.1103/PhysRevD.92.085039, e-Print: arXiv:1508.06507 [hep-th].
[337] Singer, P., *Radiative decays of light- and heavy-flavour tensor mesons*, preprint, TECHNION PHYS 82-30; TECHNION PH 82-30 (1982).
[338] Singh, L.P.S. and Hagen, C.R., *Lagrangian formulation for arbitrary spin. I. The boson case*, Phys. Rev., D9, No. 4, 898 (1974).
[339] Singh, L.P.S. and Hagen, C.R., *Lagrangian formulation for arbitrary spin. II. The fermion case*, Phys. Rev., D9, No. 4, 910 (1974).
[340] Skenderis, K. and Nieuwenhuizen, P. van, *On the Hamiltonian Approach and Path Integration for a Point Particle Minimally Coupled to Electromagnetism*, preprint, hep-th/9401024 (1994).
[341] Sokolov, A. *et al.*, *Quantum Electrodynamics*, Mir Publishers, Moskow, 1988.
[342] Srednicki, M., *Quantum Field Theory*, Cambridge Univ. Press, 2007.
[343] Strassler M., *Higgs Discovery: Is it a Higgs?*, 2010, retrieved 2013-08-06.
[344] Straumann, N., *On Pauli's invention of non-Abelian Kaluza–Klein Theory in 1953*, arXiv:gr-qc/0012054 (2000).
[345] Strumia, S., *Agravity*, JHEP 1406 (2014) 080, arXiv:1403.4226, Bibcode: 2014JHEP...06..080S, doi:10.1007/JHEP06 (2014) 080.
[346] Sudarshan, E. and Marshak, R., *Phys. Rev.*, 101, 1860 (1958).
[347] Szabo, R.J., *Quantum Field Theory on Noncommutative Spaces*, Physics Reports 378, 207-99 (2003).
[348] Taylor, M., *Generalized entanglement entropy*, July 23, 2015, 40 pp., e-Print: arXiv:1507.06410 [hep-th].
[349] Teodorescu, P.P. and Nicorovici Porumbaru, N., *Applications of group theory in mechanics and physics*, Technical Publishing House, Bucharest, 1985.
[350] 't Hooft, G., *50 Years of Yang–Mills theory*, World Scientific, ISBN 981-238-934-2, 2005.
[351] 't Hooft, G., *The Conceptual Basis of Quantum Field Theory*, in Butterfield, J., and Earman, J., eds., Philosophy of Physics, Part A, Elsevier, 661–730, 2007.
[352] Toró, T., *An. Univ. Timisoara*, VII, Nr.1, 99 (1969).
[353] Tourrenc, A., *C. R. Acad. Sci.*, 271, Nr.1, B1; Nr.2, B69 (1970).
[354] Tourrenc, A., *C. R. Acad. Sci.*, 273, Nr.5, B229 (1971).
[355] Tully, C., *Elementary Particle Physics in a Nutshell*, Princeton University Press, 2011.
[356] Tzou, K., *Nuovo Cimento*, 33, 286 (1964).
[357] Törnqvist, N.A., *The tensor mesons in the unitarized quark model: is there a glueball component in the f and f' meson?*, preprint, HU TFT 83-23 (1983).

[358] Tatomir, D., *An. st. Univ. Iasi (s.n.)*, Sec. Ib. fiz, XXII, 71, 84 (1976); XXIII, 75 (1977); XXV, 21 (1979).

[359] Tatomir, D., *Weak interactions in the presence of the gravitational field. Considerations on particle scattering in a static gravitational field, described by the Schwarzschild metric. The S-matrix for the Rarita–Schwinger gravitational field.* Communicated at National Colloquium on the Theory of Relativity, Iasi, April 27–28 (1979).

[360] Tatomir, D., *Rev. Roum. Phys.*, 25, Nr. 3, 267; 277 (1980).

[361] Tatomir, D., *Considerations on particle scattering and transformation in the presence of gravitational field.* Communicated at National Colloquium on Physics and Thechnology of Amorphous and Crystalline Materials, Iasi, June 7–8, 1980.

[362] Tatomir, D., in *Abstracts of Contributed Papers for the Discusion Groups*, 9th International Conference on General Relativity and Gravitation, 594, Fr. Schiller Univ., Jena, July 14–19 (1980).

[363] Tatomir, D., Radu, D. and Mihalache, O., *Quantum Scatterings on Static Spherically Symmetric Gravitational Background in the Weak Field Newtonian Limit*, Acta Physica Slovaca, 48, No. 1, p. 55–66 (1998).

[364] Tatomir, D., *S-Matrix for the Interaction between Gravitational and Yang–Mills Fields*, Scientific Annals of the Alexandru Ioan Cuza University of Iasi, (n.s.), section I b, Phys., Tome XXIV, p. 91 (1978).

[365] Tatomir, D., Radu, D. and Mihalache, O., *Massive vector, tensor and spin-3/2 particles gravitationally scattered on Schwarzschild background*, Studia Universitatis Babes-Bolyai, Physica 2, XXXVIII, p. 31–43 (1993).

[366] Tatomir, D., *Contributions in Quantum Study of the Particle Interaction in the Presence of Gravitation*, Ph. D. Thesis, Al. I. Cuza Univ., Iasi (1981).

[367] Tatomir, D., Radu, D. and Mihalache, O., *Scattering of the massive 3/2-spin particles in the static gravitational field described by Schwarzschild metric*, Annals of the West University of Timisoara, Vol. XXX, Physical Science Series, p. 35–42 (1993).

[368] Tatomir, D., Radu, D. and Mihalache, O., *Scattering of the massive two-spin particles in the static gravitational field described by Schwarzschild metric*, Annals of the West University of Timisoara, Vol. XXX, Physical Science Series, p. 49–55 (1993).

[369] Tatomir, D., Radu, D. and Mihalache, O., *On the scattering of the massive particles in exterior electromagnetic and gravitational fields*, Rom. J. Phys., 39, Nos. 5–6, p. 383–393 (1994).

[370] Tatomir, D. and Radu, D., *On the gravitational scattering of integer-spin particles*, Rom. J. Phys., 39, Nos. 7–8, p. 491–497 (1994).

[371] Tatomir, D. and Radu, D., *The Electron Scattering and Transformation of Photons into Gravitons in Electric Fields*, Rom. Rep. Phys., 50, Nos. 3–4, 871 (1998).

[372] Tatomir, D., Radu, D. and Sachelarie, A., *On the Scattering of Massive High-Spin Particles in the Static Gravitational Field Described by Schwarzschild Metric*, The Bulletin of the Polytechnic Institute of Iasi,

Tome XLII–XLVI, fasc. 3–4, Sect. I – Mathematics, Theoretical Mechanics, Physics, (1996).

[373] Tatomir, D. and Sachelarie, A., *The Graviton Photoproduction in External Electromagnetic Fields*, The Bulletin of the Polytechnic Institute of Iasi, Tome XLII (XLVI), f. 1–2, p. 99 (1996).

[374] Tatomir, D., Radu, D. and Agop, M., *The electron scattering and graviton photoproduction in external electromagnetic fields*, Il Nuovo Cimento, 112B, No. 5, p. 710–728 (1997).

[375] Tatomir, D., *Gravitational Scattering of Zero-Rest-Mass Particles*, Rom. J. Phys., 39, Nos. 5–6, p. 373 (1994).

[376] Tatomir, D., Radu, D. and Mihalache, O., *Gravitational scattering of the zero-rest mass spin-3/2 particles*, Rom. Rep. Phys, 48, Nos. 9–10, p. 799 (1996).

[377] Tatomir, D., Oancea, S. and Manta, V., *The Scattering of 3/2-Spin Particles in External Gravitational Field*, Tensor, N. S., 57, 44 (1996).

[378] Tatomir, D., *Scientific Annals of the Alexandru Ioan Cuza University of Iasi*, (n.s.), Sect. Ib. fiz, XXV, 21 (1979).

[379] Tatomir, D., in *10–th International Conference in G. R. G.*, Padova, July 4–9, p. 140 (1983).

[380] Tatomir, D. and Radu, D., *Scattering of the massive vector particles in the static gravitational field described by Schwarzschild metric*, Scientific Annals of Timisoara University, Vol. XXX, Physical Sciences Series, p. 43–48 (1993).

[381] Tatomir, D. and Radu, D., *Gravitational Scattering of the Massive Vector Particles in Schwarzschild Field (Proca Formalism)*, Scientific Annals of Al.I. Cuza University of Iasi, Tome XL, s.I.b. fasc. 2, Solid-state Physics – Theoretical Physics, p. 69 (1994).

[382] Tatomir, D., Radu, D. and Mihalache, O., *Gravitational Scattering of the Massive Two-Spin Particles (Gupta Formalism)*, Scientific Annals of Al.I. Cuza University of Iasi, Tome XL, s.I.b. fasc. 2, Solid-state Physics – Theoretical Physics, p. 63 (1994).

[383] Tatomir, D. and Radu, D., *The Scattering of Zero-Rest-Mass Particles in Static Gravitational Field Described by Schwarzschild Metric*, Scientific Annals of Al.I. Cuza University of Iasi, Tome XLI–XLII, s.I.b. fasc. 2, Solid-State Physics – Theoretical Physics, (1995–1996).

[384] Titeica, S., *Quantum Mechanics*, Romanian Academy Publishing House, Bucharest, 1984.

[385] Velo, G. and Zwanziger, D., *Propagation and Quantization of Rarita–Schwinger Waves in an External Electromagnetic Potential*, Phys. Rev., 186, 1337 (1969).

[386] Velo, G. and Zwanziger, D., *Noncausality and Other Defects of Interaction Lagrangians for Particles with Spin One and Higher*, Phys. Rev., 188, 2218 (1969).

[387] Visconti, A., *Théorie quantique des champs*, Gauthier-Villars, Paris, Tome I 1961, Tome II 1965.

[388] Vladimirov, Iu., *Jurn. Eksp. Teor. Fiz.*, 43, 89 (1962); 45, 251 (1963).

[389] Vladimirov, Iu., *Izvestia V. U. Zov., Fizika*, Nr. 2, 133 (1963); Nr. 6, 82 (1964).

[390] Vladimirov, Iu., Ishakov I., *Izvestia V. U. Zov., Fizika*, Nr. 8, 45 (1970).

[391] Vladimirov, Iu., in *Einsteinovskii sbornik*, Izd-vo Nauka, Moskva (1974).

[392] Vladimirova, L., *Izvestia V. U. Zov., Fizika*, Nr. 1, 134 (1971).

[393] Voronov, N., *Jurn. Eskp. Teor. Fiz.*, 64, 1889 (1973).

[394] Warren, S., *Fields*, A free text, also available from arXiv:hep-th/9912205, 2005.

[395] Weber, J., *Obsc. teor. otnosit. i gravit. volny*, Izd-vo I.L., Moskva (1962).

[396] Weber, J. and Hinds, C., *Phys. Rev.*, 128, 2414 (1962).

[397] Weber, J., in *Relativity, Groups and Topology*, Gordon and Breach, New York (1964).

[398] Weber, J., *Phys. Rev. Lett.*, 22, 1320 (1969).

[399] Weber, J., *Nature Phys. Sci.*, 240, 28 (1972).

[400] Weinberg, S., *Phys. Rev.*, 135, 1049 (1964); 138, 988 (1965); 140, 516 (1965).

[401] Weinberg, S., *Gravitation and Cosmology*, John Wiley and Sons, New York (1972).

[402] Weinberg, S., *Photons and Gravitons in S-Matrix Theory: Derivation of Charge Conservation and Equality of Gravitational and Inertial Mass*, Phys. Rev., B135, 1049 (1964).

[403] Wheeler, J., *Gravit., neitrino i Vselen.*, Izd-vo I.L., Moskva, 1962.

[404] Wheeler, J., in *Relativity, Groups and Topology*, Gordon and Breach, New York, 1964.

[405] Wigner, E.P., *On Unitary Representations of the Inhomogeneous Lorentz Group*, Ann. Math., 40, 149 (1939).

[406] Wilczek, F., *Quantum field theory*, Reviews of Modern Physics, 71: S83–S95. Also doi:10.1103/Rev.Mod.Phys.71 (1999).

[407] Wolfram, S., *MATHEMATICA, a System for Doing Mathematics by Computer*, Second Edition, Addison-Wesley Publishing Company, USA, 1991.

[408] * * *, *MATHEMATICA – User's Guide for Microsoft Windows*, Addison-Wesley Publishing Company, USA, 1991.

[409] Yang, C. and Mills, R., *Phys. Rev.*, 96, 191 (1954).

[410] Yndurain, F. J., Garcia-Martin, R. and Pelaez, J.R., *Experimental status of the $\pi\pi$ isoscalar S wave at low energy: $f_0$ (600) pole and scattering length*, Physical Review D 76 (7), 074034, arXiv:hep-ph/0701025, Bibcode: 2007PhRvD..76g4034G, doi:10.1103/Phys.Rev.D76, 074034 (2007).

[411] Zee, A., *Quantum Field Theory in a Nutshell*, Princeton University Press, ISBN 0-691-01019-6, 2003.

[412] Zeldovici, Ia. and Smorodinski, Ia., *Jurn. Eksp. Teor. Fiz.*, 41, 907 (1961).

[413] Zeldovici, Ia. and Novikov, I., *Dokl. Acad. Nauk S.S.S.R.*, 155, 1033 (1964).

[414] Zeldovici, Ia., in *Magic without Magic: J.A. Wheeler*, W.H. Freeman and Co, San Francisco (1972).

[415] Zet, Gh., *Unitary symmetries and gauge theories*, Gh. Asachi Publishing House, Iasi (1998).

[416] Zimmerman, P., *Gravitational self-force in scalar-tensor gravity*, July 14, 2015, 13 pp., Phys. Rev. D92 (2015) 6, 064051, doi:10.1103/PhysRevD.92.064051, e-Print: arXiv:1507.04076 [gr-qc].

# Author Index

# Subject Index